Student Solutions Manual

Beginning and Intermediate Algebra
An Integrated Approach

SIXTH EDITION

R. David Gustafson
Rock Valley College

Rosemary M. Karr
Collin College

Marilyn B. Massey
Collin College

Prepared by

Michael Weldon
Mount San Jacinto College

Australia • Brazil • Japan • Korea • Mexico • Singapore • Spain • United Kingdom • United States

For product information and technology assistance, contact us at
Cengage Learning Customer & Sales Support,
1-800-354-9706

For permission to use material from this text or product, submit all requests online at **www.cengage.com/permissions**
Further permissions questions can be emailed to
permissionrequest@cengage.com

ISBN-13: 978-0-538-49533-2
ISBN-10: 0-538-49533-2

Brooks/Cole
20 Davis Drive
Belmont, CA 94002-3098
USA

Cengage Learning is a leading provider of customized learning solutions with office locations around the globe, including Singapore, the United Kingdom, Australia, Mexico, Brazil, and Japan. Locate your local office at: **www.cengage.com/global**

Cengage Learning products are represented in Canada by Nelson Education, Ltd.

To learn more about Brooks/Cole, visit **www.cengage.com/brookscole**

Purchase any of our products at your local college store or at our preferred online store **www.CengageBrain.com**

Printed in the United States of America
2 3 4 5 6 7 8 14 13 12 11 10

Contents

Preface

This manual contains detailed solutions to the odd exercises of the text *Beginning and Intermediate Algebra*, sixth edition, by Gustafson, Karr, and Massey. It also contains solutions to all of the exercises in the Chapter Review and Chapter Test sections of the text.

Many of the exercises in the text may be solved using more than one method, but it is not feasible to list all possible solutions in this manual. Also, some of the exercises may have been solved in this manual using a method that differs slightly from that presented in the text. There are a few exercises in the text whose solutions may vary from person to person. Some of these solutions may not have been included in this manual. For the solution to an exercise like this, the notation "answers may vary" has been included.

Please remember that only reading a solution does not teach you how to solve a problem. To repeat a commonly used phrase, mathematics is not a spectator sport. You MUST make an honest attempt to solve each exercise in the text without using this manual first. This manual should be viewed more or less as a last resort. Above all, DO NOT simply copy the solution from this manual onto your own paper. Doing so will not help you learn how to do the exercise, nor will it help you to do better on quizzes or tests.

I would like to thank Shaun Williams of Brooks/Cole Publishing for her help and guidance.

This book is dedicated to my nieces Hannah and Mairin Welden.

May your study of this material be successful and rewarding.

Michael G. Welden

Exercise 1.1 (page 10)

1-9. Answers may vary.

11. $-|15| = -(15) = -15$

13. set

15. whole

17. integers

19. subset

21. rational

23. real

25. natural, prime

27. odd

29. $<$

31. variables

33. 7

35. parenthesis, open

37. absolute value

39. natural: $1, 2, 6, 9$

41. positive integers: $1, 2, 6, 9$

43. integers: $-3, -1, 0, 1, 2, 6, 9$

45. real: $-3, -\frac{1}{2}, -1, 0, 1, 2, \frac{5}{3}, \sqrt{7}, 3.25, 6, 9$

47. odd integers: $-3, -1, 1, 9$

49. composite: $6, 9$

51. $3 \boxed{<} 5$

53. $5 \boxed{} 3 + 2$
$5 \boxed{=} 5$

55. $25 \boxed{<} 32$

57. $5 + 7 \boxed{} 10$
$12 \boxed{>} 10$

59.

0 3 6

6 is greater than 3. 6 is to the right of 3.

61.

0 6 11

11 is greater than 6. 11 is to the right of 6.

63.

0 2

2 is greater than 0. 2 is to the right of 0.

65.

0 8

8 is greater than 0. 8 is to the right of 0.

67.

1 2 3 4 5 6 7 8

69.

1 5

71.

8

73.

10 12 14 16 18 20

75. $|36| = 36$

77. $|0| = 0$

79. $|-230| = 230$

81. $|12 - 4| = |8| = 8$

83. $4 + 5 = 9$
9: natural, odd, composite, whole

85. $15 - 15 = 0$
0: even, whole

87. $3 \cdot 8 = 24$
24: natural, even, composite, whole

89. $24 \div 8 = 3$
3: natural, odd, prime, whole

91. $3+9 \; \boxed{} \; 20-8$
$12 \; \boxed{=} \; 12$

93. $4 \cdot 2 \; \boxed{} \; 2 \cdot 4$
$8 \; \boxed{=} \; 8$

95. $8 \div 2 \; \boxed{} \; 4+3$
$4 \; \boxed{<} \; 7$

97. $45 \div 9 \; \boxed{} \; 36 \div 12$
$5 \; \boxed{>} \; 3$

99. $3+2+5 \; \boxed{} \; 5+2+3$
$10 \; \boxed{=} \; 10$

101. $7 > 3$

103. $8 \leq 8$

105. $3+4=7$

107. $\sqrt{2} \approx 1.41$

109. $3 \leq 7 \Rightarrow \boxed{7 \geq 3}$

111. $6 > 0 \Rightarrow \boxed{0 < 6}$

113. $3+8 > 8 \Rightarrow \boxed{8 < 3+8}$

115. $6-2 < 10-4 \Rightarrow \boxed{10-4 > 6-2}$

117. $2 \cdot 3 < 3 \cdot 4 \Rightarrow \boxed{3 \cdot 4 > 2 \cdot 3}$

119. $\frac{12}{4} < \frac{24}{6} \Rightarrow \boxed{\frac{24}{6} > \frac{12}{4}}$

121.

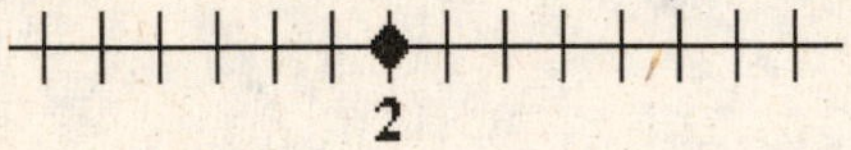

123.

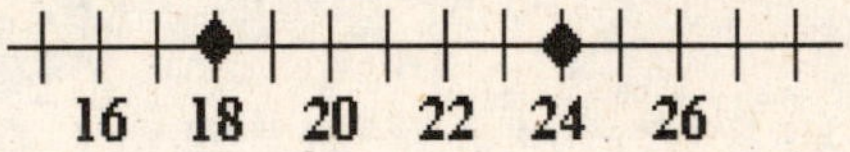

125.

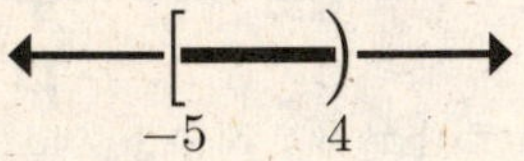

127. $|21-19| = |2| = 2$

129. If you think you have the greatest natural number, just add 1 to it to get a greater natural number.

131. The absolute value of a positive number (or 0) is equal to the positive number (or 0). The absolute value of a negative number is equal to the opposite of the negative number.

133. Answers may vary.

Exercise 1.2 (page 27)

1. $\frac{3}{6} = \frac{1 \cdot \overset{1}{\cancel{3}}}{2 \cdot \underset{1}{\cancel{3}}} = \frac{1}{2}$

3. $\frac{10}{20} = \frac{1 \cdot \overset{1}{\cancel{10}}}{2 \cdot \underset{1}{\cancel{10}}} = \frac{1}{2}$

5. $\frac{5}{6} \cdot \frac{1}{2} = \frac{5 \cdot 1}{6 \cdot 2} = \frac{5}{12}$

7. $\frac{2}{3} \div \frac{3}{2} = \frac{2}{3} \cdot \frac{2}{3} = \frac{2 \cdot 2}{3 \cdot 3} = \frac{4}{9}$

9. $\frac{4}{9} + \frac{7}{9} = \frac{4+7}{9} = \frac{11}{9}$

11. $\frac{2}{3} - \frac{1}{2} = \frac{2 \cdot \mathbf{2}}{3 \cdot \mathbf{2}} - \frac{1 \cdot \mathbf{3}}{2 \cdot \mathbf{3}} = \frac{4}{6} - \frac{3}{6}$
$= \frac{4-3}{6} = \frac{1}{6}$

13.
$$\begin{array}{r} 2.5 \\ +\;0.36 \\ \hline 2.86 \end{array}$$

SECTION 1.2

15.

$$\begin{array}{r} 0.\ 2 \\ \times\ 2.\ 5 \\ \hline 1\ 0 \\ 4\ 0\ \\ \hline 5\ 0 \end{array}$$

Put two digits to the right of the decimal point. Answer: 0.5

17. The digit in the 2nd decimal place is 4. The next digit to the right is 4. Since this digit is less than 5, round down. Leave the 4 in the 2nd decimal place, and delete all digits to the right. 3.24

19. true

21. false; 21 has factors of 3 and 7.

23. true; 8 is to the right of -2.

25. true; $|-9| = 9$, so $9 \le |-9|$.

27. $3 + 7 \boxed{=} 10$

29. $|-2| = 2$, so $|-2| \boxed{=} 2$

31. numerator

33. undefined

35. prime

37. improper

39. 1

41. multiply

43. numerators, denominator

45. least common denominator, equivalent

47. terminating, 2

49. divisor, dividend, quotient

51. $30 = 6 \cdot 5$
$= 2 \cdot 3 \cdot 5$

53. $70 = 7 \cdot 10$
$= 7 \cdot 2 \cdot 5$
$= 2 \cdot 5 \cdot 7$

55. $\frac{6}{12} = \frac{1 \cdot \cancel{6}^{1}}{2 \cdot \cancel{6}_{1}} = \frac{1}{2}$

57. $\frac{15}{20} = \frac{3 \cdot \cancel{5}^{1}}{4 \cdot \cancel{5}_{1}} = \frac{3}{4}$

59. $\frac{24}{18} = \frac{4 \cdot \cancel{6}^{1}}{3 \cdot \cancel{6}_{1}} = \frac{4}{3}$

61. $\frac{72}{64} = \frac{9 \cdot \cancel{8}^{1}}{8 \cdot \cancel{8}_{1}} = \frac{9}{8}$

63. $\frac{1}{2} \cdot \frac{3}{5} = \frac{1 \cdot 3}{2 \cdot 5} = \frac{3}{10}$

65. $\frac{4}{3} \cdot \frac{6}{5} = \frac{4 \cdot 6}{3 \cdot 5} = \frac{4 \cdot 2 \cdot \cancel{3}^{1}}{\cancel{3}_{1} \cdot 5} = \frac{8}{5}$

67. $12 \cdot \frac{5}{6} = \frac{12}{1} \cdot \frac{5}{6} = \frac{12 \cdot 5}{1 \cdot 6} = \frac{2 \cdot \cancel{6}^{1} \cdot 5}{1 \cdot \cancel{6}_{1}}$
$= \frac{10}{1} = 10$

69. $\frac{10}{21} \cdot 14 = \frac{10}{21} \cdot \frac{14}{1} = \frac{10 \cdot 14}{21 \cdot 1} = \frac{10 \cdot 2 \cdot \cancel{7}^{1}}{3 \cdot \cancel{7}_{1}}$
$= \frac{20}{3}$

71. $\frac{3}{5} \div \frac{2}{3} = \frac{3}{5} \cdot \frac{3}{2} = \frac{3 \cdot 3}{5 \cdot 2} = \frac{9}{10}$

73. $\frac{3}{4} \div \frac{6}{5} = \frac{3}{4} \cdot \frac{5}{6} = \frac{3 \cdot 5}{4 \cdot 6} = \frac{\cancel{3}^{1} \cdot 5}{4 \cdot 2 \cdot \cancel{3}_{1}} = \frac{5}{8}$

SECTION 1.2

75. $6 \div \frac{3}{14} = \frac{6}{1} \div \frac{3}{14} = \frac{6}{1} \cdot \frac{14}{3} = \frac{6 \cdot 14}{1 \cdot 3} = \frac{2 \cdot \overset{1}{\cancel{3}} \cdot 14}{1 \cdot \underset{1}{\cancel{3}}} = \frac{28}{1} = 28$

77. $\frac{42}{30} \div 7 = \frac{42}{30} \div \frac{7}{1} = \frac{42}{30} \cdot \frac{1}{7} = \frac{42 \cdot 1}{30 \cdot 7} = \frac{\overset{1}{\cancel{6}} \cdot \overset{1}{\cancel{7}} \cdot 1}{5 \cdot \underset{1}{\cancel{6}} \cdot \underset{1}{\cancel{7}}} = \frac{1}{5}$

79. $\frac{3}{5} + \frac{3}{5} = \frac{3+3}{5} = \frac{6}{5}$

81. $\frac{4}{13} - \frac{3}{13} = \frac{4-3}{13} = \frac{1}{13}$

83. $\frac{1}{6} + \frac{1}{24} = \frac{1 \cdot \mathbf{4}}{6 \cdot \mathbf{4}} + \frac{1}{24} = \frac{4}{24} + \frac{1}{24} = \frac{4+1}{24} = \frac{5}{24}$

85. $\frac{7}{10} - \frac{1}{14} = \frac{7 \cdot \mathbf{7}}{10 \cdot \mathbf{7}} - \frac{1 \cdot \mathbf{5}}{14 \cdot \mathbf{5}} = \frac{49}{70} - \frac{5}{70} = \frac{49-5}{70} = \frac{44}{70} = \frac{22 \cdot \overset{1}{\cancel{2}}}{35 \cdot \underset{1}{\cancel{2}}} = \frac{22}{35}$

87. $4\frac{3}{5} + \frac{3}{5} = \left(4 + \frac{3}{5}\right) + \frac{3}{5} = \left(\frac{20}{5} + \frac{3}{5}\right) + \frac{3}{5} = \frac{23}{5} + \frac{3}{5} = \frac{26}{5} = 5\frac{1}{5}$

89. $3\frac{1}{3} - 1\frac{2}{3} = \left(3 + \frac{1}{3}\right) - \left(1 + \frac{2}{3}\right) = \left(\frac{9}{3} + \frac{1}{3}\right) - \left(\frac{3}{3} + \frac{2}{3}\right) = \frac{10}{3} - \frac{5}{3} = \frac{5}{3} = 1\frac{2}{3}$

91. $3\frac{3}{4} - 2\frac{1}{2} = \left(3 + \frac{3}{4}\right) - \left(2 + \frac{1}{2}\right) = \left(\frac{12}{4} + \frac{3}{4}\right) - \left(\frac{8}{4} + \frac{2}{4}\right) = \frac{15}{4} - \frac{10}{4} = \frac{5}{4} = 1\frac{1}{4}$

93. $8\frac{2}{9} - 7\frac{2}{3} = \left(8 + \frac{2}{9}\right) - \left(7 + \frac{2}{3}\right) = \left(\frac{72}{9} + \frac{2}{9}\right) - \left(\frac{63}{9} + \frac{6}{9}\right) = \frac{74}{9} - \frac{69}{9} = \frac{5}{9}$

95.

$$\begin{array}{r} 0.8 \\ 5\overline{)4.0} \\ \underline{4\,0} \\ 0 \end{array}$$

$\frac{4}{5} = 0.8$, terminating

97.

$$\begin{array}{r} 0.40909 \\ 22\overline{)9.00000} \\ \underline{8\,8} \\ 20 \\ \underline{0} \\ 200 \\ \underline{198} \\ 20 \\ \underline{0} \\ 200 \\ \underline{198} \\ 2 \end{array}$$

$\frac{9}{22} = 0.4\overline{09}$ repeating

99.
$$\begin{array}{r} 23.45 \\ +\ 135.2 \\ \hline 158.65 \end{array}$$

101.
$$\begin{array}{r} 6\ \ {}^{11}\!\!\not{1}\ {}^{13} \\ 6\not{7}.\not{2}\not{3}5 \\ -\ 22.45 \\ \hline 44.785 \end{array}$$

103.
$$\begin{array}{r} 3.4 \\ \times\ 13.2 \\ \hline 68 \\ 102 \\ 34 \\ \hline 4488 \end{array}$$

Put two digits to the right of the decimal point. Answer: 44.88

105. $0.23\overline{)1.0465}$

Move decimal points 2 places right.

$$\begin{array}{r} 4.55 \\ 23.\overline{)104.65} \\ 92 \\ \hline 126 \\ 115 \\ \hline 115 \\ 115 \\ \hline 0 \end{array}$$

107. The digit in the 2nd decimal place is 6. The next digit to the right is 9. Since this digit is 5 or more, round up. Change the 6 in the 2nd decimal place to 7, and delete all digits to the right. 587.27

The digit in the 3rd decimal place is 9. The next digit to the right is 4. Since this digit is less than 5, round down. Leave the 9 in the 3rd decimal place, and delete all digits to the right. 587.269

109. The digit in the 2nd decimal place is 9. The next digit to the right is 8. Since this digit is 5 or more, round up. Change the 9 in the 2nd decimal place to 0, increase the digit in the 1st decimal place from 3 to 4, and delete all digits to the right. 6,025.40

The digit in the 3rd decimal place is 8. The next digit to the right is 2. Since this digit is less than 5, round down. Leave the 8 in the 3rd decimal place, and delete all digits to the right. 6,025.398

111. $\frac{5}{12}\cdot\frac{18}{5}=\frac{5\cdot 18}{12\cdot 5}=\frac{\overset{1}{\not{5}}\cdot 3\cdot\overset{1}{\not{6}}}{2\cdot\underset{1}{\not{6}}\cdot\underset{1}{\not{5}}}=\frac{3}{2}$

113. $\frac{17}{34}\cdot\frac{3}{6}=\frac{17\cdot 3}{34\cdot 6}=\frac{\overset{1}{\not{17}}\cdot\overset{1}{\not{3}}}{2\cdot\underset{1}{\not{17}}\cdot 2\cdot\underset{1}{\not{3}}}=\frac{1}{4}$

115. $\frac{2}{13}\div\frac{8}{13}=\frac{2}{13}\cdot\frac{13}{8}=\frac{2\cdot 13}{13\cdot 8}=\frac{\overset{1}{\not{2}}\cdot\overset{1}{\not{13}}}{\underset{1}{\not{13}}\cdot 4\cdot\underset{1}{\not{2}}}=\frac{1}{4}$

117. $\frac{21}{35}\div\frac{3}{14}=\frac{21}{35}\cdot\frac{14}{3}=\frac{21\cdot 14}{35\cdot 3}=\frac{\overset{1}{\not{7}}\cdot\overset{1}{\not{3}}\cdot 14}{5\cdot\underset{1}{\not{7}}\cdot\underset{1}{\not{3}}}=\frac{14}{5}$

119. $\frac{3}{5}+\frac{2}{3}=\frac{3\cdot\mathbf{3}}{5\cdot\mathbf{3}}+\frac{2\cdot\mathbf{5}}{3\cdot\mathbf{5}}=\frac{9}{15}+\frac{10}{15}=\frac{9+10}{15}=\frac{19}{15}$

121. $\frac{9}{4}-\frac{5}{6}=\frac{9\cdot\mathbf{3}}{4\cdot\mathbf{3}}-\frac{5\cdot\mathbf{2}}{6\cdot\mathbf{2}}=\frac{27}{12}-\frac{10}{12}=\frac{27-10}{12}=\frac{17}{12}$

SECTION 1.2

123. $3 - \frac{3}{4} = \frac{3}{1} - \frac{3}{4} = \frac{3 \cdot \mathbf{4}}{1 \cdot \mathbf{4}} - \frac{3}{4} = \frac{12}{4} - \frac{3}{4} = \frac{12-3}{4} = \frac{9}{4}$

125. $\frac{17}{3} + 4 = \frac{17}{3} + \frac{4}{1} = \frac{17}{3} + \frac{4 \cdot \mathbf{3}}{1 \cdot \mathbf{3}} = \frac{17}{3} + \frac{12}{3} = \frac{17+12}{3} = \frac{29}{3}$

Problems 127-133 are to be solved using a calculator. The keystrokes needed to solve each problem using a TI-84 graphing calculator appear in each solution. There may be other solutions. Keystrokes for other calculators may be slightly different.

127. [3] [2] [3] [.] [2] [4] [+] [2] [7] [.] [2] [5] [4] [3] [ENTER] $\{350.4943\} \Rightarrow 350.49$

129. [2] [5] [.] [2] [5] [×] [1] [3] [2] [.] [1] [7] [9] [ENTER] $\{3337.51975\} \Rightarrow 3{,}337.52$

131. [4] [.] [5] [6] [9] [4] [3] [2] [3] [÷] [.] [4] [5] [6] [ENTER] $\{10.02068487\} \Rightarrow 10.02$

133. [5] [5] [.] [7] [7] [4] [4] [3] [−] [.] [5] [6] [8] [2] [4] [5] [ENTER]
$\{55.206185\} \Rightarrow 55.21$

135. $43\frac{1}{2} - 12\frac{1}{3} = 43 + \frac{1}{2} - 12 - \frac{1}{3} = \frac{258}{6} + \frac{3}{6} - \frac{72}{6} - \frac{2}{6} = \frac{187}{6} = 31\frac{1}{6}$ acres

137. $14 \cdot 3\frac{1}{4} = \frac{14}{1} \cdot \frac{13}{4} = \frac{14 \cdot 13}{1 \cdot 4} = \frac{7 \cdot \overset{1}{\cancel{2}} \cdot 13}{1 \cdot 2 \cdot \underset{1}{\cancel{2}}} = \frac{91}{2} = 45\frac{1}{2}$ yd

139. $187.75 - 46.8 - 72.5 = \$68.45$ million

141. 34% of 36,000 = 0.34(36,000) = $12,240

143. 0.23(17,500) = 4,025 defective ⇒ 17,500 − 4025 = 13,475 acceptable units

145. 0.12(18,700,000) = 2,244,000 increase ⇒ sales = 18,700,000 + 2,244,000 = $20,944,000

147. # gallons = 15,675.2 ÷ 25.5 = 614.7137255 ⇒ cost = 614.7137255(2.87) ≈ $1,764.23

149. Area = length · width = (253.5 ft)(178.5 ft) = 45,249.75 ft^2
Drums of sealer = 45,249.75 ÷ 4,000 ≈ 11.3 ⇒ needs 12 drums; Cost = 12(97.50) = $1,170

151. Standard = 37.50(2,530) = $94,875; High-capacity = 57.35(1,670) = $95,774.50
The high-capacity order will produce the greater profit.

153. Silage per cow = 0.57(12,000) = 6840 pounds; 30(6,840) = 205,200 lb of silage

155. Regular = 1,730 + 36(107.75) = 1,730 + 3,879 = 5,609
High = 4,170 + 36(57.50) = 4,170 + 2,070 = 6,240
The high-efficiency furnace will be more expensive after 3 years.

157-161. Answers may vary.

163. No. Each proper fraction is less than 1. When a number is multiplied by a number less than 1, the result is smaller than the original number.

Exercise 1.3 (page 38)

1. $2^5 = 2 \cdot 2 \cdot 2 \cdot 2 \cdot 2 = 32$

3. $4^3 = 4 \cdot 4 \cdot 4 = 64$

5. $3(2)^3 = 3 \cdot 2 \cdot 2 \cdot 2 = 24$

7. $3 + 2 \cdot 4 = 3 + 8 = 11$

9. $4 + 2^2 \cdot 3 = 4 + 4 \cdot 3 = 4 + 12 = 16$

11. 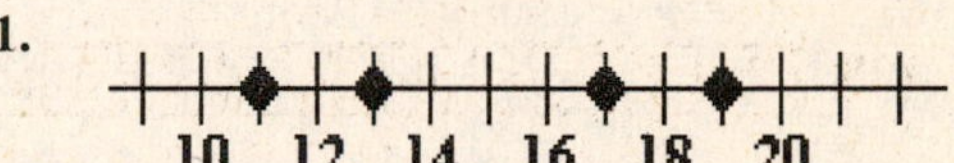

13. 17 is a prime number.

15. exponent

17. grouping

19. perimeter, circumference

21. $P = 4s$, units

23. $P = 2l + 2w$, units

25. $P = a + b + c$, units

27. $P = a + b + c + d$, units

29. $C = \pi D$, or $C = 2\pi r$, units

31. $V = lwh$, cubic units

33. $V = \frac{1}{3}Bh$, cubic units

35. $V = \frac{4}{3}\pi r^3$, cubic units

37. $4^2 = 4 \cdot 4 = 16$

39. $\left(\frac{1}{10}\right)^4 = \frac{1}{10} \cdot \frac{1}{10} \cdot \frac{1}{10} \cdot \frac{1}{10} = \frac{1}{10{,}000}$

41. $x^2 = x \cdot x$

43. $3z^4 = 3 \cdot z \cdot z \cdot z \cdot z$

45. $(5t)^2 = 5t \cdot 5t$

47. $5(2x)^3 = 5 \cdot 2x \cdot 2x \cdot 2x$

49. $4(3)^2 = 4 \cdot 9 = 36$

51. $(5 \cdot 2)^3 = 10^3 = 1{,}000$

53. $2(3^2) = 2 \cdot 9 = 18$

55. $(3 \cdot 2)^3 = 6^3 = 216$

57. $3 \cdot 5 - 4 = 15 - 4 = 11$

59. $3(5 - 4) = 3(1) = 3$

61. $2 + 3 \cdot 5 - 4 = 2 + 15 - 4 = 17 - 4 = 13$

63. $64 \div (3 + 1) = 64 \div 4 = 16$

65. $3^2 + 2(1 + 4) - 2 = 9 + 2(5) - 2 = 9 + 10 - 2 = 19 - 2 = 17$

67. $\frac{3}{5} \cdot \frac{10}{3} + \frac{1}{2} \cdot 12 = \frac{3}{5} \cdot \frac{10}{3} + \frac{1}{2} \cdot \frac{12}{1} = \frac{\overset{1}{\cancel{3}}}{\underset{1}{\cancel{5}}} \cdot \frac{2 \cdot \overset{1}{\cancel{5}}}{\underset{1}{\cancel{3}}} + \frac{1}{\underset{1}{\cancel{2}}} \cdot \frac{6 \cdot \overset{1}{\cancel{2}}}{1} = \frac{2}{1} + \frac{6}{1} = 2 + 6 = 8$

69. $\left[\frac{1}{3} - \left(\frac{1}{2}\right)^2\right]^2 = \left[\frac{1}{3} - \frac{1}{4}\right]^2 = \left[\frac{1 \cdot \mathbf{4}}{3 \cdot \mathbf{4}} - \frac{1 \cdot \mathbf{3}}{4 \cdot \mathbf{3}}\right]^2 = \left[\frac{4}{12} - \frac{3}{12}\right]^2 = \left[\frac{1}{12}\right]^2 = \frac{1}{144}$

SECTION 1.3

71. $\frac{(3+5)^2+2}{2(8-5)} = \frac{8^2+2}{2(3)} = \frac{64+2}{6} = \frac{66}{6} = 11$

73. $\frac{(5-3)^2+2}{4^2-(8+2)} = \frac{2^2+2}{16-10} = \frac{4+2}{6} = \frac{6}{6} = 1$

75. $\frac{3\cdot 7-5(3\cdot 4-11)}{4(3+2)-3^2+5} = \frac{3\cdot 7-5(12-11)}{4(5)-3^2+5} = \frac{3\cdot 7-5(1)}{4(5)-9+5} = \frac{21-5}{20-9+5} = \frac{16}{16} = 1$

77. $P = 4s = 4(4\text{ in.}) = 16\text{ in.}$

79. $P = a+b+c = 3\text{ m}+5\text{ m}+7\text{ m} = 15\text{ m}$

81. $A = s^2 = (5\text{ m})^2 = 25\text{ m}^2$

83. $A = bh = (6\text{ ft})(10\text{ ft}) = 60\text{ ft}^2$

85. $C = 2\pi r \approx 2\left(\frac{22}{7}\right)(14\text{ m}) = \frac{2}{1}\cdot\frac{22}{7}\cdot\frac{14}{1}\text{ m} = \frac{2\cdot 22\cdot 2\cdot \overset{1}{\cancel{7}}}{\underset{1}{\cancel{7}}}\text{ m} = 88\text{ m}$

87. $A = \pi r^2 \approx \frac{22}{7}(21\text{ ft})^2 = \frac{22}{7}(441\text{ ft}^2) = \frac{22}{7}\cdot\frac{441}{1}\text{ ft}^2 = \frac{22}{\underset{1}{\cancel{7}}}\cdot\frac{63\cdot\overset{1}{\cancel{7}}}{1}\text{ ft}^2 = 1{,}386\text{ ft}^2$

89. $V = \frac{1}{3}Bh = \frac{1}{3}(3\text{ cm})^2(2\text{ cm}) = \frac{1}{3}(9\text{ cm}^2)(2\text{ cm}) = (3\text{ cm}^2)(2\text{ cm}) = 6\text{ cm}^3$

91. $V = \frac{4}{3}\pi r^3 \approx \frac{4}{3}\cdot\frac{22}{7}(6\text{ m})^3 = \frac{88}{21}(216\text{ m}^3) = \frac{88}{21}\cdot\frac{216}{1}\text{ m}^3 \approx 905\text{ m}^3$

93. Cylinder: $V = Bh = \pi(4\text{ cm})^2(14\text{ cm}) \approx \frac{22}{7}\cdot\frac{16}{1}\text{ cm}^2\cdot\frac{14}{1}\text{ cm} = \frac{22\cdot 16\cdot 2\cdot\overset{1}{\cancel{7}}}{\underset{1}{\cancel{7}}\cdot 1\cdot 1}\text{ cm}^3 = 704\text{ cm}^3$

Cone: $V = \frac{1}{3}Bh = \frac{1}{3}\pi(4\text{ cm})^2(21\text{ cm}) \approx \frac{1}{3}\cdot\frac{22}{7}\cdot\frac{16}{1}\text{ cm}^2\cdot\frac{21}{1}\text{ cm} = \frac{22\cdot 16\cdot\overset{1}{\cancel{3}}\cdot\overset{1}{\cancel{7}}}{\underset{1}{\cancel{3}}\cdot\underset{1}{\cancel{7}}\cdot 1\cdot 1}\text{ cm}^3 = 352\text{ cm}^3$

Total $= 704\text{ cm}^3 + 352\text{ cm}^3 = 1{,}056\text{ cm}^3$

95. $6^2 = 6\cdot 6 = 36$

97. $3+5^2 = 3+25 = 28$

99. $(3+5)^2 = (8)^2 = 64$

101. $(7+9)\div(2\cdot 4) = 16\div 8 = 2$

103. $(5+7)\div 3\cdot 4 = 12\div 3\cdot 4 = 4\cdot 4 = 16$

105. $24\div 4\cdot 3+3 = 6\cdot 3+3 = 18+3 = 21$

107. $5^2-(7-3)^2 = 5^2-4^2 = 25-16 = 9$

109. $(2\cdot 3-4)^3 = (6-4)^3 = 2^3 = 8$

111. $\frac{2[4+2(3-1)]}{3[3(2\cdot 3-4)]} = \frac{2[4+2(2)]}{3[3(6-4)]} = \frac{2[4+4]}{3[3(2)]} = \frac{2[8]}{3[6]} = \frac{16}{18} = \frac{8}{9}$

Problems 113-115 are to be solved using a calculator. The keystrokes needed to solve each problem using a TI-84 graphing calculator appear in each solution. There may be other solutions. Keystrokes for other calculators may be slightly different.

113. [7] [.] [9] [^] [3] [ENTER] {493.039}

115. [2] [5] [.] [3] [^] [2] [ENTER] {640.09}

117. $39 = (3 \cdot 8) + (5 \cdot 3)$

119. $87 = (3 \cdot 8 + 5) \cdot 3$

121. $V = \frac{4}{3}\pi r^3 = \frac{4}{3}\pi(21.35\text{ ft})^3 \approx 40{,}764.51\text{ ft}^3$

123. $P = 4s = 4\left(30\frac{2}{5}\text{ m}\right) = 4\left(30 + \frac{2}{5}\text{ m}\right) = 4\left(\frac{150}{5} + \frac{2}{5}\text{ m}\right) = \frac{4}{1}\left(\frac{152}{5}\text{ m}\right) = \frac{608}{5}\text{ m} = 121\frac{3}{5}\text{ m}$

125. $V = lwh = (40\text{ ft})(40\text{ ft})(9\text{ ft}) = 14{,}400\text{ ft}^3$; Per student $= 14{,}400\text{ ft}^3 \div 30 = 480\text{ ft}^3$ per student

127. $f = \dfrac{rs}{(r+s)(n-1)} = \dfrac{(8)(12)}{(8+12)(1.6-1)} = \dfrac{96}{(20)(0.6)} = \dfrac{96}{12} = 8$

129. **Answers may vary.**

131. Increasing powers produce larger numbers.

Exercise 1.4 (page 48)

1. $2 + 3 = +(2 + 3) = 5$

3. $-4 + 7 = +(7 - 4) = 3$

5. $6 - 2 = 4$

7. $-5 - (-7) = -5 + (+7) = +(7 - 5) = 2$

9. $5 + 3(7 - 2) = 5 + 3(5) = 5 + 15 = 20$

11. $5 + 3(7) - 2 = 5 + 21 - 2 = 26 - 2 = 24$

13. arrows

15. unlike

17. subtract, greater

19. add, opposite

21. $4 + 8 = +(4 + 8) = 12$

23. $(-3) + (-7) = -(3 + 7) = -10$

25. $\frac{1}{5} + \left(+\frac{1}{7}\right) = \frac{7}{35} + \left(+\frac{5}{35}\right) = +\left(\frac{7}{35} + \frac{5}{35}\right) = \frac{12}{35}$

27. $44.902 + 33.098 = +(44.902 + 33.098) = 78$

29. $6 + (-4) = +(6 - 4) = 2$

31. $(-0.4) + 0.9 = +(0.9 - 0.4) = 0.5$

33. $\frac{2}{3} + \left(-\frac{1}{4}\right) = \frac{8}{12} + \left(-\frac{3}{12}\right) = +\left(\frac{8}{12} - \frac{3}{12}\right) = +\frac{5}{12}$

35. $87.63 + (-102.6) = -(102.6 - 87.63) = -14.97$

SECTION 1.4

37. $5 + [4 + (-2)] = 5 + [2] = 7$

39. $-2 + (-4 + 5) = -2 + 1 = -1$

41. $(-3 + 5) + 2 = 2 + 2 = 4$

43. $-15 + (-4 + 12) = -15 + 8 = -7$

45. $\begin{array}{r} 5 \\ + \;\; -4 \\ \hline 1 \end{array}$

47. $\begin{array}{r} -1.3 \\ + \;\; 3.5 \\ \hline 2.2 \end{array}$

49. $8 - 4 = 8 + (-4) = 4$

51. $8 - (-4) = 8 + (+4) = 12$

53. $0 - (-5) = 0 + (+5) = 5$

55. $\frac{5}{3} - \frac{7}{6} = \frac{10}{6} - \frac{7}{6} = \frac{10}{6} + \left(-\frac{7}{6}\right) = \frac{3}{6} = \frac{1}{2}$

57. $\begin{array}{r} 8 \\ - \;\; 4 \\ \hline \end{array} \Rightarrow \begin{array}{r} 8 \\ + \;\; -4 \\ \hline 4 \end{array}$

59. $\begin{array}{r} -10 \\ - \;\; -3 \\ \hline \end{array} \Rightarrow \begin{array}{r} -10 \\ + \;\; 3 \\ \hline -7 \end{array}$

61. $+3 - [(-4) - 3] = +3 - [(-4) + (-3)] = +3 - [-7] = +3 + [+7] = 10$

63. $(5 - 3) + (3 - 5) = [5 + (-3)] + [3 + (-5)] = [2] + [-2] = 0$

65. $5 - [4 + (-2) - 5] = 5 - [4 + (-2) + (-5)] = 5 - [-3] = 5 + [+3] = 8$

67. $\frac{5 - (-4)}{3 - (-6)} = \frac{5 + (+4)}{3 + (+6)} = \frac{9}{9} = 1$

69. $\left(\frac{5}{2} - 3\right) - \left(\frac{3}{2} - 5\right) = \left(\frac{5}{2} - \frac{6}{2}\right) - \left(\frac{3}{2} - \frac{10}{2}\right) = \left(-\frac{1}{2}\right) - \left(-\frac{7}{2}\right) = \left(-\frac{1}{2}\right) + \left(+\frac{7}{2}\right)$
$= \frac{6}{2} = 3$

71. $(5.2 - 2.5) - (5.25 - 5) = [5.2 + (-2.5)] - [5.25 + (-5)] = 2.7 - [0.25] = 2.7 + (-0.25)$
$= 2.45$

Problems 73-75 are to be solved using a calculator. The keystrokes needed to solve each problem using a TI-84 graphing calculator appear in each solution. There may be other solutions. Keystrokes for other calculators may be slightly different.

73. [2] [.] [3] [4] [−] [3] [.] [4] [7] [+] [.] [7] [2] [ENTER]
$\{-0.41\} \Rightarrow -0.41$

75. [2] [.] [3] [4] [x^2] [−] [3] [.] [4] [7] [x^2] [−] [.] [7] [2] [x^2] [ENTER]
$\{-7.0837\} \Rightarrow -7.08$

77. $9 + (-11) = -(11 - 9) = -2$

79. $[-4 + (-3)] + [2 + (-2)] = [-7] + [0]$
$= -7$

81. $-4+(-3+2)+(-3)=-4+(-1)+(-3)=-5+(-3)=-8$

83. $-|8+(-4)|+7=-|4|+7=-4+7=3$

85. $-5.2+|-2.5+(-4)|=-5.2+|-6.5|$
$=-5.2+6.5=1.3$

87. $-3\frac{1}{2}-5\frac{1}{4}=-\frac{7}{2}-\frac{21}{4}=-\frac{14}{4}-\frac{21}{4}=-\frac{14}{4}+\left(-\frac{21}{4}\right)=-\frac{35}{4}=-8\frac{3}{4}$

89. $-6.7-(-2.5)=-6.7+(+2.5)=-4.2$

91. $\frac{-4-2}{-[2+(-3)]}=\frac{-4+(-2)}{-[-1]}=\frac{-6}{+1}=-6$

93. $\left(\frac{3}{4}-\frac{4}{5}\right)-\left(\frac{2}{3}+\frac{1}{4}\right)=\left(\frac{15}{20}-\frac{16}{20}\right)-\left(\frac{8}{12}+\frac{3}{12}\right)=\left(-\frac{1}{20}\right)-\left(\frac{11}{12}\right)=\left(-\frac{3}{60}\right)-\left(\frac{55}{60}\right)$
$=-\left(\frac{3}{60}+\frac{55}{60}\right)$
$=-\frac{58}{60}=-\frac{29}{30}$

95. $(-575)+(+400)=-175$
She still owes $175.

97. $(+13)+(-4)=+9$

99. $(-14)+10=-4°$

101. 1700 – (–300) = 2000 years

103. (–2,300) + (1,750) + (1,875) = +1,325 m

105. 32,000 – 28,000 = 4,000 ft

107. +32 – (+27) = 5°

109. 12,153 – 23 + 57 = 12,187

111. 500 · 2 – 300 = 1000 – 300 = 700 shares

113. 437.45 + 25.17 + 37.93 + 45.26 – 17.13 – 83.44 – 22.58 = $422.66

115. 115,000 – 78 – 446 – 216 – 7,612.32 – 23,445.11 + 223 = $83,425.57

117. **Answers may vary.**

119. The answers agree if the two numbers have the same sign. The answers do not agree if the numbers have opposite signs.

Exercise 1.5 (page 56)

1. $1(-3)=-3$

3. $-2(3)(-4)=-6(-4)=24$

5. $\frac{-12}{6}=-2$

7. $\frac{3(6)}{-2}=\frac{18}{-2}=-9$

SECTION 1.5

9. $12 \div 4(-3) = 3(-3) = -9$

11. $30 \cdot 37\frac{1}{2} = \frac{30}{1} \cdot \frac{75}{2} = \frac{15 \cdot \overset{1}{\cancel{2}} \cdot 75}{1 \cdot \underset{1}{\cancel{2}}} = 1{,}125$ lb

13. $3^3 - 8(3)^2 = 27 - 8(9) = 27 - 72 = -45$

15. positive

17. positive

19. positive

21. a

23. 0

25. $(+6)(+8) = 48$

27. $(-8)(-7) = 56$

29. $(+12)(-12) = -144$

31. $(-32)(-14) = 448$

33. $(-2)(3)(4) = (-6)(4) = -24$

35. $(-2)^2 = (-2)(-2) = 4$

37. $(-4)^3 = (-4)(-4)(-4) = (+16)(-4) = -64$

39. $(3)(-4)(-6) = (-12)(-6) = 72$

41. $2 + (-1)(-3) = 2 + 3 = 5$

43. $(-1 + 2)(-3) = 1(-3) = -3$

45. $[-1 - (-3)][-1 + (-3)] = [-1 + 3][-4] = [2][-4] = -8$

47. $-1(2) + (2)(-3)^2 = -2 + 2(9) = -2 + 18 = 16$

49. $\left(\frac{1}{2}\right)(-32) = -\frac{1}{2} \cdot \frac{32}{1} = -\frac{32}{2} = -16$

51. $\left(-\frac{3}{4}\right)\left(-\frac{8}{3}\right) = +\frac{3}{4} \cdot \frac{8}{3} = \frac{24}{12} = 2$

53. $\frac{80}{-20} = -4$

55. $\frac{-110}{-55} = 2$

57. $\frac{-160}{40} = -4$

59. $\frac{320}{-16} = -20$

61. $\frac{3(4)}{-2} = \frac{12}{-2} = -6$

63. $\frac{5(-18)}{3} = \frac{-90}{3} = -30$

65. $\frac{8 - 12}{-2} = \frac{-4}{-2} = 2$

67. $\frac{20 - 25}{7 - 12} = \frac{-5}{-5} = 1$

69. $\frac{6 - 2(3)}{-3(8 - 4)} = \frac{6 - 6}{-3(4)} = \frac{0}{-12} = 0$

71. $\frac{-4(3) - 5}{3(2) - 6} = \frac{-12 - 5}{6 - 6} = \frac{-17}{0} \Rightarrow$ undefined

Problems 73-75 are to be solved using a calculator. The keystrokes needed to solve each problem using a TI-84 graphing calculator appear in each solution. There may be other solutions. Keystrokes for other calculators may be slightly different.

73. [( ] [(–)] [6] [+] [4] [×] [(–)] [3] [ )] [÷] [( ] [4] [–] [6] [ )] [ENTER] {9}

75.

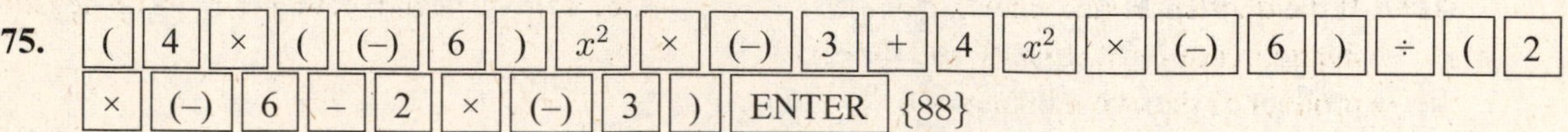

SECTION 1.5

77. $(-3)\left(-\frac{1}{3}\right) = +\frac{3}{1} \cdot \frac{1}{3} = \frac{3}{3} = 1$

79. $(-1)(2^3) = (-1)(8) = -8$

81. $(-2)(-2)(-2)(-3)(-4) = (+4)(-2)(-3)(-4) = (-8)(-3)(-4) = (+24)(-4) = -96$

83. $(2)(-5)(-6)(-7) = (-10)(-6)(-7) = (+60)(-7) = -420$

85. $(-7)^2 = (-7)(-7) = 49$

87. $-(-3)^2 = -(-3)(-3) = -(+9) = -9$

89. $(-1)^2[2-(-3)] = (-1)(-1)[2+(+3)]$
$= 1[5] = 5$

91. $(-3)(-1) - (-3)(2) = 3 - (-6) = 3 + 6$
$= 9$

93. $(-1)^3(-2)^2 + (-3)^2 = (-1)(-1)(-1)(-2)(-2) + (-3)(-3) = -4 + 9 = 5$

95. $\frac{4+(-18)}{-2} = \frac{-14}{-2} = 7$

97. $\frac{(-2)(5)(4)}{-3+1} = \frac{-40}{-2} = 20$

99. $\frac{1}{2} - \frac{2}{3} - \frac{3}{4} = \frac{6}{12} + \left(-\frac{8}{12}\right) + \left(-\frac{9}{12}\right)$
$= -\frac{11}{12}$

101. $\frac{1}{2} - \frac{2}{3} = \frac{3}{6} + \left(-\frac{4}{6}\right) = -\frac{1}{6}$

103. $\left(\frac{1}{2} - \frac{2}{3}\right)\left(\frac{1}{2} + \frac{2}{3}\right) = \left(\frac{3}{6} - \frac{4}{6}\right)\left(\frac{3}{6} + \frac{4}{6}\right) = \left(-\frac{1}{6}\right)\left(\frac{7}{6}\right) = -\frac{7}{36}$

105. $\left(\frac{1}{3} - \frac{1}{2}\right)\left(\frac{2}{3} - \frac{1}{2}\right) = \left(\frac{2}{6} - \frac{3}{6}\right)\left(\frac{4}{6} - \frac{3}{6}\right) = \left(-\frac{1}{6}\right)\left(\frac{1}{6}\right) = -\frac{1}{36}$

107. $(+2)(+3) = +6$

109. $(-350)(+15) = -5250$

111. $(+23)(-120) = -2{,}760$

113. $\frac{-18}{-3} = +6$

115. **a.** $75(-32) = -\$2400$ **b.** $57(-17) = -\$969$ **c.** $87(-12) = -\$1044$
d. $(-2400) + (-969) + (-1044) = -\$4413$

117. $\frac{(+26)+(+35)+(+17)+(-25)+(-31)+(-12)+(-24)}{7} = \frac{-14}{7} = -2$ per day

119. $613.50(18) = \$11{,}043 \Rightarrow$ enough \$

121. **Answers may vary.**

123. If the quotient is undefined, then the denominator must equal 0, and the product of the two numbers is 0.

125. If x^5 is negative, then x must be negative.

Exercise 1.6 (page 64)

1. $x + y = -2 + 3 = 1$

3. $7x + y = 7(-2) + 3 = -14 + 3 = -11$

5. $4x^2 = 4(-2)^2 = 4(4) = 16$

7. $-3x^2 = -3(-2)^2 = -3(4) = -12$

9. $0.14 \cdot 3{,}800 = 532$

11. $\dfrac{-4 + (7 - 9)}{(-9 - 7) + 4} = \dfrac{-4 + (-2)}{-16 + 4} = \dfrac{-6}{-12} = \dfrac{1}{2}$

13. sum

15. multiplication

17. algebraic

19. term, coefficient

21. $x + y$

23. $y - x$

25. xy

27. $3xy$

29. $\dfrac{y}{x}$

31. $\dfrac{3z}{4x}$

33. $x + y = (-2) + 5 = 3$

35. $xyz = (-2)(5)(-3) = (-10)(-3) = 30$

37. $\dfrac{yz}{x} = \dfrac{(5)(-3)}{-2} = \dfrac{-15}{-2} = \dfrac{15}{2}$

39. $\dfrac{3(x + z)}{y} = \dfrac{3[(-2) + (-3)]}{5} = \dfrac{3[-5]}{5} = \dfrac{-15}{5} = -3$

41. $\dfrac{x(y + z) - 25}{(x + z)^2 - y^2} = \dfrac{-2[5 + (-3)] - 25}{[-2 + (-3)]^2 - 5^2} = \dfrac{-2[2] - 25}{[-5]^2 - 25} = \dfrac{-4 - 25}{25 - 25} = \dfrac{-29}{0} \Rightarrow$ undefined

43. $\dfrac{3(x + z^2) + 4}{y(x - z)} = \dfrac{3\left[(-2) + (-3)^2\right] + 4}{5[(-2) - (-3)]} = \dfrac{3(-2 + 9) + 4}{5(1)} = \dfrac{3(7) + 4}{5} = \dfrac{21 + 4}{5} = \dfrac{25}{5} = 5$

45. $6d$: 1 term; coef. $= 6$

47. $-xy - 4t + 35$: 3 terms; coef. $= -1$

49. $3ab + bc - cd - ef$: 4 terms; coef. $= 3$

51. $-4xyz + 7xy - z$: 3 terms; coef. $= -4$

53. $3x + 4y + 2z + 2$: 4 terms; coef. $= 3$

55. $z + \dfrac{x}{y}$

57. $z - xy$

59. $\dfrac{xy}{x + z}$

61. $\dfrac{x - 4}{3y}$

63. the quotient obtained when the sum of 3 and x is divided by y

65. the product of x, y and the sum of x and y

67. the sum of x and 3

69. the quotient obtained when x is divided by y

71. the product of 2, x and y (or twice the product of x and y)

73. the quotient obtained when 5 is divided by the sum of x and y

75. $x + z = 8 + 2 = 10$

77. $y - z = 4 - 2 = 2$

79. $yz - 3 = (4)(2) - 3 = 5$

81. $\frac{xy}{z} = \frac{(8)(4)}{2} = 16$

83. 3rd term: $19x$; factors: $19, x$

85. x is common to the 1st and 3rd terms.

87. 1st term: $3xyz$; factors: $3, x, y, z$

89. 3rd term: $17xz$; factors: $17, x, z$

91. coefficients: 5, 1 and 8

93. x and y are common to the 1st and 3rd terms.

95. coefficients: 3, 1 and 25; $3 \cdot 1 \cdot 25 = 75$

97. x and y are common to the 1st and 3rd terms.

99. $c + 4$

101. **a.** $h - 20$ **b.** $c + 20$

103. $35{,}000n$

105. $500 - x$

107. $\$(3d + 5)$

109. $\frac{N(N-1)}{2} = \frac{10{,}000(10{,}000-1)}{2} = \frac{10{,}000(9{,}999)}{2} = \frac{99{,}990{,}000}{2} = 49{,}995{,}000$ comparisons

111-113. Answers may vary.

115. $37x \Rightarrow 37(2x)$
$37(2x) = 2(37x) \Rightarrow 37x$ is doubled.

Exercise 1.7 (page 73)

In exercises 1-6, there are many different possible answers. The answer listed is just an example.

1. $2(xy) = (2x)y$

3. $2(x + y) = 2x + 2y$

5. $5 - 3 = 2$
$3 - 5 = -2$

7. $x + y^2 \geq z$

9. $|x| \boxed{\geq} 0$

11. positive

13. real

15. $a + b = b + \underline{a}$

17. $(a + b) + c = a + \underline{(b + c)}$

19. $a(b + c) = ab + \underline{ac}$

21. $a \cdot 1 = \underline{a}$

23. element, multiplication

25. $a, \frac{1}{a}$, multiplicative

27. $x + y = 12 + (-2) = 10$

29. $xy = 12(-2) = -24$

31. $x^2 = 12^2 = 144$

33. $\frac{x}{y^2} = \frac{12}{(-2)^2} = \frac{12}{4} = 3$

35. $x + y = 5 + 7 = 12$
$y + x = 7 + 5 = 12$

37. $3x + 2y = 3(5) + 2(7) = 15 + 14 = 29$
$2y + 3x = 2(7) + 3(5) = 14 + 15 = 29$

39. $x(x + y) = 5(5 + 7) = 5(12) = 60$
$(x + y)x = (5 + 7)5 = (12)5 = 60$

41. $x^2(yz^2) = 5^2[7(-1)^2] = 25[7(1)] = 25[7] = \boxed{175}$
$(x^2y)z^2 = [5^2(7)](-1)^2 = [25(7)](1) = [175](1) = \boxed{175}$

43. $4(x+2) = 4x+8$

45. $2(z-3) = 2z-6$

47. $3(x+y) = 3x+3y$

49. $x(x+3) = x \cdot x + x \cdot 3 = x^2 + 3x$

51. $-x(a+b) = (-x)a + (-x)b = -ax - bx$

53. $-4(x^2+x+2) = (-4)x^2 + (-4)x + (-4)2 = -4x^2 - 4x - 8$

55. additive inverse: -2
multiplicative inverse: $\frac{1}{2}$

57. additive inverse: $-\frac{1}{3}$
multiplicative inverse: 3

59. additive inverse: 0
multiplicative inverse: none

61. additive inverse: $\frac{5}{2}$
multiplicative inverse: $-\frac{2}{5}$

63. additive inverse: 0.2
multiplicative inverse: -5

65. additive inverse: $-\frac{4}{3}$
multiplicative inverse: $\frac{3}{4}$

67. $3(x+2) = 3x + 3(2)$

69. $y^2x = xy^2$

71. $(x+y)z = (y+x)z$

73. $(xy)z = x(yz)$

75. $(x+y)+z = [2+(-3)]+1 = -1+1 = \boxed{0}$; $x+(y+z) = 2+(-3+1) = 2+(-2) = \boxed{0}$

77. $(xz)y = [2(1)](-3) = [2](-3) = \boxed{-6}$; $x(yz) = 2[-3(1)] = 2[-3] = \boxed{-6}$

79. $-5(t+2) = -5t + (-5)(2) = -5t - 10$

81. $-2a(x-a) = -2ax + (-2a)(-a)$
$= -2ax + 2a^2$

83. commutative property of addition

85. commutative property of multiplication

87. distributive property

89. commutative property of addition

91. multiplication identity property

93. additive inverse property

95. addition identity property

97. **Answers may vary.**

99. Closure for addition would not be true (odd number plus odd number equals even number).
Closure for multiplication would be true (odd number times odd number equals odd number).
There would be no additive identity (0 is an even number).
There would be a multiplicative identity, 1 (1 is an odd number).

Chapter 1 Review (page 75)

1. natural: $1, 2, 3, 4, 5$

2. prime: $2, 3, 5$

3. odd, natural: $1, 3, 5$

CHAPTER 1 REVIEW

4. composite: 4

5. integers: $-6, 0, 5$

6. rational: $-6, -\frac{2}{3}, 0, 2.6, 5$

7. prime: 5

8. real: $-6, -\frac{2}{3}, 0, \sqrt{2}, 2.6, \pi, 5$

9. even integers: $-6, 0$

10. odd integers: 5

11. irrational: $\sqrt{2}, \pi$

12. negative numbers: $-6, -\dfrac{2}{3}$

13. $-5 \ \boxed{\phantom{<}}\ 12 - 12$
$-5 \ \boxed{<}\ 0$

14. $\dfrac{24}{6} \ \boxed{\phantom{<}}\ 5$
$4 \ \boxed{<}\ 5$

15. $13 - 13 \ \boxed{}\ 5 - \dfrac{25}{5}$
$0 \ \boxed{}\ 5 - 5$
$0 \ \boxed{=}\ 0$

16. $\dfrac{21}{7} \ \boxed{}\ -33$
$3 \ \boxed{>}\ -33$

17. $-(-8) = +8$

18. $-(12 - 4) = -(8) = -8$

19.

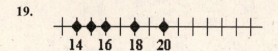

14 16 18 20

20.

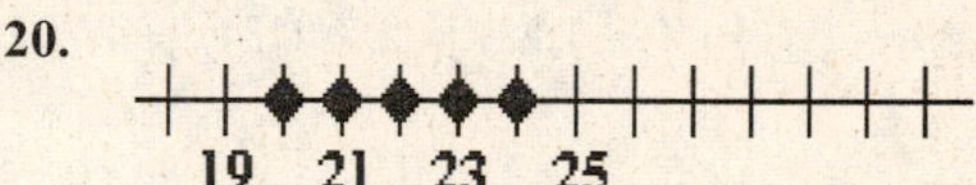

21.

-3 2

22.

-4 3

23. $|53 - 42| = |11| = 11$

24. $|-31| = 31$

25. $\dfrac{45}{27} = \dfrac{5 \cdot \cancel{9}^{1}}{3 \cdot \cancel{9}_{1}} = \dfrac{5}{3}$

26. $\dfrac{121}{11} = \dfrac{11 \cdot \cancel{11}^{1}}{1 \cdot \cancel{11}_{1}} = \dfrac{11}{1} = 11$

27. $\dfrac{31}{15} \cdot \dfrac{10}{62} = \dfrac{\cancel{31}^{1} \cdot \cancel{2}^{1} \cdot \cancel{5}^{1}}{3 \cdot \cancel{5}_{1} \cdot \cancel{2}_{1} \cdot \cancel{31}_{1}} = \dfrac{1}{3}$

28. $\dfrac{25}{36} \cdot \dfrac{12}{15} \cdot \dfrac{3}{5} = \dfrac{\cancel{5}^{1} \cdot \cancel{5}^{1} \cdot \cancel{12}^{1} \cdot \cancel{3}^{1}}{\cancel{3}_{1} \cdot \cancel{12}_{1} \cdot 3 \cdot \cancel{5}_{1} \cdot \cancel{5}_{1}} = \dfrac{1}{3}$

29. $\dfrac{18}{21} \div \dfrac{6}{7} = \dfrac{18}{21} \cdot \dfrac{7}{6} = \dfrac{\cancel{3}^{1} \cdot \cancel{6}^{1} \cdot \cancel{7}^{1}}{\cancel{3}_{1} \cdot \cancel{7}_{1} \cdot \cancel{6}_{1}} = \dfrac{1}{1} = 1$

30. $\dfrac{14}{24} \div \dfrac{7}{12} \div \dfrac{2}{5} = \dfrac{14}{24} \cdot \dfrac{12}{7} \cdot \dfrac{5}{2}$
$= \dfrac{\cancel{2}^{1} \cdot \cancel{7}^{1} \cdot \cancel{12}^{1} \cdot 5}{\cancel{2}_{1} \cdot \cancel{12}_{1} \cdot \cancel{7}_{1} \cdot 2} = \dfrac{5}{2}$

31. $\dfrac{7}{12} + \dfrac{9}{12} = \dfrac{7 + 9}{12} = \dfrac{16}{12} = \dfrac{4 \cdot \cancel{4}^{1}}{3 \cdot \cancel{4}_{1}} = \dfrac{4}{3}$

32. $\dfrac{13}{24} - \dfrac{5}{24} = \dfrac{13 - 5}{24} = \dfrac{8}{24} = \dfrac{\cancel{8}^{1}}{3 \cdot \cancel{8}_{1}} = \dfrac{1}{3}$

CHAPTER 1 REVIEW

33. $\frac{1}{3}+\frac{1}{7}=\frac{1\cdot\mathbf{7}}{3\cdot\mathbf{7}}+\frac{1\cdot\mathbf{3}}{7\cdot\mathbf{3}}=\frac{7}{21}+\frac{3}{21}$
$=\frac{7+3}{21}=\frac{10}{21}$

34. $\frac{5}{7}+\frac{4}{9}=\frac{5\cdot\mathbf{9}}{7\cdot\mathbf{9}}+\frac{4\cdot\mathbf{7}}{9\cdot\mathbf{7}}=\frac{45}{63}+\frac{28}{63}$
$=\frac{45+28}{63}=\frac{73}{63}$

35. $\frac{2}{3}-\frac{1}{7}=\frac{2\cdot\mathbf{7}}{3\cdot\mathbf{7}}-\frac{1\cdot\mathbf{3}}{7\cdot\mathbf{3}}=\frac{14}{21}-\frac{3}{21}$
$=\frac{14-3}{21}=\frac{11}{21}$

36. $\frac{4}{5}-\frac{2}{3}=\frac{4\cdot\mathbf{3}}{5\cdot\mathbf{3}}-\frac{2\cdot\mathbf{5}}{3\cdot\mathbf{5}}=\frac{12}{15}-\frac{10}{15}$
$=\frac{12-10}{15}=\frac{2}{15}$

37. $3\frac{2}{3}+5\frac{1}{4}=\frac{11}{3}+\frac{21}{4}=\frac{11\cdot\mathbf{4}}{3\cdot\mathbf{4}}+\frac{21\cdot\mathbf{3}}{4\cdot\mathbf{3}}=\frac{44}{12}+\frac{63}{12}=\frac{44+63}{12}=\frac{107}{12}=8\frac{11}{12}$

38. $7\frac{5}{12}-4\frac{1}{2}=\frac{89}{12}-\frac{9}{2}=\frac{89}{12}-\frac{9\cdot\mathbf{6}}{2\cdot\mathbf{6}}=\frac{89}{12}-\frac{54}{12}=\frac{89-54}{12}=\frac{35}{12}=2\frac{11}{12}$

39. $32.71+15.9=48.61$

40. $27.92-14.93=12.99$

41. $5.3\cdot3.5=18.55$

42. $21.83\div5.9=3.7$

43. $2.7(4.92-3.18)=2.7(1.74)\approx4.70$

44. $\frac{3.3+2.5}{0.22}=\frac{5.8}{0.22}\approx26.36$

45. $\frac{12.5}{14.7-11.2}=\frac{12.5}{3.5}\approx3.57$

46. $(3-0.7)(3.63-2)=(2.3)(1.63)\approx3.75$

47. $17\frac{1}{2}+15\frac{3}{4}=17+\frac{2}{4}+15+\frac{3}{4}=32+\frac{5}{4}=32+1\frac{1}{4}=33\frac{1}{4}$
$100-33\frac{1}{4}=100-33-\frac{1}{4}=67-\frac{1}{4}=66\frac{3}{4}$ acres left

48. avg. $=\frac{5.2+4.7+9.5+8}{4}=\frac{27.4}{4}$
$=6.85$ hours

49. $0.15(380)=57$

50. Front/Back: $2(2.7+2.7+4.2)=2(9.6)=19.2$ ft
Top/Bottom: $2(1.2+1.2+4.2)=2(6.6)=13.2$ ft
Sides: $2(1.2+2.7)=2(3.9)=7.8$ ft
TOTAL $=19.2+13.2+7.8=40.2$ ft

51. $3^4=3\cdot3\cdot3\cdot3=81$

52. $\left(\frac{2}{3}\right)^2=\frac{2}{3}\cdot\frac{2}{3}=\frac{4}{9}$

53. $(0.5)^2=(0.5)(0.5)=0.25$

54. $5^2+2^3=5\cdot5+2\cdot2\cdot2=25+8=33$

55. $3^2+4^2=9+16=25$

56. $(3+4)^2=7^2=49$

57. $A=\frac{1}{2}bh=\frac{1}{2}\left(6\frac{1}{2}\text{ ft}\right)(7\text{ ft})=\frac{1}{2}\cdot\frac{13}{2}\cdot\frac{7}{1}\text{ ft}^2=\frac{91}{4}\text{ ft}^2=22\frac{3}{4}\text{ ft}^2$

58. $V = Bh = \pi r^2 h = \pi\left(\frac{32.1}{2}\text{ ft}\right)^2(18.7\text{ ft}) = \pi(257.6025\text{ ft}^2)(18.7\text{ ft}) \approx 15{,}133.6\text{ ft}^3$

59. $5 + 3^3 = 5 + 27 = 32$

60. $7 \cdot 2 - 7 = 14 - 7 = 7$

61. $4 + (8 \div 4) = 4 + 2 = 6$

62. $(4 + 8) \div 4 = 12 \div 4 = 3$

63. $5^3 - \frac{81}{3} = 125 - 27 = 98$

64. $(5-2)^2 + 5^2 + 2^2 = 3^2 + 5^2 + 2^2 = 9 + 25 + 4 = 38$

65. $\frac{4 \cdot 3 + 3^4}{31} = \frac{12 + 81}{31} = \frac{93}{31} = 3$

66. $\frac{4}{3} \cdot \frac{9}{2} + \frac{1}{2} \cdot 18 = \frac{2 \cdot \overset{1}{\cancel{2}} \cdot 3 \cdot \overset{1}{\cancel{3}}}{\underset{1}{\cancel{3}} \cdot \underset{1}{\cancel{2}}} + \frac{1 \cdot 9 \cdot \overset{1}{\cancel{2}}}{\underset{1}{\cancel{2}} \cdot 1} = \frac{6}{1} + \frac{9}{1} = 15$

67. $8^2 - 6 = 64 - 6 = 58$

68. $(8-6)^2 = 2^2 = 4$

69. $\frac{6+8}{6-4} = \frac{14}{2} = 7$

70. $\frac{6(8) - 12}{4 + 8} = \frac{48 - 12}{12} = \frac{36}{12} = 3$

71. $2^2 + 2(3)^2 = 4 + 2(9) = 4 + 18 = 22$

72. $\frac{2^2 + 3}{2^3 - 1} = \frac{4 + 3}{8 - 1} = \frac{7}{7} = 1$

73. $(+7) + (+8) = +(7 + 8) = 15$

74. $(-25) + (-32) = -(25 + 32) = -57$

75. $(-2.7) + (-3.8) = -(2.7 + 3.8) = -6.5$

76. $\frac{1}{3} + \frac{1}{6} = \frac{2}{6} + \frac{1}{6} = \frac{3}{6} = \frac{1}{2}$

77. $(+12) + (-24) = -(24 - 12) = -12$

78. $(-44) + (+60) = +(60 - 44) = 16$

79. $3.7 + (-2.5) = +(3.7 - 2.5) = 1.2$

80. $-5.6 + (+2.06) = -(5.6 - 2.06) = -3.54$

81. $15 - (-4) = 15 + (+4) = 19$

82. $-12 - (-13) = -12 + (+13) = 1$

83. $[-5 + (-5)] - (-5) = [-10] + (+5) = -5$

84. $1 - [5 - (-3)] = 1 - [5 + (+3)] = 1 - [8] = -7$

85. $\frac{5}{6} - \left(-\frac{2}{3}\right) = \frac{5}{6} + \frac{2}{3} = \frac{5}{6} + \frac{4}{6} = \frac{9}{6} = \frac{3}{2}$

86. $\frac{2}{3} - \left(\frac{1}{3} - \frac{2}{3}\right) = \frac{2}{3} - \left(-\frac{1}{3}\right) = \frac{2}{3} + \frac{1}{3} = \frac{3}{3} = 1$

87. $\left|\frac{3}{7}-\left(-\frac{4}{7}\right)\right|=\left|\frac{3}{7}+\frac{4}{7}\right|=\left|\frac{7}{7}\right|=|1|=1$

88. $\frac{3}{7}-\left|-\frac{4}{7}\right|=\frac{3}{7}-\left(+\frac{4}{7}\right)$
$=\frac{3}{7}+\left(-\frac{4}{7}\right)=-\frac{1}{7}$

89. $(+3)(+4)=12$

90. $(-5)(-12)=60$

91. $\left(-\frac{3}{14}\right)\left(-\frac{7}{6}\right)=+\frac{3}{14}\cdot\frac{7}{6}$
$=\frac{\overset{1}{\cancel{3}}\cdot\overset{1}{\cancel{7}}}{2\cdot\underset{1}{\cancel{7}}\cdot 2\cdot\underset{1}{\cancel{3}}}=\frac{1}{4}$

92. $(3.75)(0.37)=1.3875$

93. $5(-7)=-35$

94. $(-15)(7)=-105$

95. $\left(-\frac{1}{2}\right)\left(\frac{4}{3}\right)=-\frac{1}{2}\cdot\frac{4}{3}=-\frac{1\cdot 2\cdot\overset{1}{\cancel{2}}}{\underset{1}{\cancel{2}}\cdot 3}=-\frac{2}{3}$

96. $(-12.2)(3.7)=-45.14$

97. $\frac{+25}{+5}=5$

98. $\frac{-14}{-2}=7$

99. $\frac{(-2)(-7)}{4}=\frac{+14}{4}=+\frac{7\cdot\overset{1}{\cancel{2}}}{2\cdot\underset{1}{\cancel{2}}}=\frac{7}{2}$

100. $\frac{-22.5}{-3.75}=6$

101. $\frac{-25}{5}=-5$

102. $\frac{(-3)(-4)}{(-6)}=\frac{+12}{-6}=-2$

103. $\left(\frac{-10}{2}\right)^2-(-1)^3=(-5)^2-(-1)^3$
$=25-(-1)$
$=25+1=26$

104. $\frac{[-3+(-4)]^2}{10+(-3)}=\frac{[-7]^2}{7}=\frac{49}{7}=7$

105. $\left(\frac{-3+(-3)}{3}\right)\left(\frac{-15}{5}\right)=\left(\frac{-6}{3}\right)\left(\frac{-15}{5}\right)$
$=(-2)(-3)=6$

106. $\frac{-2-(-8)}{5+(-1)}=\frac{-2+(+8)}{4}=\frac{6}{4}=\frac{3}{2}$

107. xz

108. $x+2y$

109. $2(x+y)$

110. $x-yz$

111. the product of 3, x and y

112. 5 decreased by the product of y and z

113. 5 less than the product of y and z

114. the quotient obtained when the sum of x, y and z is divided by twice their product

115. $y+z=-3+(-1)=-4$

116. $x+y=2+(-3)=-1$

117. $x + (y + z) = 2 + [-3 + (-1)]$
$= 2 + [-4] = -2$

118. $x - y = 2 - (-3) = 2 + (+3) = 5$

119. $x - (y - z) = 2 - [-3 - (-1)]$
$= 2 - [-3 + (+1)]$
$= 2 - [-2] = 2 + (+2) = 4$

120. $(x - y) - z = [2 - (-3)] - (-1)$
$= [2 + (+3)] + (+1)$
$= 5 + 1 = 6$

121. $xy = (2)(-3) = -6$

122. $yz = (-3)(-1) = 3$

123. $x(x + z) = 2[2 + (-1)] = 2[1] = 2$

124. $xyz = (2)(-3)(-1) = (-6)(-1) = 6$

125. $y^2z + x = (-3)^2(-1) + 2$
$= 9(-1) + 2 = -9 + 2 = -7$

126. $yz^3 + (xy)^2 = (-3)(-1)^3 + [2(-3)]^2$
$= (-3)(-1) + [-6]^2$
$= 3 + 36 = 39$

127. $\dfrac{xy}{z} = \dfrac{(2)(-3)}{(-1)} = \dfrac{-6}{-1} = 6$

128. $\dfrac{|xy|}{3z} = \dfrac{|2(-3)|}{3(-1)} = \dfrac{|-6|}{-3} = \dfrac{6}{-3} = -2$

129. three terms

130. 7

131. 1

132. $2 + 4 + 3 = 9$

133. closure property of addition

134. commutative property of multiplication

135. associative property of addition

136. distributive property

137. commutative property of addition

138. associative property of multiplication

139. commutative property of addition

140. multiplicative identity property

141. additive inverse property

142. additive identity property

Chapter 1 Test (page 82)

1. $31, 37, 41, 43, 47$

2. 2

3.

2 4 6 8 10

4.

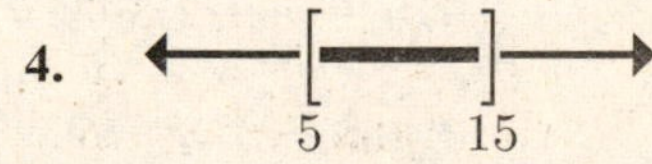

5. $-|23| = -(+23) = -23$

6. $-|7| + |-7| = -(+7) + (+7) = -7 + 7 = 0$

CHAPTER 1 TEST

7. $3(4-2) \ \square \ -2(2-5)$
$3(2) \ \square \ -2(-3)$
$6 \ \boxed{=} \ 6$

8. $1+4\cdot 3 \ \square \ -2(-7)$
$1+12 \ \square \ +14$
$13 \ \boxed{<} \ 14$

9. $25\% \text{ of } 136 \ \square \ \frac{1}{2} \text{ of } 66$
$0.25(136) \ \square \ \frac{1}{2}(66)$
$34 \ \boxed{>} \ 33$

10. $-13.7 \ \square \ -|-13.7|$
$-13.7 \ \square \ -(+13.7)$
$-13.7 \ \boxed{=} \ -13.7$

11. $\frac{26}{40} = \frac{13\cdot \cancel{2}^{1}}{20\cdot \cancel{2}_{1}} = \frac{13}{20}$

12. $\frac{7}{8}\cdot\frac{24}{21} = \frac{\cancel{7}\cdot\cancel{3}\cdot\cancel{8}}{\cancel{8}\cdot\cancel{3}\cdot\cancel{7}} = \frac{1}{1} = 1$

13. $\frac{18}{35} \div \frac{9}{14} = \frac{18}{35}\cdot\frac{14}{9} = \frac{2\cdot\cancel{9}\cdot 2\cdot\cancel{7}}{5\cdot\cancel{7}\cdot\cancel{9}} = \frac{4}{5}$

14. $\frac{24}{16} + 3 = \frac{3\cdot\cancel{8}}{2\cdot\cancel{8}} + \frac{3}{1} = \frac{3}{2} + \frac{3\cdot\mathbf{2}}{1\cdot\mathbf{2}} = \frac{3}{2} + \frac{6}{2} = \frac{3+6}{2} = \frac{9}{2}\left(\text{or } 4\frac{1}{2}\right)$

15. $\frac{17-5}{36} - \frac{2(13-5)}{12} = \frac{12}{36} - \frac{2(8)}{12} = \frac{12}{36} - \frac{16}{12} = \frac{\cancel{12}}{3\cdot\cancel{12}} - \frac{4\cdot\cancel{4}}{3\cdot\cancel{4}} = \frac{1}{3} - \frac{4}{3} = \frac{1-4}{3} = \frac{-3}{3} = -1$

16. $\frac{|-7-(-6)|}{-7-|-6|} = \frac{|-7+(+6)|}{-7-(+6)} = \frac{|-1|}{-7+(-6)} = \frac{1}{-13} = -\frac{1}{13}$

17. $0.17(457) = 77.69 \approx 77.7$

18. $A = lw = (23.56 \text{ ft})(12.8 \text{ ft}) = 301.568 \text{ ft}^2 \approx 301.57 \text{ ft}^2$

19. $A = \frac{1}{2}bh = \frac{1}{2}(16 \text{ cm})(8 \text{ cm}) = \frac{1}{2}(128 \text{ cm}^2) = 64 \text{ cm}^2$

20. $V = Bh = \pi r^2 h = \pi(7 \text{ in.})^2(10 \text{ in.}) = \pi(49 \text{ in.}^2)(10 \text{ in.}) = \pi(490 \text{ in.}^3) \approx 1{,}539 \text{ in.}^3$

21. $xy + z = (-2)(3) + 4 = -6 + 4 = -2$

22. $x(y+z) = -2(3+4) = -2(7) = -14$

23. $\frac{z+4y}{2x} = \frac{4+4(3)}{2(-2)} = \frac{4+12}{-4} = \frac{16}{-4} = -4$

24. $|x^3 - z| = |(-2)^3 - 4| = |-8-4| = |-12| = 12$

25. $x^3 + y^2 + z = (-2)^3 + (3)^2 + 4 = -8 + 9 + 4 = 5$

26. $|x| - 3|y| - 4|z| = |-2| - 3|3| - 4|4| = 2 - 3(3) - 4(4) = 2 - 9 - 16 = -23$

27. $\frac{xy}{x+y}$

28. $5y - (x+y)$

29. $x(12+12)+y(7+7)=24x+14y$

30. $\$(12a+8b)$

31. 3

32. 4 terms

33. $3(x+2)=3x+6$

34. $-p(r-t)=(-p)r+(-p)(-t)=-pr+pt$

35. 0

36. 5

37. commutative property of multiplication

38. distributive property

39. commutative property of addition

40. multiplicative inverse property

Exercise 2.1 (page 95)

1.
$$\begin{aligned} x-9&=11\\ x-9+\mathbf{9}&=11+\mathbf{9}\\ x&=20 \end{aligned}$$

3.
$$\begin{aligned} w+5&=7\\ w+5-\mathbf{5}&=7-\mathbf{5}\\ w&=2 \end{aligned}$$

5.
$$\begin{aligned} 3x&=3\\ \frac{3x}{3}&=\frac{3}{3}\\ x&=1 \end{aligned}$$

7.
$$\begin{aligned} \frac{x}{5}&=2\\ \mathbf{5}\cdot\frac{x}{5}&=\mathbf{5}\cdot 2\\ x&=10 \end{aligned}$$

9. $\frac{4}{5}+\frac{2}{3}=\frac{4\cdot 3}{5\cdot 3}+\frac{2\cdot 5}{3\cdot 5}=\frac{12}{15}+\frac{10}{15}=\frac{22}{15}$

11. $\frac{5}{9}\div\frac{3}{5}=\frac{5}{9}\cdot\frac{5}{3}=\frac{25}{27}$

13. $2+3\cdot 4=2+12=14$

15. $3+4^3(-5)=3+64(-5)=3+(-320)$
$=-317$

17. equation, expression

19. equivalent

21. equal

23. equal

25. regular price

27. 100

29. $x=2$; equation

31. $6x+7$; expression

33. $x+7=0$; equation

35. $3(x-4)$; expression

37.
$$\begin{aligned} x+2&=3\\ 1+2&\stackrel{?}{=}3\\ 3&=3 \end{aligned}$$
1 is a solution.

39.
$$\begin{aligned} a-7&=0\\ -7-7&\stackrel{?}{=}0\\ -14&\neq 0 \end{aligned}$$
-7 is not a solution.

41.
$$\begin{aligned} \frac{y}{7}&=4\\ \frac{28}{7}&\stackrel{?}{=}4\\ 4&=4 \end{aligned}$$
28 is a solution.

43.
$$\frac{x}{5} = x$$
$$\frac{0}{5} \stackrel{?}{=} 0$$
$$0 = 0$$
0 is a solution.

45.
$$3k + 5 = 5k - 1$$
$$3(3) + 5 \stackrel{?}{=} 5(3) - 1$$
$$9 + 5 \stackrel{?}{=} 15 - 1$$
$$14 = 14$$
3 is a solution.

47.
$$\frac{5+x}{10} - x = \frac{1}{2}$$
$$\frac{5+0}{10} - 0 \stackrel{?}{=} \frac{1}{2}$$
$$\frac{5}{10} - 0 \stackrel{?}{=} \frac{1}{2}$$
$$\tfrac{1}{2} = \tfrac{1}{2}$$
0 is a solution.

49.
$$y - 7 = 12$$
$$y - 7 \mathbf{+ 7} = 12 \mathbf{+ 7}$$
$$y = 19$$
$$y - 7 = 12$$
$$19 - 7 \stackrel{?}{=} 12$$
$$12 = 12$$

51.
$$a - 4 = -12$$
$$a - 4 \mathbf{+ 4} = -12 \mathbf{+ 4}$$
$$a = -8$$
$$a - 4 = -12$$
$$-8 - 4 \stackrel{?}{=} -12$$
$$-12 = -12$$

53.
$$p - 404 = 115$$
$$p - 404 \mathbf{+ 404} = 115 \mathbf{+ 404}$$
$$p = 519$$
Check:
$$p - 404 = 115$$
$$519 - 404 \stackrel{?}{=} 115$$
$$115 = 115$$

55.
$$r - \frac{1}{5} = \frac{3}{10}$$
$$r - \frac{1}{5} \mathbf{+ \frac{1}{5}} = \frac{3}{10} \mathbf{+ \frac{1}{5}}$$
$$r = \frac{3}{10} + \frac{2}{10} = \frac{5}{10} = \frac{1}{2}$$
Check:
$$r - \frac{1}{5} = \frac{3}{10}$$
$$\frac{1}{2} - \frac{1}{5} \stackrel{?}{=} \frac{3}{10}$$
$$\frac{5}{10} - \frac{2}{10} \stackrel{?}{=} \frac{3}{10}$$
$$\tfrac{3}{10} = \tfrac{3}{10}$$

57.
$$x + 7 = 13$$
$$x + 7 \mathbf{- 7} = 13 \mathbf{- 7}$$
$$x = 6$$
$$x + 7 = 13$$
$$6 + 7 \stackrel{?}{=} 13$$
$$13 = 13$$

59.
$$b + 3 = -10$$
$$b + 3 \mathbf{- 3} = -10 \mathbf{- 3}$$
$$b = -13$$
$$b + 3 = -10$$
$$-13 + 3 \stackrel{?}{=} -10$$
$$-10 = -10$$

61.
$$41 = 45 + q$$
$$41 \mathbf{- 45} = 45 \mathbf{- 45} + q$$
$$-4 = q$$
$$41 = 45 + q$$
$$41 \stackrel{?}{=} 45 + (-4)$$
$$41 = 41$$

63.
$$k + \frac{2}{3} = \frac{1}{5}$$
$$k + \frac{2}{3} \mathbf{- \frac{2}{3}} = \frac{1}{5} \mathbf{- \frac{2}{3}}$$
$$k = \frac{3}{15} - \frac{10}{15} = -\frac{7}{15}$$
Check:
$$k + \frac{2}{3} = \frac{1}{5}$$
$$-\frac{7}{15} + \frac{2}{3} \stackrel{?}{=} \frac{1}{5}$$
$$-\frac{7}{15} + \frac{10}{15} \stackrel{?}{=} \frac{1}{5}$$
$$\frac{3}{15} \stackrel{?}{=} \frac{1}{5}$$
$$\tfrac{1}{5} = \tfrac{1}{5}$$

65.
$$\frac{x}{5} = 5 \qquad \frac{x}{5} = 5$$
$$\mathbf{5} \cdot \frac{x}{5} = \mathbf{5} \cdot 5 \qquad \frac{25}{5} \stackrel{?}{=} 5$$
$$x = 25 \qquad 5 = 5$$

67.
$$\frac{b}{3} = 5 \qquad \frac{b}{3} = 5$$
$$\mathbf{3} \cdot \frac{b}{3} = \mathbf{3} \cdot 5 \qquad \frac{15}{3} \stackrel{?}{=} 5$$
$$b = 15 \qquad 5 = 5$$

69.
$$\frac{b}{3} = \frac{1}{3} \qquad \frac{b}{3} = \frac{1}{3}$$
$$\mathbf{3} \cdot \frac{b}{3} = \mathbf{3} \cdot \frac{1}{3} \qquad \frac{1}{3} = \frac{1}{3}$$
$$b = 1$$

71.
$$\frac{u}{5} = -\frac{3}{10} \qquad \frac{u}{5} = -\frac{3}{10}$$
$$\mathbf{10} \cdot \frac{u}{5} = \mathbf{10} \cdot \left(-\frac{3}{10}\right) \qquad \frac{-\frac{3}{2}}{5} \stackrel{?}{=} -\frac{3}{10}$$
$$2u = -3 \qquad \frac{-\frac{3}{2} \cdot 2}{5 \cdot 2} \stackrel{?}{=} -\frac{3}{10}$$
$$\frac{2u}{2} = \frac{-3}{2} \qquad \frac{-3}{10} = -\frac{3}{10}$$
$$u = -\frac{3}{2}$$

73.
$$6x = 18 \qquad 6x = 18$$
$$\frac{6x}{6} = \frac{18}{6} \qquad 6(3) \stackrel{?}{=} 18$$
$$x = 3 \qquad 18 = 18$$

75.
$$11x = -121 \qquad 11x = -121$$
$$\frac{11x}{11} = \frac{-121}{11} \qquad 11(-11) \stackrel{?}{=} -121$$
$$x = -11 \qquad -121 = -121$$

77.
$$-4x = 36 \qquad -4x = 36$$
$$\frac{-4x}{-4} = \frac{36}{-4} \qquad -4(-9) \stackrel{?}{=} 36$$
$$x = -9 \qquad 36 = 36$$

79.
$$4w = 108 \qquad 4w = 108$$
$$\frac{4w}{4} = \frac{108}{4} \qquad 4(27) \stackrel{?}{=} 108$$
$$w = 27 \qquad 108 = 108$$

81.
$$5x = \frac{5}{8} \qquad 5x = \frac{5}{8}$$
$$\frac{1}{5} \cdot 5x = \frac{1}{5} \cdot \frac{5}{8} \qquad 5 \cdot \frac{1}{8} \stackrel{?}{=} \frac{5}{8}$$
$$x = \frac{1}{8} \qquad \frac{5}{8} = \frac{5}{8}$$

83.
$$\frac{1}{7}w = 14 \qquad \frac{1}{7}w = 14$$
$$\mathbf{7} \cdot \frac{1}{7}w = \mathbf{7} \cdot 14 \qquad \frac{1}{7}(98) \stackrel{?}{=} 14$$
$$w = 98 \qquad 14 = 14$$

85.
$$-1.2w = -102 \qquad -1.2w = -102$$
$$\frac{-1.2w}{-1.2} = \frac{-102}{-1.2} \qquad -1.2(85) \stackrel{?}{=} -102$$
$$w = 85 \qquad -102 = -102$$

87.
$$0.25x = 1228 \qquad 0.25x = 1228$$
$$\frac{0.25x}{0.25} = \frac{1228}{0.25} \qquad 0.25(4912) \stackrel{?}{=} 1228$$
$$x = 4912 \qquad 4912 = 4912$$

89. Let r = the regular price. Then

[Sale price] = [Regular price] − [Markdown]

$$7995 = r - 1350$$
$$7995 + \mathbf{1350} = r - 1350 + \mathbf{1350}$$
$$9345 = r$$

The regular price is \$9,345.

91. Let w = the wholesale cost. Then

[Retail price] = [Wholesale cost] + [Markup]

$$175 = w + 85$$
$$175 - \mathbf{85} = w + 85 - \mathbf{85}$$
$$90 = w$$

The wholesale cost is \$90.

93.
$$\begin{aligned} rb &= \boldsymbol{a} \\ 40\% \cdot 200 &= \boldsymbol{a} \\ 0.40(200) &= \boldsymbol{a} \\ 80 &= \boldsymbol{a} \end{aligned}$$

95.
$$\begin{aligned} rb &= \boldsymbol{a} \\ 50\% \cdot 38 &= \boldsymbol{a} \\ 0.50(38) &= \boldsymbol{a} \\ 19 &= \boldsymbol{a} \end{aligned}$$

97.
$$\begin{aligned} rb &= \boldsymbol{a} \\ 15\% \cdot b &= 48 \\ 0.15b &= 48 \\ \frac{0.15b}{0.15} &= \frac{48}{0.15} \\ b &= 320 \end{aligned}$$

99.
$$\begin{aligned} rb &= \boldsymbol{a} \\ 35\% \cdot b &= 133 \\ 0.35b &= 133 \\ \frac{0.35b}{0.35} &= \frac{133}{0.35} \\ b &= 380 \end{aligned}$$

101.
$$\begin{aligned} rb &= \boldsymbol{a} \\ 28\% \cdot b &= 42 \\ 0.28b &= 42 \\ \frac{0.28b}{0.28} &= \frac{42}{0.28} \\ b &= 150 \end{aligned}$$

103.
$$\begin{aligned} rb &= \boldsymbol{a} \\ r(357.5) &= 71.5 \\ \frac{r(357.5)}{357.5} &= \frac{71.5}{357.5} \\ r &= 0.20 \\ r &= 20\% \end{aligned}$$

105.
$$\begin{aligned} p + 0.27 &= 3.57 \\ p + 0.27 - \mathbf{0.27} &= 3.57 - \mathbf{0.27} \\ p &= 3.3 \end{aligned}$$

Check:
$$\begin{aligned} p + 0.27 &= 3.57 \\ 3.3 + 0.27 &\stackrel{?}{=} 3.57 \\ 3.57 &= 3.57 \end{aligned}$$

107.
$$\begin{aligned} \frac{x}{32} &= -2 \\ \mathbf{32} \cdot \frac{x}{32} &= \mathbf{32} \cdot (-2) \\ x &= -64 \end{aligned} \qquad \begin{aligned} \frac{x}{32} &= -2 \\ \frac{-64}{32} &\stackrel{?}{=} -2 \\ -2 &= -2 \end{aligned}$$

109.
$$\begin{aligned} -57 &= b - 29 \\ -57 + \mathbf{29} &= b - 29 + \mathbf{29} \\ -28 &= b \end{aligned}$$

Check:
$$\begin{aligned} -57 &= b - 29 \\ -57 &\stackrel{?}{=} -28 - 29 \\ -57 &= -57 \end{aligned}$$

111.
$$\begin{aligned} y - 2.63 &= -8.21 \\ y - 2.63 + \mathbf{2.63} &= -8.21 + \mathbf{2.63} \\ y &= -5.58 \end{aligned}$$

Check:
$$\begin{aligned} y - 2.63 &= -8.21 \\ -5.58 - 2.63 &\stackrel{?}{=} -8.21 \\ -8.21 &= -8.21 \end{aligned}$$

113.
$$\begin{aligned} \frac{y}{-3} &= -\frac{5}{6} \\ \mathbf{-6} \cdot \frac{y}{-3} &= \mathbf{-6} \cdot \left(-\frac{5}{6}\right) \\ 2y &= 5 \\ \frac{2y}{2} &= \frac{5}{2} \\ y &= \frac{5}{2} \end{aligned} \qquad \begin{aligned} \frac{y}{-3} &= -\frac{5}{6} \\ \frac{\frac{5}{2}}{-3} &\stackrel{?}{=} -\frac{5}{6} \\ \frac{\frac{5}{2} \cdot 2}{-3 \cdot 2} &\stackrel{?}{=} -\frac{5}{6} \\ \frac{5}{-6} &= -\frac{5}{6} \end{aligned}$$

115.
$$\begin{aligned} -37 + w &= 37 \\ -37 + \mathbf{37} + w &= 37 + \mathbf{37} \\ w &= 74 \end{aligned}$$

Check:
$$\begin{aligned} -37 + w &= 37 \\ -37 + 74 &\stackrel{?}{=} 37 \\ 37 &= 37 \end{aligned}$$

117.
$$\begin{aligned} -3 &= \frac{x}{11} \\ \mathbf{11} \cdot (-3) &= \mathbf{11} \cdot \frac{x}{11} \\ -33 &= x \end{aligned} \qquad \begin{aligned} -3 &= \frac{x}{11} \\ -3 &\stackrel{?}{=} \frac{-33}{11} \\ -3 &= -3 \end{aligned}$$

119.

$$b+7=\frac{20}{3}$$
$$b+7-\mathbf{7}=\frac{20}{3}-\mathbf{7}$$
$$b=\frac{20}{3}-\frac{21}{3}=-\frac{1}{3}$$

Check:
$$b+7=\frac{20}{3}$$
$$-\frac{1}{3}+7\stackrel{?}{=}\frac{20}{3}$$
$$-\frac{1}{3}+\frac{21}{3}\stackrel{?}{=}\frac{20}{3}$$
$$\frac{20}{3}=\frac{20}{3}$$

121.

$$2x=\frac{1}{7}$$
$$\frac{1}{2}\cdot 2x=\frac{1}{2}\cdot\frac{1}{7}$$
$$x=\frac{1}{14}$$

$$2x=\frac{1}{7}$$
$$2\cdot\frac{1}{14}\stackrel{?}{=}\frac{1}{7}$$
$$\frac{2}{14}\stackrel{?}{=}\frac{1}{7}$$
$$\frac{1}{7}=\frac{1}{7}$$

123.

$$-\frac{3}{5}=x-\frac{2}{5}$$
$$-\frac{3}{5}\mathbf{+\frac{2}{5}}=x-\frac{2}{5}\mathbf{+\frac{2}{5}}$$
$$-\frac{1}{5}=x$$

Check:
$$-\frac{3}{5}=x-\frac{2}{5}$$
$$-\frac{3}{5}\stackrel{?}{=}-\frac{1}{5}-\frac{2}{5}$$
$$-\frac{3}{5}=-\frac{3}{5}$$

125.

$$\frac{1}{7}x=\frac{5}{7}$$
$$\mathbf{7}\cdot\frac{1}{7}x=\mathbf{7}\cdot\frac{5}{7}$$
$$x=5$$

$$\frac{1}{7}x=\frac{5}{7}$$
$$\frac{1}{7}(5)=\frac{5}{7}$$

127.

$$-32r=64$$
$$\frac{-32r}{-32}=\frac{64}{-32}$$
$$r=-2$$

$$-32r=64$$
$$-32(-2)\stackrel{?}{=}64$$
$$64=64$$

129.

$$18x=-9$$
$$\frac{18x}{18}=\frac{-9}{18}$$
$$x=-\frac{1}{2}$$

$$18x=-9$$
$$18\left(-\frac{1}{2}\right)\stackrel{?}{=}-9$$
$$-9=-9$$

131.

$$rb=\boldsymbol{a}$$
$$r(4)=0.32$$
$$\frac{r(4)}{4}=\frac{0.32}{4}$$
$$r=0.08$$
$$r=8\%$$

133.

$$rb=\boldsymbol{a}$$
$$r(17)=34$$
$$\frac{r(17)}{17}=\frac{34}{17}$$
$$r=2.00$$
$$r=200\%$$

135. Let b = the selling price

$$rb=a$$
$$0.05b=13.50$$
$$\frac{0.05b}{0.05}=\frac{13.50}{0.05}$$
$$b=270$$

The selling price is $270.

137. Let $r =$ the percentage for sales tax.

$$rb = a$$
$$r(12) = 0.72$$
$$\frac{r(12)}{12} = \frac{0.72}{12}$$
$$r = 0.06$$
$$r = 6\%$$

The sales tax is computed at a rate of 6%.

139.

$$A = p + i$$
$$5010 = 4750 + i$$
$$5010 - \mathbf{4750} = 4750 - \mathbf{4750} + i$$
$$260 = i$$

The deposit earned $260 in interest.

141. Let $x =$ the original number in the audience.

$$\frac{1}{3} \cdot \boxed{\text{Original audience}} = \boxed{\text{Number who left}}$$
$$\frac{1}{3}x = 78$$
$$\mathbf{3} \cdot \frac{1}{3}x = \mathbf{3} \cdot 78$$
$$x = 234$$

There were originally 234 in the audience.

143. Let $r =$ the percentage not pleased.

$$a = 9200 - 4140 = 5060$$
$$rb = a$$
$$r(9200) = 5060$$
$$\frac{r(9200)}{9200} = \frac{5060}{9200}$$
$$r = 0.55$$
$$r = 55\%$$

55% of those surveyed were not pleased.

145. Let $x =$ the number employed at the factory.

$$rb = a$$
$$0.90(x) = 2484$$
$$\frac{0.90x}{0.90} = \frac{2484}{0.90}$$
$$x = 2760$$

There are a total of 2,760 employees.

147. Let $x =$ the original number of shares owned.

$$1.5 \cdot \boxed{\text{Original \# of shares}} = \boxed{\text{\# of shares after split}}$$
$$1.5x = 555$$
$$\frac{1.5x}{1.5} = \frac{555}{1.5}$$
$$x = 370$$

The shareholder owned 370 shares.

149. Let $c =$ the original cost.

$$\boxed{\text{Original cost}} - \boxed{\text{Depreciation}} = \boxed{\text{New cost}}$$
$$c - 7500 = 10250$$
$$c - 7500 \mathbf{+ 7500} = 10250 \mathbf{+ 7500}$$
$$c = 17750$$

The original cost was $17,750.

151. Let $t =$ the tax paid.

$$\boxed{\text{Cost}} = \boxed{\text{Price}} + \boxed{\text{Tax}}$$
$$39.32 = 37.10 + t$$
$$39.32 - \mathbf{37.10} = 37.10 - \mathbf{37.10} + t$$
$$2.22 = t$$

The tax was $2.22.

153. Let $b =$ the cost of the brush. Then

$$\boxed{\text{Primer cost}} + \boxed{\text{Paint cost}} + \boxed{\text{Brush cost}} = \boxed{\text{Total cost}}$$
$$10.99 + 14.50 + b = 30.44$$
$$25.49 + b = 30.44$$
$$25.49 - \mathbf{25.49} + b = 30.44 - \mathbf{25.49}$$
$$b = 4.95$$

The cost of the brush was $4.95.

155. Let $c =$ the condominium price. Then

$$\boxed{\text{Condo price}} = \boxed{\text{House price}} - 57595$$
$$c = 202744 - 57595$$
$$c = 145149$$

The price of the condominium is $145,149.

157. Answers may vary.

159.
$$\begin{aligned} A_{\text{circle}} &= A_{\text{square}} \\ \pi r^2 &= s^2 \\ \pi(4.5)^2 &= 8^2 \\ \pi(20.25) &= 64 \\ \frac{\pi(20.25)}{20.25} &= \frac{64}{20.25} \\ \pi &\approx 3.16 \end{aligned}$$

Exercise 2.2 (page 105)

1. add 7

3. add 3

5. multiply by 3

7.
$$\begin{aligned} 7x - 7 &= 14 \\ 7x - 7 + 7 &= 14 + 7 \\ 7x &= 21 \\ \frac{7x}{7} &= \frac{21}{7} \\ x &= 3 \end{aligned}$$

9.
$$\begin{aligned} P &= 2l + 2w \\ &= 2(8.5\text{ cm}) + 2(16.5\text{ cm}) \\ &= 17\text{ cm} + 33\text{ cm} = 50\text{ cm} \end{aligned}$$

11. $A = \frac{1}{2}h(b+d) = \frac{1}{2}(8.5\text{ in.})(6.7\text{ in.} + 12.2\text{ in.}) = \frac{1}{2}(8.5\text{ in.})(18.9\text{ in.}) = \frac{1}{2}(160.65\text{ in.}^2) = 80.325\text{ in.}^2$

13. cost

15. percent

17. percent of increase

19.
$$\begin{aligned} 5x - 1 &= 4 \\ 5x - 1 + 1 &= 4 + 1 \\ 5x &= 5 \\ \frac{5x}{5} &= \frac{5}{5} \\ x &= 1 \end{aligned}$$
$$\begin{aligned} 5x - 1 &= 4 \\ 5(1) - 1 &\stackrel{?}{=} 4 \\ 5 - 1 &\stackrel{?}{=} 4 \\ 4 &= 4 \end{aligned}$$

21.
$$\begin{aligned} -6x + 2 &= 14 \\ -6x + 2 - 2 &= 14 - 2 \\ -6x &= 12 \\ \frac{-6x}{-6} &= \frac{12}{-6} \\ x &= -2 \end{aligned}$$
$$\begin{aligned} -6x + 2 &= 14 \\ -6(-2) + 2 &\stackrel{?}{=} 14 \\ 12 + 2 &\stackrel{?}{=} 14 \\ 14 &= 14 \end{aligned}$$

23.
$$\begin{aligned} 6x + 2 &= -4 \\ 6x + 2 - 2 &= -4 - 2 \\ 6x &= -6 \\ \frac{6x}{6} &= \frac{-6}{6} \\ x &= -1 \end{aligned}$$
$$\begin{aligned} 6x + 2 &= -4 \\ 6(-1) + 2 &\stackrel{?}{=} -4 \\ -6 + 2 &\stackrel{?}{=} -4 \\ -4 &= -4 \end{aligned}$$

25.
$$\begin{aligned} 3x - 8 &= 1 \\ 3x - 8 + 8 &= 1 + 8 \\ 3x &= 9 \\ \frac{3x}{3} &= \frac{9}{3} \\ x &= 3 \end{aligned}$$
$$\begin{aligned} 3x - 8 &= 1 \\ 3(3) - 8 &\stackrel{?}{=} 1 \\ 9 - 8 &\stackrel{?}{=} 1 \\ 1 &= 1 \end{aligned}$$

SECTION 2.2

27.

$$\begin{aligned} \frac{z}{9}+5&=-1 \\ \frac{z}{9}+5-5&=-1-5 \\ \frac{z}{9}&=-6 \\ 9\cdot\frac{z}{9}&=9(-6) \\ z&=-54 \end{aligned} \qquad \begin{aligned} \frac{z}{9}+5&=-1 \\ \frac{-54}{9}+5&\stackrel{?}{=}-1 \\ -6+5&\stackrel{?}{=}-1 \\ -1&=-1 \end{aligned}$$

29.

$$\begin{aligned} \frac{b}{3}+5&=2 \\ \frac{b}{3}+5-5&=2-5 \\ \frac{b}{3}&=-3 \\ 3\cdot\frac{b}{3}&=3(-3) \\ b&=-9 \end{aligned} \qquad \begin{aligned} \frac{b}{3}+5&=2 \\ \frac{-9}{3}+5&\stackrel{?}{=}2 \\ -3+5&\stackrel{?}{=}2 \\ 2&=2 \end{aligned}$$

31.

$$\begin{aligned} \frac{x}{3}-3&=-2 \\ \frac{x}{3}-3+3&=-2+3 \\ \frac{x}{3}&=1 \\ 3\cdot\frac{x}{3}&=3\cdot 1 \\ x&=3 \end{aligned} \qquad \begin{aligned} \frac{x}{3}-3&=-2 \\ \frac{3}{3}-3&\stackrel{?}{=}-2 \\ 1-3&\stackrel{?}{=}-2 \\ -2&=-2 \end{aligned}$$

33.

$$\begin{aligned} \frac{p}{11}+9&=6 \\ \frac{p}{11}+9-9&=6-9 \\ \frac{p}{11}&=-3 \\ 11\cdot\frac{p}{11}&=11(-3) \\ p&=-33 \end{aligned} \qquad \begin{aligned} \frac{p}{11}+9&=6 \\ \frac{-33}{11}+9&\stackrel{?}{=}6 \\ -3+9&\stackrel{?}{=}6 \\ 6&=6 \end{aligned}$$

35.

$$\begin{aligned} \frac{b+5}{3}&=11 \\ 3\cdot\frac{b+5}{3}&=3\cdot 11 \\ b+5&=33 \\ b+5-5&=33-5 \\ b&=28 \end{aligned} \qquad \begin{aligned} \frac{b+5}{3}&=11 \\ \frac{28+5}{3}&\stackrel{?}{=}11 \\ \frac{33}{3}&\stackrel{?}{=}11 \\ 11&=11 \end{aligned}$$

37.

$$\begin{aligned} \frac{r+7}{3}&=4 \\ 3\cdot\frac{r+7}{3}&=3\cdot 4 \\ r+7&=12 \\ r+7-7&=12-7 \\ r&=5 \end{aligned} \qquad \begin{aligned} \frac{r+7}{3}&=4 \\ \frac{5+7}{3}&\stackrel{?}{=}4 \\ \frac{12}{3}&\stackrel{?}{=}4 \\ 4&=4 \end{aligned}$$

39.

$$\begin{aligned} \frac{3x-12}{2}&=9 \\ 2\cdot\frac{3x-12}{2}&=2\cdot 9 \\ 3x-12&=18 \\ 3x-12+12&=18+12 \\ 3x&=30 \\ \frac{3x}{3}&=\frac{30}{3} \\ x&=10 \end{aligned} \qquad \begin{aligned} \frac{3x-12}{2}&=9 \\ \frac{3(10)-12}{2}&\stackrel{?}{=}9 \\ \frac{30-12}{2}&\stackrel{?}{=}9 \\ \frac{18}{2}&\stackrel{?}{=}9 \\ 9&=9 \end{aligned}$$

41.

$$\begin{aligned} \frac{5k-8}{9}&=1 \\ 9\cdot\frac{5k-8}{9}&=9\cdot 1 \\ 5k-8&=9 \\ 5k-8+8&=9+8 \\ 5k&=17 \\ \frac{5k}{5}&=\frac{17}{5} \\ k&=\frac{17}{5} \end{aligned} \qquad \begin{aligned} \frac{5k-8}{9}&=1 \\ \frac{5\cdot\frac{17}{5}-8}{9}&\stackrel{?}{=}1 \\ \frac{17-8}{9}&\stackrel{?}{=}1 \\ \frac{9}{9}&\stackrel{?}{=}1 \\ 1&=1 \end{aligned}$$

43.

$$\begin{aligned}\frac{k}{5}-\frac{1}{2}&=\frac{3}{2}\\ \frac{k}{5}-\frac{1}{2}+\frac{1}{2}&=\frac{3}{2}+\frac{1}{2}\\ \frac{k}{5}&=\frac{4}{2}\\ \frac{k}{5}&=2\\ 5\cdot\frac{k}{5}&=5\cdot 2\\ k&=10\end{aligned}$$

$$\begin{aligned}\frac{k}{5}-\frac{1}{2}&=\frac{3}{2}\\ \frac{10}{5}-\frac{1}{2}&\stackrel{?}{=}\frac{3}{2}\\ 2-\frac{1}{2}&\stackrel{?}{=}\frac{3}{2}\\ \frac{4}{2}-\frac{1}{2}&\stackrel{?}{=}\frac{3}{2}\\ \frac{3}{2}&=\frac{3}{2}\end{aligned}$$

45.

$$\begin{aligned}\frac{w}{16}+\frac{5}{4}&=1\\ \frac{w}{16}+\frac{5}{4}-\frac{5}{4}&=1-\frac{5}{4}\\ \frac{w}{16}&=\frac{4}{4}-\frac{5}{4}\\ \frac{w}{16}&=-\frac{1}{4}\\ 16\cdot\frac{w}{16}&=16\left(-\frac{1}{4}\right)\\ w&=-4\end{aligned}$$

$$\begin{aligned}\frac{w}{16}+\frac{5}{4}&=1\\ \frac{-4}{16}+\frac{5}{4}&\stackrel{?}{=}1\\ -\frac{1}{4}+\frac{5}{4}&\stackrel{?}{=}1\\ \frac{4}{4}&\stackrel{?}{=}1\\ 1&=1\end{aligned}$$

47.

$$\begin{aligned}\frac{3x}{2}-6&=9\\ \frac{3x}{2}-6+6&=9+6\\ \frac{3x}{2}&=15\\ 2\cdot\frac{3x}{2}&=2\cdot 15\\ 3x&=30\\ \frac{3x}{3}&=\frac{30}{3}\\ x&=10\end{aligned}$$

$$\begin{aligned}\frac{3x}{2}-6&=9\\ \frac{3(10)}{2}-6&\stackrel{?}{=}9\\ \frac{30}{2}-6&\stackrel{?}{=}9\\ 15-6&\stackrel{?}{=}9\\ 9&=9\end{aligned}$$

49.

$$\begin{aligned}\frac{3y}{2}+5&=11\\ \frac{3y}{2}+5-5&=11-5\\ \frac{3y}{2}&=6\\ 2\cdot\frac{3y}{2}&=2\cdot 6\\ 3y&=12\\ \frac{3y}{3}&=\frac{12}{3}\\ y&=4\end{aligned}$$

$$\begin{aligned}\frac{3y}{2}+5&=11\\ \frac{3(4)}{2}+5&\stackrel{?}{=}11\\ \frac{12}{2}+5&\stackrel{?}{=}11\\ 6+5&\stackrel{?}{=}11\\ 11&=11\end{aligned}$$

51.

$$\begin{aligned}\frac{2.4x+4.8}{1.2}&=8\\ 1.2\cdot\frac{2.4x+4.8}{1.2}&=1.2(8)\\ 2.4x+4.8&=9.6\\ 2.4x+4.8-4.8&=9.6-4.8\\ 2.4x&=4.8\\ \frac{2.4x}{2.4}&=\frac{4.8}{2.4}\\ x&=2\end{aligned}$$

$$\begin{aligned}\frac{2.4x+4.8}{1.2}&=8\\ \frac{2.4(2)+4.8}{1.2}&\stackrel{?}{=}8\\ \frac{4.8+4.8}{1.2}&\stackrel{?}{=}8\\ \frac{9.6}{1.2}&\stackrel{?}{=}8\\ 8&=8\end{aligned}$$

53.

$$\begin{aligned} \frac{2.1x-0.13}{0.8}+2.5&=0.5 \\ \frac{2.1x-0.13}{0.8}+2.5-2.5&=0.5-2.5 \\ \frac{2.1x-0.13}{0.8}&=-2 \\ 0.8\cdot\frac{2.1x-0.13}{0.8}&=0.8(-2) \\ 2.1x-0.13&=-1.6 \\ 2.1x-0.13+0.13&=-1.6+0.13 \\ 2.1x&=-1.47 \\ \frac{2.1x}{2.1}&=\frac{-1.47}{2.1} \\ x&=-0.7 \end{aligned}$$

$$\begin{aligned} \frac{2.1x-0.13}{0.8}+2.5&=0.5 \\ \frac{2.1(-0.7)-0.13}{0.8}+2.5&\stackrel{?}{=}0.5 \\ \frac{-1.47-0.13}{0.8}+2.5&\stackrel{?}{=}0.5 \\ \frac{-1.6}{0.8}+2.5&\stackrel{?}{=}0.5 \\ -2+2.5&\stackrel{?}{=}0.5 \\ 0.5&=0.5 \end{aligned}$$

55.

$$\begin{aligned} 11x+17&=-5 \\ 11x+17-17&=-5-17 \\ 11x&=-22 \\ \frac{11x}{11}&=\frac{-22}{11} \\ x&=-2 \end{aligned}$$

$$\begin{aligned} 11x+17&=-5 \\ 11(-2)+17&\stackrel{?}{=}-5 \\ -22+17&\stackrel{?}{=}-5 \\ -5&=-5 \end{aligned}$$

57.

$$\begin{aligned} 43p+72&=158 \\ 43p+72-72&=158-72 \\ 43p&=86 \\ \frac{43p}{43}&=\frac{86}{43} \\ p&=2 \end{aligned}$$

$$\begin{aligned} 43p+72&=158 \\ 43(2)+72&\stackrel{?}{=}158 \\ 86+72&\stackrel{?}{=}158 \\ 158&=158 \end{aligned}$$

59.

$$\begin{aligned} -47-21n&=58 \\ -47+47-21n&=58+47 \\ -21n&=105 \\ \frac{-21n}{-21}&=\frac{105}{-21} \\ n&=-5 \end{aligned}$$

$$\begin{aligned} -47-21n&=58 \\ -47-21(-5)&\stackrel{?}{=}58 \\ -47+105&\stackrel{?}{=}58 \\ 58&=58 \end{aligned}$$

61.
$$\begin{aligned} 2y - \frac{5}{3} &= \frac{4}{3} \\ 2y - \frac{5}{3} + \frac{5}{3} &= \frac{4}{3} + \frac{5}{3} \\ 2y &= \frac{9}{3} \\ 2y &= 3 \\ \frac{2y}{2} &= \frac{3}{2} \\ y &= \frac{3}{2} \end{aligned}$$
$$\begin{aligned} 2y - \frac{5}{3} &= \frac{4}{3} \\ 2\left(\frac{3}{2}\right) - \frac{5}{3} &\stackrel{?}{=} \frac{4}{3} \\ 3 - \frac{5}{3} &\stackrel{?}{=} \frac{4}{3} \\ \frac{9}{3} - \frac{5}{3} &\stackrel{?}{=} \frac{4}{3} \\ \frac{4}{3} &= \frac{4}{3} \end{aligned}$$

63.
$$\begin{aligned} -0.4y - 12 &= -20 \\ -0.4y - 12 + 12 &= -20 + 12 \\ -0.4y &= -8 \\ \frac{-0.4y}{-0.4} &= \frac{-8}{-0.4} \\ y &= 20 \end{aligned}$$
$$\begin{aligned} -0.4y - 12 &= -20 \\ -0.4(20) - 12 &\stackrel{?}{=} -20 \\ -8 - 12 &\stackrel{?}{=} -20 \\ -20 &= -20 \end{aligned}$$

65.
$$\begin{aligned} \frac{2x}{3} + \frac{1}{2} &= 3 \\ \frac{2x}{3} + \frac{1}{2} - \frac{1}{2} &= 3 - \frac{1}{2} \\ \frac{2x}{3} &= \frac{5}{2} \\ 3 \cdot \frac{2x}{3} &= 3 \cdot \frac{5}{2} \\ 2x &= \frac{15}{2} \\ \frac{1}{2} \cdot 2x &= \frac{1}{2} \cdot \frac{15}{2} \\ x &= \frac{15}{4} \end{aligned}$$
$$\begin{aligned} \frac{2x}{3} + \frac{1}{2} &= 3 \\ \frac{2\left(\frac{15}{4}\right)}{3} + \frac{1}{2} &\stackrel{?}{=} 3 \\ \frac{\frac{15}{2}}{3} + \frac{1}{2} &\stackrel{?}{=} 3 \\ \frac{15}{6} + \frac{1}{2} &\stackrel{?}{=} 3 \\ \frac{5}{2} + \frac{1}{2} &\stackrel{?}{=} 3 \\ \frac{6}{2} &\stackrel{?}{=} 3 \\ 3 &= 3 \end{aligned}$$

67.
$$\begin{aligned} \frac{3x}{4} - \frac{2}{5} &= 2 \\ \frac{3x}{4} - \frac{2}{5} + \frac{2}{5} &= 2 + \frac{2}{5} \\ \frac{3x}{4} &= \frac{12}{5} \\ 4 \cdot \frac{3x}{4} &= 4 \cdot \frac{12}{5} \\ 3x &= \frac{48}{5} \\ \frac{1}{3} \cdot 3x &= \frac{1}{3} \cdot \frac{48}{5} \\ x &= \frac{16}{5} \end{aligned}$$
$$\begin{aligned} \frac{3x}{4} - \frac{2}{5} &= 2 \\ \frac{3\left(\frac{16}{5}\right)}{4} - \frac{2}{5} &\stackrel{?}{=} 2 \\ \frac{\frac{48}{5}}{4} - \frac{2}{5} &\stackrel{?}{=} 2 \\ \frac{48}{20} - \frac{2}{5} &\stackrel{?}{=} 2 \\ \frac{12}{5} - \frac{2}{5} &\stackrel{?}{=} 2 \\ \frac{10}{5} &\stackrel{?}{=} 2 \\ 2 &= 2 \end{aligned}$$

69.
$$\begin{aligned} \frac{u-2}{5} &= 1 \\ 5 \cdot \frac{u-2}{5} &= 5 \cdot 1 \\ u - 2 &= 5 \\ u - 2 + 2 &= 5 + 2 \\ u &= 7 \end{aligned}$$
$$\begin{aligned} \frac{u-2}{5} &= 1 \\ \frac{7-2}{5} &\stackrel{?}{=} 1 \\ \frac{5}{5} &\stackrel{?}{=} 1 \\ 1 &= 1 \end{aligned}$$

71.
$$\begin{aligned} \frac{x-4}{4} &= -3 \\ 4 \cdot \frac{x-4}{4} &= 4(-3) \\ x - 4 &= -12 \\ x - 4 + 4 &= -12 + 4 \\ x &= -8 \end{aligned}$$
$$\begin{aligned} \frac{x-4}{4} &= -3 \\ \frac{-8-4}{4} &\stackrel{?}{=} -3 \\ \frac{-12}{4} &\stackrel{?}{=} -3 \\ -3 &= -3 \end{aligned}$$

73.

$$\begin{aligned} \frac{3z+2}{17} &= 0 \\ 17\cdot\frac{3z+2}{17} &= 17\cdot 0 \\ 3z+2 &= 0 \\ 3z+2-2 &= 0-2 \\ 3z &= -2 \\ \frac{3z}{3} &= \frac{-2}{3} \\ z &= -\frac{2}{3} \end{aligned}$$

$$\begin{aligned} \frac{3z+2}{17} &= 0 \\ \frac{3(-\frac{2}{3})+2}{17} &\stackrel{?}{=} 0 \\ \frac{-2+2}{17} &\stackrel{?}{=} 0 \\ \frac{0}{17} &\stackrel{?}{=} 0 \\ 0 &= 0 \end{aligned}$$

75.

$$\begin{aligned} \frac{17k-28}{21}+\frac{4}{3} &= 0 \\ \frac{17k-28}{21}+\frac{4}{3}-\frac{4}{3} &= 0-\frac{4}{3} \\ \frac{17k-28}{21} &= -\frac{4}{3} \\ 21\cdot\frac{17k-28}{21} &= 21\left(-\frac{4}{3}\right) \\ 17k-28 &= -28 \\ 17k-28+28 &= -28+28 \\ 17k &= 0 \\ \frac{17k}{17} &= \frac{0}{17} \\ k &= 0 \end{aligned}$$

$$\begin{aligned} \frac{17k-28}{21}+\frac{4}{3} &= 0 \\ \frac{17\cdot 0-28}{21}+\frac{4}{3} &\stackrel{?}{=} 0 \\ \frac{0-28}{21}+\frac{4}{3} &\stackrel{?}{=} 0 \\ -\frac{4}{3}+\frac{4}{3} &\stackrel{?}{=} 0 \\ 0 &= 0 \end{aligned}$$

77.

$$\begin{aligned} -\frac{x}{3}-\frac{1}{2} &= -\frac{5}{2} \\ -\frac{x}{3}-\frac{1}{2}+\frac{1}{2} &= -\frac{5}{2}+\frac{1}{2} \\ -\frac{x}{3} &= -\frac{4}{2} \\ -\frac{x}{3} &= -2 \\ -3\left(-\frac{x}{3}\right) &= -3(-2) \\ x &= 6 \end{aligned}$$

$$\begin{aligned} -\frac{x}{3}-\frac{1}{2} &= -\frac{5}{2} \\ -\frac{6}{3}-\frac{1}{2} &\stackrel{?}{=} -\frac{5}{2} \\ -2-\frac{1}{2} &\stackrel{?}{=} -\frac{5}{2} \\ -\frac{4}{2}-\frac{1}{2} &\stackrel{?}{=} -\frac{5}{2} \\ -\frac{5}{2} &= -\frac{5}{2} \end{aligned}$$

79.

$$\begin{aligned} \frac{9-5w}{15} &= \frac{2}{5} \\ 15\cdot\frac{9-5w}{15} &= 15\cdot\frac{2}{5} \\ 9-5w &= 6 \\ 9-9-5w &= 6-9 \\ -5w &= -3 \\ \frac{-5w}{-5} &= \frac{-3}{-5} \\ w &= \frac{3}{5} \end{aligned}$$

$$\begin{aligned} \frac{9-5w}{15} &= \frac{2}{5} \\ \frac{9-5(\frac{3}{5})}{15} &\stackrel{?}{=} \frac{2}{5} \\ \frac{9-3}{15} &\stackrel{?}{=} \frac{2}{5} \\ \frac{6}{15} &\stackrel{?}{=} \frac{2}{5} \\ \frac{2}{5} &= \frac{2}{5} \end{aligned}$$

81. Let x = her former rent.
Then $2x - 100$ = the new rent.

$$\boxed{\text{The new rent}} = 400$$
$$2x - 100 = 400$$
$$2x - 100 + 100 = 400 + 100$$
$$2x = 500$$
$$\frac{2x}{2} = \frac{500}{2}$$
$$x = 250$$

Her former rent was \$250.

83. Let x = the number of days.
Then $16 + 12x$ = the total cost.

$$\boxed{\text{The total cost}} = 100$$
$$16 + 12x = 100$$
$$16 - 16 + 12x = 100 - 16$$
$$12x = 84$$
$$\frac{12x}{12} = \frac{84}{12}$$
$$x = 7$$

The owner was gone for 7 days.

85. Let x = the regular price. Then $0.80x$ = the sale price.

$$\boxed{\text{Final price}} = 0.90 \cdot \boxed{\text{Sale price}}$$
$$36 = 0.90(0.80x)$$
$$36 = 0.72x$$
$$\frac{36}{0.72} = \frac{0.72x}{0.72}$$

$50 = x$; The original price was \$50.

87. For a purchase of \$100:

$$\boxed{\text{Markdown}} = \boxed{\text{Percent markdown}} \cdot \boxed{\text{Regular price}}$$
$$15 = r \cdot 100$$
$$\frac{15}{100} = \frac{r \cdot 100}{100}$$
$$0.15 = r$$
$$15\% = r$$

For a purchase of \$250:

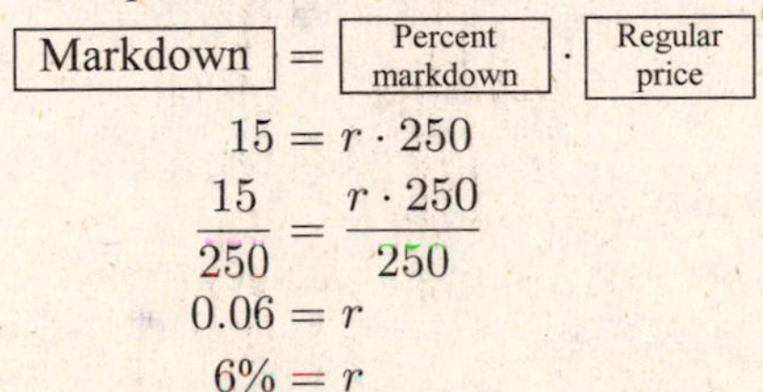

$$\boxed{\text{Markdown}} = \boxed{\text{Percent markdown}} \cdot \boxed{\text{Regular price}}$$
$$15 = r \cdot 250$$
$$\frac{15}{250} = \frac{r \cdot 250}{250}$$
$$0.06 = r$$
$$6\% = r$$

The range of the percent discount is from 6% to 15%.

89. Let x = the original number.
Then $3x - 6$ = the other number.

$$\boxed{\text{The other number}} = 9$$
$$3x - 6 = 9$$
$$3x - 6 + 6 = 9 + 6$$
$$3x = 15$$
$$\frac{3x}{3} = \frac{15}{3}$$
$$x = 5$$

The original number is 5.

91. Let x = the original number.
Then $\frac{x+7}{2}$ = the other number.

$$\boxed{\text{The other number}} = 5$$
$$\frac{x+7}{2} = 5$$
$$2 \cdot \frac{x+7}{2} = 2 \cdot 5$$
$$x + 7 = 10$$
$$x + 7 - 7 = 10 - 7$$
$$x = 3$$

The original number is 3.

93. Let $x =$ the # of minutes (after the 1st). Then $0.85 + 0.27x =$ the total cost.

$$\boxed{\text{The total cost}} = 8.41$$
$$0.85 + 0.27x = 8.41$$
$$0.85 - 0.85 + 0.27x = 8.41 - 0.85$$
$$0.27x = 7.56$$
$$\frac{0.27x}{0.27} = \frac{7.56}{0.27}$$
$$x = 28$$

She can talk for 28 minutes **after the first minute**, for a total of 29 minutes.

95. Let $x =$ the money from ticket sales.
Then $1500 + 0.20x =$ the total income.

$$\boxed{\text{The total income}} = 2980$$
$$1500 + 0.20x = 2980$$
$$1500 - 1500 + 0.20x = 2980 - 1500$$
$$0.20x = 1480$$
$$\frac{0.20x}{0.20} = \frac{1480}{0.20}$$
$$x = 7400$$

The total ticket sales were \$7,400.

97. Let $x =$ the score on the fifth exam.

$$\boxed{\text{Average score}} = 90$$
$$\frac{85 + 80 + 95 + 78 + x}{5} = 90$$
$$\frac{338 + x}{5} = 90$$
$$5 \cdot \frac{338 + x}{5} = 5 \cdot 90$$
$$338 + x = 450$$
$$338 - 338 + x = 450 - 338$$
$$x = 112$$

It is impossible to receive an A.

99. Answers may vary.

101.

$$\frac{7x + \#}{22} = \frac{1}{2}$$
$$\frac{7(1) + \#}{22} = \frac{1}{2}$$
$$\frac{7 + \#}{22} = \frac{1}{2}$$
$$22 \cdot \frac{7 + \#}{22} = 22 \cdot \frac{1}{2}$$
$$7 + \# = 11$$
$$7 - 7 + \# = 11 - 7$$
$$\# = 4$$

The original equation was $\frac{7x + 4}{22} = \frac{1}{2}$.

Exercise 2.3 (page 113)

1. $3x + 5x = (3 + 5)x = 8x$

3. $3x + 2x - 5x = (3 + 2 - 5)x = 0x = 0$

5.
$$\begin{aligned} 3(x + 2) - 3x + 6 &= 3 \cdot x + 3 \cdot 2 - 3x + 6 \\ &= 3x + 6 - 3x + 6 \\ &= 0x + 12 = 12 \end{aligned}$$

7.
$$5x = 4x + 3$$
$$5x - 4x = 4x - 4x + 3$$
$$x = 3$$

SECTION 2.3

9.
$$\begin{aligned} 3x &= 2(x+1) \\ 3x &= 2x+2 \\ 3x-2x &= 2x-2x+2 \\ x &= 2 \end{aligned}$$

11. $x^2z(y^3-z) = (-3)^2(0)\left[(-5)^3-0\right] = 0$

13. $\dfrac{x-y^2}{2y-1+x} = \dfrac{-3-(-5)^2}{2(-5)-1+(-3)} = \dfrac{-3-(+25)}{-10-1+(-3)} = \dfrac{-28}{-14} = 2$

15.
$$\begin{aligned} \frac{6}{7}-\frac{5}{8} &= \frac{6\cdot \mathbf{8}}{7\cdot \mathbf{8}}-\frac{5\cdot \mathbf{7}}{8\cdot \mathbf{7}} = \frac{48}{56}-\frac{35}{56} \\ &= \frac{48-35}{56} = \frac{13}{56} \end{aligned}$$

17. $\dfrac{6}{7}\div\dfrac{5}{8} = \dfrac{6}{7}\cdot\dfrac{8}{5} = \dfrac{48}{35}$

19. variables, like, unlike, numerical

21. identity, contradiction

23. $3x+17x = (3+17)x = 20x$

25. $8x^2-5x^2 = (8-5)x^2 = 3x^2$

27. $9x+3y \Rightarrow$ unlike terms

29.
$$\begin{aligned} 3(x+2)+4x &= 3\cdot x+3\cdot 2+4x \\ &= 3x+6+4x = 7x+6 \end{aligned}$$

31.
$$\begin{aligned} 5(z-3)+2z &= 5\cdot z-5\cdot 3+2z \\ &= 5z-15+2z = 7z-15 \end{aligned}$$

33.
$$\begin{aligned} 12(x+11)-11 &= 12\cdot x+12\cdot 11-11 \\ &= 12x+132-11 \\ &= 12x+121 \end{aligned}$$

35. $8(y+7)-2(y-3) = 8\cdot y+8\cdot 7+(-2)\cdot y+(-2)(-3) = 8y+56-2y+6 = 6y+62$

37. $2x+4(y-x)+3y = 2x+4\cdot y+4(-x)+3y = 2x+4y-4x+3y = -2x+7y$

39.
$$\begin{aligned} 9(x+11)+5(13-x) &= 0 \\ 9x+99+65-5x &= 0 \\ 4x+164 &= 0 \\ 4x+164-164 &= 0-164 \\ 4x &= -164 \\ \frac{4x}{4} &= \frac{-164}{4} \\ x &= -41 \end{aligned}$$

$$\begin{aligned} 9(x+11)+5(13-x) &= 0 \\ 9(-41+11)+5[13-(-41)] &\stackrel{?}{=} 0 \\ 9(-30)+5[13+41] &\stackrel{?}{=} 0 \\ -270+5(54) &\stackrel{?}{=} 0 \\ -270+270 &\stackrel{?}{=} 0 \\ 0 &= 0 \end{aligned}$$

41.
$$\begin{aligned} 11x+6(3-x) &= 3 \\ 11x+18-6x &= 3 \\ 5x+18 &= 3 \\ 5x+18-18 &= 3-18 \\ 5x &= -15 \\ \frac{5x}{5} &= \frac{-15}{5} \\ x &= -3 \end{aligned}$$

$$\begin{aligned} 11x+6(3-x) &= 3 \\ 11(-3)+6[3-(-3)] &\stackrel{?}{=} 3 \\ -33+6[3+3] &\stackrel{?}{=} 3 \\ -33+6[6] &\stackrel{?}{=} 3 \\ -33+36 &\stackrel{?}{=} 3 \\ 3 &= 3 \end{aligned}$$

SECTION 2.3

43.

$$\begin{aligned} 3x + 2 &= 2x \\ 3x - 3x + 2 &= 2x - 3x \\ 2 &= -x \\ -1(2) &= -1(-x) \\ -2 &= x \end{aligned}$$

$$\begin{aligned} 3x + 2 &= 2x \\ 3(-2) + 2 &\stackrel{?}{=} 2(-2) \\ -6 + 2 &\stackrel{?}{=} -4 \\ -4 &= -4 \end{aligned}$$

45.

$$\begin{aligned} 5x - 3 &= 4x \\ 5x - 5x - 3 &= 4x - 5x \\ -3 &= -x \\ -1(-3) &= -1(-x) \\ 3 &= x \end{aligned}$$

$$\begin{aligned} 5x - 3 &= 4x \\ 5(3) - 3 &\stackrel{?}{=} 4(3) \\ 15 - 3 &\stackrel{?}{=} 12 \\ 12 &= 12 \end{aligned}$$

47.

$$\begin{aligned} 9y - 3 &= 6y \\ 9y - 9y - 3 &= 6y - 9y \\ -3 &= -3y \\ \frac{-3}{-3} &= \frac{-3y}{-3} \\ 1 &= y \end{aligned}$$

$$\begin{aligned} 9y - 3 &= 6y \\ 9(1) - 3 &\stackrel{?}{=} 6(1) \\ 9 - 3 &\stackrel{?}{=} 6 \\ 6 &= 6 \end{aligned}$$

49.

$$\begin{aligned} 8y - 7 &= y \\ 8y - 8y - 7 &= y - 8y \\ -7 &= -7y \\ \frac{-7}{-7} &= \frac{-7y}{-7} \\ 1 &= y \end{aligned}$$

$$\begin{aligned} 8y - 7 &= y \\ 8(1) - 7 &\stackrel{?}{=} 1 \\ 8 - 7 &\stackrel{?}{=} 1 \\ 1 &= 1 \end{aligned}$$

51.

$$\begin{aligned} 3(a + 2) &= 4a \\ 3a + 6 &= 4a \\ 3a - 3a + 6 &= 4a - 3a \\ 6 &= a \end{aligned}$$

$$\begin{aligned} 3(a + 2) &= 4a \\ 3(6 + 2) &\stackrel{?}{=} 4(6) \\ 3(8) &\stackrel{?}{=} 24 \\ 24 &= 24 \end{aligned}$$

53.

$$\begin{aligned} 5(b + 7) &= 6b \\ 5b + 35 &= 6b \\ 5b - 5b + 35 &= 6b - 5b \\ 35 &= b \end{aligned}$$

$$\begin{aligned} 5(b + 7) &= 6b \\ 5(35 + 7) &\stackrel{?}{=} 6(35) \\ 5(42) &\stackrel{?}{=} 210 \\ 210 &= 210 \end{aligned}$$

55.

$$\begin{aligned} 2 + 3(x - 5) &= 4(x - 1) \\ 2 + 3x - 15 &= 4x - 4 \\ 3x - 13 &= 4x - 4 \\ 3x - 3x - 13 &= 4x - 3x - 4 \\ -13 &= x - 4 \\ -13 + 4 &= x - 4 + 4 \\ -9 &= x \end{aligned}$$

$$\begin{aligned} 2 + 3(x - 5) &= 4(x - 1) \\ 2 + 3(-9 - 5) &\stackrel{?}{=} 4(-9 - 1) \\ 2 + 3(-14) &\stackrel{?}{=} 4(-10) \\ 2 + (-42) &\stackrel{?}{=} -40 \\ -40 &= -40 \end{aligned}$$

57.

$$\begin{aligned} 3(a + 2) &= 2(a - 7) \\ 3a + 6 &= 2a - 14 \\ 3a - 2a + 6 &= 2a - 2a - 14 \\ a + 6 &= -14 \\ a + 6 - 6 &= -14 - 6 \\ a &= -20 \end{aligned}$$

$$\begin{aligned} 3(a + 2) &= 2(a - 7) \\ 3(-20 + 2) &\stackrel{?}{=} 2(-20 - 7) \\ 3(-18) &\stackrel{?}{=} 2(-27) \\ -54 &= -54 \end{aligned}$$

59.

$$\begin{aligned} \frac{3(t-7)}{2} &= t-6 \\ 2\cdot\frac{3(t-7)}{2} &= 2(t-6) \\ 3(t-7) &= 2(t-6) \\ 3t-21 &= 2t-12 \\ 3t-2t-21 &= 2t-2t-12 \\ t-21 &= -12 \\ t-21+21 &= -12+21 \\ t &= 9 \end{aligned}$$

$$\begin{aligned} \frac{3(t-7)}{2} &= t-6 \\ \frac{3(9-7)}{2} &\stackrel{?}{=} 9-6 \\ \frac{3(2)}{2} &\stackrel{?}{=} 9-6 \\ \frac{6}{2} &\stackrel{?}{=} 3 \\ 3 &= 3 \end{aligned}$$

61.

$$\begin{aligned} \frac{2(t-1)}{6}-2 &= \frac{t+2}{6} \\ 6\left[\frac{2(t-1)}{6}-2\right] &= 6\cdot\frac{t+2}{6} \\ 6\cdot\frac{2(t-1)}{6}-6\cdot 2 &= t+2 \\ 2(t-1)-12 &= t+2 \\ 2t-2-12 &= t+2 \\ 2t-14 &= t+2 \\ 2t-t-14 &= t-t+2 \\ t-14 &= 2 \\ t-14+14 &= 2+14 \\ t &= 16 \end{aligned}$$

$$\begin{aligned} \frac{2(t-1)}{6}-2 &= \frac{t+2}{6} \\ \frac{2(16-1)}{6}-2 &\stackrel{?}{=} \frac{16+2}{6} \\ \frac{2(15)}{6}-2 &\stackrel{?}{=} \frac{18}{6} \\ \frac{30}{6}-2 &\stackrel{?}{=} 3 \\ 5-2 &\stackrel{?}{=} 3 \\ 3 &= 3 \end{aligned}$$

63.

$$\begin{aligned} 3.1(x-2) &= 1.3x+2.8 \\ 3.1x-6.2 &= 1.3x+2.8 \\ 3.1x-1.3x-6.2 &= 1.3x-1.3x+2.8 \\ 1.8x-6.2 &= 2.8 \\ 1.8x-6.2+6.2 &= 2.8+6.2 \\ 1.8x &= 9.0 \\ \frac{1.8x}{1.8} &= \frac{9.0}{1.8} \\ x &= 5 \end{aligned}$$

$$\begin{aligned} 3.1(x-2) &= 1.3x+2.8 \\ 3.1(5-2) &\stackrel{?}{=} 1.3(5)+2.8 \\ 3.1(3) &\stackrel{?}{=} 6.5+2.8 \\ 9.3 &= 9.3 \end{aligned}$$

65.

$$\begin{aligned} 2.7(y+1) &= 0.3(3y+33) \\ 2.7y+2.7 &= 0.9y+9.9 \\ 2.7y-0.9y+2.7 &= 0.9y-0.9y+9.9 \\ 1.8y+2.7 &= 9.9 \\ 1.8y+2.7-2.7 &= 9.9-2.7 \\ 1.8y &= 7.2 \\ \frac{1.8y}{1.8} &= \frac{7.2}{1.8} \\ y &= 4 \end{aligned}$$

$$\begin{aligned} 2.7(y+1) &= 0.3(3y+33) \\ 2.7(4+1) &= 0.3[3(4)+33] \\ 2.7(5) &\stackrel{?}{=} 0.3(12+33) \\ 13.5 &\stackrel{?}{=} 0.3(45) \\ 13.5 &= 13.5 \end{aligned}$$

SECTION 2.3

67.
$$\begin{aligned} 8x+3(2-x) &= 5(x+2)-4 \\ 8x+6-3x &= 5x+10-4 \\ 5x+6 &= 5x+6 \\ 5x-5x+6 &= 5x-5x+6 \\ 6 &= 6 \end{aligned}$$
Identity, $\mathbb{R}$

69.
$$\begin{aligned} 2(s+2) &= 2(s+1)+3 \\ 2s+4 &= 2s+2+3 \\ 2s+4 &= 2s+5 \\ 2s-2s+4 &= 2s-2s+5 \\ 4 &\neq 5 \end{aligned}$$
Contradiction, $\emptyset$

71.
$$\begin{aligned} \frac{5(x+3)}{3}-x &= \frac{2(x+8)}{3} \\ 3\left[\frac{5(x+3)}{3}-x\right] &= 3\cdot\frac{2(x+8)}{3} \\ 3\cdot\frac{5(x+3)}{3}-3\cdot x &= 2(x+8) \\ 5(x+3)-3x &= 2x+16 \\ 5x+15-3x &= 2x+16 \\ 2x+15 &= 2x+16 \\ 2x-2x+15 &= 2x-2x+16 \\ 15 &\neq 16 \end{aligned}$$
Contradiction, $\emptyset$

73.
$$\begin{aligned} x+7 &= \frac{2x+6}{2}+4 \\ 2(x+7) &= 2\left[\frac{2x+6}{2}+4\right] \\ 2x+14 &= 2\cdot\frac{2x+6}{2}+2\cdot 4 \\ 2x+14 &= 2x+6+8 \\ 2x+14 &= 2x+14 \\ 2x-2x+14 &= 2x-2x+14 \\ 14 &= 14 \end{aligned}$$
Identity, $\mathbb{R}$

75. Expression:
$$\begin{aligned} (x+2)-(x-y) = 1(x+2)-1(x-y) = 1\cdot x+1\cdot 2+(-1)\cdot x+(-1)(-y) &= x+2-x+y \\ &= y+2 \end{aligned}$$

77. Equation:
$$\begin{aligned} \frac{4(2x-10)}{3} &= 2(x-4) \\ 3\cdot\frac{4(2x-10)}{3} &= 3\cdot 2(x-4) \\ 4(2x-10) &= 6(x-4) \\ 8x-40 &= 6x-24 \\ 8x-6x-40 &= 6x-6x-24 \\ 2x-40 &= -24 \\ 2x-40+40 &= -24+40 \\ 2x &= 16 \\ \frac{2x}{2} &= \frac{16}{2} \\ x &= 8 \end{aligned}$$

79. Expression:
$$2\left(4x+\frac{9}{2}\right)-3\left(x+\frac{2}{3}\right) = 2\cdot 4x+2\cdot\frac{9}{2}+(-3)\cdot x+(-3)\cdot\frac{2}{3} = 8x+9-3x-2 = 5x+7$$

81. Equation:

$$\begin{aligned}
\frac{8(5-q)}{5} &= -2q \\
5 \cdot \frac{8(5-q)}{5} &= 5(-2q) \\
8(5-q) &= -10q \\
40 - 8q &= -10q \\
40 - 8q + 8q &= -10q + 8q \\
40 &= -2q \\
\frac{40}{-2} &= \frac{-2q}{-2} \\
-20 &= q
\end{aligned}$$

83. Equation:

$$\begin{aligned}
\frac{3x+14}{2} &= x - 2 + \frac{x+18}{2} \\
2 \cdot \frac{3x+14}{2} &= 2\left[x - 2 + \frac{x+18}{2}\right] \\
3x + 14 &= 2x - 2 \cdot 2 + 2 \cdot \frac{x+18}{2} \\
3x + 14 &= 2x - 4 + x + 18 \\
3x + 14 &= 3x + 14 \\
3x - 3x + 14 &= 3x - 3x + 14 \\
14 &= 14
\end{aligned}$$

Identity, $\mathbb{R}$

85. Equation:

$$\begin{aligned}
5 - 7r &= 8r \\
5 - 7r + 7r &= 8r + 7r \\
5 &= 15r \\
\frac{5}{15} &= \frac{15r}{15} \\
\frac{1}{3} &= r
\end{aligned}$$

87. Equation:

$$\begin{aligned}
22 - 3r &= 8r \\
22 - 3r + 3r &= 8r + 3r \\
22 &= 11r \\
\frac{22}{11} &= \frac{11r}{11} \\
2 &= r
\end{aligned}$$

89. Expression: $8(x+3) - 3x = 8 \cdot x + 8 \cdot 3 - 3x = 8x + 24 - 3x = 5x + 24$

91. Equation:

$$\begin{aligned}
19.1x - 4(x + 0.3) &= -46.5 \\
19.1x - 4x - 1.2 &= -46.5 \\
15.1x - 1.2 &= -46.5 \\
15.1x - 1.2 + 1.2 &= -46.5 + 1.2 \\
15.1x &= -45.3 \\
\frac{15.1x}{15.1} &= \frac{-45.3}{15.1} \\
x &= -3
\end{aligned}$$

93. Expression: $3.2(m+1.3) - 2.5(m-7.2) = 3.2m + 4.16 - 2.5m + 18 = 0.7m + 22.16$

95. Equation:

$$\begin{aligned}
14.3(x+2) + 13.7(x-3) &= 15.5 \\
14.3x + 28.6 + 13.7x - 41.1 &= 15.5 \\
28.0x - 12.5 &= 15.5 \\
28x - 12.5 + 12.5 &= 15.5 + 12.5 \\
28x &= 28 \\
\frac{28x}{28} &= \frac{28}{28} \\
x &= 1
\end{aligned}$$

97. Equation:

$$\begin{aligned} 10x + 3(2 - x) &= 5(x + 2) - 4 \\ 10x + 6 - 3x &= 5x + 10 - 4 \\ 7x + 6 &= 5x + 6 \\ 7x - 5x + 6 &= 5x - 5x + 6 \\ 2x + 6 &= 6 \\ 2x + 6 - 6 &= 6 - 6 \\ 2x &= 0 \\ \frac{2x}{2} &= \frac{0}{2} \\ x &= 0 \end{aligned}$$

99.

$$\begin{aligned} \frac{3.7(2.3x - 2.7)}{1.5} &= 5.2(x - 1.2) \\ 1.5 \cdot \frac{3.7(2.3x - 2.7)}{1.5} &= 1.5(5.2)(x - 1.2) \\ 3.7(2.3x - 2.7) &= 7.8(x - 1.2) \\ 8.51x - 9.99 &= 7.8x - 9.36 \\ 8.51x - 7.8x - 9.99 &= 7.8x - 7.8x - 9.36 \\ 0.71x - 9.99 &= -9.36 \\ 0.71x - 9.99 + 9.99 &= -9.36 + 9.99 \\ 0.71x &= 0.63 \\ \frac{0.71x}{0.71} &= \frac{0.63}{0.71} \\ x &\approx 0.887 \approx 0.9 \end{aligned}$$

101. They are like terms because the variable parts are exactly the same.

103. $7xxy^3 = 7x^2y^3$; $5x^2yyy = 5x^2y^3$
They are like terms because the variable parts are exactly the same.

105. Let $x =$ the number.

$$\begin{aligned} x &= 2x \\ x - x &= 2x - x \\ 0 &= x; \text{ The number is 0.} \end{aligned}$$

Exercise 2.4 (page 120)

1.

$$\begin{aligned} ab + c &= 0 \\ ab + c - c &= 0 - c \\ ab &= -c \\ \frac{ab}{b} &= \frac{-c}{b} \\ a &= -\frac{c}{b} \end{aligned}$$

3.

$$\begin{aligned} a &= \frac{b}{c} \\ ac &= \frac{b}{c} \cdot c \\ ac &= b, \text{ or } b = ac \end{aligned}$$

5. $2x - 5y + 3x = 2x + 3x - 5y = (2 + 3)x - 5y = 5x - 5y$

7.

$$\begin{aligned} \frac{3}{5}(x + 5) - \frac{8}{5}(10 + x) = \frac{3}{5}x + \frac{3}{5} \cdot 5 - \frac{8}{5} \cdot 10 - \frac{8}{5}x = \frac{3}{5}x + 3 - 16 - \frac{8}{5}x &= \left(\frac{3}{5} - \frac{8}{5}\right)x - 13 \\ &= -\frac{5}{5}x - 13 \\ &= -x - 13 \end{aligned}$$

9. literal

11. isolate

13. subtract

SECTION 2.4

15.
$$E = IR$$
$$\frac{E}{R} = \frac{IR}{R}$$
$$\frac{E}{R} = I, \text{ or } I = \frac{E}{R}$$

17.
$$V = lwh$$
$$\frac{V}{lh} = \frac{lwh}{lh}$$
$$\frac{V}{lh} = w, \text{ or } w = \frac{V}{lh}$$

19.
$$K = A + 32$$
$$K - 32 = A + 32 - 32$$
$$K - 32 = A, \text{ or } A = K - 32$$

21.
$$V = \frac{1}{3}Bh$$
$$3V = 3 \cdot \frac{1}{3}Bh$$
$$3V = Bh$$
$$\frac{3V}{B} = \frac{Bh}{B}$$
$$\frac{3V}{B} = h, \text{ or } h = \frac{3V}{B}$$

23.
$$V = \frac{1}{3}\pi r^2 h$$
$$3 \cdot V = 3 \cdot \frac{1}{3}\pi r^2 h$$
$$3V = \pi r^2 h$$
$$\frac{3V}{\pi r^2} = \frac{\pi r^2 h}{\pi r^2}$$
$$\frac{3V}{\pi r^2} = h, \text{ or } h = \frac{3V}{\pi r^2}$$

25.
$$y = \frac{1}{2}(x + 2)$$
$$2y = 2 \cdot \frac{1}{2}(x + 2)$$
$$2y = x + 2$$
$$2y - 2 = x + 2 - 2$$
$$2y - 2 = x, \text{ or } x = 2y - 2$$

27.
$$A = \frac{B + 4}{8}$$
$$8A = 8 \cdot \frac{B + 4}{8}$$
$$8A = B + 4$$
$$8A - 4 = B + 4 - 4$$
$$8A - 4 = B, \text{ or } B = 8A - 4$$

29.
$$A = \frac{3}{2}(B + 5)$$
$$\frac{2}{3}A = \frac{2}{3} \cdot \frac{3}{2}(B + 5)$$
$$\frac{2}{3}A = B + 5$$
$$\frac{2}{3}A - 5 = B + 5 - 5$$
$$\frac{2}{3}A - 5 = B, \text{ or } B = \frac{2}{3}A - 5$$

31.
$$p = \frac{h}{2}(q + r)$$
$$\frac{2}{h} \cdot p = \frac{2}{h} \cdot \frac{h}{2}(q + r)$$
$$\frac{2p}{h} = q + r$$
$$\frac{2p}{h} - r = q + r - r$$
$$\frac{2p}{h} - r = q, \text{ or } q = \frac{2p}{h} - r$$

33.
$$G = 2b(r - 1)$$
$$\frac{G}{2b} = \frac{2b(r - 1)}{2b}$$
$$\frac{G}{2b} = r - 1$$
$$\frac{G}{2b} + 1 = r - 1 + 1$$
$$\frac{G}{2b} + 1 = r, \text{ or } r = \frac{G}{2b} + 1$$

35.
$$d = rt \qquad t = \frac{d}{r}$$
$$\frac{d}{r} = \frac{rt}{r} \qquad t = \frac{135}{45}$$
$$\frac{d}{r} = t \qquad t = 3$$

37.
$$\begin{aligned} P &= a + b + c & c &= P - a - b \\ P - a &= a - a + b + c & c &= 37 - 15 - 19 \\ P - a &= b + c & c &= 22 - 19 \\ P - a - b &= b - b + c & c &= 3 \\ P - a - b &= c \end{aligned}$$

39.
$$P = 4s$$
$$\frac{P}{4} = \frac{4s}{4}$$
$$\frac{P}{4} = s, \text{ or } s = \frac{P}{4}$$

41.
$$P = 2l + 2w$$
$$P - 2l = 2l - 2l + 2w$$
$$P - 2l = 2w$$
$$\frac{P - 2l}{2} = \frac{2w}{2}$$
$$\frac{P - 2l}{2} = w, \text{ or } w = \frac{P - 2l}{2}$$

43.
$$A = P + Prt$$
$$A - P = P - P + Prt$$
$$A - P = Prt$$
$$\frac{A - P}{Pr} = \frac{Prt}{Pr}$$
$$\frac{A - P}{Pr} = t, \text{ or } t = \frac{A - P}{Pr}$$

45.
$$K = \frac{wv^2}{2g}$$
$$2g \cdot K = 2g \cdot \frac{wv^2}{2g}$$
$$2gK = wv^2$$
$$\frac{2gK}{v^2} = \frac{wv^2}{v^2}$$
$$\frac{2gK}{v^2} = w, \text{ or } w = \frac{2gK}{v^2}$$

47.
$$K = \frac{wv^2}{2g}$$
$$2g \cdot K = 2g \cdot \frac{wv^2}{2g}$$
$$2gK = wv^2$$
$$\frac{2gK}{2K} = \frac{wv^2}{2K}$$
$$g = \frac{wv^2}{2K}$$

49.
$$F = \frac{GMm}{d^2}$$
$$d^2 \cdot F = d^2 \cdot \frac{GMm}{d^2}$$
$$d^2F = GMm$$
$$\frac{d^2F}{Gm} = \frac{GMm}{Gm}$$
$$\frac{d^2F}{Gm} = M, \text{ or } M = \frac{d^2F}{Gm}$$

51.

$$i = prt \qquad t = \frac{i}{pr}$$
$$\frac{i}{pr} = \frac{prt}{pr} \qquad t = \frac{12}{100(0.06)}$$
$$\frac{i}{pr} = t \qquad t = \frac{12}{6} = 2$$

53.

$$K = \frac{1}{2}h(a+b) \qquad h = \frac{2K}{a+b}$$
$$2 \cdot K = 2 \cdot \frac{1}{2}h(a+b) \qquad h = \frac{2(48)}{7+5}$$
$$2K = h(a+b) \qquad h = \frac{96}{12} = 8$$
$$\frac{2K}{a+b} = \frac{h(a+b)}{a+b}$$
$$\frac{2K}{a+b} = h$$

55.

$$V = \frac{1}{3}\pi r^2 h \qquad h = \frac{3V}{\pi r^2}$$
$$3 \cdot V = 3 \cdot \frac{1}{3}\pi r^2 h \qquad h = \frac{3(36\pi)}{\pi(6)^2}$$
$$3V = \pi r^2 h \qquad h = \frac{108\pi}{36\pi} = 3 \text{ inches}$$
$$\frac{3V}{\pi r^2} = \frac{\pi r^2 h}{\pi r^2}$$
$$\frac{3V}{\pi r^2} = h$$

57.

$$E = IR \qquad I = \frac{E}{R}$$
$$\frac{E}{R} = \frac{IR}{R} \qquad I = \frac{48}{12}$$
$$\frac{E}{R} = I \qquad I = 4 \text{ amperes}$$

59.

$$P = I^2 R \qquad R = \frac{P}{I^2}$$
$$\frac{P}{I^2} = \frac{I^2 R}{I^2} \qquad R = \frac{2700}{14^2}$$
$$\frac{P}{I^2} = R \qquad R = \frac{2700}{196} = 13.78 \text{ ohms}$$

61.

$$F = \frac{GMm}{d^2}$$
$$d^2 \cdot F = d^2 \cdot \frac{GMm}{d^2}$$
$$d^2 F = GMm$$
$$\frac{Fd^2}{GM} = \frac{GMm}{GM}$$
$$\frac{Fd^2}{GM} = m, \text{ or } m = \frac{Fd^2}{GM}$$

63.

$$L = 2D + 3.25(r+R)$$
$$L - 3.25(r+R) = 2D + 3.25(r+R) - 3.25(r+R)$$
$$L - 3.25(r+R) = 2D$$
$$\frac{L - 3.25(r+R)}{2} = \frac{2D}{2}$$
$$\frac{L - 3.25r - 3.25R}{2} = D$$

$$D = \frac{L - 3.25r - 3.25R}{2}$$
$$D = \frac{25 - 3.25(1) - 3.25(3)}{2}$$
$$D = \frac{25 - 3.25 - 9.75}{2} = \frac{12}{2} = 6 \text{ ft}$$

65.

$$C = 0.15(T - C)$$
$$C = 0.15T - 0.15C$$
$$C + 0.15C = 0.15T - 0.15C + 0.15C$$
$$1.15C = 0.15T$$
$$\frac{1.15C}{1.15} = \frac{0.15T}{1.15}$$

$C \approx 0.1304T$; The maximum contribution is about 13% of taxable income.

67. **Answers may vary.**

69. $E = mc^2$
$E = 1(300{,}000)^2 = 90{,}000{,}000{,}000$ joules

Exercise 2.5 (page 128)

1. $90° - 20° = 70°$

3. $P = 2l + 2w = 2(6 \text{ ft}) + 2(4 \text{ ft})$
$= 12 \text{ ft} + 8 \text{ ft} = 20 \text{ ft}$

5. $V = \frac{1}{3}Bh = \frac{1}{3}s^2h = \frac{1}{3}(10 \text{ cm})^2(6 \text{ cm}) = \frac{1}{3}(100 \text{ cm}^2)(6 \text{ cm}) = \frac{1}{3}(600 \text{ cm}^3) = 200 \text{ cm}^3$

7. $3(x+2) + 4(x-3) = 3 \cdot x + 3 \cdot 2 + 4 \cdot x - 4 \cdot 3 = 3x + 6 + 4x - 12 = 7x - 6$

9. $\frac{1}{2}(x+1) - \frac{1}{2}(x+4) = \frac{1}{2} \cdot x + \frac{1}{2} \cdot 1 - \frac{1}{2} \cdot x - \frac{1}{2} \cdot 4 = \frac{1}{2}x + \frac{1}{2} - \frac{1}{2}x - \frac{4}{2} = -\frac{3}{2}$

11. $A = P + Prt = 1200 + 1200(0.08)(3) = 1200 + 288 = \$1488$

13. $2l + 2w$

15. vertex

17. degrees

19. straight

21. supplementary

23. Let $x =$ the length of one part.
Then $2x =$ the length of the other part.

$$\boxed{\text{Length of first part}} + \boxed{\text{Length of second part}} = 12$$
$$x + 2x = 12$$
$$3x = 12$$
$$\frac{3x}{3} = \frac{12}{3}$$
$$x = 4$$

The parts are 4 feet and 8 feet long.

25.
$$\boxed{\text{Sum of 3 lengths}} = 30$$
$$x + 10 + 2x + x = 30$$
$$4x + 10 = 30$$
$$4x + 10 - 10 = 30 - 10$$
$$4x = 20$$
$$\frac{4x}{4} = \frac{20}{4}$$
$$x = 5$$

The sections are 15, 10, and 5 feet long.

27.
$$\boxed{\text{Sum of 3 lengths}} = 24$$
$$x + x + 4 + x + 2 = 24$$
$$3x + 6 = 24$$
$$3x + 6 - 6 = 24 - 6$$
$$3x = 18$$
$$\frac{3x}{3} = \frac{18}{3}$$
$$x = 6$$

The sections are 6, 8, and 10 feet long.

29. Let $x =$ the number of hardcover books.
Then $11x =$ the number of paperbacks.

$$\boxed{\text{Number of hardcover books}} + \boxed{\text{Number of paperbacks}} = 114{,}000$$
$$x + 11x = 114{,}000$$
$$12x = 114{,}000$$
$$\frac{12x}{12} = \frac{114{,}000}{12}$$
$$x = 9500$$

There were 9500 hardcover books and 104,500 paperbacks sold.

31.
$$\begin{aligned} x + 40° &= 50° \\ x + 40° - 40° &= 50° - 40° \\ x &= 10° \end{aligned}$$

33.
$$\begin{aligned} x + 21° &= 180° \\ x + 21° - 21° &= 180° - 21° \\ x &= 159° \end{aligned}$$

35.
$$\begin{aligned} x + 12° &= 59° \\ x + 12° - 12° &= 59° - 12° \\ x &= 47° \end{aligned}$$

37.
$$\begin{aligned} x + 63° &= 90° \\ x + 63° - 63° &= 90° - 63° \\ x &= 27° \end{aligned}$$

39. Let $x =$ the measure of the complement.
$$\begin{aligned} x + 37° &= 90° \\ x + 37° - 37° &= 90° - 37° \\ x &= 53° \end{aligned}$$

41. Let $x =$ the measure of the complement.
$$\begin{aligned} x + 40° &= 90° \\ x + 40° - 40° &= 90° - 40° \\ x &= 50° \end{aligned}$$
Let $x =$ the measure of the supplement of the complement.
$$\begin{aligned} x + 50° &= 180° \\ x + 50° - 50° &= 180° - 50° \\ x &= 130° \end{aligned}$$

43.
$$\begin{aligned} \boxed{\text{Perimeter}} &= 90 \\ 2(w + 7) + 2w &= 90 \\ 2w + 14 + 2w &= 90 \\ 4w + 14 &= 90 \\ 4w + 14 - 14 &= 90 - 14 \\ 4w &= 76 \\ \frac{4w}{4} &= \frac{76}{4} \\ w &= 19 \end{aligned}$$
The dimensions are 19 cm by 26 cm.

45. Let $w =$ the width of the picture.
Then $2w + 5 =$ the length of the picture.
$$\begin{aligned} \boxed{\text{Perimeter}} &= 112 \\ 2w + 2(2w + 5) &= 112 \\ 2w + 4w + 10 &= 112 \\ 6w + 10 &= 112 \\ 6w + 10 - 10 &= 112 - 10 \\ 6w &= 102 \\ \frac{6w}{6} &= \frac{102}{6} \\ w &= 17 \end{aligned}$$
The dimensions are 17 in. by 39 in.

47.
$$\begin{aligned} \boxed{\text{Sum of three sides}} &= 57 \\ x + x + x &= 57 \\ 3x &= 57 \\ \frac{3x}{3} &= \frac{57}{3} \\ x &= 19 \end{aligned}$$
Each side is 19 feet long.

49. Let $a =$ the measure of the vertex angle.
Then $4a =$ the measure of the other angles.
$$\begin{aligned} \boxed{\text{Sum of angle measures}} &= 180 \\ a + 4a + 4a &= 180 \\ 9a &= 180 \\ \frac{9a}{9} &= \frac{180}{9} \\ a &= 20 \end{aligned}$$
The vertex angle measures 20°.

SECTION 2.5

51. Let x = amount invested at 5%.

$$\begin{aligned} I &= PRT \\ 300 &= x(0.05)(1) \\ \frac{300}{0.05} &= \frac{0.05x}{0.05} \\ 6000 &= x\text{; He invested \$6,000.} \end{aligned}$$

53. Let x = amount in 9% fund. Then $24000 - x$ = amount in 14% fund.

$\boxed{\text{Interest at 9\%}} + \boxed{\text{Interest at 14\%}} = \boxed{\text{Total interest}}$

$$\begin{aligned} 0.09x + 0.14(24000 - x) &= 3135 \\ 9x + 14(24000 - x) &= 313500 \\ 9x + 336000 - 14x &= 313500 \\ -5x + 336000 &= 313500 \\ -5x + 336000 - 336000 &= 313500 - 336000 \\ -5x &= -22500 \\ \frac{-5x}{-5} &= \frac{-22500}{-5} \\ x &= 4500 \Rightarrow \text{\$4,500 was invested at 9\%, and \$19,500 was invested at 14\%.} \end{aligned}$$

55. Let x = amount invested in each account.

$\boxed{\text{Interest at 8\%}} + \boxed{\text{Interest at 11\%}} = \boxed{\text{Total interest}}$

$$\begin{aligned} 0.08x + 0.11x &= 712.50 \\ 8x + 11x &= 71250 \\ 19x &= 71250 \\ \frac{19x}{19} &= \frac{71250}{19} \\ x &= 3750 \end{aligned}$$

\$3,750 is invested in each account.

57. Let x = amount invested at 7%.

$\boxed{\text{Interest at 6\%}} + \boxed{\text{Interest at 7\%}} = \boxed{\text{Total interest}}$

$$\begin{aligned} 0.06(15000) + 0.07x &= 1250 \\ 6(15000) + 7x &= 125000 \\ 90000 + 7x &= 125000 \\ 90000 - 90000 + 7x &= 125000 - 90000 \\ 7x &= 35000 \\ \tfrac{7x}{7} &= \tfrac{35000}{7} \\ x &= 5000 \end{aligned}$$

\$5,000 should be invested at 7%.

59. Let r = the fund rate and $r + 0.01$ = the CD rate.
The client invests \$21,000 in CDs, and \$10,500 in the fund.

$\boxed{\text{CD interest}} = \boxed{\text{Fund interest}} + 840$

$$\begin{aligned} (r + 0.01)21000 &= r(10500) + 840 \\ 21000r + 210 &= 10500r + 840 \\ 21000r - 10500r + 210 &= 10500r - 10500r + 840 \\ 10500r + 210 &= 840 \\ 10500r + 210 - 210 &= 840 - 210 \\ 10500r &= 630 \\ \frac{10500r}{10500} &= \frac{630}{10500} \\ r &= 0.06 \Rightarrow \text{The rates are 6\% and 7\%.} \end{aligned}$$

61. **Answers may vary.**

63. Pairs of vertical angles have equal measures.

Exercise 2.6 (page 137)

1. $d = rt = 50h$

3. $0.40(12) = 4.8$ oz

5. $3 + 4(-5) = 3 + (-20) = -17$

7. $2^3 - 3^2 = 2 \cdot 2 \cdot 2 - 3 \cdot 3 = 8 - 9 = -1$

9.
$$\begin{aligned} -2x + 3 &= 9 \\ -2x + 3 - 3 &= 9 - 3 \\ -2x &= 6 \\ \frac{-2x}{-2} &= \frac{6}{-2} \\ x &= -3 \end{aligned}$$

11.
$$\begin{aligned} \frac{2}{3}p + 1 &= 5 \\ \frac{2}{3}p + 1 - 1 &= 5 - 1 \\ \frac{2}{3}p &= 4 \\ \frac{3}{2} \cdot \frac{2}{3}p &= \frac{3}{2} \cdot 4 \\ p &= \tfrac{12}{2} \\ p &= 6 \end{aligned}$$

13. $d = rt$

15. $v = pn$

17. Let t = time for cars to meet.

	r	t	d
Car 1 (A to B)	50	t	$50t$
Car 2 (B to A)	55	t	$55t$

$$\begin{aligned} \boxed{\text{Distance for car 1}} + \boxed{\text{Distance for car 2}} &= 315 \\ 50t + 55t &= 315 \\ 105t &= 315 \\ \frac{105t}{105} &= \frac{315}{105} \\ t &= 3 \end{aligned}$$

The cars meet after 3 hours.

19. Let t = days for crews to meet.

	r	t	d
Crew 1	1.5	t	$1.5t$
Crew 2	1.2	t	$1.2t$

$$\begin{aligned} \boxed{\text{Distance for crew 1}} + \boxed{\text{Distance for crew 2}} &= 9.45 \\ 1.5t + 1.2t &= 9.45 \\ 2.7t &= 9.45 \\ \frac{2.7t}{2.7} &= \frac{9.45}{2.7} \\ t &= 3.5 \end{aligned}$$

The crews meet after 3.5 days.

21. Let t = time for cars to be 715 miles apart.

	r	t	d
Car 1 (going east)	60	t	$60t$
Car 2 (going west)	50	t	$50t$

$$\begin{aligned} \boxed{\text{Distance 1}} + \boxed{\text{Distance 2}} &= 715 \\ 60t + 50t &= 715 \\ 110t &= 715 \\ \frac{110t}{110} &= \frac{715}{110} \\ t &= 6.5 \end{aligned}$$

They will be 715 miles apart after 6.5 hours.

23. Let t = time for boys to be 2 miles apart.

	r	t	d
Boy 1	3	t	$3t$
Boy 2	4	t	$4t$

$$\begin{aligned} \boxed{\text{Distance 1}} + \boxed{\text{Distance 2}} &= 2 \\ 3t + 4t &= 2 \\ 7t &= 2 \\ \frac{7t}{7} &= \frac{2}{7} \\ t &= \frac{2}{7} \end{aligned}$$

They will lose contact after $\frac{2}{7}$ hour (17 min.).

SECTION 2.6

25.

	r	t	d
Car	60 mph	t	$60t$
Bus	50 mph	$t+2$	$50(t+2)$

$\boxed{\text{Car distance}} = \boxed{\text{Bus distance}}$

$$60t = 50(t+2)$$
$$60t = 50t + 100$$
$$60t - 50t = 50t - 50t + 100$$
$$10t = 100$$
$$\frac{10t}{10} = \frac{100}{10}$$
$$t = 10$$

The car will overtake the bus after 10 hours.

27. Let t = time for cars to be 82.5 miles apart.

	r	t	d
Car 1	42	t	$42t$
Car 2	53	t	$53t$

$\boxed{\text{Distance 2}} - \boxed{\text{Distance 1}} = 82.5$

$$53t - 42t = 82.5$$
$$11t = 82.5$$
$$\frac{11t}{11} = \frac{82.5}{11}$$
$$t = 7.5$$

They will be 82.5 miles apart after 7.5 hours.

29. Let r = rate of slow train. Then $r + 20$ = rate of fast train.

	r	t	d
Slow train	r	3	$3r$
Fast train	$r+20$	3	$3(r+20)$

$\boxed{\text{Slow dist.}} + \boxed{\text{Fast dist.}} = 330$

$$3r + 3(r+20) = 330$$
$$3r + 3r + 60 = 330$$
$$6r + 60 = 330$$
$$6r + 60 - 60 = 330 - 60$$
$$6r = 270$$
$$\frac{6r}{6} = \frac{270}{6}$$
$$r = 45$$

The rates are 45 mph and 65 mph.

31. Let t = slower time. Then $5 - t$ = faster time.

	r	t	d
1st part	40	t	$40t$
2nd part	50	$5-t$	$50(5-t)$

$\boxed{\text{1st dist.}} + \boxed{\text{2nd dist.}} = 210$

$$40t + 50(5-t) = 210$$
$$40t + 250 - 50t = 210$$
$$-10t + 250 = 210$$
$$-10t + 250 - 250 = 210 - 250$$
$$-10t = -40$$
$$\frac{-10t}{-10} = \frac{-40}{-10}$$
$$t = 4$$

The car averaged 40 mph for 4 hours.

33. Let T = total number of liters of solution.

12% of the total = liters of acid

$$0.12T = 0.3$$
$$\frac{0.12T}{0.12} = \frac{0.3}{0.12}$$

$T = 2.5$; There are 2.5 liters of the solution.

35. Let $x =$ gallons of \$1.15 fuel.

$$\boxed{\text{Value of \$1.15 fuel}} + \boxed{\text{Value of \$0.85 fuel}} = \boxed{\text{Value of mixture}}$$

$$\begin{aligned} 1.15x + 0.85(20) &= 1(20 + x) \\ 115x + 85(20) &= 100(20 + x) \\ 115x + 1700 &= 2000 + 100x \\ 115x - 100x + 1700 &= 2000 + 100x - 100x \\ 15x + 1700 &= 2000 \\ 15x + 1700 - 1700 &= 2000 - 1700 \\ 15x &= 300 \\ \frac{15x}{15} &= \frac{300}{15} \\ x &= 20 \Rightarrow \text{20 gallons of the \$1.15 fuel should be used.} \end{aligned}$$

37. Let $x =$ gallons of 3% solution used.

$$\boxed{\text{Salt in 3\% solution}} + \boxed{\text{Salt in 7\% solution}} = \boxed{\text{Salt in mixture}}$$

$$\begin{aligned} 0.03x + 0.07(50) &= 0.05(x + 50) \\ 3x + 7(50) &= 5(x + 50) \\ 3x + 350 &= 5x + 250 \\ 3x - 5x + 350 &= 5x - 5x + 250 \\ -2x + 350 &= 250 \\ -2x + 350 - 350 &= 250 - 350 \\ -2x &= -100 \\ \frac{-2x}{-2} &= \frac{-100}{-2} \\ x &= 50 \end{aligned}$$

50 gallons of the 3% mixture should be used.

39. Let $x =$ ounces of water (0%) added.

$$\boxed{\text{Amt. in 10\% sol.}} + \boxed{\text{Amt. in water}} = \boxed{\text{Amt. in 8\% sol.}}$$

$$\begin{aligned} 0.10(30) + 0(x) &= 0.08(30 + x) \\ 10(30) + 0 &= 8(30 + x) \\ 300 &= 240 + 8x \\ 300 - 240 &= 240 - 240 + 8x \\ 60 &= 8x \\ \frac{60}{8} &= \frac{8x}{8} \\ 7.5 &= x \end{aligned}$$

7.5 ounces of water should be added.

41. Let $x =$ pounds of lemon drops. Then $100 - x =$ pounds of jelly beans.

$$\boxed{\text{Value of lemon drops}} + \boxed{\text{Value of jelly beans}} = \boxed{\text{Value of mixture}}$$

$$\begin{aligned} 1.90x + 1.20(100 - x) &= 1.48(100) \\ 190x + 120(100 - x) &= 148(100) \\ 190x + 12000 - 120x &= 14800 \\ 70x + 12000 &= 14800 \\ 70x + 12000 - 12000 &= 14800 - 12000 \\ 70x &= 2800 \\ \frac{70x}{70} &= \frac{2800}{70} \\ x &= 40 \Rightarrow \text{40 lb of lemon drops and 60 lb of jelly beans should be used.} \end{aligned}$$

43. Let c = cost of cashews. Then $c - 0.30$ = cost of peanuts. 20 pounds of each are used.

Value of cashews	+	Value of peanuts	=	Value of mixture

$$\begin{aligned} 20c + 20(c - 0.30) &= 1.05(40) \\ 20c + 20c - 6 &= 42 \\ 40c - 6 &= 42 \\ 40c - 6 + 6 &= 42 + 6 \\ 40c &= 48 \\ \frac{40c}{40} &= \frac{48}{40} \\ c &= 1.20 \end{aligned}$$

A bag of cashews is worth \$1.20.

45. Let x = pounds of regular coffee used.

Value of regular	+	Value of gourmet	=	Value of mixture

$$\begin{aligned} 4(x) + 7(40) &= 5(x + 40) \\ 4x + 280 &= 5x + 200 \\ 4x - 5x + 280 &= 5x - 5x + 200 \\ -x + 280 &= 200 \\ -x + 280 - 280 &= 200 - 280 \\ -x &= -80 \\ x &= 80 \end{aligned}$$

80 pounds of regular coffee should be used.

47. Let c = cost of hazelnut beans per pound.

Value of chocolate	+	Value of hazelnut	=	Value of mixture

$$\begin{aligned} 7(2) + c(5) &= 6(7) \\ 14 + 5c &= 42 \\ 14 - 14 + 5c &= 42 - 14 \\ 5c &= 28 \\ c &= \tfrac{28}{5} = \$5.60 \Rightarrow \text{The hazelnut beans cost \$5.60 per pound.} \end{aligned}$$

49. Answers may vary.

51. Answers may vary.

53. Let x and $x + 2$ represent the integers.

$$\begin{aligned} x + x + 2 &= 16 \\ 2x + 2 &= 16 \\ 2x &= 14 \\ x &= 7 \end{aligned}$$

This says that the integers are 7 and 9, but these are not **even** integers. The equation has a solution, but not the problem itself.

55. You cannot mix a 10% solution and a 20% solution and end up with a solution that has a greater concentration (30%) than either of the original solutions.

Exercise 2.7 (page 145)

1.
$$\begin{aligned} 2x &< 4 \\ \frac{2x}{2} &< \frac{4}{2} \\ x &< 2 \end{aligned}$$

3.
$$\begin{aligned} -3x &\le -6 \\ \frac{-3x}{-3} &\ge \frac{-6}{-3} \\ x &\ge 2 \end{aligned}$$

5.
$$\begin{aligned} 2x - 5 &< 7 \\ 2x - 5 + 5 &< 7 + 5 \\ 2x &< 12 \\ \frac{2x}{2} &< \frac{12}{2} \\ x &< 6 \end{aligned}$$

7. $3x^2 - 2(y^2 - x^2) = 3x^2 + (-2)y^2 - (-2)x^2 = 3x^2 - 2y^2 + 2x^2 = 5x^2 - 2y^2$

9. $\frac{1}{3}(x+6)-\frac{4}{3}(x-9)=\frac{1}{3}x+\frac{1}{3}\cdot 6-\frac{4}{3}x-\frac{4}{3}(-9)=-\frac{3}{3}x+2+12=-x+14$

11. is less than, is greater than

13. double inequality

15. inequality

17.
$$\begin{aligned} x+2 &> 5\\ x+2-2 &> 5-2\\ x &> 3 \end{aligned}$$

19.
$$\begin{aligned} 2x+9 &\le x+8\\ 2x-x+9 &\le x-x+8\\ x+9 &\le 8\\ x+9-9 &\le 8-9\\ x &\le -1 \end{aligned}$$

21.
$$\begin{aligned} 2x-3 &\le 5\\ 2x-3+3 &\le 5+3\\ 2x &\le 8\\ \frac{2x}{2} &\le \frac{8}{2}\\ x &\le 4 \end{aligned}$$

23.
$$\begin{aligned} 8x+4 &> 6x-2\\ 8x-6x+4 &> 6x-6x-2\\ 2x+4 &> -2\\ 2x+4-4 &> -2-4\\ 2x &> -6\\ \frac{2x}{2} &> \frac{-6}{2}\\ x &> -3 \end{aligned}$$

25.
$$\begin{aligned} 7x+2 &> 4x-1\\ 7x-4x+2 &> 4x-4x-1\\ 3x+2 &> -1\\ 3x+2-2 &> -1-2\\ 3x &> -3\\ \frac{3x}{3} &> \frac{-3}{3}\\ x &> -1 \end{aligned}$$

27.
$$\begin{aligned} \frac{5}{2}(7x-15)+x &\ge \frac{13}{2}x-\frac{3}{2}\\ 2\left[\frac{5}{2}(7x-15)+x\right] &\ge 2\left[\frac{13}{2}x-\frac{3}{2}\right]\\ 2\cdot\frac{5}{2}(7x-15)+2x &\ge 2\cdot\frac{13}{2}x-2\cdot\frac{3}{2}\\ 5(7x-15)+2x &\ge 13x-3\\ 35x-75+2x &\ge 13x-3\\ 37x-75 &\ge 13x-3\\ 37x-13x-75 &\ge 13x-13x-3\\ 24x-75 &\ge -3\\ 24x-75+75 &\ge -3+75\\ 24x &\ge 72\\ \frac{24x}{24} &\ge \frac{72}{24}\\ x &\ge 3 \end{aligned}$$

29.
$$\begin{aligned} -x-3 &\le 7\\ -x-3+3 &\le 7+3\\ -x &\le 10\\ -1(-x) &\ge -1(10)\\ x &\ge -10 \end{aligned}$$

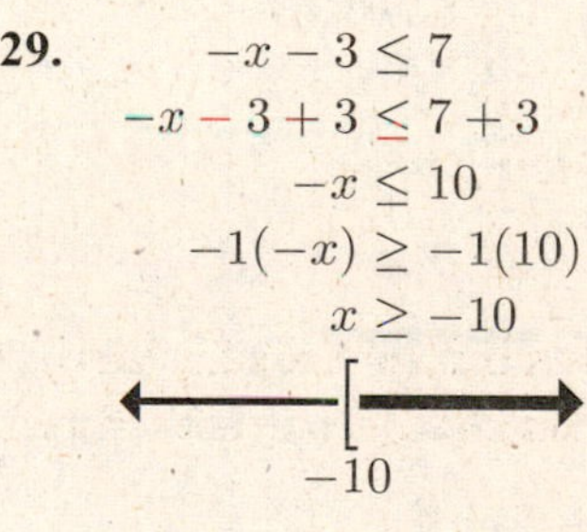

SECTION 2.7

31.

$$\begin{aligned} -3x - 5 &< 4 \\ -3x - 5 + 5 &< 4 + 5 \\ -3x &< 9 \\ \frac{-3x}{-3} &> \frac{9}{-3} \\ x &> -3 \end{aligned}$$

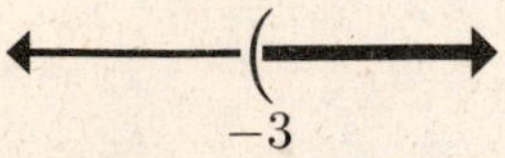

−3

33.

$$\begin{aligned} -3x - 7 &> -1 \\ -3x - 7 + 7 &> -1 + 7 \\ -3x &> 6 \\ \frac{-3x}{-3} &< \frac{6}{-3} \\ x &< -2 \end{aligned}$$

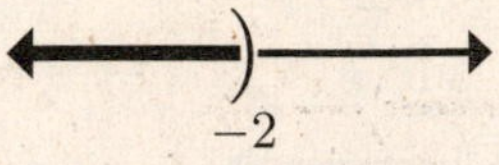

−2

35.

$$\begin{aligned} -4x + 1 &> 17 \\ -4x + 1 - 1 &> 17 - 1 \\ -4x &> 16 \\ \frac{-4x}{-4} &< \frac{16}{-4} \\ x &< -4 \end{aligned}$$

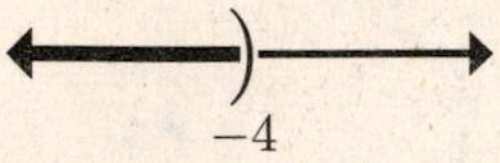

−4

37.

$$\begin{aligned} 9 - 2x &> 24 - 7x \\ 9 - 2x + 7x &> 24 - 7x + 7x \\ 9 + 5x &> 24 \\ 9 - 9 + 5x &> 24 - 9 \\ 5x &> 15 \\ \frac{5x}{5} &> \frac{15}{5} \\ x &> 3 \end{aligned}$$

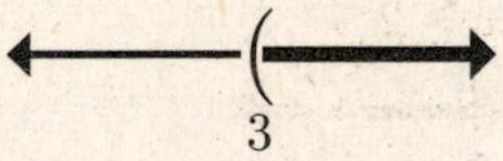

3

39.

$$\begin{aligned} 3(x - 8) &< 5x + 6 \\ 3x - 24 &< 5x + 6 \\ 3x - 5x - 24 &< 5x - 5x + 6 \\ -2x - 24 &< 6 \\ -2x - 24 + 24 &< 6 + 24 \\ -2x &< 30 \\ \frac{-2x}{-2} &> \frac{30}{-2} \\ x &> -15 \end{aligned}$$

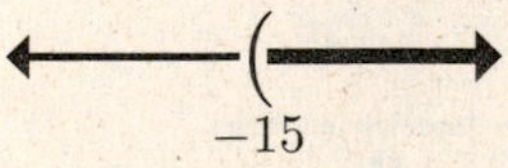

−15

41.

$$\begin{aligned} 2 < \quad x - 5 \quad &< 5 \\ 2 + 5 < x - 5 + 5 &< 5 + 5 \\ 7 < \quad x \quad &< 10 \end{aligned}$$

7 10

43.

$$\begin{aligned} -5 < \quad x + 4 \quad &\le 7 \\ -5 - 4 < x + 4 - 4 &\le 7 - 4 \\ -9 < \quad x \quad &\le 3 \end{aligned}$$

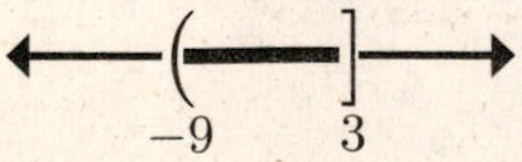

−9 3

45.

$$\begin{aligned} 0 \le \quad x + 10 \quad &\le 10 \\ 0 - 10 \le x + 10 - 10 &\le 10 - 10 \\ -10 \le \quad x \quad &\le 0 \end{aligned}$$

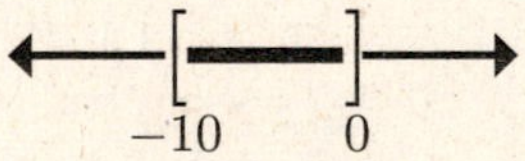

−10 0

47.

$$\begin{aligned} -6 &< 3(x+2) < 9 \\ -6 &< 3x+6 < 9 \\ -6-6 &< 3x+6-6 < 9-6 \\ -12 &< 3x < 3 \\ \frac{-12}{3} &< \frac{3x}{3} < \frac{3}{3} \\ -4 &< x < 1 \end{aligned}$$

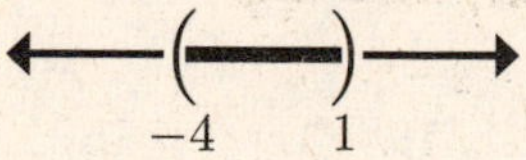

49.

$$\begin{aligned} 5+x &\geq 3 \\ 5-5+x &\geq 3-5 \\ x &\geq -2 \end{aligned}$$

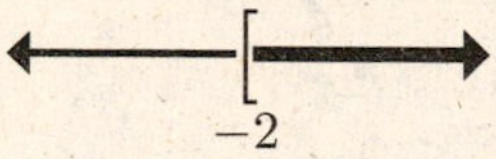

51.

$$\begin{aligned} 7-x &\leq 3x-1 \\ 7-x-3x &\leq 3x-3x-1 \\ 7-4x &\leq -1 \\ 7-7-4x &\leq -1-7 \\ -4x &\leq -8 \\ \frac{-4x}{-4} &\geq \frac{-8}{-4} \\ x &\geq 2 \end{aligned}$$

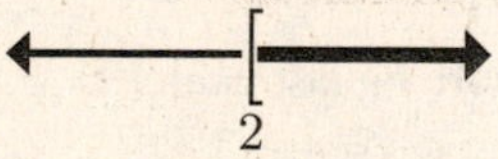

53.

$$\begin{aligned} 8(5-x) &\leq 10(8-x) \\ 40-8x &\leq 80-10x \\ 40-8x+10x &\leq 80-10x+10x \\ 40+2x &\leq 80 \\ 40-40+2x &\leq 80-40 \\ 2x &\leq 40 \\ \frac{2x}{2} &\leq \frac{40}{2} \\ x &\leq 20 \end{aligned}$$

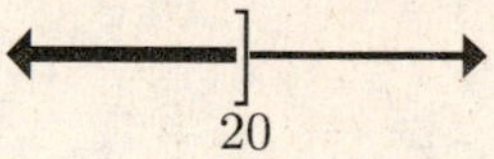

55.

$$\begin{aligned} \frac{3x-3}{2} &< 2x+2 \\ 2\cdot\frac{3x-3}{2} &< 2(2x+2) \\ 3x-3 &< 4x+4 \\ 3x-4x-3 &< 4x-4x+4 \\ -x-3 &< 4 \\ -x-3+3 &< 4+3 \\ -x &< 7 \\ -1(-x) &> -1(7) \\ x &> -7 \end{aligned}$$

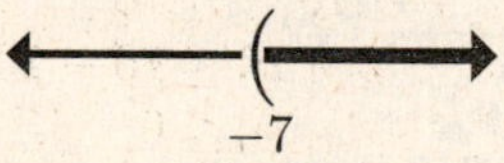

57.

$$\begin{aligned} \frac{2(x+5)}{3} &\leq 3x-6 \\ 3\cdot\frac{2(x+5)}{3} &\leq 3(3x-6) \\ 2(x+5) &\leq 9x-18 \\ 2x+10 &\leq 9x-18 \\ 2x-9x+10 &\leq 9x-9x-18 \\ -7x+10 &\leq -18 \\ -7x+10-10 &\leq -18-10 \\ -7x &\leq -28 \\ \frac{-7x}{-7} &\geq \frac{-28}{-7} \\ x &\geq 4 \end{aligned}$$

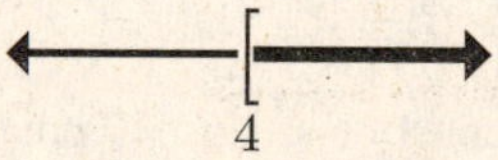

59.

$$\begin{aligned} 4 < -2x &< 10 \\ \frac{4}{-2} > \frac{-2x}{-2} &> \frac{10}{-2} \\ -2 > x &> -5 \\ -5 < x &< -2 \end{aligned}$$

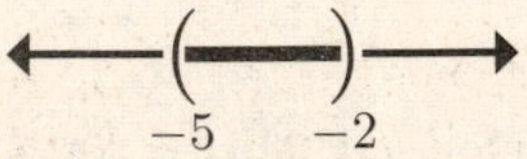

61.

$$\begin{aligned} -3 \le \frac{x}{2} &\le 5 \\ 2(-3) \le 2 \cdot \frac{x}{2} &\le 2(5) \\ -6 \le x &\le 10 \end{aligned}$$

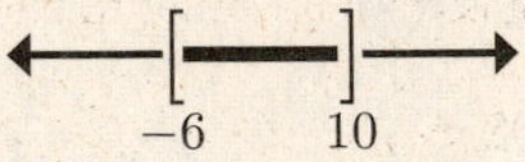

63.

$$\begin{aligned} 3 \le 2x - 1 &< 5 \\ 3 + 1 \le 2x - 1 + 1 &< 5 + 1 \\ 4 \le 2x &< 6 \\ \frac{4}{2} \le \frac{2x}{2} &< \frac{6}{2} \\ 2 \le x &< 3 \end{aligned}$$

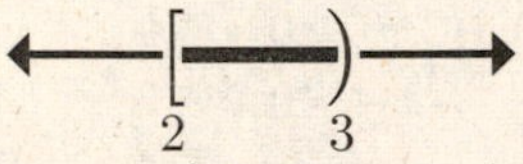

65.

$$\begin{aligned} 0 < 10 - 5x &\le 15 \\ 0 - 10 < 10 - 10 - 5x &\le 15 - 10 \\ -10 < -5x &\le 5 \\ \frac{-10}{-5} > \frac{-5x}{-5} &\ge \frac{5}{-5} \\ 2 > x &\ge -1 \\ -1 \le x &< 2 \end{aligned}$$

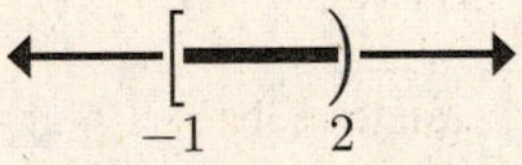

67.

$$\begin{aligned} -4 < \frac{x-2}{2} &< 6 \\ 2(-4) < 2 \cdot \frac{x-2}{2} &< 2(6) \\ -8 < x - 2 &< 12 \\ -8 + 2 < x - 2 + 2 &< 12 + 2 \\ -6 < x &< 14 \end{aligned}$$

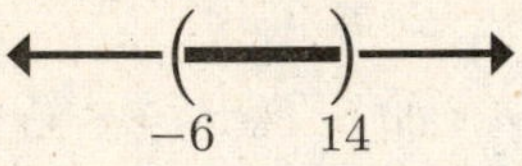

69. Let s = score on last exam.

$$\begin{aligned} \text{Average score} &\ge 80 \\ \frac{68 + 75 + 79 + s}{4} &\ge 80 \\ 4 \cdot \frac{68 + 75 + 79 + s}{4} &\ge 4(80) \\ 68 + 75 + 79 + s &\ge 320 \\ 222 + s &\ge 320 \\ 222 - 222 + s &\ge 320 - 222 \\ s &\ge 98 \end{aligned}$$

Her last test score must be at least 98%.

71. Let s = length of each side.

$$\begin{aligned} \text{Perimeter} &\ge 68 \\ 4s &\ge 68 \\ \frac{4s}{4} &\ge \frac{68}{4} \\ s &\ge 17 \end{aligned}$$

Each side must be at least 17 cm long.

73. Let r = rating of third model.

$$\begin{aligned} \text{Average rating} &\ge 21 \\ \frac{17 + 19 + r}{3} &\ge 21 \\ 3 \cdot \frac{17 + 19 + r}{3} &\ge 3(21) \\ 17 + 19 + r &\ge 63 \\ 36 + r &\ge 63 \\ 36 - 36 + r &\ge 63 - 36 \\ r &\ge 27 \end{aligned}$$

It must have a rating of at least 27 mpg.

75.
$$\begin{aligned} 470\text{ ft} \le &\text{ range in feet } \le 13{,}143\text{ ft} \\ \frac{470\text{ ft}}{5280} \le &\text{ range in miles } \le \frac{13143\text{ ft}}{5280} \\ 0.1\text{ miles} \le &\text{ range in miles } \le 2.5\text{ miles} \end{aligned}$$
The range is from 0.1 miles to 2.5 miles.

77.
$$\begin{aligned} 17500\text{ ft} < &\text{ range in feet } < 21700\text{ ft} \\ \frac{17500\text{ ft}}{5280} < &\text{ range in miles } < \frac{21700\text{ ft}}{5280} \\ 3.3\text{ miles} < &\text{ range in miles } < 4.1\text{ miles} \end{aligned}$$
The range is between 3.3 miles and 4.1 miles.

79.
$$\begin{aligned} 19^\circ < &\quad C &< 22^\circ \\ 19^\circ < &\quad \frac{5}{9}(F-32) &< 22^\circ \\ \frac{9}{5}(19^\circ) < &\quad \frac{9}{5}\cdot\frac{5}{9}(F-32) &< \frac{9}{5}(22^\circ) \\ 34.2^\circ < &\quad F-32 &< 39.6^\circ \\ 34.2^\circ + 32 < &\quad F-32+32 &< 39.6^\circ + 32 \\ 66.2^\circ < &\quad F &< 71.6^\circ \end{aligned}$$
The Fahrenheit temperature is between 66.2° and 71.6°.

81.
$$\begin{aligned} 5.9\text{ in.} < &\quad r &< 6.1\text{ in.} \\ 2\pi(5.9\text{ in.}) < &\quad 2\pi\cdot r &< 2\pi(6.1\text{ in}) \\ 2(3.14)(5.9\text{ in.}) < &\quad 2\pi r &< 2(3.14)(6.1\text{ in.}) \\ 37.052\text{ in.} < &\quad C &< 38.308\text{ in.} \end{aligned}$$
The circumference can vary between about 37.052 inches and 38.308 inches.

83.
$$\begin{aligned} 150\text{ lb} < &\text{ range in lbs } < 190\text{ lb} \\ \frac{150\text{ lb}}{2.2} < &\text{ range in kg } < \frac{190\text{ lb}}{2.2} \\ 68.18\text{ kg} < &\text{ range in kg } < 86.36\text{ kg} \end{aligned}$$
The weight is between 68.18 and 86.36 kg.

85. Let $w =$ width. Then $2w - 3 =$ length.
$$\begin{aligned} 24 < &\quad \text{perimeter} &< 48 \\ 24 < &\quad 2w + 2(2w-3) &< 48 \\ 24 < &\quad 2w + 4w - 6 &< 48 \\ 24 < &\quad 6w - 6 &< 48 \\ 24 + 6 < &\quad 6w - 6 + 6 &< 48 + 6 \\ 30 < &\quad 6w &< 54 \\ \frac{30}{6} < &\quad \frac{6w}{6} &< \frac{54}{6} \\ 5 < &\quad w &< 9 \end{aligned}$$
The width could be between 5 and 9 feet.

87. **Answers may vary.**

89. This is not correct because x could represent a negative number, and then multiplying both sides of the equation would require the inequality to change from $<$ to $>$.

Chapter 2 Review (page 149)

1.
$$\begin{aligned} 3x + 7 &= 1 \\ 3(-2) + 7 &\stackrel{?}{=} 1 \\ -6 + 7 &\stackrel{?}{=} 1 \\ 1 &= 1 \end{aligned}$$
-2 is a solution.

2.
$$\begin{aligned} 5 - 2x &= 3 \\ 5 - 2(-1) &\stackrel{?}{=} 3 \\ 5 + 2 &\stackrel{?}{=} 3 \\ 7 &\ne 3 \end{aligned}$$
-1 is not a solution.

3.
$$\begin{aligned} 2(x+3) &= x \\ 2(-3+3) &\stackrel{?}{=} -3 \\ 2(0) &\stackrel{?}{=} -3 \\ 0 &\ne -3 \end{aligned}$$
-3 is not a solution.

4. $5(3-x)=2-4x$
$5(3-13)\stackrel{?}{=}2-4(13)$
$5(-10)\stackrel{?}{=}2-52$
$-50=-50$
13 is a solution.

5. $3(x+5)=2(x-3)$
$3(-21+5)\stackrel{?}{=}2(-21-3)$
$3(-16)\stackrel{?}{=}2(-24)$
$-48=-48$
-21 is a solution.

6. $2(x-7)=x+14$
$2(0-7)\stackrel{?}{=}0+14$
$2(-7)\stackrel{?}{=}14$
$-14\neq 14$
0 is not a solution.

7. $x-7=-6$
$x-7+7=-6+7$
$x=1$

$x-7=-6$
$1-7\stackrel{?}{=}-6$
$-6=-6$

8. $y-4=5$
$y-4+4=5+4$
$y=9$

$y-4=5$
$9-4\stackrel{?}{=}5$
$5=5$

9. $p+4=20$
$p+4-4=20-4$
$p=16$

$p+4=20$
$16+4\stackrel{?}{=}20$
$20=20$

10. $x+\frac{3}{5}=\frac{3}{5}$
$x+\frac{3}{5}-\frac{3}{5}=\frac{3}{5}-\frac{3}{5}$
$x=0$

$x+\frac{3}{5}=\frac{3}{5}$
$0+\frac{3}{5}\stackrel{?}{=}\frac{3}{5}$
$\frac{3}{5}=\frac{3}{5}$

11. $y-\frac{7}{2}=\frac{1}{2}$
$y-\frac{7}{2}+\frac{7}{2}=\frac{1}{2}+\frac{7}{2}$
$y=\frac{8}{2}$
$y=4$

$y-\frac{7}{2}=\frac{1}{2}$
$4-\frac{7}{2}\stackrel{?}{=}\frac{1}{2}$
$\frac{8}{2}-\frac{7}{2}\stackrel{?}{=}\frac{1}{2}$
$\frac{1}{2}=\frac{1}{2}$

12. $z+\frac{5}{3}=-\frac{1}{3}$
$z+\frac{5}{3}-\frac{5}{3}=-\frac{1}{3}-\frac{5}{3}$
$z=-\frac{6}{3}$
$z=-2$

$z+\frac{5}{3}=-\frac{1}{3}$
$-2+\frac{5}{3}\stackrel{?}{=}-\frac{1}{3}$
$-\frac{6}{3}+\frac{5}{3}\stackrel{?}{=}-\frac{1}{3}$
$-\frac{1}{3}=-\frac{1}{3}$

13. Let $r=$ the regular price. Then

$\boxed{\text{Sale price}}=\boxed{\text{Regular price}}-\boxed{\text{Markdown}}$
$69.95=r-35.45$
$69.95+35.45=r-35.45+35.45$
$105.40=r$

The regular price is \$105.40.

14. Let $w=$ the wholesale cost. Then

$\boxed{\text{Retail price}}=\boxed{\text{Wholesale cost}}+\boxed{\text{Markup}}$
$212.95=w+115.25$
$212.95-115.25=w+115.25-115.25$
$97.70=w$

The wholesale cost is \$97.70.

15. $3x=15$
$\frac{3x}{3}=\frac{15}{3}$
$x=5$

$3x=15$
$3(5)\stackrel{?}{=}15$
$15=15$

16. $8r=-16$
$\frac{8r}{8}=\frac{-16}{8}$
$r=-2$

$8r=-16$
$8(-2)\stackrel{?}{=}-16$
$-16=-16$

17. $10z=5$
$\frac{10z}{10}=\frac{5}{10}$
$z=\frac{1}{2}$

$10z=5$
$10\left(\frac{1}{2}\right)\stackrel{?}{=}5$
$5=5$

18. $14q=21$
$\frac{14q}{14}=\frac{21}{14}$
$q=\frac{3}{2}$

$14q=21$
$14\left(\frac{3}{2}\right)\stackrel{?}{=}21$
$21=21$

19.
$$\frac{y}{3}=6 \qquad 3\cdot\frac{y}{3}=3\cdot 6 \qquad y=18$$
$$\frac{y}{3}=6 \qquad \frac{18}{3}\stackrel{?}{=}6 \qquad 6=6$$

20.
$$\frac{w}{7}=-5 \qquad 7\cdot\frac{w}{7}=7(-5) \qquad w=-35$$
$$\frac{w}{7}=-5 \qquad \frac{-35}{7}\stackrel{?}{=}-5 \qquad -5=-5$$

21.
$$\frac{a}{-7}=\frac{1}{14} \qquad -7\cdot\frac{a}{-7}=-7\left(\frac{1}{14}\right) \qquad a=-\frac{7}{14} \qquad a=-\frac{1}{2}$$
$$\frac{a}{-7}=\frac{1}{14} \qquad \frac{-\frac{1}{2}}{-7}\stackrel{?}{=}\frac{1}{14} \qquad \frac{2(-\frac{1}{2})}{2(-7)}\stackrel{?}{=}\frac{1}{14} \qquad \frac{-1}{-14}=\frac{1}{14}$$

22.
$$\frac{p}{12}=\frac{1}{2} \qquad 12\cdot\frac{p}{12}=12\left(\frac{1}{2}\right) \qquad p=\frac{12}{2} \qquad p=6$$
$$\frac{p}{12}=\frac{1}{2} \qquad \frac{6}{12}\stackrel{?}{=}\frac{1}{2} \qquad \frac{1}{2}=\frac{1}{2}$$

23.
$$rb=a \qquad 35\%\cdot 700=a \qquad 0.35(700)=a \qquad 245=a$$

24.
$$rb=a \qquad 72\%\cdot b=936 \qquad 0.72b=936 \qquad \frac{0.72b}{0.72}=\frac{936}{0.72} \qquad b=1{,}300$$

25.
$$rb=a \qquad r\cdot 2300=851 \qquad \frac{r\cdot 2300}{2300}=\frac{851}{2300} \qquad r=0.37 \qquad r=37\%$$

26.
$$rb=a \qquad r\cdot 576=72 \qquad \frac{r\cdot 576}{576}=\frac{72}{576} \qquad r=0.125 \qquad r=12.5\%$$

27.
$$5y+6=21 \qquad 5y+6-6=21-6 \qquad 5y=15 \qquad \frac{5y}{5}=\frac{15}{5} \qquad y=3$$
$$5y+6=21 \qquad 5(3)+6\stackrel{?}{=}21 \qquad 15+6\stackrel{?}{=}21 \qquad 21=21$$

28.
$$5y-9=1 \qquad 5y-9+9=1+9 \qquad 5y=10 \qquad \frac{5y}{5}=\frac{10}{5} \qquad y=2$$
$$5y-9=1 \qquad 5(2)-9\stackrel{?}{=}1 \qquad 10-9\stackrel{?}{=}1 \qquad 1=1$$

29.
$$-12z+4=-8 \qquad -12z+4-4=-8-4 \qquad -12z=-12 \qquad \frac{-12z}{-12}=\frac{-12}{-12} \qquad z=1$$
$$-12z+4=-8 \qquad -12(1)+4\stackrel{?}{=}-8 \qquad -12+4\stackrel{?}{=}-8 \qquad -8=-8$$

30.
$$17z+3=20 \qquad 17z+3-3=20-3 \qquad 17z=17 \qquad \frac{17z}{17}=\frac{17}{17} \qquad z=1$$
$$17z+3=20 \qquad 17(1)+3\stackrel{?}{=}20 \qquad 17+3\stackrel{?}{=}20 \qquad 20=20$$

31.
$$13-13p=0 \qquad 13-13-13p=0-13 \qquad -13p=-13 \qquad \frac{-13p}{-13}=\frac{-13}{-13} \qquad p=1$$
$$13-13p=0 \qquad 13-13(1)\stackrel{?}{=}0 \qquad 13-13\stackrel{?}{=}0 \qquad 0=0$$

CHAPTER 2 REVIEW

32.
$$\begin{aligned} 10 + 7p &= -4 \\ 10 - 10 + 7p &= -4 - 10 \\ 7p &= -14 \\ \frac{7p}{7} &= \frac{-14}{7} \\ p &= -2 \end{aligned} \qquad \begin{aligned} 10 + 7p &= -4 \\ 10 + 7(-2) &\stackrel{?}{=} -4 \\ 10 + (-14) &\stackrel{?}{=} -4 \\ -4 &= -4 \end{aligned}$$

33.
$$\begin{aligned} 23a - 43 &= 3 \\ 23a - 43 + 43 &= 3 + 43 \\ 23a &= 46 \\ \frac{23a}{23} &= \frac{46}{23} \\ a &= 2 \end{aligned} \qquad \begin{aligned} 23a - 43 &= 3 \\ 23(2) - 43 &\stackrel{?}{=} 3 \\ 46 - 43 &\stackrel{?}{=} 3 \\ 3 &= 3 \end{aligned}$$

34.
$$\begin{aligned} 84 - 21a &= -63 \\ 84 - 84 - 21a &= -63 - 84 \\ -21a &= -147 \\ \frac{-21a}{-21} &= \frac{-147}{-21} \\ a &= 7 \end{aligned} \qquad \begin{aligned} 84 - 21a &= -63 \\ 84 - 21(7) &\stackrel{?}{=} -63 \\ 84 - 147 &\stackrel{?}{=} -63 \\ -63 &= -63 \end{aligned}$$

35.
$$\begin{aligned} 3x + 7 &= 1 \\ 3x + 7 - 7 &= 1 - 7 \\ 3x &= -6 \\ \frac{3x}{3} &= \frac{-6}{3} \\ x &= -2 \end{aligned} \qquad \begin{aligned} 3x + 7 &= 1 \\ 3(-2) + 7 &\stackrel{?}{=} 1 \\ -6 + 7 &\stackrel{?}{=} 1 \\ 1 &= 1 \end{aligned}$$

36.
$$\begin{aligned} 7 - 9x &= 16 \\ 7 - 7 - 9x &= 16 - 7 \\ -9x &= 9 \\ \frac{-9x}{-9} &= \frac{9}{-9} \\ x &= -1 \end{aligned} \qquad \begin{aligned} 7 - 9x &= 16 \\ 7 - 9(-1) &\stackrel{?}{=} 16 \\ 7 + 9 &\stackrel{?}{=} 16 \\ 16 &= 16 \end{aligned}$$

37.
$$\begin{aligned} \frac{b+3}{4} &= 2 \\ 4 \cdot \frac{b+3}{4} &= 4 \cdot 2 \\ b + 3 &= 8 \\ b + 3 - 3 &= 8 - 3 \\ b &= 5 \end{aligned} \qquad \begin{aligned} \frac{b+3}{4} &= 2 \\ \frac{5+3}{4} &\stackrel{?}{=} 2 \\ \frac{8}{4} &\stackrel{?}{=} 2 \\ 2 &= 2 \end{aligned}$$

38.
$$\begin{aligned} \frac{b-7}{2} &= -2 \\ 2 \cdot \frac{b-7}{2} &= 2(-2) \\ b - 7 &= -4 \\ b - 7 + 7 &= -4 + 7 \\ b &= 3 \end{aligned} \qquad \begin{aligned} \frac{b-7}{2} &= -2 \\ \frac{3-7}{2} &\stackrel{?}{=} -2 \\ \frac{-4}{2} &\stackrel{?}{=} -2 \\ -2 &= -2 \end{aligned}$$

39.
$$\begin{aligned} \frac{x-8}{5} &= 1 \\ 5 \cdot \frac{x-8}{5} &= 5(1) \\ x - 8 &= 5 \\ x - 8 + 8 &= 5 + 8 \\ x &= 13 \end{aligned} \qquad \begin{aligned} \frac{x-8}{5} &= 1 \\ \frac{13-8}{5} &\stackrel{?}{=} 1 \\ \frac{5}{5} &\stackrel{?}{=} 1 \\ 1 &= 1 \end{aligned}$$

40.
$$\begin{aligned} \frac{x+10}{2} &= -1 \\ 2 \cdot \frac{x+10}{2} &= 2(-1) \\ x + 10 &= -2 \\ x + 10 - 10 &= -2 - 10 \\ x &= -12 \end{aligned} \qquad \begin{aligned} \frac{x+10}{2} &= -1 \\ \frac{-12+10}{2} &\stackrel{?}{=} -1 \\ \frac{-2}{2} &\stackrel{?}{=} -1 \\ -1 &= -1 \end{aligned}$$

41.

$$\begin{aligned} \frac{2y-2}{4} &= 2 \\ 4\cdot\frac{2y-2}{4} &= 4(2) \\ 2y-2 &= 8 \\ 2y-2+2 &= 8+2 \\ 2y &= 10 \\ \frac{2y}{2} &= \frac{10}{2} \\ y &= 5 \end{aligned}
\qquad
\begin{aligned} \frac{2y-2}{4} &= 2 \\ \frac{2(5)-2}{4} &\stackrel{?}{=} 2 \\ \frac{10-2}{4} &\stackrel{?}{=} 2 \\ \frac{8}{4} &\stackrel{?}{=} 2 \\ 2 &= 2 \end{aligned}$$

42.

$$\begin{aligned} \frac{3y+12}{11} &= 3 \\ 11\cdot\frac{3y+12}{11} &= 11(3) \\ 3y+12 &= 33 \\ 3y+12-12 &= 33-12 \\ 3y &= 21 \\ \frac{3y}{3} &= \frac{21}{3} \\ y &= 7 \end{aligned}
\qquad
\begin{aligned} \frac{3y+12}{11} &= 3 \\ \frac{3(7)+12}{11} &\stackrel{?}{=} 3 \\ \frac{21+12}{11} &\stackrel{?}{=} 3 \\ \frac{33}{11} &\stackrel{?}{=} 3 \\ 3 &= 3 \end{aligned}$$

43.

$$\begin{aligned} \frac{x}{2}+7 &= 11 \\ \frac{x}{2}+7-7 &= 11-7 \\ \frac{x}{2} &= 4 \\ 2\cdot\frac{x}{2} &= 2\cdot 4 \\ x &= 8 \end{aligned}
\qquad
\begin{aligned} \frac{x}{2}+7 &= 11 \\ \frac{8}{2}+7 &\stackrel{?}{=} 11 \\ 4+7 &\stackrel{?}{=} 11 \\ 11 &= 11 \end{aligned}$$

44.

$$\begin{aligned} \frac{r}{3}-3 &= 7 \\ \frac{r}{3}-3+3 &= 7+3 \\ \frac{r}{3} &= 10 \\ 3\cdot\frac{r}{3} &= 3\cdot 10 \\ r &= 30 \end{aligned}
\qquad
\begin{aligned} \frac{r}{3}-3 &= 7 \\ \frac{30}{3}-3 &\stackrel{?}{=} 7 \\ 10-3 &\stackrel{?}{=} 7 \\ 7 &= 7 \end{aligned}$$

45.

$$\begin{aligned} \frac{a}{2}+\frac{9}{4} &= 6 \\ \frac{a}{2}+\frac{9}{4}-\frac{9}{4} &= \frac{24}{4}-\frac{9}{4} \\ \frac{a}{2} &= \frac{15}{4} \\ 2\cdot\frac{a}{2} &= 2\cdot\frac{15}{4} \\ a &= \frac{15}{2} \end{aligned}
\qquad
\begin{aligned} \frac{a}{2}+\frac{9}{4} &= 6 \\ \frac{\frac{15}{2}}{2}+\frac{9}{4} &\stackrel{?}{=} 6 \\ \frac{2\cdot\frac{15}{2}}{2\cdot 2}+\frac{9}{4} &\stackrel{?}{=} 6 \\ \frac{15}{4}+\frac{9}{4} &\stackrel{?}{=} 6 \\ \frac{24}{4} &\stackrel{?}{=} 6 \\ 6 &= 6 \end{aligned}$$

46.

$$\begin{aligned} \frac{x}{8}-2.3 &= 3.2 \\ \frac{x}{8}-2.3+2.3 &= 3.2+2.3 \\ \frac{x}{8} &= 5.5 \\ 8\cdot\frac{x}{8} &= 8(5.5) \\ x &= 44 \end{aligned}
\qquad
\begin{aligned} \frac{x}{8}-2.3 &= 3.2 \\ \frac{44}{8}-2.3 &\stackrel{?}{=} 3.2 \\ 5.5-2.3 &\stackrel{?}{=} 3.2 \\ 3.2 &= 3.2 \end{aligned}$$

47. Let x = the regular price.

[Sale price] = [Regular price] − [Markdown]

$$\begin{aligned} 240 &= x-0.25x \\ 240 &= 0.75x \\ \frac{240}{0.75} &= \frac{0.75x}{0.75} \\ 320 &= x \end{aligned}$$

The regular price is \$320.

48. Let r = the sales tax rate.

[Total price] = [Price before tax] + [Sales tax]

$$\begin{aligned} 40.47 &= 38+38r \\ 40.47-38 &= 38-38+38r \\ 2.47 &= 38r \\ \frac{2.47}{38} &= \frac{38r}{38} \\ 0.065 &= r \end{aligned}$$

The sales tax rate is 6.5%.

49. Let r = the % increase.

$$\boxed{\text{New price}} = \boxed{\text{Original price}} + \boxed{\text{Increase}}$$

$$\begin{aligned} 1100 &= 560 + 560r \\ 1100 - 560 &= 560 - 560 + 560r \\ 540 &= 560r \\ \frac{540}{560} &= \frac{560r}{560} \\ 0.964 &= r \end{aligned}$$

The percent increase is 96.4%.

50. Let r = the % discount.

$$\boxed{\text{Markdown}} = \boxed{\text{Percent markdown}} \cdot \boxed{\text{Regular price}}$$

$$\begin{aligned} 465 - 215 &= r \cdot 465 \\ 250 &= 465r \\ \frac{250}{465} &= \frac{465r}{465} \\ 0.538 &= r \end{aligned}$$

The percent markdown is 53.8%.

51. $5x + 9x = (5 + 9)x = 14x$

52. $7a + 12a = (7 + 12)a = 19a$

53. $18b - 13b = (18 - 13)b = 5b$

54. $21x - 23x = (21 - 23)x = -2x$

55. $5y - 7y = (5 - 7)y = -2y$

56. $19x - 19 \Rightarrow$ unlike terms

57. $$\begin{aligned} 7(x + 2) + 2(x - 7) &= 7x + 14 + 2x - 14 \\ &= 7x + 2x + 14 - 14 \\ &= (7 + 2)x + 0 = 9x \end{aligned}$$

58. $$\begin{aligned} 2(3 - x) + x - 6x &= 6 - 2x + x - 6x \\ &= 6 + (-2 + 1 - 6)x \\ &= 6 - 7x \end{aligned}$$

59. $$\begin{aligned} y^2 + 3(y^2 - 2) &= y^2 + 3y^2 - 6 \\ &= (1 + 3)y^2 - 6 \\ &= 4y^2 - 6 \end{aligned}$$

60. $$\begin{aligned} 2x^2 - 2(x^2 - 2) &= 2x^2 - 2x^2 + 4 \\ &= (2 - 2)x^2 + 4 \\ &= 0x^2 + 4 = 4 \end{aligned}$$

61. $$\begin{aligned} 2x - 19 &= 2 - x \\ 2x + x - 19 &= 2 - x + x \\ 3x - 19 &= 2 \\ 3x - 19 + 19 &= 2 + 19 \\ 3x &= 21 \\ \frac{3x}{3} &= \frac{21}{3} \\ x &= 7 \end{aligned}$$

$$\begin{aligned} 2x - 19 &= 2 - x \\ 2(7) - 19 &\stackrel{?}{=} 2 - 7 \\ 14 - 19 &\stackrel{?}{=} -5 \\ -5 &= -5 \end{aligned}$$

62. $$\begin{aligned} 5b - 19 &= 2b + 20 \\ 5b - 2b - 19 &= 2b - 2b + 20 \\ 3b - 19 &= 20 \\ 3b - 19 + 19 &= 20 + 19 \\ 3b &= 39 \\ \tfrac{3b}{3} &= \tfrac{39}{3} \\ b &= 13 \end{aligned}$$

$$\begin{aligned} 5b - 19 &= 2b + 20 \\ 5(13) - 19 &\stackrel{?}{=} 2(13) + 20 \\ 65 - 19 &\stackrel{?}{=} 26 + 20 \\ 46 &= 46 \end{aligned}$$

63.

$$\begin{aligned} 3x + 20 &= 5 - 2x \\ 3x + 2x + 20 &= 5 - 2x + 2x \\ 5x + 20 &= 5 \\ 5x + 20 - 20 &= 5 - 20 \\ 5x &= -15 \\ \tfrac{5x}{5} &= \tfrac{-15}{5} \\ x &= -3 \end{aligned}$$

$$\begin{aligned} 3x + 20 &= 5 - 2x \\ 3(-3) + 20 &\stackrel{?}{=} 5 - 2(-3) \\ -9 + 20 &\stackrel{?}{=} 5 + 6 \\ 11 &= 11 \end{aligned}$$

64.

$$\begin{aligned} 0.9x + 10 &= 0.7x + 1.8 \\ 0.9x - 0.7x + 10 &= 0.7x - 0.7x + 1.8 \\ 0.2x + 10 &= 1.8 \\ 0.2x + 10 - 10 &= 1.8 - 10 \\ 0.2x &= -8.2 \\ \tfrac{0.2x}{0.2} &= \tfrac{-8.2}{0.2} \\ x &= -41 \end{aligned}$$

$$\begin{aligned} 0.9x + 10 &= 0.7x + 1.8 \\ 0.9(-41) + 10 &\stackrel{?}{=} 0.7(-41) + 1.8 \\ -36.9 + 10 &\stackrel{?}{=} -28.7 + 1.8 \\ -26.9 &= -26.9 \end{aligned}$$

65.

$$\begin{aligned} 10(p - 3) &= 3(p + 11) \\ 10p - 30 &= 3p + 33 \\ 10p - 3p - 30 &= 3p - 3p + 33 \\ 7p - 30 &= 33 \\ 7p - 30 + 30 &= 33 + 30 \\ 7p &= 63 \\ \tfrac{7p}{7} &= \tfrac{63}{7} \\ p &= 9 \end{aligned}$$

$$\begin{aligned} 10(p - 3) &= 3(p + 11) \\ 10(9 - 3) &\stackrel{?}{=} 3(9 + 11) \\ 10(6) &\stackrel{?}{=} 3(20) \\ 60 &= 60 \end{aligned}$$

66.

$$\begin{aligned} 2(5x - 7) &= 2(x - 35) \\ 10x - 14 &= 2x - 70 \\ 10x - 2x - 14 &= 2x - 2x - 70 \\ 8x - 14 &= -70 \\ 8x - 14 + 14 &= -70 + 14 \\ 8x &= -56 \\ \tfrac{8x}{8} &= \tfrac{-56}{8} \\ x &= -7 \end{aligned}$$

$$\begin{aligned} 2(5x - 7) &= 2(x - 35) \\ 2[5(-7) - 7] &\stackrel{?}{=} 2(-7 - 35) \\ 2[-35 - 7] &\stackrel{?}{=} 2(-42) \\ 2(-42) &\stackrel{?}{=} 2(-42) \\ -84 &= -84 \end{aligned}$$

67.

$$\begin{aligned} \frac{3u-6}{5} &= 3 \\ 5 \cdot \frac{3u-6}{5} &= 5(3) \\ 3u-6 &= 15 \\ 3u-6+6 &= 15+6 \\ 3u &= 21 \\ \frac{3u}{3} &= \frac{21}{3} \\ u &= 7 \end{aligned}$$

$$\begin{aligned} \frac{3u-6}{5} &= 3 \\ \frac{3(7)-6}{5} &\stackrel{?}{=} 3 \\ \frac{21-6}{5} &\stackrel{?}{=} 3 \\ \frac{15}{5} &\stackrel{?}{=} 3 \\ 3 &= 3 \end{aligned}$$

68.

$$\begin{aligned} \frac{5v-35}{3} &= -5 \\ 3 \cdot \frac{5v-35}{3} &= 3(-5) \\ 5v-35 &= -15 \\ 5v-35+35 &= -15+35 \\ 5v &= 20 \\ \frac{5v}{5} &= \frac{20}{5} \\ v &= 4 \end{aligned}$$

$$\begin{aligned} \frac{5v-35}{3} &= -5 \\ \frac{5(4)-35}{3} &\stackrel{?}{=} -5 \\ \frac{20-35}{3} &\stackrel{?}{=} -5 \\ \frac{-15}{3} &\stackrel{?}{=} -5 \\ -5 &= -5 \end{aligned}$$

69.

$$\begin{aligned} \frac{7x-28}{4} &= -21 \\ 4 \cdot \frac{7x-28}{4} &= 4(-21) \\ 7x-28 &= -84 \\ 7x-28+28 &= -84+28 \\ 7x &= -56 \\ \frac{7x}{7} &= \frac{-56}{7} \\ x &= -8 \end{aligned}$$

$$\begin{aligned} \frac{7x-28}{4} &= -21 \\ \frac{7(-8)-28}{4} &\stackrel{?}{=} -21 \\ \frac{-56-28}{4} &\stackrel{?}{=} -21 \\ \frac{-84}{4} &\stackrel{?}{=} -21 \\ -21 &= -21 \end{aligned}$$

70.

$$\begin{aligned} \frac{27+9y}{5} &= -27 \\ 5 \cdot \frac{27+9y}{5} &= 5(-27) \\ 27+9y &= -135 \\ 27-27+9y &= -135-27 \\ 9y &= -162 \\ \frac{9y}{9} &= \frac{-162}{9} \\ y &= -18 \end{aligned}$$

$$\begin{aligned} \frac{27+9y}{5} &= -27 \\ \frac{27+9(-18)}{5} &\stackrel{?}{=} -27 \\ \frac{27+(-162)}{5} &\stackrel{?}{=} -27 \\ \frac{-135}{5} &\stackrel{?}{=} -27 \\ -27 &= -27 \end{aligned}$$

71.
$$\begin{aligned} 2x-5&=x-5+x\\ 2x-5&=2x-5\\ 2x-2x-5&=2x-2x-5\\ -5&=-5 \end{aligned}$$
Identity, $\mathbb{R}$

72.
$$\begin{aligned} -3(a+1)-a&=-4a+3\\ -3a-3-a&=-4a+3\\ -4a-3&=-4a+3\\ -4a+4a-3&=-4a+4a+3\\ -3&\neq 3 \end{aligned}$$
Contradiction, $\emptyset$

73.
$$\begin{aligned} 2(x-1)+4&=4(1+x)-(2x+2)\\ 2x-2+4&=4+4x-2x-2\\ 2x+2&=2x+2\\ 2x-2x+2&=2x-2x+2\\ 2&=2 \end{aligned}$$
Identity, $\mathbb{R}$

74.
$$\begin{aligned} 3(2x+1)+3&=9(x+2)+9-3x\\ 6x+3+3&=9x+18+9-3x\\ 6x+6&=6x+27\\ 6x-6x+6&=6x-6x+27\\ 6&\neq 27 \end{aligned}$$
Contradiction, $\emptyset$

75.
$$\begin{aligned} E&=IR\\ \frac{E}{I}&=\frac{IR}{I}\\ \frac{E}{I}&=R, \text{ or } R=\frac{E}{I} \end{aligned}$$

76.
$$\begin{aligned} i&=prt\\ \frac{i}{pr}&=\frac{prt}{pr}\\ \frac{i}{pr}&=t, \text{ or } t=\frac{i}{pr} \end{aligned}$$

77.
$$\begin{aligned} P&=I^2R\\ \frac{P}{I^2}&=\frac{I^2R}{I^2}\\ \frac{P}{I^2}&=R, \text{ or } R=\frac{P}{I^2} \end{aligned}$$

78.
$$\begin{aligned} d&=rt\\ \frac{d}{t}&=\frac{rt}{t}\\ \frac{d}{t}&=r, \text{ or } r=\frac{d}{t} \end{aligned}$$

79.
$$\begin{aligned} V&=lwh\\ \frac{V}{lw}&=\frac{lwh}{lw}\\ \frac{V}{lw}&=h, \text{ or } h=\frac{V}{lw} \end{aligned}$$

80.
$$\begin{aligned} y&=mx+b\\ y-b&=mx+b-b\\ y-b&=mx\\ \frac{y-b}{x}&=\frac{mx}{x}\\ \frac{y-b}{x}&=m, \text{ or } m=\frac{y-b}{x} \end{aligned}$$

81.
$$\begin{aligned} V&=\pi r^2h\\ \frac{V}{\pi r^2}&=\frac{\pi r^2h}{\pi r^2}\\ \frac{V}{\pi r^2}&=h, \text{ or } h=\frac{V}{\pi r^2} \end{aligned}$$

82.
$$\begin{aligned} a&=2\pi rh\\ \frac{a}{2\pi h}&=\frac{2\pi rh}{2\pi h}\\ \frac{a}{2\pi h}&=r, \text{ or } r=\frac{a}{2\pi h} \end{aligned}$$

83.
$$\begin{aligned} F&=\frac{GMm}{d^2}\\ d^2\cdot F&=d^2\cdot\frac{GMm}{d^2}\\ d^2F&=GMm\\ \frac{d^2F}{Mm}&=\frac{GMm}{Mm}\\ \frac{d^2F}{Mm}&=G, \text{ or } G=\frac{d^2F}{Mm} \end{aligned}$$

84.
$$P = \frac{RT}{mV}$$
$$mV \cdot P = mV \cdot \frac{RT}{mV}$$
$$mVP = RT$$
$$\frac{mVP}{VP} = \frac{RT}{VP}$$
$$m = \frac{RT}{VP}$$

85. Let $x =$ the length of one part. Then $2x - 7 =$ the length of the other part.

$$\boxed{\text{Length of first part}} + \boxed{\text{Length of second part}} = 8$$
$$x + 2x - 7 = 8$$
$$3x - 7 = 8$$
$$3x - 7 + 7 = 8 + 7$$
$$3x = 15$$
$$\frac{3x}{3} = \frac{15}{3}$$
$$x = 5$$

One part should be cut 5 feet long.

86.
$$x + 47° = 62°$$
$$x + 47° - 47° = 62° - 47°$$
$$x = 15°$$

87.
$$x + 135° = 180°$$
$$x + 135° - 135° = 180° - 135°$$
$$x = 45°$$

88. Let $x =$ the measure of the complement.
$$x + 69° = 90°$$
$$x + 69° - 69° = 90° - 69°$$
$$x = 21°$$

89. Let $x =$ the measure of the supplement.
$$x + 69° = 180°$$
$$x + 69° - 69° = 180° - 69°$$
$$x = 111°$$

90. Let $w =$ the width of the picture. Then $2w + 3 =$ the length of the picture.
$$\boxed{\text{Perimeter}} = 84$$
$$2w + 2(2w + 3) = 84$$
$$2w + 4w + 6 = 84$$
$$6w + 6 = 84$$
$$6w + 6 - 6 = 84 - 6$$
$$6w = 78$$
$$\frac{6w}{6} = \frac{78}{6}$$
$w = 13 \Rightarrow$ The width is 13 inches.

91. Let $x =$ amount in 7% CD. Then $27000 - x =$ amount in 9% fund.
$$\boxed{\text{Interest at 7\%}} + \boxed{\text{Interest at 9\%}} = \boxed{\text{Total interest}}$$
$$0.07x + 0.09(27000 - x) = 2110$$
$$7x + 9(27000 - x) = 211000$$
$$7x + 243000 - 9x = 211000$$
$$-2x + 243000 = 211000$$
$$-2x + 243000 - 243000 = 211000 - 243000$$
$$-2x = -32000$$
$$\frac{-2x}{-2} = \frac{-32000}{-2}$$
$x = 16000 \Rightarrow$ \$16,000 is invested at 7%, and \$11,000 is invested at 9%.

92. Let t = time for friends to meet.

	r	t	d
Walk	3	t	$3t$
Bike	12	t	$12t$

$$\boxed{\text{Distance walked}} + \boxed{\text{Distance biked}} = 5$$
$$3t + 12t = 5$$
$$15t = 5$$
$$\frac{15t}{15} = \frac{5}{15}$$
$$t = \frac{1}{3}$$

They meet after $\frac{1}{3}$ hour (or 20 minutes).

93. Let t = time to be 90 miles apart.

	r	t	d
1st Team	20	t	$20t$
2nd Team	25	t	$25t$

$$\boxed{\text{Distance of 1st Team}} + \boxed{\text{Distance of 2nd Team}} = 90$$
$$20t + 25t = 90$$
$$45t = 90$$
$$\frac{45t}{45} = \frac{90}{45}$$
$$t = 2$$

They will lose contact in 2 hours.

94. Let t = time needed to be 75 miles apart.

	r	t	d
Bus	65	t	$65t$
Truck	55	t	$55t$

$$\boxed{\text{Bus Distance}} - \boxed{\text{Truck Distance}} = 75$$
$$65t - 55t = 75$$
$$10t = 75$$
$$\frac{10t}{10} = \frac{75}{10}$$
$$t = 7.5$$

They will be 75 miles apart after 7.5 hours.

95. Let x = liters of 1% butterfat milk used.

$$\boxed{\text{Butterfat in 4\% milk}} + \boxed{\text{Butterfat in 1\% milk}} = \boxed{\text{Butterfat in 2\% mixture}}$$
$$0.04(12) + 0.01x = 0.02(12 + x)$$
$$4(12) + 1x = 2(12 + x)$$
$$48 + x = 24 + 2x$$
$$x - 2x + 48 = 24 + 2x - 2x$$
$$-x + 48 = 24$$
$$-x + 48 - 48 = 24 - 48$$
$$-x = -24$$
$$x = 24$$

24 liters of the 1% milk should be used.

96. Let x = liters of 12% solution used.

$$\boxed{\text{Acid in 6\% solution}} + \boxed{\text{Acid in 12\% solution}} = \boxed{\text{Acid in 8\% solution}}$$
$$0.06(2) + 0.12x = 0.08(2 + x)$$
$$0.12 + 0.12x = 0.16 + 0.08x$$
$$12 + 12x = 16 + 8x$$
$$12 + 12x - 8x = 16 + 8x - 8x$$
$$12 + 4x = 16$$
$$12 - 12 + 4x = 16 - 12$$
$$4x = 4$$
$$\frac{4x}{4} = \frac{4}{4}$$
$$x = 1$$

1 liter of the 12% solution should be used.

97. Let x = pounds of 90¢ candy.
Then $20 - x$ = pounds of \$1.50 candy.

$$\boxed{\text{Value of 90¢ candy}} + \boxed{\text{Value of \$1.50 candy}} = \boxed{\text{Value of mixture}}$$
$$0.90x + 1.50(20 - x) = 1.20(20)$$
$$90x + 150(20 - x) = 120(20)$$
$$90x + 3000 - 150x = 2400$$
$$-60x + 3000 = 2400$$
$$-60x + 3000 - 3000 = 2400 - 3000$$
$$-60x = -600$$
$$\frac{-60x}{-60} = \frac{-600}{-60}$$
$$x = 10$$

10 lb of each should be used.

98.
$$\begin{aligned} 3x+2 &< 5 \\ 3x+2-2 &< 5-2 \\ 3x &< 3 \\ \frac{3x}{3} &< \frac{3}{3} \\ x &< 1 \end{aligned}$$

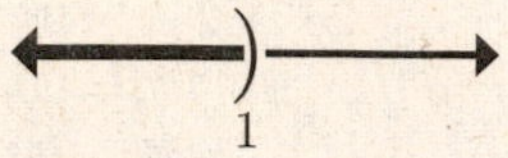

99.
$$\begin{aligned} -5x-8 &> 7 \\ -5x-8+8 &> 7+8 \\ -5x &> 15 \\ \frac{-5x}{-5} &< \frac{15}{-5} \\ x &< -3 \end{aligned}$$

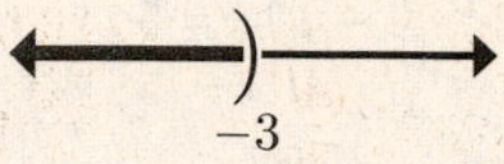

100.
$$\begin{aligned} 5x-3 &\geq 2x+9 \\ 5x-2x-3 &\geq 2x-2x+9 \\ 3x-3 &\geq 9 \\ 3x-3+3 &\geq 9+3 \\ 3x &\geq 12 \\ \frac{3x}{3} &\geq \frac{12}{3} \\ x &\geq 4 \end{aligned}$$

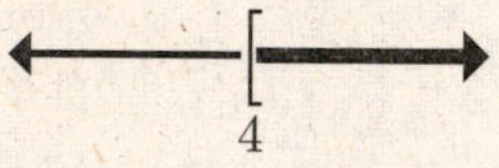

101.
$$\begin{aligned} 7x+1 &\leq 8x-5 \\ 7x-8x+1 &\leq 8x-8x-5 \\ -x+1 &\leq -5 \\ -x+1-1 &\leq -5-1 \\ -x &\leq -6 \\ -(-x) &\geq -(-6) \\ x &\geq 6 \end{aligned}$$

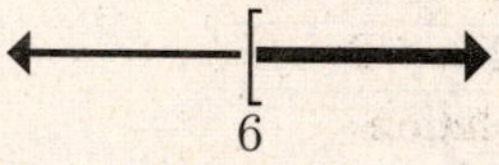

102.
$$\begin{aligned} 5(3-x) &\leq 3(x-3) \\ 15-5x &\leq 3x-9 \\ 15-5x-3x &\leq 3x-3x-9 \\ 15-8x &\leq -9 \\ 15-15-8x &\leq -9-15 \\ -8x &\leq -24 \\ \frac{-8x}{-8} &\geq \frac{-24}{-8} \\ x &\geq 3 \end{aligned}$$

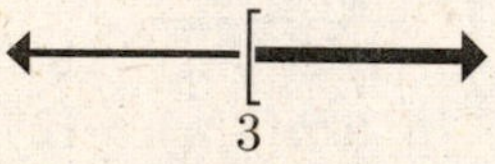

103.
$$\begin{aligned} 3(5-x) &\geq 2x \\ 15-3x &\geq 2x \\ 15-3x+3x &\geq 2x+3x \\ 15 &\geq 5x \\ \frac{15}{5} &\geq \frac{5x}{5} \\ 3 &\geq x, \text{ or } x \leq 3 \end{aligned}$$

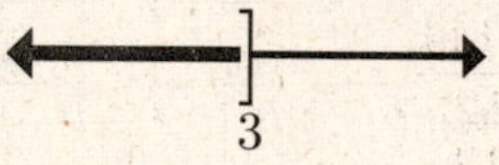

104.
$$\begin{aligned} 8 < \; & x+2 < 13 \\ 8-2 < \; & x+2-2 < 13-2 \\ 6 < \; & x < 11 \end{aligned}$$

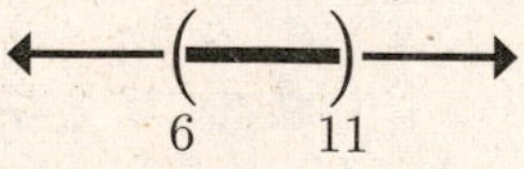

105.
$$\begin{aligned} 0 \leq \; & 2-2x < 4 \\ 0-2 \leq \; & 2-2-2x < 4-2 \\ -2 \leq \; & -2x < 2 \\ \frac{-2}{-2} \geq \; & \frac{-2x}{-2} > \frac{2}{-2} \\ 1 \geq \; & x > -1 \\ -1 < \; & x \leq 1 \end{aligned}$$

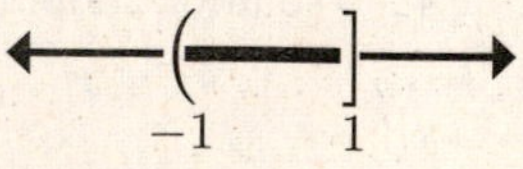

106. Let l = length. Then $l - 6$ = width.

$$\begin{aligned} 0 \text{ ft} &< \text{perimeter} \le 68 \text{ ft} \\ 0 \text{ ft} &< 2l + 2(l-6) \le 68 \text{ ft} \\ 0 \text{ ft} &< 2l + 2l - 12 \le 68 \text{ ft} \\ 0 \text{ ft} &< 4l - 12 \le 68 \text{ ft} \\ 0 + 12 \text{ ft} &< 4l - 12 + 12 \le 68 + 12 \text{ ft} \\ 12 \text{ ft} &< 4l \le 80 \text{ ft} \\ \frac{12}{4} \text{ ft} &< \frac{4l}{4} \le \frac{80}{4} \text{ ft} \\ 3 \text{ ft} &< l \le 20 \text{ ft} \end{aligned}$$

If $l = 3$, then the width is negative, which is impossible. To keep the width positive, we must have $l - 6 > 0$, which implies $l > 6$. Thus the solution is $6 \text{ ft} < l \le 20 \text{ ft}$.

Chapter 2 Test (page 154)

1.
$$\begin{aligned} 5x + 3 &= -2 \\ 5(-1) + 3 &\stackrel{?}{=} -2 \\ -5 + 3 &\stackrel{?}{=} -2 \\ -2 &= -2 \end{aligned}$$
-1 is a solution.

2.
$$\begin{aligned} 3(x+2) &= 2x \\ 3(-6+2) &\stackrel{?}{=} 2(-6) \\ 3(-4) &\stackrel{?}{=} -12 \\ -12 &= -12 \end{aligned}$$
-6 is a solution.

3.
$$\begin{aligned} -3(2-x) &= 0 \\ -3[2-(-2)] &\stackrel{?}{=} 0 \\ -3[4] &\stackrel{?}{=} 0 \\ -12 &\neq 0 \end{aligned}$$
-2 is not a solution.

4.
$$\begin{aligned} 3(x+2) &= 2x + 7 \\ 3(1+2) &\stackrel{?}{=} 2(1) + 7 \\ 3(3) &\stackrel{?}{=} 2 + 7 \\ 9 &= 9 \end{aligned}$$
1 is a solution.

5.
$$\begin{aligned} x + 17 &= -19 \\ x + 17 - 17 &= -19 - 17 \\ x &= -36 \end{aligned}$$

6.
$$\begin{aligned} a - 15 &= 32 \\ a - 15 + 15 &= 32 + 15 \\ a &= 47 \end{aligned}$$

7.
$$\begin{aligned} 12x &= -144 \\ \frac{12x}{12} &= \frac{-144}{12} \\ x &= -12 \end{aligned}$$

8.
$$\begin{aligned} \frac{x}{7} &= -1 \\ 7 \cdot \frac{x}{7} &= 7(-1) \\ x &= -7 \end{aligned}$$

9.
$$\begin{aligned} 8x + 2 &= -14 \\ 8x + 2 - 2 &= -14 - 2 \\ 8x &= -16 \\ \frac{8x}{8} &= \frac{-16}{8} \\ x &= -2 \end{aligned}$$

10.
$$\begin{aligned} 3 &= 5 - 2x \\ 3 - 5 &= 5 - 5 - 2x \\ -2 &= -2x \\ \frac{-2}{-2} &= \frac{-2x}{-2} \\ 1 &= x \end{aligned}$$

11.
$$\begin{aligned} \frac{2x-5}{3} &= 3 \\ 3 \cdot \frac{2x-5}{3} &= 3(3) \\ 2x - 5 &= 9 \\ 2x - 5 + 5 &= 9 + 5 \\ 2x &= 14 \\ \tfrac{2x}{2} &= \tfrac{14}{2} \\ x &= 7 \end{aligned}$$

12.
$$\begin{aligned} 23 - 5(x+10) &= -12 \\ 23 - 5x - 50 &= -12 \\ -5x - 27 &= -12 \\ -5x - 27 + 27 &= -12 + 27 \\ -5x &= 15 \\ \frac{-5x}{-5} &= \frac{15}{-5} \\ x &= -3 \end{aligned}$$

13. $x + 5(x-3) = x + 5x - 15 = 6x - 15$

14. $3x - 5(2-x) = 3x - 10 + 5x = 8x - 10$

CHAPTER 2 TEST

15. $-3(x+3)+3(x-3) = -3x-9+3x-9 = -3x+3x-9-9 = -18$

16. $-4(2x-5)-7(4x+1) = -8x+20-28x-7 = -36x+13$

17.
$$\begin{aligned} \frac{3x-18}{2} &= 6x \\ 2\cdot\frac{3x-18}{2} &= 2(6x) \\ 3x-18 &= 12x \\ 3x-3x-18 &= 12x-3x \\ -18 &= 9x \\ \tfrac{-18}{9} &= \tfrac{9x}{9} \\ -2 &= x \end{aligned}$$

18.
$$\begin{aligned} \frac{7}{8}(x-4) &= 5x-\frac{7}{2} \\ 8\cdot\frac{7}{8}(x-4) &= 8\left(5x-\frac{7}{2}\right) \\ 7(x-4) &= 8(5x)-8\cdot\tfrac{7}{2} \\ 7x-28 &= 40x-28 \\ 7x-40x-28 &= 40x-40x-28 \\ -33x-28 &= -28 \\ -33x-28+28 &= -28+28 \\ -33x &= 0 \\ x &= 0 \end{aligned}$$

19.
$$\begin{aligned} d &= rt \\ \frac{d}{r} &= \frac{rt}{r} \\ \frac{d}{r} &= t, \text{ or } t = \frac{d}{r} \end{aligned}$$

20.
$$\begin{aligned} P &= 2l+2w \\ P-2w &= 2l+2w-2w \\ P-2w &= 2l \\ \frac{P-2w}{2} &= \frac{2l}{2} \\ \tfrac{P-2w}{2} &= l, \text{ or } l = \tfrac{P-2w}{2} \end{aligned}$$

21.
$$\begin{aligned} A &= 2\pi rh \\ \frac{A}{2\pi r} &= \frac{2\pi rh}{2\pi r} \\ \frac{A}{2\pi r} &= h, \text{ or } h = \frac{A}{2\pi r} \end{aligned}$$

22.
$$\begin{aligned} A &= P+Prt \\ A-P &= P-P+Prt \\ A-P &= Prt \\ \frac{A-P}{Pt} &= \frac{Prt}{Pt} \\ \frac{A-P}{Pt} &= r, \text{ or } r = \frac{A-P}{Pt} \end{aligned}$$

23.
$$\begin{aligned} x+45 &= 120 \\ x+45-45 &= 120-45 \\ x &= 75^\circ \end{aligned}$$

24. Let x = the measure of the supplement.
$$\begin{aligned} x+105 &= 180 \\ x+105-105 &= 180-105 \\ x &= 75^\circ \end{aligned}$$

25. Let x = amount at 6%. Then $10000-x$ = amount at 5%.

[Interest at 6%] + [Interest at 5%] = [Total interest]

$$\begin{aligned} 0.06x+0.05(10000-x) &= 560 \\ 6x+5(10000-x) &= 56000 \\ 6x+50000-5x &= 56000 \\ x+50000 &= 56000 \\ x+50000-50000 &= 56000-50000 \end{aligned}$$

$x = 6000 \Rightarrow$ \$6,000 is invested at 6%, and \$4,000 is invested at 5%.

26. Let t = time to meet.

	r	t	d
Car	65	t	$65t$
Truck	55	t	$55t$

[Car Dist] + [Truck Dist] = 72

$$\begin{aligned} 65t + 55t &= 72 \\ 120t &= 72 \\ \tfrac{120t}{120} &= \tfrac{72}{120} \\ t &= \tfrac{3}{5} \end{aligned}$$

They meet after $\frac{3}{5}$ of an hour, or $\frac{3}{5}(60) = 36$ minutes.

27. Let x = liters of water used.

[Amt. in 10% sol.] + [Amt. in water] = [Amt. in 8% sol.]

$$\begin{aligned} 0.10(30) + 0(x) &= 0.08(30 + x) \\ 10(30) + 0 &= 8(30 + x) \\ 300 &= 240 + 8x \\ 300 - 240 &= 8x \\ 60 &= 8x \\ \tfrac{60}{8} &= \tfrac{8x}{8} \\ \tfrac{15}{2} &= x \end{aligned}$$

$7\frac{1}{2}$ liters of water should be used.

28. Let x = pounds of peanuts used.

[Value of cashews] + [Value of peanuts] = [Value of mixture]

$$\begin{aligned} 6(20) + 3x &= 4(20 + x) \\ 120 + 3x &= 80 + 4x \\ 120 + 3x - 4x &= 80 + 4x - 4x \\ -x + 120 &= 80 \\ -x + 120 - 120 &= 80 - 120 \\ -x &= -40 \\ x &= 40; \end{aligned}$$

40 pounds of peanuts should be used.

29.

$$\begin{aligned} 8x - 20 &\geq 4 \\ 8x - 20 + 20 &\geq 4 + 20 \\ 8x &\geq 24 \\ \frac{8x}{8} &\geq \frac{24}{8} \\ x &\geq 3 \end{aligned}$$

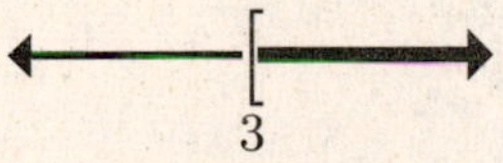

30.

$$\begin{aligned} x - 2(x + 7) &> 14 \\ x - 2x - 14 &> 14 \\ -x - 14 &> 14 \\ -x - 14 + 14 &> 14 + 14 \\ -x &> 28 \\ -(-x) &< -(28) \\ x &< -28 \end{aligned}$$

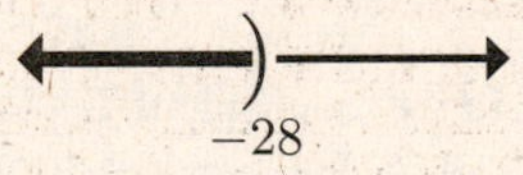

31.

$$\begin{aligned} -4 \leq\ & 2(x + 1) < 10 \\ -4 \leq\ & 2x + 2 < 10 \\ -4 - 2 \leq\ & 2x + 2 - 2 < 10 - 2 \\ -6 \leq\ & 2x < 8 \\ \frac{-6}{2} \leq\ & \frac{2x}{2} < \frac{8}{2} \\ -3 \leq\ & x < 4 \end{aligned}$$

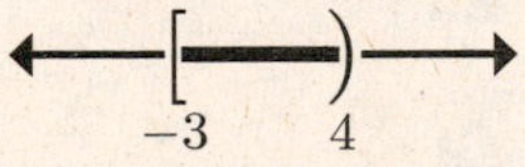

32.

$$\begin{aligned} -2 <\ & 5(x - 1) \leq 10 \\ -2 <\ & 5x - 5 \leq 10 \\ -2 + 5 <\ & 5x - 5 + 5 \leq 10 + 5 \\ 3 <\ & 5x \leq 15 \\ \frac{3}{5} <\ & \frac{5x}{5} \leq \frac{15}{5} \\ \frac{3}{5} <\ & x \leq 3 \end{aligned}$$

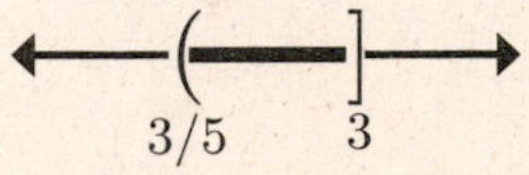

Cumulative Review Exercises (page 155)

1. $\frac{27}{9} = 3$: integer, rational number, real number

2. $-0.25 = -\frac{1}{4}$: rational number, real number

3. 2 3 4 5 6 7

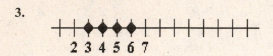

4. 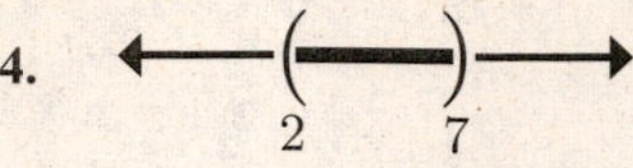

5. $\frac{|-3| - |3|}{|-3-3|} = \frac{(+3) - (+3)}{|-6|} = \frac{0}{+6} = 0$

6. $\frac{5}{7} \cdot \frac{14}{3} = \frac{5}{\overset{}{\underset{1}{\cancel{7}}}} \cdot \frac{2 \cdot \overset{1}{\cancel{7}}}{3} = \frac{10}{3}$

7. $2\frac{3}{5} + 5\frac{1}{2} = \frac{13}{5} + \frac{11}{2} = \frac{13 \cdot 2}{5 \cdot 2} + \frac{11 \cdot 5}{2 \cdot 5} = \frac{26}{10} + \frac{55}{10} = \frac{81}{10} = 8\frac{1}{10}$

8. $35.7 - 0.05 = 35.65$

9. $(3x - 2y)z = [3(-5) - 2(3)]0 = 0$

10. $\frac{x - 3y + |z|}{2 - x} = \frac{-5 - 3(3) + |0|}{2 - (-5)} = \frac{-5 - 9 + 0}{2 + 5} = \frac{-14}{7} = -2$

11. $x^2 - y^2 + z^2 = (-5)^2 - (3)^2 + 0^2$
$= 25 - 9 + 0 = 16$

12. $\frac{x}{y} + \frac{y+2}{3-z} = \frac{-5}{3} + \frac{3+2}{3-0} = -\frac{5}{3} + \frac{5}{3} = 0$

13.
$$\begin{aligned} rb &= a \\ 0.075(330) &= a \\ 24.75 &= a \end{aligned}$$

14.
$$\begin{aligned} rb &= a \\ 0.32b &= 1688 \\ \frac{0.32b}{0.32} &= \frac{1688}{0.32} \\ b &= 5{,}275 \end{aligned}$$

15. 2nd term: $5x^2y$; coefficient: 5

16. 3rd term: $37y$; factors: 37, y

17. $3x - 5x + 2y = -2x + 2y$

18. $3(x - 7) + 2(8 - x) = 3x - 21 + 16 - 2x$
$= x - 5$

19. $2x^2y^3 - 4x^2y^3 = -2x^2y^3$

20. $2(3 - x) + 5(x + 2) = 6 - 2x + 5x + 10 = 3x + 16$

CUMULATIVE REVIEW EXERCISES

21.
$$\begin{aligned} 3(x-5)+2 &= 2x \\ 3x-15+2 &= 2x \\ 3x-13 &= 2x \\ 3x-3x-13 &= 2x-3x \\ -13 &= -x \\ 13 &= x \end{aligned}$$

22.
$$\begin{aligned} \frac{x-5}{3}-5 &= 7 \\ \frac{x-5}{3}-5+5 &= 7+5 \\ \frac{x-5}{3} &= 12 \\ 3\cdot\frac{x-5}{3} &= 3(12) \\ x-5 &= 36 \\ x-5+5 &= 36+5 \\ x &= 41 \end{aligned}$$

23.
$$\begin{aligned} \frac{2x-1}{5} &= \frac{1}{2} \\ 10\cdot\frac{2x-1}{5} &= 10\cdot\frac{1}{2} \\ 2(2x-1) &= 5 \\ 4x-2 &= 5 \\ 4x-2+2 &= 5+2 \\ 4x &= 7 \\ \tfrac{4x}{4} &= \tfrac{7}{4} \\ x &= \tfrac{7}{4} \end{aligned}$$

24.
$$\begin{aligned} 2(a-3)-3(a-2) &= -a \\ 2a-6-3a+6 &= -a \\ -a &= -a \\ -a+a &= -a+a \\ 0 &= 0 \end{aligned}$$
Identity, $\mathbb{R}$

25.
$$\begin{aligned} A &= \frac{1}{2}h(b+B) \\ 2A &= 2\cdot\frac{1}{2}h(b+B) \\ 2A &= h(b+B) \\ \frac{2A}{b+B} &= \frac{h(b+B)}{b+B} \\ \frac{2A}{b+B} &= h, \text{ or } h = \frac{2A}{b+B} \end{aligned}$$

26.
$$\begin{aligned} y &= mx+b \\ y-b &= mx+b-b \\ y-b &= mx \\ \frac{y-b}{m} &= \frac{mx}{m} \\ \frac{y-b}{m} &= x, \text{ or } x = \frac{y-b}{m} \end{aligned}$$

27. Let x = Dealer's invoice.

[Price] = [Dealer's invoice] + [Markup]

$$\begin{aligned} 23499 &= x+0.03x \\ 23499 &= 1.03x \\ \frac{23499}{1.03} &= \frac{1.03x}{1.03} \\ 22814.56 &= x \end{aligned}$$

The dealer's invoice was \$22,814.56.

28. Let x = original sofa price.

[Price] = [New sofa price] + [New chair price]

$$\begin{aligned} 780 &= x-0.35x+300-0.35(300) \\ 780 &= 0.65x+300-105 \\ 780 &= 0.65x+195 \\ 780-195 &= 0.65x+195-195 \\ 585 &= 0.65x \\ \frac{585}{0.65} &= \frac{0.65x}{0.65} \\ 900 &= x \end{aligned}$$

The original price of the sofa was \$900.

CUMULATIVE REVIEW EXERCISES

29. Let $x =$ original car price.

$$\begin{aligned} 13725.25 &= x + 0.085x \\ 13725.25 &= 1.085x \\ \frac{13725.25}{1.085} &= \frac{1.085x}{1.085} \\ 12650 &= x \end{aligned}$$

The original price was \$12,650.

30. Let x and $3x =$ the unknown amounts.

$$\boxed{\text{Cement}} + \boxed{\text{Gravel}} = \boxed{\text{Concrete}}$$

$$\begin{aligned} x + 3x &= 500 \\ 4x &= 500 \\ \frac{4x}{4} &= \frac{500}{4} \\ x &= 125 \text{ lbs cement} \end{aligned}$$

31. Let x and $2x =$ the lengths.

$$\begin{aligned} \text{Total length} &= 35 \\ 14 + x + 2x &= 35 \\ 14 + 3x &= 35 \\ 14 - 14 + 3x &= 35 - 14 \\ 3x &= 21 \\ \frac{3x}{3} &= \frac{21}{3} \\ x &= 7 \text{ ft} \end{aligned}$$

The section will not span the doorway.

32. Let $x =$ the length of one panel.

$$\begin{aligned} \boxed{\text{Total length}} &= 18 \\ x + x + 3.4 &= 18 \\ 2x + 3.4 &= 18 \\ 2x + 3.4 - 3.4 &= 18 - 3.4 \\ 2x &= 14.6 \\ \frac{2x}{2} &= \frac{14.6}{2} \\ x &= 7.3 \end{aligned}$$

The lengths are 7.3 and 10.7 feet.

33. Let $n =$ number of kwh used.

$$\begin{aligned} 17.50 + 0.18n &= 43.96 \\ 17.50 - 17.50 + 0.18n &= 43.96 - 17.50 \\ 0.18n &= 26.46 \\ \frac{0.18n}{0.18} &= \frac{26.46}{0.18} \\ n &= 147 \end{aligned}$$

147 kwh were used that month.

34. Let $n =$ number of feet required.

$$\begin{aligned} 35 + 1.50n &= 162.50 \\ 35 - 35 + 1.50n &= 162.50 - 35 \\ 1.50n &= 127.50 \\ \frac{1.50n}{1.50} &= \frac{127.50}{1.50} \\ n &= 85 \end{aligned}$$

85 ft of gutters were required.

35. $4^2 - 5^2 = 16 - 25 = -9$

36. $(4 - 5)^2 = (-1)^2 = 1$

37. $5(4^3 - 2^3) = 5(64 - 8) = 5(56) = 280$

38. $$\begin{aligned} -2(5^4 - 7^3) &= -2(625 - 343) \\ &= -2(282) = -564 \end{aligned}$$

39.

$$\begin{aligned} 8(4 + x) &> 10(6 + x) \\ 32 + 8x &> 60 + 10x \\ 32 + 8x - 10x &> 60 + 10x - 10x \\ 32 - 2x &> 60 \\ 32 - 32 - 2x &> 60 - 32 \\ -2x &> 28 \\ \frac{-2x}{-2} &< \frac{28}{-2} \\ x &< -14 \end{aligned}$$

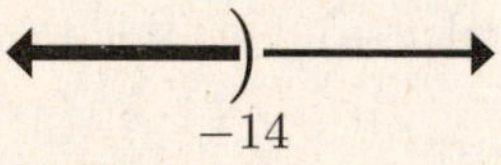

40.

$$\begin{array}{rcl} -9 < & 3(x + 2) & \le 3 \\ -9 < & 3x + 6 & \le 3 \\ -9 - 6 < & 3x + 6 - 6 & \le 3 - 6 \\ -15 < & 3x & \le -3 \\ \frac{-15}{3} < & \frac{3x}{3} & \le \frac{-3}{3} \\ -5 < & x & \le -1 \end{array}$$

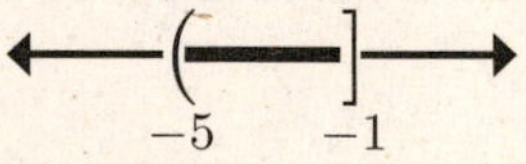

Exercise 3.1 (page 165)

1. **Answers will vary.**

3. IV

5. $-3-3(-5) = -3-(-15) = -3+15$
$= 12$

7. opposite of -8: 8

9.
$$\begin{aligned} -4x+7 &= -21 \\ -4x+7-7 &= -21-7 \\ -4x &= -28 \\ \frac{-4x}{-4} &= \frac{-28}{-4} \\ x &= 7 \end{aligned}$$

11.
$$\begin{aligned} (x+1)(x+y)^2 &= (-2+1)[-2+(-5)]^2 \\ &= (-1)(-7)^2 \\ &= -1(+49) = -49 \end{aligned}$$

13. ordered pair

15. origin

17. rectangular, Cartesian

19. coordinates

21. no

23. origin, left, up

25. II

27.

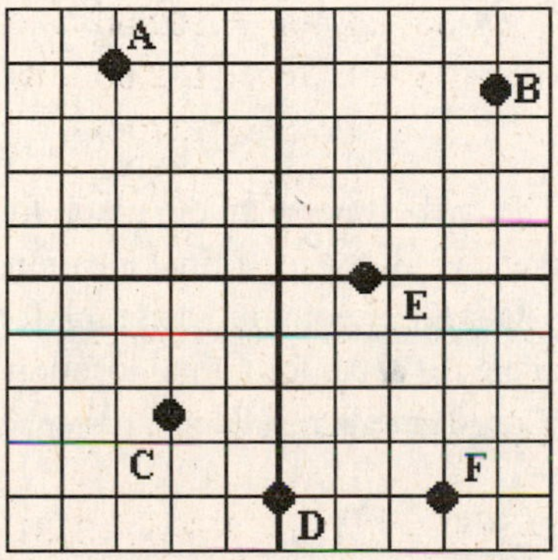

29.

x	y
4	3 or −3
0	5 or −5
−3	4 or −4
5 or −5	0
−4	−3 or 3
0	−5 or 5
3	−4 or 4

31. The point $(-10, 60)$ indicates that 10 minutes before the workout started, her heart rate was 60 beats per minute.

33. The point with an x-coordinate of 30 is $(30, 150)$, so her heart rate was 150 beats per minute one half-hour after starting.

35. The points on the graph with a y-coordinate of 100 have x-coordinates of approximately 5 and 50, so her heart rate was 100 beats per minute after about 5 and 50 minutes.

37. Before the workout, her heart rate was 60 beats per minute. After the workout, her heart rate was about 70 beats per minute, or about 10 beats per minute higher.

39. To find the charge for a 1-day rental, find the y-coordinate of the point with an x-coordinate of 1. This is the point $(1, 2)$. The charge will be \$2.

41. To find the charge for a 5-day rental, find the y-coordinate of the point with an x-coordinate of 5. This is the point $(5, 7)$. The charge will be \$7.

43. To find the cost for a 1-oz. letter, find the y-coordinate of the point with an x-coordinate of 1. This is the point $(1, 44)$. Cost $= 44¢$

To find the cost for a $2\frac{1}{2}$-oz. letter, find the y-coordinate of the point with an x-coordinate of $2\frac{1}{2}$. This is the point $\left(2\frac{1}{2}, 76\right)$. Cost $= 76¢$

45. To find the cost for a 0.75-oz. letter, find the y-coordinate of the point with an x-coordinate of 0.75. This is the point $(0.75, 44)$, with a cost of $44¢$. To find the cost for a 2.75-oz. letter, find the y-coordinate of the point with an x-coordinate of 2.75. This is the point $(2.75, 76)$, with a cost of $76¢$. The difference is $76 - 44 = 32¢$.

47.

City	Ordered Pair
Carbondale	$\left(5\frac{1}{2}, J\right)$
Champaign	$(7, D)$
Chicago	$\left(8\frac{1}{2}, B\right)$
Peoria	$\left(5\frac{3}{4}, C\right)$
Rockford	$\left(5\frac{3}{4}, A\right)$
Springfield	$\left(4\frac{1}{2}, E\right)$
St. Louis	$\left(4\frac{1}{4}, H\right)$

49. **a.** The highest point on the graph for the 60° angle is $(-2, 7)$, for a height of 7 ft. The highest point on the graph for the 30° angle is $(0, 3)$, for a height of 3 ft. The 60° angle results in a height 4 ft greater.

b. The farthest point on the graph for the 60° angle is $(3, 0)$, for a final location of 3 ft. The farthest point on the graph for the 30° angle is $(7, 0)$, for a final location of 7 ft. The 30° angle results in a distance 4 ft farther.

51.

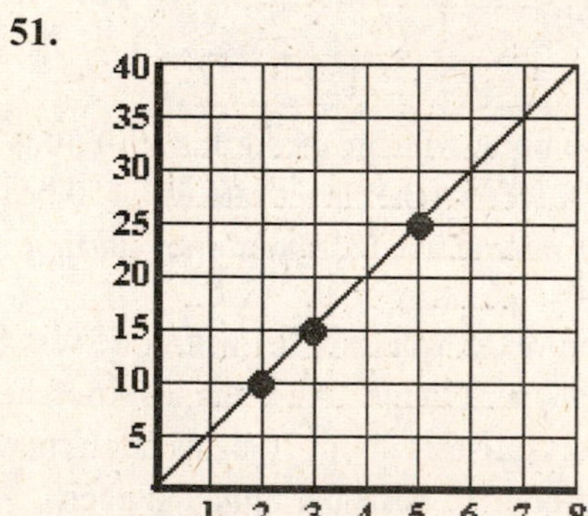

a. $(7, 35) \Rightarrow 35$ miles
b. $(4, 20) \Rightarrow 4$ gallons
c. $(6.5, 32.5) \Rightarrow 32.5$ miles

53.

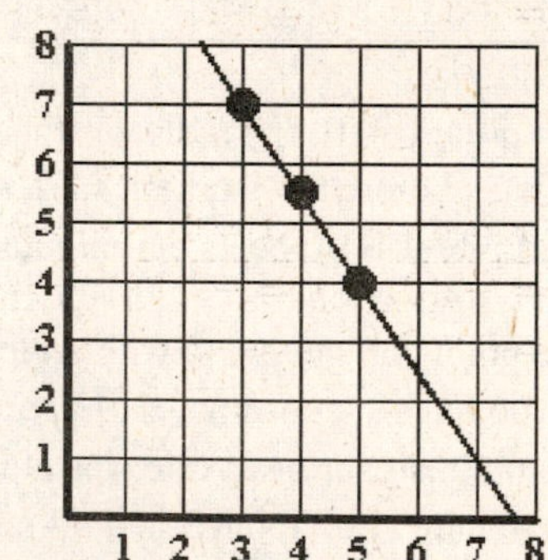

a. A 3-year old car is worth \$7000.
b. $(7, 1) \Rightarrow \$1000$
c. $(6, 2500) \Rightarrow 6$ years

55-59. Answers may vary.

Exercise 3.2 (page 180)

1. 3

3. Answers will vary. Here are three possible points: $(1, 7)$, $(2, 6)$, $(3, 5)$.

5. Vertical lines have no y-intercepts.

7.
$$\frac{x}{8} = -12$$
$$8 \cdot \frac{x}{8} = 8(-12)$$
$$x = -96$$

9. expression

11.
$$rb = a$$
$$0.005(250) = a$$
$$1.25 = a$$

13. $-2.5 - (-2.6) = -2.5 + 2.6 = 0.1$

15. two

17. independent, dependent

19. linear

21. y-intercept

23.
$$x - 2y = -4$$
$$4 - 2(4) \stackrel{?}{=} -4$$
$$4 - 8 \stackrel{?}{=} -4$$
$$-4 = -4 \Rightarrow (4, 4) \text{ is a solution.}$$

25.
$$y = \frac{2}{3}x + 5$$
$$12 \stackrel{?}{=} \frac{2}{3}(6) + 5$$
$$12 \stackrel{?}{=} 4 + 5$$
$$12 \neq 9 \Rightarrow (6, 12) \text{ is not a solution.}$$

27. $y = x - 3$

$x = 0$	$x = 1$	$x = -2$	$x = -4$
$y = 0 - 3$	$y = 1 - 3$	$y = -2 - 3$	$y = -4 - 3$
$y = -3$	$y = -2$	$y = -5$	$y = -7$

x	y	(x, y)
0	-3	$(0, -3)$
1	-2	$(1, -2)$
-2	-5	$(-2, -5)$
-4	-7	$(-4, -7)$

29. $y = -2x$

$x = 0$	$x = 1$	$x = 3$	$x = -2$
$y = -2(0)$	$y = -2(1)$	$y = -2(3)$	$y = -2(-2)$
$y = 0$	$y = -2$	$y = -6$	$y = 4$

x	y	(x, y)
0	0	$(0, 0)$
1	-2	$(1, -2)$
3	-6	$(3, -6)$
-2	4	$(-2, 4)$

SECTION 3.2

31.

$$y = 2x$$

$$\begin{array}{ccc} x = 0 & x = 1 & x = -1 \\ \hline y = 2(0) & y = 2(1) & y = 2(-1) \\ y = 0 & y = 2 & y = -2 \end{array}$$

x	y
0	0
1	2
−1	−2

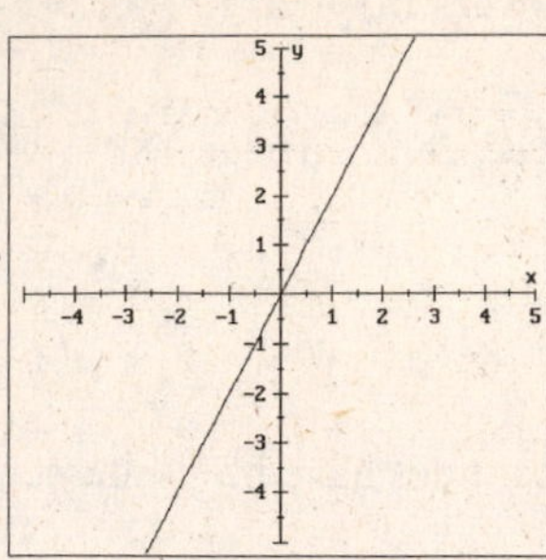

33.

$$y = 2x - 1$$

$$\begin{array}{ccc} x = 0 & x = 1 & x = -1 \\ \hline y = 2(0) - 1 & y = 2(1) - 1 & y = 2(-1) - 1 \\ y = 0 - 1 & y = 2 - 1 & y = -2 - 1 \\ y = -1 & y = 1 & y = -3 \end{array}$$

x	y
0	−1
1	1
−1	−3

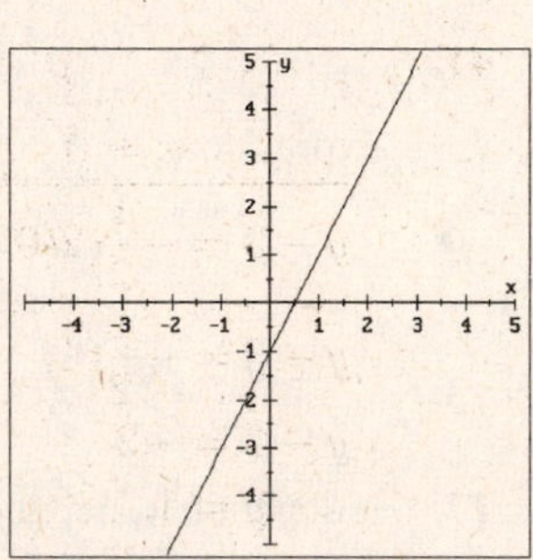

35.

$$y = 1.2x - 2$$

$$\begin{array}{ccc} x = 0 & x = 5 & x = -5 \\ \hline y = 1.2(0) - 2 & y = 1.2(5) - 2 & y = 1.2(-5) - 2 \\ y = 0 - 2 & y = 6 - 2 & y = -6 - 2 \\ y = -2 & y = 4 & y = -8 \end{array}$$

x	y
0	−2
5	4
−5	−8

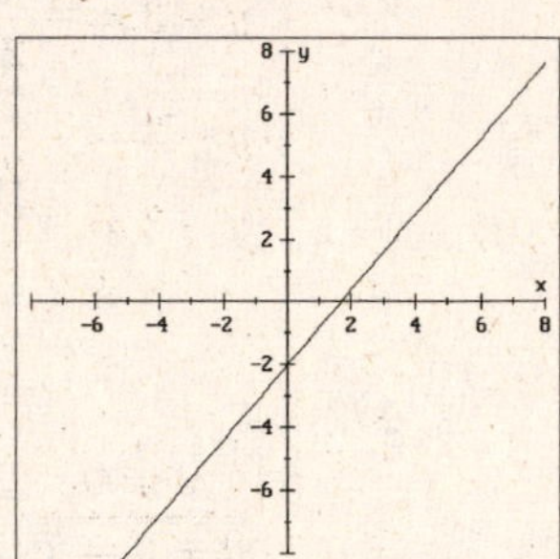

37.

$$y = 2.5x - 5$$

$$\begin{array}{cc} x = -4 & x = -2 \\ \hline y = 2.5(-4) - 5 & y = 2.5(-2) - 5 \\ y = -10 - 5 & y = -5 - 5 \\ y = -15 & y = -10 \\ x = 0 & x = 4 \\ \hline y = 2.5(0) - 5 & y = 2.5(4) - 5 \\ y = 0 - 5 & y = 10 - 5 \\ y = -5 & y = 5 \end{array}$$

x	y
−4	−15
−2	−10
0	−5
4	5

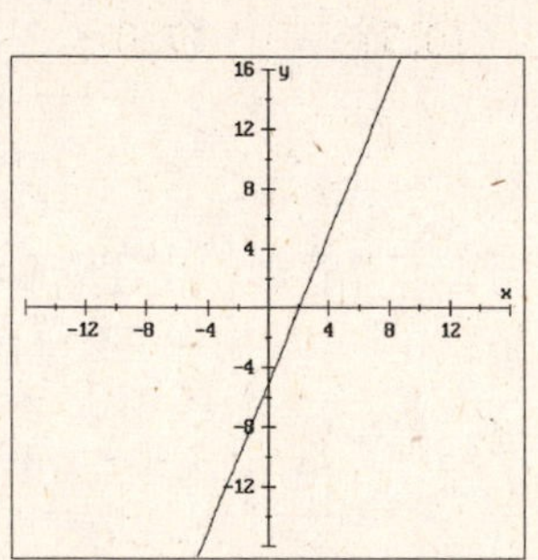

SECTION 3.2

39.

$$y = \frac{x}{2} - 2$$

$x = 0$	$x = 2$	$x = -2$
$y = \frac{0}{2} - 2$	$y = \frac{2}{2} - 2$	$y = \frac{-2}{2} - 2$
$y = 0 - 2$	$y = 1 - 2$	$y = -1 - 2$
$y = -2$	$y = -1$	$y = -3$

x	y
0	-2
2	-1
-2	-3

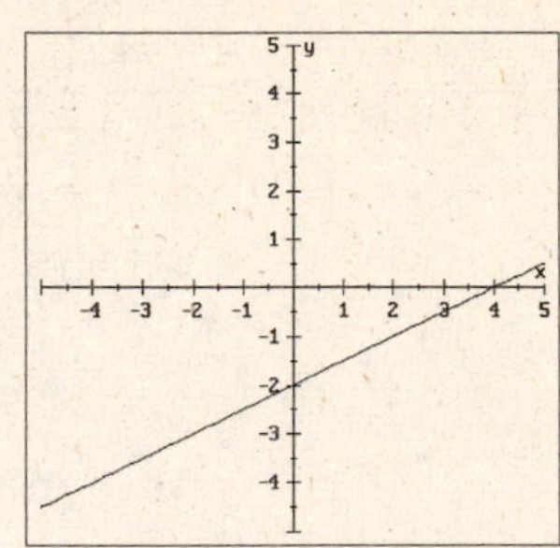

41.

$$y - 3 = -\frac{1}{2}(2x + 4)$$

$x = 0$	$x = 1$
$y - 3 = -\frac{1}{2}(2(0) + 4)$	$y - 3 = -\frac{1}{2}(2(1) + 4)$
$y - 3 = -\frac{1}{2}(4)$	$y - 3 = -\frac{1}{2}(6)$
$y - 3 = -2$	$y - 3 = -3$
$y = 1$	$y = 0$

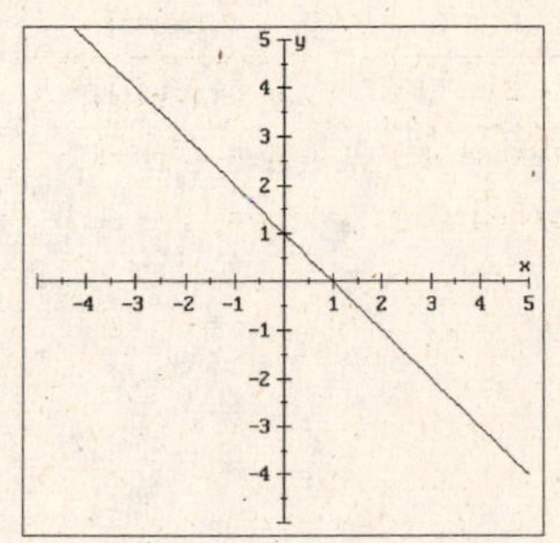

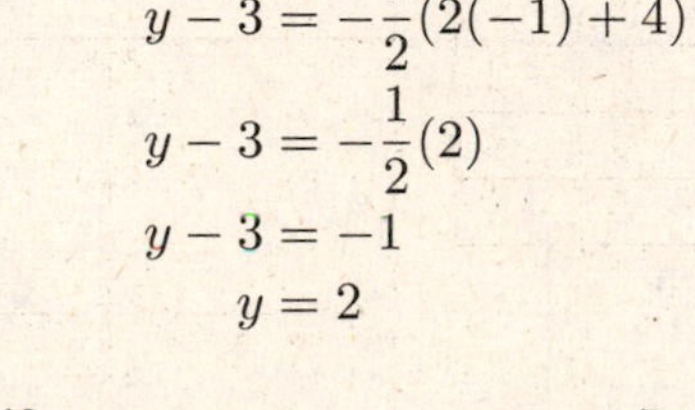

$x = -1$

$$y - 3 = -\frac{1}{2}(2(-1) + 4)$$
$$y - 3 = -\frac{1}{2}(2)$$
$$y - 3 = -1$$
$$y = 2$$

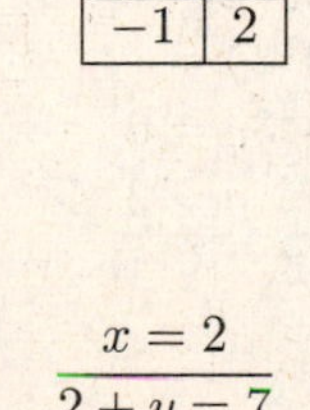

x	y
0	1
1	0
-1	2

43.

$$x + y = 7$$

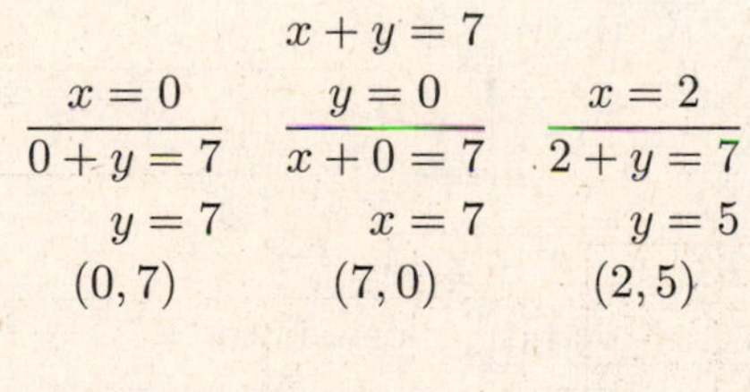

$x = 0$	$y = 0$	$x = 2$
$0 + y = 7$	$x + 0 = 7$	$2 + y = 7$
$y = 7$	$x = 7$	$y = 5$
$(0, 7)$	$(7, 0)$	$(2, 5)$

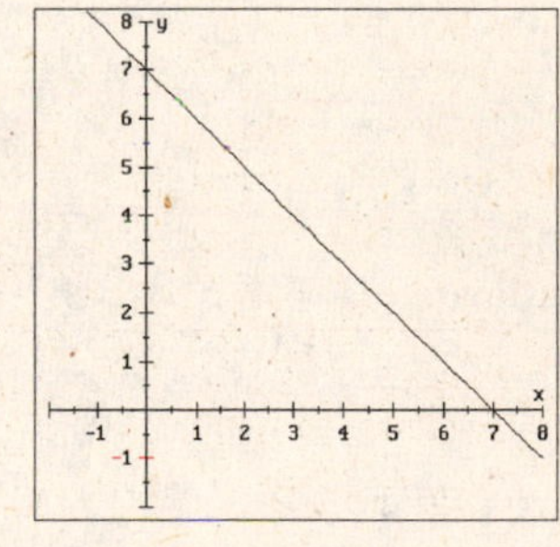

45.

$$x - y = 7$$

$x = 0$	$y = 0$	$x = 2$
$0 - y = 7$	$x - 0 = 7$	$2 - y = 7$
$-y = 7$	$x = 7$	$-y = 5$
$y = -7$	$(7, 0)$	$y = -5$
$(0, -7)$		$(2, -5)$

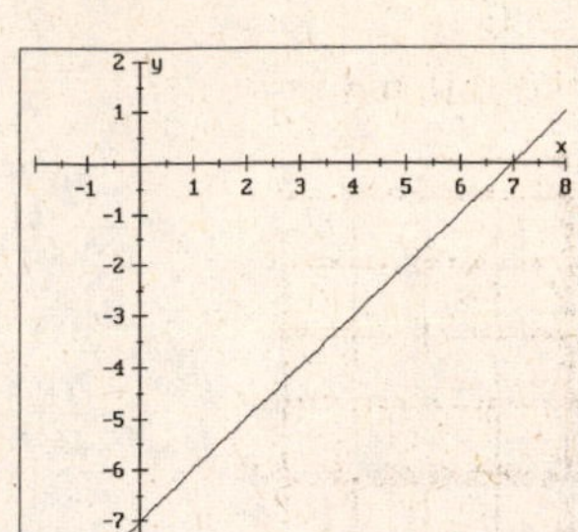

47.

$$y = -2x + 5$$
$$2x + y = 5$$

$x = 0$	$y = 0$	$x = 1$
$2(0) + y = 5$	$2x + 0 = 5$	$y = -2x + 5$
$0 + y = 5$	$2x = 5$	$y = -2(1) + 5$
$y = 5$	$x = \frac{5}{2}$	$y = 3$
$(0, 5)$	$\left(\frac{5}{2}, 0\right)$	$(1, 3)$

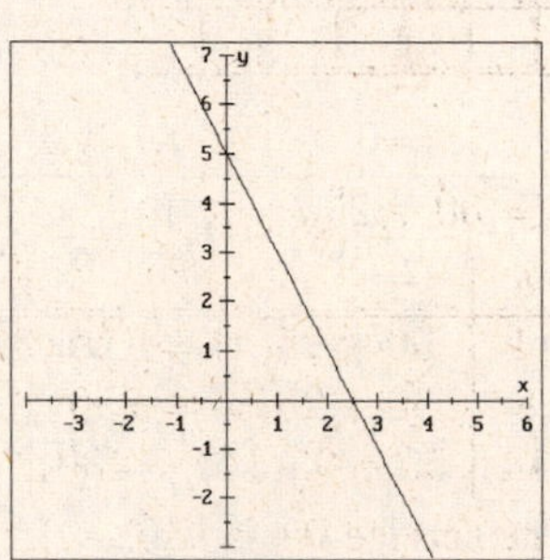

49.

$$2x + 3y = 12$$

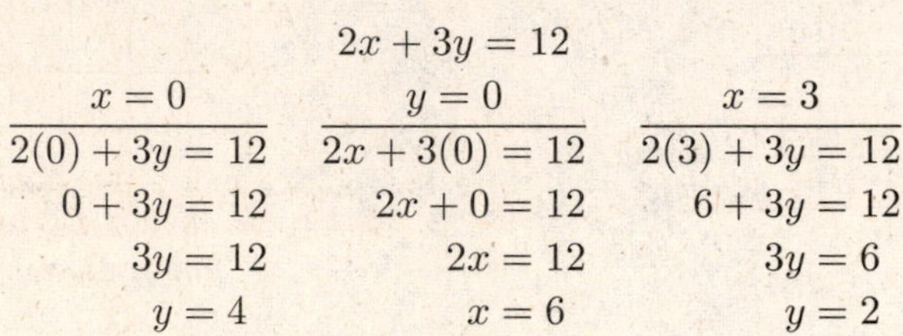

$x = 0$	$y = 0$	$x = 3$
$2(0) + 3y = 12$	$2x + 3(0) = 12$	$2(3) + 3y = 12$
$0 + 3y = 12$	$2x + 0 = 12$	$6 + 3y = 12$
$3y = 12$	$2x = 12$	$3y = 6$
$y = 4$	$x = 6$	$y = 2$
$(0, 4)$	$(6, 0)$	$(3, 2)$

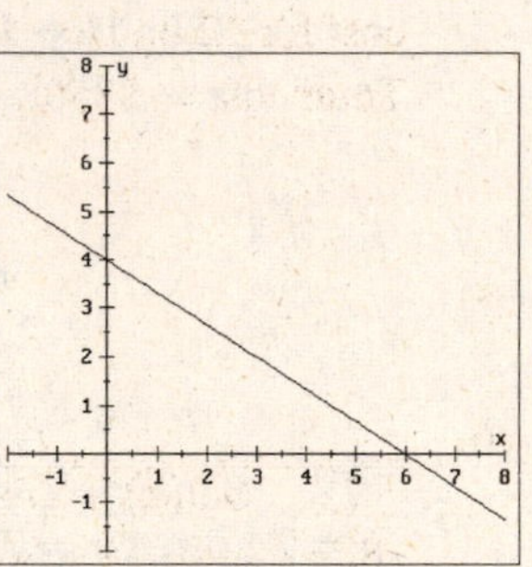

51. $y = -5$

horizontal, y-coordinate $= -5$

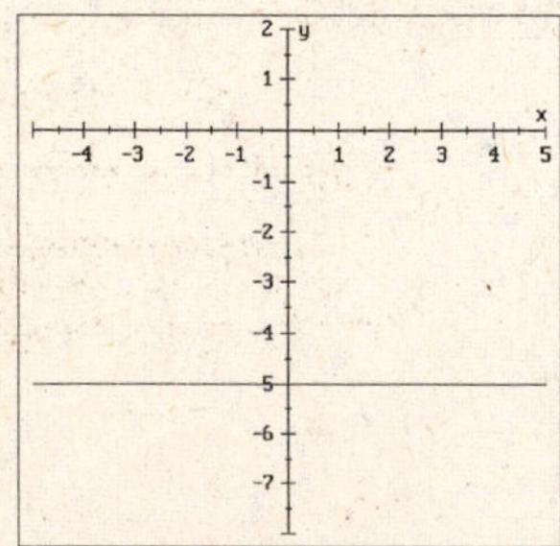

53. $x = 5$

vertical, x-coordinate $= 5$

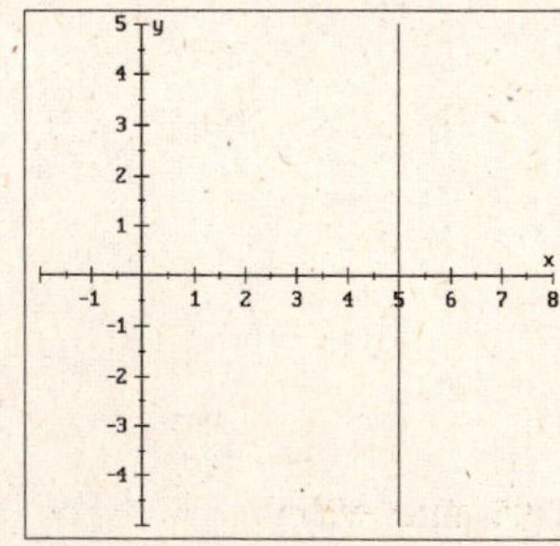

55. $y = 0$

horizontal, y-coordinate $= 0$

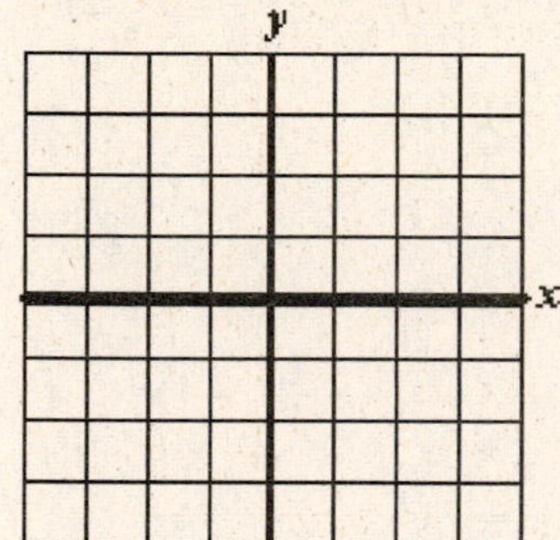

57. $2x = 5$

$x = \frac{5}{2} = 2\frac{1}{2}$

vertical, x-coordinate $= \frac{5}{2}$

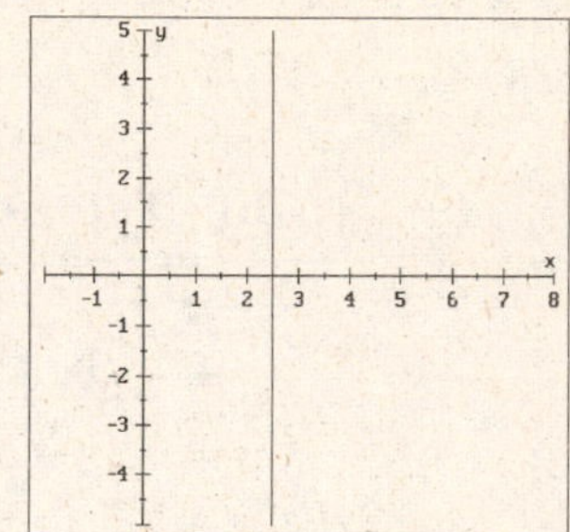

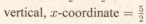

59. **a.** $c = 50 + 25u$

b.

u	c	(u, c)
4	$50 + 25(4) = 150$	$(4, 150)$
8	$50 + 25(8) = 250$	$(8, 250)$
14	$50 + 25(14) = 400$	$(14, 400)$

c. The service fee is \$50.

d. cost for 18 units = \$500

cost for 12 units = \$350

Total cost = \$850

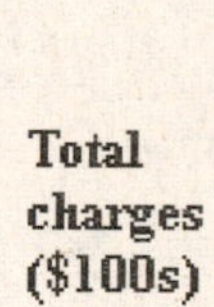

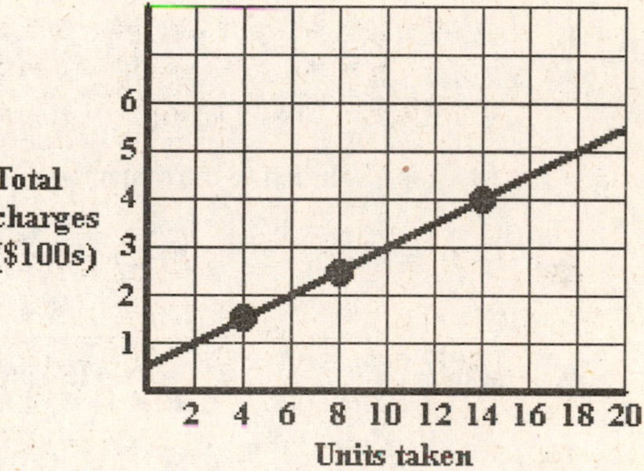

61. **a.**

r	h	(r, h)
7	$3.9(7) + 28.9 = 56.2$	$(7, 56.2)$
8.5	$3.9(8.5) + 28.9 = 62.1$	$(8.5, 62.1)$
9	$3.9(9) + 28.9 = 64.0$	$(9, 64.0)$

b. ...taller the woman is.

c. 58.1 inches tall

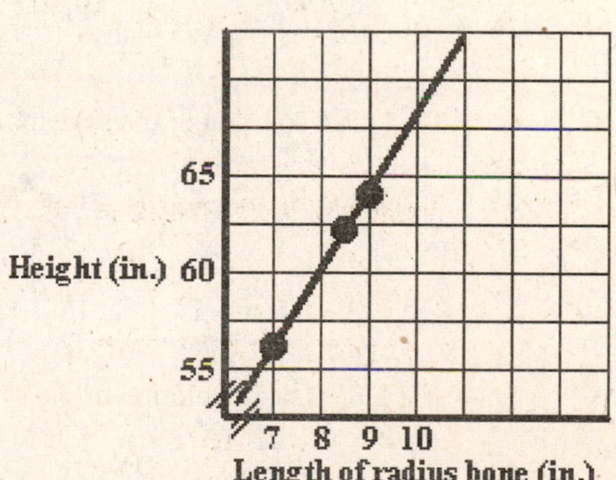

63-67. Answers may vary.

69. $x_M = \frac{a+c}{2} = \frac{5+7}{2} = \frac{12}{2} = 6$

$y_M = \frac{b+d}{2} = \frac{3+9}{2} = \frac{12}{2} = 6$

$M(6,6)$

71. $x_M = \frac{a+c}{2} = \frac{2+(-3)}{2} = \frac{-1}{2} = -\frac{1}{2}$

$y_M = \frac{b+d}{2} = \frac{-7+12}{2} = \frac{5}{2}$

$M\left(-\frac{1}{2}, \frac{5}{2}\right)$

73. $x_M = \frac{a+c}{2} = \frac{4+10}{2} = \frac{14}{2} = 7$

$y_M = \frac{b+d}{2} = \frac{6+6}{2} = \frac{12}{2} = 6$

$M(7,6)$

Exercise 3.3 (page 192)

1.
$$\begin{array}{rr} x+y=5 & x-y=1 \\ 3+2 \stackrel{?}{=} 5 & 3-2 \stackrel{?}{=} 1 \\ 5=5 & 1=1 \end{array}$$
$(3,2)$ is a solution to the system.

3.
$$\begin{array}{rr} x+y=5 & x-y=2 \\ 4+1 \stackrel{?}{=} 5 & 4-1 \stackrel{?}{=} 2 \\ 5=5 & 3\neq 2 \end{array}$$
$(4,1)$ is not a solution to the system.

5. $(-2)^4 = (-2)(-2)(-2)(-2) = 16$

7. $3x - x^2 = 3(-3) - (-3)^2 = -9 - (+9)$
$= -9 - 9 = -18$

9. system

11. independent

13. inconsistent; $\emptyset$

15.
$$\begin{array}{rr} x+y=2 & 2x-y=1 \\ 1+1 \stackrel{?}{=} 2 & 2(1)-1 \stackrel{?}{=} 1 \\ 2=2 & 2-1 \stackrel{?}{=} 1 \\ & 1=1 \end{array}$$
$(1,1)$ is a solution to the system.

17.
$$\begin{array}{rr} 2x+y=4 & x+y=1 \\ 2(3)+(-2) \stackrel{?}{=} 4 & 3+(-2) \stackrel{?}{=} 1 \\ 6+(-2) \stackrel{?}{=} 4 & 1=1 \\ 4=4 & \end{array}$$
$(3,-2)$ is a solution to the system.

19.
$$\begin{array}{rr} 2x-3y=-7 & 4x-5y=25 \\ 2(4)-3(5) \stackrel{?}{=} -7 & 4(4)-5(5) \stackrel{?}{=} 25 \\ 8-15 \stackrel{?}{=} -7 & 16-25 \stackrel{?}{=} 25 \\ -7=-7 & -9\neq 25 \end{array}$$
$(4,5)$ is not a solution to the system.

21.
$$\begin{array}{rr} 4x+5y=-23 & -3x+2y=0 \\ 4(-2)+5(-3) \stackrel{?}{=} -23 & -3(-2)+2(-3) \stackrel{?}{=} 0 \\ -8+(-15) \stackrel{?}{=} -23 & 6+(-6) \stackrel{?}{=} 0 \\ -23=-23 & 0=0 \end{array}$$
$(-2,-3)$ is a solution to the system.

23.

$$\begin{array}{rr} 2x + y = 4 & 4x - 3y = 11 \\ 2\left(\frac{1}{2}\right) + 3 \stackrel{?}{=} 4 & 4\left(\frac{1}{2}\right) - 3(3) \stackrel{?}{=} 11 \\ 1 + 3 \stackrel{?}{=} 4 & 2 - 9 \stackrel{?}{=} 11 \\ 4 = 4 & -7 \neq 11 \end{array}$$

$\left(\frac{1}{2}, 3\right)$ is not a solution to the system.

25.

$$\begin{array}{rr} 5x - 4y = -6 & 8y = 10x + 12 \\ 5\left(-\frac{2}{5}\right) - 4\left(\frac{1}{4}\right) \stackrel{?}{=} -6 & 8\left(\frac{1}{4}\right) \stackrel{?}{=} 10\left(-\frac{2}{5}\right) + 12 \\ -2 - 1 \stackrel{?}{=} -6 & 2 \stackrel{?}{=} -4 + 12 \\ -3 \neq -6 & 2 \neq 8 \end{array}$$

$\left(-\frac{2}{5}, \frac{1}{4}\right)$ is not a solution to the system.

27. $x + y = 2$ $\quad x - y = 0$

x	y
0	2
1	1

x	y
0	0
1	1

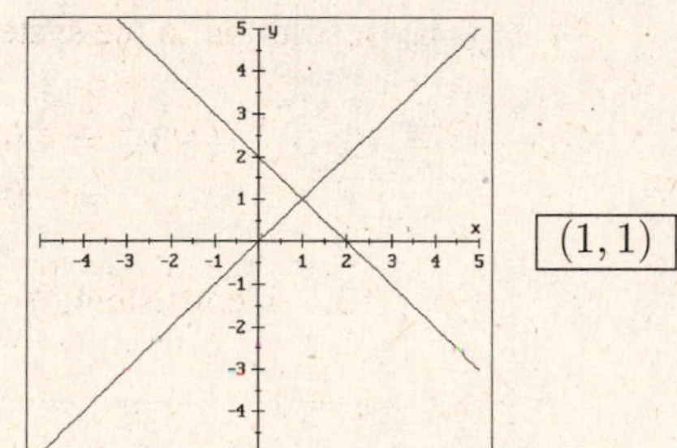

$(1, 1)$

29. $x + y = 2$ $\quad x - y = 4$

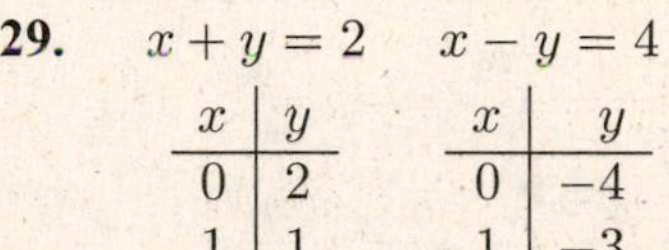

x	y
0	2
1	1

x	y
0	-4
1	-3

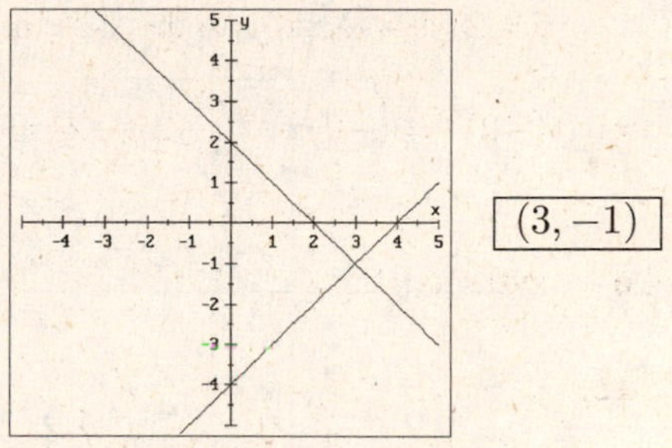

$(3, -1)$

31. $y = 2x$ $\quad x + y = 0$

x	y
0	0
1	2

x	y
2	-2
1	-1

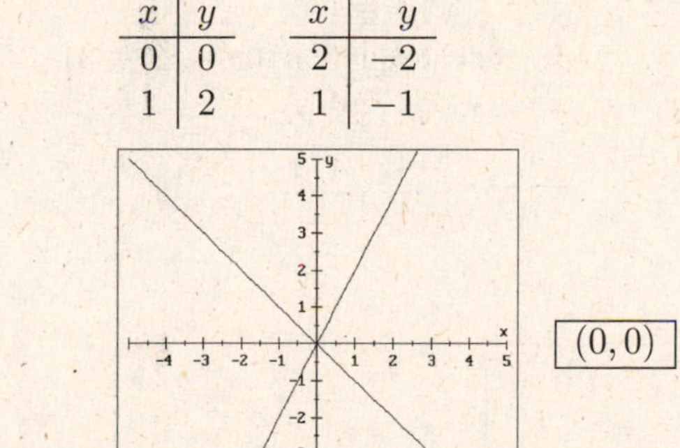

$(0, 0)$

33. $3x + 2y = -8$ $\quad 2x - 3y = -1$

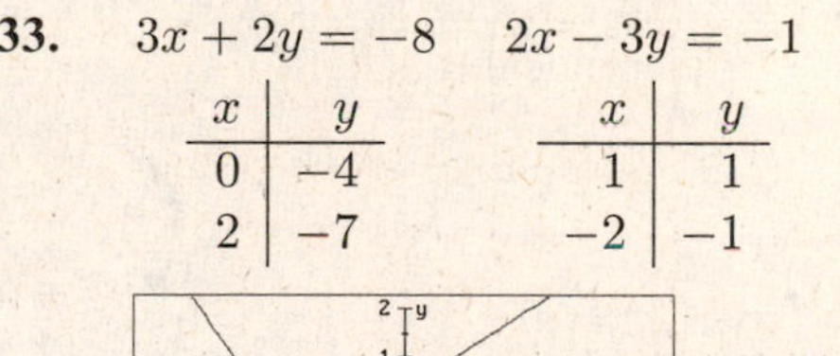

x	y
0	-4
2	-7

x	y
1	1
-2	-1

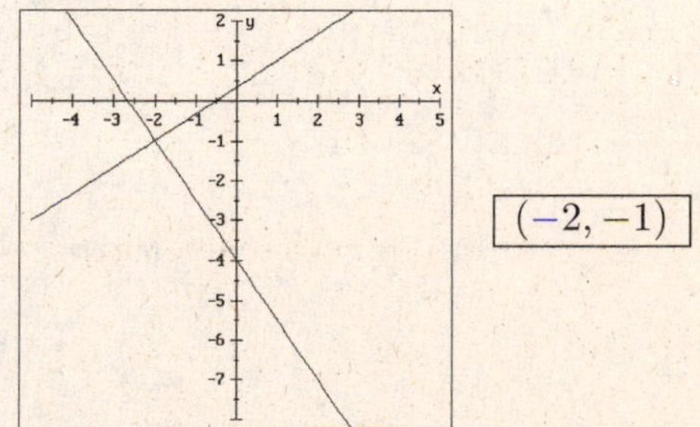

$(-2, -1)$

35. $3x - 6y = 18$ $\quad x = 2y + 3$

x	y
6	0
2	-2

x	y
3	0
5	1

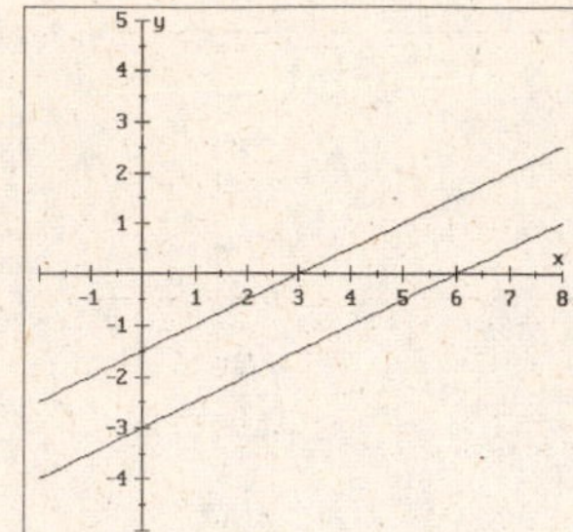

inconsistent system; solution set: $\emptyset$

37. $y = x$ $\quad x - y = 7$

x	y
0	0
2	2

x	y
4	-3
3	-4

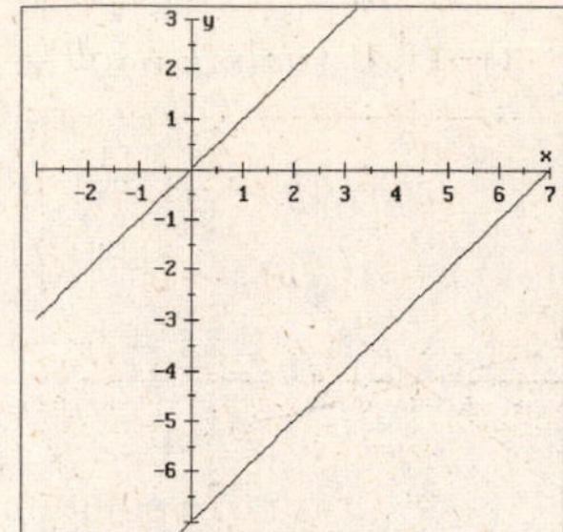

inconsistent system; solution set: $\emptyset$

39. $4x - 2y = 8$ $\quad y = 2x - 4$

x	y
0	-4
1	-2

x	y
0	-4
1	-2

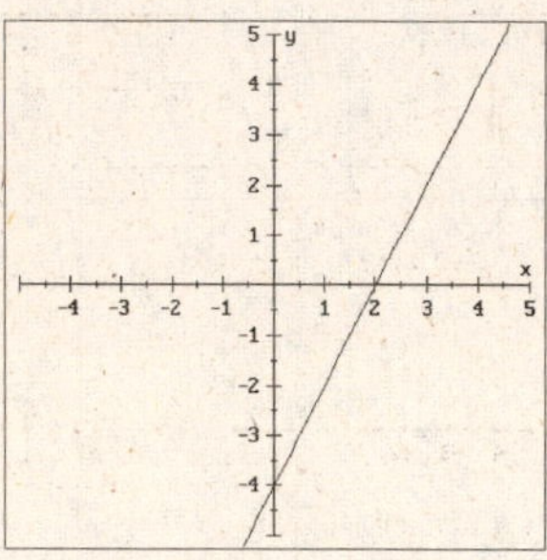

dependent equations
$(x, 2x - 4)$

41. $6x + 3y = 9$ $\quad y + 2x = 3$

$y = -2x + 3$

x	y
0	3
2	-1

x	y
0	3
2	-1

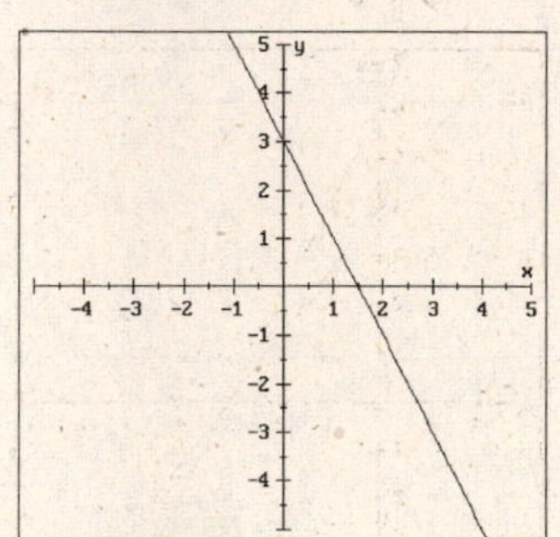

dependent equations $(x, -2x + 3)$

43. $x + 2y = -4$ $\quad x - \frac{1}{2}y = 6$

$2x - y = 12$

x	y
0	-2
2	-3

x	y
4	-4
6	0

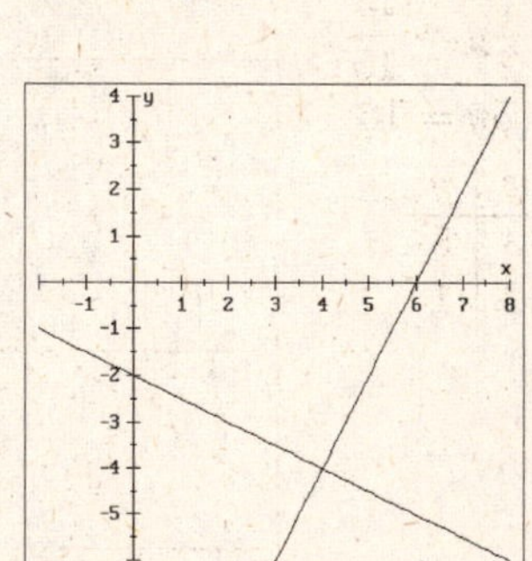

$\boxed{(4, -4)}$

SECTION 3.3

45. $-\frac{3}{4}x + y = 3$ $\quad$ $\frac{1}{4}x + y = -1$

$-3x + 4y = 12$ $\quad$ $x + 4y = -4$

x	y
0	3
−4	0

x	y
0	−1
−4	0

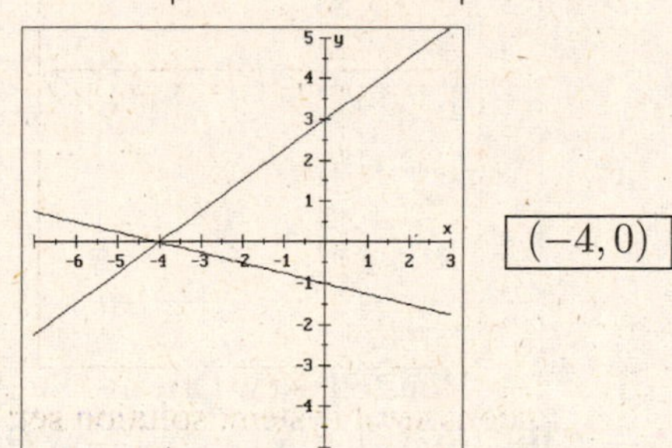

$(-4, 0)$

47. $2x - 3y = -18$ $\quad$ $3x + 2y = -1$

x	y
−3	4
−6	2

x	y
1	−2
−1	1

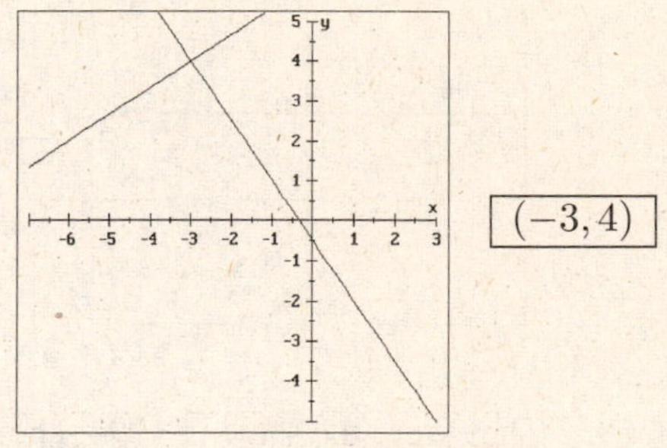

$(-3, 4)$

49. $4x = 3(4 - y)$ $\quad$ $2y = 4(3 - x)$

$4x = 12 - 3y$ $\quad$ $2y = 12 - 4x$

x	y
0	4
3	0

x	y
3	0
0	6

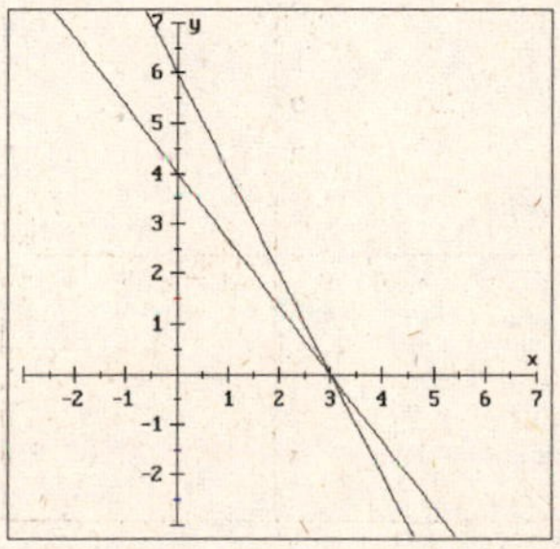

$(3, 0)$

51. $\frac{1}{2}x + \frac{1}{4}y = 0$ $\quad$ $\frac{1}{4}x - \frac{3}{8}y = -2$

$2x + y = 0$ $\quad$ $2x - 3y = -16$

x	y
0	0
1	−2

x	y
−2	4
−5	2

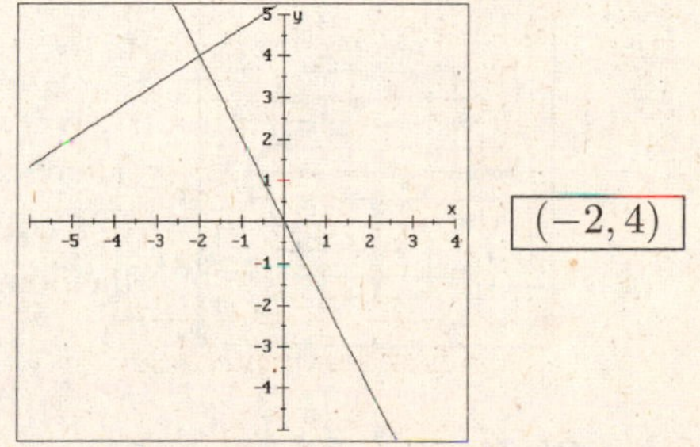

$(-2, 4)$

53. $\frac{1}{3}x - \frac{1}{2}y = \frac{1}{6}$ $\quad$ $\frac{2}{5}x + \frac{1}{2}y = \frac{13}{10}$

$2x - 3y = 1$ $\quad$ $4x + 5y = 13$

x	y
5	3
2	1

x	y
−3	5
2	1

$(2, 1)$

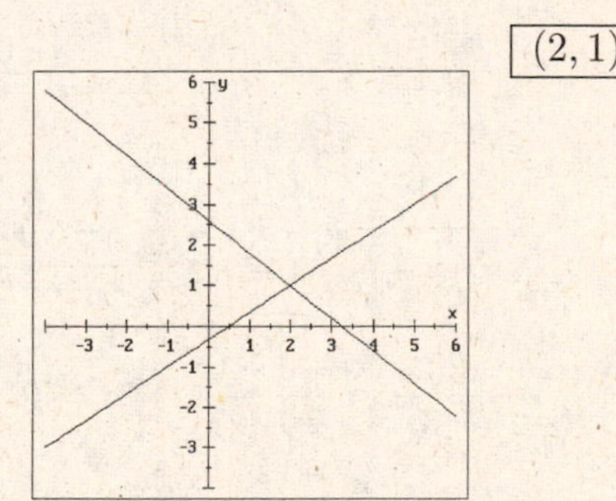

55. $\begin{cases} y = 4 - x \\ y = 2 + x \end{cases}$

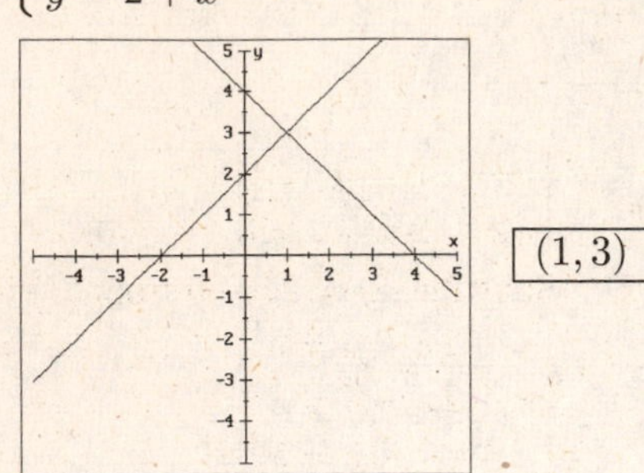

$\boxed{(1,3)}$

57. $\begin{cases} 3x - 6y = 4 & \Rightarrow y = \frac{1}{2}x - \frac{2}{3} \\ 2x + y = 1 & \Rightarrow y = -2x + 1 \end{cases}$

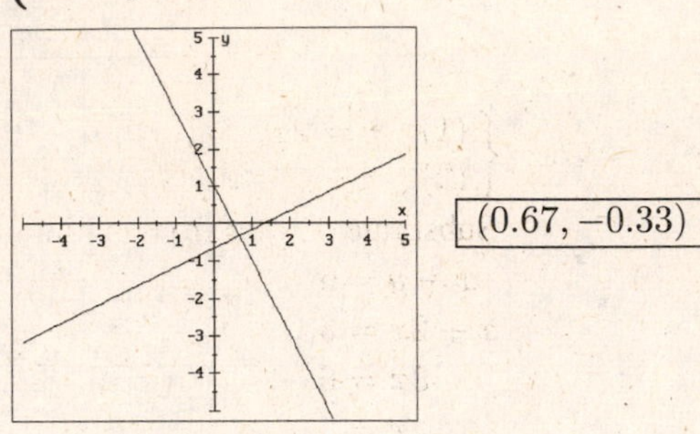

$\boxed{(0.67, -0.33)}$

59. They were the same in 1994; about 4100

61. **a.** Houston, New Orleans, St. Augustine
b. St. Louis, Memphis, New Orleans
c. New Orleans

63.

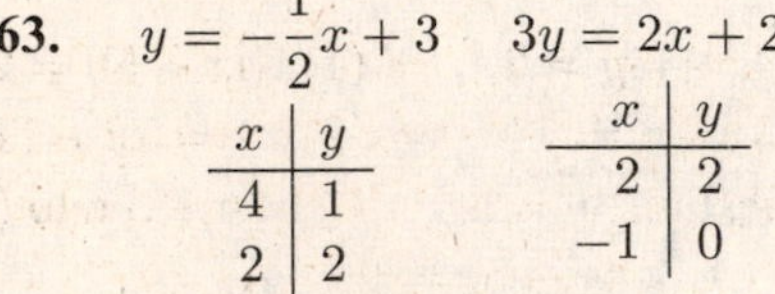

$y = -\frac{1}{2}x + 3$

x	y
4	1
2	2

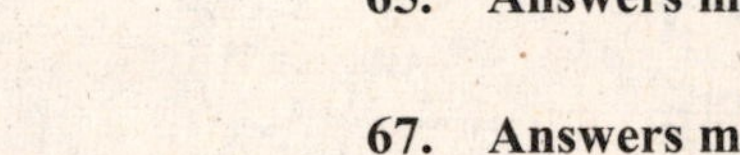

$3y = 2x + 2$

x	y
2	2
−1	0

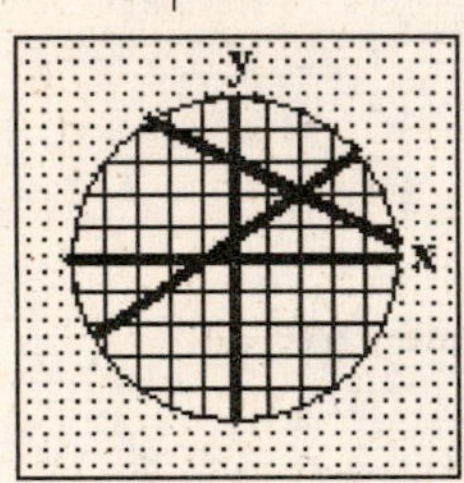

There is a chance of a collision at the point (2, 2).

65. **Answers may vary.**

67. **Answers may vary.**

Exercise 3.4 (page 201)

1. $y = x + 1$
$y = (2z) + 1$
$y = 2z + 1$

3. $y = x + 1$
$y = (3t + 2) + 1$
$y = 3t + 3$

5. $y^2 - x^2 = (3)^2 - (-2)^2 = 9 - 4 = 5$

7. $\frac{3x - 2y}{2x + y} = \frac{3(-2) - 2(3)}{2(-2) + 3} = \frac{-6 - 6}{-4 + 3}$
$= \frac{-12}{-1} = 12$

9. $-x(3y - 4) = -(-2)[3(3) - 4]$
$= -(-2)(5) = 10$

SECTION 3.4

11. y, terms

13.
$$\begin{aligned} 2(x-6) &= 2x-15 \\ 2x-12 &= 2x-15 \\ -12 &\neq -15 \end{aligned}$$
solution set: $\emptyset$

15. infinitely many

17. $\begin{cases} (1) \quad y = 2x \\ (2) \quad x + y = 6 \end{cases}$

Substitute $y = 2x$ from (1) into (2):
$$\begin{aligned} x + y &= 6 \\ x + 2x &= 6 \\ 3x &= 6 \\ x &= 2 \end{aligned}$$
Substitute and solve for y:
$y = 2x$
$y = 2(2) = 4$
$\boxed{x = 2, y = 4}$

19. $\begin{cases} (1) \quad y = 2x - 6 \\ (2) \quad 2x + y = 6 \end{cases}$

Substitute $y = 2x - 6$ from (1) into (2):
$$\begin{aligned} 2x + y &= 6 \\ 2x + 2x - 6 &= 6 \\ 4x &= 12 \\ x &= 3 \end{aligned}$$
Substitute and solve for y:
$y = 2x - 6$
$y = 2(3) - 6 = 6 - 6 = 0$
$\boxed{x = 3, y = 0}$

21. $\begin{cases} (1) \quad y = 2x + 5 \\ (2) \quad x + 2y = -5 \end{cases}$

Substitute $y = 2x + 5$ from (1) into (2):
$$\begin{aligned} x + 2y &= -5 \\ x + 2(2x + 5) &= -5 \\ x + 4x + 10 &= -5 \\ 5x &= -15 \\ x &= -3 \end{aligned}$$
Substitute and solve for y:
$y = 2x + 5$
$y = 2(-3) + 5 = -6 + 5 = -1$
$\boxed{x = -3, y = -1}$

23. $\begin{cases} 4x + 5y = 2 & \Rightarrow (1) \quad 4x + 5y = 2 \\ 3x - y = 11 & \Rightarrow (2) \quad y = 3x - 11 \end{cases}$

Substitute $y = 3x - 11$ from (2) into (1):
$$\begin{aligned} 4x + 5y &= 2 \\ 4x + 5(3x - 11) &= 2 \\ 4x + 15x - 55 &= 2 \\ 19x &= 57 \\ x &= 3 \end{aligned}$$
Substitute and solve for y:
$y = 3x - 11$
$y = 3(3) - 11 = 9 - 11 = -2$
$\boxed{x = 3, y = -2}$

25. $\begin{cases} 2x + y = 0 & \Rightarrow (1) \quad y = -2x \\ 3x + 2y = 1 & \Rightarrow (2) \quad 3x + 2y = 1 \end{cases}$

Substitute $y = -2x$ from (1) into (2):
$$\begin{aligned} 3x + 2y &= 1 \\ 3x + 2(-2x) &= 1 \\ 3x - 4x &= 1 \\ -x &= 1 \\ x &= -1 \end{aligned}$$
Substitute and solve for y:
$y = -2x$
$y = -2(-1) = 2$
$\boxed{x = -1, y = 2}$

27. $\begin{cases} (1) \quad 2x + 3y = 5 & \Rightarrow 2x = 5 - 3y \quad \Rightarrow x = \dfrac{5-3y}{2} \\ (2) \quad 3x + 2y = 5 \end{cases}$

Substitute $x = \dfrac{5-3y}{2}$ from (1) into (2):

$$\begin{aligned} 3x + 2y &= 5 \\ 3 \cdot \frac{5-3y}{2} + 2y &= 5 \\ 3(5-3y) + 4y &= 10 \\ 15 - 9y + 4y &= 10 \\ -5y &= -5 \\ y &= 1 \end{aligned}$$

Substitute and solve for x:

$$x = \frac{5-3y}{2}$$
$$x = \frac{5-3(1)}{2}$$
$$x = \frac{5-3}{2} = \frac{2}{2} = 1$$

Solution: $\boxed{x = 1, y = 1}$

29. $\begin{cases} (1) \quad 2x + 5y = -2 & \Rightarrow 2x = -2 - 5y \quad \Rightarrow x = \dfrac{-2-5y}{2} \\ (2) \quad 4x + 3y = 10 \end{cases}$

Substitute $x = \dfrac{-2-5y}{2}$ from (1) into (2):

$$\begin{aligned} 4x + 3y &= 10 \\ 4 \cdot \frac{-2-5y}{2} + 3y &= 10 \\ 4(-2-5y) + 6y &= 20 \\ -8 - 20y + 6y &= 20 \\ -14y &= 28 \\ y &= -2 \end{aligned}$$

Substitute and solve for x:

$$x = \frac{-2-5y}{2}$$
$$x = \frac{-2-5(-2)}{2}$$
$$x = \frac{-2+10}{2} = \frac{8}{2} = 4$$

Solution: $\boxed{x = 4, y = -2}$

31. $\begin{cases} (1) \quad 2a = 3b - 13 \\ (2) \quad b = 2a + 7 \end{cases}$

Substitute $b = 2a + 7$ from (2) into (1):

$$\begin{aligned} 2a &= 3b - 13 \\ 2a &= 3(2a+7) - 13 \\ 2a &= 6a + 21 - 13 \\ -4a &= 8 \\ a &= -2 \end{aligned}$$

Substitute and solve for b:

$b = 2a + 7$

$b = 2(-2) + 7 = -4 + 7 = 3$

$\boxed{a = -2, b = 3}$

33. $\begin{cases} (1) \quad 3(x-1)+3=8+2y \quad \Rightarrow 3x-3+3=8+2y \quad \Rightarrow 3x=2y+8 \quad \Rightarrow x=\dfrac{2y+8}{3} \\ (2) \quad 2(x+1)=4+3y \quad \Rightarrow 2x+2=4+3y \quad \Rightarrow 2x-3y=2 \end{cases}$

Substitute $x=\dfrac{2y+8}{3}$ from (1) into (2):

$$\begin{aligned} 2x-3y&=2 \\ 2\cdot\frac{2y+8}{3}-3y&=2 \\ 2(2y+8)-9y&=6 \\ 4y+16-9y&=6 \\ -5y&=-10 \\ y&=2 \end{aligned}$$

Substitute and solve for x:

$$x=\frac{2y+8}{3}$$
$$x=\frac{2(2)+8}{3}$$
$$x=\frac{4+8}{3}=\frac{12}{3}=4$$

Solution: $\boxed{x=4, y=2}$

35. $\begin{cases} (1) \quad 6a=5(3+b+a)-a \;\Rightarrow 6a=15+5b+5a-a \;\Rightarrow 2a=5b+15 \;\Rightarrow a=\dfrac{5b+15}{2} \\ (2) \quad 3(a-b)+4b=5(1+b) \;\Rightarrow 3a-3b+4b=5+5b \;\Rightarrow 3a-4b=5 \end{cases}$

Substitute $a=\dfrac{5b+15}{2}$ from (1) into (2):

$$\begin{aligned} 3a-4b&=5 \\ 3\cdot\frac{5b+15}{2}-4b&=5 \\ 3(5b+15)-8b&=10 \\ 15b+45-8b&=10 \\ 7b&=-35 \\ b&=-5 \end{aligned}$$

Substitute and solve for a:

$$a=\frac{5b+15}{2}$$
$$a=\frac{5(-5)+15}{2}$$
$$a=\frac{-25+15}{2}=\frac{-10}{2}=-5$$

Solution: $\boxed{a=-5, b=-5}$

37. $\begin{cases} 8y=15-4x & \Rightarrow (1) \quad 8y=15-4x \\ x+2y=4 & \Rightarrow (2) \quad x=-2y+4 \end{cases}$

Substitute $x=-2y+4$ from (2) into (1):

$$8y=15-4x$$
$$8y=15-4(-2y+4)$$
$$8y=15+8y-16$$
$$0\neq-1$$

$\boxed{\text{inconsistent system, } \emptyset}$

39. $\begin{cases} (1) \quad a=\dfrac{3}{2}b+5 \\ (2) \quad 2a-3b=8 \end{cases}$

Substitute $a=\dfrac{3}{2}b+5$ from (1) into (2):

$$\begin{aligned} 2a-3b&=8 \\ 2\left(\tfrac{3}{2}b+5\right)-3b&=8 \\ 3b+10-3b&=8 \\ 10&\neq8 \end{aligned}$$

$\boxed{\text{inconsistent system, } \emptyset}$

41. $\begin{cases} 9x = 3y + 12 & \Rightarrow (1) \quad 9x = 3y + 12 \\ 4 = 3x - y & \Rightarrow (2) \quad y = 3x - 4 \end{cases}$

Substitute $y = 3x - 4$ from (2) into (1):

$$\begin{aligned} 9x &= 3y + 12 \\ 9x &= 3(3x - 4) + 12 \\ 9x &= 9x - 12 + 12 \\ 0 &= 0 \end{aligned}$$

$\boxed{\text{dependent equations; } (x, 3x - 4)}$

43. $\begin{cases} (1) \quad 3a + 6b = -15 \\ (2) \quad a = -2b - 5 & \Rightarrow b = \dfrac{a-5}{2} \end{cases}$

Substitute $b = \dfrac{a+5}{2}$ from (1) into (2):

$$\begin{aligned} 3a + 6b &= -15 \\ 3a + 6 \cdot \tfrac{a-5}{2} &= -15 \\ 3a + 3(a - 5) &= -15 \\ -15 &= -15 \end{aligned}$$

$\boxed{\text{dependent equations; } \left(a, \tfrac{a+5}{2}\right)}$

45. $\begin{cases} 2x + y = 4 & \Rightarrow (1) \quad y = -2x + 4 \\ 4x + y = 5 & \Rightarrow (2) \quad 4x + y = 5 \end{cases}$

Substitute $y = -2x + 4$ from (1) into (2):

$$\begin{aligned} 4x + y &= 5 \\ 4x + (-2x + 4) &= 5 \\ 4x - 2x + 4 &= 5 \\ 2x &= 1 \\ x &= \frac{1}{2} \end{aligned}$$

Substitute and solve for y:

$$\begin{aligned} y &= -2x + 4 \\ y &= -2\left(\frac{1}{2}\right) + 4 = -1 + 4 = 3 \end{aligned}$$

$\boxed{x = \tfrac{1}{2}, y = 3}$

47. $\begin{cases} r + 3s = 9 & \Rightarrow (1) \quad r = -3s + 9 \\ 3r + 2s = 13 & \Rightarrow (2) \quad 3r + 2s = 13 \end{cases}$

Substitute $r = -3s + 9$ from (1) into (2):

$$\begin{aligned} 3r + 2s &= 13 \\ 3(-3s + 9) + 2s &= 13 \\ -9s + 27 + 2s &= 13 \\ -7s &= -14 \\ s &= 2 \end{aligned}$$

Substitute and solve for r:

$$\begin{aligned} r &= -3s + 9 \\ r &= -3(2) + 9 = -6 + 9 = 3 \end{aligned}$$

$\boxed{r = 3, s = 2}$

49. $\begin{cases} (1) \quad y - x = 3x & \Rightarrow y = 4x \\ (2) \quad 2(x + y) = 14 - y & \Rightarrow 2x + 2y = 14 - y & \Rightarrow 2x + 3y = 14 \end{cases}$

Substitute $y = 4x$ from (1) into (2):

$$\begin{aligned} 2x + 3y &= 14 \\ 2x + 3(4x) &= 14 \\ 2x + 12x &= 14 \\ 14x &= 14 \\ x &= 1 \end{aligned}$$

Substitute and solve for y:

$$\begin{aligned} y &= 4x \\ y &= 4(1) = 4 \end{aligned}$$

Solution:

$\boxed{x = 1, y = 4}$

51. $\begin{cases} 3x + 4y = -7 & \Rightarrow (1) \quad 3x + 4y = -7 \\ 2y - x = -1 & \Rightarrow (2) \quad x = 2y + 1 \end{cases}$

Substitute $x = 2y + 1$ from (2) into (1):

$$\begin{aligned} 3x + 4y &= -7 \\ 3(2y + 1) + 4y &= -7 \\ 6y + 3 + 4y &= -7 \\ 10y &= -10 \\ y &= -1 \end{aligned}$$

Substitute and solve for x:

$$\begin{aligned} x &= 2y + 1 \\ x &= 2(-1) + 1 = -2 + 1 = -1 \end{aligned}$$

Solution:

$\boxed{x = -1, y = -1}$

53. $\begin{cases} (1) \quad 2x - 3y = -3 & \Rightarrow 2x = 3y - 3 \quad \Rightarrow x = \dfrac{3y-3}{2} \\ (2) \quad 3x + 5y = -14 \end{cases}$

Substitute $x = \dfrac{3y-3}{2}$ from (1) into (2):

$$\begin{aligned} 3x + 5y &= -14 \\ 3 \cdot \frac{3y-3}{2} + 5y &= -14 \\ 3(3y-3) + 10y &= -28 \\ 9y - 9 + 10y &= -28 \\ 19y &= -19 \\ y &= -1 \end{aligned}$$

Substitute and solve for x:

$$\begin{aligned} x &= \frac{3y-3}{2} \\ x &= \frac{3(-1)-3}{2} \\ x &= \frac{-3-3}{2} = \frac{-6}{2} = -3 \end{aligned}$$

Solution: $\boxed{x = -3, y = -1}$

55. $\begin{cases} (1) \quad 7x - 2y = -1 & \Rightarrow 7x = 2y - 1 \quad \Rightarrow x = \dfrac{2y-1}{7} \\ (2) \quad -5x + 2y = -1 \end{cases}$

Substitute $x = \dfrac{2y-1}{7}$ from (1) into (2):

$$\begin{aligned} -5x + 2y &= -1 \\ -5 \cdot \frac{2y-1}{7} + 2y &= -1 \\ -5(2y-1) + 14y &= -7 \\ -10y + 5 + 14y &= -7 \\ 4y &= -12 \\ y &= -3 \end{aligned}$$

Substitute and solve for x:

$$\begin{aligned} x &= \frac{2y-1}{7} \\ x &= \frac{2(-3)-1}{7} \\ x &= \frac{-6-1}{7} = \frac{-7}{7} = -1 \end{aligned}$$

Solution: $\boxed{x = -1, y = -3}$

57. $\begin{cases} (1) \quad 2a + 3b = 2 & \Rightarrow 2a = 2 - 3b \quad \Rightarrow a = \dfrac{2-3b}{2} \\ (2) \quad 8a - 3b = 3 \end{cases}$

Substitute $a = \dfrac{2-3b}{2}$ from (1) into (2):

$$\begin{aligned} 8a - 3b &= 3 \\ 8 \cdot \frac{2-3b}{2} - 3b &= 3 \\ 4(2-3b) - 3b &= 3 \\ 8 - 12b - 3b &= 3 \\ -15b &= -5 \\ b &= \frac{1}{3} \end{aligned}$$

Substitute and solve for a:

$$\begin{aligned} a &= \frac{2-3b}{2} \\ a &= \frac{2-3(\frac{1}{3})}{2} \\ a &= \frac{2-1}{2} = \frac{1}{2} \end{aligned}$$

Solution: $\boxed{a = \frac{1}{2}, b = \frac{1}{3}}$

59. $\begin{cases} (1) & \frac{1}{2}x + \frac{1}{2}y = -1 & \Rightarrow x + y = -2 & \Rightarrow x = -y - 2 \\ (2) & \frac{1}{3}x - \frac{1}{2}y = -4 & \Rightarrow 2x - 3y = -24 \end{cases}$

Substitute $x = -y - 2$ from (1) into (2):

$$\begin{aligned} 2x - 3y &= -24 \\ 2(-y-2) - 3y &= -24 \\ -2y - 4 - 3y &= -24 \\ -5y &= -20 \\ y &= 4 \end{aligned}$$

Substitute and solve for x:

$x = -y - 2$

$x = -4 - 2 = -6$

Solution: $\boxed{x = -6, y = 4}$

61. $\begin{cases} (1) & 5x = \frac{1}{2}y - 1 & \Rightarrow 10x = y - 2 \\ (2) & \frac{1}{4}y = 10x - 1 & \Rightarrow y = 40x - 4 \end{cases}$

Substitute $y = 40x - 4$ from (2) into (1):

$$\begin{aligned} 10x &= y - 2 \\ 10x &= 40x - 4 - 2 \\ -30x &= -6 \\ x &= \frac{1}{5} \end{aligned}$$

Substitute and solve for y:

$y = 40x - 4$

$y = 40\left(\frac{1}{5}\right) - 4$

$y = 8 - 4 = 4$

Solution: $\boxed{x = \frac{1}{5}, y = 4}$

63. $\begin{cases} (1) & \frac{6x-1}{3} - \frac{5}{3} = \frac{3y+1}{2} & \Rightarrow [\times 6] \Rightarrow & 2(6x-1) - 2(5) = 3(3y+1) \\ & & & 12x - 2 - 10 = 9y + 3 \quad \Rightarrow 12x = 9y + 15 \\ (2) & \frac{1+5y}{4} + \frac{x+3}{4} = \frac{17}{2} & \Rightarrow [\times 4] \Rightarrow & 1 + 5y + x + 3 = 2(17) \quad \Rightarrow x = 30 - 5y \end{cases}$

Substitute $x = 30 - 5y$ from (2) into (1):

$$\begin{aligned} 12x &= 9y + 15 \\ 12(30 - 5y) &= 9y + 15 \\ 360 - 60y &= 9y + 15 \\ -69y &= -345 \\ y &= 5 \end{aligned}$$

Substitute and solve for x:

$x = 30 - 5y$

$x = 30 - 5(5)$

$x = 30 - 25 = 5$

Solution: $\boxed{x = 5, y = 5}$

65. $\begin{cases} (1) & x + y = 90 \\ (2) & y = 2x \end{cases}$

Substitute $y = 2x$ from (2) into (1):

$$\begin{aligned} x + y &= 90 \\ x + 2x &= 90 \\ 3x &= 90 \\ x &= 30 \end{aligned}$$

Substitute and solve for y:

$y = 2x$

$y = 2(30) = 60$

$\boxed{x = 30°, y = 60°}$

67. Answers may vary.

69. Answers may vary.

Exercise 3.5 (page 208)

1.
$$\begin{aligned} x + y &= 1 \\ x - y &= 1 \\ \hline 2x \quad &= 2 \\ x \quad &= 1 \end{aligned}$$

3.
$$\begin{aligned} -x + \ y &= 3 \\ x + \ y &= 3 \\ \hline 2y &= 6 \\ y &= 3 \end{aligned}$$

5.
$$\begin{aligned} 8(3x - 5) - 12 &= 4(2x + 3) \\ 24x - 40 - 12 &= 8x + 12 \\ 24x - 52 &= 8x + 12 \\ 24x - 8x - 52 &= 8x - 8x + 12 \\ 16x - 52 + 52 &= 12 + 52 \\ 16x &= 64 \\ \frac{16x}{16} &= \frac{64}{16} \\ x &= 4 \end{aligned}$$

7.
$$\begin{aligned} x - 2 &= \frac{x + 2}{3} \\ 3(x - 2) &= 3 \cdot \frac{x + 2}{3} \\ 3x - 6 &= x + 2 \\ 3x - x - 6 &= x - x + 2 \\ 2x - 6 + 6 &= 2 + 6 \\ 2x &= 8 \\ \frac{2x}{2} &= \frac{8}{2} \\ x &= 4 \end{aligned}$$

9.
$$\begin{aligned} 7x - 9 &\le 5 \\ 7x - 9 + 9 &\le 5 + 9 \\ 7x &\le 14 \\ \frac{7x}{7} &\le \frac{14}{7} \\ x &\le 2 \end{aligned}$$

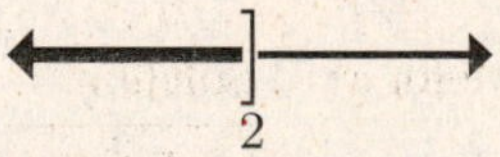

11. coefficient

13. general

15. 15

17.
$$\begin{aligned} x + y &= \ \ 5 \\ x - y &= -3 \\ \hline 2x \quad &= \ \ 2 \\ x \quad &= \ \ 1 \end{aligned}$$

Substitute and solve for y:

$$\begin{aligned} x + y &= 5 \\ 1 + y &= 5 \\ y &= 4 \quad \boxed{(1, 4)} \end{aligned}$$

19.
$$\begin{aligned} x - y &= -5 \\ x + y &= \ \ 1 \\ \hline 2x \quad &= -4 \\ x \quad &= -2 \end{aligned}$$

Substitute and solve for y:

$$\begin{aligned} x + y &= 1 \\ -2 + y &= 1 \\ y &= 3 \quad \boxed{(-2, 3)} \end{aligned}$$

21.
$$\begin{aligned} 2x + \ y &= -1 \\ -2x + \ y &= \ \ 3 \\ \hline 2y &= \ \ 2 \\ y &= \ \ 1 \end{aligned}$$

Substitute and solve for x:

$$\begin{aligned} 2x + y &= -1 \\ 2x + 1 &= -1 \\ 2x &= -2 \\ x &= -1 \end{aligned}$$

Solution:

$$\boxed{(-1, 1)}$$

23. $\begin{aligned} 2x - 3y &= -11 \\ 3x + 3y &= 21 \\ \hline 5x &= 10 \\ x &= 2 \end{aligned}$

Substitute and solve for y:

$$\begin{aligned} 3x + 3y &= 21 \\ 3(2) + 3y &= 21 \\ 6 + 3y &= 21 \\ 3y &= 15 \\ y &= 5 \end{aligned}$$

Solution: $\boxed{(2, 5)}$

25. $\begin{aligned} x + y &= 5 \Rightarrow \times (-1) \\ x + 2y &= 8 \end{aligned}$ $\quad$ $\begin{aligned} -x - y &= -5 \\ x + 2y &= 8 \\ \hline y &= 3 \end{aligned}$ $\quad$ $\begin{aligned} x + 2y &= 8 \\ x + 2(3) &= 8 \\ x + 6 &= 8 \\ x &= 2 \end{aligned}$

Solution: $\boxed{(2, 3)}$

27. $\begin{aligned} 2x + y &= 4 \Rightarrow \times (-1) \\ 2x + 3y &= 0 \end{aligned}$ $\quad$ $\begin{aligned} -2x - y &= -4 \\ 2x + 3y &= 0 \\ \hline 2y &= -4 \\ y &= -2 \end{aligned}$ $\quad$ $\begin{aligned} 2x + y &= 4 \\ 2x + (-2) &= 4 \\ 2x &= 6 \\ x &= 3 \end{aligned}$

Solution: $\boxed{(3, -2)}$

29. $\begin{aligned} 3x + 29 &= 5y \Rightarrow 3x - 5y = -29 \Rightarrow \times (-1) \\ 4y - 34 &= -3x \Rightarrow 3x + 4y = 34 \end{aligned}$ $\quad$ $\begin{aligned} -3x + 5y &= 29 \\ 3x + 4y &= 34 \\ \hline 9y &= 63 \\ y &= 7 \end{aligned}$

$$\begin{aligned} 3x + 4y &= 34 \\ 3x + 4(7) &= 34 \\ 3x + 28 &= 34 \\ 3x &= 6 \\ x &= 2 \end{aligned}$$

Solution: $\boxed{(2, 7)}$

31. $\begin{aligned} 2x + y &= 10 \Rightarrow \times (-2) \\ x + 2y &= 10 \end{aligned}$ $\quad$ $\begin{aligned} -4x - 2y &= -20 \\ x + 2y &= 10 \\ \hline -3x &= -10 \\ x &= \tfrac{10}{3} \end{aligned}$ $\quad$ $\begin{aligned} 2x + y &= 10 \\ 2\left(\tfrac{10}{3}\right) + y &= 10 \\ \tfrac{20}{3} + y &= \tfrac{30}{3} \\ y &= \tfrac{10}{3} \end{aligned}$

Solution: $\boxed{\left(\frac{10}{3}, \frac{10}{3}\right)}$

33. $\begin{aligned} 3x + 2y &= 0 \Rightarrow \times (3) \\ 2x - 3y &= -13 \Rightarrow \times (2) \end{aligned}$ $\quad$ $\begin{aligned} 9x + 6y &= 0 \\ 4x - 6y &= -26 \\ \hline 13x &= -26 \\ x &= -2 \end{aligned}$ $\quad$ $\begin{aligned} 3x + 2y &= 0 \\ 3(-2) + 2y &= 0 \\ -6 + 2y &= 0 \\ 2y &= 6 \\ y &= 3 \end{aligned}$

Solution: $\boxed{(-2, 3)}$

35. $\begin{aligned} 4x + 5y &= -20 \Rightarrow \times (4) \\ 5x - 4y &= -25 \Rightarrow \times (5) \end{aligned}$ $\quad$ $\begin{aligned} 16x + 20y &= -80 \\ 25x - 20y &= -125 \\ \hline 41x &= -205 \\ x &= -5 \end{aligned}$ $\quad$ $\begin{aligned} 4x + 5y &= -20 \\ 4(-5) + 5y &= -20 \\ -20 + 5y &= -20 \\ 5y &= 0 \\ y &= 0 \end{aligned}$

Solution: $\boxed{(-5, 0)}$

SECTION 3.5

37.

$$\begin{array}{l} 6x = -3y \\ \underline{5y = 2x + 12} \end{array} \Rightarrow \begin{array}{l} 6x + 3y = 0 \\ \underline{-2x + 5y = 12} \Rightarrow \times (3) \end{array} \quad \begin{array}{r} 6x + 3y = 0 \\ \underline{-6x + 15y = 36} \\ 18y = 36 \\ y = 2 \end{array} \quad \begin{array}{r} 6x = -3y \\ 6x = -3(2) \\ 6x = -6 \\ x = -1 \end{array}$$

Solution: $\boxed{(-1, 2)}$

39.

$$\begin{array}{l} 3x - 2y = -1 \Rightarrow \times (3) \\ \underline{2x + 3y = -5} \Rightarrow \times (2) \end{array} \quad \begin{array}{r} 9x - 6y = -3 \\ \underline{4x + 6y = -10} \\ 13x = -13 \\ x = -1 \end{array} \quad \begin{array}{r} 2x + 3y = -5 \\ 2(-1) + 3y = -5 \\ -2 + 3y = -5 \\ 3y = -3 \\ y = -1 \end{array}$$

Solution: $\boxed{(-1, -1)}$

41.

$$\begin{array}{l} \frac{3}{5}x + \frac{4}{5}y = 1 \Rightarrow \times (5) \\ \underline{-\frac{1}{4}x + \frac{3}{8}y = 1} \Rightarrow \times (8) \end{array} \quad \begin{array}{l} 3x + 4y = 5 \Rightarrow \times (2) \\ \underline{-2x + 3y = 8} \Rightarrow \times (3) \end{array} \quad \begin{array}{r} 6x + 8y = 10 \\ \underline{-6x + 9y = 24} \\ 17y = 34 \\ y = 2 \end{array} \quad \begin{array}{r} 3x + 4y = 5 \\ 3x + 4(2) = 5 \\ 3x + 8 = 5 \\ 3x = -3 \\ x = -1 \end{array}$$

Solution: $\boxed{(-1, 2)}$

43.

$$\begin{array}{l} \frac{3}{5}x + y = 1 \Rightarrow \times (5) \\ \underline{\frac{4}{5}x - y = -1} \Rightarrow \times (5) \end{array} \quad \begin{array}{r} 3x + 5y = 5 \\ \underline{4x - 5y = -5} \\ 7x = 0 \\ x = 0 \end{array} \quad \begin{array}{r} 3x + 5y = 5 \\ 3(0) + 5y = 5 \\ 5y = 5 \\ y = 1 \end{array}$$

Solution: $\boxed{(0, 1)}$

45.

$$\begin{array}{l} 2x = 3(y - 2) \\ \underline{2(x + 4) = 3y} \end{array} \Rightarrow \begin{array}{l} 2x = 3y - 6 \\ \underline{2x + 8 = 3y} \end{array} \Rightarrow \begin{array}{l} 2x - 3y = -6 \Rightarrow \times (-1) \\ \underline{2x - 3y = -8} \end{array} \quad \begin{array}{r} -2x + 3y = 6 \\ \underline{2x - 3y = -8} \\ 0 \neq -2 \end{array}$$

$\boxed{\text{solution set: } \emptyset}$

47.

$$\begin{array}{l} 4x = 3(4 - y) \\ \underline{3y = 4(2 - x)} \end{array} \Rightarrow \begin{array}{l} 4x = 12 - 3y \\ \underline{3y = 8 - 4x} \end{array} \Rightarrow \begin{array}{l} 4x + 3y = 12 \\ \underline{4x + 3y = 8} \Rightarrow \times (-1) \end{array} \quad \begin{array}{r} 4x + 3y = 12 \\ \underline{-4x - 3y = -8} \\ 0 \neq 4 \end{array}$$

$\boxed{\text{solution set: } \emptyset}$

49.

$$\begin{array}{l} 3(x - 2) = 4y \\ \underline{2(2y + 3) = 3x} \end{array} \Rightarrow \begin{array}{l} 3x - 6 = 4y \\ \underline{4y + 6 = 3x} \end{array} \Rightarrow \begin{array}{r} 3x - 4y = 6 \\ \underline{-3x + 4y = -6} \\ 0 = 0 \end{array}$$

$$\begin{aligned} 4y + 6 &= 3x \\ 4y &= 3x - 6 \\ y &= \frac{3x - 6}{4} \end{aligned} \Rightarrow \boxed{\text{infinitely many solutions of the form } \left(x, \frac{3x-6}{4}\right)}$$

51. $\dfrac{3(x-2y)=12 \quad\Rightarrow}{x=2(y+2)\quad\Rightarrow}$ $\dfrac{3x-6y=12}{x=2y+4}\Rightarrow$ $\dfrac{3x-6y=12}{x-2y=4}\Rightarrow \times(-3)$ $\dfrac{3x-6y=12}{-3x+6y=-12}$ $0=0$

$$\begin{aligned}3x-6y&=12\\-6y&=-3x+12\\6y&=3x-12\\y&=\frac{3x-12}{6}\\y&=\frac{3(x-4)}{6}=\frac{x-4}{2}\end{aligned}\Rightarrow \boxed{\text{infinitely many solutions of the form } \left(x, \tfrac{x-4}{2}\right)}$$

53.
$$\begin{aligned}2x+\ y&=-2\\-2x-3y&=-6\\\hline -2y&=-8\\y&=4\end{aligned}$$

Substitute and solve for x:

$$\begin{aligned}2x+y&=-2\\2x+4&=-2\\2x&=-6\\x&=-3\quad\boxed{(-3,4)}\end{aligned}$$

55.
$$\begin{aligned}4x+3y&=24\\4x-3y&=-24\\\hline 8x&=0\\x&=0\end{aligned}$$

Substitute and solve for y:

$$\begin{aligned}4x+3y&=24\\4(0)+3y&=24\\0+3y&=24\\3y&=24\\y&=8\quad\boxed{(0,8)}\end{aligned}$$

57. $\dfrac{5(x-1)=8-3(y+2)}{4(x+2)-7=3(2-y)}$ $\Rightarrow$ $\dfrac{5x-5=8-3y-6}{4x+8-7=6-3y}$ $\Rightarrow$ $\dfrac{5x+3y=7}{4x+3y=5}$

$\dfrac{5x+3y=7\Rightarrow\times(-1)}{4x+3y=5}$

$$\begin{aligned}-5x-3y&=-7\\4x+3y&=5\\\hline -x&=-2\\x&=2\end{aligned}$$

$$\begin{aligned}5x+3y&=7\\5(2)+3y&=7\\10+3y&=7\\3y&=-3\\y&=-1\end{aligned}$$

Solution: $\boxed{(2,-1)}$

59. $\dfrac{2x+3y=2\Rightarrow\times(-2)}{4x-9y=-1}$

$$\begin{aligned}-4x-6y&=-4\\4x-9y&=-1\\\hline -15y&=-5\\y&=\tfrac{-5}{-15}=\tfrac{1}{3}\end{aligned}$$

$$\begin{aligned}2x+3y&=2\\2x+3\left(\tfrac{1}{3}\right)&=2\\2x+1&=2\\2x&=1\\x&=\tfrac{1}{2}\end{aligned}$$

Solution: $\boxed{\left(\dfrac{1}{2},\dfrac{1}{3}\right)}$

61.
$$\begin{aligned} 4(2x-y)&=18 &&\Rightarrow 8x-4y=18 &&\Rightarrow 8x-4y=18\\ 3(x-3)&=2y-1 &&\Rightarrow 3x-9=2y-1 &&\Rightarrow 3x-2y=8 \end{aligned}$$

$$\begin{aligned} 8x-4y&=18 & 8x-4y&=18 \\ 3x-2y&=8 \Rightarrow \times(-2) & -6x+4y&=-16 \\ & & 2x&=2 \\ & & x&=1 \end{aligned}$$

$$\begin{aligned} 8x-4y&=18\\ 8(1)-4y&=18\\ 8-4y&=18\\ -4y&=10\\ y&=-\tfrac{10}{4}=-\tfrac{5}{2} \end{aligned}$$

Solution: $\boxed{(1,-\frac{5}{2})}$

63.
$$\begin{aligned} \frac{x}{2}-\frac{y}{3}&=-2 \Rightarrow \times(6) & 3x-2y&=-12 \Rightarrow & 3x-2y&=-12\\ \frac{2x-3}{2}+\frac{6y+1}{3}&=\frac{17}{6} \Rightarrow \times(6) & 3(2x-3)+2(6y+1)&=17 \Rightarrow & 6x-9+12y+2&=17 \end{aligned}$$

$$\begin{aligned} 3x-2y&=-12 \Rightarrow \times(-2) & -6x+4y&=24\\ 6x+12y&=24 & 6x+12y&=24\\ & & 16y&=48\\ & & y&=3 \end{aligned}$$

$$\begin{aligned} 3x-2y&=-12\\ 3x-2(3)&=-12\\ 3x-6&=-12\\ 3x&=-6\\ x&=-2 \end{aligned}$$

Solution: $\boxed{(-2,3)}$

65.
$$\begin{aligned} \frac{x-3}{2}+\frac{y+5}{3}&=\frac{11}{6} \Rightarrow \times(6) & 3(x-3)+2(y+5)&=11 \Rightarrow 3x-9+2y+10=11\\ \frac{x+3}{3}-\frac{5}{12}&=\frac{y+3}{4} \Rightarrow \times(12) & 4(x+3)-5&=3(y+3) \Rightarrow 4x+12-5=3y+9 \end{aligned}$$

$$\begin{aligned} 3x+2y&=10 \Rightarrow \times(3) & 9x+6y&=30\\ 4x-3y&=2 \Rightarrow \times(2) & 8x-6y&=4\\ & & 17x&=34\\ & & x&=2 \end{aligned}$$

$$\begin{aligned} 3x+2y&=10\\ 3(2)+2y&=10\\ 6+2y&=10\\ 2y&=4\\ y&=2 \end{aligned}$$

Solution: $\boxed{(2,2)}$

67.
$$\begin{aligned} 2(x+y)&=10 \Rightarrow 2x+2y=10 \Rightarrow \times(5) & 10x+10y&=50\\ 5(x-y)&=5 \Rightarrow 5x-5y=5 \Rightarrow \times(2) & 10x-10y&=10\\ & & 20x&=60\\ & & x&=3 \end{aligned}$$

$$\begin{aligned} 2x+2y&=10\\ 2(3)+2y&=10\\ 6+2y&=10\\ 2y&=4\\ y&=2 \end{aligned}$$

Solution: $\boxed{x=3,\ y=2}$

69. **Answers may vary.**

71. $(1,4)$

Exercise 3.6 (page 218)

1. $2x$

3. $2x+3y$

5. $\$(3x+2y)$

SECTION 3.6

7. $x < 4$

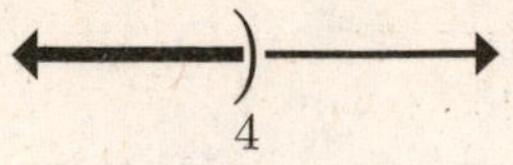

9. $-1 < x \le 2$

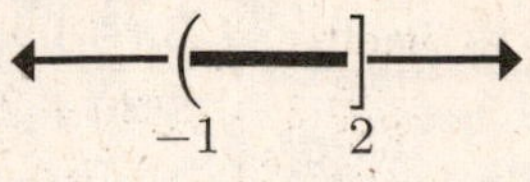

11. $8 \cdot 8 \cdot 8 \cdot c = 8^3c$

13. $a \cdot a \cdot b \cdot b = a^2b^2$

15. variable

17. system

19. Let $x =$ President's salary.
Let $y =$ Vice President's salary.

$$\begin{cases} (1) & x = y + 207400 \\ (2) & x + y = 592600 \end{cases}$$

Substitute $x = y + 207400$ into (2):

$$\begin{aligned} x + y &= 592600 \\ y + 207400 + y &= 592600 \\ 2y &= 385200 \\ y &= 192600 \end{aligned}$$

Substitute; solve for x:

$$\begin{aligned} x &= y + 207400 \\ x &= 192600 + 207400 \\ x &= 400000 \end{aligned}$$

The president makes $400,000 and the vice president makes $192,600.

21. Let $x =$ older amount.
Let $y =$ younger amount.

$$\begin{cases} (1) & x = 2y + 10000 \\ (2) & x + y = 497500 \end{cases}$$

Substitute $x = 2y + 10000$ into (2):

$$\begin{aligned} x + y &= 497500 \\ 2y + 10000 + y &= 497500 \\ 3y &= 487500 \\ y &= 162500 \end{aligned}$$

The younger son will get $162,500.

23. Let $x =$ the length of one piece.
Let $y =$ the other length.

$$\begin{cases} (1) & x = y + 5 \\ (2) & x + y = 25 \end{cases}$$

Substitute $x = y + 5$ into (2):

$$\begin{aligned} x + y &= 25 \\ y + 5 + y &= 25 \\ 2y &= 20 \\ y &= 10 \end{aligned}$$

Substitute; solve for x:

$$\begin{aligned} x &= y + 5 \\ x &= 10 + 5 = 15 \end{aligned}$$

The lengths are 10 ft and 15 ft.

25. Let $x =$ cost of a catcher's mitt and $y =$ cost of an outfielder's glove.

$$\begin{aligned} x + 10y &= 239.50 \Rightarrow \times(-1) \\ x + 5y &= 134.50 \end{aligned}$$

$$\begin{aligned} -x - 10y &= -239.50 \\ x + 5y &= 134.50 \\ \hline -5y &= -105 \\ y &= 21 \end{aligned}$$

$$\begin{aligned} x + 5y &= 134.50 \\ x + 5(21) &= 134.50 \\ x + 105 &= 134.50 \\ x &= 29.50 \end{aligned}$$

A catcher's mitt costs $29.50, while an outfielder's glove costs $21.

27. Let $w =$ the width.
Let $l =$ the length.

$$\begin{cases} (1) & l = w + 5 \\ (2) & 2l + 2w = 110 \end{cases}$$

Substitute $l = w + 5$ into (2):

$$\begin{aligned} 2l + 2w &= 110 \\ 2(w + 5) + 2w &= 110 \\ 2w + 10 + 2w &= 110 \\ 4w &= 100 \\ w &= 25 \end{aligned}$$

Substitute; solve for l:

$$\begin{aligned} l &= w + 5 \\ l &= 25 + 5 = 30 \end{aligned}$$

The dimensions are 25 ft by 30 ft.

SECTION 3.6

29. Let x = the length. Let y = the width.

$$\begin{cases} (1) & x = 2y + 2 \\ (2) & 2x + 2y = 34 \end{cases}$$

Substitute $x = 2y + 2$ into (2):

$$\begin{aligned} 2x + 2y &= 34 \\ 2(2y + 2) + 2y &= 34 \\ 4y + 4 + 2y &= 34 \\ 6y &= 30 \\ y &= 5 \end{aligned}$$

Substitute; solve for x:

$x = 2y + 2$

$x = 2(5) + 2 = 12$

The area is $(5)(12) = 60$ ft^2.

31. Let c = the cost. Let n = number of years.

$$\begin{cases} (1) & c = 2250 + 412n \\ (2) & c = 1715 + 466n \end{cases}$$

Substitute:

$$\begin{aligned} 2250 + 412n &= 1715 + 466n \\ -54n &= -535 \\ n &\approx 9.9 \end{aligned}$$

The break point is about 9.9 years.

33. Since 7 years is less than the break point, choose the furnace with the lower up-front cost (80+).

35. Let x = Bill's principal and y = Janette's principal.

$$\begin{aligned} x + y &= 5000 \\ 0.05x + 0.07y &= 310 \Rightarrow \times (100) \end{aligned} \qquad \begin{aligned} x + y &= 5000 \Rightarrow \times(-5) \\ 5x + 7y &= 31000 \end{aligned} \qquad \begin{aligned} -5x - 5y &= -25000 \\ 5x + 7y &= 31000 \\ \hline 2y &= 6000 \\ y &= 3000 \end{aligned}$$

$$\begin{aligned} x + y &= 5000 \\ x + 3000 &= 5000 \\ x &= 2000 \end{aligned}$$

Bill invested \$2000.

37. Let x = number of student tickets and y = number of nonstudent tickets.

$$\begin{aligned} x + y &= 350 \Rightarrow \times (-1) \\ x + 2y &= 450 \end{aligned} \qquad \begin{aligned} -x - y &= -350 \\ x + 2y &= 450 \\ \hline y &= 100 \end{aligned} \qquad \begin{aligned} x + y &= 350 \\ x + 100 &= 350 \\ x &= 250 \end{aligned}$$

There were 250 student tickets sold.

39. Let b = speed of boat in still water and c = speed of current.

	d	r	t	Equation $(d = r \cdot t)$
Downstream	24	$b + c$	2	$24 = (b + c) \cdot 2 \Rightarrow 2b + 2c = 24$
Upstream	24	$b - c$	3	$24 = (b - c) \cdot 3 \Rightarrow 3b - 3c = 24$

$$\begin{aligned} 2b + 2c &= 24 \Rightarrow \times (3) \\ 3b - 3c &= 24 \Rightarrow \times (2) \end{aligned} \qquad \begin{aligned} 6b + 6c &= 72 \\ 6b - 6c &= 48 \\ \hline 12b &= 120 \\ b &= 10 \end{aligned}$$

The speed of the boat is 10 mph in still water.

SECTION 3.6

41. Let $p =$ airspeed of plane and $w =$ speed of wind.

	d	r	t	Equation $(d = r \cdot t)$
With wind	600	$p+w$	2	$600 = (p+w) \cdot 2 \Rightarrow 2p+2w = 600$
Against wind	600	$p-w$	3	$600 = (p-w) \cdot 3 \Rightarrow 3p-3w = 600$

$$\begin{aligned} 2p+2w &= 600 \Rightarrow \times\ (3) \quad 6p+6w = 1800 \\ 3p-3w &= 600 \Rightarrow \times\ (2) \quad 6p-6w = 1200 \\ &\qquad\qquad\quad 12p = 3000 \\ &\qquad\qquad\quad p = 250 \end{aligned}$$

$$\begin{aligned} 2p+2w &= 600 \\ 2(250)+2w &= 600 \\ 500+2w &= 600 \\ 2w &= 100 \\ w &= 50 \end{aligned}$$

The speed of the wind is 50 mph.

43. Let $x =$ liters of first solution and $y =$ liters of second solution.

	Fractional part that is alcohol	Number of liters of solution	Number of liters of alcohol
First solution	0.40	x	$0.40x$
Second solution	0.55	y	$0.55y$
Final solution	0.50	15	$0.50(15) = 7.5$

$$\begin{aligned} x + y &= 15 \qquad\qquad\quad\ \ x+y = 15 \Rightarrow \times(-40) \quad -40x - 40y = -600 \\ 0.40x + 0.55y &= 7.5 \Rightarrow \times(100) \quad 40x+55y = 750 \qquad\qquad\quad 40x+55y = 750 \\ &\qquad\qquad\qquad\qquad\qquad\qquad\qquad\qquad\qquad\qquad 15y = 150 \\ &\qquad\qquad\qquad\qquad\qquad\qquad\qquad\qquad\qquad\qquad y = 10 \end{aligned}$$

$$\begin{aligned} x+y &= 15 \\ x+10 &= 15 \\ x &= 5 \end{aligned}$$

The chemist should use 5 L of the 40% solution and 10 L of the 55% solution.

45. Let $x =$ pounds of peanuts and $y =$ pounds of cashews.

	Cost per pound	Number of pounds	Total value
Peanuts	3	x	$3x$
Cashews	6	y	$6y$
Mixture	4	48	$4(48) = 192$

$$\begin{aligned} x + y &= 48 \Rightarrow \times\ (-3) \quad -3x-3y = -144 \\ 3x+6y &= 192 \qquad\qquad\qquad\ \ 3x+6y = 192 \\ &\qquad\qquad\qquad\qquad\quad\ \ 3y = 48 \\ &\qquad\qquad\qquad\qquad\quad\ \ y = 16 \end{aligned}$$

$$\begin{aligned} x+y &= 48 \\ x+16 &= 48 \\ x &= 32 \end{aligned}$$

The merchant should use 32 pounds of peanuts and 16 pounds of cashews.

47. Let x and y represent the integers.

$$\begin{cases} (1) & x = 2y \\ (2) & x+y = 96 \end{cases}$$

Substitute $x = 2y$ into (2):

$$\begin{aligned} x+y &= 96 \\ 2y+y &= 96 \\ 3y &= 96 \\ y &= 32 \end{aligned}$$

Substitute; solve for x:

$$\begin{aligned} x &= 2y \\ x &= 2(32) = 64 \end{aligned}$$

The integers are 32 and 64.

SECTION 3.6

49. Let x and y represent the integers.

$$\begin{array}{l} 3x + y = 29 \Rightarrow \times\,(-2) \\ \underline{x + 2y = 18} \end{array} \quad \begin{array}{rcl} -6x - 2y &=& -58 \\ \underline{x + 2y} &\underline{=}& \underline{18} \\ -5x &=& -40 \\ x &=& 8 \end{array} \quad \begin{array}{r} x + 2y = 18 \\ 8 + 2y = 18 \\ 2y = 10 \\ y = 5 \end{array}$$

The integers are 5 and 8.

51. Let $x =$ cost of contact cleaner and $y =$ cost of soaking solution.

$$\begin{array}{l} 2x + 3y = 29.40 \Rightarrow \times(-2) \\ \underline{3x + 2y = 28.60 \Rightarrow \times(3)} \end{array} \quad \begin{array}{rcl} -4x - 6y &=& -58.80 \\ \underline{9x + 6y} &\underline{=}& \underline{85.80} \\ 5x &=& 27 \\ x &=& 5.40 \end{array} \quad \begin{array}{r} 2x + 3y = 29.40 \\ 2(5.40) + 3y = 29.40 \\ 10.80 + 3y = 29.40 \\ 3y = 18.60 \\ y = 6.20 \end{array}$$

The contact cleaner costs \$5.40, while the soaking solution costs \$6.20.

53. Let $x =$ the width.
Let $y =$ the length.

$$\begin{cases} (1) & x = y - 26 \\ (2) & 2x + 2y = 332 \end{cases}$$

Substitute $x = y - 26$ into (2):

$$\begin{aligned} 2x + 2y &= 332 \\ 2(y - 26) + 2y &= 332 \\ 2y - 52 + 2y &= 332 \\ 4y &= 384 \\ y &= 96 \end{aligned}$$

Substitute; solve for x:

$x = y - 26$

$x = 96 - 26 = 70$

The area is $(96)(70) = 6{,}720$ ft^2.

55. Let $x =$ pounds of rye and $y =$ pounds of bluegrass.

$$\begin{cases} (1) & y = 2x \\ (2) & x + y = 100 \end{cases}$$

Substitute $y = 2x$ from (1) into (2):

$$\begin{aligned} x + y &= 100 \\ x + 2x &= 100 \\ 3x &= 100 \\ x &= 33\tfrac{1}{3} \end{aligned}$$

Substitute and solve for y:

$y = 2x$

$y = 2\left(33\tfrac{1}{3}\right) = 66\tfrac{2}{3}$

Add 15 pounds to $66\tfrac{2}{3}$.

He used $81\tfrac{2}{3}$ pounds of bluegrass.

57. Let $x =$ deaths from cancer.
Let $y =$ deaths from diabetes.

$$\begin{cases} (1) & x = 7y \\ (2) & x + y = 308000 \end{cases}$$

Substitute $x = 7y$ into (2):

$$\begin{aligned} x + y &= 308000 \\ 7y + y &= 308000 \\ 8y &= 308000 \\ y &= 38500 \end{aligned}$$

Substitute; solve for x:

$x = 7y$

$x = 7(38500) = 269500$

38,500 die from diabetes and 269,500 die from cancer.

59. Let $r =$ the lower rate. Then $r + 0.02 =$ the higher rate.

$$\boxed{\text{Interest on \$950}} + \boxed{\text{Interest on \$1,200}} = \boxed{\text{Total interest}}$$

$$\begin{aligned} 950r + 1200(r + 0.02) &= 88.50 \\ 950r + 1200r + 24 &= 88.50 \\ 2150r &= 64.50 \\ r &= 0.03 \Rightarrow \text{The lower rate is 3\%.} \end{aligned}$$

61. $\begin{cases} (1) & p = -\frac{1}{2}q + 1300 \\ (2) & p = \frac{1}{3}q + \frac{1400}{3} \end{cases}$

Substitute:

$$-\tfrac{1}{2}q + 1300 = \tfrac{1}{3}q + \tfrac{1400}{3}$$
$$6\left(-\tfrac{1}{2}q + 1300\right) = 6\left(\tfrac{1}{3}q + \tfrac{1400}{3}\right)$$
$$-3q + 7800 = 2q + 2800$$
$$-5q = -5000$$
$$q = 1000$$

Substitute $q = 1000$ and solve for p:

$$p = -\tfrac{1}{2}q + 1300$$
$$= -\tfrac{1}{2}(1000) + 1300$$
$$= -500 + 1300 = 800$$

The equilibrium price is \$800.

63. Answers may vary.

Exercise 3.7 (page 230)

1. $y > 3x + 2$
$0 \overset{?}{>} 3(0) + 2$
$0 \overset{?}{>} 0 + 2$
$0 \not> 2$
$(0, 0)$ is not a solution.

3. $y > 3x + 2$
$4 \overset{?}{>} 3(-2) + 2$
$4 \overset{?}{>} -6 + 2$
$4 > -4$
$(-2, 4)$ is a solution.

5. $y \le \frac{1}{2}x - 1$
$0 \overset{?}{\le} \frac{1}{2}(0) - 1$
$0 \overset{?}{\le} 0 - 1$
$0 \not\le -1$
$(0, 0)$ is not a solution.

7. $y \le \frac{1}{2}x - 1$
$3 \overset{?}{\le} \frac{1}{2}(4) - 1$
$3 \overset{?}{\le} 2 - 1$
$3 \not\le 1$
$(4, 3)$ is not a solution.

9. $3x + 5 = 14$
$3x = 9$
$x = 3$

11. $A = P + Prt$
$A - P = Prt$
$\frac{A - P}{Pr} = \frac{Prt}{Pr}$
$\frac{A - P}{Pr} = t$, or $t = \frac{A - P}{Pr}$

13. $2a + 5(a - 3) = 2a + 5a - 15 = 7a - 15$

15. $4(b - a) + 3b + 2a = 4b - 4a + 3b + 2a$
$= -2a + 7b$

17. inequality

19. boundary

21. inequalities

23. doubly shaded

25. **a.** does not include boundary **b.** does include boundary

27. **a.**
$5x - 3y \ge 0$
$5(1) - 3(1) \overset{?}{\ge} 0$
$5 - 3 \overset{?}{\ge} 0$
$2 \ge 0$
$(1, 1)$ is a solution.

b.
$5x - 3y \ge 0$
$5(-2) - 3(-3) \overset{?}{\ge} 0$
$-10 + 9 \overset{?}{\ge} 0$
$-1 \not\ge 0$
$(-2, -3)$ is not a solution.

SECTION 3.7

27. c.
$$5x - 3y \geq 0$$
$$5(0) - 3(0) \overset{?}{\geq} 0$$
$$0 - 0 \overset{?}{\geq} 0$$
$$0 \geq 0$$
$(0, 0)$ is a solution.

d.
$$5x - 3y \geq 0$$
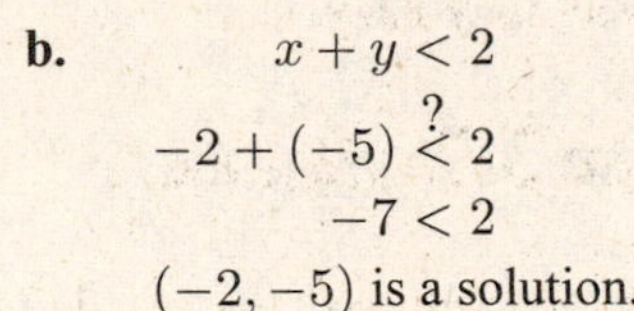
$$5\left(\frac{1}{5}\right) - 3\left(\frac{4}{3}\right) \overset{?}{\geq} 0$$
$$1 - 4 \overset{?}{\geq} 0$$
$$-3 \ngeq 0$$
$\left(\frac{1}{5}, \frac{4}{3}\right)$ is not a solution.

29. a.
$$x + y < 2$$
$$2 + 1 \overset{?}{<} 2$$
$$3 \nless 2$$
$(2, 1)$ is not a solution.

b.
$$x + y < 2$$
$$-2 + (-5) \overset{?}{<} 2$$
$$-7 < 2$$
$(-2, -5)$ is a solution.

c.
$$x + y < 2$$
$$-0.1 + 0.3 \overset{?}{<} 2$$
$$0.2 < 2$$
$(-0.1, 0.3)$ is a solution.

d.
$$x + y < 2$$
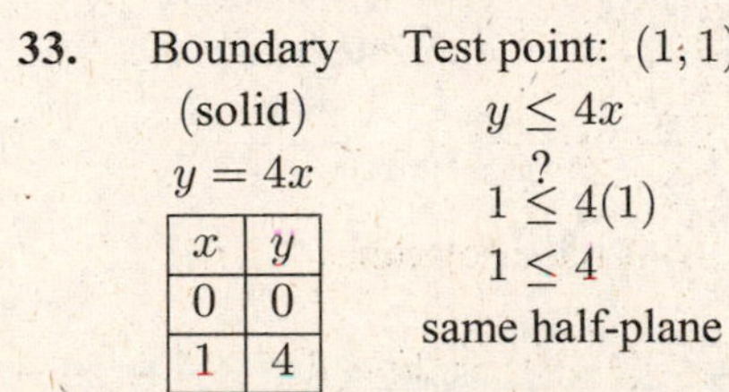
$$-3 + \tfrac{3}{4} \overset{?}{<} 2$$
$$-\tfrac{12}{4} + \tfrac{3}{4} \overset{?}{<} 2$$
$$-\tfrac{9}{4} < 2$$
$\left(-3, \tfrac{3}{4}\right)$ is a solution.

31. Boundary (solid) $y = x + 2$

x	y
0	2
2	4

Test point: $(0, 0)$
$$y \leq x + 2$$
$$0 \overset{?}{\leq} 0 + 2$$
$$0 \leq 2$$
same half-plane

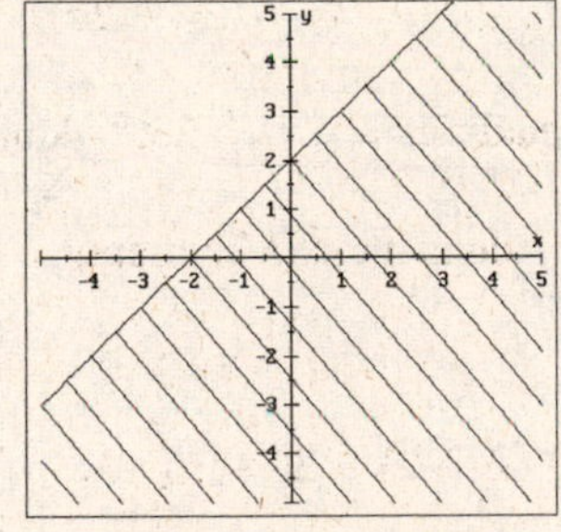

33. Boundary (solid) $y = 4x$

x	y
0	0
1	4

Test point: $(1, 1)$
$$y \leq 4x$$
$$1 \overset{?}{\leq} 4(1)$$
$$1 \leq 4$$
same half-plane

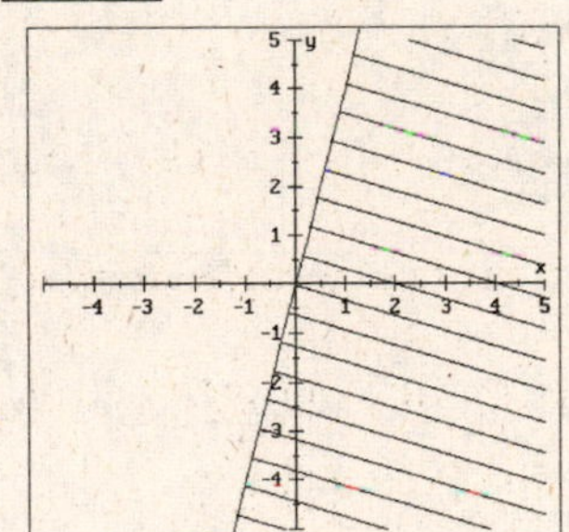

35. Boundary (dotted) $y = x - 3$

x	y
0	-3
3	0

Test point: $(0, 0)$

$y > x - 3$

$0 \overset{?}{>} 0 - 3$

$0 > -3$

same half-plane

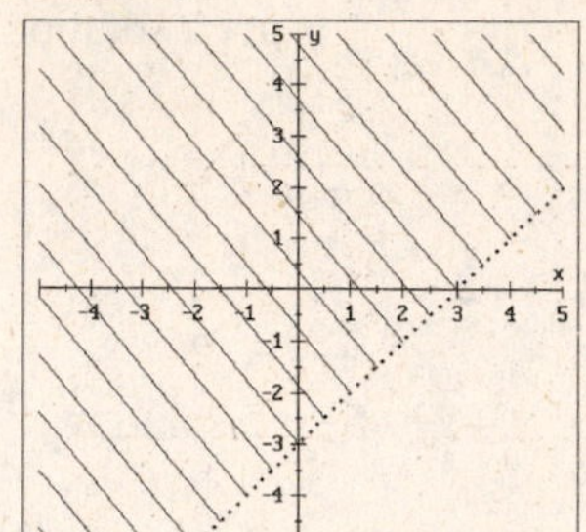

37. Boundary (dotted) $y = 2x - 4$

x	y
0	-4
2	0

Test point: $(0, 0)$

$y > 2x - 4$

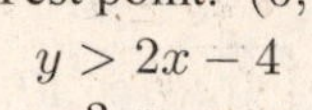

$0 \overset{?}{>} 2(0) - 4$

$0 \overset{?}{>} 0 - 4$

$0 > -4$

same half-plane

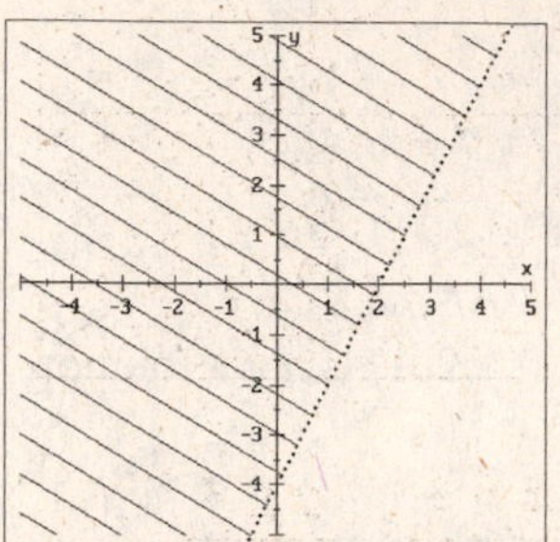

39. Boundary (solid) $y = 2x$

x	y
0	0
2	4

Test point: $(1, 1)$

$y \geq 2x$

$1 \overset{?}{\geq} 2(1)$

$1 \not\geq 2$

opposite half-plane

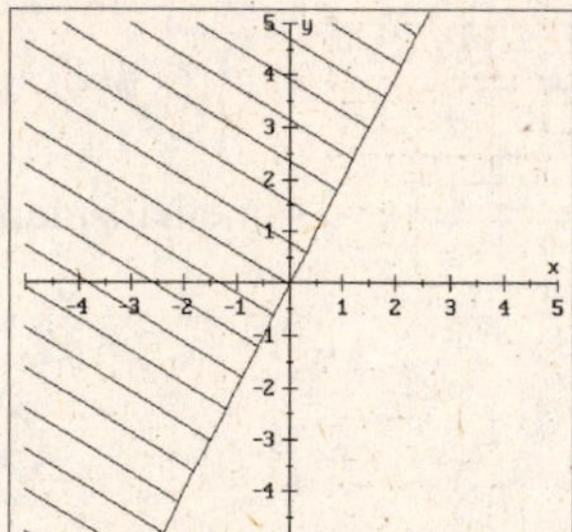

41. Boundary (dotted) $x = 2$

x	y
2	0
2	4

Test point: $(0, 0)$

$x < 2$

$0 < 2$

same half-plane

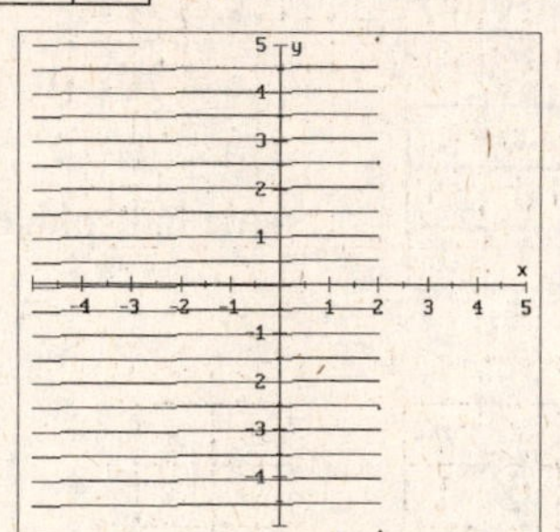

43. $x + 2y \leq 3$ $2x - y \geq 1$

x	y
1	1
3	0

x	y
0	-1
2	3

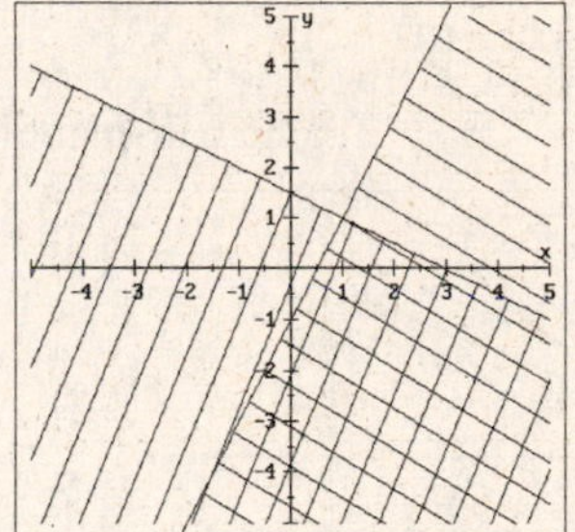

45. $x + y < -1$ $x - y > -1$

x	y
0	-1
-1	0

x	y
0	1
-1	0

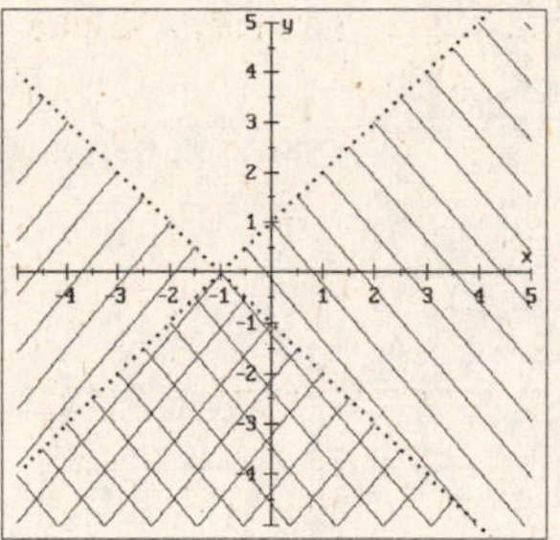

47. $x > 2$ $y \leq 3$

x	y
2	0
2	2

x	y
0	3
1	3

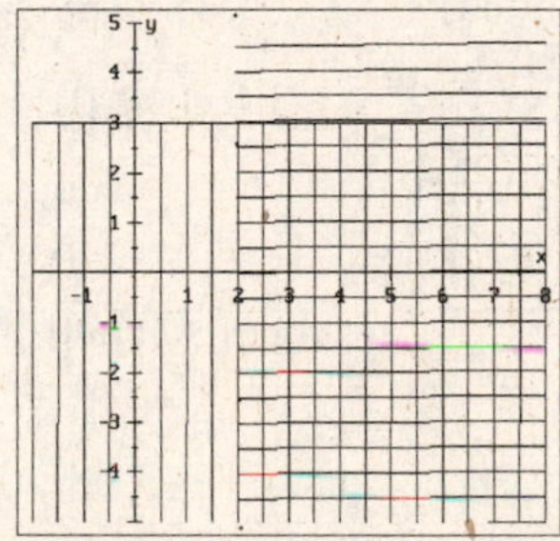

49. $x \leq 0$ $y < 0$

x	y
0	0
0	2

x	y
0	0
1	0

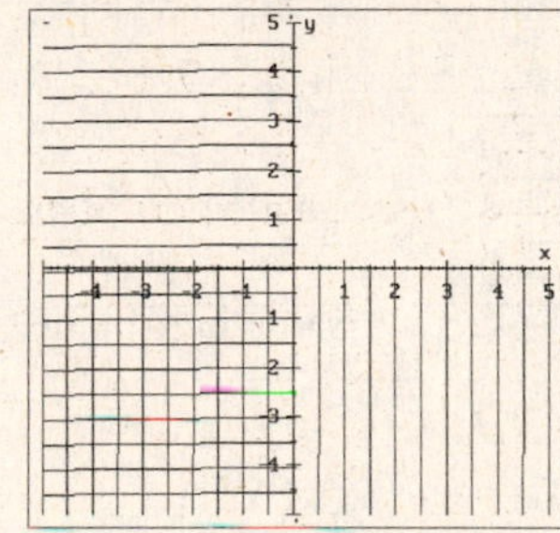

51. $x + y < 1$ $x + y > 3$

x	y
0	1
1	0

x	y
0	3
3	0

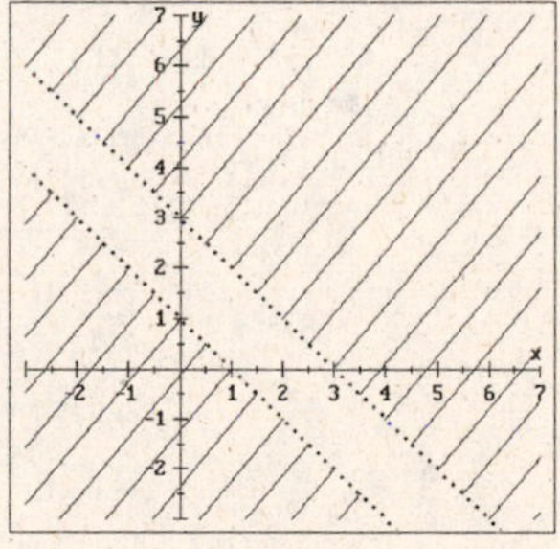

solution set: $\emptyset$

53. $y \leq -\frac{4}{3}x - 2$ $4x + 3y > 15$

x	y
0	-2
-3	2

x	y
0	5
3	1

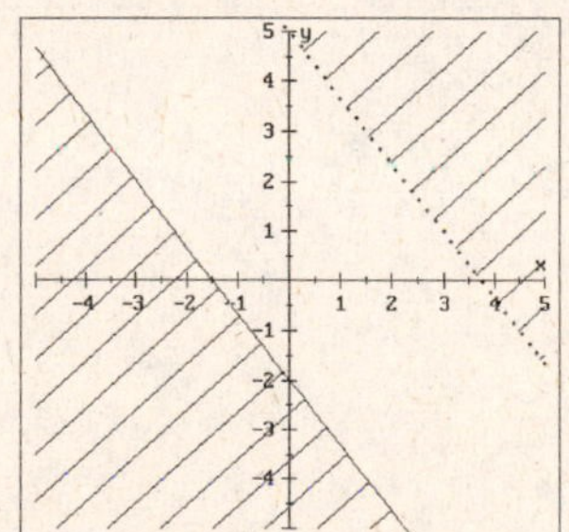

solution set: $\emptyset$

55. Boundary (solid)

$x - 2y = 4$

x	y
0	-2
4	0

Test point: $(0, 0)$

$x - 2y \leq 4$

$0 - 2(0) \overset{?}{\leq} 4$

$0 - 0 \overset{?}{\leq} 4$

$0 \leq 4$

same half-plane

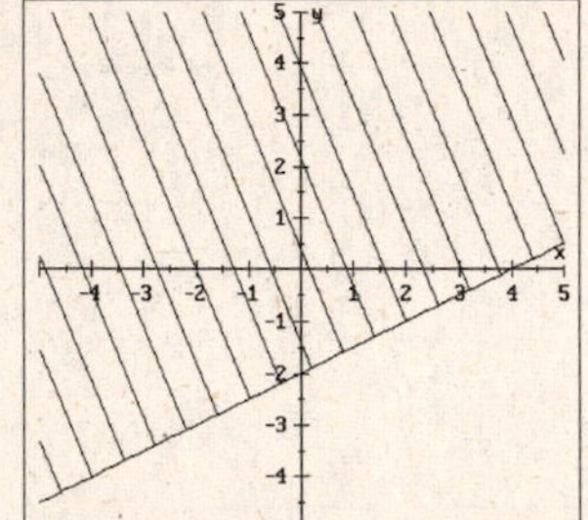

57. Boundary (dotted)

$y = 2 - 3x$

x	y
0	2
1	-1

Test point: $(0, 0)$

$y < 2 - 3x$

$0 \overset{?}{<} 2 - 3(0)$

$0 \overset{?}{<} 2 - 0$

$0 < 2$

same half-plane

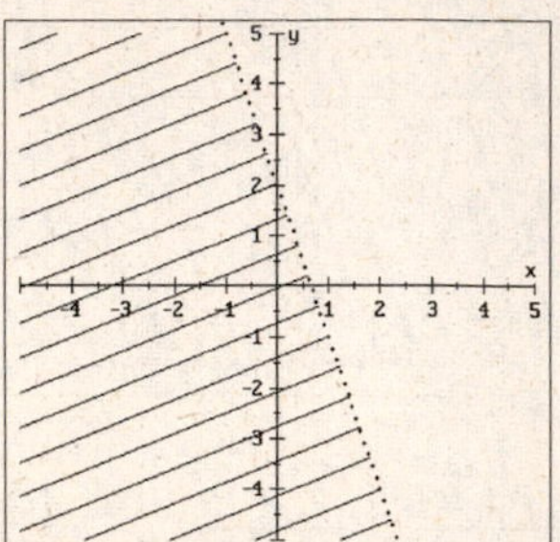

59. Boundary (dotted)

$2y - x = 8$

x	y
0	4
-8	0

Test point: $(0, 0)$

$2y - x < 8$

$2(0) - 0 \overset{?}{<} 8$

$0 - 0 \overset{?}{<} 8$

$0 < 8$

same half-plane

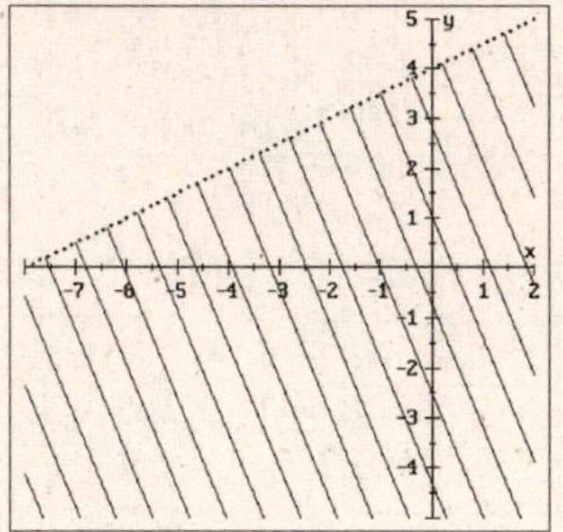

61. Boundary (dotted)

$3x - 4y = 12$

x	y
0	-3
4	0

Test point: $(0, 0)$

$3x - 4y > 12$

$3(0) - 4(0) \overset{?}{>} 12$

$0 - 0 \overset{?}{>} 12$

$0 \not> 12$

opposite half-plane

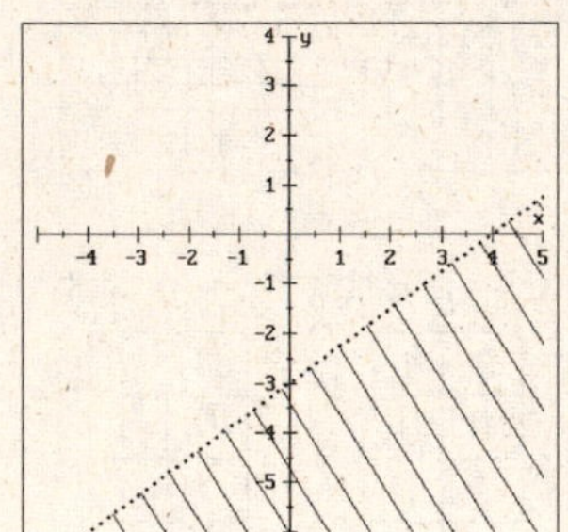

63. Boundary (solid)

$5x + 4y = 20$

x	y
0	5
4	0

Test point: $(0, 0)$

$$5x + 4y \geq 20$$
$$5(0) + 4(0) \overset{?}{\geq} 20$$
$$0 + 0 \overset{?}{\geq} 20$$
$$0 \not\geq 20$$

opposite half-plane

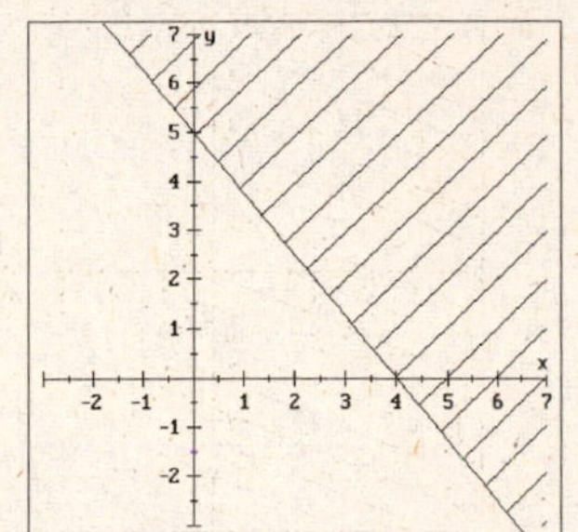

65.

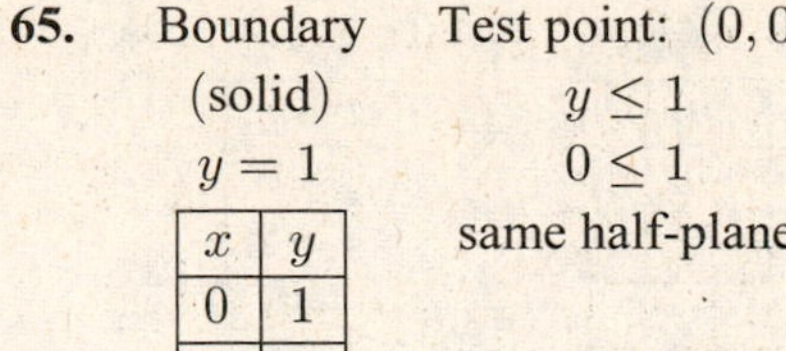

Boundary (solid)

$y = 1$

x	y
0	1
4	1

Test point: $(0, 0)$

$$y \leq 1$$
$$0 \leq 1$$

same half-plane

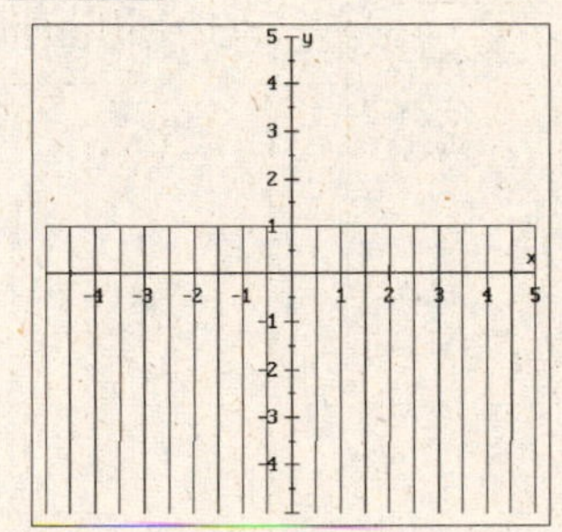

67. $2x - y < 4$

x	y
0	-4
2	0

$x + y \geq -1$

x	y
0	-1
-1	0

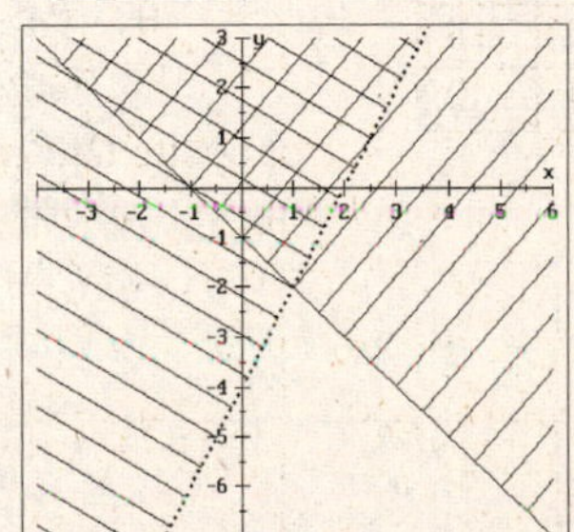

69. $3x + 4y > -7$

x	y
-1	-1
-5	2

$2x - 3y \geq 1$

x	y
2	1
-1	-1

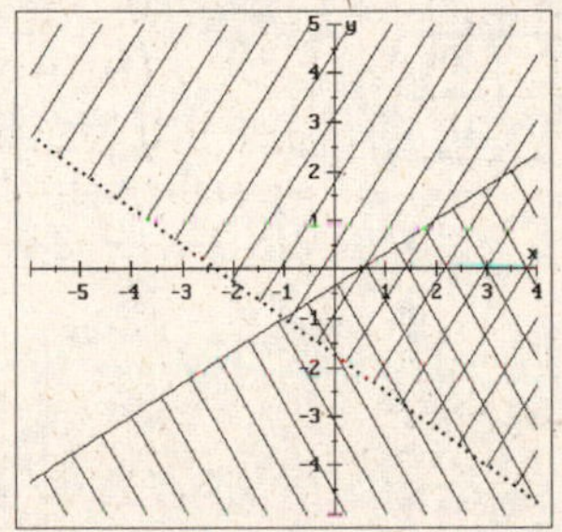

SECTION 3.7

71. $2x - 4y > -6$ $3x + y \geq 5$

x	y
1	2
-3	0

x	y
0	5
1	2

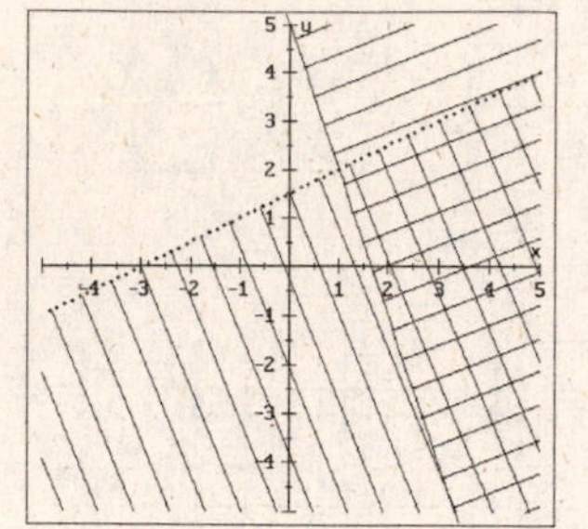

73. $\frac{x}{2} + \frac{y}{3} \geq 2$ $\frac{x}{2} - \frac{y}{2} < -1$

$3x + 2y \geq 12$ $x - y < -2$

x	y
0	6
4	0

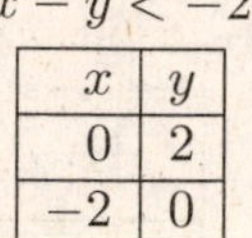

x	y
0	2
-2	0

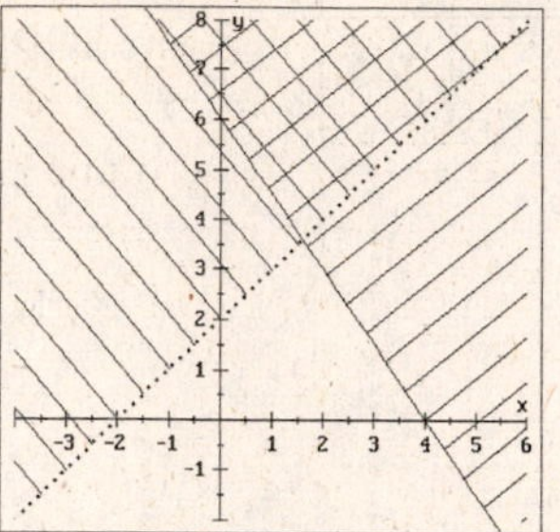

75.

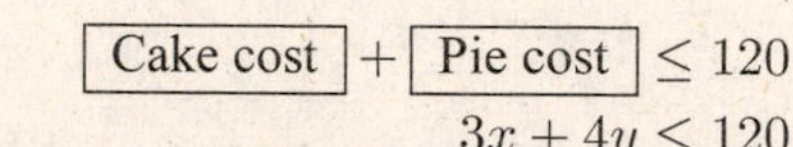

$\boxed{\text{Cake cost}} + \boxed{\text{Pie cost}} \leq 120$

$3x + 4y \leq 120$

Boundary (solid)

$3x + 4y = 120$

x	y
0	30
40	0

Test point: $(0, 0)$

$3x + 4y \leq 120$

$3(0) + 4(0) \overset{?}{\leq} 120$

$0 + 0 \overset{?}{\leq} 120$

$0 \leq 120$

same half-plane

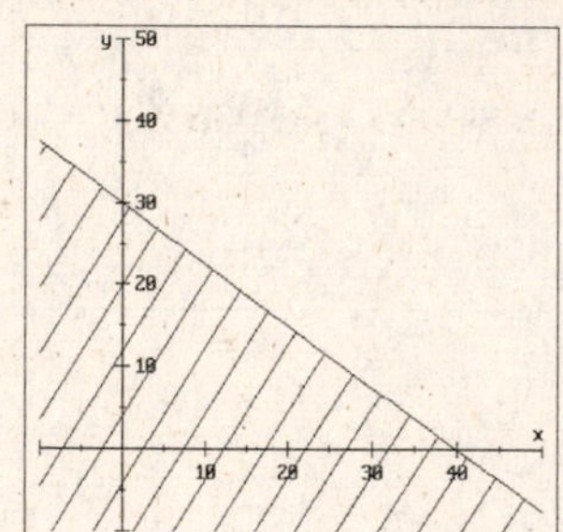

Solutions: $(10, 10), (20, 10), (10, 20)$

77. $\boxed{\text{Leather}} + \boxed{\text{Nylon}} \geq 4400$

$100x + 88y \geq 4400$

Boundary (solid)

x	y
0	50
44	0

Test point: $(0, 0)$

$100x + 88y \geq 4400$

$100(0) + 88(0) \overset{?}{\geq} 4400$

$0 + 0 \overset{?}{\geq} 4400$

$0 \ngeq 4400$

opposite half-plane

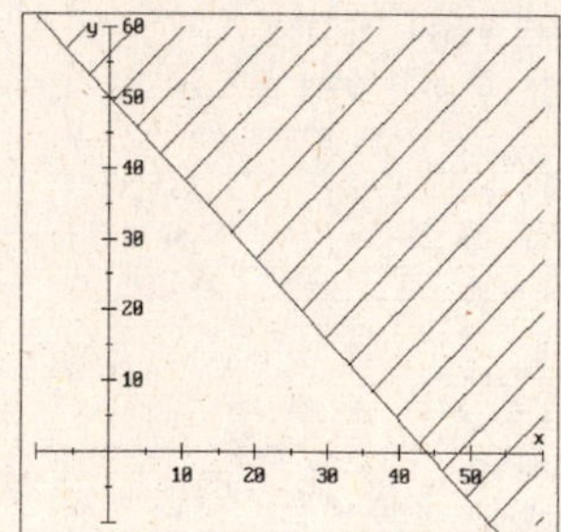

Solutions: $(50, 50), (30, 40), (40, 40)$

79. $\boxed{\text{R. stock}} + \boxed{\text{M. stock}} \le 8000$

$40x + 50y \le 8000$

Boundary (solid)

$40x + 50y = 8000$

x	y
0	160
200	0

Test point: $(0, 0)$

$40x + 50y \le 8000$

$40(0) + 50(0) \overset{?}{\le} 8000$

$0 + 0 \overset{?}{\le} 8000$

$0 \le 8000$

same half-plane

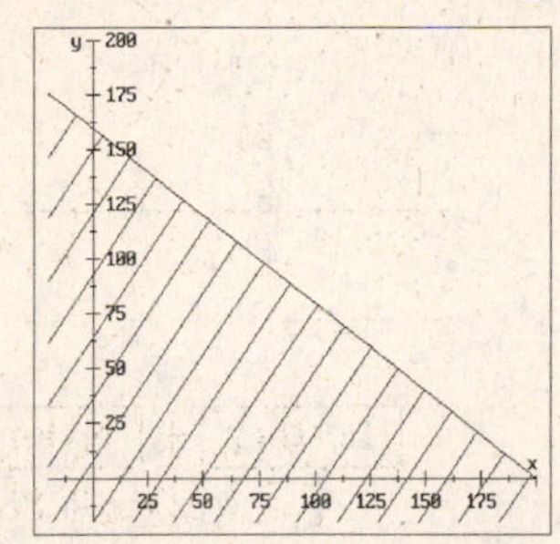

Solutions: $(80, 40), (80, 80), (120, 40)$

81. $10x + 15y \ge 30 \qquad 10x + 15y \le 60$

x	y
0	2
3	0

x	y
0	4
6	0

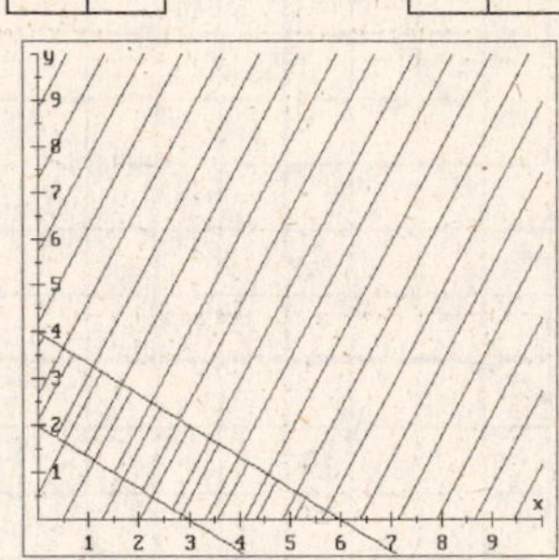

1 \$10 CD and 2 \$15 CDs

4 \$10 CDs and 1 \$15 CD

83. $150x + 100y \le 900 \qquad y > x$

x	y
0	9
6	0

x	y
0	0
1	1

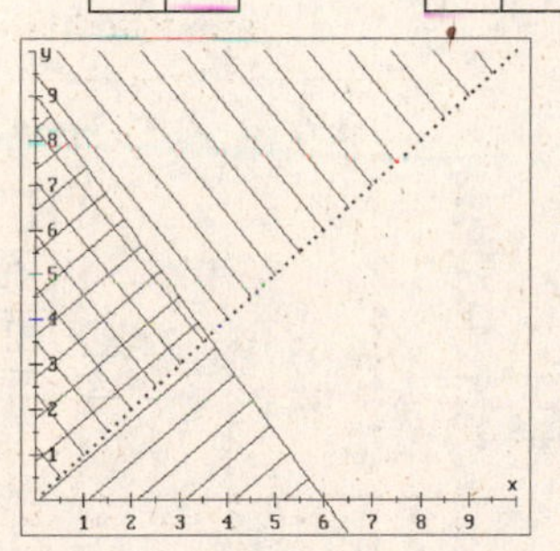

2 desk chairs and 4 side chairs

1 desk chair and 5 side chairs

85. Answers may vary.

87. Answers may vary.

89. Answers may vary.

91. Answers may vary.

Chapter 3 Review (page 236)

1-6.

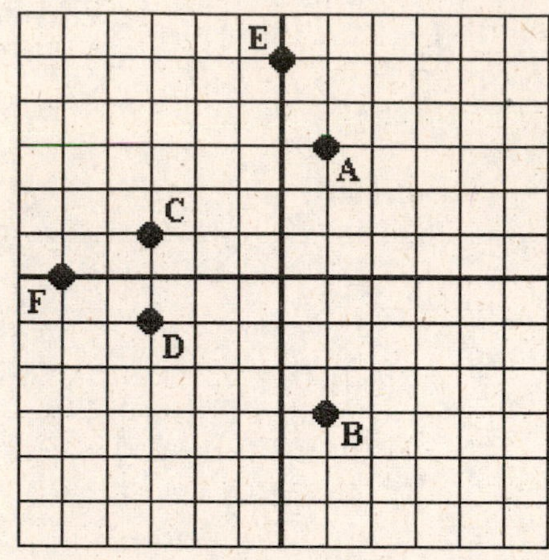

7. $(3, 1)$
8. $(-4, 5)$
9. $(-3, -4)$
10. $(2, -3)$
11. $(0, 0)$
12. $(0, 4)$
13. $(-5, 0)$
14. $(0, -3)$

15.
$$3x - 4y = 12$$
$$3(2) - 4(1) \stackrel{?}{=} 12$$
$$6 - 4 \stackrel{?}{=} 12$$
$$2 \neq 12$$
$(2, 1)$ is not a solution.

16.
$$3x - 4y = 12$$
$$3(3) - 4\left(-\frac{3}{4}\right) \stackrel{?}{=} 12$$
$$9 + 3 \stackrel{?}{=} 12$$
$$12 = 12$$
$\left(3, -\frac{3}{4}\right)$ is a solution.

17. $y = x - 5$

x	y
0	-5
2	-3

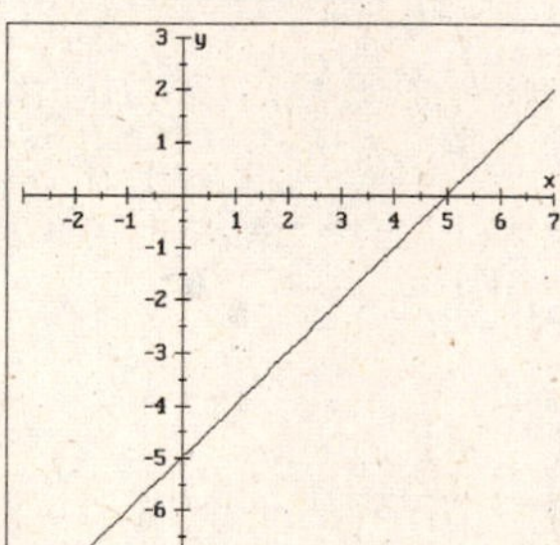

18. $y = 2x + 1$

x	y
0	1
2	5

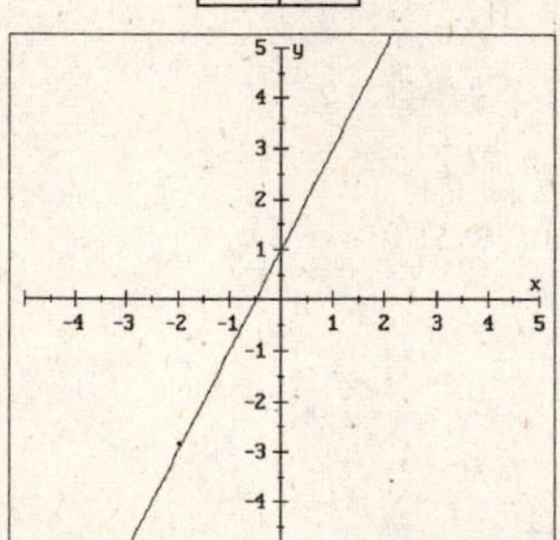

19. $y = \frac{x}{2} + 2$

x	y
0	2
2	3

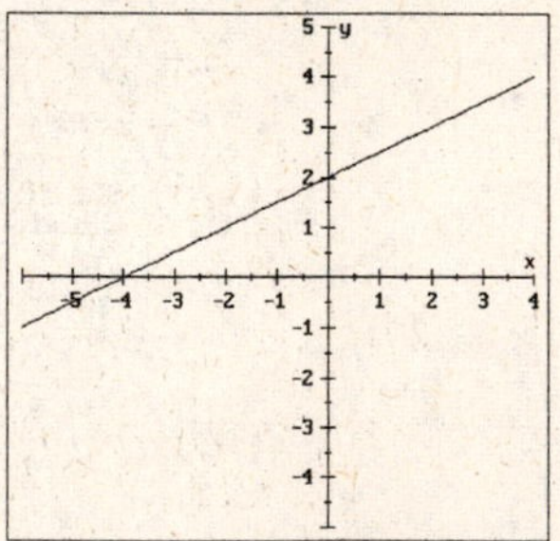

20. $y = 3$

x	y
0	3
2	3

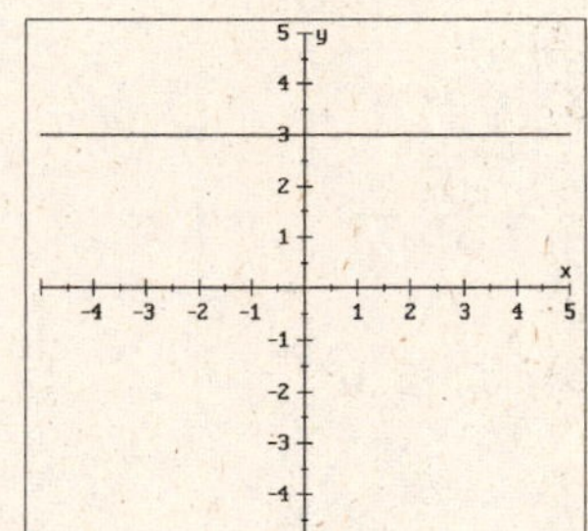

21. $x + y = 4$

x	y
0	4
4	0

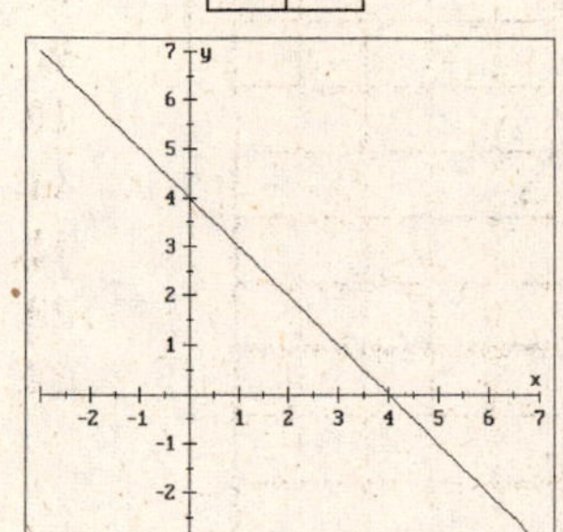

22. $x - y = -3$

x	y
0	3
-3	0

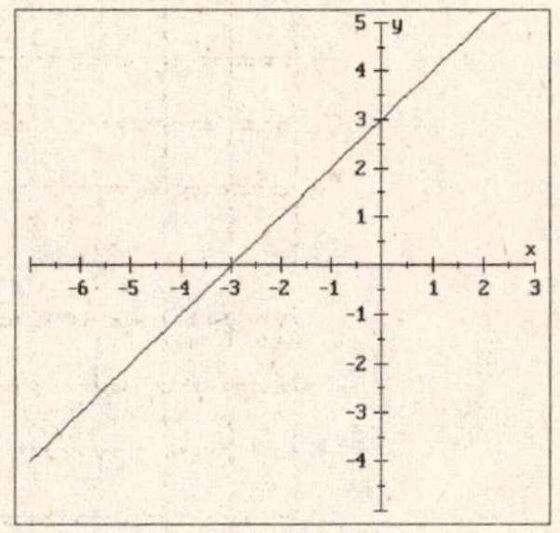

23. $3x + 5y = 15$

x	y
0	3
5	0

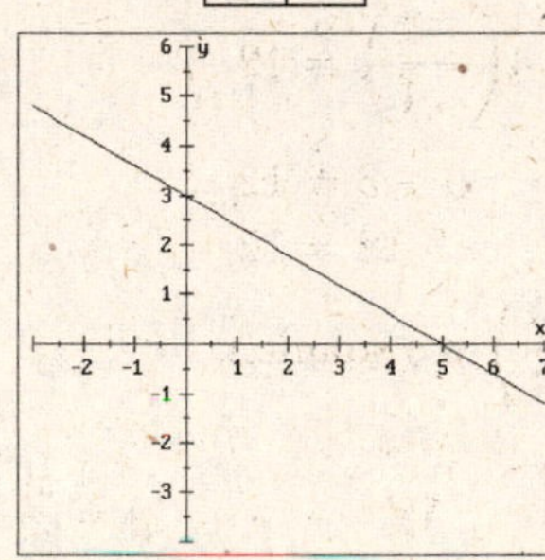

24. $7x - 4y = 28$

x	y
0	-7
4	0

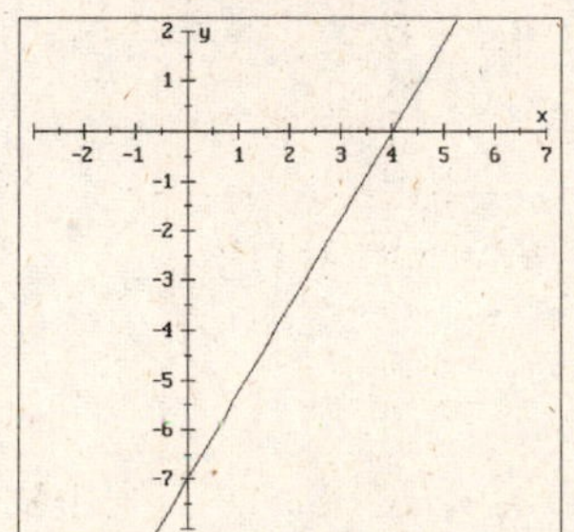

25.

$$\begin{aligned} 3x - y &= -2 \\ 3(1) - 5 &\stackrel{?}{=} -2 \\ 3 - 5 &\stackrel{?}{=} -2 \\ -2 &= -2 \end{aligned} \qquad \begin{aligned} 2x + 3y &= 17 \\ 2(1) + 3(5) &\stackrel{?}{=} 17 \\ 2 + 15 &\stackrel{?}{=} 17 \\ 17 &= 17 \end{aligned}$$

$(1, 5)$ is a solution to the system.

26.

$$\begin{aligned} 5x + 3y &= 2 \\ 5(-2) + 3(4) &\stackrel{?}{=} 2 \\ -10 + 12 &\stackrel{?}{=} 2 \\ 2 &= 2 \end{aligned} \qquad \begin{aligned} -3x + 2y &= 16 \\ -3(-2) + 2(4) &\stackrel{?}{=} 16 \\ 6 + 8 &\stackrel{?}{=} 16 \\ 14 &\neq 16 \end{aligned}$$

$(-2, 4)$ is not a solution to the system.

27.

$$\begin{aligned} 2x + 4y &= 30 \\ 2(14) + 4\left(\frac{1}{2}\right) &\stackrel{?}{=} 30 \\ 28 + 2 &\stackrel{?}{=} 30 \\ 30 &= 30 \end{aligned} \qquad \begin{aligned} \frac{x}{4} - y &= 3 \\ \frac{14}{4} - \frac{1}{2} &\stackrel{?}{=} 3 \\ \frac{7}{2} - \frac{1}{2} &\stackrel{?}{=} 3 \\ \frac{6}{2} &= 3 \end{aligned}$$

$(14, \frac{1}{2})$ is a solution to the system.

28.

$$\begin{aligned} 4x - 6y &= 18 \\ 4\left(\frac{7}{2}\right) - 6\left(-\frac{2}{3}\right) &\stackrel{?}{=} 18 \\ 14 + 4 &\stackrel{?}{=} 18 \\ 18 &= 18 \end{aligned} \qquad \begin{aligned} \frac{x}{3} + \frac{y}{2} &= \frac{5}{6} \\ \frac{\frac{7}{2}}{3} + \frac{-\frac{2}{3}}{2} &\stackrel{?}{=} \frac{5}{6} \\ \frac{7}{6} - \frac{2}{6} &\stackrel{?}{=} \frac{5}{6} \\ \frac{5}{6} &= \frac{5}{6} \end{aligned}$$

$\left(\frac{7}{2}, -\frac{2}{3}\right)$ is a solution to the system.

29. $x + y = 7$ $\quad 2x - y = 5$

x	y
0	7
7	0

x	y
2	-1
3	1

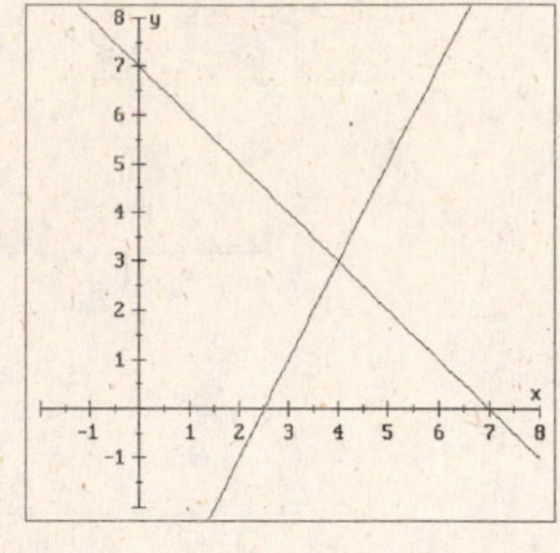

$\boxed{(4,3)}$

30. $\frac{x}{3} + \frac{y}{5} = -1$ $\quad x - 3y = -3$

x	y
0	-5
-3	0

x	y
0	1
-3	0

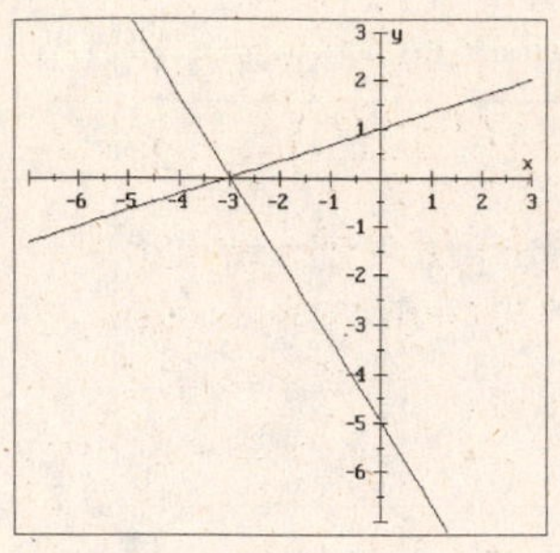

$\boxed{(-3,0)}$

31. $3x + 6y = 6$ $\quad x + 2y = 2$

x	y
0	1
2	0

$2y = -x + 2$

$y = -\frac{1}{2}x + 1$

x	y
0	1
2	0

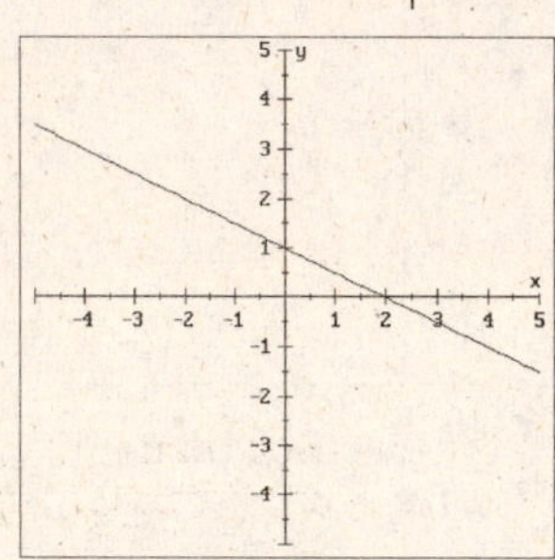

dependent equations; $\left(x, -\frac{1}{2}x + 1\right)$

32. $6x + 3y = 12$ $\quad 2x + y = 2$

x	y
0	4
2	0

x	y
0	2
1	0

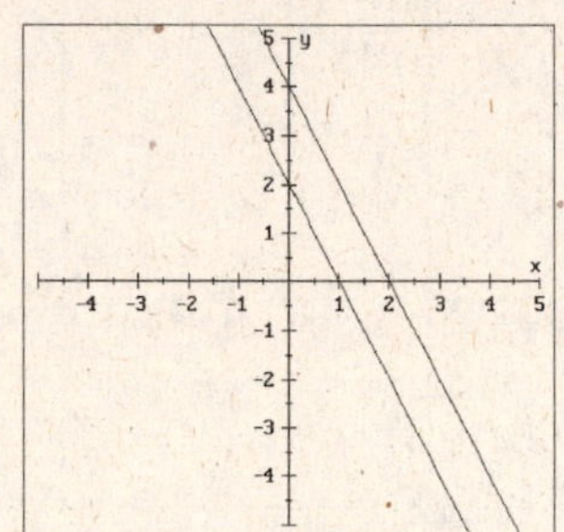

inconsistent system
solution set: $\emptyset$

33. $\begin{cases} (1) \quad x = 3y + 5 \\ (2) \quad 5x - 4y = 3 \end{cases}$

Substitute $x = 3y + 5$ from (1):

$$\begin{aligned} 5x - 4y &= 3 \\ 5(3y + 5) - 4y &= 3 \\ 15y + 25 - 4y &= 3 \\ 11y &= -22 \\ y &= -2 \end{aligned}$$

Substitute and solve for x:

$$\begin{aligned} x &= 3y + 5 \\ x &= 3(-2) + 5 \\ x &= -6 + 5 = -1 \end{aligned}$$

Solution:

$\boxed{x = -1, y = -2}$

34. $\begin{cases} 3x - \frac{2y}{5} = 2(x-2) & \Rightarrow 15x - 2y = 10(x-2) & \Rightarrow 5x - 2y = -20 \ (1) \\ 2x - 3 = 3 - 2y & \Rightarrow 2x + 2y = 6 & \Rightarrow 2x + 2y = 6 \ (2) \end{cases}$

Substitute $y = \frac{6-2x}{2}$ from (2):

$$\begin{aligned} 5x - 2y &= -20 \\ 5x - 2\left(\frac{6-2x}{2}\right) &= -20 \\ 5x - (6 - 2x) &= -20 \\ 5x - 6 + 2x &= -20 \\ 7x &= -14 \\ x &= -2 \end{aligned}$$

Substitute and solve for y:

$$\begin{aligned} y &= \frac{6-2x}{2} \\ y &= \frac{6-2(-2)}{2} \\ y &= \frac{6+4}{2} = \frac{10}{2} = 5 \end{aligned}$$

Solution: $\boxed{x = -2, y = 5}$

35. $\begin{cases} 8x + 5y = 3 \ (1) \\ 5x - 8y = 13 \ (2) \end{cases}$

Substitute $y = \frac{3-8x}{5}$ from (1):

$$\begin{aligned} 5x - 8y &= 13 \\ 5x - 8\left(\frac{3-8x}{5}\right) &= 13 \\ 25x - 8(3 - 8x) &= 65 \\ 25x - 24 + 64x &= 65 \\ 89x &= 89 \\ x &= 1 \end{aligned}$$

Substitute and solve for y:

$$\begin{aligned} y &= \frac{3-8x}{5} \\ y &= \frac{3-8(1)}{5} \\ y &= \frac{3-8}{5} = \frac{-5}{5} = -1 \end{aligned}$$

Solution: $\boxed{x = 1, y = -1}$

36. $\begin{cases} 6(x+2) = y - 1 & \Rightarrow 6x + 12 = y - 1 & \Rightarrow 6x - y = -13 \ (1) \\ 5(y-1) = x + 2 & \Rightarrow 5y - 5 = x + 2 & \Rightarrow -x + 5y = 7 \ (2) \end{cases}$

Substitute $y = 6x + 13$ from (1):

$$\begin{aligned} -x + 5y &= 7 \\ -x + 5(6x + 13) &= 7 \\ -x + 30x + 65 &= 7 \\ 29x &= -58 \\ x &= -2 \end{aligned}$$

Substitute and solve for y:

$$\begin{aligned} y &= 6x + 13 \\ y &= 6(-2) + 13 \\ y &= -12 + 13 = 1 \end{aligned}$$

Solution: $\boxed{x = -2, y = 1}$

37.

$$\begin{array}{rcl} 2x + y &=& 1 \\ 5x - y &=& 20 \\ \hline 7x &=& 21 \\ x &=& 3 \end{array}$$

$$\begin{aligned} 2x + y &= 1 \\ 2(3) + y &= 1 \\ 6 + y &= 1 \\ y &= -5 \end{aligned}$$

Solution: $\boxed{(3, -5)}$

38. $x + 8y = 7 \Rightarrow \times\ (-1)$
$x - 4y = 1$

$$\begin{array}{rcl} -x - 8y &=& -7 \\ x - 4y &=& 1 \\ \hline -12y &=& -6 \\ y &=& \frac{1}{2} \end{array}$$

$$\begin{aligned} x + 8y &= 7 \\ x + 8\left(\tfrac{1}{2}\right) &= 7 \\ x + 4 &= 7 \\ x &= 3 \end{aligned}$$

Solution: $\boxed{\left(3, \tfrac{1}{2}\right)}$

39. $\begin{aligned} 5x + y &= 2 \Rightarrow \times(-2) \\ 3x + 2y &= 11 \end{aligned}$ $\quad \begin{aligned} -10x - 2y &= -4 \\ 3x + 2y &= 11 \\ \hline -7x &= 7 \\ x &= -1 \end{aligned}$ $\quad \begin{aligned} 5x + y &= 2 \\ 5(-1) + y &= 2 \\ -5 + y &= 2 \\ y &= 7 \end{aligned}$ Solution: $\boxed{(-1, 7)}$

40. $\begin{aligned} x + y &= 3 \Rightarrow x + y = 3 \Rightarrow \times(-1) \\ 3x &= 2 - y \Rightarrow 3x + y = 2 \end{aligned}$ $\quad \begin{aligned} -x - y &= -3 \\ 3x + y &= 2 \\ \hline 2x &= -1 \\ x &= -\tfrac{1}{2} \end{aligned}$ $\quad \begin{aligned} x + y &= 3 \\ -\tfrac{1}{2} + y &= \tfrac{6}{2} \\ y &= \tfrac{7}{2} \end{aligned}$ Solution: $\boxed{\left(-\tfrac{1}{2}, \tfrac{7}{2}\right)}$

41. $\begin{aligned} 11x + 3y &= 27 \Rightarrow \times(4) \\ 8x + 4y &= 36 \Rightarrow \times(-3) \end{aligned}$ $\quad \begin{aligned} 44x + 12y &= 108 \\ -24x - 12y &= -108 \\ \hline 20x &= 0 \\ x &= 0 \end{aligned}$ $\quad \begin{aligned} 11x + 3y &= 27 \\ 11(0) + 3y &= 27 \\ 0 + 3y &= 27 \\ 3y &= 27 \\ y &= 9 \end{aligned}$ Solution: $\boxed{(0, 9)}$

42. $\begin{aligned} 9x + 3y &= 5 \Rightarrow 9x + 3y = 5 \\ 3x &= 4 - y \Rightarrow 3x + y = 4 \Rightarrow \times(-3) \end{aligned}$ $\quad \begin{aligned} 9x + 3y &= 5 \\ -9x - 3y &= -12 \\ \hline 0 &\neq -7 \end{aligned}$ $\boxed{\text{inconsistent system; } \emptyset}$

43. $\begin{aligned} 9x + 3y &= 5 \\ 3x + y &= \tfrac{5}{3} \Rightarrow \times(-3) \end{aligned}$ $\quad \begin{aligned} 9x + 3y &= 5 \\ -9x - 3y &= -5 \\ \hline 0 &= 0 \end{aligned}$ $\boxed{\text{dependent equations; } \left(x, -3x + \tfrac{5}{3}\right)}$

44. $\begin{aligned} \frac{x}{3} + \frac{y+2}{2} &= 1 \Rightarrow \times(6) \quad 2x + 3(y+2) = 6 \Rightarrow \quad 2x + 3y + 6 = 6 \\ \frac{x+8}{8} + \frac{y-3}{3} &= 0 \Rightarrow \times(24) \quad 3(x+8) + 8(y-3) = 0 \Rightarrow 3x + 24 + 8y - 24 = 0 \end{aligned}$

$\begin{aligned} 2x + 3y &= 0 \Rightarrow \times(3) \\ 3x + 8y &= 0 \Rightarrow \times(-2) \end{aligned}$ $\quad \begin{aligned} 6x + 9y &= 0 \\ -6x - 16y &= 0 \\ \hline -7y &= 0 \\ y &= 0 \end{aligned}$ $\quad \begin{aligned} 2x + 3y &= 0 \\ 2x + 3(0) &= 0 \\ 2x &= 0 \\ x &= 0 \end{aligned}$ Solution: $\boxed{(0, 0)}$

45. Let x and y represent the integers.

$\begin{cases} (1) & x = 5y \\ (2) & x + y = 18 \end{cases}$

Substitute $x = 5y$ into (2):

$\begin{aligned} x + y &= 18 \\ 5y + y &= 18 \\ 6y &= 18 \\ y &= 3 \end{aligned}$

Substitute; solve for x:

$x = 5y$

$x = 5(3) = 15$

The integers are 3 and 15.

46. Let $w =$ the width.
Let $l =$ the length.
$\begin{cases} (1) & l = 3w \\ (2) & 2l + 2w = 24 \end{cases}$

Substitute $l = 3w$ into (2):
$$\begin{aligned} 2l + 2w &= 24 \\ 2(3w) + 2w &= 24 \\ 6w + 2w &= 24 \\ 8w &= 24 \\ w &= 3 \end{aligned}$$

Substitute; solve for l:
$l = 3w$
$l = 3(3) = 9$
The dimensions are 3 ft by 9 ft.

47. Let $x =$ the cost of a grapefruit.
Let $y =$ the cost of an orange.
$\begin{cases} (1) & x = y + 15 \\ (2) & x + y = 85 \end{cases}$

Substitute $x = y + 15$ into (2):
$$\begin{aligned} x + y &= 85 \\ y + 15 + y &= 85 \\ 2y &= 70 \\ y &= 35 \end{aligned}$$

Substitute; solve for x:
$x = y + 15$
$x = 35 + 15 = 50$
A grapefruit costs 50¢.

48. Let $x =$ the electric bill.
Let $y =$ the gas bill.
$\begin{cases} (1) & x = y - 23 \\ (2) & x + y = 109 \end{cases}$

Substitute $x = y - 23$ into (2):
$$\begin{aligned} x + y &= 109 \\ y - 23 + y &= 109 \\ 2y &= 132 \\ y &= 66 \end{aligned}$$

The gas bill was \$66.

49. Let $x =$ cost of a gallon of milk and $y =$ cost of one dozen eggs.

$$\begin{aligned} 2x + 3y = 6.80 &\Rightarrow \times(-2) & -4x - 6y &= -13.60 \\ 3x + 2y = 7.35 &\Rightarrow \times(3) & 9x + 6y &= 22.05 \\ & & 5x &= 8.45 \\ & & x &= 1.69 \end{aligned}$$

A gallon of milk costs \$1.69.

50. Let $x =$ principal at 10% and $y =$ principal at 6%.

$$\begin{aligned} x + y &= 3000 & x + y = 3000 &\Rightarrow \times(-6) & -6x - 6y &= -18000 \\ 0.10x + 0.06y &= 270 \Rightarrow \times(100) & 10x + 6y = 27000 & & 10x + 6y &= 27000 \\ & & & & 4x &= 9000 \\ & & & & x &= 2250 \end{aligned}$$

$$\begin{aligned} x + y &= 3000 \\ 2250 + y &= 3000 \\ y &= 750 \end{aligned}$$

He invested \$750 at the 6% rate.

51. Let $b =$ speed of boat in still water and $c =$ speed of current.

	d	r	t	Equation $(d = r \cdot t)$
Downstream	56	$b + c$	4	$56 = (b + c) \cdot 4 \Rightarrow 4b + 4c = 56$
Upstream	56	$b - c$	7	$56 = (b - c) \cdot 7 \Rightarrow 7b - 7c = 56$

$$\begin{aligned} 4b + 4c = 56 &\Rightarrow \times(7) & 28b + 28c &= 392 \\ 7b - 7c = 56 &\Rightarrow \times(4) & 28b - 28c &= 224 \\ & & 56b &= 616 \\ & & b &= 11 \end{aligned}$$

$$\begin{aligned} 4b + 4c &= 56 \\ 4(11) + 4c &= 56 \\ 44 + 4c &= 56 \\ 4c &= 12 \\ c &= 3 \end{aligned}$$

The speed of the current is 3 mph.

52. Let x = liters of first solution and y = milliliters of second solution.

	Fractional part that is saline	Number of liters of solution	Number of milliliters of saline
First solution	0.1	x	$0.1x$
Second solution	0.6	y	$0.6y$
Final solution	0.3	50	$0.3(50) = 15$

$$\begin{aligned} x + \quad y &= 50 \\ 0.1x + 0.6y &= 15 \Rightarrow \times(10) \end{aligned} \qquad \begin{aligned} x + \ y &= \ 50 \Rightarrow \times(-1) \\ x + 6y &= 150 \end{aligned} \qquad \begin{aligned} -x - \ y &= -50 \\ x + 6y &= \ 150 \\ \hline 5y &= \ 100 \\ y &= \ 20 \end{aligned}$$

$$\begin{aligned} x + y &= 50 \\ x + 20 &= 50 \\ x &= 30 \end{aligned}$$

The chemist should use 30 mL of the 10% solution and 20 mL of the 60% solution.

53. Boundary (solid) $y = x + 2$

x	y
0	2
−2	0

Test point: $(0, 0)$

$y \geq x + 2$

$0 \overset{?}{\geq} 0 + 2$

$0 \not\geq 2$

opposite half-plane

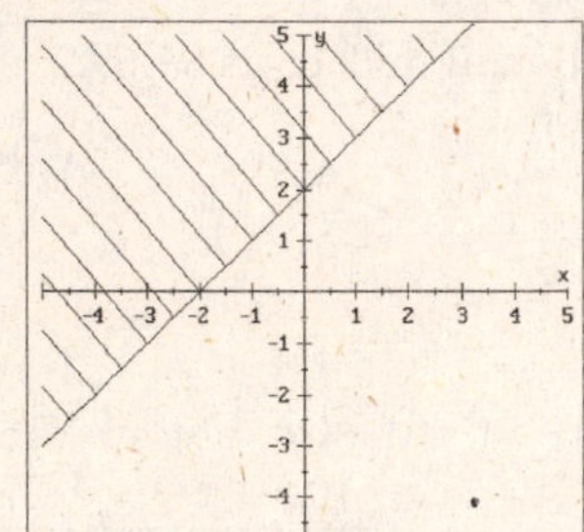

54. Boundary (dotted) $x = 3$

x	y
3	0
3	4

Test point: $(0, 0)$

$x < 3$

$0 < 3$

same half-plane

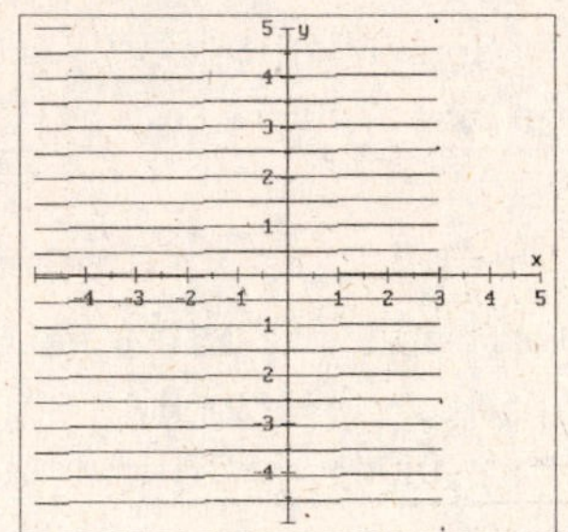

55. $5x + 3y < 15$

x	y
0	5
3	0

$3x - y > 3$

x	y
0	−3
1	0

56. $5x - 3y \geq 5$

x	y
1	0
4	5

$3x + 2y \geq 3$

x	y
1	0
3	−3

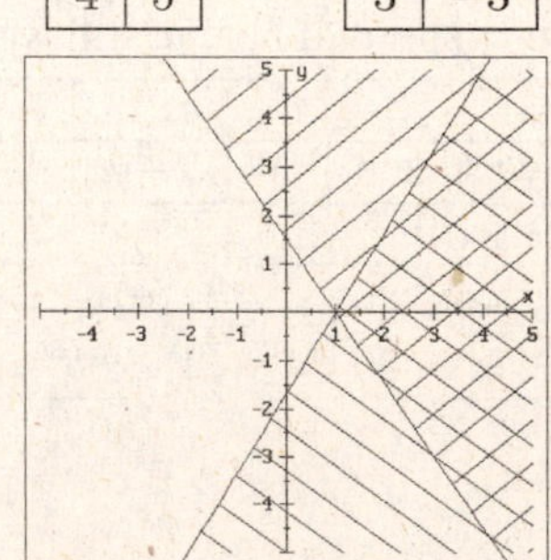

57. $x \geq 3y$ $y < 3x$

x	y
0	0
3	1

x	y
0	0
1	3

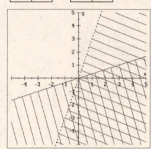

58. $x \geq 0$ $x \leq 3$

x	y
0	0
0	4

x	y
3	1
3	0

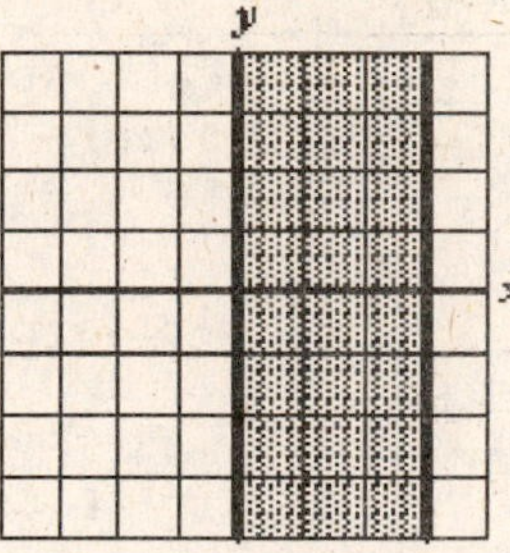

59. $10x + 20y \geq 40$ $10x + 20y \leq 60$

x	y
0	2
4	0

x	y
0	3
6	0

3 shirts and 1 pair of pants
1 shirt and 2 pairs of pants

Chapter 3 Test (page 243)

1. $$y = \frac{x}{2} + 1$$

x	y
0	1
2	3

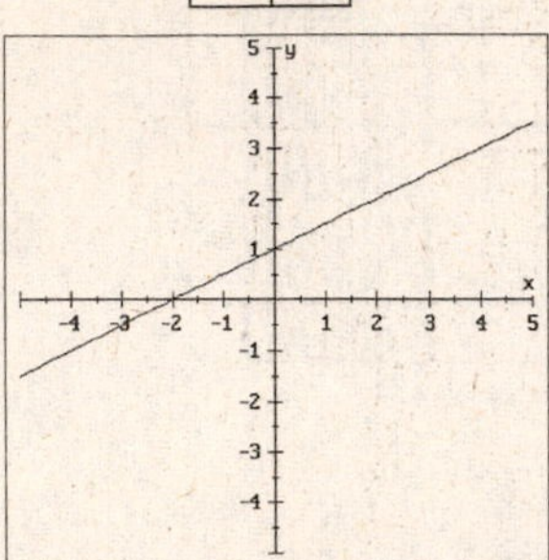

2. $$\begin{aligned} 2(x+1) - y &= 4 \\ 2x + 2 - y &= 4 \\ 2x - y &= 2 \end{aligned}$$

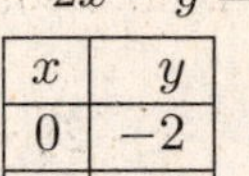

x	y
0	-2
1	0

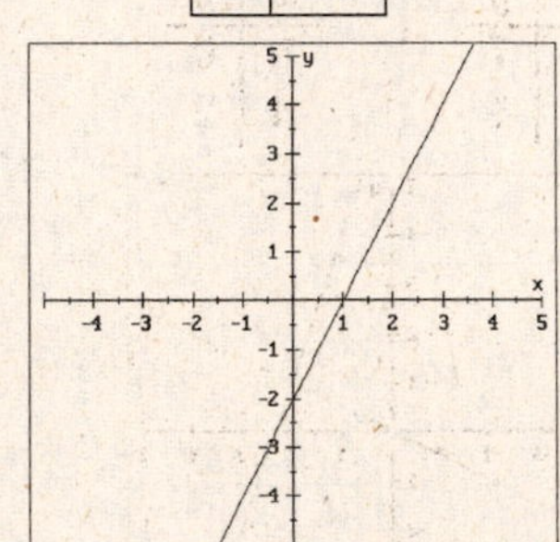

3. $$x = 1$$

x	y
1	2
1	0

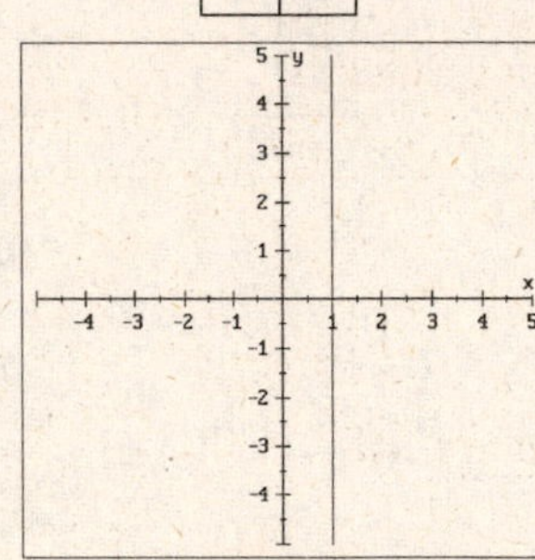

4. $$\begin{aligned} 2y &= 8 \\ y &= 4 \end{aligned}$$

x	y
0	4
2	4

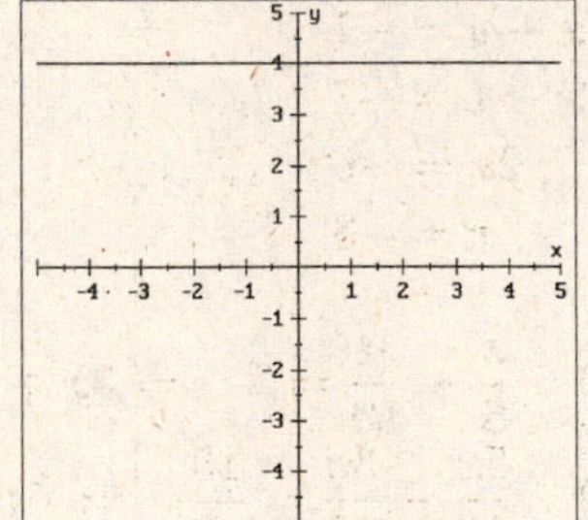

5. $$\begin{aligned} 3x - 2y &= 12 & 2x + 3y &= -5 \\ 3(2) - 2(-3) &\stackrel{?}{=} 12 & 2(2) + 3(-3) &\stackrel{?}{=} -5 \\ 6 + 6 &\stackrel{?}{=} 12 & 4 - 9 &\stackrel{?}{=} -5 \\ 12 &= 12 & -5 &= -5 \end{aligned}$$

$(2, -3)$ is a solution to the system.

6.
$$\begin{array}{rr} 4x + y = -9 & 2x - 3y = -7 \\ 4(-2) + (-1) \stackrel{?}{=} -9 & 2(-2) - 3(-1) \stackrel{?}{=} -7 \\ -8 - 1 \stackrel{?}{=} -9 & -4 + 3 \stackrel{?}{=} -7 \\ -9 = -9 & -1 \neq -7 \end{array}$$
$(-2, -1)$ is not a solution to the system.

7. $3x + y = 7$ $\quad x - 2y = 0$

x	y
0	7
2	1

x	y
0	0
4	2

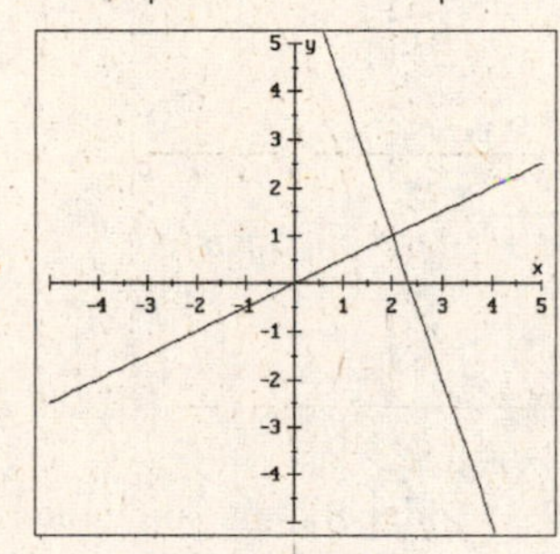

$\boxed{(2,1)}$

8. $x + \frac{y}{2} = 1$ $\quad y = 1 - 3x$

x	y
0	2
1	0

x	y
0	1
1	-2

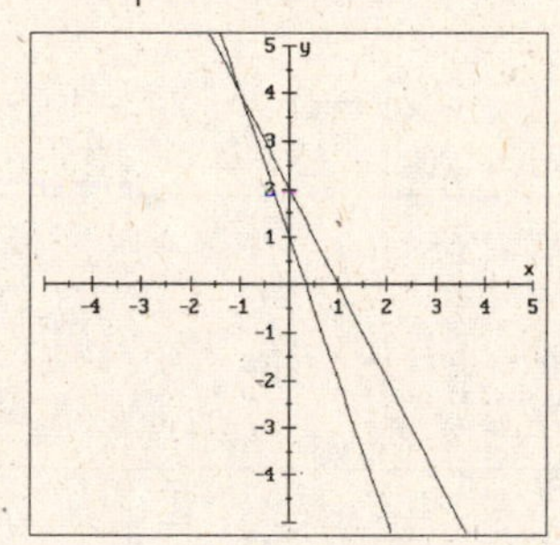

$\boxed{(-1,4)}$

9. $\begin{cases} (1) & y = x - 1 \\ (2) & x + y = -7 \end{cases}$

Substitute $y = x - 1$ from (1):
$$\begin{aligned} x + y &= -7 \\ x + x - 1 &= -7 \\ 2x &= -6 \\ x &= -3 \end{aligned}$$

Substitute and solve for y:
$$\begin{aligned} y &= x - 1 \\ y &= -3 - 1 \\ y &= -4 \end{aligned}$$

Solution: $\boxed{x = -3, y = -4}$

10. $\begin{cases} (1) & \frac{x}{6} + \frac{y}{10} = 3 \quad \Rightarrow 5x + 3y = 90 \quad \Rightarrow x = \frac{90 - 3y}{5} \\ (2) & \frac{5x}{16} - \frac{3y}{16} = \frac{15}{8} \quad \Rightarrow 5x - 3y = 30 \end{cases}$

Substitute $x = \frac{90 - 3y}{5}$ from (1) into (2):
$$\begin{aligned} 5x - 3y &= 30 \\ 5\left(\frac{90 - 3y}{5}\right) - 3y &= 30 \\ 90 - 3y - 3y &= 30 \\ -6y &= -60 \\ y &= 10 \end{aligned}$$

Substitute and solve for x:
$$\begin{aligned} x &= \frac{90 - 3y}{5} \\ x &= \frac{90 - 3(10)}{5} \\ x &= \frac{90 - 30}{5} = \frac{60}{5} = 12 \end{aligned}$$

Solution: $\boxed{x = 12, y = 10}$

CHAPTER 3 TEST

11.

$$\begin{array}{rl} 3x - y = & 2 \\ 2x + y = & 8 \\ \hline 5x \quad = & 10 \\ x \quad = & 2 \end{array} \qquad \begin{aligned} 2x + y &= 8 \\ 2(2) + y &= 8 \\ 4 + y &= 8 \\ y &= 4 \end{aligned}$$

Solution: $\boxed{(2,4)}$

12.

$$\begin{array}{rl} 4x + \ 3 = -3y & \\ \dfrac{-x}{7} + \dfrac{4y}{21} = \ 1 \Rightarrow \times(21) \end{array} \qquad \begin{array}{rl} 4x + 3y = -3 \Rightarrow \times(3) \\ -3x + 4y = \ 21 \Rightarrow \times(4) \end{array} \qquad \begin{array}{rl} 12x + \ 9y = & -9 \\ -12x + 16y = & 84 \\ \hline 25y = & 75 \\ y = & 3 \end{array} \qquad \begin{aligned} 4x + 3y &= -3 \\ 4x + 3(3) &= -3 \\ 4x + 9 &= -3 \\ 4x &= -12 \\ x &= -3 \end{aligned}$$

$\boxed{(-3,3)}$

13.

$$\begin{array}{rl} 2x + 3(y-2) = 0 \Rightarrow 2x + 3y - 6 = 0 \Rightarrow & 2x + 3y = 6 \\ -3y = 2(x-4) \Rightarrow \quad -3y = 2x - 8 \Rightarrow & -2x - 3y = -8 \\ \hline & 0 \neq -2 \end{array}$$

$\Rightarrow$ inconsistent

14.

$$\begin{array}{rl} \dfrac{x}{3} + y - 4 = 0 \Rightarrow x + 3y - 12 = 0 \Rightarrow & x + 3y = 12 \\ -3y = x - 12 \Rightarrow -x - 3y = -12 \Rightarrow & -x - 3y = -12 \\ \hline & 0 = 0 \end{array}$$

$\Rightarrow$ dependent (and consistent)

15. Let x and y represent the numbers.

$$\begin{cases} (1) \quad x = 3y + 2 \\ (2) \quad x + y = -18 \end{cases}$$

Substitute $x = 3y + 2$.

$$\begin{aligned} x + y &= -18 \\ 3y + 2 + y &= -18 \\ 4y &= -20 \\ y &= -5 \end{aligned}$$

Substitute; solve for x:

$$\begin{aligned} x &= 3y + 2 \\ x &= 3(-5) + 2 \\ x &= -15 + 2 = -13 \end{aligned}$$

The numbers are -5 and -13.
The product is $(-5)(-13) = 65$.

16. Let $x =$ number of adult tickets and $y =$ number of child tickets.

$$\begin{array}{rl} x + \quad y = & 7 \Rightarrow \times(-14) \\ 21x + 14y = & 119 \end{array} \qquad \begin{array}{rl} -14x - 14y = & -98 \\ 21x + 14y = & 119 \\ \hline 7x \quad = & 21 \\ x \quad = & 3 \end{array}$$

He bought 3 adult tickets.

17. Let $x =$ principal at 3% and $y =$ principal at 4%.

$$\begin{array}{rl} x + \quad y = & 10000 \\ 0.03x + 0.04y = & 340 \Rightarrow \times(100) \end{array} \qquad \begin{array}{rl} x + \ y = & 10000 \Rightarrow \times(-4) \\ 3x + 4y = & 34000 \end{array} \qquad \begin{array}{rl} -4x - 4y = & -40000 \\ 3x + 4y = & 34000 \\ \hline -x \quad = & -6000 \\ x \quad = & 6000 \end{array}$$

$$\begin{aligned} x + y &= 10000 \\ 6000 + y &= 10000 \\ y &= 4000 \end{aligned}$$

She invested \$4000 at the 4% rate.

18. Let b = speed of boat in still water and c = speed of current.

	d	r	t	Equation $(d = r \cdot t)$
Downstream	8	$b+c$	2	$8 = (b+c) \cdot 2 \Rightarrow 2b + 2c = 8$
Upstream	8	$b-c$	4	$8 = (b-c) \cdot 4 \Rightarrow 4b - 4c = 8$

$$\begin{array}{l} 2b + 2c = 8 \Rightarrow \times\ (2) \\ \underline{4b - 4c = 8 \Rightarrow} \end{array} \quad \begin{array}{rl} 4b + 4c &= 16 \\ \underline{4b - 4c} &\underline{= \ \ 8} \\ 8b &= 24 \\ b &= 3 \end{array} \quad \begin{array}{r} 2b + 2c = 8 \\ 2(3) + 2c = 8 \\ 6 + 2c = 8 \\ 2c = 2 \\ c = 1 \end{array}$$

The speed of the current is 1 mph.

19. $x + y < 3$ $\quad x - y < 1$

x	y
0	3
3	0

x	y
0	-1
1	0

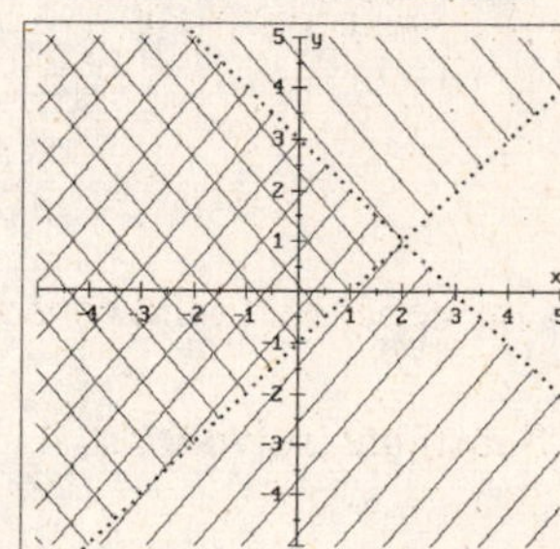

20. $2x + 3y \le 6$ $\quad x \ge 2$

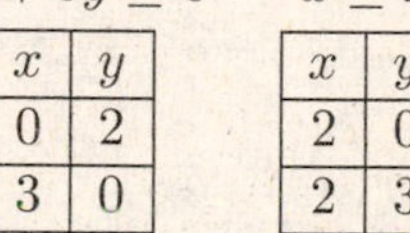

x	y
0	2
3	0

x	y
2	0
2	3

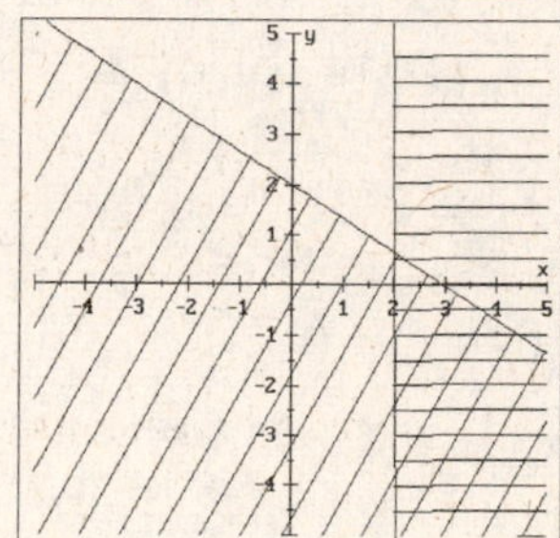

Exercise 4.1 (page 252)

1. Base: x
Exponent: 3

3. Base: b
Exponent: c

5. $6^2 = 6 \cdot 6$
$= 36$

7. $2^3 + 1^3 = 8 + 1$
$= 9$

9. −3 −1 1 3

11. the product of 3 and the sum of x and y

13. $|2x| + 3$

15. base; -5; 3

17. $(3x)(3x)(3x)(3x)$

19. $y \cdot y \cdot y \cdot y \cdot y$

21. $x^n y^n$

23. a^{bc}

25. Answers may vary.

27. Base: 4
Exponent: 3

29. Base: x
Exponent: 5

31. Base: $2y$
Exponent: 3

33. Base: x
Exponent: 4

35. Base: x
Exponent: 1

37. Base: x
Exponent: 3

39. $5^4 = 5 \cdot 5 \cdot 5 \cdot 5 = 625$

41. $2^2 + 3^2 = 4 + 9 = 13$

43. $5^4 - 4^3 = 625 - 64 = 561$

SECTION 4.1

45. $-5(3^4+4^3)=-5(81+64)=-5(145)=-725$

47. $5^3=5\cdot 5\cdot 5$

49. $x^7=x\cdot x\cdot x\cdot x\cdot x\cdot x\cdot x$

51. $-4x^5=-4\cdot x\cdot x\cdot x\cdot x\cdot x$

53. $(3t)^5=(3t)(3t)(3t)(3t)(3t)$

55. $2\cdot 2\cdot 2=2^3$

57. $x\cdot x\cdot x\cdot x=x^4$

59. $(2x)(2x)(2x)=(2x)^3$

61. $-4\cdot t\cdot t\cdot t\cdot t=-4t^4$

63. $x^4x^3=x^{4+3}=x^7$

65. $x^5x^5=x^{5+5}=x^{10}$

67. $a^3a^4a^5=a^{3+4+5}=a^{12}$

69. $y^3(y^2y^4)=y^3y^{2+4}=y^3y^6=y^{3+6}=y^9$

71. $4x^2(3x^5)=4\cdot 3x^2x^5=12x^{2+5}=12x^7$

73. $(-y^2)(4y^3)=-1\cdot 4y^2y^3=-4y^{2+3}=-4y^5$

75. $(3^2)^4=3^{2\cdot 4}=3^8$

77. $(y^5)^3=y^{5\cdot 3}=y^{15}$

79. $(x^2x^3)^5=(x^{2+3})^5=(x^5)^5=x^{5\cdot 5}=x^{25}$

81. $(a^2a^7)^3=(a^{2+7})^3=(a^9)^3=a^{9\cdot 3}=a^{27}$

83. $(x^5)^2(x^7)^3=x^{5\cdot 2}x^{7\cdot 3}=x^{10}x^{21}=x^{10+21}=x^{31}$

85. $(r^3r^2)^4(r^3r^5)^2=(r^{3+2})^4(r^{3+5})^2=(r^5)^4(r^8)^2=r^{5\cdot 4}r^{8\cdot 2}=r^{20}r^{16}=r^{20+16}=r^{36}$

87. $(xy)^3=x^3y^3$

89. $(r^3s^2)^2=(r^3)^2(s^2)^2=r^{3\cdot 2}s^{2\cdot 2}=r^6s^4$

91. $\left(4ab^2\right)^2=4^2a^2\left(b^2\right)^2=16a^2b^{2\cdot 2}=16a^2b^4$

93. $\left(-2r^2s^3t\right)^3=(-2)^3\left(r^2\right)^3\left(s^3\right)^3t^3=-8r^{2\cdot 3}s^{3\cdot 3}t^3=-8r^6s^9t^3$

95. $\left(\frac{a}{b}\right)^3=\frac{a^3}{b^3}$

97. $\left(\frac{x^2}{y^3}\right)^5=\frac{(x^2)^5}{(y^3)^5}=\frac{x^{2\cdot 5}}{y^{3\cdot 5}}=\frac{x^{10}}{y^{15}}$

99. $\frac{x^5}{x^3}=x^{5-3}=x^2$

101. $\frac{y^3y^4}{yy^2}=\frac{y^{3+4}}{y^{1+2}}=\frac{y^7}{y^3}=y^{7-3}=y^4$

103. $\frac{12a^2a^3a^4}{4(a^4)^2}=\frac{12a^{2+3+4}}{4a^{4\cdot 2}}=\frac{12a^9}{4a^8}=\frac{12}{4}a^{9-8}=3a^1=3a$

105. $\frac{(ab^2)^3}{(ab)^2}=\frac{a^3(b^2)^3}{a^2b^2}=\frac{a^3b^{2\cdot 3}}{a^2b^2}=\frac{a^3b^6}{a^2b^2}=a^{3-2}b^{6-2}=a^1b^4=ab^4$

107. $tt^2=t^{1+2}=t^3$

109. $6x^3\left(-x^2\right)\left(-x^4\right)=6(-1)(-1)x^3x^2x^4=6x^{3+2+4}=6x^9$

111. $(a^3)^7=a^{3\cdot 7}=a^{21}$

113. $(3zz^2z^3)^5 = (3z^{1+2+3})^5 = (3z^6)^5 = 3^5(z^6)^5 = 243z^{6\cdot 5} = 243z^{30}$

115. $(s^3)^3(s^2)^2(s^5)^4 = s^{3\cdot 3}s^{2\cdot 2}s^{5\cdot 4} = s^9s^4s^{20} = s^{9+4+20} = s^{33}$

117. $\left(\frac{-2a}{b}\right)^5 = \frac{(-2a)^5}{b^5} = \frac{(-2)^5a^5}{b^5} = \frac{-32a^5}{b^5}$

119. $\left(\frac{b^2}{3a}\right)^3 = \frac{(b^2)^3}{(3a)^3} = \frac{b^{2\cdot 3}}{3^3a^3} = \frac{b^6}{27a^3}$

121. $\frac{17(x^4y^3)^8}{34(x^5y^2)^4} = \frac{17(x^4)^8(y^3)^8}{34(x^5)^4(y^2)^4} = \frac{17x^{4\cdot 8}y^{3\cdot 8}}{34x^{5\cdot 4}y^{2\cdot 4}} = \frac{17x^{32}y^{24}}{34x^{20}y^8} = \frac{17}{34}x^{32-20}y^{24-8} = \frac{1}{2}x^{12}y^{16} = \frac{x^{12}y^{16}}{2}$

123. $\left(\frac{y^3y}{2yy^2}\right)^3 = \left(\frac{y^{3+1}}{2y^{1+2}}\right)^3 = \left(\frac{y^4}{2y^3}\right)^3 = \left(\frac{1}{2}y^{4-3}\right)^3 = \left(\frac{1}{2}y^1\right)^3 = \left(\frac{1}{2}\right)^3(y^1)^3 = \frac{1}{8}y^{1\cdot 3} = \frac{y^3}{8}$

125. $\left(\frac{-2r^3r^3}{3r^4r}\right)^3 = \left(\frac{-2r^{3+3}}{3r^{4+1}}\right)^3 = \left(\frac{-2r^6}{3r^5}\right)^3 = \left(\frac{-2}{3}r^{6-5}\right)^3 = \frac{(-2)^3}{3^3}(r^1)^3 = \frac{-8}{27}r^{1\cdot 3} = -\frac{8r^3}{27}$

127. $\frac{20(r^4s^3)^4}{6(rs^3)^3} = \frac{20(r^4)^4(s^3)^4}{6r^3(s^3)^3} = \frac{20r^{4\cdot 4}s^{3\cdot 4}}{6r^3s^{3\cdot 3}} = \frac{20r^{16}s^{12}}{6r^3s^9} = \frac{20}{6}r^{16-3}s^{12-9} = \frac{10}{3}r^{13}s^3 = \frac{10r^{13}s^3}{3}$

129.

Bounce	1	2	3	4
Height	$\frac{1}{2}(32)$	$\frac{1}{2}\left[\frac{1}{2}(32)\right] = 32\left(\frac{1}{2}\right)^2$	$\frac{1}{2}\left[32\left(\frac{1}{2}\right)^2\right] = 32\left(\frac{1}{2}\right)^3$	$\frac{1}{2}\left[32\left(\frac{1}{2}\right)^3\right] = 32\left(\frac{1}{2}\right)^4$

$32\left(\frac{1}{2}\right)^4 = 32\left(\frac{1^4}{2^4}\right) = 32\left(\frac{1}{16}\right) = \frac{32}{16} = 2 \text{ ft}$

131.

Years	7	14	21	28
Value	2,000	4,000	8,000	16,000

133. $A = P(1+r)^t = 8000(1+0.06)^{30} = 8000(1.06)^{30} \approx 8000(5.743491) \approx \$45{,}947.93$

135. Answers may vary.

137. No. $2^3 = 8, 3^2 = 9 \Rightarrow 2^3 \neq 3^2$

Exercise 4.2 (page 259)

1. $2^{-1} = \frac{1}{2^1} = \frac{1}{2}$

3. $\left(\frac{1}{2}\right)^{-1} = \frac{1}{\left(\frac{1}{2}\right)^1} = \frac{1}{\frac{1}{2}} = 2$

5. $x^{-1}x^2 = x^{-1+2} = x^1 = x$

7. $\frac{x^5x^2}{x^7} = \frac{x^{5+2}}{x^7} = \frac{x^7}{x^7} = x^{7-7} = x^0 = 1$

9. $\frac{3a^2+4b+8}{a+2b^2} = \frac{3(-2)^2+4(3)+8}{-2+2(3)^2} = \frac{3(4)+12+8}{-2+2(9)} = \frac{12+12+8}{-2+18} = \frac{32}{16} = 2$

SECTION 4.2

11.
$$\begin{aligned}5\left(x-\frac{1}{2}\right)&=\frac{7}{2}\\5x-\frac{5}{2}&=\frac{7}{2}\\5x-\frac{5}{2}+\frac{5}{2}&=\frac{7}{2}+\frac{5}{2}\\5x&=\frac{12}{2}\\5x&=6\\\frac{5x}{5}&=\frac{6}{5}\\x&=\frac{6}{5}\end{aligned}$$

13.
$$\begin{aligned}P&=L+\frac{s}{f}i\\P-L&=\frac{s}{f}i\\f(P-L)&=f\cdot\frac{s}{f}i\\f(P-L)&=si\\\frac{f(P-L)}{i}&=\frac{si}{i}\\\frac{f(P-L)}{i}&=s\text{, or } s=\frac{f(P-L)}{i}\end{aligned}$$

15. $1, \dfrac{1}{x^n}$

17. $\dfrac{1}{8^2}$

19. $14^0 = 1$

21. $2x^0 = 2\cdot x^0 = 2\cdot 1 = 2$

23. $\left(\dfrac{a^2b^3}{ab^4}\right)^0 = 1$

25. $\dfrac{8^y}{8^y} = 8^{y-y} = 8^0 = 1$

27. $5^{-4} = \dfrac{1}{5^4} = \dfrac{1}{625}$

29. $x^{-2} = \dfrac{1}{x^2}$

31. $(2y)^{-4} = \dfrac{1}{(2y)^4} = \dfrac{1}{2^4y^4} = \dfrac{1}{16y^4}$

33. $(-5p)^{-3} = \dfrac{1}{(-5p)^3} = \dfrac{1}{(-5)^3p^3} = -\dfrac{1}{125p^3}$

35. $\left(y^2y^4\right)^{-2} = \left(y^6\right)^{-2} = y^{-12} = \dfrac{1}{y^{12}}$

37. $3x^{-3} = 3\cdot x^{-3} = 3\cdot\dfrac{1}{x^3} = \dfrac{3}{x^3}$

39. $\dfrac{y^4}{y^5} = y^{4-5} = y^{-1} = \dfrac{1}{y^1} = \dfrac{1}{y}$

41. $\dfrac{p^3}{p^6} = p^{3-6} = p^{-3} = \dfrac{1}{p^3}$

43.
$$\begin{aligned}\frac{x^{-2}x^{-3}}{x^{-10}} = \frac{x^{-2+(-3)}}{x^{-10}} = \frac{x^{-5}}{x^{-10}} &= x^{-5-(-10)}\\&= x^5\end{aligned}$$

45. $\dfrac{15a^3b^8}{3a^4b^4} = \dfrac{15}{3}a^{3-4}b^{8-4} = 5a^{-1}b^4 = \dfrac{5b^4}{a}$

47. $\left(a^{-4}\right)^3 = a^{-4\cdot 3} = a^{-12} = \dfrac{1}{a^{12}}$

49.
$$\begin{aligned}\left(\frac{a^3}{a^{-4}}\right)^{-2} = \left(a^{3-(-4)}\right)^{-2} = \left(a^7\right)^{-2} &= a^{-14}\\&= \frac{1}{a^{14}}\end{aligned}$$

51. $\left(\dfrac{6a^2b^3}{2ab^2}\right)^{-2} = \left(\dfrac{6}{2}a^{2-1}b^{3-2}\right)^{-2} = \left(3a^1b^1\right)^{-2} = \dfrac{1}{(3ab)^2} = \dfrac{1}{3^2a^2b^2} = \dfrac{1}{9a^2b^2}$

53. $\left(\dfrac{18a^2b^3c^{-4}}{3a^{-1}b^2c}\right)^{-3} = \left(\dfrac{18}{3}a^{2-(-1)}b^{3-2}c^{-4-1}\right)^{-3} = \left(6a^3b^1c^{-5}\right)^{-3} = 6^{-3}\left(a^3\right)^{-3}\left(b^1\right)^{-3}\left(c^{-5}\right)^{-3}$

$= \dfrac{1}{6^3}a^{-9}b^{-3}c^{15} = \dfrac{c^{15}}{216a^9b^3}$

55. $x^{2m}x^m = x^{2m+m} = x^{3m}$

57. $\dfrac{x^{3n}}{x^{6n}} = x^{3n-6n} = x^{-3n} = \dfrac{1}{x^{3n}}$

59. $y^{3m+2}y^{-m} = y^{3m+2+(-m)} = y^{2m+2}$

61. $\left(y^{n+2}\right)^2 = y^{(n+2)\cdot 2} = y^{2n+4}$

63. $2^5 \cdot 2^{-2} = 2^{5+(-2)} = 2^3 = 8$

65. $4^{-3} \cdot 4^{-2} \cdot 4^5 = 4^{-3+(-2)+5} = 4^0 = 1$

67. $\dfrac{3^5 \cdot 3^{-2}}{3^3} = \dfrac{3^{5+(-2)}}{3^3} = \dfrac{3^3}{3^3} = 3^{3-3}$

$= 3^0 = 1$

69. $\dfrac{2^5 \cdot 2^7}{2^6 \cdot 2^{-3}} = \dfrac{2^{5+7}}{2^{6+(-3)}} = \dfrac{2^{12}}{2^3} = 2^{12-3}$

$= 2^9 = 512$

71. $(-x)^0 = 1$

73. $\dfrac{x^0 - 5x^0}{2x^0} = \dfrac{1-5(1)}{2(1)} = \dfrac{1-5}{2} = \dfrac{-4}{2} = -2$

75. $b^{-5} = \dfrac{1}{b^5}$

77. $u^{2m}v^{3n}u^{3m}v^{-3n} = u^{2m+3m}v^{3n+(-3n)}$

$= u^{5m}v^0 = u^{5m}$

79. $\left(x^{3-2n}\right)^{-4} = x^{(3-2n)(-4)} = x^{-12+8n} = x^{8n-12}$

81. $\left(y^{2-n}\right)^{-4} = y^{(2-n)(-4)} = y^{-8+4n} = y^{4n-8}$

83. $\dfrac{y^{3m}}{y^{2m}} = y^{3m-2m} = y^m$

85. $(4t)^{-3} = \dfrac{1}{(4t)^3} = \dfrac{1}{4^3t^3} = \dfrac{1}{64t^3}$

87. $\left(ab^2\right)^{-3} = \dfrac{1}{\left(ab^2\right)^3} = \dfrac{1}{a^3\left(b^2\right)^3} = \dfrac{1}{a^3b^6}$

89. $\left(x^2y\right)^{-2} = \dfrac{1}{\left(x^2y\right)^2} = \dfrac{1}{\left(x^2\right)^2y^2} = \dfrac{1}{x^4y^2}$

91. $\dfrac{\left(r^2\right)^3}{\left(r^3\right)^4} = \dfrac{r^6}{r^{12}} = r^{6-12} = r^{-6} = \dfrac{1}{r^6}$

93. $\dfrac{y^4y^3}{y^4y^{-2}} = \dfrac{y^7}{y^2} = y^{7-2} = y^5$

95. $\dfrac{a^4a^{-2}}{a^2a^0} = \dfrac{a^2}{a^2} = a^{2-2} = a^0 = 1$

97. $\left(ab^2\right)^{-2} = \dfrac{1}{\left(ab^2\right)^2} = \dfrac{1}{a^2\left(b^2\right)^2} = \dfrac{1}{a^2b^4}$

99. $\left(x^2y\right)^{-3} = \dfrac{1}{\left(x^2y\right)^3} = \dfrac{1}{\left(x^2\right)^3y^3} = \dfrac{1}{x^6y^3}$

101. $\left(x^{-4}x^3\right)^3 = \left(x^{-1}\right)^3 = x^{-3} = \dfrac{1}{x^3}$

103. $\left(y^3y^{-2}\right)^{-2} = \left(y^1\right)^{-2} = y^{-2} = \dfrac{1}{y^2}$

105. $\left(a^{-2}b^{-3}\right)^{-4} = \left(a^{-2}\right)^{-4}\left(b^{-3}\right)^{-4} = a^8b^{12}$

107. $\left(-2x^3y^{-2}\right)^{-5} = (-2)^{-5}\left(x^3\right)^{-5}\left(y^{-2}\right)^{-5} = \dfrac{1}{(-2)^5}x^{-15}y^{10} = -\dfrac{y^{10}}{32x^{15}}$

SECTION 4.2

109. $\left(\frac{b^5}{b^{-2}}\right)^{-2} = \left(b^{5-(-2)}\right)^{-2} = \left(b^7\right)^{-2} = b^{-14} = \frac{1}{b^{14}}$

111. $\left(\frac{4x^2}{3x^{-5}}\right)^4 = \left(\frac{4}{3}x^{2-(-5)}\right)^4 = \left(\frac{4}{3}x^7\right)^4 = \left(\frac{4}{3}\right)^4 \left(x^7\right)^4 = \frac{256}{81}x^{28} = \frac{256x^{28}}{81}$

113. $\left(\frac{12y^3z^{-2}}{3y^{-4}z^3}\right)^2 = \left(\frac{12}{3}y^{3-(-4)}z^{-2-3}\right)^2 = \left(4y^7z^{-5}\right)^2 = 4^2\left(y^7\right)^2\left(z^{-5}\right)^2 = 16y^{14}z^{-10} = \frac{16y^{14}}{z^{10}}$

115. $\left(\frac{2x^3y^{-2}}{4xy^2}\right)^7 = \left(\frac{2}{4}x^{3-1}y^{-2-2}\right)^7 = \left(\frac{1}{2}x^2y^{-4}\right)^7 = \left(\frac{1}{2}\right)^7\left(x^2\right)^7\left(y^{-4}\right)^7 = \frac{1}{128}x^{14}y^{-28} = \frac{x^{14}}{128y^{28}}$

117. $\left(\frac{14u^{-2}v^3}{21u^{-3}v}\right)^4 = \left(\frac{14}{21}u^{-2-(-3)}v^{3-1}\right)^4 = \left(\frac{2}{3}u^1v^2\right)^4 = \left(\frac{2}{3}\right)^4\left(u^1\right)^4\left(v^2\right)^4 = \frac{16}{81}u^4v^8 = \frac{16u^4v^8}{81}$

119. $\frac{\left(2x^{-2}y\right)^{-3}}{\left(4x^2y^{-1}\right)^3} = \frac{2^{-3}\left(x^{-2}\right)^{-3}y^{-3}}{4^3\left(x^2\right)^3\left(y^{-1}\right)^3} = \frac{x^6y^{-3}}{2^3\cdot 4^3x^6y^{-3}} = \frac{1}{8\cdot 64}x^{6-6}y^{-3-(-3)} = \frac{1}{512}x^0y^0 = \frac{1}{512}$

121. $\frac{\left(17x^5y^{-5}z\right)^{-3}}{\left(17x^{-5}y^3z^2\right)^{-4}} = \frac{17^{-3}\left(x^5\right)^{-3}\left(y^{-5}\right)^{-3}z^{-3}}{17^{-4}\left(x^{-5}\right)^{-4}\left(y^3\right)^{-4}\left(z^2\right)^{-4}} = \frac{17^{-3}x^{-15}y^{15}z^{-3}}{17^{-4}x^{20}y^{-12}z^{-8}}$

$= 17^{-3-(-4)}x^{-15-20}y^{15-(-12)}z^{-3-(-8)}$

$= 17^1x^{-35}y^{27}z^5 = \frac{17y^{27}z^5}{x^{35}}$

123. $\frac{x^{3n}}{x^{6n}} = x^{3n-6n} = x^{-3n} = \frac{1}{x^{3n}}$

Problems 125-129 are to be solved using a calculator. The keystrokes needed to solve each problem using a TI-84 graphing calculator appear in each solution. There may be other solutions. Keystrokes for other calculators may be slightly different.

125. $P = A(1+i)^{-n} = 100{,}000(1+0.07)^{-40}$

[1] [0] [0] [0] [0] [0] [(] [1] [+] [.] [0] [7] [)] [^] [(-)] [4] [0] [=]

The original deposit should be \$6,678.04.

127. $P = A(1+i)^{-n} = 100{,}000(1+0.09)^{-40}$

[1] [0] [0] [0] [0] [0] [(] [1] [+] [.] [0] [9] [)] [^] [(-)] [4] [0] [=]

The original deposit should be \$3,183.76.

129. $P = A(1+i)^{-n} = 1{,}000{,}000(1+0.08)^{-60}$

[1] [0] [0] [0] [0] [0] [0] [(] [1] [+] [.] [0] [8] [)] [^] [(-)] [6] [0] [=]

The original deposit should be \$9,875.85.

131. Answers may vary.

133. If $x > 1$, then x raised to a negative power is less than x. If $x = 1$, then x raised to a negative power is equal to x. If $0 < x < 1$, then x raised to a negative power is greater than x.

Exercise 4.3 (page 266)

1. $3.72 \times 10^2 = 372$, so 3.72×10^2 is larger.

3. $3.72 \times 10^3 = 3{,}720$; $4.72 \times 10^3 = 4{,}720$
4.72×10^3 is larger.

5. $3.72 \times 10^{-1} = 0.372$
$4.72 \times 10^{-2} = 0.0472$; 3.72×10^{-1} is larger.

7. $-5y^{55} = -5(-1)^{55} = -5(-1) = 5$

9. commutative property of addition

11.
$$\begin{aligned} 3(x-4) - 6 &= 0 \\ 3x - 12 - 6 &= 0 \\ 3x - 18 &= 0 \\ 3x &= 18 \\ x &= 6 \end{aligned}$$

13. scientific notation

15. $23{,}000 = 2.3 \times 10^4$

17. $1{,}700{,}000 = 1.7 \times 10^6$

19. $0.062 = 6.2 \times 10^{-2}$

21. $0.00000275 = 2.75 \times 10^{-6}$

23.
$$\begin{aligned} 42.5 \times 10^2 &= 4.25 \times 10^1 \times 10^2 \\ &= 4.25 \times 10^3 \end{aligned}$$

25.
$$\begin{aligned} 0.25 \times 10^{-2} &= 2.5 \times 10^{-1} \times 10^{-2} \\ &= 2.5 \times 10^{-3} \end{aligned}$$

27. $2.3 \times 10^2 = 230$

29. $8.12 \times 10^5 = 812{,}000$

31. $1.15 \times 10^{-3} = 0.00115$

33. $9.76 \times 10^{-4} = 0.000976$

35. $(3.4 \times 10^2)(2.1 \times 10^3) = (3.4)(2.1) \times 10^2 \times 10^3 = 7.14 \times 10^5 = 714{,}000$

37. $$\frac{9.3 \times 10^2}{3.1 \times 10^{-2}} = \frac{9.3}{3.1} \times \frac{10^2}{10^{-2}} = 3 \times 10^4 = 30{,}000$$

39. $$\frac{96{,}000}{(12{,}000)(0.00004)} = \frac{9.6 \times 10^4}{(1.2 \times 10^4)(4 \times 10^{-5})} = \frac{9.6 \times 10^4}{4.8 \times 10^{-1}} = \frac{9.6}{4.8} \times \frac{10^4}{10^{-1}} = 2 \times 10^5 = 200{,}000$$

41. $$\begin{aligned} \frac{2{,}475}{(132{,}000{,}000)(0.25)} = \frac{2.475 \times 10^3}{(1.32 \times 10^8)(2.5 \times 10^{-1})} = \frac{2.475 \times 10^3}{3.3 \times 10^7} = \frac{2.475}{3.3} \times \frac{10^3}{10^7} &= 0.75 \times 10^{-4} \\ &= 0.000075 \end{aligned}$$

43. $0.0000051 = 5.1 \times 10^{-6}$

45. $257{,}000{,}000 = 2.57 \times 10^8$

47. $$\frac{2.4 \times 10^2}{6 \times 10^{23}} = \frac{2.4}{6} \times \frac{10^2}{10^{23}} = 0.4 \times 10^{-21} = 4 \times 10^{-1} \times 10^{-21} = 4 \times 10^{-22}$$

SECTION 4.3

49. $25 \times 10^6 = 25{,}000{,}000$

51. $0.51 \times 10^{-3} = 0.00051$

53. $25{,}700{,}000{,}000{,}000 = 2.57 \times 10^{13}$ mi

55. $1.14 \times 10^8 = 114{,}000{,}000$ mi

57. $0.00622 = 6.22 \times 10^{-3}$ mi

59. 3.6×10^7 mi $\times$ 5280 ft/mi $= 19008 \times 10^7$ ft $= 1.9008 \times 10^{11}$ ft

61. 2.617×10^{11} bar $\times$ 42 gal/bar $= 109.914 \times 10^{11}$ gal $= 1.09914 \times 10^{13}$ gallons

63. $\dfrac{3.3 \times 10^4 \text{ cm}}{\text{sec}} = \dfrac{3.3 \times 10^4 \text{ cm}}{\text{sec}} \cdot \dfrac{1 \text{ m}}{100 \text{ cm}} \cdot \dfrac{1 \text{ km}}{1{,}000 \text{ m}} = \dfrac{3.3 \times 10^4}{10^5} \text{ km/sec} = 3.3 \times 10^{-1} \text{ km/sec}$

65. x-rays, visible light, infrared

67. 1.5×10^{-4} in.; $25{,}000{,}000{,}000{,}000 = 2.5 \times 10^{13}$

69. Answers may vary.

71. Answers may vary.

Exercise 4.4 (page 277)

For exercises 1-7, answers may vary. The answers below are only examples of correct answers.

1. $x + 2$

3. $x^2 + 2x + 1$

5. $x^3 + 4$

7. 5

9.
$$\begin{aligned} 5(u-5)+9 &= 2(u+4) \\ 5u - 25 + 9 &= 2u + 8 \\ 5u - 16 &= 2u + 8 \\ 3u - 16 &= 8 \\ 3u &= 24 \\ u &= 8 \end{aligned}$$

11.
$$\begin{aligned} -4(3y+2) &\le 28 \\ -12y - 8 &\le 28 \\ -12y &\le 36 \\ \tfrac{-12y}{-12} &\ge \tfrac{36}{-12} \\ y &\ge -3 \end{aligned}$$

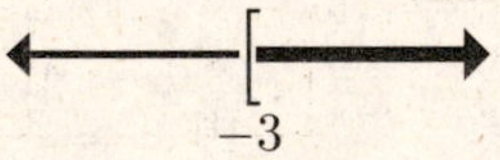

13. $(x^2x^4)^3 = (x^6)^3 = x^{18}$

15. $\left(\dfrac{y^2y^5}{y^4}\right)^3 = \left(\dfrac{y^7}{y^4}\right)^3 = (y^3)^3 = y^9$

17. algebraic

19. monomial; binomial; trinomial

21. sum

23. polynomial

25. cubic

27. descending; 5

29. function

31. yes

33. no; fractional exponent on x

35. binomial

37. trinomial

39. monomial

41. binomial

SECTION 4.4

43. 4th

45. 3rd

47. 8th
$(3+5=8)$

49. 6th
$(3+1+2=6)$

51. $5x-3=5(2)-3=10-3=7$

53. $5x-3=5(-1)-3=-5-3=-8$

55. $-x^2-4=-(0)^2-4=-0-4=-4$

57. $-x^2-4=-(-1)^2-4=-1-4=-5$

59.

x	x^2-3
-2	$(-2)^2-3=4-3=1$
-1	$(-1)^2-3=1-3=-2$
0	$(0)^2-3=0-3=-3$
1	$(1)^2-3=1-3=-2$
2	$(2)^2-3=4-3=1$

61.

x	x^3+2
-2	$(-2)^3+2=-8+2=-6$
-1	$(-1)^3+2=-1+2=1$
0	$(0)^3+2=0+2=2$
1	$(1)^3+2=1+2=3$
2	$(2)^3+2=8+2=10$

63. $f(0)=5(0)+1=0+1=1$

65. $f\left(-\frac{1}{2}\right)=5\left(-\frac{1}{2}\right)+1=-\frac{5}{2}+\frac{2}{2}=-\frac{3}{2}$

67.
$$\begin{aligned} f(x)&=26\\ 5x+1&=26\\ 5x&=25\\ x&=5 \end{aligned}$$

69.
$$\begin{aligned} f(x)&=-21\\ 5x+1&=-21\\ 5x&=-22\\ x&=-\tfrac{22}{5} \end{aligned}$$

71. $f(x)=x^2-1$

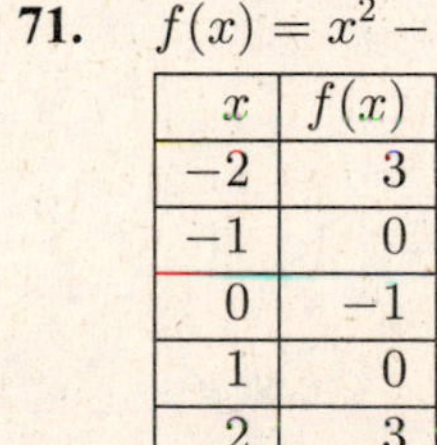

x	$f(x)$
-2	3
-1	0
0	-1
1	0
2	3

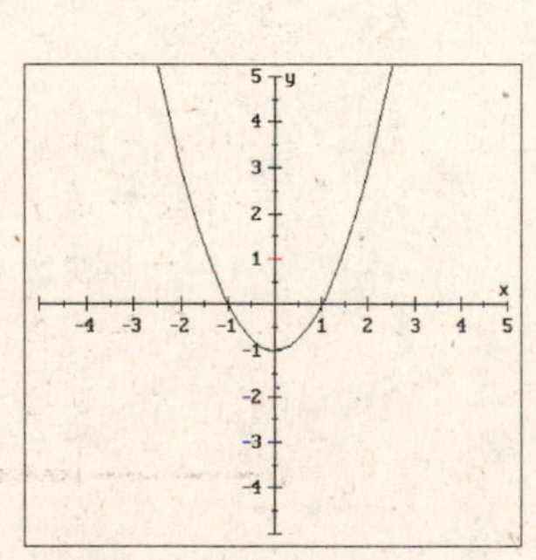

73. $f(x)=x^3+2$

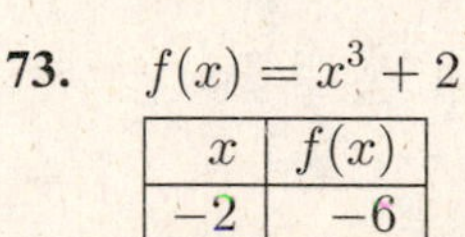

x	$f(x)$
-2	-6
-1	1
0	2
1	3
2	10

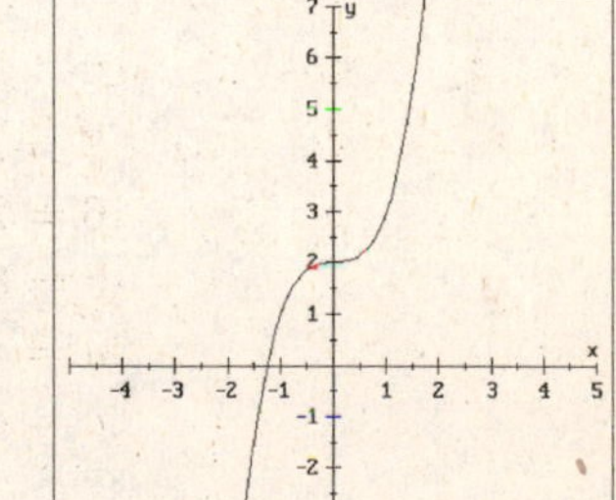

75. trinomial

77. none of these
(4 terms)

79. binomial

81. 12th

83. 0th

85. $f(0)=(0)^2-2(0)+3=0-0+3=3$

87. $f(-2)=(-2)^2-2(-2)+3$
$=4+4+3=11$

89. $f(0.5)=(0.5)^2-2(0.5)+3$
$=0.25-1+3=2.25$

91. $h = -16t^2 + 64t$
$h = -16(2)^2 + 64(2)$
$h = -16(4) + 128$
$h = -64 + 128 = 64$ ft

93. $f(d) = -0.08d^2 + 100d$
$f(815) = -0.08(815)^2 + 100(815)$
$= -0.08(664225) + 81500$
$= -53138 + 81500$
$= \$28{,}362$

95. $d = 0.04v^2 + 0.9v$
$d = 0.04(30)^2 + 0.9(30)$
$d = 0.04(900) + 27$
$d = 36 + 27 = 63$ ft

97. **Answers may vary.**

99. There are many possible polynomials. One is $2x - 2$.

Exercise 4.5 (page 284)

1. $x^3 + 3x^3 = 4x^3$

3. $(x + 3y) - (x + y) = x + 3y - x - y$
$= 2y$

5. $(2x - y^2) - (2x + y^2) = 2x - y^2 - 2x - y^2$
$= -2y^2$

7. $3x^2 + 2y + x^2 - y = 4x^2 + y$

9. $ab + cd = (3)(-2) + (-1)(2)$
$= -6 + (-2) = -8$

11. $a(b + c) = 3[-2 + (-1)] = 3(-3) = -9$

13. $-4(2x - 9) \geq 12$
$-8x + 36 \geq 12$
$-8x \geq -24$
$\frac{-8x}{-8} \leq \frac{-24}{-8}$
$x \leq 3$

3

15. monomial

17. coefficients; variables

19. like terms

21. like terms; $3y + 4y = 7y$

23. unlike terms

25. like terms
$3x^3 + 4x^3 + 6x^3 = 13x^3$

27. like terms; $-5x^3y^2 + 13x^3y^2 = 8x^3y^2$

29. like terms
$-23t^6 + 32t^6 + 56t^6 = 65t^6$

31. unlike terms

33. $4y + 5y = 9y$

35. $-2x + 3x = 1x = x$

37. $-8t^2 + 4t^2 = -4t^2$

39. $15x^2 + 10x^2 = 25x^2$

41. $-18a - 3a = -21a$

43. $32u^3 - 16u^3 = 16u^3$

SECTION 4.5

45. $18x^5y^2 - 11x^5y^2 = 7x^5y^2$

47. $22ab^2 - 30ab^2 = -8ab^2$

49. $(3x+7)+(4x-3) = 3x+7+4x-3 = 7x+4$

51. $(2x+3y+z)+(5x-10y+z) = 2x+3y+z+5x-10y+z = 7x-7y+2z$

53.
$$\begin{array}{rr} & 3x^2+4x+\ 5 \\ + & \underline{2x^2-3x+\ 6} \\ & 5x^2+\ x+11 \end{array}$$

55.
$$\begin{array}{rr} & 2x^3-3x^2+4x-7 \\ + & \underline{-9x^3-4x^2-5x+6} \\ & -7x^3-7x^2-\ x-1 \end{array}$$

57. $(4a+3)-(2a-4) = 4a+3-2a+4 = 2a+7$

59. $(3a^2-2a+4)-(a^2-3a+7) = 3a^2-2a+4-a^2+3a-7 = 2a^2+a-3$

61.
$$\begin{array}{rr} & 3x^2+4x-5 \\ - & \underline{-2x^2-2x+3} \\ & 3x^2+4x-5 \\ + & \underline{2x^2+2x-3} \\ & 5x^2+6x-8 \end{array}$$

63.
$$\begin{array}{rr} & 4x^3+4x^2-3x+10 \\ - & \underline{5x^3-2x^2-4x-\ 4} \\ & 4x^3+4x^2-3x+10 \\ + & \underline{-5x^3+2x^2+4x+\ 4} \\ & -x^3+6x^2+\ x+14 \end{array}$$

65. $(-8x-3y)-(11x+y) = -8x-3y-11x-y = -19x-4y$

67. $(2x^2-3x+1)-(4x^2-3x+2) = 2x^2-3x+1-4x^2+3x-2 = -2x^2-1$

69. $2(x+3)+4(x-2) = 2x+6+4x-8 = 6x-2$

71. $-2(x^2+7x-1)-3(x^2-2x+7) = -2x^2-14x+2-3x^2+6x-21 = -5x^2-8x-19$

73. $2(x^2-5x-4)-3(x^2-5x-4)+6(x^2-5x-4)$
$= 2x^2-10x-8-3x^2+15x+12+6x^2-30x-24 = 5x^2-25x-20$

75. $2(2y^2-2y+2)-4(3y^2-4y-1)+4(y^3-y^2-y)$
$= 4y^2-4y+4-12y^2+16y+4+4y^3-4y^2-4y = 4y^3-12y^2+8y+8$

77. $3rst+4rst+7rst = 14rst$

79. $-4a^2bc+5a^2bc-7a^2bc = -6a^2bc$

81. $(3x)^2-4x^2+10x^2 = 9x^2-4x^2+10x^2 = 15x^2$

83. $5x^2y^2+2(xy)^2-(3x^2)y^2 = 5x^2y^2+2x^2y^2-3x^2y^2 = 4x^2y^2$

85. $(-3x^2y)^4+(4x^4y^2)^2-2x^8y^4 = 81x^8y^4+16x^8y^4-2x^8y^4 = 95x^8y^4$

87. $2(x+3)+3(x+3) = 2x+6+3x+9 = 5x+15$

89. $-8(x-y)+11(x-y) = -8x+8y+11x-11y = 3x-3y$

SECTION 4.5

91. $(4c^2 + 3c - 2) + (3c^2 + 4c + 2) = 4c^2 + 3c - 2 + 3c^2 + 4c + 2 = 7c^2 + 7c$

93.
$$\begin{array}{rr} & -3x^2y + 4xy + 25y^2 \\ + & 5x^2y - 3xy - 12y^2 \\ \hline & 2x^2y + \ \ xy + 13y^2 \end{array}$$

95.
$$\begin{array}{rr} & -2x^2y^2 - \ \ 4xy + 12y^2 \\ - & 10x^2y^2 + \ \ 9xy - 24y^2 \\ \hline & -2x^2y^2 - \ \ 4xy + 12y^2 \\ + & -10x^2y^2 - \ \ 9xy + 24y^2 \\ \hline & -12x^2y^2 - 13xy + 36y^2 \end{array}$$

97.
$$\begin{aligned} 2(a^2b^2 - ab) - 3(ab + 2ab^2) + (b^2 - ab + a^2b^2) &= 2a^2b^2 - 2ab - 3ab - 6ab^2 + b^2 - ab + a^2b^2 \\ &= 3a^2b^2 - 6ab + b^2 - 6ab^2 \end{aligned}$$

99.
$$\begin{aligned} &-4(x^2y^2 + xy^3 + xy^2z) - 2(x^2y^2 - 4xy^2z) - 2(8xy^3 - y) \\ &\qquad = -4x^2y^2 - 4xy^3 - 4xy^2z - 2x^2y^2 + 8xy^2z - 16xy^3 + 2y \\ &\qquad = -6x^2y^2 + 4xy^2z - 20xy^3 + 2y \end{aligned}$$

101.
$$\begin{aligned} &[(2x^2 - 3x + 4) + (3x^2 - 2)] + (x^2 + x - 3) \\ &\qquad = [2x^2 - 3x + 4 + 3x^2 - 2] + x^2 + x - 3 \\ &\qquad = 5x^2 - 3x + 2 + x^2 + x - 3 = 6x^2 - 2x - 1 \end{aligned}$$

103.
$$\begin{aligned} &[(3t^3 + t^2) + (-t^3 + 6t - 3)] - (t^3 - 2t^2 + 2) \\ &\qquad = [3t^3 + t^2 - t^3 + 6t - 3] - t^3 + 2t^2 - 2 \\ &\qquad = 2t^3 + t^2 + 6t - 3 - t^3 + 2t^2 - 2 = t^3 + 3t^2 + 6t - 5 \end{aligned}$$

105.
$$\begin{aligned} &[(-2x^2 - 7x + 1) + (-4x^2 + 8x - 1)] + (3x^2 + 4x - 7) \\ &\qquad = [-2x^2 - 7x + 1 - 4x^2 + 8x - 1] + 3x^2 + 4x - 7 \\ &\qquad = -6x^2 + x + 3x^2 + 4x - 7 = -3x^2 + 5x - 7 \end{aligned}$$

107.
$$\begin{aligned} y &= 900x + 105{,}000 \\ y &= 900(10) + 105{,}000 \\ y &= 9{,}000 + 105{,}000 = \$114{,}000 \end{aligned}$$

109.
$$\begin{aligned} y &= 1{,}000x + 120{,}000 \\ y &= 1{,}000(12) + 120{,}000 \\ y &= 12{,}000 + 120{,}000 = \$132{,}000 \end{aligned}$$

111. a.
$$\begin{aligned} y &= 900x + 105{,}000 & y &= 1{,}000x + 120{,}000 \\ y &= 900(20) + 105{,}000 & y &= 1{,}000(20) + 120{,}000 \\ y &= 18{,}000 + 105{,}000 = \$123{,}000 & y &= 20{,}000 + 120{,}000 = \$140{,}000 \end{aligned}$$
$$\$123{,}000 + \$140{,}000 = \$263{,}000$$

111. b. $y = 1{,}900x + 225{,}000 = 1{,}900(20) + 225{,}000 = 38{,}000 + 225{,}000 = \$263{,}000$

113. $y = -1{,}100x + 6{,}600$

115. $y = (-1{,}100x + 6{,}600) + (-1{,}700x + 9{,}200) = -2{,}800x + 15{,}800$

117. Answers may vary.

119. $P(x+h)+P(x)=[3(x+h)-5]+[3x-5]=3x+3h-5+3x-5=6x+3h-10$

121. $P(x)-Q(x)=\left(x^{23}+5x^2+73\right)-\left(x^{23}+4x^2+73\right)=x^{23}+5x^2+73-x^{23}-4x^2-73=x^2$
$P(7)-Q(7)=7^2=49$

Exercise 4.6 (page 294)

1. $$\begin{aligned}2x^2(3x-1)&=2x^2(3x)+2x^2(-1)\\&=6x^3-2x^2\end{aligned}$$

3. $$\begin{aligned}7xy(x+y)&=7xy(x)+7xy(y)\\&=7x^2y+7xy^2\end{aligned}$$

5. $(x+3)(x+2)=x(x)+x(2)+3(x)+3(2)=x^2+2x+3x+6=x^2+5x+6$

7. $(2x+3)(x+2)=2x(x)+2x(2)+3(x)+3(2)=2x^2+4x+3x+6=2x^2+7x+6$

9. $(x+3)^2=(x+3)(x+3)=x(x)+x(3)+3(x)+3(3)=x^2+3x+3x+9=x^2+6x+9$

11. distributive property

13. commutative property of multiplication

15. $$\begin{aligned}\tfrac{5}{3}(5y+6)-10&=0\\3\left[\tfrac{5}{3}(5y+6)-10\right]&=3(0)\\3\cdot\tfrac{5}{3}(5y+6)-3\cdot10&=0\\5(5y+6)-30&=0\\25y+30-30&=0\\25y&=0\\y&=0\end{aligned}$$

17. monomial

19. special products

21. $(2x)(3x)=6x^2$

23. $(5)(3x)=15x$

25. $(3x^2)(4x^3)=(3)(4)x^2x^3=12x^5$

27. $(-5t^3)(2t^4)=(-5)(2)t^3t^4=-10t^7$

29. $$\begin{aligned}\left(2x^2y^3\right)\left(3x^3y^2\right)&=(2)(3)x^2x^3y^3y^2\\&=6x^5y^5\end{aligned}$$

31. $$\begin{aligned}\left(3b^2\right)(-2b)\left(4b^3\right)&=(3)(-2)(4)b^2bb^3\\&=-24b^6\end{aligned}$$

33. $\left(a^2b^3c\right)^5=\left(a^2\right)^5\left(b^3\right)^5c^5=a^{10}b^{15}c^5$

35. $(a^3b^2c)(abc^3)^2=(a^3b^2c)\left[a^2b^2(c^3)^2\right]=(a^3b^2c)(a^2b^2c^6)=a^3a^2b^2b^2cc^6=a^5b^4c^7$

37. $3(x+4)=3(x)+3(4)=3x+12$

39. $-4(t+7)=-4(t)+(-4)(7)=-4t-28$

41. $3x(x-2)=3x(x)-3x(2)=3x^2-6x$

43. $$\begin{aligned}-2x^2\left(3x^2-x\right)&=-2x^2\left(3x^2\right)-\left(-2x^2\right)(x)\\&=-6x^4+2x^3\end{aligned}$$

45. $3xy(x+y)=3xy(x)+3xy(y)=3x^2y+3xy^2$

47. $2x^2(3x^2+4x-7)=2x^2(3x^2)+2x^2(4x)-2x^2(7)=6x^4+8x^3-14x^2$

SECTION 4.6

49. $\frac{1}{4}x^2(8x^5-4)=\frac{1}{4}x^2(8x^5)-\frac{1}{4}x^2(4)=2x^7-x^2$

51. $-\frac{2}{3}r^2t^2(9r-3t)=-\frac{2}{3}r^2t^2(9r)-\left(-\frac{2}{3}r^2t^2\right)(3t)=-6r^3t^2+2r^2t^3$

53. $(a+4)(a+5)=a(a)+a(5)+4(a)+4(5)=a^2+5a+4a+20=a^2+9a+20$

55. $(3x-2)(x+4)=3x(x)+3x(4)+(-2)(x)+(-2)(4)=3x^2+12x-2x-8=3x^2+10x-8$

57. $(2a+4)(3a-5)=2a(3a)-2a(5)+4(3a)-4(5)=6a^2-10a+12a-20=6a^2+2a-20$

59. $$\begin{aligned}(3x-5)(2x+1)=3x(2x)+3x(1)+(-5)(2x)+(-5)(1)&=6x^2+3x-10x-5\\&=6x^2-7x-5\end{aligned}$$

61. $(2s+3t)(3s-t)=2s(3s)-2s(t)+3t(3s)-3t(t)=6s^2-2st+9st-3t^2=6s^2+7st-3t^2$

63. $(u+v)(u+2t)=u(u)+u(2t)+v(u)+v(2t)=u^2+2tu+uv+2tv$

65. $(x+y)(x+z)=x(x)+x(z)+y(x)+y(z)=x^2+xz+xy+yz$

67. $$\begin{aligned}2(x-4)(x+1)=2[x(x)+x(1)+(-4)(x)+(-4)(1)]&=2\left[x^2+x-4x-4\right]\\&=2\left[x^2-3x-4\right]=2x^2-6x-8\end{aligned}$$

69. $$\begin{aligned}3a(a+b)(a-b)=3a[a(a)-a(b)+b(a)-b(b)]&=3a\left[a^2-ab+ab-b^2\right]\\&=3a\left[a^2-b^2\right]=3a^3-3ab^2\end{aligned}$$

71. $(3xy)(-2x^2y^3)(x+y)=(-6x^3y^4)(x+y)=-6x^3y^4(x)+(-6x^3y^4)(y)=-6x^4y^4-6x^3y^5$

73. $$\begin{aligned}2t(t+2)+3t(t-5)=[2t(t)+2t(2)]+[3t(t)-3t(5)]&=\left[2t^2+4t\right]+\left[3t^2-15t\right]\\&=2t^2+4t+3t^2-15t=5t^2-11t\end{aligned}$$

75. $(x+5)^2=(x+5)(x+5)=x(x)+x(5)+5(x)+5(5)=x^2+5x+5x+25=x^2+10x+25$

77. $(x-4)^2=(x-4)(x-4)=x(x)+x(-4)-4(x)-4(-4)=x^2-4x-4x+16=x^2-8x+16$

79. $$\begin{aligned}(2s+1)^2=(2s+1)(2s+1)=2s(2s)+2s(1)+1(2s)+1(1)&=4s^2+2s+2s+1\\&=4s^2+4s+1\end{aligned}$$

81. $$\begin{aligned}(x-2y)^2=(x-2y)(x-2y)&=x(x)-x(2y)+(-2y)(x)-(-2y)(2y)\\&=x^2-2xy-2xy+4y^2=x^2-4xy+4y^2\end{aligned}$$

83. $(r+4)(r-4)=r(r)-r(4)+4(r)-4(4)=r^2-4r+4r-16=r^2-16$

85. $(4x+5)(4x-5)=4x(4x)-4x(5)+5(4x)-5(5)=16x^2-20x+20x-25=16x^2-25$

SECTION 4.6

87. $(2x+1)(x^2+3x-1) = (2x+1)(x^2) + (2x+1)(3x) - (2x+1)(1)$
$= 2x(x^2) + 1(x^2) + 2x(3x) + 1(3x) - 2x(1) - 1(1)$
$= 2x^3 + x^2 + 6x^2 + 3x - 2x - 1 = 2x^3 + 7x^2 + x - 1$

89. $(4t+3)(t^2+2t+3) = 4t(t^2) + 4t(2t) + 4t(3) + 3(t^2) + 3(2t) + 3(3)$
$= 4t^3 + 8t^2 + 12t + 3t^2 + 6t + 9 = 4t^3 + 11t^2 + 18t + 9$

91.
$$\begin{array}{r} 4x + 3 \\ \underline{x + 2} \\ 4x^2 + 3x \\ \underline{8x + 6} \\ 4x^2 + 11x + 6 \end{array}$$

93.
$$\begin{array}{r} 4x - 2y \\ \underline{3x + 5y} \\ 12x^2 - 6xy \\ \underline{20xy - 10y^2} \\ 12x^2 + 14xy - 10y^2 \end{array}$$

95.
$$\begin{aligned} (s-4)(s+1) &= s^2+5 \\ s^2+s-4s-4 &= s^2+5 \\ s^2-3s-4 &= s^2+5 \\ -3s-4 &= 5 \\ -3s &= 9 \\ s &= -3 \end{aligned}$$

97.
$$\begin{aligned} z(z+2) &= (z+4)(z-4) \\ z^2+2z &= z^2-4z+4z-16 \\ z^2+2z &= z^2-16 \\ 2z &= -16 \\ z &= -8 \end{aligned}$$

99.
$$\begin{aligned} (x+4)(x-4) &= (x-2)(x+6) \\ x^2-4x+4x-16 &= x^2+6x-2x-12 \\ x^2-16 &= x^2+4x-12 \\ -16 &= 4x-12 \\ -4 &= 4x \\ -1 &= x \end{aligned}$$

101.
$$\begin{aligned} (a-3)^2 &= (a+3)^2 \\ (a-3)(a-3) &= (a+3)(a+3) \\ a^2-3a-3a+9 &= a^2+3a+3a+9 \\ a^2-6a+9 &= a^2+6a+9 \\ -6a+9 &= 6a+9 \\ -6a &= 6a \\ 12a &= 0 \\ a &= 0 \end{aligned}$$

103. $(x^2y^3)^5 = (x^2)^5(y^3)^5 = x^{10}y^{15}$

105. $(x^5y^2)^3 = (x^5)^3(y^2)^3 = x^{15}y^6$

107. $(x^2y^5)(x^2z^5)(-3y^2z^3) = (1)(1)(-3)x^2x^2y^5y^2z^5x^3 = -3x^4y^7z^8$

109. $(x+3)(2x-3) = x(2x) - x(3) + 3(2x) - 3(3) = 2x^2 - 3x + 6x - 9 = 2x^2 + 3x - 9$

111. $(t-3)(t-3) = t(t) - t(3) + (-3)(t) - (-3)(3) = t^2 - 3t - 3t + 9 = t^2 - 6t + 9$

113. $(3x-5)(2x+1) = 3x(2x) + 3x(1) + (-5)(2x) + (-5)(1) = 6x^2 + 3x - 10x - 5$
$= 6x^2 - 7x - 5$

115. $(-2r-3s)(2r+7s) = (-2r)(2r) + (-2r)(7s) + (-3s)(2r) + (-3s)(7s)$
$= -4r^2 - 14rs - 6rs - 21s^2 = -4r^2 - 20rs - 21s^2$

SECTION 4.6

117. $(2a-3b)^2 = (2a-3b)(2a-3b) = 2a(2a) - 2a(3b) + (-3b)(2a) - (-3b)(3b)$
$= 4a^2 - 6ab - 6ab + 9b^2 = 4a^2 - 12ab + 9b^2$

119. $(4x+5y)(4x-5y) = 4x(4x) - 4x(5y) + 5y(4x) - 5y(5y)$
$= 16x^2 - 20xy + 20xy - 25y^2 = 16x^2 - 25y^2$

121.
$$\begin{array}{r} x^2 + x + 1 \\ x - 1 \\ \hline x^3 + x^2 + x \phantom{{}-1} \\ -x^2 - x - 1 \\ \hline x^3 \phantom{{}+x^2+x} - 1 \end{array}$$

123. $(-3x+y)(x^2 - 8xy + 16y^2)$
$= -3x(x^2) - (-3x)(8xy) + (-3x)(16y^2) + y(x^2) - y(8xy) + y(16y^2)$
$= -3x^3 + 24x^2y - 48xy^2 + x^2y - 8xy^2 + 16y^3 = -3x^3 + 25x^2y - 56xy^2 + 16y^3$

125. $(x-2y)(x^2 + 2xy + 4y^2)$
$= x(x^2) + x(2xy) + x(4y^2) + (-2y)(x^2) + (-2y)(2xy) + (-2y)(4y^2)$
$= x^3 + 2x^2y + 4xy^2 - 2x^2y - 4xy^2 - 8y^3 = x^3 - 8y^3$

127. $3xy(x+y) - 2x(xy-x) = [3xy(x) + 3xy(y)] - [(2x)(xy) - (2x)(x)]$
$= [3x^2y + 3xy^2] - [2x^2y - 2x^2]$
$= 3x^2y + 3xy^2 - 2x^2y + 2x^2 = x^2y + 3xy^2 + 2x^2$

129. $(x+y)(x-y) + x(x+y) = [x(x) - x(y) + y(x) - y(y)] + [x(x) + x(y)]$
$= [x^2 - xy + xy - y^2] + [x^2 + xy]$
$= [x^2 - y^2] + [x^2 + xy] = x^2 - y^2 + x^2 + xy = 2x^2 + xy - y^2$

131.
$$\begin{aligned} 7s^2 + (s-3)(2s+1) &= (3s-1)^2 \\ 7s^2 + (s-3)(2s+1) &= (3s-1)(3s-1) \\ 7s^2 + 2s^2 + s - 6s - 3 &= 9s^2 - 3s - 3s + 1 \\ 9s^2 - 5s - 3 &= 9s^2 - 6s + 1 \\ -5s - 3 &= -6s + 1 \\ -5s &= -6s + 4 \\ s &= 4 \end{aligned}$$

133. $(x-3)^2 - (x+3)^2 = [(x-3)(x-3)] - [(x+3)(x+3)]$
$= [x(x) - x(3) + (-3)(x) - (-3)(3)] - [x(x) + x(3) + 3(x) + 3(3)]$
$= [x^2 - 3x - 3x + 9] - [x^2 + 3x + 3x + 9]$
$= [x^2 - 6x + 9] - [x^2 + 6x + 9] = x^2 - 6x + 9 - x^2 - 6x - 9 = -12x$

135. $(3x+4)(2x-2)-(2x+1)(x+3)$
$= [3x(2x)-3x(2)+4(2x)-4(2)]-[2x(x)+2x(3)+1(x)+1(3)]$
$= \left[6x^2-6x+8x-8\right]-\left[2x^2+6x+x+3\right]$
$= \left[6x^2+2x-8\right]-\left[2x^2+7x+3\right] = 6x^2+2x-8-2x^2-7x-3 = 4x^2-5x-11$

137. $(b+2)(b-2)+2b(b+1) = [b(b)-b(2)+2(b)-2(2)]+[2b(b)+2b(1)]$
$= \left[b^2-2b+2b-4\right]+\left[2b^2+2b\right]$
$= \left[b^2-4\right]+\left[2b^2+2b\right] = b^2-4+2b^2+2b = 3b^2+2b-4$

139. Let $r =$ the smaller radius.
Then $r+3 =$ the larger radius.

$$\boxed{\text{Larger area}} - \boxed{\text{Smaller area}} = 15\pi$$
$$\pi(r+3)^2-\pi r^2 = 15\pi$$
$$\pi(r+3)(r+3)-\pi r^2 = 15\pi$$
$$\pi\left(r^2+3r+3r+9\right)-\pi r^2 = 15\pi$$
$$\pi\left(r^2+6r+9\right)-\pi r^2 = 15\pi$$
$$\pi r^2+6\pi r+9\pi-\pi r^2 = 15\pi$$
$$6\pi r+9\pi = 15\pi$$
$$6\pi r = 6\pi$$
$$\frac{6\pi r}{6\pi} = \frac{6\pi}{6\pi}$$
$$r = 1$$

The larger radius $= 1+3 = 4$ m.

141. Let $s =$ the side of the softball field. Then
$s+30 =$ the side of the baseball field.

$$\boxed{\text{Larger area}} - \boxed{\text{Smaller area}} = 4500$$
$$(s+30)^2-s^2 = 4500$$
$$(s+30)(s+30)-s^2 = 4500$$
$$s^2+30s+30s+900-s^2 = 4500$$
$$60s+900 = 4500$$
$$60s = 3600$$
$$s = 60$$

The baseball field has a side of
$60+30 = 90$ feet.

143. **Answers may vary.**

145. **Answers may vary.**

Exercise 4.7 (page 301)

1. $\dfrac{4x^3y}{2xy} = \dfrac{4}{2}x^{3-1}y^{1-1} = 2x^2y^0 = 2x^2$

3. $\dfrac{35ab^2c^3}{7abc} = \dfrac{35}{7}a^{1-1}b^{2-1}c^{3-1} = 5a^0b^1c^2 = 5bc^2$

5. $\dfrac{(x+y)+(x-y)}{2x} = \dfrac{x+y+x-y}{2x} = \dfrac{2x}{2x} = 1$

7. binomial

9. none of these

11. 2

13. $\dfrac{5}{15} = \dfrac{5(1)}{5(3)} = \dfrac{1}{3}$

15. $\dfrac{-125}{75} = -\dfrac{25(5)}{25(3)} = -\dfrac{5}{3}$

17. $\dfrac{120}{160} = \dfrac{40(3)}{40(4)} = \dfrac{3}{4}$

19. $\dfrac{-3612}{-3612} = 1$

21. $\dfrac{-90}{360} = -\dfrac{90(1)}{90(4)} = -\dfrac{1}{4}$

23. $\dfrac{5880}{2660} = \dfrac{140(42)}{140(19)} = \dfrac{42}{19}$

25. polynomial

26. monomial

27. two

29. $\dfrac{a}{b}$

31. $\dfrac{xy}{yz} = \dfrac{x}{z}$

33. $\dfrac{r^3s^2}{rs^3} = r^{3-1}s^{2-3} = r^2s^{-1} = \dfrac{r^2}{s}$

35. $\dfrac{8x^3y^2}{4xy^3} = \dfrac{8}{4}x^{3-1}y^{2-3} = 2x^2y^{-1} = \dfrac{2x^2}{y}$

37. $\dfrac{12u^5v}{-4u^2v^3} = \dfrac{12}{-4}u^{5-2}v^{1-3} = -3u^3v^{-2}$
$= -\dfrac{3u^3}{v^2}$

39. $\dfrac{6x+9y}{3xy} = \dfrac{6x}{3xy} + \dfrac{9y}{3xy} = \dfrac{2}{y} + \dfrac{3}{x}$

41. $\dfrac{xy+6}{3y} = \dfrac{xy}{3y} + \dfrac{6}{3y} = \dfrac{x}{3} + \dfrac{2}{y}$

43. $\dfrac{5x-10y}{25xy} = \dfrac{5x}{25xy} - \dfrac{10y}{25xy} = \dfrac{1}{5y} - \dfrac{2}{5x}$

45. $\dfrac{3x^2+6y^3}{3x^2y^2} = \dfrac{3x^2}{3x^2y^2} + \dfrac{6y^3}{3x^2y^2} = \dfrac{1}{y^2} + \dfrac{2y}{x^2}$

47. $\dfrac{4x-2y+8z}{4xy} = \dfrac{4x}{4xy} - \dfrac{2y}{4xy} + \dfrac{8z}{4xy}$
$= \dfrac{1}{y} - \dfrac{1}{2x} + \dfrac{2z}{xy}$

49. $\dfrac{12x^3y^2-8x^2y-4x}{4xy} = \dfrac{12x^3y^2}{4xy} - \dfrac{8x^2y}{4xy} - \dfrac{4x}{4xy} = 3x^2y - 2x - \dfrac{1}{y}$

51. $\dfrac{-25x^2y+30xy^2-5xy}{-5xy} = \dfrac{-25x^2y}{-5xy} + \dfrac{30xy^2}{-5xy} - \dfrac{5xy}{-5xy} = 5x - 6y + 1$

53. $\dfrac{15a^3b^2-10a^2b^3}{5a^2b^2} = \dfrac{15a^3b^2}{5a^2b^2} - \dfrac{10a^2b^3}{5a^2b^2} = 3a - 2b$

55. $\dfrac{5x(4x-2y)}{2y} = \dfrac{20x^2-10xy}{2y} = \dfrac{20x^2}{2y} - \dfrac{10xy}{2y} = \dfrac{10x^2}{y} - 5x$

57. $\dfrac{(-2x)^3+(3x^2)^2}{6x^2} = \dfrac{-8x^3+9x^4}{6x^2} = \dfrac{-8x^3}{6x^2} + \dfrac{9x^4}{6x^2} = -\dfrac{4x}{3} + \dfrac{3x^2}{2}$

59. $\dfrac{4x^2y^2-2(x^2y^2+xy)}{2xy} = \dfrac{4x^2y^2-2x^2y^2-2xy}{2xy} = \dfrac{2x^2y^2-2xy}{2xy} = \dfrac{2x^2y^2}{2xy} - \dfrac{2xy}{2xy} = xy - 1$

61. $\dfrac{(a+b)^2-(a-b)^2}{2ab} = \dfrac{(a+b)(a+b)-(a-b)(a-b)}{2ab}$
$= \dfrac{[a^2+ab+ab+b^2]-[a^2-ab-ab+b^2]}{2ab}$
$= \dfrac{[a^2+2ab+b^2]-[a^2-2ab+b^2]}{2ab}$
$= \dfrac{a^2+2ab+b^2-a^2+2ab-b^2}{2ab} = \dfrac{4ab}{2ab} = 2$

SECTION 4.7

63. $\dfrac{-16r^3y^2}{-4r^2y^4} = \dfrac{-16}{-4}r^{3-2}y^{2-4} = 4ry^{-2}$
$= \dfrac{4r}{y^2}$

65. $\dfrac{-65rs^2t}{15r^2s^3t} = \dfrac{-65}{15}r^{1-2}s^{2-3}t^{1-1}$
$= -\dfrac{13}{3}r^{-1}s^{-1}t^0 = -\dfrac{13}{3rs}$

67. $\dfrac{x^2x^3}{xy^6} = \dfrac{x^5}{xy^6} = \dfrac{x^{5-1}}{y^6} = \dfrac{x^4}{y^6}$

69. $\dfrac{(a^3b^4)^3}{ab^4} = \dfrac{a^9b^{12}}{ab^4} = a^{9-1}b^{12-4} = a^8b^8$

71. $\dfrac{15(r^2s^3)^2}{-5(rs^5)^3} = \dfrac{15r^4s^6}{-5r^3s^{15}} = \dfrac{15}{-5}r^{4-3}s^{6-15} = -3r^1s^{-9} = -\dfrac{3r}{s^9}$

73. $\dfrac{-32(x^3y)^3}{128(x^2y^2)^3} = \dfrac{-32x^9y^3}{128x^6y^6} = \dfrac{-32}{128}x^{9-6}y^{3-6} = -\dfrac{1}{4}x^3y^{-3} = -\dfrac{x^3}{4y^3}$

75. $\dfrac{(5a^2b)^3}{(2a^2b^2)^3} = \dfrac{125a^6b^3}{8a^6b^6} = \dfrac{125}{8}a^{6-6}b^{3-6} = \dfrac{125}{8}a^0b^{-3} = \dfrac{125}{8b^3}$

77. $\dfrac{-(3x^3y^4)^3}{-(9x^4y^5)^2} = \dfrac{-27x^9y^{12}}{-81x^8y^{10}} = \dfrac{-27}{-81}x^{9-8}y^{12-10} = \dfrac{1}{3}xy^2 = \dfrac{xy^2}{3}$

79. $\dfrac{(a^2a^3)^4}{(a^4)^3} = \dfrac{(a^5)^4}{a^{12}} = \dfrac{a^{20}}{a^{12}} = a^{20-12} = a^8$

81. $\dfrac{(z^3z^{-4})^3}{(z^{-3})^2} = \dfrac{(z^{-1})^3}{z^{-6}} = \dfrac{z^{-3}}{z^{-6}} = z^{-3-(-6)} = z^3$

83. $\dfrac{(a^2b)^3(ab^2)^2}{(3a^3b^2)^4} = \dfrac{(a^6b^3)(a^2b^4)}{81a^{12}b^8} = \dfrac{a^8b^7}{81a^{12}b^8} = \dfrac{1}{81}a^{8-12}b^{7-8} = \dfrac{1}{81}a^{-4}b^{-1} = \dfrac{1}{81a^4b}$

85. $\dfrac{(3x-y)(2x-3y)}{6xy} = \dfrac{6x^2-9xy-2xy+3y^2}{6xy} = \dfrac{6x^2-11xy+3y^2}{6xy} = \dfrac{6x^2}{6xy} - \dfrac{11xy}{6xy} + \dfrac{3y^2}{6xy}$
$= \dfrac{x}{y} - \dfrac{11}{6} + \dfrac{y}{2x}$

87. $l = \dfrac{P-2w}{2}$
$l = \dfrac{P}{2} - \dfrac{2w}{2}$
$l = \dfrac{P}{2} - w$
They are the same.

89. $C = \dfrac{0.15x+12}{x}$
$C = \dfrac{0.15x}{x} + \dfrac{12}{x}$
$C = 0.15 + \dfrac{12}{x}$
They are the same.

91. **Answers may vary.**

93. $\dfrac{x^{500}-x^{499}}{x^{499}} = \dfrac{x^{500}}{x^{499}} - \dfrac{x^{499}}{x^{499}} = x-1$
Let $x = 501$: $x - 1 = 501 - 1 = 500$

Exercise 4.8 (page 309)

1.
$$\begin{array}{rr} & 2+\frac{3}{x} \\ x & \overline{)\,2x+3} \\ & \underline{2x\quad\;\;} \\ & 3 \end{array}$$

3.
$$\begin{array}{rr} & 2+\frac{1}{x+1} \\ x+1 & \overline{)\,2x+3} \\ & \underline{2x+2} \\ & 1 \end{array}$$

5.
$$\begin{array}{rr} & x \\ x+1 & \overline{)\,x^2+x} \\ & \underline{x^2+x} \\ & 0 \end{array}$$

7. $21, 22, 24, 25, 26, 27, 28$

9. $|a-b| = |-2-3| = |-5| = 5$

11. $-|a^2-b^2| = -\left|(-2)^2-3^2\right| = -|4-9| = -|-5| = -(+5) = -5$

13. $3(2x^2-4x+5)+2(x^2+3x-7) = 6x^2-12x+15+2x^2+6x-14 = 8x^2-6x+1$

15. divisor; dividend

17. remainder

19. $4x^3-2x^2+7x+6$

21. $6x^4-x^3+2x^2+9x$

23. $0x^3$ and $0x$

25.
$$\begin{array}{rr} & x+2 \\ x+2 & \overline{)\,x^2+4x+4} \\ & \underline{x^2+2x\qquad} \\ & 2x+4 \\ & \underline{2x+4} \\ & 0 \end{array}$$

27.
$$\begin{array}{rr} & x+2 \\ x+5 & \overline{)\,x^2+7x+10} \\ & \underline{x^2+5x\qquad} \\ & 2x+10 \\ & \underline{2x+10} \\ & 0 \end{array}$$

29.
$$\begin{array}{rr} & x-3 \\ x-2 & \overline{)\,x^2-5x+6} \\ & \underline{x^2-2x\qquad} \\ & -3x+6 \\ & \underline{-3x+6} \\ & 0 \end{array}$$

31.
$$\begin{array}{rr} & a+5 \\ a-4 & \overline{)\,a^2+a-20} \\ & \underline{a^2-4a\qquad} \\ & 5a-20 \\ & \underline{5a-20} \\ & 0 \end{array}$$

33.
$$\begin{array}{rr} & 3a-2 \\ 2a+3 & \overline{)\,6a^2+5a-6} \\ & \underline{6a^2+9a\qquad} \\ & -4a-6 \\ & \underline{-4a-6} \\ & 0 \end{array}$$

35.
$$\begin{array}{rr} & b+3 \\ 3b+2 & \overline{)\,3b^2+11b+6} \\ & \underline{3b^2+2b\qquad} \\ & 9b+6 \\ & \underline{9b+6} \\ & 0 \end{array}$$

37.
$$\begin{array}{rr} & x+1+\frac{-1}{2x+3} \\ 2x+3 & \overline{)\,2x^2+5x+2} \\ & \underline{2x^2+3x\qquad} \\ & 2x+2 \\ & \underline{2x+3} \\ & -1 \end{array}$$

39.
$$\begin{array}{rr} & 2x+2+\frac{-3}{2x+1} \\ 2x+1 & \overline{)\,4x^2+6x-1} \\ & \underline{4x^2+2x\qquad} \\ & 4x-1 \\ & \underline{4x+2} \\ & -3 \end{array}$$

41.
$$\begin{array}{rr} & a+b \\ a+b & \overline{)\,a^2+2ab+b^2} \\ & \underline{a^2+ab\qquad} \\ & ab+b^2 \\ & \underline{ab+b^2} \\ & 0 \end{array}$$

43.
$$\begin{array}{rr} & 2x-y \\ x+2y & \overline{)\,2x^2+3xy-2y^2} \\ & \underline{2x^2+4xy\qquad} \\ & -xy-2y^2 \\ & \underline{-xy-2y^2} \\ & 0 \end{array}$$

45.
$$\begin{array}{rr} & x-3y \\ 2x-y & \overline{)\,2x^2-7xy+3y^2} \\ & \underline{2x^2-xy\qquad} \\ & -6xy+3y^2 \\ & \underline{-6xy+3y^2} \\ & 0 \end{array}$$

47.
$$\begin{array}{rr} & a+2b \\ a+b & \overline{)\,a^2+3ab+2b^2} \\ & \underline{a^2+ab\qquad} \\ & 2ab+2b^2 \\ & \underline{2ab+2b^2} \\ & 0 \end{array}$$

SECTION 4.8

49.
$$\begin{array}{r} 2x+\ \ 1 \\ 5x+3\,\overline{)\,10x^2+11x+3} \\ \underline{10x^2+\ \ 6x\qquad} \\ 5x+3 \\ \underline{5x+3} \\ 0 \end{array}$$

51.
$$\begin{array}{r} x-\ \ 7 \\ 2x+4\,\overline{)\,2x^2-10x-28} \\ \underline{2x^2+\ \ 4x\qquad} \\ -14x-28 \\ \underline{-14x-28} \\ 0 \end{array}$$

53.
$$\begin{array}{r} x+\ 1 \\ x-1\,\overline{)\,x^2+0x-1} \\ \underline{x^2-\ \ x\qquad} \\ x-1 \\ \underline{x-1} \\ 0 \end{array}$$

55.
$$\begin{array}{r} 2x-\ \ 3 \\ 2x+3\,\overline{)\,4x^2+0x-9} \\ \underline{4x^2+6x\qquad} \\ -6x-9 \\ \underline{-6x-9} \\ 0 \end{array}$$

57.
$$\begin{array}{r} x-\ \ \ y \\ x+y\,\overline{)\,x^2+0xy-y^2} \\ \underline{x^2+\ \ xy\qquad} \\ -\ \ xy-y^2 \\ \underline{-\ \ xy-y^2} \\ 0 \end{array}$$

59.
$$\begin{array}{r} x^2+\ \ 2x+\ \ 4 \\ x-2\,\overline{)\,x^3+0x^2+0x-8} \\ \underline{x^3-2x^2\qquad\qquad} \\ 2x^2+0x\qquad \\ \underline{2x^2-4x\qquad} \\ 4x-8 \\ \underline{4x-8} \\ 0 \end{array}$$

61.
$$\begin{array}{r} x^2+\ \ \ xy+\ \ \ \ y^2 \\ x-y\,\overline{)\,x^3+0x^2y+0xy^2-y^3} \\ \underline{x^3-\ \ x^2y\qquad\qquad\qquad} \\ x^2y+0xy^2\qquad \\ \underline{x^2y-\ \ xy^2\qquad} \\ xy^2-y^3 \\ \underline{xy^2-y^3} \\ 0 \end{array}$$

63.
$$\begin{array}{r} a^2-\ \ 3a+\ \ 10+\ \frac{-30}{a+3} \\ a+3\,\overline{)\,a^3+0a^2+\ \ \ \ \ a+\ \ \ \ \ 0\quad} \\ \underline{a^3+3a^2\qquad\qquad\qquad\quad} \\ -3a^2+\ \ \ \ a\qquad\quad \\ \underline{-3a^2-\ \ 9a\qquad\quad} \\ 10a+\ \ \ \ 0\quad \\ \underline{10a+\ \ 30\quad} \\ -30\quad \end{array}$$

65.
$$\begin{array}{r} 3x+\ \ 2y \\ 2x-y\,\overline{)\,6x^2+\ \ xy-2y^2} \\ \underline{6x^2-3xy\qquad} \\ 4xy-2y^2 \\ \underline{4xy-2y^2} \\ 0 \end{array}$$

67.
$$\begin{array}{r} x+\ \ \ \ 5y \\ 3x-2y\,\overline{)\,3x^2+13xy-10y^2} \\ \underline{3x^2-\ \ 2xy\qquad} \\ 15xy-10y^2 \\ \underline{15xy-10y^2} \\ 0 \end{array}$$

69.
$$\begin{array}{r} x-\ \ \ \ 5y \\ 4x+y\,\overline{)\,4x^2-19xy-5y^2} \\ \underline{4x^2+\ \ \ \ xy\qquad} \\ -20xy-5y^2 \\ \underline{-20xy-5y^2} \\ 0 \end{array}$$

71.
$$\begin{array}{r} x^2+\ \ 2x-\ \ 1 \\ 2x+3\,\overline{)\,2x^3+7x^2+4x-3} \\ \underline{2x^3+3x^2\qquad\qquad} \\ 4x^2+4x\qquad \\ \underline{4x^2+6x\qquad} \\ -2x-3 \\ \underline{-2x-3} \\ 0 \end{array}$$

73.
$$\begin{array}{r} 2x^2+\ \ \ 2x+\ \ 1 \\ 3x+2\,\overline{)\,6x^3+10x^2+7x+2} \\ \underline{6x^3+\ \ 4x^2\qquad\qquad} \\ 6x^2+7x\qquad \\ \underline{6x^2+4x\qquad} \\ 3x+2 \\ \underline{3x+2} \\ 0 \end{array}$$

75.
$$\begin{array}{r} x^2+\ \ \ \ xy+\ \ \ \ y^2 \\ 2x+y\,\overline{)\,2x^3+3x^2y+3xy^2+y^3} \\ \underline{2x^3+\ \ x^2y\qquad\qquad\qquad} \\ 2x^2y+3xy^2\qquad \\ \underline{2x^2y+\ \ xy^2\qquad} \\ 2xy^2+y^3 \\ \underline{2xy^2+y^3} \\ 0 \end{array}$$

77.
$$\begin{array}{r|l} & x^2+2x+1 \\ \hline x+1 & x^3+3x^2+3x+1 \\ & \underline{x^3+x^2} \\ & 2x^2+3x \\ & \underline{2x^2+2x} \\ & x+1 \\ & \underline{x+1} \\ & 0 \end{array}$$

79.
$$\begin{array}{r|l} & x^2+2x-1+\frac{6}{2x+3} \\ \hline 2x+3 & 2x^3+7x^2+4x+3 \\ & \underline{2x^3+3x^2} \\ & 4x^2+4x \\ & \underline{4x^2+6x} \\ & -2x+3 \\ & \underline{-2x-3} \\ & 6 \end{array}$$

81.
$$\begin{array}{r|l} & 2x^2+8x+14+\frac{31}{x-2} \\ \hline x-2 & 2x^3+4x^2-2x+3 \\ & \underline{2x^3-4x^2} \\ & 8x^2-2x \\ & \underline{8x^2-16x} \\ & 14x+3 \\ & \underline{14x-28} \\ & 31 \end{array}$$

83.
$$\begin{array}{r|l} & 3y^2+6y-9+\frac{7}{2y+3} \\ \hline 2y+3 & 6y^3+21y^2+0y-20 \\ & \underline{6y^3+9y^2} \\ & 12y^2+0y \\ & \underline{12y^2+18y} \\ & -18y-20 \\ & \underline{-18y-27} \\ & 7 \end{array}$$

85. **Answers may vary.**

87. x^2-2x is added to the dividend when it should be subtracted.

Chapter 4 Review (page 312)

1. $(-3x)^4=(-3x)(-3x)(-3x)(-3x)$

2. $\left(\frac{1}{2}pq\right)^3=\left(\frac{1}{2}pq\right)\left(\frac{1}{2}pq\right)\left(\frac{1}{2}pq\right)$

3. $5^3=5\cdot5\cdot5=125$

4. $3^5=3\cdot3\cdot3\cdot3\cdot3=243$

5. $(-8)^2=(-8)(-8)=64$

6. $-8^2=-1\cdot8\cdot8=-64$

7. $3^2+2^2=9+4=13$

8. $(3+2)^2=(5)^2=25$

9. $x^3x^2=x^{3+2}=x^5$

10. $x^2x^7=x^{2+7}=x^9$

11. $(y^7)^3=y^{7\cdot3}=y^{21}$

12. $(x^{21})^2=x^{21\cdot2}=x^{42}$

13. $(ab)^3=a^3b^3$

14. $(3x)^4=3^4x^4=81x^4$

15. $b^3b^4b^5=b^{3+4+5}=b^{12}$

16. $-z^2(z^3y^2)=-z^{2+3}y^2=-y^2z^5$

17. $(16s)^2s=16^2s^2s=256s^3$

18. $-3y(y^5)=-3y^6$

19. $(x^2x^3)^3=(x^5)^3=x^{15}$

20. $(2x^2y)^2=2^2(x^2)^2y^2=4x^4y^2$

21. $\dfrac{x^7}{x^3}=x^{7-3}=x^4$

22. $\left(\dfrac{x^2y}{xy^2}\right)^2=\dfrac{x^4y^2}{x^2y^4}=x^2y^{-2}=\dfrac{x^2}{y^2}$

CHAPTER 4 REVIEW

23. $\dfrac{8(y^2x)^2}{4(yx^2)^2} = \dfrac{8y^4x^2}{4y^2x^4} = \dfrac{8}{4}y^2x^{-2} = \dfrac{2y^2}{x^2}$

24. $\dfrac{(5y^2z^3)^3}{25(yz)^5} = \dfrac{125y^6z^9}{25y^5z^5} = 5yz^4$

25. $x^0 = 1$

26. $(3x^2y^2)^0 = 1$

27. $(3x^0)^2 = (3 \cdot 1)^2 = 3^2 = 9$

28. $(3x^2y^0)^2 = (3x^2 \cdot 1)^2 = (3x^2)^2 = 9x^4$

29. $x^{-3} = \dfrac{1}{x^3}$

30. $x^{-2}x^3 = x^1 = x$

31. $y^4y^{-3} = y^1 = y$

32. $\dfrac{x^3}{x^{-7}} = x^{3-(-7)} = x^{10}$

33. $\left(x^{-3}x^4\right)^{-2} = \left(x^1\right)^{-2} = x^{-2} = \dfrac{1}{x^2}$

34. $\left(a^{-2}b\right)^{-3} = a^6b^{-3} = \dfrac{a^6}{b^3}$

35. $\left(\dfrac{x^2}{x}\right)^{-5} = (x)^{-5} = \dfrac{1}{x^5}$

36. $\left(\dfrac{15z^4}{5z^3}\right)^{-2} = (3z)^{-2} = \dfrac{1}{(3z)^2} = \dfrac{1}{9z^2}$

37. $728 = 7.28 \times 10^2$

38. $9{,}370 = 9.37 \times 10^3$

39. $0.0136 = 1.36 \times 10^{-2}$

40. $0.00942 = 9.42 \times 10^{-3}$

41. $7.73 = 7.73 \times 10^0$

42. $753 \times 10^3 = 7.53 \times 10^2 \times 10^3$
$= 7.53 \times 10^5$

43. $0.018 \times 10^{-2} = 1.8 \times 10^{-2} \times 10^{-2}$
$= 1.8 \times 10^{-4}$

44. $600 \times 10^2 = 6.00 \times 10^2 \times 10^2$
$= 6 \times 10^4$

45. $7.26 \times 10^5 = 726{,}000$

46. $3.91 \times 10^{-4} = 0.000391$

47. $2.68 \times 10^0 = 2.68$

48. $5.76 \times 10^1 = 57.6$

49. $739 \times 10^{-2} = 7.39$

50. $0.437 \times 10^{-3} = 0.000437$

51. $\dfrac{(0.00012)(0.00004)}{0.00000016} = \dfrac{(1.2 \times 10^{-4})(4 \times 10^{-5})}{1.6 \times 10^{-7}} = \dfrac{4.8 \times 10^{-9}}{1.6 \times 10^{-7}} = 3 \times 10^{-2} = 0.03$

52. $\dfrac{(4{,}800)(20{,}000)}{600{,}000} = \dfrac{(4.8 \times 10^3)(2 \times 10^4)}{6 \times 10^5} = \dfrac{9.6 \times 10^7}{6 \times 10^5} = 1.6 \times 10^2 = 160$

53. $13x^7$: monomial, degree = 7

54. $5^3x + x^2$: binomial, degree = 2

55. $-3x^5 + x - 1$: trinomial, degree = 5

56. $9xy + 21x^3y^2$: binomial,
degree = 3 + 2 = 5

57. $3x + 2 = 3(3) + 2 = 9 + 2 = 11$

58. $3x + 2 = 3(0) + 2 = 0 + 2 = 2$

59. $3x + 2 = 3(-2) + 2 = -6 + 2 = -4$

60. $3x + 2 = 3\left(\frac{2}{3}\right) + 2 = 2 + 2 = 4$

61. $5x^4 - x = 5(3)^4 - 3 = 5(81) - 3 = 405 - 3 = 402$

62. $5x^4 - x = 5(0)^4 - 0 = 5(0) - 0 = 0 - 0 = 0$

63. $5x^4 - x = 5(-2)^4 - (-2) = 5(16) + 2 = 80 + 2 = 82$

64. $5x^4 - x = 5(-0.3)^4 - (-0.3) = 5(0.0081) + 0.3 = 0.0405 + 3 = 0.3405$

65. $f(0) = 0^2 - 4 = 0 - 4 = -4$

66. $f(5) = 5^2 - 4 = 25 - 4 = 21$

67. $f(-2) = (-2)^2 - 4 = 4 - 4 = 0$

68. $f\left(\frac{1}{2}\right) = \left(\frac{1}{2}\right)^2 - 4 = \frac{1}{4} - 4 = -\frac{15}{4}$

69. $f(x) = x^2 - 5$

x	-2	-1	0	1	2
$f(x)$	-1	-4	-5	-4	-1

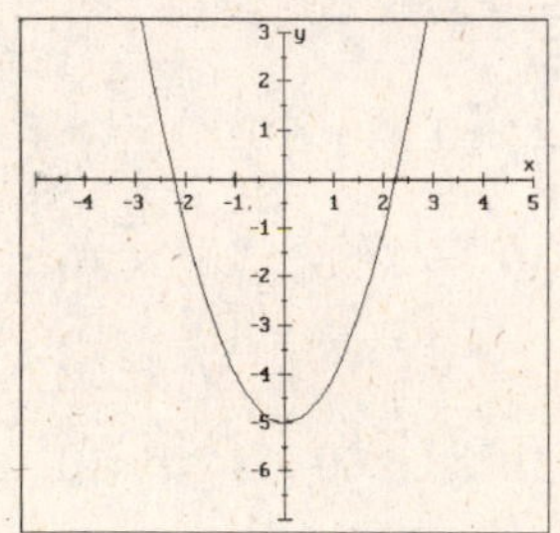

70. $f(x) = x^3 - 2$

x	-2	-1	0	1	2
$f(x)$	-10	-3	-2	-1	6

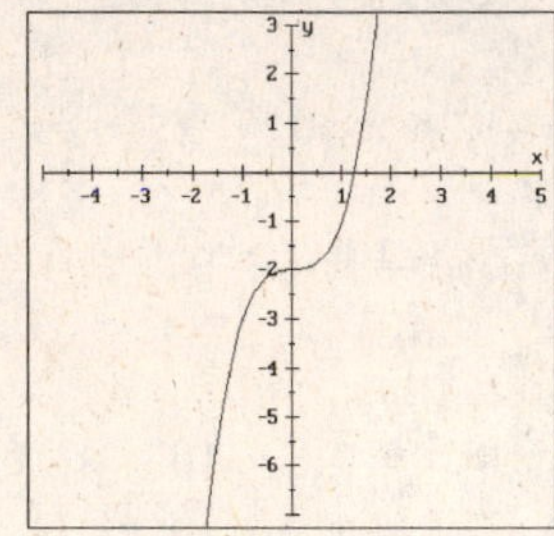

71. $3x + 5x - x = 7x$

72. $3x + 2y$: not like terms, so not possible

73. $(xy)^2 + 3x^2y^2 = x^2y^2 + 3x^2y^2 = 4x^2y^2$

74. $-2x^2yz + 3yx^2z = -2x^2yz + 3x^2yz = x^2yz$

75. $(3x^2 + 2x) + (5x^2 - 8x) = 3x^2 + 2x + 5x^2 - 8x = 8x^2 - 6x$

76. $(7a^2 + 2a - 5) - (3a^2 - 2a + 1) = 7a^2 + 2a - 5 - 3a^2 + 2a - 1 = 4a^2 + 4a - 6$

77. $3(9x^2 + 3x + 7) - 2(11x^2 - 5x + 9) = 27x^2 + 9x + 21 - 22x^2 + 10x - 18 = 5x^2 + 19x + 3$

78. $4(4x^3 + 2x^2 - 3x - 8) - 5(2x^3 - 3x + 8) = 16x^3 + 8x^2 - 12x - 32 - 10x^3 + 15x - 40 = 6x^3 + 8x^2 + 3x - 72$

79. $(2x^2y^3)(5xy^2) = 10x^3y^5$

80. $(xyz^3)(x^3z)^2 = xyz^3x^6z^2 = x^7yz^5$

81. $5(x + 3) = 5x + 15$

82. $3(2x + 4) = 6x + 12$

CHAPTER 4 REVIEW

83. $x^2(3x^2-5)=3x^4-5x^2$

84. $2y^2(y^2+5y)=2y^4+10y^3$

85. $-x^2y(y^2-xy)=-x^2y^3+x^3y^2$

86. $-3xy(xy-x)=-3x^2y^2+3x^2y$

87. $\begin{aligned}(x+3)(x+2)&=x^2+2x+3x+6\\&=x^2+5x+6\end{aligned}$

88. $\begin{aligned}(2x+1)(x-1)&=2x^2-2x+x-1\\&=2x^2-x-1\end{aligned}$

89. $\begin{aligned}(3a-3)(2a+2)&=6a^2+6a-6a-6\\&=6a^2-6\end{aligned}$

90. $\begin{aligned}6(a-1)(a+1)&=6\left(a^2+a-a-1\right)\\&=6\left(a^2-1\right)=6a^2-6\end{aligned}$

91. $\begin{aligned}(a-b)(2a+b)&=2a^2+ab-2ab-b^2\\&=2a^2-ab-b^2\end{aligned}$

92. $\begin{aligned}(3x-y)(2x+y)&=6x^2+3xy-2xy-y^2\\&=6x^2+xy-y^2\end{aligned}$

93. $\begin{aligned}(x+3)(x+3)&=x^2+3x+3x+9\\&=x^2+6x+9\end{aligned}$

94. $\begin{aligned}(x+5)(x-5)&=x^2-5x+5x-25\\&=x^2-25\end{aligned}$

95. $\begin{aligned}(y-2)(y+2)&=y^2+2y-2y-4\\&=y^2-4\end{aligned}$

96. $\begin{aligned}(x+4)^2&=(x+4)(x+4)\\&=x^2+4x+4x+16\\&=x^2+8x+16\end{aligned}$

97. $(x-3)^2=(x-3)(x-3)=x^2-3x-3x+9=x^2-6x+9$

98. $(y-1)^2=(y-1)(y-1)=y^2-y-y+1=y^2-2y+1$

99. $(2y+1)^2=(2y+1)(2y+1)=4y^2+2y+2y+1=4y^2+4y+1$

100. $(y^2+1)(y^2-1)=y^4-y^2+y^2-1=y^4-1$

101. $(3x+1)\left(x^2+2x+1\right)=3x^3+6x^2+3x+x^2+2x+1=3x^3+7x^2+5x+1$

102. $(2a-3)\left(4a^2+6a+9\right)=8a^3+12a^2+18a-12a^2-18a-27=8a^3-27$

103. $\begin{aligned}x^2+3&=x(x+3)\\x^2+3&=x^2+3x\\3&=3x\\1&=x\end{aligned}$

104. $\begin{aligned}x^2+x&=(x+1)(x+2)\\x^2+x&=x^2+2x+x+2\\x^2+x&=x^2+3x+2\\x&=3x+2\\-2x&=2\\x&=-1\end{aligned}$

105.
$$\begin{aligned}(x+2)(x-5) &= (x-4)(x-1)\\ x^2-5x+2x-10 &= x^2-x-4x+4\\ x^2-3x-10 &= x^2-5x+4\\ -3x-10 &= -5x+4\\ -3x &= -5x+14\\ 2x &= 14\\ x &= 7\end{aligned}$$

106.
$$\begin{aligned}(x-1)(x-2) &= (x-3)(x+1)\\ x^2-2x-x+2 &= x^2+x-3x-3\\ x^2-3x+2 &= x^2-2x-3\\ -3x+2 &= -2x-3\\ -3x &= -2x-5\\ -x &= -5\\ x &= 5\end{aligned}$$

107.
$$\begin{aligned}x^2+x(x+2) &= x(2x+1)+1\\ x^2+x^2+2x &= 2x^2+x+1\\ 2x^2+2x &= 2x^2+x+1\\ 2x &= x+1\\ x &= 1\end{aligned}$$

108.
$$\begin{aligned}(x+5)(3x+1) &= x^2+(2x-1)(x-5)\\ 3x^2+16x+5 &= 3x^2-11x+5\\ 16x+5 &= -11x+5\\ 16x &= -11x\\ 27x &= 0\\ x &= 0\end{aligned}$$

109. $$\frac{3x+6y}{2xy}=\frac{3x}{2xy}+\frac{6y}{2xy}=\frac{3}{2y}+\frac{3}{x}$$

110. $$\frac{14xy-21x}{7xy}=\frac{14xy}{7xy}-\frac{21x}{7xy}=2-\frac{3}{y}$$

111. $$\frac{15a^2bc+20ab^2c-25abc^2}{-5abc}=\frac{15a^2bc}{-5abc}+\frac{20ab^2c}{-5abc}-\frac{25abc^2}{-5abc}=-3a-4b+5c$$

112.
$$\begin{aligned}\frac{(x+y)^2+(x-y)^2}{-2xy} &= \frac{(x+y)(x+y)+(x-y)(x-y)}{-2xy}\\ &= \frac{x^2+xy+xy+y^2+x^2-xy-xy+y^2}{-2xy}\\ &= \frac{2x^2+2y^2}{-2xy}=\frac{2x^2}{-2xy}+\frac{2y^2}{-2xy}=-\frac{x}{y}-\frac{y}{x}\end{aligned}$$

113.
$$\begin{array}{r|l} & x + 1 + \frac{3}{x+2} \\ \hline x+2 & x^2+3x+5 \\ & \underline{x^2+2x} \\ & x+5 \\ & \underline{x+2} \\ & 3\end{array}$$

114.
$$\begin{array}{r|l} & x - 5 \\ \hline x-1 & x^2-6x+5 \\ & \underline{x^2-x} \\ & -5x+5 \\ & \underline{-5x+5} \\ & 0\end{array}$$

115.
$$\begin{array}{r|l} & 2x + 1 \\ \hline x+3 & 2x^2+7x+3 \\ & \underline{2x^2+6x} \\ & x+3 \\ & \underline{x+3} \\ & 0\end{array}$$

116.
$$\begin{array}{r|l} & x + 5 + \frac{3}{3x-1} \\ \hline 3x-1 & 3x^2+14x-2 \\ & \underline{3x^2-x} \\ & 15x-2 \\ & \underline{15x-5} \\ & 3\end{array}$$

117.

$$\begin{array}{r} 3x^2 + 2x + 1 + \frac{2}{2x-1} \\ 2x-1 \overline{)\,6x^3 + x^2 + 0x + 1} \\ \underline{6x^3 - 3x^2} \\ 4x^2 + 0x \\ \underline{4x^2 - 2x} \\ 2x + 1 \\ \underline{2x - 1} \\ 2 \end{array}$$

118.

$$\begin{array}{r} 3x^2 - x - 4 \\ 3x+1 \overline{)\,9x^3 + 0x^2 - 13x - 4} \\ \underline{9x^3 + 3x^2} \\ -3x^2 - 13x \\ \underline{-3x^2 - x} \\ -12x - 4 \\ \underline{-12x - 4} \\ 0 \end{array}$$

Chapter 4 Test (page 316)

1. $2xxxyyyy = 2x^3y^4$

2. $3^2 + 5^3 = 9 + 125 = 134$

3. $y^2(yy^3) = y^2y^4 = y^6$

4. $(-3b^2)(2b^3)(-b^2) = (-3)(2)(-1)b^2b^3b^2 = 6b^7$

5. $(2x^3)^5(x^2)^3 = 32x^{15}x^6 = 32x^{21}$

6. $(2rr^2r^3)^3 = (2r^6)^3 = 8r^{18}$

7. $3x^0 = 3(1) = 3$

8. $2y^{-5}y^2 = 2y^{-3} = \dfrac{2}{y^3}$

9. $\dfrac{y^2}{yy^{-2}} = \dfrac{y^2}{y^{-1}} = y^3$

10. $\left(\dfrac{a^2b^{-1}}{4a^3b^{-2}}\right)^{-3} = \left(\dfrac{1}{4}a^{-1}b^1\right)^{-3} = \left(\dfrac{1}{4}\right)^{-3}a^3b^{-3} = \dfrac{64a^3}{b^3}$

11. $28{,}000 = 2.8 \times 10^4$

12. $0.0025 = 2.5 \times 10^{-3}$

13. $7.4 \times 10^3 = 7{,}400$

14. $9.3 \times 10^{-5} = 0.000093$

15. $3x^2 + 2$: binomial

16. $2 + 3 + 5 = 10$th degree

17. $x^2 + x - 2 = (-2)^2 + (-2) - 2$
$= 4 - 2 - 2 = 0$

18. $f(x) = x^2 + 2$

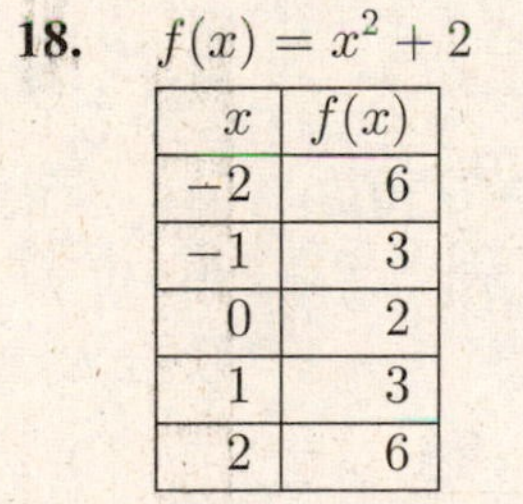

x	$f(x)$
-2	6
-1	3
0	2
1	3
2	6

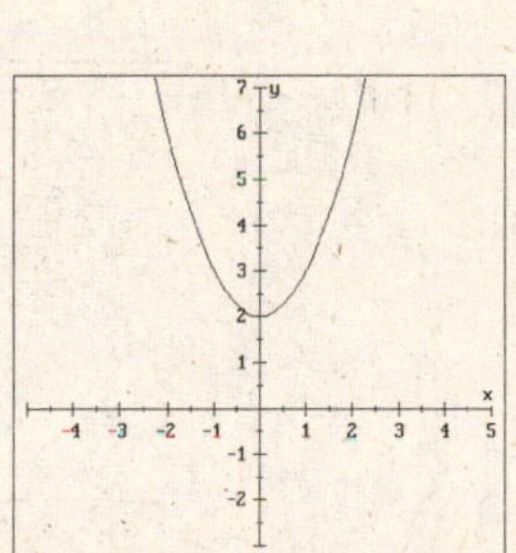

19. $-6(x - y) + 2(x + y) - 3(x + 2y) = -6x + 6y + 2x + 2y - 3x - 6y = -7x + 2y$

20. $-2(x^2 + 3x - 1) - 3(x^2 - x + 2) + 5(x^2 + 2) = -2x^2 - 6x + 2 - 3x^2 + 3x - 6 + 5x^2 + 10$
$= -3x + 6$

CHAPTER 4 TEST

21.
$$\begin{array}{r} 3x^3 + 4x^2 - x - 7 \\ + \quad \underline{2x^3 - 2x^2 + 3x + 2} \\ 5x^3 + 2x^2 + 2x - 5 \end{array}$$

22.
$$\begin{array}{r} 2x^2 - 7x + 3 \\ - \quad \underline{3x^2 - 2x - 1} \\ \end{array} \Rightarrow \begin{array}{r} 2x^2 - 7x + 3 \\ + \quad \underline{-3x^2 + 2x + 1} \\ -x^2 - 5x + 4 \end{array}$$

23. $(-2x^3)(2x^2y) = -4x^5y$

24. $3y^2(y^2 - 2y + 3) = 3y^4 - 6y^3 + 9y^2$

25. $(2x - 5)(3x + 4) = 6x^2 + 8x - 15x - 20 = 6x^2 - 7x - 20$

26. $(2x - 3)(x^2 - 2x + 4) = 2x^3 - 4x^2 + 8x - 3x^2 + 6x - 12 = 2x^3 - 7x^2 + 14x - 12$

27. $\dfrac{8x^2y^3z^4}{16x^3y^2z^4} = \dfrac{1}{2}x^{-1}y^1z^0 = \dfrac{y}{2x}$

28.
$$\begin{aligned} \frac{6a^2 - 12b^2}{24ab} &= \frac{6a^2}{24ab} - \frac{12b^2}{24ab} \\ &= \frac{a}{4b} - \frac{b}{2a} \end{aligned}$$

29.
$$\begin{array}{r} x - 2 \\ 2x + 3\overline{\big)\, 2x^2 - x - 6} \\ \underline{2x^2 + 3x} \\ -4x - 6 \\ \underline{-4x - 6} \\ 0 \end{array}$$

30.
$$\begin{aligned} (a + 2)^2 &= (a - 3)^2 \\ (a + 2)(a + 2) &= (a - 3)(a - 3) \\ a^2 + 4a + 4 &= a^2 - 6a + 9 \\ 4a + 4 &= -6a + 9 \\ 10a &= 5 \\ a &= \frac{5}{10} = \frac{1}{2} \end{aligned}$$

Cumulative Review Exercises (page 317)

1. $5 + 3 \cdot 2 = 5 + 6 = 11$

2. $3 \cdot 5^2 - 4 = 3 \cdot 25 - 4 = 75 - 4 = 71$

3. $\dfrac{3x - y}{xy} = \dfrac{3(2) - (-5)}{2(-5)} = \dfrac{6 + 5}{-10} = -\dfrac{11}{10}$

4. $\dfrac{x^2 - y^2}{x + y} = \dfrac{2^2 - (-5)^2}{2 + (-5)} = \dfrac{4 - 25}{-3} = \dfrac{-21}{-3} = 7$

5.
$$\begin{aligned} \frac{4}{5}x + 6 &= 18 \\ \frac{4}{5}x &= 12 \\ 5 \cdot \frac{4}{5}x &= 5(12) \\ 4x &= 60 \\ x &= 15 \end{aligned}$$

6.
$$\begin{aligned} x - 2 &= \frac{x + 2}{3} \\ 3(x - 2) &= 3 \cdot \frac{x + 2}{3} \\ 3x - 6 &= x + 2 \\ 2x - 6 &= 2 \\ 2x &= 8 \\ x &= 4 \end{aligned}$$

7.
$$\begin{aligned} 2(5x + 2) &= 3(3x - 2) \\ 10x + 4 &= 9x - 6 \\ x + 4 &= -6 \\ x &= -10 \end{aligned}$$

8.
$$\begin{aligned} 4(y + 1) &= -2(4 - y) \\ 4y + 4 &= -8 + 2y \\ 2y + 4 &= -8 \\ 2y &= -12 \\ y &= -6 \end{aligned}$$

9.
$$\begin{aligned} 5x - 3 &> 7 \\ 5x &> 10 \\ x &> 2 \end{aligned}$$

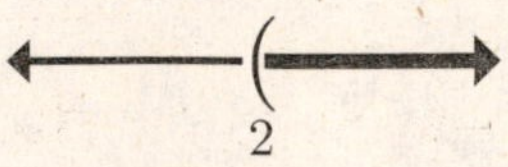

10.
$$\begin{aligned} 7x - 9 &< 5 \\ 7x &< 14 \\ x &< 2 \end{aligned}$$

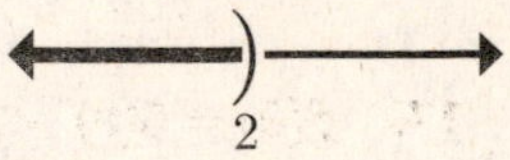

CUMULATIVE REVIEW EXERCISES

11.
$$\begin{aligned} -2 < -x+3 &< 5 \\ -5 < \quad -x \quad &< 2 \\ \frac{-5}{-1} > \frac{-x}{-1} &> \frac{2}{-1} \\ 5 > \quad x \quad &> -2 \\ -2 < \quad x \quad &< 5 \end{aligned}$$

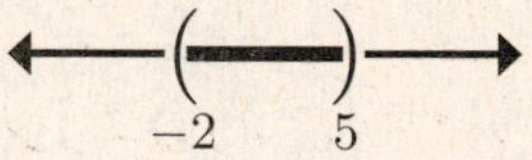

12.
$$\begin{aligned} 0 \le \frac{4-x}{3} &\le 2 \\ 3(0) \le 3\cdot\frac{4-x}{3} &\le 3(2) \\ 0 \le 4-x &\le 6 \\ -4 \le -x &\le 2 \\ \frac{-4}{-1} \ge \frac{-x}{-1} &\ge \frac{2}{-1} \\ 4 \ge x &\ge -2 \\ -2 \le x &\le 4 \end{aligned}$$

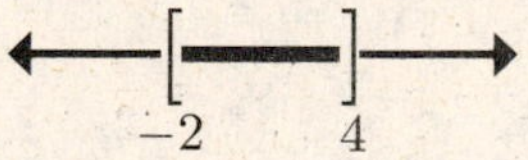

13.
$$\begin{aligned} A &= p+prt \\ A-p &= p-p+prt \\ A-p &= prt \\ \frac{A-p}{pt} &= \frac{prt}{pt} \\ \frac{A-p}{pt} &= r, \text{ or } r = \frac{A-p}{pt} \end{aligned}$$

14.
$$\begin{aligned} A &= \frac{1}{2}bh \\ 2A &= 2\cdot\frac{1}{2}bh \\ 2A &= bh \\ \frac{2A}{b} &= \frac{bh}{b} \\ \frac{2A}{b} &= h, \text{ or } h = \frac{2A}{b} \end{aligned}$$

15. $3x - 4y = 12$

$x = 0$	$y = 0$
$3(0) - 4y = 12$	$3x - 4(0) = 12$
$0 - 4y = 12$	$3x - 0 = 12$
$-4y = 12$	$3x = 12$
$y = -3$	$x = 4$

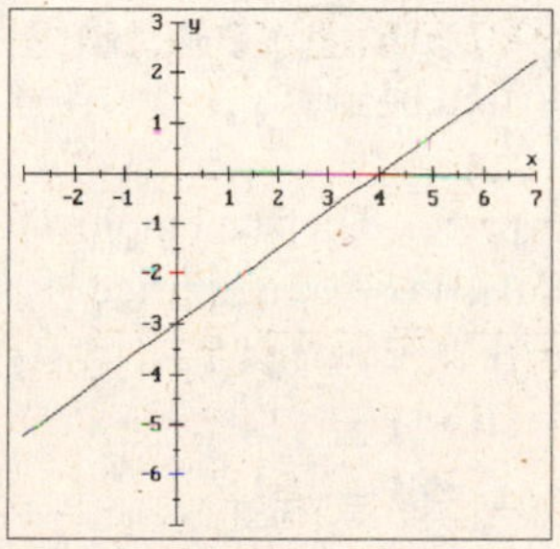

16. $y - 2 = \frac{1}{2}(x - 4)$

$x = 0$	$x = 4$
$y - 2 = \frac{1}{2}(0-4)$	$y - 2 = \frac{1}{2}(4-4)$
$y - 2 = \frac{1}{2}(-4)$	$y - 2 = \frac{1}{2}(0)$
$y - 2 = -2$	$y - 2 = 0$
$y = 0$	$y = 2$

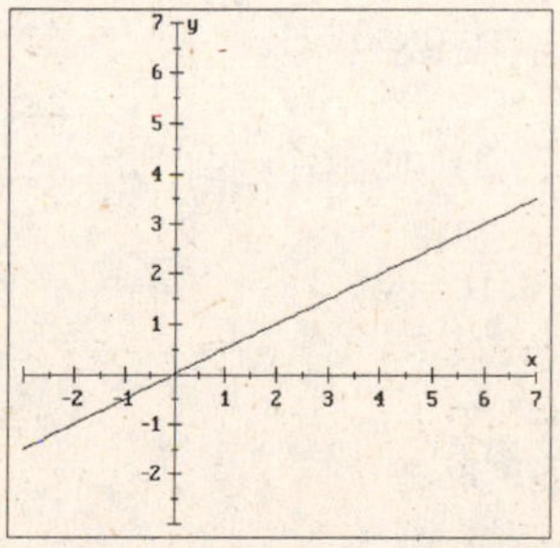

17. $f(0) = 5(0) - 2 = 0 - 2 = -2$

18. $f(3) = 5(3) - 2 = 15 - 2 = 13$

19. $f(-2) = 5(-2) - 2 = -10 - 2 = -12$

20. $f\left(\frac{1}{5}\right) = 5\left(\frac{1}{5}\right) - 2 = 1 - 2 = -1$

21. $(y^3y^5)y^6 = y^8y^6 = y^{14}$

22. $\frac{x^3y^4}{x^2y^3} = x^{3-2}y^{4-3} = x^1y^1 = xy$

23. $\frac{a^4b^{-3}}{a^{-3}b^3} = a^{4-(-3)}b^{-3-3} = a^7b^{-6} = \frac{a^7}{b^6}$

24. $\left(\frac{-x^{-2}y^3}{x^{-3}y^2}\right)^2 = \left(-x^{-2-(-3)}y^{3-2}\right)^2$
$= \left(-x^1y^1\right)^2 = x^2y^2$

25. $(3x^2 + 2x - 7) - (2x^2 - 2x + 7) = 3x^2 + 2x - 7 - 2x^2 + 2x - 7 = x^2 + 4x - 14$

26. $(3x - 7)(2x + 8) = 6x^2 + 24x - 14x - 56 = 6x^2 + 10x - 56$

27. $(x - 2)(x^2 + 2x + 4) = x^3 + 2x^2 + 4x - 2x^2 - 4x - 8 = x^3 - 8$

28.
$$\begin{array}{r|l} & 2x + 1 \\ \hline x - 3 & 2x^2 - 5x - 3 \\ & \underline{2x^2 - 6x} \\ & \quad\quad x - 3 \\ & \quad\quad \underline{x - 3} \\ & \quad\quad\quad 0 \end{array}$$

29. $(1.6 \times 10^2)(3 \times 10^{16}) = 4.8 \times 10^{18}$ m

30. $A = 2lw + 2wd + 2ld$
$202 = 2(9)(5) + 2(5)d + 2(9)d$
$202 = 90 + 28d$
$112 = 28d \Rightarrow d = 4$ inches

31. $A = \pi(R + r)(R - r)$
$= \pi(17 + 3)(17 - 3)$
$= \pi(20)(14) = 280\pi \approx 879.6$ in.^{2}

32. Let $r =$ the regular price. Then an employee can purchase the TV for $0.75r$.
Purchase price + Sales tax = Total price
$0.75r + 0.08(0.75r) = 414.72$
$0.81r = 414.72$
$\frac{0.81r}{0.81} = \frac{414.72}{0.81}$
$r = 512 \Rightarrow$ The regular price is \$512.

Exercise 5.1 (page 327)

1. $36 = 4 \cdot 9 = (2 \cdot 2) \cdot (3 \cdot 3) = 2^2 \cdot 3^2$

3. $81 = 9 \cdot 9 = (3 \cdot 3) \cdot (3 \cdot 3) = 3^4$

5. $3 = 3; 6 = 2 \cdot 3; 9 = 3 \cdot 3$
GCF = 3

7. $a(x + 3) = a \cdot (x + 3)$
$3(x + 3) = 3 \cdot (x + 3)$; GCF $= x + 3$

9. $15xy + 10 = 5y(3x + 2)$

11. $a(x + 3) - 3(x + 3) = (x + 3)(a - 3)$

SECTION 5.1

13. $3x - 2(x+1) = 5$
$3x - 2x - 2 = 5$
$x = 7$

15. $\dfrac{2x-7}{5} = 3$
$5 \cdot \dfrac{2x-7}{5} = 5(3)$
$2x - 7 = 15$
$2x = 22$
$x = 11$

17. prime-factored

19. largest (or greatest)

21. grouping

23. $12 = 2 \cdot 6 = 2 \cdot (2 \cdot 3) = 2^2 \cdot 3$

25. $15 = 3 \cdot 5$

27. $40 = 4 \cdot 10 = (2 \cdot 2) \cdot (2 \cdot 5) = 2^3 \cdot 5$

29. $98 = 2 \cdot 49 = 2 \cdot (7 \cdot 7) = 2 \cdot 7^2$

31. $225 = 15 \cdot 15 = (3 \cdot 5) \cdot (3 \cdot 5) = 3^2 \cdot 5^2$

33. $288 = 24 \cdot 12 = (4 \cdot 6) \cdot (4 \cdot 3)$
$= (2^2 \cdot 2 \cdot 3) \cdot (2^2 \cdot 3)$
$= 2^5 \cdot 3^2$

35. $5xy^2 = 5 \cdot x \cdot y \cdot y$; $10xy = 2 \cdot 5 \cdot x \cdot y$
GCF $= 5xy$

37. $6x^2y^2 = 2 \cdot 3 \cdot x \cdot x \cdot y \cdot y$; $12xyz = 2 \cdot 2 \cdot 3 \cdot x \cdot y \cdot z$; $18xy^2z^3 = 2 \cdot 3 \cdot 3 \cdot x \cdot y \cdot y \cdot z \cdot z \cdot z$
GCF $= 2 \cdot 3 \cdot xy = 6xy$

39. $4a + 12 = \boxed{4}(a+3)$

41. $r^4 + r^2 = r^2\left(\boxed{r^2} + 1\right)$

43. $3x + 6 = 3(x+2)$

45. $4x - 8 = 4(x-2)$

47. $xy - xz = x(y - z)$

49. $t^3 + 2t^2 = t^2(t+2)$

51. $a^3b^3z^3 - a^2b^3z^2 = a^2b^3z^2(az - 1)$

53. $24x^2y^3z^4 + 8xy^2z^3 = 8xy^2z^3(3xyz + 1)$

55. $3x + 3y - 6z = 3(x + y - 2z)$

57. $ab + ac - ad = a(b + c - d)$

59. $4y^2 + 8y - 2xy = 2y(2y + 4 - x)$

61. $12r^2 - 3rs + 9r^2s^2 = 3r(4r - s + 3rs^2)$

63. $abx - ab^2x + abx^2 = abx(1 - b + x)$

65. $4x^2y^2z^2 - 6xy^2z^2 + 12xyz^2 = 2xyz^2(2xy - 3y + 6)$

67. $-x - 2 = -(x+2)$

69. $-a - b = -(a+b)$

71. $-2x + 5y = -(2x - 5y)$

73. $-2a + 3b = -(2a - 3b)$

75. $-3xy + 2z + 5w = -(3xy - 2z - 5w)$

77. $-3ab - 5ac + 9bc = -(3ab + 5ac - 9bc)$

79. $-3x^2y - 6xy^2 = -3xy(x + 2y)$

81. $-4a^2b^3 + 12a^3b^2 = -4a^2b^2(b - 3a)$

SECTION 5.1

83. $-8a^5b^2 - 8a^3b^4 = -8a^3b^2(a^2 + b^2)$

85. $-4a^2b^2c^2 + 14a^2b^2c - 10ab^2c^2 = -2ab^2c(2ac - 7a + 5c)$

87. $a(x+y) + b(x+y) = (x+y)\boxed{(a+b)}$ **89.** $p(m-n) - q(m-n) = \boxed{(m-n)}(p-q)$

91. $(x+y)2 + (x+y)b = (x+y)(2+b)$

93. $(x-3)^2 + (x-3) = (x-3)(x-3) + 1(x-3) = (x-3)(x-3+1) = (x-3)(x-2)$

95. $x(y+1) - 5(y+1) = (y+1)(x-5)$

97. $(3t+5)^2 - (3t+5) = (3t+5)(3t+5) - 1(3t+5) = (3t+5)(3t+5-1) = (3t+5)(3t+4)$

99. $2x + 2y + ax + ay = 2(x+y) + a(x+y) = (x+y)(2+a)$

101. $9p - 9q + mp - mq = 9(p-q) + m(p-q) = (p-q)(9+m)$

103. $ax + bx - a - b = x(a+b) - 1(a+b) = (a+b)(x-1)$

105. $x(a-b) + y(b-a) = x(a-b) - y(a-b) = (a-b)(x-y)$

107. $4y^2 + 8y - 2xy = 2y(2y + \boxed{4} - \boxed{x})$ **109.** $r^4 + r^2 = r^2(r^2+1)$

111. $12uvw^3 - 18uv^2w^2 = 6uvw^2(2w - 3v)$

113. $70a^3b^2c^2 + 49a^2b^3c^3 - 21a^2b^2c^2 = 7a^2b^2c^2(10a + 7bc - 3)$

115. $-3m - 4n + 1 = -(3m + 4n - 1)$

117. $-14a^6b^6 + 49a^2b^3 - 21ab = -7ab(2a^5b^5 - 7ab^2 + 3)$

119. $-5a^2b^3c + 15a^3b^4c^2 - 25a^4b^3c = -5a^2b^3c(1 - 3abc + 5a^2)$

121. $3(r-2s) - x(r-2s) = (r-2s)(3-x)$

123. $3x(a+b+c) - 2y(a+b+c) = (a+b+c)(3x-2y)$

125. $14x^2y(r+2s-t) - 21xy(r+2s-t) = (r+2s-t)(14x^2y - 21xy) = 7xy(r+2s-t)(2x-3)$

127. $(x+3)(x+1) - y(x+1) = (x+1)(x+3-y)$

129. $(3x-y)(x^2-2) + (x^2-2) = (3x-y)(x^2-2) + 1(x^2-2) = (x^2-2)(3x-y+1)$

131. $3x(c-3d) + 6y(c-3d) = (c-3d)(3x+6y) = 3(c-3d)(x+2y)$

133. $xr + xs + yr + ys = x(r+s) + y(r+s) = (r+s)(x+y)$

135. $2ax + 2bx + 3a + 3b = 2x(a+b) + 3(a+b) = (a+b)(2x+3)$

137. $2ab + 2ac + 3b + 3c = 2a(b+c) + 3(b+c) = (b+c)(2a+3)$

139. $3tv - 9tw + uv - 3uw = 3t(v-3w) + u(v-3w) = (v-3w)(3t+u)$

141. $9mp + 3mq - 3np - nq = 3m(3p+q) - n(3p+q) = (3p+q)(3m-n)$

143. $ax^3 + bx^3 + 2ax^2y + 2bx^2y = x^2[ax + bx + 2ay + 2by] = x^2[x(a+b) + 2y(a+b)]$
$= x^2(a+b)(x+2y)$

145. $4a^2b + 12a^2 - 8ab - 24a = 4a[ab + 3a - 2b - 6] = 4a[a(b+3) - 2(b+3)]$
$= 4a(b+3)(a-2)$

147. $2x^2 + 2xy - 3x - 3y = 2x(x+y) - 3(x+y) = (x+y)(2x-3)$

149. $x^3 + 2x^2 + x + 2 = x^2(x+2) + 1(x+2) = (x+2)(x^2+1)$

151. $x^3y - x^2y - xy^2 + y^2 = y[x^3 - x^2 - xy + y] = y[x^2(x-1) - y(x-1)]$
$= y(x-1)(x^2-y)$

153. $2r - bs - 2s + br = 2r - 2s + br - bs = 2(r-s) + b(r-s) = (r-s)(2+b)$

155. $ax + by + bx + ay = ax + ay + bx + by = a(x+y) + b(x+y) = (x+y)(a+b)$

157. $ac + bd - ad - bc = ac - bc - ad + bd = c(a-b) - d(a-b) = (a-b)(c-d)$

159. $ar^2 - brs + ars - br^2 = r[ar - bs + as - br] = r[ar - br + as - bs] = r[r(a-b) + s(a-b)]$
$= r(a-b)(r+s)$

161. $ba + 3 + a + 3b = ba + a + 3b + 3 = a(b+1) + 3(b+1) = (b+1)(a+3)$

163. $pr + qs - ps - qr = pr - qr - ps + qs = r(p-q) - s(p-q) = (p-q)(r-s)$

165. Answers may vary.

167. Answers may vary.

169. $ax + ay + bx + by = a(x+y) + b(x+y) = (x+y)(a+b)$
$ax + bx + ay + by = x(a+b) + y(a+b) = (a+b)(x+y)$

Exercise 5.2 (page 333)

1. $x^2 - 9 = x^2 - 3^2 = (x+3)(x-3)$

3. $z^2 - 4 = z^2 - 2^2 = (z+2)(z-2)$

SECTION 5.2

5. $25 - t^2 = 5^2 - t^2 = (5+t)(5-t)$

7. $100 - y^2 = 10^2 - y^2 = (10+y)(10-y)$

9.
$$\begin{aligned} \frac{p}{w} + \frac{v^2}{2g} + h &= k \\ \frac{p}{w} &= k - h - \frac{v^2}{2g} \\ w \cdot \frac{p}{w} &= w\left(k - h - \frac{v^2}{2g}\right) \\ p &= w\left(k - h - \frac{v^2}{2g}\right) \end{aligned}$$

11. difference of two squares

13. prime

15. $x^2 - 9 = (x+3)\boxed{(x-3)}$

17. $4m^2 - 9n^2 = (2m+3n)\boxed{(2m-3n)}$

19. $x^2 - 16 = x^2 - 4^2 = (x+4)(x-4)$

21. $y^2 - 49 = y^2 - 7^2 = (y+7)(y-7)$

23. $4y^2 - 49 = (2y)^2 - 7^2 = (2y+7)(2y-7)$

25. $9x^2 - y^2 = (3x)^2 - y^2 = (3x+y)(3x-y)$

27.
$$\begin{aligned} 25t^2 - 36u^2 &= (5t)^2 - (6u)^2 \\ &= (5t+6u)(5t-6u) \end{aligned}$$

29.
$$\begin{aligned} 16a^2 - 25b^2 &= (4a)^2 - (5b)^2 \\ &= (4a+5b)(4a-5b) \end{aligned}$$

31. $8x^2 - 32y^2 = 8[x^2 - 4y^2] = 8\left[x^2 - (2y)^2\right] = 8(x+2y)(x-2y)$

33. $2a^2 - 8y^2 = 2[a^2 - 4y^2] = 2\left[a^2 - (2y)^2\right] = 2(a+2y)(a-2y)$

35. $3r^2 - 12s^2 = 3[r^2 - 4s^2] = 3\left[r^2 - (2s)^2\right] = 3(r+2s)(r-2s)$

37. $x^3 - xy^2 = x(x^2 - y^2) = x(x+y)(x-y)$

39.
$$\begin{aligned} a^4 - 16 &= \left(a^2+4\right)\left(a^2-4\right) \\ &= \left(a^2+4\right)(a+2)(a-2) \end{aligned}$$

41.
$$\begin{aligned} a^4 - b^4 &= \left(a^2+b^2\right)\left(a^2-b^2\right) \\ &= \left(a^2+b^2\right)(a+b)(a-b) \end{aligned}$$

43. $2x^4 - 2y^4 = 2(x^4 - y^4) = 2(x^2+y^2)(x^2-y^2) = 2(x^2+y^2)(x+y)(x-y)$

45. $a^4b - b^5 = b(a^4 - b^4) = b(a^2+b^2)(a^2-b^2) = b(a^2+b^2)(a+b)(a-b)$

47. $2x^4y - 512y^5 = 2y(x^4 - 256y^4) = 2y(x^2+16y^2)(x^2-16y^2) = 2y(x^2+16y^2)(x+4y)(x-4y)$

49. $a^3 - 9a + 3a^2 - 27 = a(a^2-9) + 3(a^2-9) = (a^2-9)(a+3) = (a+3)(a-3)(a+3)$

51.
$$\begin{aligned} a^4 - 4b^2 &= \left(a^2\right)^2 - (2b)^2 \\ &= \left(a^2+2b\right)\left(a^2-2b\right) \end{aligned}$$

53. $a^2 + b^2$: sum of squares $\Rightarrow$ prime

55. $49y^2 - 225z^4 = (7y)^2 - (15z^2)^2 = (7y+15z^2)(7y-15z^2)$

57. $196x^4 - 169y^2 = (14x^2)^2 - (13y)^2 = (14x^2+13y)(14x^2-13y)$

SECTION 5.2

59. $4a^2x - 9b^2x = x[4a^2 - 9b^2] = x\left[(2a)^2 - (3b)^2\right] = x(2a+3b)(2a-3b)$

61. $3m^3 - 3mn^2 = 3m(m^2 - n^2) = 3m(m+n)(m-n)$

63. $4x^4 - x^2y^2 = x^2(4x^2 - y^2) = x^2\left[(2x)^2 - y^2\right] = x^2(2x+y)(2x-y)$

65. $2a^3b - 242ab^3 = 2ab[a^2 - 121b^2] = 2ab\left[a^2 - (11b)^2\right] = 2ab(a+11b)(a-11b)$

67. $x^4 - 81 = (x^2+9)(x^2-9) = (x^2+9)(x+3)(x-3)$

69. $81r^4 - 256s^4 = (9r^2+16s^2)(9r^2-16s^2) = (9r^2+16s^2)(3r+4s)(3r-4s)$

71. $a^4 - b^8 = (a^2+b^4)(a^2-b^4) = (a^2+b^4)(a+b^2)(a-b^2)$

73. $x^8 - y^8 = (x^4+y^4)(x^4-y^4) = (x^4+y^4)(x^2+y^2)(x^2-y^2) = (x^4+y^4)(x^2+y^2)(x+y)(x-y)$

75. $$\begin{aligned} 48m^4n - 243n^5 = 3n\left(16m^4 - 81n^4\right) &= 3n\left(4m^2+9n^2\right)\left(4m^2-9n^2\right) \\ &= 3n\left(4m^2+9n^2\right)(2m+3n)(2m-3n) \end{aligned}$$

77. $$\begin{aligned} 2p^{10}q - 32p^2q^5 = 2p^2q\left(p^8 - 16q^4\right) &= 2p^2q\left(p^4+4q^2\right)\left(p^4-4q^2\right) \\ &= 2p^2q\left(p^4+4q^2\right)\left(p^2+2q\right)\left(p^2-2q\right) \end{aligned}$$

79. $2x^9y + 2xy^9 = 2xy(x^8+y^8)$

81. $$\begin{aligned} a^6b^2 - a^2b^6c^4 = a^2b^2\left(a^4 - b^4c^4\right) &= a^2b^2\left(a^2+b^2c^2\right)\left(a^2-b^2c^2\right) \\ &= a^2b^2\left(a^2+b^2c^2\right)(a+bc)(a-bc) \end{aligned}$$

83. $a^2b^7 - 625a^2b^3 = a^2b^3(b^4-625) = a^2b^3(b^2+25)(b^2-25) = a^2b^3(b^2+25)(b+5)(b-5)$

85. $$\begin{aligned} 243r^5s - 48rs^5 = 3rs\left(81r^4 - 16s^4\right) &= 3rs\left(9r^2+4s^2\right)\left(9r^2-4s^2\right) \\ &= 3rs\left(9r^2+4s^2\right)(3r+2s)(3r-2s) \end{aligned}$$

87. $16(x-y)^2 - 9 = [4(x-y)]^2 - 3^2 = [4(x-y)+3][4(x-y)-3] = (4x-4y+3)(4x-4y-3)$

89. $b^3 - 25b - 2b^2 + 50 = b(b^2-25) - 2(b^2-25) = (b^2-25)(b-2) = (b+5)(b-5)(b-2)$

91. $a^3 - 49a + 2a^2 - 98 = a(a^2-49) + 2(a^2-49) = (a^2-49)(a+2) = (a+7)(a-7)(a+2)$

93. $$\begin{aligned} 3m^3 - 3mn^2 + 3am^2 - 3an^2 = 3\left[m^3 - mn^2 + am^2 - an^2\right] &= 3\left[m\left(m^2-n^2\right) + a\left(m^2-n^2\right)\right] \\ &= 3\left(m^2-n^2\right)(m+a) \\ &= 3(m+n)(m-n)(m+a) \end{aligned}$$

95. $2m^3n^2 - 32mn^2 + 8m^2 - 128 = 2\left[m^3n^2 - 16mn^2 + 4m^2 - 64\right]$
$= 2\left[mn^2\left(m^2 - 16\right) + 4\left(m^2 - 16\right)\right]$
$= 2\left(m^2 - 16\right)\left(mn^2 + 4\right) = 2(m+4)(m-4)\left(mn^2 + 4\right)$

97. **Answers may vary.**

99. $399 \cdot 401 = (400-1)(400+1) = 400^2 - 1^2 = 160{,}000 - 1 = 159{,}999$

Exercise 5.3 (page 343)

1. $x^2 + 5x + 4 = (x+1)\left(x + \boxed{4}\right)$

3. $x^2 + x - 6 = \left(x \boxed{-} 2\right)\left(x + \boxed{3}\right)$

5. $x^2 + 5x - 6 = \left(x + \boxed{6}\right)\left(x - \boxed{1}\right)$

7. $x - 3 > 5$
$x > 8$

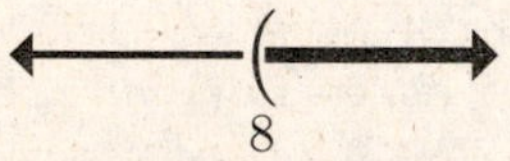

9. $-3x - 5 \geq 4$
$-3x \geq 9$
$\frac{-3x}{-3} \leq \frac{9}{-3}$
$x \leq -3$

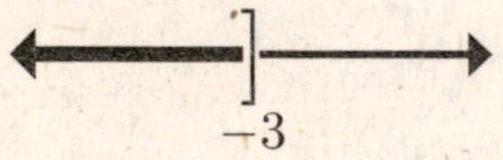

11. $\frac{3(x-1)}{4} < 12$
$4 \cdot \frac{3x-3}{4} < 4(12)$
$3x - 3 < 48$
$3x < 51$
$x < 17$

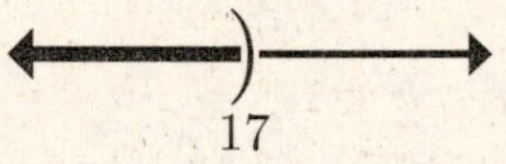

13. $-2 < x \leq 4$

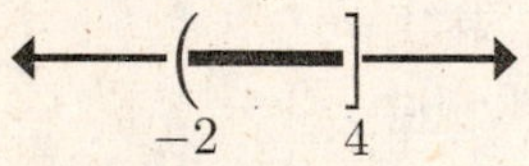

15. $x^2 + 2xy + y^2 = (x+y)^2$

17. $y^2 + 6y + 8 = \left(y + \boxed{4}\right)\left(y + \boxed{2}\right)$

19. $x^2 - xy - 2y^2 = \left(x + \boxed{y}\right)\left(x - \boxed{2y}\right)$

21. $x^2 + 3x + 2 = (x+2)(x+1)$

23. $z^2 + 12z + 11 = (z+1)(z+11)$

25. $t^2 - 9t + 14 = (t-2)(t-7)$

27. $p^2 - 6p + 5 = (p-5)(p-1)$

29. $a^2 + 6a - 16 = (a+8)(a-2)$

31. $s^2 + 11s - 26 = (s+13)(s-2)$

33. $c^2 + 4c - 5 = (c+5)(c-1)$

35. $t^2 - 5t - 50 = (t-10)(t+5)$

37. $a^2 - 4a - 5 = (a-5)(a+1)$

39. $y^2 - y - 30 = (y-6)(y+5)$

41. $m^2 + 3mn - 10n^2 = (m+5n)(m-2n)$

43. $a^2 - 4ab - 12b^2 = (a-6b)(a+2b)$

SECTION 5.3

45. $a^2 + 10ab + 9b^2 = (a + 9b)(a + b)$

47. $m^2 - 11mn + 10n^2 = (m - 10n)(m - n)$

49. $-x^2 - 7x - 10 = -(x^2 + 7x + 10)$
$= -(x + 5)(x + 2)$

51. $-y^2 - 2y + 15 = -(y^2 + 2y - 15)$
$= -(y + 5)(y - 3)$

53. $-t^2 - 15t + 34 = -(t^2 + 15t - 34)$
$= -(t + 17)(t - 2)$

55. $-r^2 + 14r - 40 = -(r^2 - 14r + 40)$
$= -(r - 4)(r - 10)$

57. $u^2 + 10u + 15$: prime

59. $r^2 - 9r - 12$: prime

61. $2x^2 + 10x + 12 = 2(x^2 + 5x + 6)$
$= 2(x + 2)(x + 3)$

63. $3y^3 - 21y^2 + 18y = 3y(y^2 - 7y + 6)$
$= 3y(y - 6)(y - 1)$

65. $3z^2 - 15tz + 12t^2 = 3(z^2 - 5tz + 4t^2) = 3(z - 4t)(z - t)$

67. $-4x^2y - 4x^3 + 24xy^2 = -4x(xy + x^2 - 6y^2) = -4x(x^2 + xy - 6y^2) = -4x(x + 3y)(x - 2y)$

69. $x^2 + 4x + 4 - y^2 = (x + 2)(x + 2) - y^2 = (x + 2)^2 - y^2 = (x + 2 + y)(x + 2 - y)$

71. $b^2 - 6b + 9 - c^2 = (b - 3)(b - 3) - c^2 = (b - 3)^2 - c^2 = (b - 3 + c)(b - 3 - c)$

73. $x^2 + 3x + 2$: $a = 1, b = 3, c = 2 \Rightarrow$ key # $= ac = 1(2) = 2$.
Find two factors of 2 whose sum is $b = 3$: 1 and 2.
Rewrite and factor: $x^2 + 3x + 2 = x^2 + 1x + 2x + 2$
$= x(x + 1) + 2(x + 1) = (x + 1)(x + 2)$

75. $t^2 - 9t + 14$: $a = 1, b = -9, c = 14 \Rightarrow$ key # $= ac = 1(14) = 14$.
Find two factors of 14 whose sum is $b = -9$: -7 and -2.
Rewrite and factor: $t^2 - 9t + 14 = t^2 - 7t - 2t + 14$
$= t(t - 7) - 2(t - 7) = (t - 7)(t - 2)$

77. $a^2 + 6a - 16$: $a = 1, b = 6, c = -16 \Rightarrow$ key # $= ac = 1(-16) = -16$.
Find two factors of -16 whose sum is $b = 6$: 8 and -2.
Rewrite and factor: $a^2 + 6a - 16 = a^2 + 8a - 2a - 16$
$= a(a + 8) - 2(a + 8) = (a + 8)(a - 2)$

79. $y^2 - y - 30$: $a = 1, b = -1, c = -30 \Rightarrow$ key # $= ac = 1(-30) = -30$.
Find two factors of -30 whose sum is $b = -1$: 5 and -6.
Rewrite and factor: $y^2 - y - 30 = y^2 + 5y - 6y - 30$
$= y(y + 5) - 6(y + 5) = (y + 5)(y - 6)$

81. $x^2 + 6x + 9 = (x + 3)(x + 3)$
$= (x + 3)^2$

83. $y^2 - 8y + 16 = (y - 4)(y - 4)$
$= (y - 4)^2$

85. $u^2 - 18u + 81 = (u-9)(u-9)$
$= (u-9)^2$

87. $x^2 + 4xy + 4y^2 = (x+2y)(x+2y)$
$= (x+2y)^2$

89. $4 - 5x + x^2 = x^2 - 5x + 4$
$= (x-4)(x-1)$

91. $10y + 9 + y^2 = y^2 + 10y + 9$
$= (y+9)(y+1)$

93. $-r^2 + 2s^2 + rs = -r^2 + rs + 2s^2$
$= -(r^2 - rs - 2s^2)$
$= -(r-2s)(r+s)$

95. $4rx + r^2 + 3x^2 = r^2 + 4rx + 3x^2$
$= (r+3x)(r+x)$

97. $-3ab + a^2 + 2b^2 = a^2 - 3ab + 2b^2$
$= (a-2b)(a-b)$

99. $-a^2 - 4ab - 3b^2 = -(a^2 + 4ab + 3b^2)$
$= -(a+3b)(a+b)$

101. $-x^2 + 6xy + 7y^2 = -(x^2 - 6xy - 7y^2)$
$= -(x-7y)(x+y)$

103. $3y^3 + 6y^2 + 3y = 3y(y^2 + 2y + 1)$
$= 3y(y+1)(y+1)$
$= 3y(y+1)^2$

105. $12xy + 4x^2y - 72y = 4y(3x + x^2 - 18) = 4y(x^2 + 3x - 18) = 4y(x+6)(x-3)$

107. $y^2 + 2yz + z^2 = (y+z)(y+z)$
$= (y+z)^2$

109. $t^2 + 20t + 100 = (t+10)(t+10)$
$= (t+10)^2$

111. $r^2 - 10rs + 25s^2 = (r-5s)(r-5s) = (r-5s)^2$

113. $a^2 + 2ab + b^2 - 4 = (a+b)(a+b) - 4 = (a+b)^2 - 2^2 = (a+b+2)(a+b-2)$

115. $b^2 - y^2 - 4y - 4 = b^2 - (y^2 + 4y + 4) = b^2 - (y+2)(y+2) = b^2 - (y+2)^2$
$= [b + (y+2)][b - (y+2)]$
$= (b+y+2)(b-y-2)$

117. **Answers may vary.**

119. Both answers check. Both may be factored more completely:
$(2x+6)(x+7) = 2(x+3)(x+7)$
$(x+3)(2x+14) = (x+3) \cdot 2(x+7)$

Exercise 5.4 (page 352)

1. $2x^2 + 5x + 3 = (\boxed{2}\,x + \boxed{3})(x+1)$

3. $6x^2 + 5x - 1 = (x \boxed{+} 1)(6x \boxed{-} 1)$

SECTION 5.4

5. $4x^2+4x-3=\left(2x+\boxed{3}\right)\left(2x-\boxed{1}\right)$

7.
$$\begin{aligned} l &= f+(n-1)d \\ l &= f+nd-d \\ l-f+d &= nd \\ \frac{l-f+d}{d} &= \frac{nd}{d} \\ \frac{l-f+d}{d} &= n \end{aligned}$$

9. descending

11. opposites

13. $6x^2+7x+2=(2x+1)\left(3x+\boxed{2}\right)$

15. $6x^2+x-2=\left(3x+\boxed{2}\right)\left(2x-\boxed{1}\right)$

17. $12x^2-7xy+y^2=\left(3x-\boxed{y}\right)\left(4x-\boxed{y}\right)$

19. $3a^2+10a+3=(3a+1)(a+3)$

21. $3a^2+13a+4=(3a+1)(a+4)$

23. $6b^2-5b+1=(2b-1)(3b-1)$

25. $2y^2-7y+3=(2y-1)(y-3)$

27. $5t^2+13t+6=(5t+3)(t+2)$

29. $16m^2-14m+3=(8m-3)(2m-1)$

31. $3a^2-4a-4=(3a+2)(a-2)$

33. $2x^2-3x-2=(2x+1)(x-2)$

35. $2m^2+5m-12=(2m-3)(m+4)$

37. $6y^2+y-2=(2y-1)(3y+2)$

39. $2x^2+3xy+y^2=(2x+y)(x+y)$

41. $3x^2-4xy+y^2=(3x-y)(x-y)$

43. $2u^2+uv-3v^2=(2u+3v)(u-v)$

45. $6p^2-pq-2q^2=(2p+q)(3p-2q)$

47.
$$\begin{aligned} -26x+6x^2-20 &= 6x^2-26x-20 \\ &= 2\left(3x^2-13x-10\right) \\ &= 2(3x+2)(x-5) \end{aligned}$$

49.
$$\begin{aligned} 15+8a^2-26a &= 8a^2-26a+15 \\ &= (4a-3)(2a-5) \end{aligned}$$

51. $12x^2+10y^2-23xy=12x^2-23xy+10y^2=(3x-2y)(4x-5y)$

53. $-21mn-10n^2+10m^2=10m^2-21mn-10n^2=(5m+2n)(2m-5n)$

55. $4z^2+13z+3$: $a=4, b=13, c=3 \Rightarrow$ key # $=ac=4(3)=12$.
Find two factors of 12 whose sum is $b=13$: 12 and 1.
Rewrite and factor: $4z^2+13z+3=4z^2+12z+1z+3$
$$=4z(z+3)+1(z+3)=(z+3)(4z+1)$$

57. $4x^2+8x+3$: $a=4, b=8, c=3 \Rightarrow$ key # $=ac=4(3)=12$.
Find two factors of 12 whose sum is $b=8$: 6 and 2.
Rewrite and factor: $4x^2+8x+3=4x^2+6x+2x+3$
$$=2x(2x+3)+1(2x+3)=(2x+3)(2x+1)$$

SECTION 5.4

59. $10u^2 - 13u - 3$: $a = 10, b = -13, c = -3 \Rightarrow$ key # $= ac = 10(-3) = -30$.
Find two factors of -30 whose sum is $b = -13$: -15 and 2.
Rewrite and factor: $10u^2 - 13u - 3 = 10u^2 - 15u + 2u - 3$
$= 5u(2u - 3) + 1(2u - 3) = (2u - 3)(5u + 1)$

61. $10y^2 - 3y - 1$: $a = 10, b = -3, c = -1 \Rightarrow$ key # $= ac = 10(-1) = -10$.
Find two factors of -10 whose sum is $b = -3$: -5 and 2.
Rewrite and factor: $10y^2 - 3y - 1 = 10y^2 - 5y + 2y - 1$
$= 5y(2y - 1) + 1(2y - 1) = (2y - 1)(5y + 1)$

63. $9x^2 - 12x + 4 = (3x - 2)(3x - 2) = (3x - 2)^2$

65. $25x^2 + 30x + 9 = (5x + 3)(5x + 3) = (5x + 3)^2$

67. $4x^2 + 12x + 9 = (2x + 3)(2x + 3) = (2x + 3)^2$

69. $9x^2 + 12x + 4 = (3x + 2)(3x + 2) = (3x + 2)^2$

71. $4x^2 + 4xy + y^2 - 16 = \left(4x^2 + 4xy + y^2\right) - 16 = (2x + y)(2x + y) - 16$
$= (2x + y)^2 - 4^2 = (2x + y + 4)(2x + y - 4)$

73. $9 - a^2 - 4ab - 4b^2 = 9 - \left(a^2 + 4ab + 4b^2\right) = 9 - (a + 2b)(a + 2b)$
$= 3^2 - (a + 2b)^2$
$= (3 + (a + 2b))(3 - (a + 2b))$
$= (3 + a + 2b)(3 - a - 2b)$

75. $4a^2 - 15ab + 9b^2 = (4a - 3b)(a - 3b)$

77. $2a^2 + 3b^2 + 5ab = 2a^2 + 5ab + 3b^2 = (2a + 3b)(a + b)$

79. $pq + 6p^2 - q^2 = 6p^2 + pq - q^2 = (2p + q)(3p - q)$

81. $b^2 + 4a^2 + 16ab = 4a^2 + 16ab + b^2$
prime

83. $-12y^2 - 12 + 25y = -\left(12y^2 - 25y + 12\right) = -(3y - 4)(4y - 3)$

85. $3x^2 + 6 + x = 3x^2 + x + 6 \Rightarrow$ prime

87. $16x^2 - 8xy + y^2 = (4x - y)(4x - y) = (4x - y)^2$

89. $4x^2 + 8xy + 3y^2 = (2x + y)(2x + 3y)$

91. $4x^2 + 10x - 6 = 2\left(2x^2 + 5x - 3\right) = 2(2x - 1)(x + 3)$

93. $y^3 + 13y^2 + 12y = y\left(y^2 + 13y + 12\right) = y(y + 12)(y + 1)$

95. $6x^3 - 15x^2 - 9x = 3x\left(2x^2 - 5x - 3\right) = 3x(2x + 1)(x - 3)$

97. $30r^5 + 63r^4 - 30r^3 = 3r^3(10r^2 + 21r - 10) = 3r^3(5r - 2)(2r + 5)$

99. $4a^2 - 4ab - 8b^2 = 4(a^2 - ab - 2b^2)$
$= 4(a - 2b)(a + b)$

101. $8x^2 - 12xy - 8y^2 = 4(2x^2 - 3xy - 2y^2)$
$= 4(2x + y)(x - 2y)$

103. $4a^2 - 4ab + b^2 = (2a - b)(2a - b)$

105. $-16m^3n - 20m^2n^2 - 6mn^3 = -2mn(8m^2 + 10mn + 3n^2) = -2mn(4m + 3n)(2m + n)$

107. $-28u^3v^3 + 26u^2v^4 - 6uv^5 = -2uv^3(14u^2 - 13uv + 3v^2) = -2uv^3(7u - 3v)(2u - v)$

109. $9p^2 + 1 + 6p - q^2 = 9p^2 + 6p + 1 - q^2 = (3p + 1)(3p + 1) - q^2 = (3p + 1)^2 - q^2$
$= (3p + 1 + q)(3p + 1 - q)$

111. Answers may vary.

113. $(6x + 1)(x + 6) = 6x^2 + 37x + 6; (6x - 1)(x - 6) = 6x^2 - 37x + 6 \Rightarrow b = \pm 37$
$(6x + 2)(x + 3) = 6x^2 + 20x + 6; (6x - 2)(x - 3) = 6x^2 - 20x + 6 \Rightarrow b = \pm 20$
$(6x + 3)(x + 2) = 6x^2 + 15x + 6; (6x - 3)(x - 2) = 6x^2 - 15x + 6 \Rightarrow b = \pm 15$
$(6x + 6)(x + 1) = 6x^2 + 12x + 6; (6x - 6)(x - 1) = 6x^2 - 12x + 6 \Rightarrow b = \pm 12$
$(2x + 1)(3x + 6) = 6x^2 + 15x + 6; (2x - 1)(3x - 6) = 6x^2 - 15x + 6 \Rightarrow b = \pm 15$
$(2x + 2)(3x + 3) = 6x^2 + 12x + 6; (2x - 2)(3x - 3) = 6x^2 - 12x + 6 \Rightarrow b = \pm 12$
$(2x + 3)(3x + 2) = 6x^2 + 13x + 6; (2x - 3)(3x - 2) = 6x^2 - 13x + 6 \Rightarrow b = \pm 13$
$(2x + 6)(3x + 1) = 6x^2 + 20x + 6; (2x - 6)(3x - 1) = 6x^2 - 20x + 6 \Rightarrow b = \pm 20$
$b = \pm 12, \pm 13, \pm 15, \pm 20, \pm 37$

Exercise 5.5 (page 358)

1. $x^3 - y^3 = (x - y)(x^2 + xy + y^2)$

3. $a^3 + 8 = a^3 + 2^3$
$= (a + 2)(a^2 - 2a + 2^2)$
$= (a + 2)(a^2 - 2a + 4)$

5. $1 + 8x^3 = 1^3 + (2x)^3$
$= (1 + 2x)(1^2 - 1 \cdot 2x + (2x)^2)$
$= (1 + 2x)(1 - 2x + 4x^2)$

7. $x^3y^3 + 1 = (xy)^3 + 1^3$
$= (xy + 1)((xy)^2 - 1xy + 1^2)$
$= (xy + 1)(x^2y^2 - xy + 1)$

9. $1 \times 10^{-13} = 0.0000000000001$ cm

11. sum of two cubes

13. $x^3 + y^3 = (x + y)\boxed{(x^2 - xy + y^2)}$

15. $y^3 + 1 = y^3 + 1^3$
$= (y + 1)(y^2 - 1y + 1^2)$
$= (y + 1)(y^2 - y + 1)$

17. $8 + x^3 = 2^3 + x^3$
$= (2 + x)(2^2 - 2x + x^2)$
$= (2 + x)(4 - 2x + x^2)$

SECTION 5.5

19. $m^3 + n^3 = (m+n)(m^2 - mn + n^2)$

21. $$\begin{aligned} 8u^3 + w^3 &= (2u)^3 + w^3 \\ &= (2u+w)\left((2u)^2 - 2uw + w^2\right) \\ &= (2u+w)\left(4u^2 - 2uw + w^2\right) \end{aligned}$$

23. $$\begin{aligned} x^3 - 8 &= x^3 - 2^3 \\ &= (x-2)\left(x^2 + 2x + 2^2\right) \\ &= (x-2)\left(x^2 + 2x + 4\right) \end{aligned}$$

25. $s^3 - t^3 = (s-t)(s^2 + st + t^2)$

27. $$\begin{aligned} 125p^3 - q^3 &= (5p)^3 - q^3 \\ &= (5p-q)\left((5p)^2 + 5pq + q^2\right) \\ &= (5p-q)\left(25p^2 + 5pq + q^2\right) \end{aligned}$$

29. $$\begin{aligned} 27a^3 - b^3 &= (3a)^3 - b^3 \\ &= (3a-b)\left((3a)^2 + 3ab + b^2\right) \\ &= (3a-b)\left(9a^2 + 3ab + b^2\right) \end{aligned}$$

31. $$\begin{aligned} 2x^3 + 54 &= 2\left(x^3 + 27\right) \\ &= 2\left(x^3 + 3^3\right) \\ &= 2(x+3)\left(x^2 - 3x + 3^2\right) \\ &= 2(x+3)\left(x^2 - 3x + 9\right) \end{aligned}$$

33. $$\begin{aligned} -x^3 + 216 &= -\left(x^3 - 216\right) \\ &= -\left(x^3 - 6^3\right) \\ &= -(x-6)\left(x^2 + 6x + 6^2\right) \\ &= -(x-6)\left(x^2 + 6x + 36\right) \end{aligned}$$

35. $$\begin{aligned} 64m^3x - 8n^3x = 8x\left(8m^3 - n^3\right) = 8x\left((2m)^3 - n^3\right) &= 8x(2m-n)\left((2m)^2 + 2mn + n^2\right) \\ &= 8x(2m-n)\left(4m^2 + 2mn + n^2\right) \end{aligned}$$

37. $$\begin{aligned} x^4y + 216xy^4 = xy\left(x^3 + 216y^3\right) = xy\left(x^3 + (6y)^3\right) &= xy(x+6y)\left(x^2 - 6xy + (6y)^2\right) \\ &= xy(x+6y)\left(x^2 - 6xy + 36y^2\right) \end{aligned}$$

39. $$\begin{aligned} x^6 - 1 = \left(x^3\right)^2 - 1^2 = \left(x^3 + 1\right)\left(x^3 - 1\right) &= \left(x^3 + 1^3\right)\left(x^3 - 1^3\right) \\ &= (x+1)\left(x^2 - 1x + 1^2\right)(x-1)\left(x^2 + 1x + 1^2\right) \\ &= (x+1)\left(x^2 - x + 1\right)(x-1)\left(x^2 + x + 1\right) \end{aligned}$$

41. $$\begin{aligned} x^{12} - y^6 = \left(x^6\right)^2 - \left(y^3\right)^2 &= \left(x^6 + y^3\right)\left(x^6 - y^3\right) \\ &= \left(\left(x^2\right)^3 + y^3\right)\left(\left(x^2\right)^3 - y^3\right) \\ &= \left(x^2 + y\right)\left(\left(x^2\right)^2 - x^2y + y^2\right)\left(x^2 - y\right)\left(\left(x^2\right)^2 + x^2y + y^2\right) \\ &= \left(x^2 + y\right)\left(x^4 - x^2y + y^2\right)\left(x^2 - y\right)\left(x^4 + x^2y + y^2\right) \end{aligned}$$

43. $125 + b^3 = 5^3 + b^3 = (5+b)(5^2 - 5b + b^2) = (5+b)(25 - 5b + b^2)$

45. $27x^3 + 125 = (3x)^3 + 5^3 = (3x+5)\left((3x)^2 - (3x)5 + 5^2\right) = (3x+5)(9x^2 - 15x + 25)$

47. $$\begin{aligned} 64x^3 + 27y^3 = (4x)^3 + (3y)^3 &= (4x+3y)\left((4x)^2 - (4x)(3y) + (3y)^2\right) \\ &= (4x+3y)\left(16x^2 - 12xy + 9y^2\right) \end{aligned}$$

49. $a^3 + b^6 = a^3 + (b^2)^3 = (a + b^2)\left(a^2 - ab^2 + (b^2)^2\right) = (a + b^2)(a^2 - ab^2 + b^4)$

51. $x^3 - y^9 = x^3 - (y^3)^3 = (x - y^3)\left(x^2 + xy^3 + (y^3)^2\right) = (x - y^3)(x^2 + xy^3 + y^6)$

53. $$\begin{aligned} 4m^5n + 500m^2n^4 = 4m^2n(m^3 + 125n^3) &= 4m^2n(m^3 + (5n)^3) \\ &= 4m^2n(m + 5n)(m^2 - 5mn + (5n)^2) \\ &= 4m^2n(m + 5n)(m^2 - 5mn + 25n^2) \end{aligned}$$

55. $$\begin{aligned} 216a^4b^4 - 1000ab^7 = 8ab^4(27a^3 - 125b^3) &= 8ab^4((3a)^3 - (5b)^3) \\ &= 8ab^4(3a - 5b)((3a)^2 + (3a)(5b) + (5b)^2) \\ &= 8ab^4(3a - 5b)(9a^2 + 15ab + 25b^2) \end{aligned}$$

57. $$\begin{aligned} x^{10}y^2 - xy^5 = xy^2(x^9 - y^3) = xy^2\left((x^3)^3 - y^3\right) &= xy^2(x^3 - y)\left((x^3)^2 + x^3y + y^2\right) \\ &= xy^2(x^3 - y)(x^6 + x^3y + y^2) \end{aligned}$$

59. $$\begin{aligned} 24m^5n - 3m^2n^4 = 3m^2n(8m^3 - n^3) &= 3m^2n((2m)^3 - n^3) \\ &= 3m^2n(2m - n)((2m)^2 + 2mn + n^2) \\ &= 3m^2n(2m - n)(4m^2 + 2mn + n^2) \end{aligned}$$

61. $$\begin{aligned} x(8a^3 - b^3) + 4(8a^3 - b^3) = (8a^3 - b^3)(x + 4) &= ((2a)^3 - b^3)(x + 4) \\ &= (2a - b)((2a)^2 + 2ab + b^2)(x + 4) \\ &= (2a - b)(4a^2 + 2ab + b^2)(x + 4) \end{aligned}$$

63. $$\begin{aligned} (a^3x + b^3x) - (a^3y + b^3y) = x(a^3 + b^3) - y(a^3 + b^3) &= (a^3 + b^3)(x - y) \\ &= (a + b)(a^2 - ab + b^2)(x - y) \end{aligned}$$

65. $$\begin{aligned} (x^4 + xy^3) - (x^3y + y^4) = x(x^3 + y^3) - y(x^3 + y^3) &= (x^3 + y^3)(x - y) \\ &= (x + y)(x^2 - xy + y^2)(x - y) \end{aligned}$$

67. $$\begin{aligned} z^3(y^2 - 4) + 8(y^2 - 4) = (y^2 - 4)(z^3 + 8) &= (y^2 - 2^2)(z^3 + 2^3) \\ &= (y + 2)(y - 2)(z + 2)(z^2 - 2z + 4) \end{aligned}$$

69. **Answers may vary.**

71. $a^3 - b^3 = 11^3 - 7^3 = 1331 - 343 = \boxed{988}$
$(a - b)(a^2 + ab + b^2) = (11 - 7)(11^2 + 11 \cdot 7 + 7^2) = 4(121 + 77 + 49) = 4(247) = \boxed{988}$

Exercise 5.6 (page 362)

1. $2x^2 - 4x$: common factor $(2x)$

3. $125 + r^3s^3 = 5^3 + (rs)^3$: sum of 2 cubes

SECTION 5.6

5. $x^2 + 4$: none (prime)

7. $25r^2 - s^4 = (5r)^2 - (s^2)^2$
difference of 2 squares

9. $$\begin{aligned} 2(t-5)+t &= 3(2-t) \\ 2t-10+t &= 6-3t \\ 3t-10 &= 6-3t \\ 6t &= 16 \\ t &= \frac{16}{6} = \frac{8}{3} \end{aligned}$$

11. $$\begin{aligned} 5-3(t+1) &= t+2 \\ 5-3t-3 &= t+2 \\ -3t+2 &= t+2 \\ -4t &= 0 \\ t &= \frac{0}{-4} = 0 \end{aligned}$$

13. factors

15. binomials

17. $6x + 3 = 3(2x + 1)$

19. $x^2 - 6x - 7 = (x - 7)(x + 1)$

21. $6t^2 + 7t - 3 = (2t + 3)(3t - 1)$

23. $t^2 - 2t + 1 = (t - 1)(t - 1) = (t - 1)^2$

25. $2x^2 - 32 = 2(x^2 - 16) = 2(x^2 - 4^2) = 2(x + 4)(x - 4)$

27. $x^2 + 7x + 1 \Rightarrow$ prime

29. $-2x^5 + 128x^2 = -2x^2(x^3 - 64) = -2x^2(x^3 - 4^3) = -2x^2(x - 4)(x^2 + 4x + 16)$

31. $14t^3 - 40t^2 + 6t^4 = 2t^2(7t - 20 + 3t^2) = 2t^2(3t^2 + 7t - 20) = 2t^2(3t - 5)(t + 4)$

33. $6x^2 - x - 16 \Rightarrow$ prime

35. $$\begin{aligned} 6a^3 + 35a^2 - 6a &= a(6a^2 + 35a - 6) \\ &= a(6a - 1)(a + 6) \end{aligned}$$

37. $16x^2 - 40x^3 + 25x^4 = x^2(16 - 40x + 25x^2) = x^2(25x^2 - 40x + 16) = x^2(5x - 4)(5x - 4)$

39. $$\begin{aligned} -84x^2 - 147x - 12x^3 = -12x^3 - 84x^2 - 147x = -3x(4x^2 + 28x + 49) &= -3x(2x + 7)(2x + 7) \\ &= -3x(2x + 7)^2 \end{aligned}$$

41. $$\begin{aligned} 8x^6 - 8 = 8(x^6 - 1) = 8\left((x^3)^2 - 1^2\right) &= 8(x^3 + 1)(x^3 - 1) \\ &= 8(x + 1)(x^2 - x + 1)(x - 1)(x^2 + x + 1) \end{aligned}$$

43. $$\begin{aligned} 5x^3 - 5x^5 + 25x^2 &= -5x^5 + 5x^3 + 25x^2 \\ &= -5x^2(x^3 - x - 5) \end{aligned}$$

45. $9x^2 + 12x + 16$: prime

47. $$\begin{aligned} 2ab^2 + 8ab - 24a &= 2a(b^2 + 4b - 12) \\ &= 2a(b + 6)(b - 2) \end{aligned}$$

49. $-8p^3q^7 - 4p^2q^3 = -4p^2q^3(2pq^4 + 1)$

51. $4a^2 - 4ab + b^2 - 9 = (4a^2 - 4ab + b^2) - 9 = (2a - b)^2 - 3^2 = (2a - b + 3)(2a - b - 3)$

53. $a^3 + b^3 = (a + b)(a^2 - ab + b^2)$

55. $x^2y^2 - 2x^2 - y^2 + 2 = x^2(y^2 - 2) - 1(y^2 - 2) = (y^2 - 2)(x^2 - 1) = (y^2 - 2)(x^2 - 1^2)$
$= (y^2 - 2)(x + 1)(x - 1)$

57. $a^2 + 2ab + b^2 - y^2 = (a + b)(a + b) - y^2 = (a + b)^2 - y^2 = (a + b + y)(a + b - y)$

59. $a^2(x - a) - b^2(x - a) = (x - a)(a^2 - b^2) = (x - a)(a + b)(a - b)$

61. $8p^6 - 27q^6 = (2p^2)^3 - (3q^2)^3 = (2p^2 - 3q^2)\left((2p^2)^2 + (2p^2)(3q^2) + (3q^2)^2\right)$
$= (2p^2 - 3q^2)(4p^4 + 6p^2q^2 + 9q^4)$

63. $125p^3 - 64y^3 = (5p)^3 - (4y)^3 = (5p - 4y)((5p)^2 + (5p)(4y) + (4y)^2)$
$= (5p - 4y)(25p^2 + 20py + 16y^2)$

65. $-16x^4y^2z + 24x^5y^3z^4 - 15x^2y^3z^7 = -x^2y^2z(16x^2 - 24x^3yz^3 + 15yz^6)$

67. $81p^4 - 16q^4 = (9p^2)^2 - (4q^2)^2 = (9p^2 + 4q^2)(9p^2 - 4q^2) = (9p^2 + 4q^2)(3p + 2q)(3p - 2q)$

69. $54x^3 + 250y^6 = 2(27x^3 + 125y^6) = 2\left((3x)^3 + (5y^2)^3\right)$
$= 2(3x + 5y^2)\left((3x)^2 - (3x)(5y^2) + (5y^2)^2\right)$
$= 2(3x + 5y^2)(9x^2 - 15xy^2 + 25y^4)$

71. $x^5 - x^3y^2 + x^2y^3 - y^5 = x^3(x^2 - y^2) + y^3(x^2 - y^2) = (x^2 - y^2)(x^3 + y^3)$
$= (x + y)(x - y)(x + y)(x^2 - xy + y^2)$

73. $2a^2c - 2b^2c + 4a^2d - 4b^2d = 2(a^2c - b^2c + 2a^2d - 2b^2d)$
$= 2(c(a^2 - b^2) + 2d(a^2 - b^2))$
$= 2(a^2 - b^2)(c + 2d) = 2(a + b)(a - b)(c + 2d)$

75. Answers may vary.

77. Answers may vary.

Exercise 5.7 (page 368)

1.
$(x - 8)(x - 7) = 0$
$x - 8 = 0$ **or** $x - 7 = 0$
$x = 8$ $\quad x = 7$

3.
$x^2 + 7x = 0$
$x(x + 7) = 0$
$x = 0$ **or** $x + 7 = 0$
$x = 0$ $\quad x = -7$

5.
$x^2 - 2x + 1 = 0$
$(x - 1)(x - 1) = 0$
$x - 1 = 0$ **or** $x - 1 = 0$
$x = 1$ $\quad x = 1$

7. $u^3u^2u^4 = u^{3+2+4} = u^9$

SECTION 5.7

9. $\dfrac{a^3b^4}{a^2b^5} = a^1b^{-1} = \dfrac{a}{b}$

11. quadratic

13. second

15.
$$(x-2)(x+3)=0$$
$$x-2=0 \quad \textbf{or} \quad x+3=0$$
$$x=2 \qquad\qquad x=-3$$

17.
$$(x-4)(x+1)=0$$
$$x-4=0 \quad \textbf{or} \quad x+1=0$$
$$x=4 \qquad\qquad x=-1$$

19.
$$(2x-5)(3x+6)=0$$
$$2x-5=0 \quad \textbf{or} \quad 3x+6=0$$
$$2x=5 \qquad\qquad 3x=-6$$
$$x=\tfrac{5}{2} \qquad\qquad x=-2$$

21.
$$(x-1)(x+2)(x-3)=0$$
$$x-1=0 \ \textbf{or}\ x+2=0 \quad \textbf{or}\ x-3=0$$
$$x=1 \qquad\qquad x=-2 \qquad\qquad x=3$$

23.
$$x^2-3x=0$$
$$x(x-3)=0$$
$$x=0 \quad \textbf{or} \quad x-3=0$$
$$x=0 \qquad\qquad x=3$$

25.
$$5x^2+7x=0$$
$$x(5x+7)=0$$
$$x=0 \quad \textbf{or} \quad 5x+7=0$$
$$x=0 \qquad\qquad 5x=-7$$
$$x=0 \qquad\qquad x=-\tfrac{7}{5}$$

27.
$$x^2-7x=0$$
$$x(x-7)=0$$
$$x=0 \quad \textbf{or} \quad x-7=0$$
$$x=0 \qquad\qquad x=7$$

29.
$$3x^2+8x=0$$
$$x(3x+8)=0$$
$$x=0 \quad \textbf{or} \quad 3x+8=0$$
$$x=0 \qquad\qquad 3x=-8$$
$$x=0 \qquad\qquad x=-\tfrac{8}{3}$$

31.
$$x^2-25=0$$
$$(x+5)(x-5)=0$$
$$x+5=0 \quad \textbf{or} \quad x-5=0$$
$$x=-5 \qquad\qquad x=5$$

33.
$$9y^2-4=0$$
$$(3y+2)(3y-2)=0$$
$$3y+2=0 \quad \textbf{or} \quad 3y-2=0$$
$$3y=-2 \qquad\qquad 3y=2$$
$$y=-\tfrac{2}{3} \qquad\qquad y=\tfrac{2}{3}$$

35.
$$x^2=49$$
$$x^2-49=0$$
$$(x+7)(x-7)=0$$
$$x+7=0 \quad \textbf{or} \quad x-7=0$$
$$x=-7 \qquad\qquad x=7$$

37.
$$4x^2=81$$
$$4x^2-81=0$$
$$(2x+9)(2x-9)=0$$
$$2x+9=0 \quad \textbf{or} \quad 2x-9=0$$
$$2x=-9 \qquad\qquad 2x=9$$
$$x=-\tfrac{9}{2} \qquad\qquad x=\tfrac{9}{2}$$

39.
$$x^2-13x+12=0$$
$$(x-1)(x-12)=0$$
$$x-1=0 \quad \textbf{or} \quad x-12=0$$
$$x=1 \qquad\qquad x=12$$

41.
$$x^2-2x-15=0$$
$$(x+3)(x-5)=0$$
$$x+3=0 \quad \textbf{or} \quad x-5=0$$
$$x=-3 \qquad\qquad x=5$$

SECTION 5.7

43.
$$\begin{aligned} 6x^2 + x &= 2 \\ 6x^2 + x - 2 &= 0 \\ (2x-1)(3x+2) &= 0 \end{aligned}$$
$$2x - 1 = 0 \quad \textbf{or} \quad 3x + 2 = 0$$
$$2x = 1 \qquad 3x = -2$$
$$x = \tfrac{1}{2} \qquad x = -\tfrac{2}{3}$$

45.
$$\begin{aligned} 2x^2 - 5x &= -2 \\ 2x^2 - 5x + 2 &= 0 \\ (2x-1)(x-2) &= 0 \end{aligned}$$
$$2x - 1 = 0 \quad \textbf{or} \quad x - 2 = 0$$
$$2x = 1 \qquad x = 2$$
$$x = \tfrac{1}{2} \qquad x = 2$$

47.
$$(x-1)\left(x^2+5x+6\right) = 0$$
$$(x-1)(x+2)(x+3) = 0$$
$$x - 1 = 0 \ \textbf{or} \ x + 2 = 0 \quad \textbf{or} \ x + 3 = 0$$
$$x = 1 \qquad x = -2 \qquad x = -3$$

49.
$$(x+3)\left(x^2+2x-15\right) = 0$$
$$(x+3)(x+5)(x-3) = 0$$
$$x + 3 = 0 \quad \textbf{or} \ x + 5 = 0 \quad \textbf{or} \ x - 3 = 0$$
$$x = -3 \qquad x = -5 \qquad x = 3$$

51.
$$x^3 + 3x^2 + 2x = 0$$
$$x\left(x^2+3x+2\right) = 0$$
$$x(x+2)(x+1) = 0$$
$$x = 0 \quad \textbf{or} \quad x + 2 = 0 \quad \textbf{or} \quad x + 1 = 0$$
$$x = 0 \qquad x = -2 \qquad x = -1$$

53.
$$x^3 - 27x - 6x^2 = 0$$
$$x\left(x^2 - 27 - 6x\right) = 0$$
$$x\left(x^2 - 6x - 27\right) = 0$$
$$x(x-9)(x+3) = 0$$
$$x = 0 \quad \textbf{or} \quad x - 9 = 0 \quad \textbf{or} \quad x + 3 = 0$$
$$x = 0 \qquad x = 9 \qquad x = -3$$

55.
$$\begin{aligned} 6x^3 + 20x^2 &= -6x \\ 6x^3 + 20x^2 + 6x &= 0 \\ 2x\left(3x^2+10x+3\right) &= 0 \\ 2x(3x+1)(x+3) &= 0 \end{aligned}$$
$$2x = 0 \quad \textbf{or} \quad 3x + 1 = 0 \quad \textbf{or} \quad x + 3 = 0$$
$$x = 0 \qquad 3x = -1 \qquad x = -3$$
$$x = 0 \qquad x = -\tfrac{1}{3} \qquad x = -3$$

57.
$$\begin{aligned} x^3 + 7x^2 &= x^2 - 9x \\ x^3 + 6x^2 + 9x &= 0 \\ x\left(x^2+6x+9\right) &= 0 \\ x(x+3)(x+3) &= 0 \end{aligned}$$
$$x = 0 \ \textbf{or} \ x + 3 = 0 \quad \textbf{or} \ x + 3 = 0$$
$$x = 0 \qquad x = -3 \qquad x = -3$$

59.
$$8x^2 - 16x = 0$$
$$8x(x-2) = 0$$
$$8x = 0 \quad \textbf{or} \quad x - 2 = 0$$
$$x = 0 \qquad x = 2$$

61.
$$10x^2 + 2x = 0$$
$$2x(5x+1) = 0$$
$$2x = 0 \quad \textbf{or} \quad 5x + 1 = 0$$
$$x = 0 \qquad 5x = -1$$
$$x = 0 \qquad x = -\tfrac{1}{5}$$

63.
$$y^2 - 49 = 0$$
$$(y+7)(y-7) = 0$$
$$y + 7 = 0 \quad \textbf{or} \quad y - 7 = 0$$
$$y = -7 \qquad y = 7$$

65.
$$4x^2 - 1 = 0$$
$$(2x+1)(2x-1) = 0$$
$$2x + 1 = 0 \quad \textbf{or} \quad 2x - 1 = 0$$
$$2x = -1 \qquad 2x = 1$$
$$x = -\tfrac{1}{2} \qquad x = \tfrac{1}{2}$$

67.
$$\begin{aligned} x^2 - 4x - 21 &= 0 \\ (x+3)(x-7) &= 0 \\ x+3 = 0 \quad &\textbf{or} \quad x-7=0 \\ x = -3 \quad & \quad\quad x = 7 \end{aligned}$$

69.
$$\begin{aligned} x^2 + 8 - 9x &= 0 \\ x^2 - 9x + 8 &= 0 \\ (x-8)(x-1) &= 0 \\ x-8=0 \quad &\textbf{or} \quad x-1=0 \\ x=8 \quad & \quad\quad x=1 \end{aligned}$$

71.
$$\begin{aligned} a^2 + 8a &= -15 \\ a^2 + 8a + 15 &= 0 \\ (a+3)(a+5) &= 0 \\ a+3=0 \quad &\textbf{or} \quad a+5=0 \\ a=-3 \quad & \quad\quad a=-5 \end{aligned}$$

73.
$$\begin{aligned} 2y - 8 &= -y^2 \\ y^2 + 2y - 8 &= 0 \\ (y+4)(y-2) &= 0 \\ y+4=0 \quad &\textbf{or} \quad y-2=0 \\ y=-4 \quad & \quad\quad y=2 \end{aligned}$$

75.
$$\begin{aligned} 2x^2 + x - 3 &= 0 \\ (2x+3)(x-1) &= 0 \\ 2x+3=0 \quad &\textbf{or} \quad x-1=0 \\ 2x=-3 \quad & \quad\quad x=1 \\ x=-\tfrac{3}{2} \quad & \quad\quad x=1 \end{aligned}$$

77.
$$\begin{aligned} 14m^2 + 23m + 3 &= 0 \\ (7m+1)(2m+3) &= 0 \\ 7m+1=0 \quad &\textbf{or} \quad 2m+3=0 \\ 7m=-1 \quad & \quad\quad 2m=-3 \\ m=-\tfrac{1}{7} \quad & \quad\quad m=-\tfrac{3}{2} \end{aligned}$$

79.
$$\begin{aligned} (x-5)\big(2x^2+x-3\big) &= 0 \\ (x-5)(2x+3)(x-1) &= 0 \\ x-5=0 \ \textbf{or}\ 2x+3=0 \quad &\textbf{or}\ x-1=0 \\ x=5 \quad\quad 2x=-3 \quad & \quad\quad x=1 \\ x=-\tfrac{3}{2} & \end{aligned}$$

81.
$$\begin{aligned} \big(p^2-81\big)(p+2) &= 0 \\ (p+9)(p-9)(p+2) &= 0 \\ p+9=0 \quad \textbf{or}\ p-9=0 \ &\textbf{or}\ p+2=0 \\ p=-9 \quad\quad p=9 \quad & \quad\quad p=-2 \end{aligned}$$

83.
$$\begin{aligned} 3x^2 - 8x &= 3 \\ 3x^2 - 8x - 3 &= 0 \\ (3x+1)(x-3) &= 0 \\ 3x+1=0 \quad &\textbf{or} \quad x-3=0 \\ 3x=-1 \quad & \quad\quad x=3 \\ x=-\tfrac{1}{3} \quad & \quad\quad x=3 \end{aligned}$$

85.
$$\begin{aligned} 15x^2 - 2 &= 7x \\ 15x^2 - 7x - 2 &= 0 \\ (3x-2)(5x+1) &= 0 \\ 3x-2=0 \quad &\textbf{or} \quad 5x+1=0 \\ 3x=2 \quad & \quad\quad 5x=-1 \\ x=\tfrac{2}{3} \quad & \quad\quad x=-\tfrac{1}{5} \end{aligned}$$

87.
$$\begin{aligned} x(6x+5) &= 6 \\ 6x^2 + 5x &= 6 \\ 6x^2 + 5x - 6 &= 0 \\ (3x-2)(2x+3) &= 0 \\ 3x-2=0 \quad &\textbf{or} \quad 2x+3=0 \\ 3x=2 \quad & \quad\quad 2x=-3 \\ x=\tfrac{2}{3} \quad & \quad\quad x=-\tfrac{3}{2} \end{aligned}$$

89.
$$\begin{aligned} (x+1)(8x+1) &= 18x \\ 8x^2 + 9x + 1 &= 18x \\ 8x^2 - 9x + 1 &= 0 \\ (x-1)(8x-1) &= 0 \\ x-1=0 \quad &\textbf{or} \quad 8x-1=0 \\ x=1 \quad & \quad\quad 8x=1 \\ x=1 \quad & \quad\quad x=\tfrac{1}{8} \end{aligned}$$

91.
$$x^3 + 1.3x^2 - 0.3x = 0$$
$$10x^3 + 13x^2 - 3x = 0$$
$$x(10x^2 + 13x - 3) = 0$$
$$x(5x - 1)(2x + 3) = 0$$
$x = 0$ **or** $5x - 1 = 0$ **or** $2x + 3 = 0$
$x = 0$ $\quad 5x = 1$ $\quad 2x = -3$
$x = 0$ $\quad x = \frac{1}{5}$ $\quad x = -\frac{3}{2}$

93. **Answers may vary.**

95. **Answers may vary.**

97.
$$3a^2 + 9a - 2a - 6 = 0$$
$$3a(a + 3) - 2(a + 3) = 0$$
$$(a + 3)(3a - 2) = 0$$
$a + 3 = 0$ **or** $3a - 2 = 0$
$a = -3$ $\quad 3a = 2$
$a = \frac{2}{3}$

$$3a^2 + 9a - 2a - 6 = 0$$
$$3a^2 + 7a - 6 = 0$$
$$(a + 3)(3a - 2) = 0$$
$a + 3 = 0$ **or** $3a - 2 = 0$
$a = -3$ $\quad 3a = 2$
$a = \frac{2}{3}$

Exercise 5.8 (page 373)

1. $A = lw$

3. $A = s^2$

5. $P = 2l + 2w$

7.
$$-2(5x + 2) = 3(2 - 3x)$$
$$-10x - 4 = 6 - 9x$$
$$-4 = 6 + x$$
$$-10 = x$$

9. Let $w =$ the width and $3w =$ the length.
$$\text{Perimeter} = 120$$
$$2(w) + 2(3w) = 120$$
$$2w + 6w = 120$$
$$8w = 120$$
$$w = 15$$
The area $= 15(45) = 675$ cm^2.

11. analyze

13. Let $x =$ the first integer. Then $x + 2 =$ the other integer.
$$x(x + 2) = 35$$
$$x^2 + 2x = 35$$
$$x^2 + 2x - 35 = 0$$
$$(x + 7)(x - 5) = 0$$
$x + 7 = 0$ **or** $x - 5 = 0$
$x = -7$ $\quad x = 5$
The integers are -7 and -5, or 5 and 7.

15. Let $x =$ the integer.
$$x^2 + 4 = 10x - 5$$
$$x^2 - 10x + 9 = 0$$
$$(x - 9)(x - 1) = 0$$
$x - 9 = 0$ **or** $x - 1 = 0$
$x = 9$ $\quad x = 1$
The integer is 1 or 9.

17. Let $v = 144$ and $h = 0$:

$$h = vt - 16t^2$$
$$0 = 144t - 16t^2$$
$$0 = 16t(9 - t)$$
$$16t = 0 \quad \textbf{or} \quad 9 - t = 0$$
$$t = 0 \qquad\qquad 9 = t$$

Since $t = 0$ is when the object was first thrown, it will hit the ground in 9 seconds.

19. Let $v = 220$ and $h = 600$:

$$h = vt - 16t^2$$
$$600 = 220t - 16t^2$$
$$16t^2 - 220t + 600 = 0$$
$$4\left(4t^2 - 55t + 150\right) = 0$$
$$4(4t - 15)(t - 10) = 0$$
$$4t - 15 = 0 \quad \textbf{or} \quad t - 10 = 0$$
$$4t = 15 \qquad\qquad t = 10$$
$$t = \tfrac{15}{4} \qquad\qquad t = 10$$

The cannonball will be at a height of 600 feet after $\frac{15}{4}$ seconds and after 10 seconds.

21.

$$h = -16t^2 + 64$$
$$0 = -16t^2 + 64$$
$$16t^2 - 64 = 0$$
$$16\left(t^2 - 4\right) = 0$$
$$16(t + 2)(t - 2) = 0$$
$$t + 2 = 0 \quad \textbf{or} \quad t - 2 = 0$$
$$t = -2 \qquad\qquad t = 2$$

The value of $t = -2$ does not make sense, so the dive lasts 2 seconds.

23.

$$75^2 = h^2 + 72^2$$
$$5625 = h^2 + 5184$$
$$0 = h^2 - 441$$
$$0 = (h + 21)(h - 21)$$
$$h + 21 = 0 \quad \textbf{or} \quad h - 21 = 0$$
$$h = -21 \qquad\qquad h = 21$$

The value of $h = -21$ does not make sense, so the camper started at a height of 21 feet.

25. Area = Length · Width

$$36 = (2w + 1)w$$
$$36 = 2w^2 + w$$
$$0 = 2w^2 + w - 36$$
$$0 = (2w + 9)(w - 4)$$
$$2w + 9 = 0 \quad \textbf{or} \quad w - 4 = 0$$
$$2w = -9 \qquad\qquad w = 4$$
$$w = -\tfrac{9}{2} \qquad\qquad w = 4$$

Since the answer $w = -\frac{9}{2}$ does not make sense, the dimensions are 4 m by 9 m.

27. Let $w =$ the width and $w + 2 =$ the length.

Area = Length · Width

$$143 = (w + 2)w$$
$$143 = w^2 + 2w$$
$$0 = w^2 + 2w - 143$$
$$0 = (w + 13)(w - 11)$$
$$w + 13 = 0 \quad \textbf{or} \quad w - 11 = 0$$
$$w = -13 \qquad\qquad w = 11$$

Since the answer $w = -13$ does not make sense, the dimensions are 11 ft by 13 ft and the perimeter is 48 ft.

SECTION 5.8

29. Let $b =$ the base and $5b - 2 =$ the height.

$$A = \frac{1}{2}bh$$
$$36 = \frac{1}{2}b(5b - 2)$$
$$72 = b(5b - 2)$$
$$72 = 5b^2 - 2b$$
$$0 = 5b^2 - 2b - 72$$
$$0 = (5b + 18)(b - 4)$$
$$5b + 18 = 0 \quad \textbf{or} \quad b - 4 = 0$$
$$5b = -18 \qquad b = 4$$
$$h = -\tfrac{18}{5} \qquad b = 4$$

Since the answer $b = -\frac{18}{5}$ does not make sense, the base is 4 in. and the height is 18 in.

31. Let $x =$ the base and the height.

$$\boxed{\text{Base}} + \boxed{\text{Height}} = \boxed{\text{Area}} - 6$$
$$x + x = \tfrac{1}{2}(x)(x) - 6$$
$$2x = \tfrac{1}{2}x^2 - 6$$
$$0 = \tfrac{1}{2}x^2 - 2x - 6$$
$$0 = x^2 - 4x - 12$$
$$0 = (x - 6)(x + 2)$$
$$x - 6 = 0 \quad \textbf{or} \quad x + 2 = 0$$
$$x = 6 \qquad x = -2$$

The answer $x = -2$ does not make sense, so the area is $\frac{1}{2}(6)(6) = 18$ square units.

33.

$$\boxed{\text{Large area}} - \boxed{\text{Small area}} = \boxed{\text{Border area}}$$
$$(10 + 2w)(25 + 2w) - (10)(25) = 74$$
$$250 + 70w + 4w^2 - 250 = 74$$
$$4w^2 + 70w - 74 = 0$$
$$2(2w^2 + 35w - 37) = 0$$
$$2(2w + 37)(w - 1) = 0$$
$$2w + 37 = 0 \quad \textbf{or} \quad w - 1 = 0$$
$$2w = -37 \qquad w = 1$$
$$w = -\tfrac{37}{2} \qquad w = 1$$

Since the answer $w = -\frac{37}{2}$ does not make sense, the width should be 1 meter.

35. Let $w =$ the width and $2w + 1 =$ the height.

$$V = lwh$$
$$210 = 10w(2w + 1)$$
$$210 = 20w^2 + 10w$$
$$0 = 20w^2 + 10w - 210$$
$$0 = 10(2w^2 + w - 21)$$
$$0 = 10(2w + 7)(w - 3)$$
$$2w + 7 = 0 \quad \textbf{or} \quad w - 3 = 0$$
$$2w = -7 \qquad w = 3$$
$$w = -\tfrac{7}{2} \qquad w = 3$$

Since the answer $w = -\frac{7}{2}$ does not make sense, the width is 3 cm.

37. Let $x =$ one edge and $x - 3 =$ the other.

$$V = \tfrac{Bh}{3}$$
$$84 = \tfrac{x(x-3)(9)}{3}$$
$$252 = 9x^2 - 27x$$
$$0 = 9x^2 - 27x - 252$$
$$0 = 9(x + 4)(x - 7)$$
$$x + 4 = 0 \quad \textbf{or} \quad x - 7 = 0$$
$$x = -4 \qquad x = 7$$

Since the answer $x = -4$ does not make sense, the base is 7 cm by 4 cm.

39.

$$C = \tfrac{1}{2}(n^2 - n)$$
$$66 = \tfrac{1}{2}(n^2 - n)$$
$$2(66) = 2 \cdot \tfrac{1}{2}(n^2 - n)$$
$$132 = n^2 - n$$
$$0 = n^2 - n - 132$$
$$0 = (n + 11)(n - 12)$$
$$n + 11 = 0 \quad \textbf{or} \quad n - 12 = 0$$
$$n = -11 \qquad n = 12$$

The value of $n = -11$ does not make sense, so 12 telephones are needed.

41. Area #1 $= \pi r^2 = \pi(20 \text{ m})^2 = 400\pi \text{ m}^2$
Area #2 $= \pi r^2 = \pi(21 \text{ m})^2 = 441\pi \text{ m}^2$
Total Area $= 400\pi + 441\pi \text{ m}^2 = 841\pi \text{ m}^2$

43.
$$\begin{aligned} a^2 + b^2 &= c^2 \\ x^2 + (x+2)^2 &= (x+4)^2 \\ x^2 + x^2 + 4x + 4 &= x^2 + 8x + 16 \\ 2x^2 + 4x + 4 &= x^2 + 8x + 16 \\ x^2 - 4x - 12 &= 0 \\ (x-6)(x+2) &= 0 \\ x - 6 = 0 \quad &\textbf{or} \quad x + 2 = 0 \\ x = 6 \quad & \quad\quad x = -2 \end{aligned}$$
The value of $x = -2$ does not make sense, so $x = 6$, and the height was 16 ft.

45. **Answers may vary.**

47. Let $w =$ the width and $w + 2 =$ the length.
$$\begin{aligned} \text{Area} &= 18 \\ w(w+2) &= 18 \\ w^2 + 2w &= 18 \\ w^2 + 2w - 18 &= 0 \end{aligned}$$
This is a prime trinomial, so it cannot be factored in order to solve.

Chapter 5 Review (page 377)

1. $35 = 5 \cdot 7$

2. $45 = 9 \cdot 5 = 3^2 \cdot 5$

3. $96 = 12 \cdot 8 = 4 \cdot 3 \cdot 2^3 = 2^2 \cdot 3 \cdot 2^3 = 2^5 \cdot 3$

4. $102 = 2 \cdot 51 = 2 \cdot 3 \cdot 17$

5. $87 = 3 \cdot 29$

6. $99 = 9 \cdot 11 = 3^2 \cdot 11$

7. $2{,}050 = 50 \cdot 41 = 25 \cdot 2 \cdot 41 = 2 \cdot 5^2 \cdot 41$

8. $4{,}096 = 64 \cdot 64 = 2^6 \cdot 2^6 = 2^{12}$

9. $3x + 9y = 3(x + 3y)$

10. $5ax^2 + 15a = 5a(x^2 + 3)$

11. $7x^2 + 14x = 7x(x + 2)$

12. $3x^2 - 3x = 3x(x - 1)$

13. $2x^3 + 4x^2 - 8x = 2x(x^2 + 2x - 4)$

14. $ax + ay - az = a(x + y - z)$

15. $ax + ay - a = a(x + y - 1)$

16. $x^2yz + xy^2z = xyz(x + y)$

17. $(x + y)a + (x + y)b = (x + y)(a + b)$

18. $(x + y)^2 + (x + y) = (x + y)(x + y) + 1(x + y) = (x + y)(x + y + 1)$

19. $2x^2(x + 2) + 6x(x + 2) = (x + 2)(2x^2 + 6x) = (x + 2)(2x)(x + 3) = 2x(x + 2)(x + 3)$

20. $3x(y+z)-9x(y+z)^2 = 3x[(y+z)-3(y+z)(y+z)] = 3x(y+z)(1-3(y+z))$
$= 3x(y+z)(1-3y-3z)$

21. $3p+9q+ap+3aq = 3(p+3q)+a(p+3q) = (p+3q)(3+a)$

22. $ar-2as+7r-14s = a(r-2s)+7(r-2s) = (r-2s)(a+7)$

23. $x^2+ax+bx+ab = x(x+a)+b(x+a)$
$= (x+a)(x+b)$

24. $xy+2x-2y-4 = x(y+2)-2(y+2)$
$= (y+2)(x-2)$

25. $xa+yb+ya+xb = xa+ya+xb+yb = a(x+y)+b(x+y) = (x+y)(a+b)$

26. $x^3-4x^2+3x-12 = x^2(x-4)+3(x-4) = (x-4)(x^2+3)$

27. $x^2-9 = x^2-3^2 = (x+3)(x-3)$

28. $x^2y^2-16 = (xy+4)(xy-4)$

29. $(x+2)^2-y^2 = (x+2+y)(x+2-y)$

30. $z^2-(x+y)^2 = [z+(x+y)][z-(x+y)] = (z+x+y)(z-x-y)$

31. $6x^2y-24y^3 = 6y(x^2-4y^2) = 6y\left[x^2-(2y)^2\right] = 6y(x+2y)(x-2y)$

32. $(x+y)^2-z^2 = [(x+y)+z][(x+y)-z] = (x+y+z)(x+y-z)$

33. $x^2+10x+21 = (x+3)(x+7)$

34. $x^2+4x-21 = (x+7)(x-3)$

35. $x^2+2x-24 = (x+6)(x-4)$

36. $x^2-4x-12 = (x-6)(x+2)$

37. $2x^2-5x-3 = (2x+1)(x-3)$

38. $3x^2-14x-5 = (3x+1)(x-5)$

39. $6x^2+7x-3 = (2x+3)(3x-1)$

40. $6x^2+3x-3 = 3\left(2x^2+x-1\right)$
$= 3(2x-1)(x+1)$

41. $6x^3+17x^2-3x = x\left(6x^2+17x-3\right)$
$= x(x+3)(6x-1)$

42. $4x^3-5x^2-6x = x\left(4x^2-5x-6\right)$
$= x(x-2)(4x+3)$

43. $12x-4x^3-2x^2 = -4x^3-2x^2+12x = -2x(2x^2+x-6) = -2x(2x-3)(x+2)$

44. $-4a^3+4a^2b+24ab^2 = -4a(a^2-ab-6b^2) = -4a(a-3b)(a+2b)$

45. $c^3-27 = c^3-3^3 = (c-3)(c^2+3c+9)$

46. $d^3+8 = d^3+2^3 = (d+2)(d^2-2d+4)$

47. $2x^3+54 = 2\left(x^3+27\right)$
$= 2\left(x^3+3^3\right)$
$= 2(x+3)\left(x^2-3x+9\right)$

48. $2ab^4-2ab = 2ab\left(b^3-1\right)$
$= 2ab\left(b^3-1^3\right)$
$= 2ab(b-1)\left(b^2+b+1\right)$

49. $3x^2y - xy^2 - 6xy + 2y^2 = y\left(3x^2 - xy - 6x + 2y\right) = y[x(3x - y) - 2(3x - y)]$
$= y(3x - y)(x - 2)$

50. $5x^2 + 10x - 15xy - 30y = 5\left(x^2 + 2x - 3xy - 6y\right) = 5[x(x + 2) - 3y(x + 2)]$
$= 5(x + 2)(x - 3y)$

51. $2a^2x + 2abx + a^3 + a^2b = a\left(2ax + 2bx + a^2 + ab\right) = a[2x(a + b) + a(a + b)]$
$= a(a + b)(2x + a)$

52. $x^2 + 2ax + a^2 - y^2 = \left(x^2 + 2ax + a^2\right) - y^2 = (x + a)(x + a) - y^2$
$= (x + a)^2 - y^2 = (x + a + y)(x + a - y)$

53. $x^2 - 4 + bx + 2b = x^2 - 2^2 + bx + 2b = (x + 2)(x - 2) + b(x + 2) = (x + 2)(x - 2 + b)$

54. $ax^6 - ay^6 = a\left(x^6 - y^6\right) = a\left[\left(x^3\right)^2 - \left(y^3\right)^2\right] = a\left(x^3 + y^3\right)\left(x^3 - y^3\right)$
$= a(x + y)\left(x^2 - xy + y^2\right)(x - y)\left(x^2 + xy + y^2\right)$

55. $x^2 + 2x = 0$
$x(x + 2) = 0$
$x = 0$ **or** $x + 2 = 0$
$x = 0$ $\quad x = -2$

56. $2x^2 - 6x = 0$
$2x(x - 3) = 0$
$2x = 0$ **or** $x - 3 = 0$
$x = 0$ $\quad x = 3$

57. $3x^2 = 2x$
$3x^2 - 2x = 0$
$x(3x - 2) = 0$
$x = 0$ **or** $3x - 2 = 0$
$3x = 2$
$x = \frac{2}{3}$

58. $5x^2 + 25x = 0$
$5x(x + 5) = 0$
$5x = 0$ **or** $x + 5 = 0$
$x = \frac{0}{5}$ $\quad x = -5$
$x = 0$

59. $x^2 - 9 = 0$
$(x + 3)(x - 3) = 0$
$x + 3 = 0$ **or** $x - 3 = 0$
$x = -3$ $\quad x = 3$

60. $x^2 - 25 = 0$
$(x + 5)(x - 5) = 0$
$x + 5 = 0$ **or** $x - 5 = 0$
$x = -5$ $\quad x = 5$

61. $a^2 - 7a + 12 = 0$
$(a - 3)(a - 4) = 0$
$a - 3 = 0$ **or** $a - 4 = 0$
$a = 3$ $\quad a = 4$

62. $x^2 - 2x - 15 = 0$
$(x + 3)(x - 5) = 0$
$x + 3 = 0$ **or** $x - 5 = 0$
$x = -3$ $\quad x = 5$

63.
$$\begin{aligned} 2x - x^2 + 24 &= 0 \\ x^2 - 2x - 24 &= 0 \\ (x+4)(x-6) &= 0 \end{aligned}$$
$$x + 4 = 0 \quad \textbf{or} \quad x - 6 = 0$$
$$x = -4 \qquad\qquad x = 6$$

64.
$$\begin{aligned} 16 + x^2 - 10x &= 0 \\ x^2 - 10x + 16 &= 0 \\ (x-2)(x-8) &= 0 \end{aligned}$$
$$x - 2 = 0 \quad \textbf{or} \quad x - 8 = 0$$
$$x = 2 \qquad\qquad x = 8$$

65.
$$\begin{aligned} 2x^2 - 5x - 3 &= 0 \\ (2x+1)(x-3) &= 0 \end{aligned}$$
$$2x + 1 = 0 \quad \textbf{or} \quad x - 3 = 0$$
$$2x = -1 \qquad\qquad x = 3$$
$$x = -\tfrac{1}{2} \qquad\qquad x = 3$$

66.
$$\begin{aligned} 2x^2 + x - 3 &= 0 \\ (2x+3)(x-1) &= 0 \end{aligned}$$
$$2x + 3 = 0 \quad \textbf{or} \quad x - 1 = 0$$
$$2x = -3 \qquad\qquad x = 1$$
$$x = -\tfrac{3}{2} \qquad\qquad x = 1$$

67.
$$\begin{aligned} 4x^2 &= 1 \\ 4x^2 - 1 &= 0 \\ (2x+1)(2x-1) &= 0 \end{aligned}$$
$$2x + 1 = 0 \quad \textbf{or} \quad 2x - 1 = 0$$
$$2x = -1 \qquad\qquad 2x = 1$$
$$x = -\tfrac{1}{2} \qquad\qquad x = \tfrac{1}{2}$$

68.
$$\begin{aligned} 9x^2 &= 4 \\ 9x^2 - 4 &= 0 \\ (3x+2)(3x-2) &= 0 \end{aligned}$$
$$3x + 2 = 0 \quad \textbf{or} \quad 3x - 2 = 0$$
$$3x = -2 \qquad\qquad 3x = 2$$
$$x = -\tfrac{2}{3} \qquad\qquad x = \tfrac{2}{3}$$

69.
$$\begin{aligned} x^3 - 7x^2 + 12x &= 0 \\ x\left(x^2 - 7x + 12\right) &= 0 \\ x(x-3)(x-4) &= 0 \end{aligned}$$
$$x = 0 \quad \textbf{or} \quad x - 3 = 0 \quad \textbf{or} \quad x - 4 = 0$$
$$x = 0 \qquad\qquad x = 3 \qquad\qquad x = 4$$

70.
$$\begin{aligned} x^3 + 5x^2 + 6x &= 0 \\ x\left(x^2 + 5x + 6\right) &= 0 \\ x(x+2)(x+3) &= 0 \end{aligned}$$
$$x = 0 \quad \textbf{or} \quad x + 2 = 0 \quad \textbf{or} \quad x + 3 = 0$$
$$x = 0 \qquad\qquad x = -2 \qquad\qquad x = -3$$

71.
$$\begin{aligned} 2x^3 + 5x^2 &= 3x \\ 2x^3 + 5x^2 - 3x &= 0 \\ x\left(2x^2 + 5x - 3\right) &= 0 \\ x(2x-1)(x+3) &= 0 \end{aligned}$$
$$x = 0 \quad \textbf{or} \quad 2x - 1 = 0 \quad \textbf{or} \quad x + 3 = 0$$
$$x = 0 \qquad\qquad 2x = 1 \qquad\qquad x = -3$$
$$x = 0 \qquad\qquad x = \tfrac{1}{2} \qquad\qquad x = -3$$

72.
$$\begin{aligned} 3x^3 - 2x &= x^2 \\ 3x^3 - x^2 - 2x &= 0 \\ x\left(3x^2 - x - 2\right) &= 0 \\ x(3x+2)(x-1) &= 0 \end{aligned}$$
$$x = 0 \quad \textbf{or} \quad 3x + 2 = 0 \quad \textbf{or} \quad x - 1 = 0$$
$$x = 0 \qquad\qquad 3x = -2 \qquad\qquad x = 1$$
$$x = 0 \qquad\qquad x = -\tfrac{2}{3} \qquad\qquad x = 1$$

73. Let $x =$ one number. Then $12 - x =$ the other number.
$$\begin{aligned} x(12 - x) &= 35 \\ 12x - x^2 &= 35 \\ 0 &= x^2 - 12x + 35 \\ 0 &= (x-5)(x-7) \end{aligned}$$
$$x - 5 = 0 \quad \textbf{or} \quad x - 7 = 0$$
$$x = 5 \qquad\qquad x = 7$$
The numbers are 5 and 7.

74. Let $x =$ the positive number.
$$\begin{aligned} 3x^2 + 5x &= 2 \\ 3x^2 + 5x - 2 &= 0 \\ (3x-1)(x+2) &= 0 \end{aligned}$$
$$3x - 1 = 0 \quad \textbf{or} \quad x + 2 = 0$$
$$3x = 1 \qquad\qquad x = -2$$
$$x = \tfrac{1}{3} \qquad\qquad x = -2$$
The answer of -2 is not positive, so the number is $\frac{1}{3}$.

75. Let $w =$ the width and $w + 2 =$ the length.
Area = Length · Width

$$\begin{aligned} 48 &= (w+2)w \\ 0 &= w^2 + 2w - 48 \\ 0 &= (w+8)(w-6) \\ w+8 = 0 \quad &\textbf{or} \quad w-6=0 \\ w = -8 \quad & \quad\quad w = 6 \end{aligned}$$

Since the answer $w = -8$ does not make sense, the dimensions are 6 ft by 8 ft.

76. Let $w =$ the width and $2w + 3 =$ the length.
Area = Length · Width

$$\begin{aligned} 27 &= (2w+3)w \\ 0 &= 2w^2 + 3w - 27 \\ 0 &= (2w+9)(w-3) \\ 2w+9 = 0 \quad &\textbf{or} \quad w-3=0 \\ 2w = -9 \quad & \quad\quad w = 3 \\ w = -\tfrac{9}{2} \quad & \quad\quad w = 3 \end{aligned}$$

Since the answer $w = -\frac{9}{2}$ does not make sense, the dimensions are 3 ft by 9 ft.

77. Let $w =$ the width and $w + 3 =$ the length.

$$\begin{aligned} \text{Area} &= \text{Perimeter} \\ (w+3)w &= 2w + 2(w+3) \\ w^2 + 3w &= 2w + 2w + 6 \\ w^2 + 3w &= 4w + 6 \\ w^2 - w - 6 &= 0 \\ (w+2)(w-3) &= 0 \\ w+2 = 0 \quad &\textbf{or} \quad w-3=0 \\ w = -2 \quad & \quad\quad w = 3 \end{aligned}$$

Since the answer $w = -2$ does not make sense, the dimensions are 3 ft by 6 ft.

78. Let $x =$ the base and $x + 1 =$ the height.
Area $= \frac{1}{2}bh$

$$\begin{aligned} 21 &= \tfrac{1}{2}x(x+1) \\ 42 &= x(x+1) \\ 42 &= x^2 + x \\ 0 &= x^2 + x - 42 \\ 0 &= (x+7)(x-6) \\ x+7 = 0 \quad &\textbf{or} \quad x-6=0 \\ x = -7 \quad & \quad\quad x = 6 \end{aligned}$$

Since the answer $x = -7$ does not make sense, the base is 6 ft and the height is 7 ft.

Chapter 5 Test (page 381)

1. $196 = 14 \cdot 14 = 2 \cdot 7 \cdot 2 \cdot 7 = 2^2 \cdot 7^2$

2. $111 = 3 \cdot 37$

3. $60ab^2c^3 + 30a^3b^2c - 25a = 5a(12b^2c^3 + 6a^2b^2c - 5)$

4. $3x^2(a+b) - 6xy(a+b) = 3x[x(a+b) - 2y(a+b)] = 3x(a+b)(x-2y)$

5.
$$\begin{aligned} ax + ay + bx + by &= a(x+y) + b(x+y) \\ &= (x+y)(a+b) \end{aligned}$$

6. $x^2 - 25 = x^2 - 5^2 = (x+5)(x-5)$

7. $3a^2 - 27b^2 = 3(a^2 - 9b^2) = 3\left(a^2 - (3b)^2\right) = 3(a+3b)(a-3b)$

8.
$$\begin{aligned} 16x^4 - 81y^4 = \left(4x^2\right)^2 - \left(9y^2\right)^2 &= \left(4x^2 + 9y^2\right)\left(4x^2 - 9y^2\right) \\ &= \left(4x^2 + 9y^2\right)\left((2x)^2 - (3y)^2\right) \\ &= \left(4x^2 + 9y^2\right)(2x+3y)(2x-3y) \end{aligned}$$

9. $x^2 + 4x + 3 = (x+3)(x+1)$

10. $x^2 - 9x - 22 = (x-11)(x+2)$

CHAPTER 5 TEST

11. $x^2+10xy+9y^2=(x+y)(x+9y)$

12. $6x^2-30xy+24y^2=6(x^2-5xy+4y^2)$
$=6(x-4y)(x-y)$

13. $3x^2+13x+4=(3x+1)(x+4)$

14. $2a^2+5a-12=(2a-3)(a+4)$

15. $2x^2+3xy-2y^2=(2x-y)(x+2y)$

16. $12-25x+12x^2=12x^2-25x+12$
$=(4x-3)(3x-4)$

17. $12a^2+6ab-36b^2=6(2a^2+ab-6b^2)$
$=6(2a-3b)(a+2b)$

18. $x^3-64=x^3-4^3=(x-4)(x^2+4x+16)$

19. $216+8a^3=8(27+a^3)=8(3^3+a^3)=8(3+a)(9-3a+a^2)$

20. $x^9z^3-y^3z^6=z^3(x^9-y^3z^3)=z^3\left((x^3)^3-(yz)^3\right)=z^3(x^3-yz)\left((x^3)^2+x^3yz+(yz)^2\right)$
$=z^3(x^3-yz)(x^6+x^3yz+y^2z^2)$

21.
$x^2+3x=0$
$x(x+3)=0$
$x=0$ **or** $x+3=0$
$x=0$ $\quad$ $x=-3$

22.
$2x^2+5x+3=0$
$(2x+3)(x+1)=0$
$2x+3=0$ **or** $x+1=0$
$2x=-3$ $\quad$ $x=-1$
$x=-\frac{3}{2}$ $\quad$ $x=-1$

23.
$9y^2-81=0$
$9(y^2-9)=0$
$9(y+3)(y-3)=0$
$y+3=0$ **or** $y-3=0$
$y=-3$ $\quad$ $y=3$

24.
$-3(y-6)+2=y^2+2$
$-3y+18+2=y^2+2$
$-3y+20=y^2+2$
$0=y^2+3y-18$
$0=(y+6)(y-3)$
$y+6=0$ **or** $y-3=0$
$y=-6$ $\quad$ $y=3$

25.
$10x^2-13x=9$
$10x^2-13x-9=0$
$(2x+1)(5x-9)=0$
$2x+1=0$ **or** $5x-9=0$
$2x=-1$ $\quad$ $5x=9$
$x=-\frac{1}{2}$ $\quad$ $x=\frac{9}{5}$

26.
$10x^2-x=9$
$10x^2-x-9=0$
$(10x+9)(x-1)=0$
$10x+9=0$ **or** $x-1=0$
$10x=-9$ $\quad$ $x=1$
$x=-\frac{9}{10}$ $\quad$ $x=1$

27.
$$10x^2 + 43x = 9$$
$$10x^2 + 43x - 9 = 0$$
$$(2x + 9)(5x - 1) = 0$$
$$2x + 9 = 0 \quad \textbf{or} \quad 5x - 1 = 0$$
$$2x = -9 \qquad 5x = 1$$
$$x = -\tfrac{9}{2} \qquad x = \tfrac{1}{5}$$

28.
$$10x^2 - 89x = 9$$
$$10x^2 - 89x - 9 = 0$$
$$(10x + 1)(x - 9) = 0$$
$$10x + 1 = 0 \quad \textbf{or} \quad x - 9 = 0$$
$$10x = -1 \qquad x = 9$$
$$x = -\tfrac{1}{10} \qquad x = 9$$

29. Let $v = 192$ and $h = 0$:
$$h = vt - 16t^2$$
$$0 = 192t - 16t^2$$
$$0 = 16t(12 - t)$$
$$16t = 0 \quad \textbf{or} \quad 12 - t = 0$$
$$t = 0 \qquad 12 = t$$
Since $t = 0$ is when the cannonball was fired, it will hit the ground in 12 seconds.

30. Let $h =$ the height and $h + 2 =$ the base.
$$A = \tfrac{1}{2}bh$$
$$40 = \tfrac{1}{2}(h + 2)h$$
$$80 = (h + 2)h$$
$$0 = h^2 + 2h - 80$$
$$0 = (h + 10)(h - 8)$$
$$h + 10 = 0 \quad \textbf{or} \quad h - 8 = 0$$
$$h = -10 \qquad h = 8$$
Since the answer $h = -10$ does not make sense, the height is 8 m and the base is 10 m.

Exercise 6.1 (page 390)

1. $\dfrac{14}{21} = \dfrac{2 \cdot 7}{3 \cdot 7} = \dfrac{2}{3}$

3. $\dfrac{xyz}{wxy} = \dfrac{z \cdot xy}{w \cdot xy} = \dfrac{z}{w}$

5. $\dfrac{6x^2y}{6xy^2} = \dfrac{6xy \cdot x}{6xy \cdot y} = \dfrac{x}{y}$

7. $\dfrac{x + y}{y + x} = \dfrac{x + y}{x + y} = 1$

9. If a, b and c are real numbers, then $(a + b) + c = a + (b + c)$.

11. 0 is the additive identity.

13 $\frac{5}{3}$ is the additive inverse of $-\frac{5}{3}$.

15. numerator

17. 0

19. negatives (or opposites)

21. $\dfrac{a}{b}$

23. factor; common

25.
$$y - 2 = 0$$
$$y = 2$$

27.
$$3a - 2 = 0$$
$$3a = 2$$
$$a = \tfrac{2}{3}$$

29.
$$x^2 - x - 2 = 0$$
$$(x + 1)(x - 2) = 0$$
$$x + 1 = 0 \quad \textbf{or} \quad x - 2 = 0$$
$$x = -1 \qquad x = 2$$

31.
$$2m^2 - m - 3 = 0$$
$$(2m - 3)(m + 1) = 0$$
$$2m - 3 = 0 \quad \textbf{or} \quad m + 1 = 0$$
$$2m = 3 \qquad m = -1$$
$$m = \tfrac{3}{2} \qquad m = -1$$

33. $\dfrac{28}{35} = \dfrac{7 \cdot 4}{7 \cdot 5} = \dfrac{4}{5}$

35. $\dfrac{4x}{2} = \dfrac{2 \cdot 2x}{2 \cdot 1} = 2x$

37. $\dfrac{-6x}{18} = -\dfrac{6 \cdot x}{6 \cdot 3} = -\dfrac{x}{3}$

39. $\dfrac{2x^2}{3y} \Rightarrow$ simplest form

41. $\dfrac{(3+4)a}{24-3} = \dfrac{7a}{21} = \dfrac{7 \cdot a}{7 \cdot 3} = \dfrac{a}{3}$

43. $\dfrac{x+3}{3(x+3)} = \dfrac{1(x+3)}{3(x+3)} = \dfrac{1}{3}$

45. $\dfrac{2(x+7)}{x+7} = \dfrac{2(x+7)}{1(x+7)} = 2$

47. $\dfrac{x^2+3x}{2x+6} = \dfrac{x(x+3)}{2(x+3)} = \dfrac{x}{2}$

49. $\dfrac{x^2+3x+2}{x^2+x-2} = \dfrac{(x+2)(x+1)}{(x+2)(x-1)} = \dfrac{x+1}{x-1}$

51. $\dfrac{x^2-8x+15}{x^2-x-6} = \dfrac{(x-5)(x-3)}{(x+2)(x-3)} = \dfrac{x-5}{x+2}$

53. $\dfrac{2x^2-8x}{x^2-6x+8} = \dfrac{2x(x-4)}{(x-2)(x-4)} = \dfrac{2x}{x-2}$

55. $\dfrac{2a^3-16}{2a^2+4a+8} = \dfrac{2(a^3-8)}{2(a^2+2a+4)} = \dfrac{a^3-2^3}{a^2+2a+4} = \dfrac{(a-2)(a^2+2a+2^2)}{a^2+2a+4}$

$= \dfrac{(a-2)(a^2+2a+4)}{a^2+2a+4} = a-2$

57. $\dfrac{4(x+3)+4}{3(x+2)+6} = \dfrac{4x+12+4}{3x+6+6} = \dfrac{4x+16}{3x+12} = \dfrac{4(x+4)}{3(x+4)} = \dfrac{4}{3}$

59. $\dfrac{x^2+5x+4}{2(x+3)-(x+2)} = \dfrac{x^2+5x+4}{2x+6-x-2} = \dfrac{x^2+5x+4}{x+4} = \dfrac{(x+1)(x+4)}{x+4} = x+1$

61. $\dfrac{x^3+1}{ax+a+x+1} = \dfrac{x^3+1^3}{a(x+1)+1(x+1)} = \dfrac{(x+1)(x^2-1x+1^2)}{(x+1)(a+1)} = \dfrac{x^2-x+1}{a+1}$

63. $\dfrac{ab+b+2a+2}{ab+a+b+1} = \dfrac{b(a+1)+2(a+1)}{a(b+1)+1(b+1)} = \dfrac{(a+1)(b+2)}{(b+1)(a+1)} = \dfrac{b+2}{b+1}$

65. $\dfrac{x-7}{7-x} = \dfrac{-(7-x)}{7-x} = -1$

67. $\dfrac{6x-3y}{3y-6x} = \dfrac{-(3y-6x)}{3y-6x} = -1$

69. $\dfrac{45}{9a} = \dfrac{9 \cdot 5}{9 \cdot a} = \dfrac{5}{a}$

71. $\dfrac{7+3}{5z} = \dfrac{10}{5z} = \dfrac{5 \cdot 2}{5 \cdot z} = \dfrac{2}{z}$

73. $\dfrac{15x^2y}{5xy^2} = \dfrac{5xy \cdot 3x}{5xy \cdot y} = \dfrac{3x}{y}$

75. $\dfrac{x^2+3x+2}{x^3+x^2} = \dfrac{(x+2)(x+1)}{x^2(x+1)} = \dfrac{x+2}{x^2}$

77. $\dfrac{14xz^2}{7x^2z^2} = \dfrac{7xz^2 \cdot 2}{7xz^2 \cdot x} = \dfrac{2}{x}$

79. $\dfrac{3x+15}{x^2-25} = \dfrac{3(x+5)}{(x+5)(x-5)} = \dfrac{3}{x-5}$

81. $\dfrac{a+b-c}{c-a-b} = \dfrac{-(c-a-b)}{c-a-b} = -1$

83. $\dfrac{6a-6b+6c}{9a-9b+9c} = \dfrac{6(a-b+c)}{9(a-b+c)} = \dfrac{6}{9} = \dfrac{2}{3}$

85. $\frac{3x+3y}{x^2+xy}=\frac{3(x+y)}{x(x+y)}=\frac{3}{x}$

87. $\frac{2x^2-8}{x^2-3x+2}=\frac{2(x^2-4)}{(x-1)(x-2)}=\frac{2(x+2)(x-2)}{(x-1)(x-2)}=\frac{2(x+2)}{x-1}$

89. $\frac{x^2-2x-15}{x^2+2x-15}=\frac{(x-5)(x+3)}{(x+5)(x-3)}$

91. $\frac{15x-3x^2}{25y-5xy}=\frac{3x(5-x)}{5y(5-x)}=\frac{3x}{5y}$

93. $\frac{4+2(x-5)}{3x-5(x-2)}=\frac{4+2x-10}{3x-5x+10}=\frac{2x-6}{-2x+10}=\frac{2(x-3)}{-2(x-5)}=\frac{x-3}{5-x}$

95. $\frac{x^3+1}{x^2-x+1}=\frac{x^3+1^3}{x^2-x+1}=\frac{(x+1)(x^2-1x+1^2)}{x^2-x+1}=\frac{(x+1)(x^2-x+1)}{x^2-x+1}=x+1$

97. $\frac{xy+3y+3x+9}{x^2-9}=\frac{y(x+3)+3(x+3)}{(x+3)(x-3)}=\frac{(x+3)(y+3)}{(x+3)(x-3)}=\frac{y+3}{x-3}$

99. **Answers may vary.**

101. $\frac{x-3}{5-x}=\frac{x-3}{-(-5+x)}=-\frac{x-3}{x-5}$

Exercise 6.2 (page 399)

1. $\frac{x}{2}\cdot\frac{3}{x}=\frac{x\cdot 3}{2\cdot x}=\frac{3}{2}$

3. $\frac{5}{x+7}\cdot(x+7)=\frac{5}{x+7}\cdot\frac{x+7}{1}$
$=\frac{5(x+7)}{1(x+7)}=5$

5. $\frac{3}{4}\div 3=\frac{3}{4}\div\frac{3}{1}=\frac{3}{4}\cdot\frac{1}{3}=\frac{3\cdot 1}{4\cdot 3}=\frac{1}{4}$

7. $2x^3y^2\left(-3x^2y^4z\right)=-6x^5y^6z$

9. $(3y)^{-4}=\frac{1}{(3y)^4}=\frac{1}{81y^4}$

11. $\frac{x^{3m}}{x^{4m}}=x^{-m}=\frac{1}{x^m}$

13. $-4(y^3-4y^2+3y-2)+6(-2y^2+4)-4(-2y^3-y)$
$=-4y^3+16y^2-12y+8-12y^2+24+8y^3+4y$
$=4y^3+4y^2-8y+32$

15. numerator

17. numerators; denominators

19. 1

21. divisor; multiply

23. $\frac{5}{7}\cdot\frac{9}{13}=\frac{5\cdot 9}{7\cdot 13}=\frac{45}{91}$

25. $\frac{2y}{z}\cdot\frac{z}{3}=\frac{2y\cdot z}{z\cdot 3}=\frac{2y}{3}$

SECTION 6.2

27. $\dfrac{4x}{3y} \cdot \dfrac{3y}{7x} = \dfrac{4 \cdot x \cdot 3 \cdot y}{3 \cdot y \cdot 7 \cdot x} = \dfrac{4}{7}$

29. $\dfrac{-2xy}{x^2} \cdot \dfrac{3xy}{2} = -\dfrac{6x^2y^2}{2x^2} = -3y^2$

31. $\dfrac{ab^2}{a^2b} \cdot \dfrac{b^2c^2}{abc} \cdot \dfrac{abc^2}{a^3c^2} = \dfrac{a^2b^5c^4}{a^6b^2c^3} = \dfrac{b^3c}{a^4}$

33. $\dfrac{z+7}{7} \cdot \dfrac{z+2}{z} = \dfrac{(z+7)(z+2)}{7z}$

35. $\dfrac{(x+1)^2}{x+1} \cdot \dfrac{x+2}{x+1} = \dfrac{(x+2)(x+1)^2}{(x+1)^2}$
$= x+2$

37. $\dfrac{3y-9}{y-3} \cdot \dfrac{y}{3y^2} = \dfrac{3(y-3)}{y-3} \cdot \dfrac{y}{3y^2} = \dfrac{3y(y-3)}{3y^2(y-3)}$
$= \frac{1}{y}$

39. $\dfrac{7y-14}{y-2} \cdot \dfrac{x^2}{7x} = \dfrac{7(y-2)}{y-2} \cdot \dfrac{x^2}{7x} = \dfrac{7x^2(y-2)}{7x(y-2)} = x$

41. $\dfrac{abc^2}{a+1} \cdot \dfrac{c}{a^2b^2} \cdot \dfrac{a^2+a}{ac} = \dfrac{abc^2}{a+1} \cdot \dfrac{c}{a^2b^2} \cdot \dfrac{a(a+1)}{ac} = \dfrac{a^2bc^3(a+1)}{a^3b^2c(a+1)} = \dfrac{c^2}{ab}$

43. $\dfrac{5z-10}{z+2} \cdot \dfrac{3}{3z-6} = \dfrac{5(z-2)}{z+2} \cdot \dfrac{3}{3(z-2)} = \dfrac{15(z-2)}{3(z+2)(z-2)} = \dfrac{5}{z+2}$

45. $\dfrac{z^2+4z-5}{5z-5} \cdot \dfrac{5z}{z+5} = \dfrac{(z+5)(z-1)}{5(z-1)} \cdot \dfrac{5z}{z+5} = \dfrac{5z(z+5)(z-1)}{5(z-1)(z+5)} = z$

47. $\dfrac{3x^2+5x+2}{x^2-9} \cdot \dfrac{x-3}{x^2-4} \cdot \dfrac{x^2+5x+6}{6x+4} = \dfrac{(3x+2)(x+1)}{(x+3)(x-3)} \cdot \dfrac{x-3}{(x+2)(x-2)} \cdot \dfrac{(x+2)(x+3)}{2(3x+2)}$
$= \dfrac{(3x+2)(x+1)(x-3)(x+2)(x+3)}{2(x+3)(x-3)(x+2)(x-2)(3x+2)} = \dfrac{x+1}{2(x-2)}$

49. $\dfrac{a^2-ab+b^2}{a^3+b^3} \cdot \dfrac{ac+ad+bc+bd}{c^2-d^2} = \dfrac{a^2-ab+b^2}{(a+b)(a^2-ab+b^2)} \cdot \dfrac{a(c+d)+b(c+d)}{(c+d)(c-d)}$
$= \dfrac{a^2-ab+b^2}{(a+b)(a^2-ab+b^2)} \cdot \dfrac{(c+d)(a+b)}{(c+d)(c-d)}$
$= \dfrac{(a^2-ab+b^2)(c+d)(a+b)}{(a+b)(a^2-ab+b^2)(c+d)(c-d)} = \dfrac{1}{c-d}$

51. $\dfrac{1}{3} \div \dfrac{1}{2} = \dfrac{1}{3} \cdot \dfrac{2}{1} = \dfrac{2}{3}$

53. $\dfrac{21}{14} \div \dfrac{5}{2} = \dfrac{21}{14} \cdot \dfrac{2}{5} = \dfrac{3 \cdot 7 \cdot 2}{2 \cdot 7 \cdot 5} = \dfrac{3}{5}$

55. $\dfrac{x^2y}{3xy} \div \dfrac{xy^2}{6y} = \dfrac{x^2y}{3xy} \cdot \dfrac{6y}{xy^2} = \dfrac{6x^2y^2}{3x^2y^3} = \dfrac{2}{y}$

57. $\dfrac{x+2}{3x} \div \dfrac{x+2}{2} = \dfrac{x+2}{3x} \cdot \dfrac{2}{x+2}$
$= \dfrac{2(x+2)}{3x(x+2)} = \dfrac{2}{3x}$

59. $\dfrac{x^2-4}{3x+6} \div \dfrac{x-2}{x+2} = \dfrac{x^2-4}{3x+6} \cdot \dfrac{x+2}{x-2} = \dfrac{(x+2)(x-2)(x+2)}{3(x+2)(x-2)} = \dfrac{x+2}{3}$

SECTION 6.2

61. $\dfrac{y(y+2)}{y^2(y-3)} \div \dfrac{y^2(y+2)}{(y-3)^2} = \dfrac{y(y+2)}{y^2(y-3)} \cdot \dfrac{(y-3)^2}{y^2(y+2)} = \dfrac{y(y+2)(y-3)^2}{y^4(y-3)(y+2)} = \dfrac{y-3}{y^3}$

63. $\dfrac{5x^2+13x-6}{x+3} \div \dfrac{5x^2-17x+6}{x-2} = \dfrac{5x^2+13x-6}{x+3} \cdot \dfrac{x-2}{5x^2-17x+6}$
$= \dfrac{(5x-2)(x+3)(x-2)}{(x+3)(5x-2)(x-3)} = \dfrac{x-2}{x-3}$

65. $\dfrac{ab+4a+2b+8}{b^2+4b+16} \div \dfrac{b^2-16}{b^3-64} = \dfrac{ab+4a+2b+8}{b^2+4b+16} \cdot \dfrac{b^3-64}{b^2-16}$
$= \dfrac{a(b+4)+2(b+4)}{b^2+4b+16} \cdot \dfrac{b^3-4^3}{(b+4)(b-4)}$
$= \dfrac{(b+4)(a+2)}{b^2+4b+16} \cdot \dfrac{(b-4)(b^2+4b+16)}{(b+4)(b-4)}$
$= \dfrac{(b+4)(a+2)(b-4)(b^2+4b+16)}{(b^2+4b+16)(b+4)(b-4)} = a+2$

67. $\dfrac{x-5}{2x-8} \cdot (x-4) = \dfrac{x-5}{2(x-4)} \cdot \dfrac{x-4}{1} = \dfrac{(x-5)(x-4)}{2(x-4)} = \dfrac{x-5}{2}$

69. $\dfrac{3x+9}{x+1} \div (x+3) = \dfrac{3(x+3)}{x+1} \div \dfrac{x+3}{1} = \dfrac{3(x+3)}{x+1} \cdot \dfrac{1}{x+3} = \dfrac{3(x+3)}{(x+1)(x+3)} = \dfrac{3}{x+1}$

71. $\dfrac{x}{3} \cdot \dfrac{9}{4} \div \dfrac{x^2}{6} = \dfrac{x}{3} \cdot \dfrac{9}{4} \cdot \dfrac{6}{x^2}$
$= \dfrac{x \cdot 3 \cdot 3 \cdot 2 \cdot 3}{3 \cdot 2 \cdot 2 \cdot x \cdot x} = \dfrac{9}{2x}$

73. $\dfrac{x^2}{18} \div \dfrac{x^3}{6} \div \dfrac{12}{x^2} = \dfrac{x^2}{18} \cdot \dfrac{6}{x^3} \cdot \dfrac{x^2}{12}$
$= \dfrac{6x^4}{216x^3} = \dfrac{x}{36}$

75. $\dfrac{2}{3x-3} \div \dfrac{2x+2}{x-1} \cdot \dfrac{5}{x+1} = \dfrac{2}{3x-3} \cdot \dfrac{x-1}{2x+2} \cdot \dfrac{5}{x+1} = \dfrac{10(x-1)}{3(x-1)(2)(x+1)(x+1)}$
$= \dfrac{5}{3(x+1)^2}$

77. $\dfrac{x^2+x-6}{x^2-4} \cdot \dfrac{x^2+2x}{x-2} \div \dfrac{x^2+3x}{x+2} = \dfrac{x^2+x-6}{x^2-4} \cdot \dfrac{x^2+2x}{x-2} \cdot \dfrac{x+2}{x^2+3x}$
$= \dfrac{(x+3)(x-2)(x)(x+2)(x+2)}{(x+2)(x-2)(x-2)(x)(x+3)} = \dfrac{x+2}{x-2}$

79. $\dfrac{25}{35} \cdot \dfrac{-21}{55} = \dfrac{-5 \cdot 5 \cdot 3 \cdot 7}{7 \cdot 5 \cdot 5 \cdot 11} = -\dfrac{3}{11}$

81. $\dfrac{2}{3} \cdot \dfrac{15}{2} \cdot \dfrac{1}{7} = \dfrac{2 \cdot 3 \cdot 5}{3 \cdot 2 \cdot 7} = \dfrac{5}{7}$

83. $\dfrac{2}{y} \div \dfrac{4}{3} = \dfrac{2}{y} \cdot \dfrac{3}{4} = \dfrac{2 \cdot 3}{y \cdot 2 \cdot 2} = \dfrac{3}{2y}$

85. $\dfrac{3x}{2} \div \dfrac{x}{2} = \dfrac{3x}{2} \cdot \dfrac{2}{x} = \dfrac{3 \cdot x \cdot 2}{2 \cdot x} = 3$

87. $\dfrac{7z}{9z} \cdot \dfrac{4z}{2z} = \dfrac{7 \cdot z \cdot 2 \cdot 2 \cdot z}{9 \cdot z \cdot 2 \cdot z} = \dfrac{14}{9}$

89. $\dfrac{2x^2y}{3xy} \cdot \dfrac{3xy^2}{2} = \dfrac{6x^3y^3}{6xy} = x^2y^2$

SECTION 6.2

91. $\dfrac{8x^2y^2}{4x^2} \cdot \dfrac{2xy}{2y} = \dfrac{16x^3y^3}{8x^2y} = 2xy^2$

93. $\dfrac{10r^2st^3}{6rs^2} \cdot \dfrac{3r^3t}{2rst} \cdot \dfrac{2s^3t^4}{5s^2t^3} = \dfrac{60r^5s^4t^8}{60r^2s^5t^4} = \dfrac{r^3t^4}{s}$

95. $\dfrac{3x}{y} \div \dfrac{2x}{4} = \dfrac{3x}{y} \cdot \dfrac{4}{2x} = \dfrac{3 \cdot x \cdot 2 \cdot 2}{y \cdot 2 \cdot x} = \dfrac{6}{y}$

97. $\dfrac{4x}{3x} \div \dfrac{2y}{9y} = \dfrac{4x}{3x} \cdot \dfrac{9y}{2y} = \dfrac{2 \cdot 2 \cdot x \cdot 3 \cdot 3 \cdot y}{3 \cdot x \cdot 2 \cdot y}$
$= 6$

99. $\dfrac{x^2}{3} \div \dfrac{2x}{4} = \dfrac{x^2}{3} \cdot \dfrac{4}{2x} = \dfrac{x \cdot x \cdot 2 \cdot 2}{3 \cdot 2 \cdot x} = \dfrac{2x}{3}$

101. $\dfrac{x-2}{2} \cdot \dfrac{2x}{x-2} = \dfrac{2x(x-2)}{2(x-2)} = x$

103. $\dfrac{x+5}{5} \cdot \dfrac{x}{x+5} = \dfrac{x(x+5)}{5(x+5)} = \dfrac{x}{5}$

105. $\dfrac{(z-2)^2}{3z^2} \div \dfrac{z-2}{6z} = \dfrac{(z-2)^2}{3z^2} \cdot \dfrac{6z}{z-2} = \dfrac{6z(z-2)^2}{3z^2(z-2)} = \dfrac{2(z-2)}{z}$

107.
$$\begin{aligned}\frac{m^2-2m-3}{2m+4} \cdot \frac{m^2-4}{m^2+3m+2} &= \frac{(m-3)(m+1)}{2(m+2)} \cdot \frac{(m+2)(m-2)}{(m+2)(m+1)} \\ &= \frac{(m-3)(m+1)(m+2)(m-2)}{2(m+2)(m+2)(m+1)} = \frac{(m-3)(m-2)}{2(m+2)}\end{aligned}$$

109.
$$\begin{aligned}\frac{x^2-y^2}{y^2-xy} \cdot \frac{yx^3-y^4}{ax+ay+bx+by} &= \frac{(x+y)(x-y)}{y(y-x)} \cdot \frac{y(x^3-y^3)}{a(x+y)+b(x+y)} \\ &= -\frac{(x+y)}{y} \cdot \frac{y(x-y)(x^2+xy+y^2)}{(x+y)(a+b)} \\ &= -\frac{(x+y) \cdot y(x-y)(x^2+xy+y^2)}{y(x+y)(a+b)} \\ &= -\frac{(x-y)(x^2+xy+y^2)}{a+b}\end{aligned}$$

111. $\dfrac{x^2-1}{3x-3} \div \dfrac{x+1}{3} = \dfrac{x^2-1}{3x-3} \cdot \dfrac{3}{x+1} = \dfrac{(x+1)(x-1)3}{3(x-1)(x+1)} = 1$

113.
$$\begin{aligned}\frac{2x^2+8x-42}{x-3} \div \frac{2x^2+14x}{x^2+5x} &= \frac{2(x^2+4x-21)}{x-3} \cdot \frac{x^2+5x}{2x^2+14x} \\ &= \frac{2(x+7)(x-3)(x)(x+5)}{(x-3)(2x)(x+7)} = x+5\end{aligned}$$

115.
$$\begin{aligned}\frac{x^2+7xy+12y^2}{x^2+2xy-8y^2} \cdot \frac{x^2-xy-2y^2}{x^2+4xy+3y^2} &= \frac{(x+4y)(x+3y)}{(x+4y)(x-2y)} \cdot \frac{(x-2y)(x+y)}{(x+3y)(x+y)} \\ &= \frac{(x+4y)(x+3y)(x-2y)(x+y)}{(x+4y)(x-2y)(x+3y)(x+y)} = 1\end{aligned}$$

117. $\dfrac{p^3-p^2q+pq^2}{mp-mq+np-nq}\div\dfrac{q^3+p^3}{q^2-p^2}=\dfrac{p^3-p^2q+pq^2}{mp-mq+np-nq}\cdot\dfrac{q^2-p^2}{q^3+p^3}$

$$=\frac{p(p^2-pq+q^2)}{m(p-q)+n(p-q)}\cdot\frac{(q+p)(q-p)}{(q+p)(q^2-pq+p^2)}$$

$$=\frac{p(p^2-pq+q^2)}{(p-q)(m+n)}\cdot\frac{(q+p)(q-p)}{(q+p)(q^2-pq+p^2)}$$

$$=\frac{p(p^2-pq+q^2)(q+p)(q-p)}{(p-q)(m+n)(q+p)(q^2-pq+p^2)}=-\frac{p}{m+n}$$

119. $\dfrac{x^2-1}{x^2-9}\cdot\dfrac{x+3}{x+2}\div\dfrac{5}{x+2}=\dfrac{x^2-1}{x^2-9}\cdot\dfrac{x+3}{x+2}\cdot\dfrac{x+2}{5}=\dfrac{(x+1)(x-1)(x+3)(x+2)}{5(x+3)(x-3)(x+2)}$

$$=\frac{(x+1)(x-1)}{5(x-3)}$$

121. $\dfrac{x-x^2}{x^2-4}\left(\dfrac{2x+4}{x+2}\div\dfrac{5}{x+2}\right)=\dfrac{x-x^2}{x^2-4}\left(\dfrac{2x+4}{x+2}\cdot\dfrac{x+2}{5}\right)=\dfrac{x-x^2}{x^2-4}\left(\dfrac{2(x+2)(x+2)}{5(x+2)}\right)$

$$=\frac{x(1-x)}{(x+2)(x-2)}\cdot\frac{2(x+2)}{5}$$

$$=\frac{2x(1-x)}{5(x-2)}$$

123. $\dfrac{y^2}{x+1}\cdot\dfrac{x^2+2x+1}{x^2-1}\div\dfrac{3y}{xy-y}=\dfrac{y^2}{x+1}\cdot\dfrac{(x+1)(x+1)}{(x+1)(x-1)}\cdot\dfrac{xy-y}{3y}=\dfrac{y^2(x+1)(x+1)(y)(x-1)}{(x+1)(x+1)(x-1)(3y)}$

$$=\frac{y^2}{3}$$

125. **Answers may vary.**

127. **Answers may vary.**

129. You always get the original value of x after simplifying.

Exercise 6.3 (page 411)

1. $\dfrac{6}{12}=\dfrac{1\cdot 6}{2\cdot 6}=\dfrac{1}{2}$; equal

3. not equal

5. $\dfrac{3x}{9}=\dfrac{3\cdot x}{3\cdot 3}=\dfrac{x}{3}$; equal

7. $\dfrac{5x}{3x}=\dfrac{x\cdot 5}{x\cdot 3}=\dfrac{5}{3}$; equal

9. $49=7\cdot 7=7^2$

11. $136=4\cdot 34=2\cdot 2\cdot 2\cdot 17=2^3\cdot 17$

13. $102=2\cdot 51=2\cdot 3\cdot 17$

15. $144=16\cdot 9=2\cdot 2\cdot 2\cdot 2\cdot 3\cdot 3=2^4\cdot 3^2$

17. LCD

19. numerators; common denominator

SECTION 6.3

21. $\frac{1}{3}+\frac{1}{3}=\frac{1+1}{3}=\frac{2}{3}$

23. $\frac{2}{9}+\frac{1}{9}=\frac{2+1}{9}=\frac{3}{9}=\frac{1}{3}$

25. $\frac{2x}{y}+\frac{2x}{y}=\frac{2x+2x}{y}=\frac{4x}{y}$

27. $\frac{3x-5}{x-2}+\frac{6x-13}{x-2}=\frac{3x-5+6x-13}{x-2}=\frac{9x-18}{x-2}=\frac{9(x-2)}{x-2}=9$

29. $\frac{35}{72}-\frac{44}{72}=\frac{35-44}{72}=\frac{-9}{72}=-\frac{1}{8}$

31. $\frac{2x}{y}-\frac{x}{y}=\frac{2x-x}{y}=\frac{x}{y}$

33. $\frac{6x-5}{3xy}-\frac{3x-5}{3xy}=\frac{6x-5-(3x-5)}{3xy}=\frac{6x-5-3x+5}{3xy}=\frac{3x}{3xy}=\frac{1}{y}$

35. $\frac{3y-2}{y+3}-\frac{2y-5}{y+3}=\frac{3y-2-(2y-5)}{y+3}=\frac{3y-2-2y+5}{y+3}=\frac{y+3}{y+3}=1$

37. $\frac{13x}{15}+\frac{12x}{15}-\frac{5x}{15}=\frac{13x+12x-5x}{15}=\frac{20x}{15}=\frac{4x}{3}$

39. $\frac{x+1}{x-2}-\frac{2(x-3)}{x-2}+\frac{3(x+1)}{x-2}=\frac{x+1-2(x-3)+3(x+1)}{x-2}=\frac{x+1-2x+6+3x+3}{x-2}$

$=\frac{2x+10}{x-2}=\frac{2(x+5)}{x-2}$

41. $\frac{25}{4}=\frac{25\cdot 5}{4\cdot 5}=\frac{125}{20}$

43. $\frac{8}{x}=\frac{8\cdot xy}{x\cdot xy}=\frac{8xy}{x^2y}$

45. $\frac{3x}{x+1}=\frac{3x(x+1)}{(x+1)(x+1)}=\frac{3x(x+1)}{(x+1)^2}$

47. $\frac{2y}{x}=\frac{2y(x+1)}{x(x+1)}=\frac{2y(x+1)}{x^2+x}$

49. $\frac{z}{z-1}=\frac{z(z+1)}{(z-1)(z+1)}=\frac{z(z+1)}{z^2-1}$

51. $\frac{2}{x+1}=\frac{2(x+2)}{(x+1)(x+2)}=\frac{2(x+2)}{x^2+3x+2}$

53. $2x=2\cdot x$
$6x=2\cdot 3\cdot x$
$\text{LCD}=2\cdot 3\cdot x=6x$

55. $3x=3\cdot x$
$6y=2\cdot 3\cdot y$
$9xy=3^2\cdot x\cdot y$
$\text{LCD}=2\cdot 3^2\cdot x\cdot y=18xy$

57. $x^2-1=(x+1)(x-1)$
$x+1=x+1$
$\text{LCD}=(x+1)(x-1)=x^2-1$

59. $x^2+6x=x(x+6)$
$x+6=x+6$
$x=x$
$\text{LCD}=x(x+6)=x^2+6x$

61. $\frac{1}{2}+\frac{2}{3}=\frac{1\cdot 3}{2\cdot 3}+\frac{2\cdot 2}{3\cdot 2}=\frac{3}{6}+\frac{4}{6}=\frac{7}{6}$

63. $\frac{2y}{9}+\frac{y}{3}=\frac{2y}{9}+\frac{y\cdot 3}{3\cdot 3}=\frac{2y}{9}+\frac{3y}{9}=\frac{5y}{9}$

65. $$\frac{x-1}{x}+\frac{y+1}{y}=\frac{(x-1)y}{x\cdot y}+\frac{(y+1)x}{y\cdot x}=\frac{xy-y}{xy}+\frac{xy+x}{xy}=\frac{2xy+x-y}{xy}$$

67. $$\frac{x+1}{x-1}+\frac{x-1}{x+1}=\frac{(x+1)(x+1)}{(x-1)(x+1)}+\frac{(x-1)(x-1)}{(x+1)(x-1)}=\frac{x^2+2x+1}{(x+1)(x-1)}+\frac{x^2-2x+1}{(x+1)(x-1)}$$
$$=\frac{2x^2+2}{(x+1)(x-1)}$$

69. $$\frac{x+3}{x^2}+\frac{x+5}{2x}=\frac{(x+3)2}{x^2\cdot 2}+\frac{(x+5)x}{2x\cdot x}=\frac{2x+6}{2x^2}+\frac{x^2+5x}{2x^2}=\frac{x^2+7x+6}{2x^2}$$

71. $$\frac{x}{x+1}+\frac{x-1}{x}=\frac{x\cdot x}{(x+1)x}+\frac{(x-1)(x+1)}{x(x+1)}=\frac{x^2}{x(x+1)}+\frac{x^2-1}{x(x+1)}=\frac{2x^2-1}{x(x+1)}$$

73. $$\frac{2}{3}-\frac{5}{6}=\frac{2\cdot 2}{3\cdot 2}-\frac{5}{6}=\frac{4}{6}-\frac{5}{6}=-\frac{1}{6}$$

75. $$\frac{21x}{14}-\frac{5x}{21}=\frac{21x\cdot 3}{14\cdot 3}-\frac{5x\cdot 2}{21\cdot 2}$$
$$=\frac{63x}{42}-\frac{10x}{42}=\frac{53x}{42}$$

77. $$\frac{x+5}{xy}-\frac{x-1}{x^2y}=\frac{(x+5)x}{xy\cdot x}-\frac{x-1}{x^2y}=\frac{x^2+5x}{x^2y}-\frac{x-1}{x^2y}=\frac{x^2+4x+1}{x^2y}$$

79. $$\frac{x}{x-2}+\frac{4+2x}{x^2-4}=\frac{x}{x-2}+\frac{2(x+2)}{(x+2)(x-2)}=\frac{x}{x-2}+\frac{2}{x-2}=\frac{x+2}{x-2}$$

81. $$\frac{x+1}{x+2}-\frac{x^2+1}{x^2-x-6}=\frac{x+1}{x+2}-\frac{x^2+1}{(x+2)(x-3)}=\frac{(x+1)(x-3)}{(x+2)(x-3)}-\frac{x^2+1}{(x+2)(x-3)}$$
$$=\frac{x^2-3x+x-3-x^2-1}{(x+2)(x-3)}$$
$$=\frac{-2x-4}{(x+2)(x-3)}=\frac{-2(x+2)}{(x+2)(x-3)}=-\frac{2}{x-3}$$

83. $$\frac{y+3}{y-1}-\frac{y+4}{1-y}=\frac{y+3}{y-1}+\frac{y+4}{y-1}=\frac{2y+7}{y-1}$$

85. $$\frac{2x}{x^2-3x+2}+\frac{2x}{x-1}-\frac{x}{x-2}=\frac{2x}{(x-2)(x-1)}+\frac{2x}{x-1}-\frac{x}{x-2}$$
$$=\frac{2x}{(x-2)(x-1)}+\frac{2x(x-2)}{(x-1)(x-2)}-\frac{x(x-1)}{(x-2)(x-1)}$$
$$=\frac{2x+2x(x-2)-x(x-1)}{(x-2)(x-1)}$$
$$=\frac{2x+2x^2-4x-x^2+x}{(x-2)(x-1)}$$
$$=\frac{x^2-x}{(x-2)(x-1)}=\frac{x(x-1)}{(x-2)(x-1)}=\frac{x}{x-2}$$

SECTION 6.3

87. $\dfrac{a}{a-1}-\dfrac{2}{a+2}+\dfrac{3(a-2)}{a^2+a-2}=\dfrac{a}{a-1}-\dfrac{2}{a+2}+\dfrac{3(a-2)}{(a+2)(a-1)}$

$$=\frac{a(a+2)}{(a-1)(a+2)}-\frac{2(a-1)}{(a+2)(a-1)}+\frac{3(a-2)}{(a+2)(a-1)}$$

$$=\frac{a(a+2)-2(a-1)+3(a-2)}{(a+2)(a-1)}$$

$$=\frac{a^2+2a-2a+2+3a-6}{(a+2)(a-1)}$$

$$=\frac{a^2+3a-4}{(a+2)(a-1)}=\frac{(a+4)(a-1)}{(a+2)(a-1)}=\frac{a+4}{a+2}$$

89. $x^2-x-6=(x+2)(x-3)$

$x^2-9=(x+3)(x-3)$

$\text{LCD}=(x+2)(x-3)(x+3)$

91. $\dfrac{4}{7y}+\dfrac{10}{7y}=\dfrac{4+10}{7y}=\dfrac{14}{7y}=\dfrac{2}{y}$

93. $\dfrac{y+2}{5z}+\dfrac{y+4}{5z}=\dfrac{y+2+y+4}{5z}=\dfrac{2y+6}{5z}$

95. $\dfrac{9y}{3x}-\dfrac{6y}{3x}=\dfrac{9y-6y}{3x}=\dfrac{3y}{3x}=\dfrac{y}{x}$

97. $\dfrac{x}{3y}+\dfrac{2x}{3y}-\dfrac{x}{3y}=\dfrac{x+2x-x}{3y}=\dfrac{2x}{3y}$

99. $14+\dfrac{10}{y^2}=\dfrac{14}{1}+\dfrac{10}{y^2}=\dfrac{14\cdot y^2}{1\cdot y^2}+\dfrac{10}{y^2}=\dfrac{14y^2}{y^2}+\dfrac{10}{y^2}=\dfrac{14y^2+10}{y^2}$

101. $\dfrac{3x}{y+2}-\dfrac{3y}{y+2}+\dfrac{x+y}{y+2}=\dfrac{3x-3y+x+y}{y+2}=\dfrac{4x-2y}{y+2}=\dfrac{2(2x-y)}{y+2}$

103. $\dfrac{-a}{3a^2-27}+\dfrac{1}{3a+9}=\dfrac{-a}{3(a^2-9)}+\dfrac{1}{3(a+3)}=\dfrac{-a}{3(a+3)(a-3)}+\dfrac{1}{3(a+3)}$

$$=\frac{-a}{3(a+3)(a-3)}+\frac{1(a-3)}{3(a+3)(a-3)}$$

$$=\frac{-a+1(a-3)}{3(a+3)(a-3)}$$

$$=\frac{-a+a-3}{3(a+3)(a-3)}$$

$$=\frac{-3}{3(a+3)(a-3)}=\frac{-1}{(a+3)(a-3)}$$

105. Answers may vary.

107. Answers may vary.

109. The subtraction needs to be distributed in the numerator of the 2nd fraction.

111. $\dfrac{a}{b}+\dfrac{c}{d}=\dfrac{a(d)}{b(d)}+\dfrac{c(b)}{d(b)}$

$$=\frac{ad}{bd}+\frac{bc}{bd}=\frac{ad+bc}{bd}$$

Exercise 6.4 (page 419)

1. $\dfrac{\frac{2}{3}}{\frac{1}{2}} = \dfrac{2}{3} \div \dfrac{1}{2} = \dfrac{2}{3} \cdot \dfrac{2}{1} = \dfrac{4}{3}$

3. $\dfrac{\frac{1}{2}}{2} = \dfrac{1}{2} \div 2 = \dfrac{1}{2} \div \dfrac{2}{1} = \dfrac{1}{2} \cdot \dfrac{1}{2} = \dfrac{1}{4}$

5. $t^3t^4t^2 = t^{3+4+2} = t^9$

7. $-2r(r^3)^2 = -2rr^6 = -2r^7$

9. $\left(\dfrac{3r}{4r^3}\right)^4 = \left(\dfrac{3}{4r^2}\right)^4 = \dfrac{3^4}{(4r^2)^4} = \dfrac{81}{256r^8}$

11. $\left(\dfrac{6r^{-2}}{2r^3}\right)^{-2} = \left(3r^{-5}\right)^{-2} = \dfrac{r^{10}}{9}$

13. complex fraction

15. single; divide

17. $\dfrac{\frac{2}{3}}{\frac{3}{4}} = \dfrac{2}{3} \div \dfrac{3}{4} = \dfrac{2}{3} \cdot \dfrac{4}{3} = \dfrac{8}{9}$

19. $\dfrac{\frac{4}{5}}{\frac{32}{15}} = \dfrac{4}{5} \div \dfrac{32}{15} = \dfrac{4}{5} \cdot \dfrac{15}{32} = \dfrac{3}{8}$

21. $\dfrac{\frac{x}{y}}{\frac{1}{x}} = \dfrac{x}{y} \div \dfrac{1}{x} = \dfrac{x}{y} \cdot \dfrac{x}{1} = \dfrac{x^2}{y}$

23. $\dfrac{\frac{5t^2}{9x^2}}{\frac{3t}{x^2t}} = \dfrac{5t^2}{9x^2} \div \dfrac{3t}{x^2t} = \dfrac{5t^2}{9x^2} \cdot \dfrac{x^2t}{3t}$
$= \dfrac{5x^2t^3}{27x^2t} = \dfrac{5t^2}{27}$

25. $\dfrac{\frac{2}{3}+1}{\frac{1}{3}+1} = \dfrac{(\frac{2}{3}+1)3}{(\frac{1}{3}+1)3} = \dfrac{\frac{2}{3}\cdot 3 + 1\cdot 3}{\frac{1}{3}\cdot 3 + 1 \cdot 3}$
$= \dfrac{2+3}{1+3} = \dfrac{5}{4}$

27. $\dfrac{\frac{1}{2}+\frac{3}{4}}{\frac{3}{2}+\frac{1}{4}} = \dfrac{(\frac{1}{2}+\frac{3}{4})4}{(\frac{3}{2}+\frac{1}{4})4} = \dfrac{\frac{1}{2}\cdot 4 + \frac{3}{4}\cdot 4}{\frac{3}{2}\cdot 4 + \frac{1}{4}\cdot 4}$
$= \dfrac{2+3}{6+1} = \dfrac{5}{7}$

29. $\dfrac{\frac{1}{y}+3}{\frac{3}{y}-2} = \dfrac{\left(\frac{1}{y}+3\right)y}{\left(\frac{3}{y}-2\right)y} = \dfrac{\frac{1}{y}\cdot y + 3 \cdot y}{\frac{3}{y}\cdot y - 2 \cdot y} = \dfrac{1+3y}{3-2y}$

31. $\dfrac{\frac{2}{x}+2}{\frac{4}{x}+2} = \dfrac{(\frac{2}{x}+2)x}{(\frac{4}{x}+2)x} = \dfrac{\frac{2}{x}\cdot x + 2\cdot x}{\frac{4}{x}\cdot x + 2 \cdot x} = \dfrac{2+2x}{4+2x} = \dfrac{2(1+x)}{2(2+x)} = \dfrac{1+x}{2+x}$

33. $\dfrac{\frac{3y}{x}-y}{y-\frac{y}{x}} = \dfrac{(\frac{3y}{x}-y)x}{(y-\frac{y}{x})x} = \dfrac{\frac{3y}{x}\cdot x - y \cdot x}{y \cdot x - \frac{y}{x}\cdot x} = \dfrac{3y-xy}{xy-y} = \dfrac{y(3-x)}{y(x-1)} = \dfrac{3-x}{x-1}$

35. $\dfrac{\frac{2}{a+2}+1}{\frac{3}{a+2}} = \dfrac{\left(\frac{2}{a+2}+1\right)(a+2)}{\left(\frac{3}{a+2}\right)(a+2)} = \dfrac{\frac{2}{a+2}(a+2)+1(a+2)}{3} = \dfrac{2+a+2}{3} = \dfrac{a+4}{3}$

37. $\dfrac{\frac{1}{x+1}}{1+\frac{1}{x+1}} = \dfrac{\left(\frac{1}{x+1}\right)(x+1)}{\left(1+\frac{1}{x+1}\right)(x+1)} = \dfrac{\frac{1}{x+1}(x+1)}{1(x+1)+\frac{1}{x+1}(x+1)} = \dfrac{1}{x+1+1} = \dfrac{1}{x+2}$

SECTION 6.4

39. $$\frac{\frac{x}{x+2}}{\frac{x}{x+2}+x}=\frac{\left(\frac{x}{x+2}\right)(x+2)}{\left(\frac{x}{x+2}+x\right)(x+2)}=\frac{\frac{x}{x+2}(x+2)}{\frac{x}{x+2}(x+2)+x(x+2)}=\frac{x}{x+x^2+2x}=\frac{x}{x^2+3x}$$
$$=\frac{x}{x(x+3)}=\frac{1}{x+3}$$

41. $$\frac{1}{\frac{1}{x}+\frac{1}{y}}=\frac{1(xy)}{\left(\frac{1}{x}+\frac{1}{y}\right)xy}=\frac{xy}{\frac{1}{x}\cdot xy+\frac{1}{y}\cdot xy}=\frac{xy}{y+x}$$

43. $$\frac{\frac{2}{x}}{\frac{2}{y}-\frac{4}{x}}=\frac{\frac{2}{x}(xy)}{\left(\frac{2}{y}-\frac{4}{x}\right)(xy)}=\frac{\frac{2}{x}(xy)}{\frac{2}{y}(xy)-\frac{4}{x}(xy)}=\frac{2y}{2x-4y}=\frac{2y}{2(x-2y)}=\frac{y}{x-2y}$$

45. $$\frac{\frac{3}{x}+\frac{2x}{y}}{\frac{4}{x}}=\frac{\left(\frac{3}{x}+\frac{2x}{y}\right)(xy)}{\frac{4}{x}(xy)}=\frac{\frac{3}{x}(xy)+\frac{2x}{y}(xy)}{4y}=\frac{3y+2x^2}{4y}$$

47. $$\frac{3+\frac{3}{x-1}}{3-\frac{3}{x}}=\frac{\left(3+\frac{3}{x-1}\right)(x)(x-1)}{\left(3-\frac{3}{x}\right)(x)(x-1)}=\frac{3x(x-1)+\frac{3}{x-1}(x)(x-1)}{3x(x-1)-\frac{3}{x}(x)(x-1)}=\frac{3x^2-3x+3x}{3x^2-3x-3(x-1)}$$
$$=\frac{3x^2}{3x^2-6x+3}$$
$$=\frac{3x^2}{3(x^2-2x+1)}$$
$$=\frac{x^2}{x^2-2x+1}=\frac{x^2}{(x-1)^2}$$

49. $$\frac{x^{-2}}{y^{-1}}=\frac{\frac{1}{x^2}}{\frac{1}{y}}=\frac{1}{x^2}\div\frac{1}{y}=\frac{1}{x^2}\cdot\frac{y}{1}=\frac{y}{x^2}$$

51. $$\frac{y^{-2}+1}{y^{-2}-1}=\frac{\frac{1}{y^2}+1}{\frac{1}{y^2}-1}=\frac{\left(\frac{1}{y^2}+1\right)y^2}{\left(\frac{1}{y^2}-1\right)y^2}=\frac{\frac{1}{y^2}\cdot y^2+1y^2}{\frac{1}{y^2}\cdot y^2-1y^2}=\frac{1+y^2}{1-y^2}$$

53. $$\frac{a^{-2}+a}{a+1}=\frac{\frac{1}{a^2}+a}{a+1}=\frac{\left(\frac{1}{a^2}+a\right)a^2}{(a+1)a^2}=\frac{1+a^3}{a^3+a^2}=\frac{(1+a)(1-a+a^2)}{a^2(a+1)}=\frac{a^2-a+1}{a^2}$$

55. $$\frac{2x^{-1}+4x^{-2}}{2x^{-2}+x^{-1}}=\frac{\frac{2}{x}+\frac{4}{x^2}}{\frac{2}{x^2}+\frac{1}{x}}=\frac{\left(\frac{2}{x}+\frac{4}{x^2}\right)x^2}{\left(\frac{2}{x^2}+\frac{1}{x}\right)x^2}=\frac{2x+4}{2+x}=\frac{2(x+2)}{x+2}=2$$

57. $$\frac{\frac{y}{x-1}}{\frac{y}{x}}=\frac{y}{x-1}\div\frac{y}{x}=\frac{y}{x-1}\cdot\frac{x}{y}=\frac{x}{x-1}$$

59. $$\frac{\frac{3}{x}+\frac{4}{x+1}}{\frac{2}{x+1}-\frac{3}{x}}=\frac{\left(\frac{3}{x}+\frac{4}{x+1}\right)(x)(x+1)}{\left(\frac{2}{x+1}-\frac{3}{x}\right)(x)(x+1)}=\frac{\frac{3}{x}(x)(x+1)+\frac{4}{x+1}(x)(x+1)}{\frac{2}{x+1}(x)(x+1)-\frac{3}{x}(x)(x+1)}$$
$$=\frac{3(x+1)+4x}{2x-3(x+1)}=\frac{3x+3+4x}{2x-3x-3}=\frac{7x+3}{-x-3}$$

61. $$\frac{\frac{2}{x}-\frac{3}{x+1}}{\frac{2}{x+1}-\frac{3}{x}}=\frac{\left(\frac{2}{x}-\frac{3}{x+1}\right)(x)(x+1)}{\left(\frac{2}{x+1}-\frac{3}{x}\right)(x)(x+1)}=\frac{2(x+1)-3x}{2x-3(x+1)}$$
$$=\frac{2x+2-3x}{2x-3x-3}=\frac{-x+2}{-x-3}=\frac{-(x-2)}{-(x+3)}=\frac{x-2}{x+3}$$

63. $$\frac{\frac{m}{m+2}-\frac{2}{m-1}}{\frac{3}{m+2}+\frac{m}{m-1}}=\frac{\left(\frac{m}{m+2}-\frac{2}{m-1}\right)(m+2)(m-1)}{\left(\frac{3}{m+2}+\frac{m}{m-1}\right)(m+2)(m-1)}=\frac{\frac{m}{m+2}(m+2)(m-1)-\frac{2}{m-1}(m+2)(m-1)}{\frac{3}{m+2}(m+2)(m-1)+\frac{m}{m-1}(m+2)(m-1)}$$
$$=\frac{m(m-1)-2(m+2)}{3(m-1)+m(m+2)}$$
$$=\frac{m^2-m-2m-4}{3m-3+m^2+2m}$$
$$=\frac{m^2-3m-4}{m^2+5m-3}=\frac{(m-4)(m+1)}{m^2+5m-3}$$

65. $$\frac{\frac{2}{x+2}}{\frac{3}{x-3}+\frac{1}{x}}=\frac{\left(\frac{2}{x+2}\right)(x)(x+2)(x-3)}{\left(\frac{3}{x-3}+\frac{1}{x}\right)(x)(x+2)(x-3)}=\frac{2x(x-3)}{\frac{3}{x-3}(x)(x+2)(x-3)+\frac{1}{x}(x)(x+2)(x-3)}$$
$$=\frac{2x^2-6x}{3x(x+2)+(x+2)(x-3)}$$
$$=\frac{2x^2-6x}{3x^2+6x+x^2-3x+2x-6}$$
$$=\frac{2x^2-6x}{4x^2+5x-6}=\frac{2x(x-3)}{(4x-3)(x+2)}$$

67. $$\frac{\frac{1}{x}+\frac{2}{x+1}}{\frac{2}{x-1}-\frac{1}{x}}=\frac{\left(\frac{1}{x}+\frac{2}{x+1}\right)(x)(x+1)(x-1)}{\left(\frac{2}{x-1}-\frac{1}{x}\right)(x)(x+1)(x-1)}=\frac{\frac{1}{x}(x)(x+1)(x-1)+\frac{2}{x+1}(x)(x+1)(x-1)}{\frac{2}{x-1}(x)(x+1)(x-1)-\frac{1}{x}(x)(x+1)(x-1)}$$
$$=\frac{(x+1)(x-1)+2x(x-1)}{2x(x+1)-(x+1)(x-1)}$$
$$=\frac{x^2-1+2x^2-2x}{2x^2+2x-(x^2-1)}$$
$$=\frac{3x^2-2x-1}{2x^2+2x-x^2+1}$$
$$=\frac{3x^2-2x-1}{x^2+2x+1}=\frac{(3x+1)(x-1)}{(x+1)^2}$$

69. $$\frac{\frac{1}{y^2+y}-\frac{1}{xy+x}}{\frac{1}{xy+x}-\frac{1}{y^2+y}}=\frac{\frac{1}{y(y+1)}-\frac{1}{x(y+1)}}{\frac{1}{x(y+1)}-\frac{1}{y(y+1)}}=\frac{\left(\frac{1}{y(y+1)}-\frac{1}{x(y+1)}\right)(x)(y)(y+1)}{\left(\frac{1}{x(y+1)}-\frac{1}{y(y+1)}\right)(x)(y)(y+1)}$$
$$=\frac{1x-1y}{1y-1x}=\frac{x-y}{-(x-y)}=-1$$

71. $$\frac{1-25y^{-2}}{1+10y^{-1}+25y^{-2}}=\frac{1-\frac{25}{y^2}}{1+\frac{10}{y}+\frac{25}{y^2}}=\frac{\left(1-\frac{25}{y^2}\right)y^2}{\left(1+\frac{10}{y}+\frac{25}{y^2}\right)y^2}=\frac{y^2-25}{y^2+10y+25}$$
$$=\frac{(y+5)(y-5)}{(y+5)(y+5)}=\frac{y-5}{y+5}$$

73. Answers may vary.

75. Answers may vary.

Exercise 6.5 (page 426)

1. multiply by 10

3. multiply by 9

5. $x^2+4x=x(x+4)$

7. $2x^2+x-3=(2x+3)(x-1)$

9. $x^4-16=(x^2)^2-4^2=(x^2+4)(x^2-4)=(x^2+4)(x+2)(x-2)$

11. extraneous

13. LCD

15. xy

17.
$$\begin{aligned}\frac{x}{2}+4&=\frac{3x}{2}\\ 2\left(\frac{x}{2}+4\right)&=2\left(\frac{3x}{2}\right)\\ x+8&=3x\\ 8&=2x\\ 4&=x\end{aligned}$$

19.
$$\begin{aligned}\frac{2y}{5}-8&=\frac{4y}{5}\\ 5\left(\frac{2y}{5}-8\right)&=5\left(\frac{4y}{5}\right)\\ 2y-40&=4y\\ -40&=2y\\ -20&=y\end{aligned}$$

21.
$$\begin{aligned}\frac{x}{3}+1&=\frac{x}{2}\\ 6\left(\frac{x}{3}+1\right)&=6\left(\frac{x}{2}\right)\\ 2x+6&=3x\\ 6&=x\end{aligned}$$

23.
$$\begin{aligned}\frac{x}{5}-\frac{x}{3}&=-8\\ 15\left(\frac{x}{5}-\frac{x}{3}\right)&=15(-8)\\ 3x-5x&=-120\\ -2x&=-120\\ x&=60\end{aligned}$$

25.
$$\begin{aligned}\frac{3a}{2}+\frac{a}{3}&=-22\\ 6\left(\frac{3a}{2}+\frac{a}{3}\right)&=6(-22)\\ 9a+2a&=-132\\ 11a&=-132\\ a&=-12\end{aligned}$$

27.
$$\begin{aligned}\frac{x-3}{3}+2x&=-1\\ 3\left(\frac{x-3}{3}+2x\right)&=3(-1)\\ x-3+6x&=-3\\ 7x&=0\\ x&=0\end{aligned}$$

SECTION 6.5

29.
$$\begin{aligned} \frac{z-3}{2} &= z+2 \\ 2\left(\frac{z-3}{2}\right) &= 2(z+2) \\ z-3 &= 2z+4 \\ -7 &= z \end{aligned}$$

31.
$$\begin{aligned} \frac{5(x+1)}{8} &= x+1 \\ 8\left(\frac{5(x+1)}{8}\right) &= 8(x+1) \\ 5(x+1) &= 8(x+1) \\ 5x+5 &= 8x+8 \\ -3x &= 3 \\ x &= -1 \end{aligned}$$

33.
$$\begin{aligned} \frac{a^2}{a+2} - \frac{4}{a+2} &= a \\ (a+2)\left(\frac{a^2}{a+2} - \frac{4}{a+2}\right) &= (a+2)a \\ a^2 - 4 &= a^2 + 2a \\ -4 &= 2a \\ -2 &= a \end{aligned}$$

-2 is extraneous $\Rightarrow$ no solutions

35.
$$\begin{aligned} \frac{x}{x-5} - \frac{5}{x-5} &= 3 \\ (x-5)\left(\frac{x}{x-5} - \frac{5}{x-5}\right) &= (x-5)3 \\ x-5 &= 3x-15 \\ -2x &= -10 \\ x &= 5 \end{aligned}$$

5 is extraneous $\Rightarrow$ no solutions

37.
$$\begin{aligned} \frac{3}{x} + 2 &= 3 \\ x\left(\frac{3}{x} + 2\right) &= x(3) \\ 3 + 2x &= 3x \\ 3 &= x \end{aligned}$$

39.
$$\begin{aligned} \frac{5}{a} - \frac{4}{a} &= 8 + \frac{1}{a} \\ a\left(\frac{5}{a} - \frac{4}{a}\right) &= a\left(8 + \frac{1}{a}\right) \\ 5 - 4 &= 8a + 1 \\ 0 &= 8a \\ 0 &= a \end{aligned}$$

0 is extraneous. $\Rightarrow$ no solutions

41.
$$\begin{aligned} \frac{2}{y+1} + 5 &= \frac{12}{y+1} \\ (y+1)\left(\frac{2}{y+1} + 5\right) &= (y+1)\left(\frac{12}{y+1}\right) \\ 2 + 5(y+1) &= 12 \\ 2 + 5y + 5 &= 12 \\ 5y &= 5 \\ y &= 1 \end{aligned}$$

43.
$$\begin{aligned} \frac{1}{x-1} + \frac{3}{x-1} &= 1 \\ (x-1)\left(\frac{1}{x-1} + \frac{3}{x-1}\right) &= (x-1)(1) \\ 1 + 3 &= x - 1 \\ 5 &= x \end{aligned}$$

45.

$$
\begin{aligned}
\frac{u}{u-1}+\frac{1}{u} &= \frac{u^2+1}{u^2-u} \\
\frac{u}{u-1}+\frac{1}{u} &= \frac{u^2+1}{u(u-1)} \\
u(u-1)\left(\frac{u}{u-1}+\frac{1}{u}\right) &= u(u-1)\left(\frac{u^2+1}{u(u-1)}\right) \\
u^2+u-1 &= u^2+1 \\
u-1 &= 1 \\
u &= 2
\end{aligned}
$$

47.

$$
\begin{aligned}
\frac{3}{x-2}+\frac{1}{x} &= \frac{2(3x+2)}{x^2-2x} \\
\frac{3}{x-2}+\frac{1}{x} &= \frac{6x+4}{x(x-2)} \\
x(x-2)\left(\frac{3}{x-2}+\frac{1}{x}\right) &= x(x-2)\left(\frac{6x+4}{x(x-2)}\right) \\
3x+x-2 &= 6x+4 \\
4x-2 &= 6x+4 \\
-2x &= 6 \\
x &= -3
\end{aligned}
$$

49.

$$
\begin{aligned}
\frac{-5}{s^2+s-2}+\frac{3}{s+2} &= \frac{1}{s-1} \\
\frac{-5}{(s+2)(s-1)}+\frac{3}{s+2} &= \frac{1}{s-1} \\
(s+2)(s-1)\left(\frac{-5}{(s+2)(s-1)}+\frac{3}{s+2}\right) &= (s+2)(s-1)\left(\frac{1}{s-1}\right) \\
-5+3(s-1) &= s+2 \\
-5+3s-3 &= s+2 \\
2s &= 10 \\
s &= 5
\end{aligned}
$$

51.

$$
\begin{aligned}
\frac{3y}{3y-6}+\frac{8}{y^2-4} &= \frac{2y}{2y+4} \\
\frac{3y}{3(y-2)}+\frac{8}{(y+2)(y-2)} &= \frac{2y}{2(y+2)} \\
\frac{y}{y-2}+\frac{8}{(y+2)(y-2)} &= \frac{y}{y+2} \\
(y+2)(y-2)\left(\frac{y}{y-2}+\frac{8}{(y+2)(y-2)}\right) &= (y+2)(y-2)\left(\frac{y}{y+2}\right) \\
y(y+2)+8 &= y(y-2) \\
y^2+2y+8 &= y^2-2y \\
8 &= -4y \\
-2 &= y: \quad -2 \text{ is extraneous} \Rightarrow \text{no solutions}
\end{aligned}
$$

53.

$$\begin{aligned}
\frac{n}{n^2-9}+\frac{n+8}{n+3}&=\frac{n-8}{n-3}\\
\frac{n}{(n+3)(n-3)}+\frac{n+8}{n+3}&=\frac{n-8}{n-3}\\
(n+3)(n-3)\left(\frac{n}{(n+3)(n-3)}+\frac{n+8}{n+3}\right)&=(n+3)(n-3)\left(\frac{n-8}{n-3}\right)\\
n+(n-3)(n+8)&=(n+3)(n-8)\\
n+n^2+5n-24&=n^2-5n-24\\
11n&=0\\
n&=0
\end{aligned}$$

55.

$$\begin{aligned}
\frac{b+2}{b+3}+1&=\frac{-7}{b-5}\\
(b+3)(b-5)\left(\frac{b+2}{b+3}+1\right)&=(b+3)(b-5)\left(\frac{-7}{b-5}\right)\\
(b-5)(b+2)+(b+3)(b-5)&=-7(b+3)\\
b^2-3b-10+b^2-2b-15&=-7b-21\\
2b^2+2b-4&=0\\
2\left(b^2+b-2\right)&=0\\
2(b+2)(b-1)&=0
\end{aligned}$$

$b+2=0$ **or** $b-1=0$

$b=-2$ $\quad$ $b=1$

57.

$$\begin{aligned}
y+\frac{2}{3}&=\frac{2y-12}{3y-9}\\
y+\frac{2}{3}&=\frac{2(y-6)}{3(y-3)}\\
3(y-3)\left(y+\frac{2}{3}\right)&=3(y-3)\left(\frac{2y-12}{3(y-3)}\right)\\
3y(y-3)+2(y-3)&=2y-12\\
3y^2-9y+2y-6&=2y-12\\
3y^2-9y+6&=0\\
3\left(y^2-3y+2\right)&=0\\
3(y-2)(y-1)&=0
\end{aligned}$$

$y-2=0$ **or** $y-1=0$

$y=2$ $\quad$ $y=1$

59.

$$\frac{3}{5x-20}+\frac{4}{5}=\frac{3}{5x-20}-\frac{x}{5}$$
$$\frac{3}{5(x-4)}+\frac{4}{5}=\frac{3}{5(x-4)}-\frac{x}{5}$$
$$5(x-4)\left(\frac{3}{5(x-4)}+\frac{4}{5}\right)=5(x-4)\left(\frac{3}{5(x-4)}-\frac{x}{5}\right)$$
$$3+4(x-4)=3-x(x-4)$$
$$3+4x-16=3-x^2+4x$$
$$x^2-16=0$$
$$(x+4)(x-4)=0$$

$x+4=0$ **or** $x-4=0$ $\Rightarrow x=4$ is extraneous, so the only solution is $x=-4$.
$x=-4$ $\quad x=4$

61.

$$\frac{1}{a}+\frac{1}{b}=1$$
$$ab\left(\frac{1}{a}+\frac{1}{b}\right)=ab(1)$$
$$b+a=ab$$
$$b=ab-a$$
$$b=a(b-1)$$
$$\frac{b}{b-1}=a$$

63.

$$\frac{a}{b}+\frac{c}{d}=1$$
$$bd\left(\frac{a}{b}+\frac{c}{d}\right)=bd(1)$$
$$ad+bc=bd$$
$$bc-bd=-ad$$
$$b(c-d)=-ad$$
$$b=\frac{-ad}{c-d}=\frac{ad}{d-c}$$

65.

$$\frac{c-4}{4}=\frac{c+4}{8}$$
$$8\left(\frac{c-4}{4}\right)=8\left(\frac{c+4}{8}\right)$$
$$2(c-4)=c+4$$
$$2c-8=c+4$$
$$c=12$$

67.

$$\frac{x+1}{3}+\frac{x-1}{5}=\frac{2}{15}$$
$$15\left(\frac{x+1}{3}+\frac{x-1}{5}\right)=15\left(\frac{2}{15}\right)$$
$$5(x+1)+3(x-1)=2$$
$$5x+5+3x-3=2$$
$$8x=0$$
$$x=0$$

69.

$$\frac{3x-1}{6}-\frac{x+3}{2}=\frac{3x+4}{3}$$
$$12\left(\frac{3x-1}{6}-\frac{x+3}{2}\right)=12\left(\frac{3x+4}{3}\right)$$
$$2(3x-1)-6(x+3)=4(3x+4)$$
$$6x-2-6x-18=12x+16$$
$$-36=12x$$
$$-3=x$$

71.

$$\frac{3r}{2}-\frac{3}{r}=\frac{3r}{2}+3$$
$$2r\left(\frac{3r}{2}-\frac{3}{r}\right)=2r\left(\frac{3r}{2}+3\right)$$
$$3r^2-6=3r^2+6r$$
$$-6=6r$$
$$-1=r$$

73.

$$\begin{aligned}\frac{1}{3}+\frac{2}{x-3}&=1\\3(x-3)\left(\frac{1}{3}+\frac{2}{x-3}\right)&=3(x-3)(1)\\x-3+3(2)&=3x-9\\x-3+6&=3x-9\\12&=2x\\6&=x\end{aligned}$$

75.

$$\begin{aligned}\frac{5}{4y+12}-\frac{3}{4}&=\frac{5}{4y+12}-\frac{y}{4}\\\frac{5}{4(y+3)}-\frac{3}{4}&=\frac{5}{4(y+3)}-\frac{y}{4}\\4(y+3)\left(\frac{5}{4(y+3)}-\frac{3}{4}\right)&=4(y+3)\left(\frac{5}{4(y+3)}-\frac{y}{4}\right)\\5-3(y+3)&=5-y(y+3)\\5-3y-9&=5-y^2-3y\\y^2-9&=0\\(y+3)(y-3)&=0\end{aligned}$$

$y+3=0$ **or** $y-3=0$ $\Rightarrow y=-3$ is extraneous, so the only solution is $y=3$.

$y=-3$ $\quad y=3$

77.

$$\begin{aligned}\frac{z-4}{z-3}&=\frac{z+2}{z+1}\\(z-3)(z+1)\left(\frac{z-4}{z-3}\right)&=(z-3)(z+1)\left(\frac{z+2}{z+1}\right)\\(z+1)(z-4)&=(z-3)(z+2)\\z^2-3z-4&=z^2-z-6\\-2z&=-2\\z&=1\end{aligned}$$

79.

$$\begin{aligned}\frac{1}{f}&=\frac{1}{d_1}+\frac{1}{d_2}\\fd_1d_2\cdot\frac{1}{f}&=fd_1d_2\left(\frac{1}{d_1}+\frac{1}{d_2}\right)\\d_1d_2&=fd_2+fd_1\\d_1d_2&=f(d_2+d_1)\\\frac{d_1d_2}{d_2+d_1}&=f\end{aligned}$$

81. **Answers may vary.**

83.
$$x = \frac{1}{x}$$
$$x(x) = x\left(\frac{1}{x}\right)$$
$$x^2 = 1$$
$$x^2 - 1 = 0$$
$$(x+1)(x-1) = 0$$
$$x + 1 = 0 \quad \textbf{or} \quad x - 1 = 0$$
$$x = -1 \qquad\qquad x = 1$$
The numbers 1 and -1 are equal to their own reciprocals.

Exercise 6.6 (page 432)

1. $i = pr$

3. $C = qd$

5.
$$x^2 - 5x - 6 = 0$$
$$(x-6)(x+1) = 0$$
$$x - 6 = 0 \quad \textbf{or} \quad x + 1 = 0$$
$$x = 6 \qquad\qquad x = -1$$

7.
$$(t+2)\left(t^2 + 7t + 12\right) = 0$$
$$(t+2)(t+4)(t+3) = 0$$
$$t + 2 = 0 \quad \textbf{or} \quad t + 4 = 0 \quad \textbf{or} \quad t + 3 = 0$$
$$t = -2 \qquad\qquad t = -4 \qquad\qquad t = -3$$

9.
$$y^3 - y^2 = 0$$
$$y^2(y-1) = 0$$
$$y = 0 \quad \textbf{or} \quad y = 0 \quad \textbf{or} \quad y - 1 = 0$$
$$y = 0 \qquad y = 0 \qquad\qquad y = 1$$

11.
$$2(y-4) = -y^2$$
$$2y - 8 = -y^2$$
$$y^2 + 2y - 8 = 0$$
$$(y-2)(y+4) = 0$$
$$y - 2 = 0 \quad \textbf{or} \quad y + 4 = 0$$
$$y = 2 \qquad\qquad y = -4$$

13. **Answers may vary.**

15. Let $x =$ the number.
$$\frac{2(3)}{4+x} = 1$$
$$(4+x)\left(\frac{6}{4+x}\right) = (4+x)(1)$$
$$6 = 4 + x$$
$2 = x \Rightarrow$ The number is 2.

17. Let $x =$ the number.

$$\frac{3+x}{4+2x} = \frac{4}{7}$$
$$7(4+2x)\left(\frac{3+x}{4+2x}\right) = 7(4+2x)\left(\frac{4}{7}\right)$$
$$7(x+3) = 4(4+2x)$$
$$7x + 21 = 16 + 8x$$
$$5 = x$$

The number is 5.

19. Let $x =$ minutes for both to grade the set.

$$\boxed{\text{Teacher in 1 minute}} + \boxed{\text{Aide in 1 minute}} = \boxed{\text{Total in 1 minute}}$$
$$\frac{1}{45} + \frac{1}{90} = \frac{1}{x}$$
$$90x\left(\frac{1}{45} + \frac{1}{90}\right) = 90x\left(\frac{1}{x}\right)$$
$$2x + x = 90$$
$$3x = 90$$
$$x = 30$$

It will take them 30 minutes to grade the set of quizzes.

21. Let $x =$ hours for both pipes to fill the pool.

$$\boxed{\text{1st pipe in 1 hour}} + \boxed{\text{2nd pipe in 1 hour}} = \boxed{\text{Total in 1 hour}}$$
$$\frac{1}{5} + \frac{1}{4} = \frac{1}{x}$$
$$20x\left(\frac{1}{5} + \frac{1}{4}\right) = 20x\left(\frac{1}{x}\right)$$
$$4x + 5x = 20$$
$$9x = 20$$
$$x = \frac{20}{9}$$

The pool can be filled in $2\frac{2}{9}$ hours.

23. Let $r =$ the rate for the heron.

	d	r	t
Goose	120	$r+10$	$\frac{120}{r+10}$
Heron	80	r	$\frac{80}{r}$

$$\boxed{\text{Time for goose}} = \boxed{\text{Time for heron}}$$
$$\frac{120}{r+10} = \frac{80}{r}$$
$$r(r+10)\left(\frac{120}{r+10}\right) = r(r+10)\left(\frac{80}{r}\right)$$
$$120r = 80(r+10)$$
$$120r = 80r + 800$$
$$40r = 800$$
$$20 = r$$

The heron's speed is 20 mph, while the goose's speed is 30 mph.

25. Let $r =$ the rate of the plane.

	d	r	t
Plane	300	r	$\frac{300}{r}$
Car	120	$r-90$	$\frac{120}{r-90}$

$$\boxed{\text{Time for plane}} = \boxed{\text{Time for car}}$$
$$\frac{300}{r} = \frac{120}{r-90}$$
$$r(r-90)\left(\frac{300}{r}\right) = r(r-90)\left(\frac{120}{r-90}\right)$$
$$300(r-90) = 120r$$
$$300r - 27000 = 120r$$
$$180r = 27000$$
$$r = 150$$

The plane travels at 150 miles per hour.

27. Let $r =$ the lower interest rate.

	I	P	r
Lower rate CD	175	$\frac{175}{r}$	r
Higher rate CD	200	$\frac{200}{r+.01}$	$r+.01$

$$\boxed{\text{Lower rate principal}} = \boxed{\text{Higher rate principal}}$$

$$\frac{175}{r} = \frac{200}{r+.01}$$

$$r(r+.01)\left(\frac{175}{r}\right) = r(r+.01)\left(\frac{200}{r+.01}\right)$$

$$175(r+.01) = 200r$$

$$175r + 1.75 = 200r$$

$$1.75 = 25r$$

$0.07 = r \Rightarrow$ The rates are 7% and 8%.

29. Let $r =$ the lower interest rate.

	I	P	r
Lower rate CD	225	$\frac{225}{r}$	r
Higher rate CD	450	$\frac{450}{r+.03}$	$r+.03$

$$\boxed{\text{Lower rate principal}} = \boxed{\text{Higher rate principal}}$$

$$\frac{225}{r} = \frac{450}{r+.03}$$

$$r(r+.03)\left(\frac{225}{r}\right) = r(r+.03)\left(\frac{450}{r+.03}\right)$$

$$225(r+.03) = 450r$$

$$225r + 6.75 = 450r$$

$$6.75 = 225r$$

$0.03 = r \Rightarrow$ The higher rate is 6%.

31. Let $x =$ the number. $\Rightarrow x + \frac{1}{x} = \frac{13}{6}$

$$6x\left(x + \frac{1}{x}\right) = 6x\left(\frac{13}{6}\right)$$

$$6x^2 + 6 = 13x$$

$$6x^2 - 13x + 6 = 0$$

$$(2x-3)(3x-2) = 0$$

$2x - 3 = 0$ **or** $3x - 2 = 0$

$2x = 3$ $\quad$ $3x = 2$

$x = \frac{3}{2}$ $\quad$ $x = \frac{2}{3}$

The numbers are $\frac{3}{2}$ and $\frac{2}{3}$.

33. Let $x =$ hours for pool to fill with drain open.

$$\boxed{\text{Pipe in 1 hour}} - \boxed{\text{Drain in 1 hour}} = \boxed{\text{Total in 1 hour}}$$

$$\frac{1}{4} - \frac{1}{8} = \frac{1}{x}$$

$$8x\left(\frac{1}{4} - \frac{1}{8}\right) = 8x\left(\frac{1}{x}\right)$$

$$2x - x = 8$$

$$x = 8$$

The pool can be filled in 8 hours.

35. Let $r =$ the speed of the current.

	d	r	t
Downstream	22	$18+r$	$\frac{22}{18+r}$
Upstream	14	$18-r$	$\frac{14}{18-r}$

$$\boxed{\text{Time downstream}} = \boxed{\text{Time upstream}}$$

$$\frac{22}{18+r} = \frac{14}{18-r}$$

continued on next page...

35. continued

$$(18+r)(18-r)\left(\frac{22}{18+r}\right)=(18+r)(18-r)\left(\frac{14}{18-r}\right)$$
$$22(18-r)=14(18+r)$$
$$396-22r=252+14r$$
$$144=36r$$
$$4=r\Rightarrow \text{The current has a speed of 4 miles per hour.}$$

37. Let $x=$ the number bought.

Then each cost $\frac{120}{x}$.

$$\boxed{\text{Number bought}}\cdot\boxed{\text{Amount charged for each}}=120$$
$$(x+10)\cdot\left(\frac{120}{x}-1\right)=120$$
$$120-x+\frac{1200}{x}-10=120$$
$$\frac{1200}{x}-x-10=0$$
$$x\left(\frac{1200}{x}-x-10\right)=0$$
$$-x^2-10x+1200=0$$
$$-(x^2+10x-1200)=0$$
$$-(x+40)(x-30)=0$$
$$x+40=0 \quad\textbf{or}\quad x-30=0$$
$$x=-40 \qquad\qquad x=30$$

The store can buy 30 at the regular price.

39. Let $r=$ the still-water speed.

	d	r	t
Upstream	60	$r-5$	$\frac{60}{r-5}$
Downstream	60	$r+5$	$\frac{60}{r+5}$

$$\boxed{\text{Time upstream}}+\boxed{\text{Time downstream}}=5$$
$$\frac{60}{r-5}+\frac{60}{r+5}=5$$
$$(r+5)(r-5)\left(\frac{60}{r-5}+\frac{60}{r+5}\right)=5(r+5)(r-5)$$
$$60(r+5)+60(r-5)=5(r^2-25)$$
$$60r+300+60r-300=5r^2-125$$
$$0=5r^2-120r-125$$
$$0=5(r^2-24r-25)$$
$$0=5(r+1)(r-25)$$
$$r+1=0 \quad\textbf{or}\quad r-25=0$$
$$r=-1 \qquad\qquad r=25$$

The still-water speed should be 25 miles per hour.

41. Let $x=$ the number who contributed.

$$\boxed{\text{Original share}}=\boxed{\text{Share with more workers}}+2$$
$$\frac{35}{x}=\frac{35}{x+2}+2$$
$$x(x+2)\left(\frac{35}{x}\right)=x(x+2)\left(\frac{35}{x+2}+2\right)$$
$$35(x+2)=35x+2x(x+2)$$
$$35x+70=35x+2x^2+4x$$
$$35x+70=35x+2x^2+4x$$
$$0=2x^2+4x-70$$
$$0=2(x^2+2x-35)$$
$$0=2(x+7)(x-5)$$
$$x+7=0 \quad\textbf{or}\quad x-5=0$$
$$x=-7 \qquad\qquad x=5$$

Since the answer cannot be negative, 5 workers must have contributed.

43. Let $r =$ the slower speed.

	d	r	t
G ⇒ P	512	$r+20$	$\frac{512}{r+20}$
P ⇒ U	528	r	$\frac{528}{r}$

$$\boxed{\text{Time to Poland}} + 4 = \boxed{\text{Time to Ukraine}}$$

$$\frac{512}{r+20} + 4 = \frac{528}{r}$$

$$r(r+20)\left(\frac{512}{r+20} + 4\right) = r(r+20)\left(\frac{528}{r}\right)$$

$$r(r+20)\left(\frac{512}{r+20} + 4\right) = r(r+20)\left(\frac{528}{r}\right)$$

$$r(512) + 4r(r+20) = 528(r+20)$$

$$512r + 4r^2 + 80r = 528r + 10{,}560$$

$$4r^2 + 64r - 10{,}560 = 0$$

$$4\left(r^2 + 16r - 2640\right) = 0$$

$$(r+60)(r-44) = 0$$

$$r + 60 = 0 \quad \textbf{or} \quad r - 44 = 0$$

$$r = -60 \qquad r = 44$$

The speeds are 44 and 64 miles per hour.

45. Answers may vary.

47. Answers may vary.

Exercise 6.7 (page 439)

1. $\frac{5}{8}$

3. $\frac{3}{9} = \frac{1}{3}$

5.
$$2x + 4 = 38$$
$$2x = 34$$
$$x = 17$$

7.
$$3(x+2) = 24$$
$$3x + 6 = 24$$
$$3x = 18$$
$$x = 6$$

9. $2x + 6 = 2(x+3)$

11. $2x^2 - x - 6 = (2x+3)(x-2)$

13. quotient

15. equal

17. Answers may vary.

19. $\frac{5}{7}$

21. $\frac{17}{34} = \frac{1}{2}$

23. $\frac{22}{33} = \frac{2}{3}$

25. $\frac{7}{24.5} = \frac{14}{49} = \frac{2}{7}$

27. $\frac{4\text{ oz}}{12\text{ oz}} = \frac{1}{3}$

29. $\frac{12\text{ min}}{1\text{ hr}} = \frac{12\text{ min}}{60\text{ min}} = \frac{1}{5}$

31. $\frac{3\text{ days}}{1\text{ week}} = \frac{3\text{ days}}{7\text{ days}} = \frac{3}{7}$

33. $\frac{18\text{ months}}{2\text{ years}} = \frac{18\text{ months}}{24\text{ months}} = \frac{3}{4}$

35. $\frac{125}{2000} = \frac{1}{16}$

37. $750 + 652 + 188 + 125 + 110 = \$1{,}825$

39. $\frac{110}{1825} = \frac{22}{365}$

41. $\frac{\$53.55}{17\text{ gal}} = \$3.15/\text{gal}$

43. $\frac{89¢}{6 \text{ oz}} \approx 14.873¢/\text{oz}; \frac{119¢}{8 \text{ oz}} = 14.875¢/\text{oz}$
The 6-oz can is the better buy.

45. $\frac{\$337.50}{27 \text{ hr}} = \$12.50/\text{hr}$

47. $\frac{325 \text{ mi}}{5 \text{ hr}} = 65 \text{ mi/hr}$

49. $\frac{84¢}{12 \text{ oz}} = 7¢/\text{oz}$

51. $\frac{345 \text{ mi}}{6 \text{ hr}} = 57.5 \frac{\text{mi}}{\text{hr}}; \frac{376 \text{ mi}}{6.2 \text{ hr}} \approx 60.6 \frac{\text{mi}}{\text{hr}}$
The truck travels faster.

53. $\frac{1235 \text{ mi}}{51.3 \text{ gal}} \approx 24 \frac{\text{mi}}{\text{gal}}; \frac{1456 \text{ mi}}{55.78 \text{ gal}} \approx 26 \frac{\text{mi}}{\text{gal}}$; The 2nd car had the better mpg rating.

55. $\frac{11880 \text{ gal}}{27 \text{ min}} = 440 \text{ gal/min}$

57. $995 + 1245 + 1680 + 4580 + 225 = \$8,725$

59. $\frac{1680}{8725} = \frac{336}{1745}$

61. **Answers may vary.**

63. $\frac{17}{19} = \frac{17 \cdot 21}{19 \cdot 21} = \frac{357}{399}; \frac{19}{21} = \frac{19 \cdot 19}{21 \cdot 19} = \frac{361}{399}; \frac{19}{21}$ is larger.

Exercise 6.8 (page 449)

1.
$$\frac{3}{5} \stackrel{?}{=} \frac{6}{10}$$
$$3(10) \stackrel{?}{=} 5(6)$$
$$30 = 30$$
proportion

3.
$$\frac{1}{2} \stackrel{?}{=} \frac{1}{4}$$
$$1(4) \stackrel{?}{=} 2(1)$$
$$4 \neq 2$$
not a proportion

5. $\frac{9}{10} = 0.9 = 90\%$

7. $33\frac{1}{3}\% = \frac{100}{3}\% = \frac{100}{3} \cdot \frac{1}{100} = \frac{1}{3}$

9.
$$rb = a$$
$$0.30(1600) = a$$
$$480 = a$$

11.
$$rb = a$$
$$0.25(98) = a$$
$$24.50 = a$$
$$98 - 24.50 = \$73.50$$

13. proportion; ratios

15. means

17. similar

19. ad; bc

21. triangle

23.
$$\frac{9}{7} \stackrel{?}{=} \frac{81}{70}$$
$$9 \cdot 70 \stackrel{?}{=} 7 \cdot 81$$
$$630 \neq 567$$
not a proportion

25.
$$\frac{-7}{3} \stackrel{?}{=} \frac{14}{-6}$$
$$-7(-6) \stackrel{?}{=} 3 \cdot 14$$
$$42 = 42$$
proportion

27.
$$\frac{9}{19} \stackrel{?}{=} \frac{38}{80}$$
$$9 \cdot 80 \stackrel{?}{=} 19 \cdot 38$$
$$720 \neq 722$$
not a proportion

SECTION 6.8

29.
$$\frac{10.4}{3.6} \stackrel{?}{=} \frac{41.6}{14.4}$$
$$10.4(14.4) \stackrel{?}{=} 3.6(41.6)$$
$$149.76 = 149.76$$
proportion

31.
$$\frac{6}{10} \stackrel{?}{=} \frac{15}{25}$$
$$6 \cdot 25 \stackrel{?}{=} 10 \cdot 15$$
$$150 = 150$$
proportional

33.
$$\frac{3}{7} \stackrel{?}{=} \frac{4}{8}$$
$$3 \cdot 8 \stackrel{?}{=} 7 \cdot 4$$
$$24 \neq 28$$
not proportional

35.
$$\frac{2}{3} = \frac{x}{6}$$
$$2(6) = 3x$$
$$12 = 3x$$
$$4 = x$$

37.
$$\frac{5}{10} = \frac{3}{c}$$
$$5c = 10(3)$$
$$5c = 30$$
$$c = 6$$

39.
$$\frac{x+1}{5} = \frac{3}{15}$$
$$15(x+1) = 5(3)$$
$$15x + 15 = 15$$
$$15x = 0$$
$$x = 0$$

41.
$$\frac{x+3}{12} = \frac{-7}{6}$$
$$6(x+3) = 12(-7)$$
$$6x + 18 = -84$$
$$6x = -102$$
$$x = -17$$

43.
$$\frac{-6}{x} = \frac{8}{4}$$
$$-6(4) = 8x$$
$$-24 = 8x$$
$$-3 = x$$

45.
$$\frac{x}{3} = \frac{9}{3}$$
$$x(3) = 3(9)$$
$$3x = 27$$
$$x = 9$$

47.
$$\frac{4-x}{13} = \frac{11}{26}$$
$$26(4-x) = 13(11)$$
$$104 - 26x = 143$$
$$-26x = 39$$
$$x = -\frac{39}{26} = -\frac{3}{2}$$

49.
$$\frac{2x+1}{18} = \frac{14}{3}$$
$$3(2x+1) = 18(14)$$
$$6x + 3 = 252$$
$$6x = 249$$
$$x = \frac{249}{6} = \frac{83}{2}$$

51.
$$\frac{3p-2}{12} = \frac{p+1}{3}$$
$$3(3p-2) = 12(p+1)$$
$$9p - 6 = 12p + 12$$
$$-3p = 18$$
$$p = -6$$

53. Let c = the cost of 51 pints of yogurt.
$$\frac{\text{3 pints}}{\text{51 pints}} = \frac{\text{cost of 3 pints}}{\text{cost of 51 pints}}$$
$$\frac{3}{51} = \frac{1}{c}$$
$$c \cdot 3 = 51 \cdot 1$$
$$3c = 51$$
$$c = 17$$
The 51 pints will cost \$17.

55. Let c = the cost of 39 packets.
$$\frac{\text{3 packets}}{\text{39 packets}} = \frac{\text{cost of 3 packets}}{\text{cost of 39 packets}}$$
$$\frac{3}{39} = \frac{50}{c}$$
$$c \cdot 3 = 39 \cdot 50$$
$$3c = 1{,}950$$
$$c = 650 \text{ cents} = \$6.50$$
The 39 packets will cost \$6.50.

57. Let $d =$ the # of drops of pure essence needed with 56 drops of alcohol.

$$\frac{\text{3 pure}}{\text{7 alcohol}} = \frac{\text{drops of pure}}{\text{56 drops of alcohol}}$$
$$\frac{3}{7} = \frac{d}{56}$$
$$3 \cdot 56 = 7d$$
$$168 = 7d$$
$$24 = d$$

24 drops of pure essence will be needed.

59. Let $f =$ the cups of flour needed for 12 dozen cookies.

$$\frac{1\frac{1}{4}\text{ C flour}}{3\frac{1}{2}\text{ dozen}} = \frac{\text{C flour}}{\text{12 dozen}}$$
$$\frac{1\frac{1}{4}}{3\frac{1}{2}} = \frac{f}{12}$$
$$1\frac{1}{4} \cdot 12 = 3\frac{1}{2}f$$
$$\frac{5}{4} \cdot 12 = \frac{7}{2}f$$
$$15 = \frac{7}{2}f$$
$$30 = 7f$$
$$\frac{30}{7} = f$$
$$4\frac{2}{7} = f$$

$4\frac{2}{7}\left(\approx 4\frac{1}{4}\right)$ cups of flour will be needed.

61. Let $n =$ the number defective.

$$\frac{\text{5 defective}}{\text{100 parts}} = \frac{\text{number defective}}{\text{940 parts}}$$
$$\frac{5}{100} = \frac{n}{940}$$
$$5 \cdot 940 = 100n$$
$$4700 = 100n$$
$$47 = n$$

There will be 47 defective parts.

63. Let $g =$ gallons of gas needed for 315 miles.

$$\frac{\text{42 miles}}{\text{315 miles}} = \frac{\text{gas for 42 miles}}{\text{gas for 315 miles}}$$
$$\frac{42}{315} = \frac{1}{g}$$
$$g \cdot 42 = 315 \cdot 1$$
$$42g = 315$$
$$g = \frac{315}{42} = \frac{15}{2} = 7\frac{1}{2}$$

$7\frac{1}{2}$ gallons of gas are needed to go 315 miles.

65. Let $p =$ the amount he was paid last week.

$$\frac{\text{30 hours}}{\text{40 hours}} = \frac{\text{pay for 30 hours}}{\text{pay for 40 hours}}$$
$$\frac{30}{40} = \frac{p}{412}$$
$$412 \cdot 30 = 40 \cdot p$$
$$12{,}360 = 40p$$
$$309 = p$$

He was paid \$309 last week.

67. Let $l =$ the length of a real engine.

$$\frac{\text{87 feet}}{\text{1 foot}} = \frac{\text{the length of a real engine}}{\text{the length of a model engine}}$$
$$\frac{87}{1} = \frac{l}{9}$$
$$9 \cdot 87 = 1 \cdot l$$
$$783 = l$$

A real engine is 783 inches (65′ 3″)long.

69. Let $l =$ the width of the house if it were real.

$$\frac{\text{1 inch}}{\text{8 inches}} = \frac{\text{the width of the doll house}}{\text{the width of a real house}}$$
$$\frac{1}{8} = \frac{36}{l}$$
$$l \cdot 1 = 8 \cdot 36$$
$$l = 288$$

It would be 288 in. wide if it were real.
[288 in. = 24 ft]

71. Let $x =$ the amount of oil to be added.

$$\frac{50}{1} = \frac{\text{ounces of gasoline}}{\text{ounces of oil}}$$
$$\frac{50}{1} = \frac{6 \cdot 128}{x}$$
$$\frac{50}{1} = \frac{768}{x}$$
$$x \cdot 50 = 1 \cdot 768$$
$$50x = 768$$
$$x = \frac{768}{50} \approx 15.36 \text{ ounces}$$

The directions are close.

73. $$\frac{\text{height of man}}{\text{man's shadow}} = \frac{\text{height of tree}}{\text{tree's shadow}}$$
$$\frac{6}{4} = \frac{h}{26}$$
$$6 \cdot 26 = 4h$$
$$156 = 4h$$
$$39 = h$$

The tree is 39 feet tall.

75. $$\frac{75}{32} = \frac{w}{20}$$
$$75 \cdot 20 = 32w$$
$$1500 = 32w$$
$$\frac{1500}{32} = w, \text{ or } w = 46\frac{7}{8}$$

The river is $46\frac{7}{8}$ feet wide.

77. $$\frac{x}{1350} = \frac{5}{1}$$
$$1 \cdot x = 1350 \cdot 5$$
$$x = 6750$$

The plane will lose 6,750 feet in altitude.

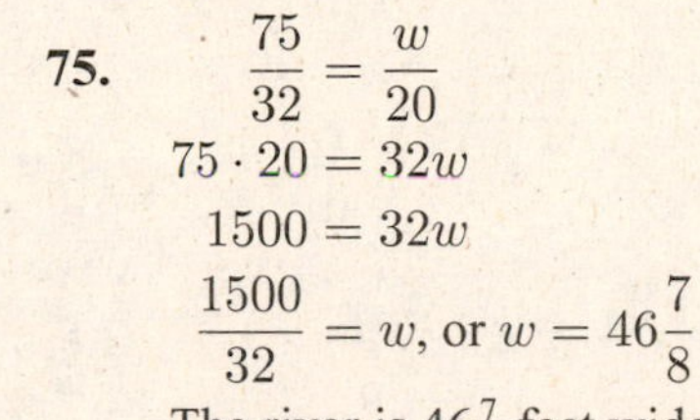

79. $$\frac{x}{750} = \frac{52800}{2500}$$
$$2500 \cdot x = 750 \cdot 52800$$
$$2500x = 39{,}600{,}000$$
$$x = 15{,}840$$

The road will rise 15,840 ft.

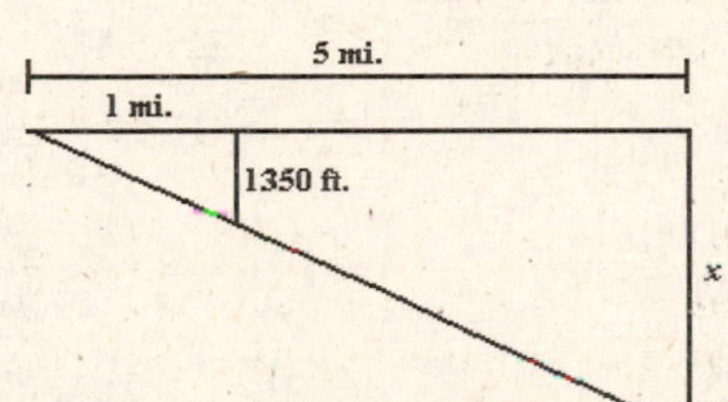

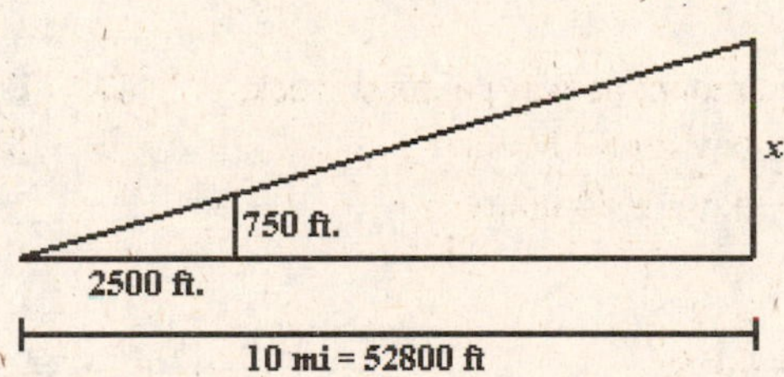

81. **Answers may vary.**

83. If $\frac{a}{b} = \frac{c}{d}$, then $ad = bc$. Thus, $\frac{a}{b} = \frac{a(b+d)}{b(b+d)} = \frac{ab+ad}{b(b+d)} = \frac{ab+bc}{b(b+d)} = \frac{b(a+c)}{b(b+d)} = \frac{a+c}{b+d}$.

Chapter 6 Review (page 453)

1.
$$\begin{aligned}(x+3)(x-3)&=0\\ x+3=0 \quad &\text{or} \quad x-3=0\\ x=-3 \quad &\qquad\quad x=3\end{aligned}$$

2.
$$\begin{aligned}x^2+x-6&=0\\ (x+3)(x-2)&=0\\ x+3=0 \quad &\text{or} \quad x-2=0\\ x=-3 \quad &\qquad\quad x=2\end{aligned}$$

3. $\frac{10}{25}=\frac{5\cdot 2}{5\cdot 5}=\frac{2}{5}$

4. $\frac{-12}{18}=-\frac{6\cdot 2}{6\cdot 3}=-\frac{2}{3}$

5. $\frac{-51}{153}=-\frac{51\cdot 1}{51\cdot 3}=-\frac{1}{3}$

6. $\frac{105}{45}=\frac{15\cdot 7}{15\cdot 3}=\frac{7}{3}$

7. $\frac{3x^2}{6x^3}=\frac{1}{2x}$

8. $\frac{5xy^2}{2x^2y^2}=\frac{5}{2x}$

9. $\frac{x^2}{x^2+x}=\frac{x^2}{x(x+1)}=\frac{x}{x+1}$

10. $\frac{x+2}{x^2+2x}=\frac{x+2}{x(x+2)}=\frac{1}{x}$

11. $\frac{6xy}{3xy}=2$

12. $\frac{8x^2y}{2x(4xy)}=\frac{8x^2y}{8x^2y}=1$

13. $\frac{3p-2}{2-3p}=-1$

14. $\frac{x^2-x-56}{x^2-5x-24}=\frac{(x-8)(x+7)}{(x-8)(x+3)}=\frac{x+7}{x+3}$

15. $\frac{2x^2-16x}{2x^2-18x+16}=\frac{2x(x-8)}{2(x^2-9x+8)}=\frac{2x(x-8)}{2(x-8)(x-1)}=\frac{x}{x-1}$

16. $\frac{a^2+2a+ab+2b}{a^2+2ab+b^2}=\frac{a(a+2)+b(a+2)}{(a+b)(a+b)}=\frac{(a+2)(a+b)}{(a+b)(a+b)}=\frac{a+2}{a+b}$

17. $\frac{3xy}{2x}\cdot\frac{4x}{2y^2}=\frac{12x^2y}{4xy^2}=\frac{3x}{y}$

18. $\frac{3x}{x^2-x}\cdot\frac{2x-2}{x^2}=\frac{3x(2x-2)}{(x^2-x)x^2}=\frac{6x(x-1)}{x^3(x-1)}$
$=\frac{6}{x^2}$

19. $\frac{x^2+3x+2}{x^2+2x}\cdot\frac{x}{x+1}=\frac{(x^2+3x+2)x}{(x^2+2x)(x+1)}=\frac{x(x+2)(x+1)}{x(x+2)(x+1)}=1$

20. $\frac{x^2+x}{3x-15}\cdot\frac{6x-30}{x^2+2x+1}=\frac{(x^2+x)(6x-30)}{(3x-15)(x^2+2x+1)}=\frac{x(x+1)\cdot 6(x-5)}{3(x-5)\cdot(x+1)(x+1)}=\frac{2x}{x+1}$

21. $\frac{3x^2}{5x^2y}\div\frac{6x}{15xy^2}=\frac{3x^2}{5x^2y}\cdot\frac{15xy^2}{6x}=\frac{45x^3y^2}{30x^3y}=\frac{3y}{2}$

22. $\frac{x^2+5x}{x^2+4x-5}\div\frac{x^2}{x-1}=\frac{x^2+5x}{x^2+4x-5}\cdot\frac{x-1}{x^2}=\frac{x(x+5)(x-1)}{x^2(x+5)(x-1)}=\frac{1}{x}$

23. $\dfrac{x^2-x-6}{2x-1} \div \dfrac{x^2-2x-3}{2x^2+x-1} = \dfrac{x^2-x-6}{2x-1} \cdot \dfrac{2x^2+x-1}{x^2-2x-3} = \dfrac{(x-3)(x+2)(2x-1)(x+1)}{(2x-1)(x-3)(x+1)}$
$= x+2$

24. $\dfrac{x^2-3x}{x^2-x-6} \div \dfrac{x^2-x}{x^2+x-2} = \dfrac{x^2-3x}{x^2-x-6} \cdot \dfrac{x^2+x-2}{x^2-x} = \dfrac{x(x-3)\cdot(x+2)(x-1)}{(x+2)(x-3)\cdot x(x-1)} = 1$

25. $\dfrac{x^2+4x+4}{x^2+x-6}\left(\dfrac{x-2}{x-1} \div \dfrac{x+2}{x^2+2x-3}\right) = \dfrac{x^2+4x+4}{x^2+x-6}\left(\dfrac{x-2}{x-1} \cdot \dfrac{x^2+2x-3}{x+2}\right)$
$= \dfrac{(x+2)(x+2)}{(x+3)(x-2)}\left(\dfrac{(x-2)(x+3)(x-1)}{(x-1)(x+2)}\right)$
$= \dfrac{(x+2)(x+2)(x-2)(x+3)(x-1)}{(x+3)(x-2)(x-1)(x+2)} = x+2$

26. $\dfrac{x}{x+y} + \dfrac{y}{x+y} = \dfrac{x+y}{x+y} = 1$

27. $\dfrac{3x}{x-7} - \dfrac{x-2}{x-7} = \dfrac{3x-x+2}{x-7} = \dfrac{2x+2}{x-7}$

28. $\dfrac{x}{x-1} + \dfrac{1}{x} = \dfrac{(x)(x)}{(x-1)(x)} + \dfrac{1(x-1)}{x(x-1)} = \dfrac{x^2}{x(x-1)} + \dfrac{x-1}{x(x-1)} = \dfrac{x^2+x-1}{x(x-1)}$

29. $\dfrac{1}{7} - \dfrac{1}{x} = \dfrac{1(x)}{7(x)} - \dfrac{1(7)}{x(7)} = \dfrac{x}{7x} - \dfrac{7}{7x} = \dfrac{x-7}{7x}$

30. $\dfrac{3}{x+1} - \dfrac{2}{x} = \dfrac{3(x)}{(x+1)(x)} - \dfrac{2(x+1)}{x(x+1)} = \dfrac{3x}{x(x+1)} - \dfrac{2x+2}{x(x+1)} = \dfrac{3x-2x-2}{x(x+1)} = \dfrac{x-2}{x(x+1)}$

31. $\dfrac{x+2}{2x} - \dfrac{2-x}{x^2} = \dfrac{(x+2)x}{2x(x)} - \dfrac{(2-x)2}{x^2(2)} = \dfrac{x^2+2x}{2x^2} - \dfrac{4-2x}{2x^2} = \dfrac{x^2+2x-4+2x}{2x^2}$
$= \dfrac{x^2+4x-4}{2x^2}$

32. $\dfrac{x}{x+2} + \dfrac{3}{x} - \dfrac{4}{x^2+2x} = \dfrac{x}{x+2} + \dfrac{3}{x} - \dfrac{4}{x(x+2)} = \dfrac{x(x)}{x(x+2)} + \dfrac{3(x+2)}{x(x+2)} - \dfrac{4}{x(x+2)}$
$= \dfrac{x^2}{x(x+2)} + \dfrac{3x+6}{x(x+2)} - \dfrac{4}{x(x+2)}$
$= \dfrac{x^2+3x+2}{x(x+2)} = \dfrac{(x+2)(x+1)}{x(x+2)} = \dfrac{x+1}{x}$

CHAPTER 6 REVIEW

33. $$\frac{2}{x-1}-\frac{3}{x+1}+\frac{x-5}{x^2-1}=\frac{2}{x-1}-\frac{3}{x+1}+\frac{x-5}{(x+1)(x-1)}$$
$$=\frac{2(x+1)}{(x-1)(x+1)}-\frac{3(x-1)}{(x+1)(x-1)}+\frac{x-5}{(x+1)(x-1)}$$
$$=\frac{2(x+1)-3(x-1)+x-5}{(x+1)(x-1)}$$
$$=\frac{2x+2-3x+3+x-5}{(x+1)(x-1)}=\frac{0}{(x+1)(x-1)}=0$$

34. $$\frac{\frac{3}{2}}{\frac{2}{3}}=\frac{3}{2}\div\frac{2}{3}=\frac{3}{2}\cdot\frac{3}{2}=\frac{9}{4}$$

35. $$\frac{\frac{3}{2}+1}{\frac{2}{3}+1}=\frac{\left(\frac{3}{2}+1\right)6}{\left(\frac{2}{3}+1\right)6}=\frac{\frac{3}{2}\cdot 6+1\cdot 6}{\frac{2}{3}\cdot 6+1\cdot 6}=\frac{9+6}{4+6}=\frac{15}{10}=\frac{3}{2}$$

36. $$\frac{\frac{1}{x}+1}{\frac{1}{x}-1}=\frac{\left(\frac{1}{x}+1\right)x}{\left(\frac{1}{x}-1\right)x}=\frac{\frac{1}{x}\cdot x+1\cdot x}{\frac{1}{x}\cdot x-1\cdot x}$$
$$=\frac{1+x}{1-x}$$

37. $$\frac{1+\frac{3}{x}}{2-\frac{1}{x^2}}=\frac{\left(1+\frac{3}{x}\right)x^2}{\left(2-\frac{1}{x^2}\right)x^2}=\frac{x^2+3x}{2x^2-1}$$

38. $$\frac{\frac{2}{x-1}+\frac{x-1}{x+1}}{\frac{1}{x^2-1}}=\frac{\frac{2}{x-1}+\frac{x-1}{x+1}}{\frac{1}{(x+1)(x-1)}}=\frac{\left(\frac{2}{x-1}+\frac{x-1}{x+1}\right)(x+1)(x-1)}{\left(\frac{1}{(x+1)(x-1)}\right)(x+1)(x-1)}=\frac{2(x+1)+(x-1)(x-1)}{1}$$
$$=2x+2+x^2-2x+1=x^2+3$$

39. $$\frac{\frac{a}{b}+c}{\frac{b}{a}+c}=\frac{\left(\frac{a}{b}+c\right)ab}{\left(\frac{b}{a}+c\right)ab}=\frac{a^2+abc}{b^2+abc}=\frac{a(a+bc)}{b(b+ac)}$$

40. $$\frac{3}{x}=\frac{2}{x-1}$$
$$x(x-1)\left(\frac{3}{x}\right)=x(x-1)\left(\frac{2}{x-1}\right)$$
$$3(x-1)=2x$$
$$3x-3=2x$$
$x=3\Rightarrow$ The answer checks.

41. $$\frac{5}{x+4}=\frac{3}{x+2}$$
$$(x+4)(x+2)\left(\frac{5}{x+4}\right)=(x+4)(x+2)\left(\frac{3}{x+2}\right)$$
$$5(x+2)=3(x+4)$$
$$5x+10=3x+12$$
$$2x=2$$
$x=1\Rightarrow$ The answer checks.

42.
$$\begin{aligned}
\frac{2}{3x}+\frac{1}{x}&=\frac{5}{9}\\
9x\left(\frac{2}{3x}+\frac{1}{x}\right)&=9x\left(\frac{5}{9}\right)\\
6+9&=5x\\
15&=5x\\
3&=x \Rightarrow \text{The answer checks.}
\end{aligned}$$

43.
$$\begin{aligned}
\frac{2x}{x+4}&=\frac{3}{x-1}\\
(x+4)(x-1)\left(\frac{2x}{x+4}\right)&=(x+4)(x-1)\left(\frac{3}{x-1}\right)\\
2x(x-1)&=3(x+4)\\
2x^2-2x&=3x+12\\
2x^2-5x-12&=0\\
(2x+3)(x-4)&=0
\end{aligned}$$
$$\begin{aligned}
2x+3=0 \quad &\textbf{or} \quad x-4=0 \Rightarrow \text{Both answers check.}\\
2x=-3 \quad &\quad\quad x=4\\
x=-\tfrac{3}{2} \quad &\quad\quad x=4
\end{aligned}$$

44.
$$\begin{aligned}
\frac{2}{x-1}+\frac{3}{x+4}&=\frac{-5}{x^2+3x-4}\\
\frac{2}{x-1}+\frac{3}{x+4}&=\frac{-5}{(x+4)(x-1)}\\
(x-1)(x+4)\left(\frac{2}{x-1}+\frac{3}{x+4}\right)&=(x-1)(x+4)\left(\frac{-5}{(x+4)(x-1)}\right)\\
2(x+4)+3(x-1)&=-5\\
2x+8+3x-3&=-5\\
5x&=-10\\
x&=-2 \Rightarrow \text{The answer checks.}
\end{aligned}$$

45.
$$\begin{aligned}
\frac{4}{x+2}-\frac{3}{x+3}&=\frac{6}{x^2+5x+6}\\
\frac{4}{x+2}-\frac{3}{x+3}&=\frac{6}{(x+2)(x+3)}\\
(x+2)(x+3)\left(\frac{4}{x+2}-\frac{3}{x+3}\right)&=(x+2)(x+3)\left(\frac{6}{(x+2)(x+3)}\right)\\
4(x+3)-3(x+2)&=6\\
4x+12-3x-6&=6\\
x&=0 \Rightarrow \text{The answer checks.}
\end{aligned}$$

46.
$$\begin{aligned} \frac{1}{r} &= \frac{1}{r_1} + \frac{1}{r_2} \\ rr_1r_2 \cdot \frac{1}{r} &= rr_1r_2\left(\frac{1}{r_1} + \frac{1}{r_2}\right) \\ r_1r_2 &= rr_2 + rr_1 \\ r_1r_2 - rr_1 &= rr_2 \\ r_1(r_2 - r) &= rr_2 \\ r_1 &= \frac{rr_2}{r_2 - r} \end{aligned}$$

47.
$$\begin{aligned} E &= 1 - \frac{T_2}{T_1} \\ T_1E &= T_1\left(1 - \frac{T_2}{T_1}\right) \\ T_1E &= T_1 - T_2 \\ T_1E - T_1 &= -T_2 \\ T_1(E - 1) &= -T_2 \\ T_1 &= -\frac{T_2}{E - 1} = \frac{T_2}{1 - E} \end{aligned}$$

48.
$$\begin{aligned} H &= \frac{RB}{R + B} \\ (R + B)H &= (R + B)\left(\frac{RB}{R + B}\right) \\ RH + BH &= RB \\ BH &= RB - RH \\ BH &= R(B - H) \\ \frac{BH}{B - H} &= R \end{aligned}$$

49. Let x = hours for both pipes to empty.

[1st pipe in 1 hour] + [2nd pipe in 1 hour] = [Total in 1 hour]

$$\begin{aligned} \frac{1}{18} + \frac{1}{20} &= \frac{1}{x} \\ 180x\left(\frac{1}{18} + \frac{1}{20}\right) &= 180x\left(\frac{1}{x}\right) \\ 10x + 9x &= 180 \\ 19x &= 180 \\ x &= \frac{180}{19} \end{aligned}$$

It can be emptied in $9\frac{9}{19}$ hours.

50. Let x = days for both working together.

[Painter in 1 day] + [Owner in 1 day] = [Total in 1 day]

$$\begin{aligned} \frac{1}{10} + \frac{1}{14} &= \frac{1}{x} \\ 70x\left(\frac{1}{10} + \frac{1}{14}\right) &= 70x\left(\frac{1}{x}\right) \\ 7x + 5x &= 70 \\ 12x &= 70 \\ x &= \frac{70}{12} = \frac{35}{6} \end{aligned}$$

It can be painted in $5\frac{5}{6}$ days.

51. Let r = the rate at which he jogs.

	d	r	t
Jogs	10	r	$\frac{10}{r}$
Rides	30	$r + 10$	$\frac{30}{r + 10}$

[Time he jogs] = [Time he rides]

$$\begin{aligned} \frac{10}{r} &= \frac{30}{r + 10} \\ r(r + 10)\left(\frac{10}{r}\right) &= r(r + 10)\left(\frac{30}{r + 10}\right) \\ 10(r + 10) &= 30r \\ 10r + 100 &= 30r \\ 100 &= 20r \\ 5 &= r \end{aligned}$$

He jogs 5 miles per hour.

52. Let $r =$ the speed of the wind.

	d	r	t
Downwind	400	$360 + r$	$\frac{400}{360 + r}$
Upwind	320	$360 - r$	$\frac{320}{360 - r}$

$$\boxed{\text{Time downwind}} = \boxed{\text{Time upwind}}$$
$$\frac{400}{360 + r} = \frac{320}{360 - r}$$
$$(360 + r)(360 - r)\left(\frac{400}{360 + r}\right) = (360 + r)(360 - r)\left(\frac{320}{360 - r}\right)$$
$$400(360 - r) = 320(360 + r)$$
$$144000 - 400r = 115200 + 320r$$
$$28800 = 720r$$
$40 = r \Rightarrow$ The wind has a speed of 40 miles per hour.

53. $\frac{3}{6} = \frac{1}{2}$

54. $\frac{12x}{15x} = \frac{4}{5}$

55. $\frac{2 \text{ ft}}{1 \text{ yd}} = \frac{2 \text{ ft}}{3 \text{ ft}} = \frac{2}{3}$

56. $\frac{5 \text{ pt}}{3 \text{ qt}} = \frac{5 \text{ pt}}{6 \text{ pt}} = \frac{5}{6}$

57. $\frac{\$8.79}{3 \text{ lb}} = \$2.93/\text{lb}$

58. $\frac{2275 \text{ kwh}}{4 \text{ weeks}} = 568.75 \text{ kwh/week}$

59.
$$\frac{4}{7} \stackrel{?}{=} \frac{20}{34}$$
$$4 \cdot 34 \stackrel{?}{=} 7 \cdot 20$$
$$136 \neq 140$$
not a proportion

60.
$$\frac{5}{7} \stackrel{?}{=} \frac{30}{42}$$
$$5 \cdot 42 \stackrel{?}{=} 7 \cdot 30$$
$$210 = 210$$
proportion

61.
$$\frac{3}{x} = \frac{6}{9}$$
$$3(9) = 6x$$
$$27 = 6x$$
$$\tfrac{27}{6} = x$$
$$\tfrac{9}{2} = x$$

62.
$$\frac{x}{3} = \frac{x}{5}$$
$$5x = 3x$$
$$2x = 0$$
$$x = 0$$

63.
$$\frac{x - 2}{5} = \frac{x}{7}$$
$$7(x - 2) = 5x$$
$$7x - 14 = 5x$$
$$2x = 14$$
$$x = 7$$

64.
$$\frac{4x - 1}{18} = \frac{x}{6}$$
$$6(4x - 1) = 18x$$
$$24x - 6 = 18x$$
$$6x = 6$$
$$x = 1$$

65. Let $h =$ the height of the pole.
$$\frac{\text{height of pole}}{\text{shadow of pole}} = \frac{\text{height of man}}{\text{shadow of man}}$$
$$\frac{h}{12} = \frac{6}{3.6}$$
$$3.6h = 12(6)$$
$$3.6h = 72$$
$$h = \frac{72}{3.6} = 20 \text{ ft}$$

Chapter 6 Test (page 459)

1. $\dfrac{48x^2y}{54xy^2} = \dfrac{6 \cdot 8x^2y}{6 \cdot 9xy^2} = \dfrac{8x}{9y}$

2. $\dfrac{2x^2 - x - 3}{4x^2 - 9} = \dfrac{(2x-3)(x+1)}{(2x+3)(2x-3)} = \dfrac{x+1}{2x+3}$

3. $\dfrac{3(x+2)-3}{2x-4-(x-5)} = \dfrac{3x+6-3}{2x-4-x+5}$
$= \dfrac{3x+3}{x+1} = \dfrac{3(x+1)}{x+1} = 3$

4. $\dfrac{12x^2y}{15xyz} \cdot \dfrac{25y^2z}{16xt} = \dfrac{3 \cdot 4 \cdot 5 \cdot 5x^2y^3z}{3 \cdot 5 \cdot 4 \cdot 4x^2ytz} = \dfrac{5y^2}{4t}$

5. $\dfrac{x^2+3x+2}{3x+9} \cdot \dfrac{x+3}{x^2-4} = \dfrac{(x+2)(x+1)}{3(x+3)} \cdot \dfrac{x+3}{(x+2)(x-2)} = \dfrac{(x+2)(x+1)(x+3)}{3(x+3)(x+2)(x-2)} = \dfrac{x+1}{3(x-2)}$

6. $\dfrac{8x^2y}{25xt} \div \dfrac{16x^2y^3}{30xyt^3} = \dfrac{8x^2y}{25xt} \cdot \dfrac{30xyt^3}{16x^2y^3} = \dfrac{4 \cdot 2 \cdot 5 \cdot 2 \cdot 3x^3y^2t^3}{5 \cdot 5 \cdot 4 \cdot 4x^3y^3t} = \dfrac{3t^2}{5y}$

7. $\dfrac{x^2-x}{3x^2+6x} \div \dfrac{3x-3}{3x^3+6x^2} = \dfrac{x^2-x}{3x^2+6x} \cdot \dfrac{3x^3+6x^2}{3x-3} = \dfrac{x(x-1)}{3x(x+2)} \cdot \dfrac{3x^2(x+2)}{3(x-1)}$
$= \dfrac{3x^3(x-1)(x+2)}{9x(x-1)(x+2)} = \dfrac{x^2}{3}$

8. $\dfrac{x^2+xy}{x-y} \cdot \dfrac{x^2-y^2}{x^2-2x} \div \dfrac{x^2+2xy+y^2}{x^2-4} = \dfrac{x^2+xy}{x-y} \cdot \dfrac{x^2-y^2}{x^2-2x} \cdot \dfrac{x^2-4}{x^2+2xy+y^2}$
$= \dfrac{x(x+y)}{x-y} \cdot \dfrac{(x+y)(x-y)}{x(x-2)} \cdot \dfrac{(x+2)(x-2)}{(x+y)(x+y)}$
$= \dfrac{x(x+y)(x+y)(x-y)(x+2)(x-2)}{x(x-y)(x-2)(x+y)(x+y)} = x+2$

9. $\dfrac{5x-4}{x-1} + \dfrac{5x+3}{x-1} = \dfrac{5x-4+5x+3}{x-1} = \dfrac{10x-1}{x-1}$

10. $\dfrac{3y+7}{2y+3} - \dfrac{3(y-2)}{2y+3} = \dfrac{3y+7-3(y-2)}{2y+3} = \dfrac{3y+7-3y+6}{2y+3} = \dfrac{13}{2y+3}$

11. $\dfrac{x+1}{x} + \dfrac{x-1}{x+1} = \dfrac{(x+1)(x+1)}{x(x+1)} + \dfrac{x(x-1)}{x(x+1)} = \dfrac{(x+1)(x+1)+x(x-1)}{x(x+1)}$
$= \dfrac{x^2+2x+1+x^2-x}{x(x+1)} = \dfrac{2x^2+x+1}{x(x+1)}$

12. $\dfrac{5x}{x-2} - 3 = \dfrac{5x}{x-2} - \dfrac{3}{1} = \dfrac{5x}{x-2} - \dfrac{3(x-2)}{x-2} = \dfrac{5x-3(x-2)}{x-2} = \dfrac{5x-3x+6}{x-2} = \dfrac{2x+6}{x-2}$

13. $\dfrac{\frac{8x^2}{xy^3}}{\frac{4y^3}{x^2y^3}} = \dfrac{8x^2}{xy^3} \div \dfrac{4y^3}{x^2y^3} = \dfrac{8x^2}{xy^3} \cdot \dfrac{x^2y^3}{4y^3} = \dfrac{8x^4y^3}{4xy^6} = \dfrac{2x^3}{y^3}$

CHAPTER 6 TEST

14. $$\frac{1+\frac{y}{x}}{\frac{y}{x}-1}=\frac{\left(1+\frac{y}{x}\right)x}{\left(\frac{y}{x}-1\right)x}=\frac{1\cdot x+\frac{y}{x}\cdot x}{\frac{y}{x}\cdot x-1\cdot x}$$
$$=\frac{x+y}{y-x}$$

15. $$\frac{x}{10}-\frac{1}{2}=\frac{x}{5}$$
$$10\left(\frac{x}{10}-\frac{1}{2}\right)=10\left(\frac{x}{5}\right)$$
$$x-5=2x$$
$$-5=x$$

16. $$3x-\frac{2(x+3)}{3}=16-\frac{x+2}{2}$$
$$6\left(3x-\frac{2x+6}{3}\right)=6\left(16-\frac{x+2}{2}\right)$$
$$18x-2(2x+6)=96-3(x+2)$$
$$18x-4x-12=96-3x-6$$
$$17x=102$$
$$x=6$$

17. $$\frac{7}{x+4}-\frac{1}{2}=\frac{3}{x+4}$$
$$2(x+4)\left(\frac{7}{x+4}-\frac{1}{2}\right)=2(x+4)\left(\frac{3}{x+4}\right)$$
$$2(7)-(x+4)=2(3)$$
$$14-x-4=6$$
$$-x=-4$$
$$x=4$$

The answer checks.

18. $$H=\frac{RB}{R+B}$$
$$(R+B)H=(R+B)\left(\frac{RB}{R+B}\right)$$
$$RH+BH=RB$$
$$RH=RB-BH$$
$$RH=B(R-H)$$
$$\frac{RH}{R-H}=B$$

19. Let $x=$ hours working together.

$$\boxed{\text{1st worker in 1 hour}}+\boxed{\text{2nd worker in 1 hour}}=\boxed{\text{Total in 1 hour}}$$
$$\frac{1}{7}+\frac{1}{9}=\frac{1}{x}$$
$$63x\left(\frac{1}{7}+\frac{1}{9}\right)=63x\left(\frac{1}{x}\right)$$
$$9x+7x=63$$
$$16x=63$$
$$x=\frac{63}{16}$$

They can finish in $3\frac{15}{16}$ hours.

20. Let $r=$ the speed of the current.

	d	r	t
Downstream	28	$23+r$	$\frac{28}{23+r}$
Upstream	18	$23-r$	$\frac{18}{23-r}$

$$\boxed{\text{Time downstream}}=\boxed{\text{Time upstream}}$$
$$\frac{28}{23+r}=\frac{18}{23-r}$$
$$(23+r)(23-r)\left(\frac{28}{23+r}\right)=(23+r)(23-r)\left(\frac{18}{23-r}\right)$$
$$28(23-r)=18(23+r)$$
$$644-28r=414+18r$$
$$230=46r$$
$$5=r$$

The current has a speed of 5 miles per hour.

21. $$\begin{aligned} \frac{575}{\frac{1}{2}} &= \frac{x}{7} \\ 7(575) &= \frac{1}{2}x \\ 4025 &= \frac{1}{2}x \\ 2(4025) &= 2\left(\frac{1}{2}x\right) \\ 8050 &= x \Rightarrow \text{The plane will lose 8,050 feet of altitude.} \end{aligned}$$

22. $\frac{6 \text{ ft}}{3 \text{ yd}} = \frac{6 \text{ ft}}{9 \text{ ft}} = \frac{2}{3}$

23. $$\begin{aligned} \frac{3xy}{5xy} &\stackrel{?}{=} \frac{3xt}{5xt} \\ 3xy \cdot 5xt &\stackrel{?}{=} 5xy \cdot 3xt \\ 15x^2yt &= 15x^2yt \Rightarrow \text{proportion} \end{aligned}$$

24. $$\begin{aligned} \frac{y}{y-1} &= \frac{y-2}{y} \\ y(y) &= (y-1)(y-2) \\ y^2 &= y^2 - 3y + 2 \\ 3y &= 2 \\ y &= \frac{2}{3} \end{aligned}$$

25. Let h = the height of the tree.
$$\begin{aligned} \frac{\text{height of tree}}{\text{shadow of tree}} &= \frac{\text{height of man}}{\text{shadow of man}} \\ \frac{h}{30} &= \frac{6}{4} \\ 4h &= 30(6) \\ 4h &= 180 \\ h &= 45 \text{ ft} \end{aligned}$$

Cumulative Review Exercises (page 460)

1. $x^2x^5 = x^{2+5} = x^7$

2. $(x^2)^5 = x^{2 \cdot 5} = x^{10}$

3. $\frac{x^5}{x^2} = x^{5-2} = x^3$

4. $(3x^5)^0 = 1$

5. $(3x^2 - 2x) + (6x^3 - 3x^2 - 1) = 3x^2 - 2x + 6x^3 - 3x^2 - 1 = 6x^3 - 2x - 1$

6. $(4x^3 - 2x) - (2x^3 - 2x^2 - 3x + 1) = 4x^3 - 2x - 2x^3 + 2x^2 + 3x - 1 = 2x^3 + 2x^2 + x - 1$

7. $3(5x^2 - 4x + 3) + 2(-x^2 + 2x - 4) = 15x^2 - 12x + 9 - 2x^2 + 4x - 8 = 13x^2 - 8x + 1$

8. $4(3x^2 - 4x - 1) - 2(-2x^2 + 4x - 3) = 12x^2 - 16x - 4 + 4x^2 - 8x + 6 = 16x^2 - 24x + 2$

9. $(3x^3y^2)(-4x^2y^3) = 3(-4)x^3x^2y^2y^3 = -12x^5y^5$

10. $-5x^2(7x^3 - 2x^2 - 2) = -35x^5 + 10x^4 + 10x^2$

11. $(3x + 1)(2x + 4) = 6x^2 + 12x + 2x + 4 = 6x^2 + 14x + 4$

CUMULATIVE REVIEW EXERCISES

12. $(5x - 4y)(3x + 2y) = 15x^2 + 10xy - 12xy - 8y^2 = 15x^2 - 2xy - 8y^2$

13.
$$\begin{array}{r|l} & x + 4 \\ \hline x+3 & x^2 + 7x + 12 \\ & \underline{x^2 + 3x} \\ & \quad\quad 4x + 12 \\ & \quad\quad \underline{4x + 12} \\ & \quad\quad\quad\quad 0 \end{array}$$

14.
$$\begin{array}{r|l} & x^2 + x + 1 \\ \hline 2x-3 & 2x^3 - x^2 - x - 3 \\ & \underline{2x^3 - 3x^2} \\ & \quad\quad 2x^2 - x \\ & \quad\quad \underline{2x^2 - 3x} \\ & \quad\quad\quad\quad 2x - 3 \\ & \quad\quad\quad\quad \underline{2x - 3} \\ & \quad\quad\quad\quad\quad\quad 0 \end{array}$$

15. $3x^2y - 6xy^2 = 3xy(x - 2y)$

16. $3(a + b) + x(a + b) = (a + b)(3 + x)$

17.
$$\begin{aligned} 2a + 2b + ab + b^2 &= 2(a + b) + b(a + b) \\ &= (a + b)(2 + b) \end{aligned}$$

18.
$$\begin{aligned} 25p^4 - 16q^2 &= \left(5p^2\right)^2 - (4q)^2 \\ &= \left(5p^2 + 4q\right)\left(5p^2 - 4q\right) \end{aligned}$$

19. $x^2 - 11x - 12 = (x - 12)(x + 1)$

20. $x^2 - xy - 6y^2 = (x - 3y)(x + 2y)$

21. $6a^2 - 7a - 20 = (2a - 5)(3a + 4)$

22. $8m^2 - 10mn - 3n^2 = (4m + n)(2m - 3n)$

23. $p^3 - 27q^3 = p^3 - (3q)^3 = (p - 3q)\left(p^2 + p(3q) + (3q)^2\right) = (p - 3q)(p^2 + 3pq + 9q^2)$

24.
$$\begin{aligned} 8r^3 + 64s^3 = 8\left(r^3 + 8s^3\right) = 8\left[r^3 + (2s)^3\right] &= 8(r + 2s)\left(r^2 - r \cdot 2s + (2s)^2\right) \\ &= 8(r + 2s)\left(r^2 - 2rs + 4s^2\right) \end{aligned}$$

25.
$$\begin{aligned} \frac{4}{5}x + 6 &= 18 \\ \frac{4}{5}x &= 12 \\ 5 \cdot \frac{4}{5}x &= 5(12) \\ 4x &= 60 \\ x &= 15 \end{aligned}$$

26.
$$\begin{aligned} 5 - \frac{x+2}{3} &= 7 - x \\ 3\left(5 - \frac{x+2}{3}\right) &= 3(7 - x) \\ 15 - (x + 2) &= 21 - 3x \\ 15 - x - 2 &= 21 - 3x \\ 2x &= 8 \\ x &= 4 \end{aligned}$$

27.
$$\begin{aligned} 6x^2 - x - 2 &= 0 \\ (2x + 1)(3x - 2) &= 0 \end{aligned}$$
$$\begin{array}{rcl} 2x + 1 = 0 & \textbf{or} & 3x - 2 = 0 \\ 2x = -1 & & 3x = 2 \\ x = -\frac{1}{2} & & x = \frac{2}{3} \end{array}$$

28.
$$\begin{aligned} 5x^2 &= 10x \\ 5x^2 - 10x &= 0 \\ 5x(x - 2) &= 0 \end{aligned}$$
$$\begin{array}{rcl} 5x = 0 & \textbf{or} & x - 2 = 0 \\ x = 0 & & x = 2 \end{array}$$

29.
$$\begin{aligned} x^2 + 3x + 2 &= 0 \\ (x + 1)(x + 2) &= 0 \end{aligned}$$
$$\begin{array}{rcl} x + 1 = 0 & \textbf{or} & x + 2 = 0 \\ x = -1 & & x = -2 \end{array}$$

30.
$$\begin{aligned} 2y^2 + 5y - 12 &= 0 \\ (2y - 3)(y + 4) &= 0 \end{aligned}$$
$$\begin{array}{rcl} 2y - 3 = 0 & \textbf{or} & y + 4 = 0 \\ 2y = 3 & & y = -4 \\ y = \frac{3}{2} & & y = -4 \end{array}$$

CUMULATIVE REVIEW EXERCISES

31. $5x - 3 > 7$

$5x > 10$

$x > 2$

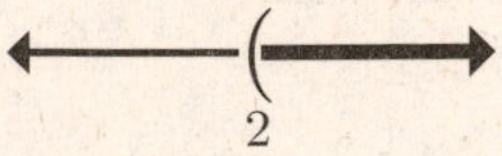

32. $7x - 9 < 5$

$7x < 14$

$x < 2$

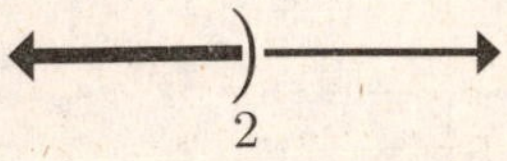

33.
$$\begin{aligned} -2 < -x + 3 &< 5 \\ -5 < -x &< 2 \\ \frac{-5}{-1} > \frac{-x}{-1} &> \frac{2}{-1} \\ 5 > x &> -2 \\ -2 < x &< 5 \end{aligned}$$

34.
$$\begin{aligned} 0 \le \frac{4-x}{3} &\le 2 \\ 3(0) \le 3 \cdot \frac{4-x}{3} &\le 3(2) \\ 0 \le 4 - x &\le 6 \\ -4 \le -x &\le 2 \\ \frac{-4}{-1} \ge \frac{-x}{-1} &\ge \frac{2}{-1} \\ 4 \ge x &\ge -2 \\ -2 \le x &\le 4 \end{aligned}$$

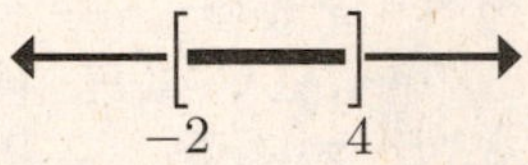

35. $4x - 3y = 12$

x	y
0	−4
3	0

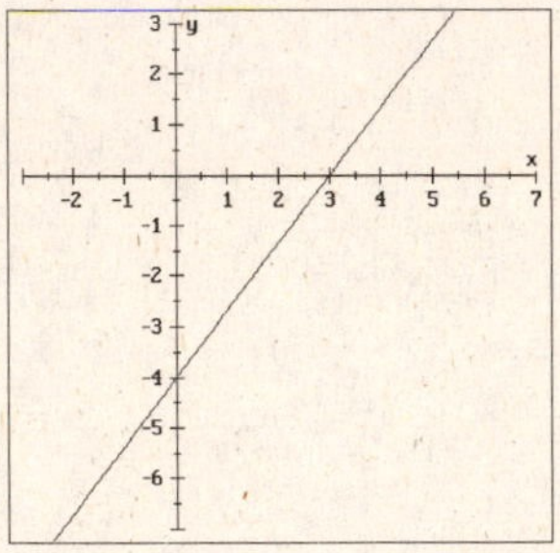

36. $3x + 4y = 4y + 12$

$3x = 12$

$x = 4$

x	y
4	0
4	2

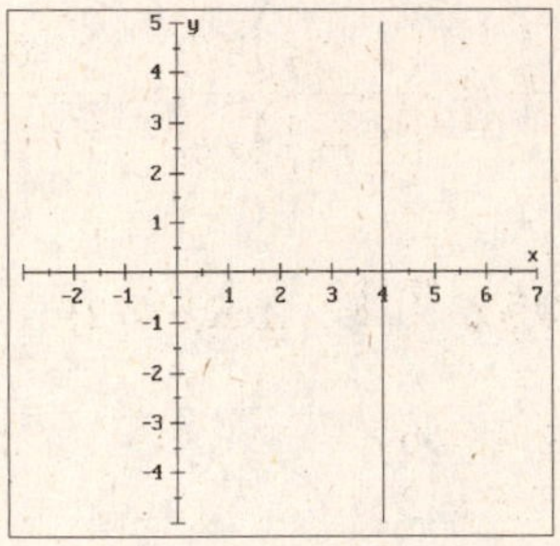

37. $f(0) = 2(0)^2 - 3 = 0 - 3 = -3$

38. $f(3) = 2(3)^2 - 3 = 2(9) - 3 = 18 - 3 = 15$

39. $f(-2) = 2(-2)^2 - 3 = 2(4) - 3$
$= 8 - 3 = 5$

40. $f(2x) = 2(2x)^2 - 3 = 2(4x^2) - 3$
$= 8x^2 - 3$

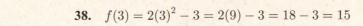

41. $$\frac{x^2 + 2x + 1}{x^2 - 1} = \frac{(x+1)(x+1)}{(x+1)(x-1)} = \frac{x+1}{x-1}$$

42. $$\frac{x^2 + 2x - 15}{x^2 + 3x - 10} = \frac{(x+5)(x-3)}{(x+5)(x-2)} = \frac{x-3}{x-2}$$

43. $$\frac{x^2 + x - 6}{5x - 5} \cdot \frac{5x - 10}{x + 3} = \frac{(x+3)(x-2) \cdot 5(x-2)}{5(x-1) \cdot (x+3)} = \frac{(x-2)^2}{x-1}$$

44. $\frac{p^2-p-6}{3p-9} \div \frac{p^2+6p+9}{p^2-9} = \frac{p^2-p-6}{3p-9} \cdot \frac{p^2-9}{p^2+6p+9} = \frac{(p-3)(p+2)\cdot(p+3)(p-3)}{3(p-3)\cdot(p+3)(p+3)}$
$= \frac{(p+2)(p-3)}{3(p+3)}$

45. $\frac{3x}{x+2} + \frac{5x}{x+2} - \frac{7x-2}{x+2} = \frac{3x+5x-(7x-2)}{x+2} = \frac{3x+5x-7x+2}{x+2} = \frac{x+2}{x+2} = 1$

46. $\frac{x-1}{x+1} + \frac{x+1}{x-1} = \frac{(x-1)(x-1)}{(x+1)(x-1)} + \frac{(x+1)(x+1)}{(x-1)(x+1)} = \frac{x^2-2x+1}{(x+1)(x-1)} + \frac{x^2+2x+1}{(x+1)(x-1)}$
$= \frac{2x^2+2}{(x+1)(x-1)}$

47. $\frac{a+1}{2a+4} - \frac{a^2}{2a^2-8} = \frac{a+1}{2(a+2)} - \frac{a^2}{2(a+2)(a-2)} = \frac{(a+1)(a-2)}{2(a+2)(a-2)} - \frac{a^2}{2(a+2)(a-2)}$
$= \frac{a^2-2a+a-2-a^2}{2(a+2)(a-2)}$
$= \frac{-a-2}{2(a+2)(a-2)}$
$= \frac{-(a+2)}{2(a+2)(a-2)} = -\frac{1}{2(a-2)}$

48. $\frac{\frac{1}{x}+\frac{1}{y}}{\frac{1}{x}-\frac{1}{y}} = \frac{\left(\frac{1}{x}+\frac{1}{y}\right)xy}{\left(\frac{1}{x}-\frac{1}{y}\right)xy} = \frac{\frac{1}{x}\cdot xy + \frac{1}{y}\cdot xy}{\frac{1}{x}\cdot xy - \frac{1}{y}\cdot xy} = \frac{y+x}{y-x}$

Exercise 7.1 (page 472)

1. $2x+4=6$
$2x=2$
$x=1$

3. $2x<4$
$x<2$

5. $-3x>12$
$\frac{-3x}{-3} < \frac{12}{-3}$
$x<-4$

7. $\left(\frac{t^3t^5t^{-6}}{t^2t^{-4}}\right)^{-3} = \left(\frac{t^2t^{-4}}{t^3t^5t^{-6}}\right)^3 = \left(\frac{t^{-2}}{t^2}\right)^3 = \left(\frac{1}{t^4}\right)^3 = \frac{1}{t^{12}}$

9. Let $x =$ the number of pies made.
$\boxed{\text{Expenses}} = \boxed{\text{Income}}$
$1200 + 3.40x = 5.95x$
$1200 = 2.55x$
$470.59 = x \Rightarrow$ He must sell at least 471 pies.

11. equation

13. multiplied; divided

15. contradiction

SECTION 7.1

17. is greater than

19. half-open

21. positive

23.
$$\begin{aligned} 2x+1&=13\\ 2x+1-1&=13-1\\ 2x&=12\\ \tfrac{2x}{2}&=\tfrac{12}{2}\\ x&=6 \end{aligned}$$

25.
$$\begin{aligned} 3(x+1)&=15\\ 3x+3&=15\\ 3x+3-3&=15-3\\ 3x&=12\\ \tfrac{3x}{3}&=\tfrac{12}{3}\\ x&=4 \end{aligned}$$

27.
$$\begin{aligned} 2r-5&=1-r\\ 2r+r-5&=1-r+r\\ 3r-5+5&=1+5\\ 3r&=6\\ \tfrac{3r}{3}&=\tfrac{6}{3}\\ r&=2 \end{aligned}$$

29.
$$\begin{aligned} 3(2y-4)-6&=3y\\ 6y-12-6&=3y\\ 6y-18&=3y\\ 6y-3y-18&=3y-3y\\ 3y-18+18&=0+18\\ 3y&=18\\ \tfrac{3y}{3}&=\tfrac{18}{3}\\ y&=6 \end{aligned}$$

31.
$$\begin{aligned} 5(5-a)&=37-2a\\ 25-5a&=37-2a\\ 25-5a+2a&=37-2a+2a\\ 25-25-3a&=37-25\\ -3a&=12\\ \tfrac{-3a}{-3}&=\tfrac{12}{-3}\\ a&=-4 \end{aligned}$$

33.
$$\begin{aligned} 4(y+1)&=-2(4-y)\\ 4y+4&=-8+2y\\ 4y-2y+4&=-8+2y-2y\\ 2y+4-4&=-8-4\\ 2y&=-12\\ \tfrac{2y}{2}&=\tfrac{-12}{2}\\ y&=-6 \end{aligned}$$

35.
$$\begin{aligned} \frac{x}{2}-\frac{x}{3}&=4\\ 6\left(\frac{x}{2}-\frac{x}{3}\right)&=6(4)\\ 6\left(\frac{x}{2}\right)-6\left(\frac{x}{3}\right)&=24\\ 3x-2x&=24\\ x&=24 \end{aligned}$$

37.
$$\begin{aligned} \frac{x}{6}+1&=\frac{x}{3}\\ 6\left(\frac{x}{6}+1\right)&=6\left(\frac{x}{3}\right)\\ 6\left(\frac{x}{6}\right)+6(1)&=2x\\ x+6&=2x\\ x-2x+6&=2x-2x\\ -x+6-6&=0-6\\ -x&=-6\\ x&=6 \end{aligned}$$

39.
$$\begin{aligned}
\frac{a+1}{3}+\frac{a-1}{5}&=\frac{2}{15}\\
15\left(\frac{a+1}{3}+\frac{a-1}{5}\right)&=15\left(\frac{2}{15}\right)\\
15\left(\frac{a+1}{3}\right)+15\left(\frac{a-1}{5}\right)&=2\\
5(a+1)+3(a-1)&=2\\
5a+5+3a-3&=2\\
8a+2-2&=2-2\\
8a&=0\\
\tfrac{8a}{8}&=\tfrac{0}{8}\\
a&=0
\end{aligned}$$

41.
$$\begin{aligned}
\frac{2z+3}{3}+\frac{3z-4}{6}&=\frac{z-2}{2}\\
6\left(\frac{2z+3}{3}+\frac{3z-4}{6}\right)&=6\cdot\frac{z-2}{2}\\
6\left(\frac{2z+3}{3}\right)+6\left(\frac{3z-4}{6}\right)&=3(z-2)\\
2(2z+3)+3z-4&=3z-6\\
4z+6+3z-4&=3z-6\\
7z+2&=3z-6\\
7z-3z+2&=3z-3z-6\\
4z+2-2&=-6-2\\
4z&=-8\\
\tfrac{4z}{4}&=\tfrac{-8}{4}\\
z&=-2
\end{aligned}$$

43.
$$\begin{aligned}
4(2-3t)+6t&=-6t+8\\
8-12t+6t&=-6t+8\\
8-6t&=-6t+8\\
-6t+8&=-6t+8
\end{aligned}$$
Identity; $\mathbb{R}$, $(-\infty,\infty)$

45.
$$\begin{aligned}
3(x-4)+6&=-2(x+4)+5x\\
3x-12+6&=-2x-8+5x\\
3x-6&=3x-8\\
3x-3x-6&=3x-3x-8\\
-6&\neq-8
\end{aligned}$$
Contradiction; $\emptyset$

47.
$$\begin{aligned}
V&=\frac{1}{3}Bh\\
3V&=3\cdot\frac{1}{3}Bh\\
3V&=Bh\\
\frac{3V}{h}&=\frac{Bh}{h}\\
\frac{3V}{h}&=B,\text{ or } B=\frac{3V}{h}
\end{aligned}$$

49.
$$\begin{aligned}
p&=2l+2w\\
p-2l&=2l-2l+2w\\
p-2l&=2w\\
\frac{p-2l}{2}&=\frac{2w}{2}\\
\frac{p-2l}{2}&=w,\text{ or } w=\frac{p-2l}{2}
\end{aligned}$$

51.
$$\begin{aligned}
z&=\frac{x-\mu}{\sigma}\\
\sigma z&=\sigma\cdot\frac{x-\mu}{\sigma}\\
\sigma z&=x-\mu\\
\sigma z+\mu&=x,\text{ or } x=z\sigma+\mu
\end{aligned}$$

53.
$$\begin{aligned}
y&=mx+b\\
y-b&=mx+b-b\\
y-b&=mx\\
\frac{y-b}{m}&=\frac{mx}{m}\\
\frac{y-b}{m}&=x,\text{ or } x=\frac{y-b}{m}
\end{aligned}$$

55.
$$\begin{aligned} 5x - 3 &> 7 \\ 5x - 3 + 3 &> 7 + 3 \\ 5x &> 10 \\ \tfrac{5x}{5} &> \tfrac{10}{5} \\ x &> 2 \end{aligned}$$
solution set: $(2, \infty)$

2

57.
$$\begin{aligned} -3x - 1 &\le 5 \\ -3x - 1 + 1 &\le 5 + 1 \\ -3x &\le 6 \\ \tfrac{-3x}{-3} &\ge \tfrac{6}{-3} \\ x &\ge -2 \end{aligned}$$
solution set: $[-2, \infty)$

-2

59.
$$\begin{aligned} -3(a + 2) &> 2(a + 1) \\ -3a - 6 &> 2a + 2 \\ -3a - 2a - 6 &> 2a - 2a + 2 \\ -5a - 6 + 6 &> 2 + 6 \\ -5a &> 8 \\ \tfrac{-5a}{-5} &< \tfrac{8}{-5} \\ a &< -\tfrac{8}{5} \end{aligned}$$
solution set: $\left(-\infty, -\frac{8}{5}\right)$

$-\frac{8}{5}$

61.

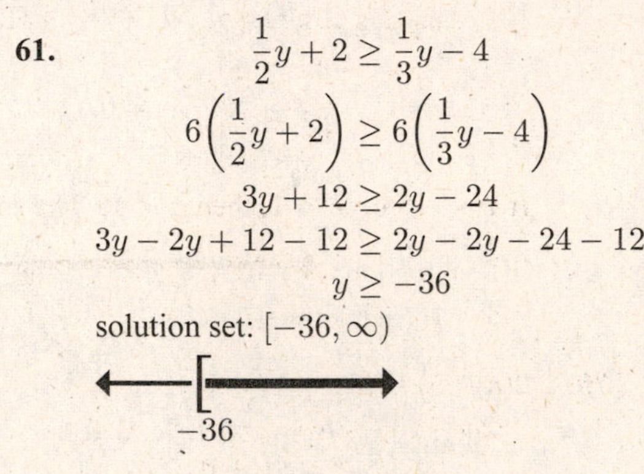

$$\begin{aligned} \frac{1}{2}y + 2 &\ge \frac{1}{3}y - 4 \\ 6\left(\frac{1}{2}y + 2\right) &\ge 6\left(\frac{1}{3}y - 4\right) \\ 3y + 12 &\ge 2y - 24 \\ 3y - 2y + 12 - 12 &\ge 2y - 2y - 24 - 12 \\ y &\ge -36 \end{aligned}$$
solution set: $[-36, \infty)$

-36

63.
$$\begin{aligned} -2 < -b + 3 &< 5 \\ -2 - 3 < -b + 3 - 3 &< 5 - 3 \\ -5 < -b &< 2 \\ \tfrac{-5}{-1} > \tfrac{-b}{-1} &> \tfrac{2}{-1} \end{aligned}$$
$5 > b > -2$, or $-2 < b < 5$

solution set: $(-2, 5)$

-2 5

65.
$$\begin{aligned} 15 > 2x - 7 &> 9 \\ 15 + 7 > 2x - 7 + 7 &> 9 + 7 \\ 22 > 2x &> 16 \\ \tfrac{22}{2} > \tfrac{2x}{2} &> \tfrac{16}{2} \end{aligned}$$
$11 > x > 8$, or $8 < x < 11$

solution set: $(8, 11)$

8 11

67.
$$\begin{aligned} -6 < -3(x - 4) &\le 24 \\ \frac{-6}{-3} > \frac{-3(x - 4)}{-3} &\ge \frac{24}{-3} \\ 2 > x - 4 &\ge -8 \\ 2 + 4 > x - 4 + 4 &\ge -8 + 4 \end{aligned}$$
$6 > x \ge -4$, or $-4 \le x < 6$

solution set: $[-4, 6)$

-4 6

69. $0 \ge \frac{1}{2}x - 4 > 6$

This inequality indicates that $0 \ge 6$ (by the transitive property). Since this is not possible, there is no solution to the inequality. $\emptyset$

71.
$$\begin{aligned} 3x+2 &< 8 \\ 3x+2-2 &< 8-2 \\ 3x &< 6 \\ x &< 2 \end{aligned} \qquad \text{or} \qquad \begin{aligned} 2x-3 &> 11 \\ 2x-3+3 &> 11+3 \\ 2x &> 14 \\ x &> 7 \end{aligned}$$

If $x < 2$ **or** $x > 7$, then the solution set is $(-\infty, 2) \cup (7, \infty)$.

73.
$$\begin{aligned} -4(x+2) &\ge 12 \\ -4x-8+8 &\ge 12+8 \\ -4x &\ge 20 \\ \tfrac{-4x}{-4} &\le \tfrac{20}{-4} \\ x &\le -5 \end{aligned} \qquad \text{or} \qquad \begin{aligned} 3x+8 &< 11 \\ 3x+8-8 &< 11-8 \\ 3x &< 3 \\ x &< 1 \end{aligned}$$

If $x \le -5$ **or** $x < 1$, then $x < 1$. The solution set is $(-\infty, 1)$.

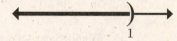

75.
$$\begin{aligned} 2(a-5)-(3a+1) &= 0 \\ 2a-10-3a-1 &= 0 \\ -a-11+11 &= 0+11 \\ -a &= 11 \\ a &= -11 \end{aligned}$$

77.
$$\begin{aligned} y(y+2) &= (y+1)^2-1 \\ y^2+2y &= (y+1)(y+1)-1 \\ y^2+2y &= y^2+2y+1-1 \\ y^2+2y &= y^2+2y \end{aligned}$$
Identity; $\mathbb{R}$, $(-\infty, \infty)$

79.
$$\begin{aligned} 8-9y &\ge -y \\ 8-9y+y &\ge -y+y \\ 8-8-8y &\ge 0-8 \\ -8y &\ge -8 \\ \tfrac{-8y}{-8} &\le \tfrac{-8}{-8} \\ y &\le 1 \end{aligned}$$

solution set: $(-\infty, 1]$

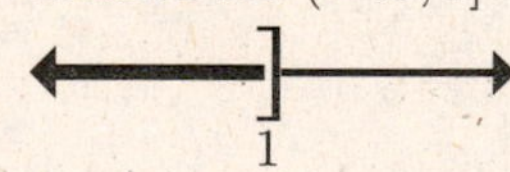

81.
$$\begin{aligned} 5(x-2) &\ge 0 \\ 5x-10+10 &\ge 0+10 \\ 5x &\ge 10 \\ x &\ge 2 \end{aligned} \qquad \text{and} \qquad \begin{aligned} -3x &< 9 \\ \tfrac{-3x}{-3} &> \tfrac{9}{-3} \\ x &> -3 \end{aligned}$$

If $x \ge 2$ **and** $x > -3$, then $x \ge 2$. The solution set is $[2, \infty)$.

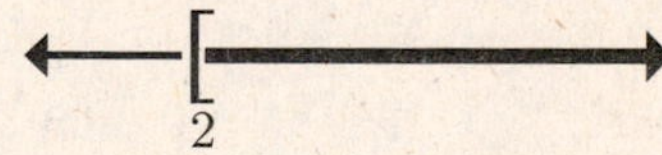

83.

$$-6 \le \frac{1}{3}a + 1 < 0$$
$$3(-6) \le 3\left(\frac{1}{3}a + 1\right) < 3(0)$$
$$-18 \le a + 3 < 0$$
$$-18 - 3 \le a + 3 - 3 < 0 - 3$$
$$-21 \le a < -3$$

solution set: $[-21, -3)$

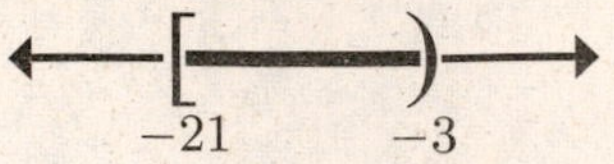

85.

$$P = L + \frac{s}{f}i$$
$$P - L = \frac{s}{f}i$$
$$f(P - L) = f \cdot \frac{s}{f}i$$
$$f(P - L) = si$$
$$\frac{f(P - L)}{i} = \frac{si}{i}$$
$$\frac{f(P - L)}{i} = s, \text{ or } s = \frac{f(P - L)}{i}$$

87. Let $w =$ the width. Then $2w =$ the length.

$\boxed{\text{Perimeter of garden}} = 72$

$$w + 2w + w + 2w = 72$$
$$6w = 72$$
$$w = 12$$

The dimension are 12 m by 24 m.

89.

$\boxed{\text{Total amount of fence}} = 150$

$$x + x + x + x + x + 5 + x + 5 + x = 150$$
$$7x + 10 = 150$$
$$7x = 140$$
$$x = 20$$

The dimensions are 20 feet by 45 feet.

91. Let $x =$ the length of the shorter piece.
Then $2x + 1 =$ the length of the other piece.

$\boxed{\text{shorter length}} + \boxed{\text{other length}} = \boxed{\text{total length}}$

$$x + 2x + 1 = 22$$
$$3x + 1 = 22$$
$$3x = 21$$
$$x = 7$$

The pieces have lengths of 7 feet and 15 feet.

93.

Price < 42
Price $-$ Cost $< 42 -$ Cost
Profit $< 42 - 27$
Profit < 15

95. Let $x =$ the # of compact discs bought.

$$175 + 8.50x \le 330$$
$$8.50x \le 155$$
$$x \le 18.24$$

He can buy at most 18 discs.

97. **Answers may vary.**

99. The 4 was not distributed correctly in the 2nd line. It should be $4x + 12 = 16$.

Exercise 7.2 (page 480)

1. $|-5| = +5 = 5$

3. $-|-6| = -(+6) = -6$

5.

$$|x| = 8$$
$$x = 8 \quad \textbf{or} \quad x = -8$$

7.

$$|x - 5| = 0$$
$$x - 5 = 0 \quad \textbf{or} \quad x - 5 = -0$$
$$x = 5 \qquad\qquad x = 5$$

SECTION 7.2

9. $$\begin{aligned} 3(2a-1) &= 2a \\ 6a-3 &= 2a \\ 4a &= 3 \\ a &= \frac{3}{4} \end{aligned}$$

11. $$\begin{aligned} \frac{5x}{2}-1 &= \frac{x}{3}+12 \\ 6\left(\frac{5x}{2}-1\right) &= 6\left(\frac{x}{3}+12\right) \\ 15x-6 &= 2x+72 \\ 13x &= 78 \\ x &= 6 \end{aligned}$$

13. x

15. 0

17. $a=-b$

19. $|8|=8$

21. $|-12|=12$

23. $-|2|=-2$

25. $-|-30|=-30$

27. $-(-|50|)=-(-50)=50$

29. $|\pi-4|=-(\pi-4)=4-\pi$ (since $\pi-4$ is less than zero.)

31. $|2|=2;\ |5|=5;$
$|2|$ is smaller.

33. $|5|=5;\ |-8|=8;$
$|5|$ is smaller.

35. $|-2|=2;\ |10|=10;$
$|-2|$ is smaller.

37. $|-3|=3;\ -|-4|=-4;$
$-|-4|$ is smaller.

39. $-|-5|=-5; -|-7|=-7;$
$-|-7|$ is smaller.

41. $-x<0$ if $x>0$; $|x+1|>0$
$-x$ is smaller.

43. $$\begin{aligned} |x| &= 8 \\ x=8 \quad &\textbf{or} \quad x=-8 \end{aligned}$$

45. $$\begin{aligned} &|x-3|=6 \\ x-3=6 \quad &\textbf{or} \quad x-3=-6 \\ x=9 \quad & \quad\quad x=-3 \end{aligned}$$

47. $$\begin{aligned} &|2x-3|=5 \\ 2x-3=5 \quad &\textbf{or} \quad 2x-3=-5 \\ 2x=8 \quad & \quad\quad 2x=-2 \\ x=4 \quad & \quad\quad x=-1 \end{aligned}$$

49. $$\begin{aligned} &|3x+2|=16 \\ 3x+2=16 \quad &\textbf{or} \quad 3x+2=-16 \\ 3x=14 \quad & \quad\quad 3x=-18 \\ x=\tfrac{14}{3} \quad & \quad\quad x=-6 \end{aligned}$$

51. $$\begin{aligned} &|x+3|+7=10 \\ &|x+3|=3 \\ x+3=3 \quad &\textbf{or} \quad x+3=-3 \\ x=0 \quad & \quad\quad x=-6 \end{aligned}$$

53. $$\begin{aligned} &|0.3x-3|-2=7 \\ &|0.3x-3|=9 \\ 0.3x-3=9 \quad &\textbf{or} \quad 0.3x-3=-9 \\ 0.3x=12 \quad & \quad\quad 0.3x=-6 \\ x=40 \quad & \quad\quad x=-20 \end{aligned}$$

55. $$\left|\tfrac{7}{2}x+3\right|=-5$$
Since an absolute value cannot be negative, there is no solution. $\emptyset$

57. $$\begin{aligned} &|3x+24|=0 \\ 3x+24=0 \quad &\textbf{or} \quad 3x+24=-0 \\ 3x=-24 \quad & \quad\quad 3x=-24 \\ x=-8 \quad & \quad\quad x=-8 \end{aligned}$$

SECTION 7.2

59.

$$|2x+1| = |3x+3|$$

$$\begin{aligned} 2x+1 &= 3x+3 \quad \textbf{or} \quad & 2x+1 &= -(3x+3) \\ -x &= 2 & 2x+1 &= -3x-3 \\ x &= -2 & 5x &= -4 \\ & & x &= -\tfrac{4}{5} \end{aligned}$$

61.

$$|3x-1| = |x+5|$$

$$\begin{aligned} 3x-1 &= x+5 \quad \textbf{or} \quad & 3x-1 &= -(x+5) \\ 2x &= 6 & 3x-1 &= -x-5 \\ x &= 3 & 4x &= -4 \\ & & x &= -1 \end{aligned}$$

63.

$$|2-x| = |3x+2|$$

$$\begin{aligned} 2-x &= 3x+2 \quad \textbf{or} \quad & 2-x &= -(3x+2) \\ -4x &= 0 & 2-x &= -3x-2 \\ x &= 0 & 2x &= -4 \\ & & x &= -2 \end{aligned}$$

65.

$$\left|\frac{x}{2}+2\right| = \left|\frac{x}{2}-2\right|$$

$$\begin{aligned} \frac{x}{2}+2 &= \frac{x}{2}-2 \quad \textbf{or} \quad & \frac{x}{2}+2 &= -\left(\frac{x}{2}-2\right) \\ 0 &= -4 & \frac{x}{2}+2 &= -\frac{x}{2}+2 \\ & & x &= 0 \end{aligned}$$

(no solution from this part)

67.

$$\left|\frac{x}{2}-1\right| = 3$$

$$\begin{aligned} \frac{x}{2}-1 &= 3 \quad \textbf{or} \quad & \frac{x}{2}-1 &= -3 \\ \frac{x}{2} &= 4 & \frac{x}{2} &= -2 \\ x &= 8 & x &= -4 \end{aligned}$$

69.

$$|3-4x| = 5$$

$$\begin{aligned} 3-4x &= 5 \quad \textbf{or} \quad & 3-4x &= -5 \\ -4x &= 2 & -4x &= -8 \\ x &= -\tfrac{1}{2} & x &= 2 \end{aligned}$$

71.

$$\left|\frac{3}{5}x-4\right| - 2 = -2$$

$$\left|\frac{3}{5}x-4\right| = 0$$

$$\begin{aligned} \frac{3}{5}x-4 &= 0 \quad \textbf{or} \quad & \frac{3}{5}x-4 &= -0 \\ \frac{3}{5}x &= 4 & \frac{3}{5}x &= 4 \\ x &= \tfrac{20}{3} & x &= \tfrac{20}{3} \end{aligned}$$

73.

$$\left|x+\tfrac{1}{3}\right| = |x-3|$$

$$\begin{aligned} x+\tfrac{1}{3} &= x-3 \quad \textbf{or} \quad & x+\tfrac{1}{3} &= -(x-3) \\ 3x+1 &= 3x-9 & x+\tfrac{1}{3} &= -x+3 \\ 0 &= -10 & 3x+1 &= -3x+9 \\ & & 6x &= 8 \\ & & x &= \tfrac{4}{3} \end{aligned}$$

(no solution from this part)

75. $|3x+7| = -|8x-2|$
Since an absolute value cannot be negative, there is no solution. $\emptyset$

77. $\left|\frac{3x+48}{3}\right| = 12$

$\frac{3x+48}{3} = 12$ **or** $\frac{3x+48}{3} = -12$
$3x+48 = 36$ $\quad$ $3x+48 = -36$
$3x = -12$ $\quad$ $3x = -84$
$x = -4$ $\quad$ $x = -28$

79. Graph $y = |0.75x + 0.12|$ and $y = 12.3$ and find the x-coordinate(s) of any point(s) of intersection:

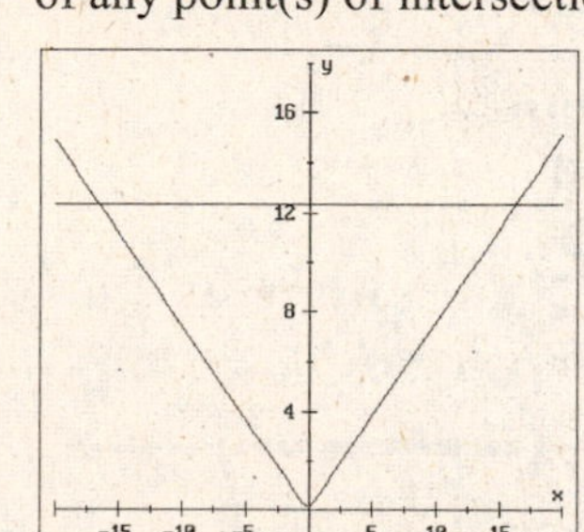

solutions: $-16.6, 16.2$

81. **Answers may vary.**

83. $|x| + k = 0$
$|x| = -k$
If this equation has exactly two solutions, then $-k > 0$, or $k < 0$.

85. **Answers may vary.**

87. **Answers may vary.**

Exercise 7.3 (page 486)

1. $|x| < 8$
$-8 < x < 8$

3. $|x| \geq 4$
$x \leq -4$ **or** $x \geq 4$

5. $|x+1| < 2$
$-2 < x+1 < 2$
$-3 < x < 1$

7. $A = p + prt$
$A - p = prt$
$\frac{A-p}{pr} = t$, or $t = \frac{A-p}{pr}$

9. $P = 2w + 2l$
$P - 2w = 2l$
$\frac{P-2w}{2} = l$, or $l = \frac{P-2w}{2}$

11. $-k < x < k$

13. $x < -k$ or $x > k$

SECTION 7.3

15. $|2x| < 8$
$-8 < 2x < 8$
$\frac{-8}{2} < \frac{2x}{2} < \frac{8}{2}$
$-4 < x < 4$
solution set: $(-4, 4)$

17. $|x + 9| \leq 12$
$-12 \leq x + 9 \leq 12$
$-12 - 9 \leq x + 9 - 9 \leq 12 - 9$
$-21 \leq x \leq 3$
solution set: $[-21, 3]$

19. $|3x - 2| \leq 10$
$-10 \leq 3x - 2 \leq 10$
$-10 + 2 \leq 3x - 2 + 2 \leq 10 + 2$
$-8 \leq 3x \leq 12$
$\frac{-8}{3} \leq \frac{3x}{3} \leq \frac{12}{3}$
$-\frac{8}{3} \leq x \leq 4$
solution set: $\left[-\frac{8}{3}, 4\right]$

21. $|3 - 2x| < 7$
$-7 < 3 - 2x < 7$
$-7 - 3 < 3 - 3 - 2x < 7 - 3$
$-10 < -2x < 4$
$\frac{-10}{-2} > \frac{-2x}{-2} > \frac{4}{-2}$
$5 > x > -2$, or $-2 < x < 5$
solution set: $(-2, 5)$

23. $|5x| > 5$
$5x < -5$ **or** $5x > 5$
$x < -1$ $\quad x > 1$
solution set: $(-\infty, -1) \cup (1, \infty)$

25. $|x - 12| > 24$
$x - 12 < -24$ **or** $x - 12 > 24$
$x < -12$ $\quad x > 36$
solution set: $(-\infty, -12) \cup (36, \infty)$

27. $|3x + 2| > 14$
$3x + 2 < -14$ **or** $3x + 2 > 14$
$3x < -16$ $\quad 3x > 12$
$x < -\frac{16}{3}$ $\quad x > 4$
solution set: $\left(-\infty, -\frac{16}{3}\right) \cup (4, \infty)$

29. $|2 - 3x| \geq 8$
$2 - 3x \leq -8$ **or** $2 - 3x \geq 8$
$-3x \leq -10$ $\quad -3x \geq 6$
$x \geq \frac{10}{3}$ $\quad x \leq -2$
solution set: $(-\infty, -2] \cup \left[\frac{10}{3}, \infty\right)$

31. $\left|\frac{1}{3}x + 7\right| + 5 > 6$
$\left|\frac{1}{3}x + 7\right| > 1$
$\frac{1}{3}x + 7 < -1$ **or** $\frac{1}{3}x + 7 > 1$
$x + 21 < -3$ $\quad x + 21 > 3$
$x < -24$ $\quad x > -18$
solution set: $(-\infty, -24) \cup (-18, \infty)$

33. $-|2x - 3| < -7$
$|2x - 3| > 7$
$2x - 3 < -7$ **or** $2x - 3 > 7$
$2x < -4$ $\quad 2x > 10$
$x < -2$ $\quad x > 5$
solution set: $(-\infty, -2) \cup (5, \infty)$

SECTION 7.3

35.
$$-|5x-1|+2<0$$
$$-|5x-1|<-2$$
$$|5x-1|>2$$
$5x-1<-2$ **or** $5x-1>2$

$5x<-1$ $\quad$ $5x>3$

$x<-\frac{1}{5}$ $\quad$ $x>\frac{3}{5}$

solution set: $\left(-\infty,-\frac{1}{5}\right)\cup\left(\frac{3}{5},\infty\right)$

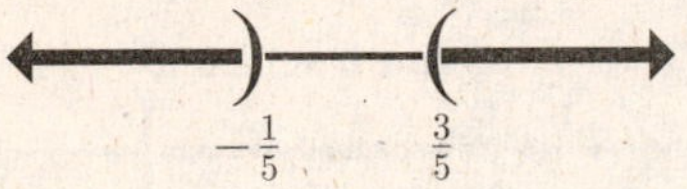

$-\frac{1}{5}$ $\quad$ $\frac{3}{5}$

37.
$$\left|\frac{x-2}{3}\right|>4$$
$\frac{x-2}{3}<-4$ **or** $\frac{x-2}{3}>4$

$x-2<-12$ $\quad$ $x-2>12$

$x<-10$ $\quad$ $x>14$

solution set: $(-\infty,-10)\cup(14,\infty)$

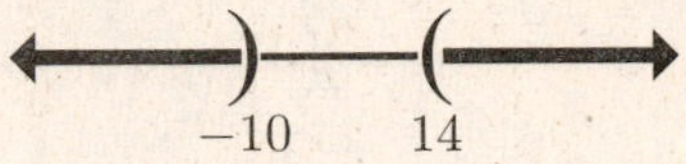

-10 $\quad$ 14

39.
$$-2|3x-4|<16$$
$$|3x-4|>-8$$
Since an absolute value is always at least 0, this inequality is true for all real numbers.

solution set: $(-\infty,\infty)$

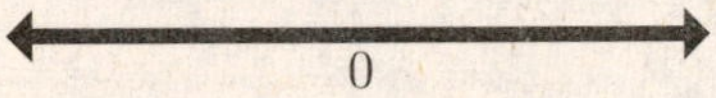

0

41. $|5x-1|+4\leq 0$

$|5x-1|\leq -4$

Since an absolute value can never be negative, this inequality has no solution. $\emptyset$

43.
$$|2x+1|+2\leq 2$$
$$|2x+1|\leq 0$$
Since an absolute value can never be less than zero, the only solution for this inequality is when the absolute value is equal to 0.
$$|2x+1|=0$$
$2x+1=0$ **or** $2x+1=-0$

$2x=-1$ $\quad$ $2x=-1$

$x=-\frac{1}{2}$ $\quad$ $x=-\frac{1}{2}$

solution set: $\left[-\frac{1}{2},-\frac{1}{2}\right]$

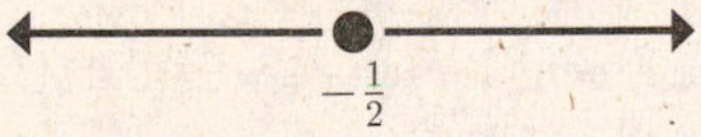

$-\frac{1}{2}$

45.
$$\left|3\left(\tfrac{x+4}{4}\right)\right|>0$$
$3\left(\frac{x+4}{4}\right)<-0$ **or** $3\left(\frac{x+4}{4}\right)>0$

$\frac{4}{3}\cdot 3\left(\frac{x+4}{4}\right)<\frac{4}{3}(-0)$ $\quad$ $\frac{4}{3}\cdot 3\left(\frac{x+4}{4}\right)>\frac{4}{3}\cdot 0$

$x+4<0$ $\quad$ $x+4>0$

$x<-4$ $\quad$ $x>-4$

solution set: $(-\infty,-4)\cup(-4,\infty)$

-4

47.
$$3|2x+5|\geq 9$$
$$|2x+5|\geq 3$$
$2x+5\leq -3$ **or** $2x+5\geq 3$

$2x\leq -8$ $\quad$ $2x\geq -2$

$x\leq -4$ $\quad$ $x\geq -1$

solution set: $(-\infty,-4]\cup[-1,\infty)$

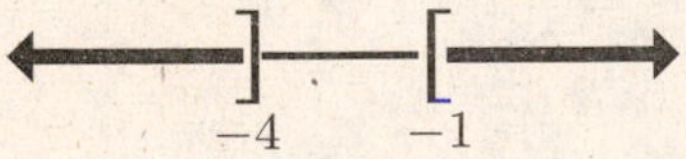

-4 $\quad$ -1

49.
$$\left|\tfrac{1}{7}x+1\right|\leq 0$$
Since an absolute value can never be less than zero, the only solution for this inequality is when the absolute value is equal to 0.
$$\left|\tfrac{1}{7}x+1\right|=0$$
$\frac{1}{7}x+1=0$ **or** $\frac{1}{7}x+1=-0$

$\frac{1}{7}x=-1$ $\quad$ $\frac{1}{7}x=-1$

$x=-7$ $\quad$ $x=-7$

solution set: $[-7,-7]$

-7

51. $\left|\frac{1}{5}x - 5\right| + 4 > 4$
$\left|\frac{1}{5}x - 5\right| > 0$
$\frac{1}{5}x - 5 < -0$ **or** $\frac{1}{5}x - 5 > 0$
$\frac{1}{5}x < 5$ $\quad \frac{1}{5}x > 5$
$x < 25$ $\quad x > 25$
solution set: $(-\infty, 25) \cup (25, \infty)$

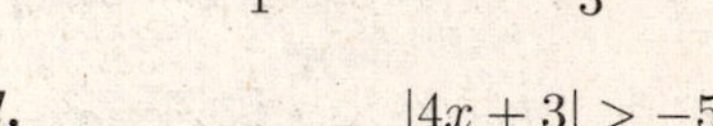

53. $3\left|\frac{1}{3}(x-2)\right| + 2 \leq 3$
$\left|\frac{1}{3}(x-2)\right| \leq \frac{1}{3}$
$-\frac{1}{3} \leq \frac{x-2}{3} \leq \frac{1}{3}$
$-1 \leq x - 2 \leq 1$
$1 \leq x \leq 3$
solution set: $[1, 3]$

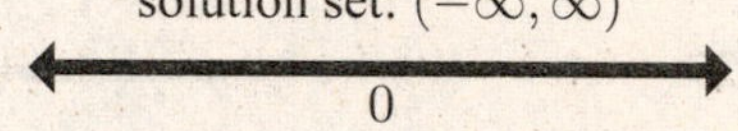

55. $|3x - 2| + 2 \geq 0$
$|3x - 2| \geq -2$
Since an absolute value is always at least 0, this inequality is true for all real numbers.
solution set: $(-\infty, \infty)$

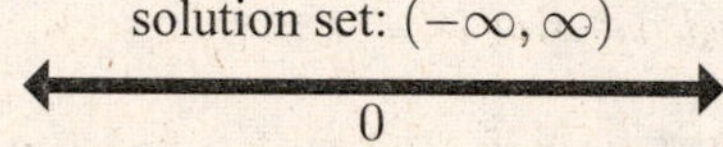

57. $|4x + 3| > -5$
Since an absolute value is always at least 0, this inequality is true for all real numbers.
solution set: $(-\infty, \infty)$

0

59. Find the x-coordinates of all points on the graph of $y = |0.5x + 0.7|$ that are below points on the graph of $y = 2.6$:

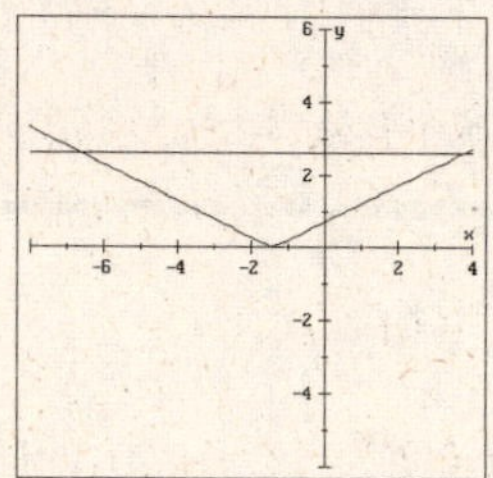

solution set: $(-6.6, 3.8)$

61. Find the x-coordinates of all points on the graph of $y = |2.15x - 3.05|$ that are above points on the graph of $y = 3.8$:

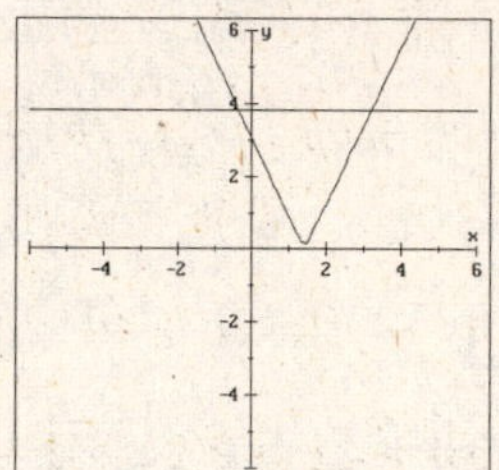

solution set: $(-\infty, -0.4) \cup (3.2, \infty)$

63. **Answers may vary.**

65. $|x| + |y| > |x + y|$ when x and y have different signs.

Exercise 7.4 (page 498)

1. $3xy^2 - 6x^2y = \mathbf{3xy} \cdot y - \mathbf{3xy} \cdot 2x = \mathbf{3xy}(y - 2x)$

3. $x^2 + 5x - 6 = (x + 6)(x - 1)$

5. $a^3 + 8 = a^3 + 2^3$
$= (a + 2)(a^2 - 2a + 2^2)$
$= (a + 2)(a^2 - 2a + 4)$

7. 1.1×10^3 ft/sec $= 1{,}100$ ft/sec

SECTION 7.4

9.
$$\begin{aligned} \frac{2}{3}(5t-3) &= 38 \\ 3 \cdot \frac{2}{3}(5t-3) &= 3(38) \\ 2(5t-3) &= 114 \\ 10t-6 &= 114 \\ 10t &= 120 \\ t &= 12 \end{aligned}$$

11. $ab + ac$

13. perfect; key number

15. $x^2 - xy + y^2$

17. $2x + 8 = \mathbf{2} \cdot x + \mathbf{2} \cdot 4 = \mathbf{2}(x+4)$

19. $2x^2 - 6x = \mathbf{2x} \cdot x - \mathbf{2x} \cdot 3 = \mathbf{2x}(x-3)$

21. $15x^2y - 10x^2y^2 = \mathbf{5x^2y} \cdot 3 - \mathbf{5x^2y} \cdot 2y = \mathbf{5x^2y}(3-2y)$

23. $13ab^2c^3 - 26a^3b^2c = \mathbf{13ab^2c} \cdot c^2 - \mathbf{13ab^2c} \cdot 2a^2 = \mathbf{13ab^2c}(c^2 - 2a^2)$

25. $27z^3 + 12z^2 + 3z = \mathbf{3z} \cdot 9z^2 + \mathbf{3z} \cdot 4z + \mathbf{3z} \cdot 1 = \mathbf{3z}(9z^2 + 4z + 1)$

27. $24s^3 - 12s^2t + 6st^2 = \mathbf{6s}(4s^2) - \mathbf{6s}(2st) + \mathbf{6s}(t^2) = \mathbf{6s}(4s^2 - 2st + t^2)$

29.
$$\begin{aligned} -3a - 6 &= (\mathbf{-3})(a) + (\mathbf{-3})(2) \\ &= \mathbf{-3}(a+2) \end{aligned}$$

31.
$$\begin{aligned} -6x^2 - 3xy &= (\mathbf{-3x})(2x) + (\mathbf{-3x})(y) \\ &= \mathbf{-3x}(2x+y) \end{aligned}$$

33.
$$\begin{aligned} x^{n+2} + x^{n+3} &= x^2(x^{n+2-2} + x^{n+3-2}) \\ &= x^2(x^n + x^{n+1}) \end{aligned}$$

35.
$$\begin{aligned} 2y^{n+2} - 3y^{n+3} &= y^n(2y^{n+2-n} - 3y^{n+3-n}) \\ &= y^n(2y^2 - 3y^3) \end{aligned}$$

37.
$$\begin{aligned} ax + bx + ay + by &= x(a+b) + y(a+b) \\ &= (a+b)(x+y) \end{aligned}$$

39.
$$\begin{aligned} x^2 + yx + 2x + 2y &= x(x+y) + 2(x+y) \\ &= (x+y)(x+2) \end{aligned}$$

41. $3c - cd + 3d - c^2 = 3c + 3d - c^2 - cd = 3(c+d) - c(c+d) = (c+d)(3-c)$

43. $4y + 8 + 2xy + 4x = 2(2y + 4 + xy + 2x) = 2[2(y+2) + x(y+2)] = 2(y+2)(2+x)$

45.
$$\begin{aligned} r_1r_2 &= rr_2 + rr_1 \\ r_1r_2 - rr_1 &= rr_2 \\ r_1(r_2 - r) &= rr_2 \\ r_1 &= \frac{rr_2}{r_2 - r} \end{aligned}$$

47.
$$\begin{aligned} S(1-r) &= a - lr \\ S - Sr &= a - lr \\ S - a &= Sr - lr \\ S - a &= r(S - l) \\ \frac{S-a}{S-l} &= r, \text{ or } r = \frac{S-a}{S-l} \end{aligned}$$

49. $x^2 - 4 = x^2 - 2^2 = (x+2)(x-2)$

51. $9y^2 - 64 = (3y)^2 - 8^2 = (3y+8)(3y-8)$

53.
$$\begin{aligned} 81a^4 - 49b^2 &= (9a^2)^2 - (7b)^2 \\ &= (9a^2 + 7b)(9a^2 - 7b) \end{aligned}$$

55.
$$\begin{aligned} (x+y)^2 - z^2 &= [(x+y)+z][(x+y)-z] \\ &= (x+y+z)(x+y-z) \end{aligned}$$

57. $x^4 - y^4 = (x^2+y^2)(x^2-y^2) = (x^2+y^2)(x+y)(x-y)$

SECTION 7.4

59. $2x^2 - 288 = 2(x^2 - 144)$
$= 2(x + 12)(x - 12)$

61. $a^2 - 3ab - 4b^2 = (a - 4b)(a + b)$

63. $x^2 + 5x + 6 = (x + 2)(x + 3)$

65. $a^2 + 5a - 52 \Rightarrow$ prime

67. $-a^2 + 4a + 32 = -(a^2 - 4a - 32)$
$= -(a - 8)(a + 4)$

69. $3x^2 + 12x - 63 = 3(x^2 + 4x - 21)$
$= 3(x + 7)(x - 3)$

71. $-3x^2 + 15x - 18 = -3(x^2 - 5x + 6)$
$= -3(x - 2)(x - 3)$

73. $2x^2 - 11x + 5 = (2x - 1)(x - 5)$

75. $6y^2 + 7y + 2 = (2y + 1)(3y + 2)$

77. $8a^2 + 6a - 9 = (4a - 3)(2a + 3)$

79. $2y^2 + yt - 6t^2 = (2y - 3t)(y + 2t)$

81. $5x^2 + 4x + 1 \Rightarrow$ prime

83. $8x^2 - 10x + 3 = (4x - 3)(2x - 1)$

85. $3x^3 - 10x^2 + 3x = x(3x^2 - 10x + 3)$
$= x(3x - 1)(x - 3)$

87. $a^2b^2 - 13ab^2 + 22b^2 = b^2(a^2 - 13a + 22)$
$= b^2(a - 11)(a - 2)$

89. $-3a^2 + ab + 2b^2 = -(3a^2 - ab - 2b^2)$
$= -(3a + 2b)(a - b)$

91. $-4x^3 - 9x + 12x^2 = -x(4x^2 - 12x + 9) = -x(2x - 3)(2x - 3) = -x(2x - 3)^2$

93. $x^{2n} + 2x^n + 1 = (x^n + 1)(x^n + 1)$
$= (x^n + 1)^2$

95. $2a^{6n} - 3a^{3n} - 2 = (2a^{3n} + 1)(a^{3n} - 2)$

97. $x^{4n} + 2x^{2n}y^{2n} + y^{4n} = (x^{2n} + y^{2n})(x^{2n} + y^{2n}) = (x^{2n} + y^{2n})^2$

99. $6x^{2n} + 7x^n - 3 = (3x^n - 1)(2x^n + 3)$

101. $x^2 + 4x + 4 - y^2 = (x^2 + 4x + 4) - y^2 = (x + 2)(x + 2) - y^2 = (x + 2)^2 - y^2$
$= [(x + 2) + y][(x + 2) - y]$
$= (x + 2 + y)(x + 2 - y)$

103. $x^2 + 2x + 1 - 9z^2 = (x^2 + 2x + 1) - 9z^2 = (x + 1)(x + 1) - 9z^2$
$= (x + 1)^2 - (3z)^2$
$= [(x + 1) + 3z][(x + 1) - 3z]$
$= (x + 1 + 3z)(x + 1 - 3z)$

105. $y^3 + 1 = y^3 + 1^3$
$= (y + 1)(y^2 - 1y + 1^2)$
$= (y + 1)(y^2 - y + 1)$

107. $a^3 - 27 = a^3 - 3^3$
$= (a - 3)(a^2 + 3a + 3^2)$
$= (a - 3)(a^2 + 3a + 9)$

SECTION 7.4

109. $8 + x^3 = 2^3 + x^3$
$= (2 + x)(2^2 - 2x + x^2)$
$= (2 + x)(4 - 2x + x^2)$

111. $2x^3 + 54 = 2(x^3 + 27)$
$= 2(x^3 + 3^3)$
$= 2(x + 3)(x^2 - 3x + 3^2)$
$= 2(x + 3)(x^2 - 3x + 9)$

113. $s^3 - t^3 = (s - t)(s^2 + st + t^2)$

115. $27x^3 + y^3 = (3x)^3 + y^3$
$= (3x + y)((3x)^2 - 3xy + y^2)$
$= (3x + y)(9x^2 - 3xy + y^2)$

117. $a^3 + 8b^3 = a^3 + (2b)^3$
$= (a + 2b)(a^2 - 2ab + (2b)^2)$
$= (a + 2b)(a^2 - 2ab + 4b^2)$

119. $27x^3 - 125y^3 = (3x)^3 - (5y)^3 = (3x - 5y)((3x)^2 + (3x)(5y) + (5y)^2)$
$= (3x - 5y)(9x^2 + 15xy + 25y^2)$

121. $x^6 - 1 = (x^3)^2 - 1^2 = (x^3 + 1)(x^3 - 1) = (x^3 + 1^3)(x^3 - 1^3)$
$= (x + 1)(x^2 - 1x + 1^2)(x - 1)(x^2 + 1x + 1^2)$
$= (x + 1)(x^2 - x + 1)(x - 1)(x^2 + x + 1)$

123. $x^{12} - y^6 = (x^6)^2 - (y^3)^2 = (x^6 + y^3)(x^6 - y^3)$
$= \left((x^2)^3 + y^3\right)\left((x^2)^3 - y^3\right)$
$= (x^2 + y)\left((x^2)^2 - x^2y + y^2\right)(x^2 - y)\left((x^2)^2 + x^2y + y^2\right)$
$= (x^2 + y)(x^4 - x^2y + y^2)(x^2 - y)(x^4 + x^2y + y^2)$

125. $2x^3 - 32x = 2x(x^2 - 16)$
$= 2x(x + 4)(x - 4)$

127. $x^4 + 8x^2 + 15 = (x^2 + 5)(x^2 + 3)$

129. $y^4 - 13y^2 + 30 = (y^2 - 10)(y^2 - 3)$

131. $a^6 - b^3 = (a^2)^3 - b^3$
$= (a^2 - b)\left((a^2)^2 + a^2b + b^2\right)$
$= (a^2 - b)(a^4 + a^2b + b^2)$

133. $x^9 + y^6 = (x^3)^3 + (y^2)^3 = (x^3 + y^2)\left((x^3)^2 - x^3y^2 + (y^2)^2\right)$
$= (x^3 + y^2)(x^6 - x^3y^2 + y^4)$

135. $-63u^3v^6z^9 + 28u^2v^7z^2 - 21u^3v^3z^4 = -7u^2v^3z^2(9uv^3z^7 - 4v^4 + 3uz^2)$

137. $a^4 - 13a^2 + 36 = (a^2 - 4)(a^2 - 9) = (a + 2)(a - 2)(a + 3)(a - 3)$

SECTION 7.4

139. $c^2 - 4a^2 + 4ab - b^2 = c^2 - (4a^2 - 4ab + b^2) = c^2 - (2a - b)(2a - b)$
$$= c^2 - (2a - b)^2$$
$$= [c + (2a - b)][c - (2a - b)]$$
$$= (c + 2a - b)(c - 2a + b)$$

141. $-x^3 + 216 = -(x^3 - 216) = -(x^3 - 6^3) = -(x - 6)(x^2 + 6x + 6^2) = -(x - 6)(x^2 + 6x + 36)$

143. $64m^3x - 8n^3x = 8x(8m^3 - n^3) = 8x((2m)^3 - n^3) = 8x(2m - n)((2m)^2 + 2mn + n^2)$
$$= 8x(2m - n)(4m^2 + 2mn + n^2)$$

145. $x^4y + 216xy^4 = xy(x^3 + 216y^3) = xy(x^3 + (6y)^3) = xy(x + 6y)(x^2 - 6xy + (6y)^2)$
$$= xy(x + 6y)(x^2 - 6xy + 36y^2)$$

147. $81r^4s^2 - 24rs^5 = 3rs^2(27r^3 - 8s^3) = 3rs^2((3r)^3 - (2s)^3)$
$$= 3rs^2(3r - 2s)((3r)^2 + (3r)(2s) + (2s)^2)$$
$$= 3rs^2(3r - 2s)(9r^2 + 6rs + 4s^2)$$

149. $125a^6b^2 + 64a^3b^5 = a^3b^2(125a^3 + 64b^3) = a^3b^2((5a)^3 + (4b)^3)$
$$= a^3b^2(5a + 4b)((5a)^2 - (5a)(4b) + (4b)^2)$$
$$= a^3b^2(5a + 4b)(25a^2 - 20ab + 16b^2)$$

151. $(m^3 + 8n^3) + (m^3x + 8n^3x) = 1(m^3 + 8n^3) + x(m^3 + 8n^3)$
$$= (m^3 + 8n^3)(1 + x)$$
$$= (m + 2n)(m^2 - 2mn + (2n)^2)(1 + x)$$
$$= (m + 2n)(m^2 - 2mn + 4n^2)(1 + x)$$

153. $(a^4 + 27a) - (a^3b + 27b) = a(a^3 + 27) - b(a^3 + 27) = (a^3 + 27)(a - b)$
$$= (a^3 + 3^3)(a - b)$$
$$= (a + 3)(a^2 - 3a + 3^2)(a - b)$$
$$= (a + 3)(a^2 - 3a + 9)(a - b)$$

155. $y^3(y^2 - 1) - 27(y^2 - 1) = (y^2 - 1)(y^3 - 27) = (y^2 - 1^2)(y^3 - 3^3)$
$$= (y + 1)(y - 1)(y - 3)(y^2 + 3y + 9)$$

157. $2mp^4 + 16mpq^3 = 2mp(p^3 + 8q^3) = 2mp(p^3 + (2q)^3) = 2mp(p + 2q)(p^2 - 2pq + (2q)^2)$
$$= 2mp(p + 2q)(p^2 - 2pq + 4q^2)$$

159. $3(x^3 + y^3) - z(x^3 + y^3) = (x^3 + y^3)(3 - z) = (x + y)(x^2 - xy + y^2)(3 - z)$

161. $a^2 - b^2 + a + b = (a^2 - b^2) + (a + b) = (a + b)(a - b) + 1(a + b)$
$$= (a + b)(a - b + 1)$$

163. $2x+y+4x^2-y^2=(2x+y)+(4x^2-y^2)=(2x+y)(1)+(2x+y)(2x-y)$
$=(2x+y)(1+2x-y)$

165. Answers may vary.

167. Answers may vary.

169. ax^2+ax+a: $a=a, b=a, c=a \Rightarrow b^2-4ac=a^2-4a(a)=a^2-4a^2=-3a^2$.
$-3a^2$ is a nonnegative perfect square only for $a=0$. The test does not work. The test actually determines the factorability of x^2+x+1. NOTE: $ax^2+ax+a=a(x^2+x+1)$.

171. $m-2n+m^2-4n^2=(m-2n)+(m^2-4n^2)=(m-2n)(1)+(m+2n)(m-2n)$
$=(m-2n)(1+m+2n)$

Exercise 7.5 (page 511)

1. $\frac{4}{6}=\frac{2}{3}$

3. $-\frac{25}{30}=-\frac{5}{6}$

5. $\frac{x^2}{xy}=\frac{x}{y}$

7. $\frac{x-2}{2-x}=\frac{x-2}{-(x-2)}=-1$

9. 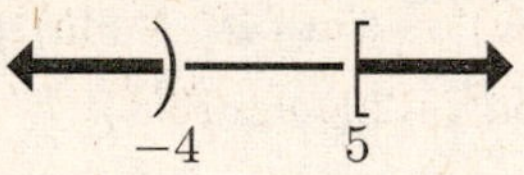

11.
$$P=2l+2w$$
$$P-2l=2w$$
$$\frac{P-2l}{2}=\frac{2w}{2}$$
$$\frac{P-2l}{2}=w, \text{ or } w=\frac{P-2l}{2}$$

13.
$$a^4-13a^2+36=0$$
$$(a^2-4)(a^2-9)=36$$
$$(a+2)(a-2)(a+3)(a-3)=36$$
$a+2=0$ **or** $a-2=0$ **or** $a+3=0$ **or** $a-3=0$
$a=-2$ $\quad a=2$ $\quad a=-3$ $\quad a=3$

15. $\frac{a}{b}$

17. $\frac{ad}{bc}$

19. $\frac{12x^3}{3x}=\frac{12}{3}\cdot\frac{x^3}{x}=4x^2$

21. $\frac{-24x^3y^4}{18x^4y^3}=-\frac{24}{18}\cdot\frac{x^3y^4}{x^4y^3}=-\frac{4y}{3x}$

23. $\frac{9y^2(y-z)}{21y(y-z)^2}=\frac{3y}{7(y-z)}$

25. $\frac{(a-b)(b-c)(c-d)}{(c-d)(b-c)(a-b)}=1$

27. $\frac{12-3x^2}{x^2-x-2}=\frac{3(4-x^2)}{(x-2)(x+1)}=\frac{3(2+x)(2-x)}{(x-2)(x+1)}=\frac{-3(x+2)(x-2)}{(x-2)(x+1)}=\frac{-3(x+2)}{x+1}$

SECTION 7.5

29. $\dfrac{x^3+8}{x^2-2x+4} = \dfrac{(x+2)(x^2-2x+4)}{x^2-2x+4}$
$= x+2$

31. $\dfrac{x^2+2x+1}{x^2+4x+3} = \dfrac{(x+1)(x+1)}{(x+1)(x+3)} = \dfrac{x+1}{x+3}$

33. $\dfrac{4x^2+24x+32}{16x^2+8x-48} = \dfrac{4(x^2+6x+8)}{8(2x^2+x-6)} = \dfrac{4(x+4)(x+2)}{8(2x-3)(x+2)} = \dfrac{x+4}{2(2x-3)}$

35. $\dfrac{x+y}{x^2-y^2} = \dfrac{x+y}{(x+y)(x-y)} = \dfrac{1}{x-y}$

37. $\dfrac{3m-6n}{3n-6m} = \dfrac{3(m-2n)}{3(n-2m)} = \dfrac{m-2n}{n-2m}$

39. $\dfrac{3x^2-3y^2}{x^2+2y+2x+yx} = \dfrac{3(x^2-y^2)}{x^2+2x+yx+2y} = \dfrac{3(x^2-y^2)}{x(x+2)+y(x+2)} = \dfrac{3(x+y)(x-y)}{(x+2)(x+y)}$
$= \dfrac{3(x-y)}{x+2}$

41. $\dfrac{x-y}{x^3-y^3-x+y} = \dfrac{x-y}{(x^3-y^3)-(x-y)} = \dfrac{x-y}{(x-y)(x^2+xy+y^2)-1(x-y)}$
$= \dfrac{x-y}{(x-y)(x^2+xy+y^2-1)}$
$= \dfrac{1}{x^2+xy+y^2-1}$

43. $\dfrac{x^2y^2}{cd} \cdot \dfrac{c^{-2}d^2}{x} = x^{2-1}y^2c^{-2-1}d^{2-1} = xy^2c^{-3}d = \dfrac{xy^2d}{c^3}$

45. $\dfrac{x^2+2x+1}{x} \cdot \dfrac{x^2-x}{x^2-1} = \dfrac{(x+1)(x+1)}{x} \cdot \dfrac{x(x-1)}{(x+1)(x-1)} = x+1$

47. $\dfrac{2x^2-x-3}{x^2-1} \cdot \dfrac{x^2+x-2}{2x^2+x-6} = \dfrac{(2x-3)(x+1)}{(x+1)(x-1)} \cdot \dfrac{(x+2)(x-1)}{(2x-3)(x+2)} = 1$

49. $\dfrac{3t^2-t-2}{6t^2-5t-6} \cdot \dfrac{4t^2-9}{2t^2+5t+3} = \dfrac{(3t+2)(t-1)}{(3t+2)(2t-3)} \cdot \dfrac{(2t+3)(2t-3)}{(2t+3)(t+1)} = \dfrac{t-1}{t+1}$

51. $\dfrac{-x^2y^{-2}}{x^{-1}y^{-3}} \div \dfrac{x^{-3}y^2}{x^4y^{-1}} = \dfrac{-x^2y^{-2}}{x^{-1}y^{-3}} \cdot \dfrac{x^4y^{-1}}{x^{-3}y^2} = \dfrac{-x^6y^{-3}}{x^{-4}y^{-1}} = -x^{6-(-4)}y^{-3-(-1)} = -x^{10}y^{-2} = -\dfrac{x^{10}}{y^2}$

53. $\dfrac{x^2-16}{x^2-25} \div \dfrac{x+4}{x-5} = \dfrac{x^2-16}{x^2-25} \cdot \dfrac{x-5}{x+4} = \dfrac{(x+4)(x-4)}{(x+5)(x-5)} \cdot \dfrac{x-5}{x+4} = \dfrac{x-4}{x+5}$

55. $\dfrac{a^2+2a-35}{12x} \div \dfrac{ax-3x}{a^2+4a-21} = \dfrac{a^2+2a-35}{12x} \cdot \dfrac{a^2+4a-21}{ax-3x}$
$= \dfrac{(a+7)(a-5)}{12x} \cdot \dfrac{(a+7)(a-3)}{x(a-3)} = \dfrac{(a+7)^2(a-5)}{12x^2}$

SECTION 7.5

57. $$(2x^2-15x+25)\div\frac{2x^2-3x-5}{x+1}=\frac{2x^2-15x+25}{1}\cdot\frac{x+1}{2x^2-3x-5}=\frac{(2x-5)(x-5)}{1}\cdot\frac{x+1}{(2x-5)(x+1)}=x-5$$

59. $$\frac{3x^2y^2}{6x^3y}\cdot\frac{-4x^7y^{-2}}{18x^{-2}y}\div\frac{36x}{18y^{-2}}=\frac{3x^2y^2}{6x^3y}\cdot\frac{-4x^7y^{-2}}{18x^{-2}y}\cdot\frac{18y^{-2}}{36x}=\frac{-2^3\cdot3^3x^9y^{-2}}{2^4\cdot3^5x^2y^2}=-\frac{x^7}{18y^4}$$

61. $$(4x+12)\cdot\frac{x^2}{2x-6}\div\frac{2}{x-3}=\frac{4x+12}{1}\cdot\frac{x^2}{2x-6}\cdot\frac{x-3}{2}=\frac{4(x+3)}{1}\cdot\frac{x^2}{2(x-3)}\cdot\frac{x-3}{2}=x^2(x+3)$$

63. $$\frac{2x^2-2x-4}{x^2+2x-8}\cdot\frac{3x^2+15x}{x+1}\div\frac{4x^2-100}{x^2-x-20}=\frac{2x^2-2x-4}{x^2+2x-8}\cdot\frac{3x^2+15x}{x+1}\cdot\frac{x^2-x-20}{4x^2-100}=\frac{2(x-2)(x+1)}{(x+4)(x-2)}\cdot\frac{3x(x+5)}{x+1}\cdot\frac{(x-5)(x+4)}{4(x+5)(x-5)}=\frac{3x}{2}$$

65. $$\frac{2x^2+5x-3}{x^2+2x-3}\div\left(\frac{x^2+2x-35}{x^2-6x+5}\div\frac{x^2-9x+14}{2x^2-5x+2}\right)$$
$$=\frac{2x^2+5x-3}{x^2+2x-3}\div\left(\frac{x^2+2x-35}{x^2-6x+5}\cdot\frac{2x^2-5x+2}{x^2-9x+14}\right)$$
$$=\frac{2x^2+5x-3}{x^2+2x-3}\div\left(\frac{(x+7)(x-5)}{(x-5)(x-1)}\cdot\frac{(2x-1)(x-2)}{(x-7)(x-2)}\right)$$
$$=\frac{2x^2+5x-3}{x^2+2x-3}\div\frac{(x+7)(2x-1)}{(x-1)(x-7)}=\frac{(2x-1)(x+3)}{(x+3)(x-1)}\cdot\frac{(x-1)(x-7)}{(x+7)(2x-1)}=\frac{x-7}{x+7}$$

67. $$\frac{x}{x+4}+\frac{5}{x+4}=\frac{x+5}{x+4}$$

69. $$\frac{5x}{x+1}+\frac{3}{x+1}-\frac{2x}{x+1}=\frac{3x+3}{x+1}=\frac{3(x+1)}{x+1}=3$$

71. $$\frac{4y}{y-4}-\frac{16}{y-4}=\frac{4y-16}{y-4}=\frac{4(y-4)}{y-4}=4$$

73. $$\frac{3(x^2+x)}{x^2-5x+6}+\frac{-3(x^2-x)}{x^2-5x+6}=\frac{3x^2+3x}{x^2-5x+6}+\frac{-3x^2+3x}{x^2-5x+6}=\frac{6x}{(x-3)(x-2)}$$

75. $$\frac{a+b}{3}+\frac{a-b}{7}=\frac{(a+b)7}{3(7)}+\frac{(a-b)3}{7(3)}=\frac{7a+7b}{21}+\frac{3a-3b}{21}=\frac{10a+4b}{21}$$

77. $$\frac{a}{2}+\frac{2a}{5}=\frac{a\cdot5}{2\cdot5}+\frac{2a\cdot2}{5\cdot2}=\frac{5a}{10}+\frac{4a}{10}=\frac{9a}{10}$$

SECTION 7.5

79. $\dfrac{3}{4x}+\dfrac{2}{3x}=\dfrac{3\cdot 3}{4x\cdot 3}+\dfrac{2\cdot 4}{3x\cdot 4}=\dfrac{9}{12x}+\dfrac{8}{12x}=\dfrac{17}{12x}$

81. $\dfrac{3}{x+2}+\dfrac{5}{x-4}=\dfrac{3(x-4)}{(x+2)(x-4)}+\dfrac{5(x+2)}{(x-4)(x+2)}=\dfrac{3x-12}{(x+2)(x-4)}+\dfrac{5x+10}{(x+2)(x-4)}$

$=\dfrac{8x-2}{(x+2)(x-4)}=\dfrac{2(4x-1)}{(x+2)(x-4)}$

83. $x+\dfrac{1}{x}=\dfrac{x}{1}+\dfrac{1}{x}=\dfrac{x(x)}{1(x)}+\dfrac{1}{x}=\dfrac{x^2}{x}+\dfrac{1}{x}=\dfrac{x^2+1}{x}$

85. $\dfrac{2}{a+4}-\dfrac{6}{a+3}=\dfrac{2(a+3)}{(a+4)(a+3)}-\dfrac{6(a+4)}{(a+3)(a+4)}=\dfrac{2a+6}{(a+4)(a+3)}-\dfrac{6a+24}{(a+4)(a+3)}$

$=\dfrac{-4a-18}{(a+4)(a+3)}=-\dfrac{2(2a+9)}{(a+4)(a+3)}$

87. $\dfrac{x}{x^2+5x+6}+\dfrac{x}{x^2-4}=\dfrac{x}{(x+2)(x+3)}+\dfrac{x}{(x+2)(x-2)}$

$=\dfrac{x(x-2)}{(x+2)(x+3)(x-2)}+\dfrac{x(x+3)}{(x+2)(x-2)(x+3)}$

$=\dfrac{x^2-2x}{(x+2)(x+3)(x-2)}+\dfrac{x^2+3x}{(x+2)(x+3)(x-2)}$

$=\dfrac{2x^2+x}{(x+2)(x+3)(x-2)}$

89. $\dfrac{8}{x^2-9}+\dfrac{2}{x-3}-\dfrac{6}{x}=\dfrac{8}{(x+3)(x-3)}+\dfrac{2}{x-3}-\dfrac{6}{x}$

$=\dfrac{8x}{x(x+3)(x-3)}+\dfrac{2x(x+3)}{x(x+3)(x-3)}-\dfrac{6(x+3)(x-3)}{x(x+3)(x-3)}$

$=\dfrac{8x}{x(x+3)(x-3)}+\dfrac{2x^2+6x}{x(x+3)(x-3)}-\dfrac{6(x^2-9)}{x(x+3)(x-3)}$

$=\dfrac{8x}{x(x+3)(x-3)}+\dfrac{2x^2+6x}{x(x+3)(x-3)}-\dfrac{6x^2-54}{x(x+3)(x-3)}$

$=\dfrac{-4x^2+14x+54}{x(x+3)(x-3)}$

91. $1+x-\dfrac{x}{x-5}=\dfrac{x+1}{1}-\dfrac{x}{x-5}=\dfrac{(x+1)(x-5)}{1(x-5)}-\dfrac{x}{x-5}$

$=\dfrac{x^2-4x-5}{x-5}-\dfrac{x}{x-5}=\dfrac{x^2-5x-5}{x-5}$

93. $\dfrac{3}{x+1}-\dfrac{2}{x-1}+\dfrac{x+3}{x^2-1}=\dfrac{3}{x+1}-\dfrac{2}{x-1}+\dfrac{x+3}{(x+1)(x-1)}$
$=\dfrac{3(x-1)}{(x+1)(x-1)}-\dfrac{2(x+1)}{(x+1)(x-1)}+\dfrac{x+3}{(x+1)(x-1)}$
$=\dfrac{3x-3}{(x+1)(x-1)}-\dfrac{2x+2}{(x+1)(x-1)}+\dfrac{x+3}{(x+1)(x-1)}$
$=\dfrac{2x-2}{(x+1)(x-1)}=\dfrac{2(x-1)}{(x+1)(x-1)}=\dfrac{2}{x+1}$

95. $\dfrac{\frac{4x}{y}}{\frac{6xz}{y^2}}=\dfrac{4x}{y}\div\dfrac{6xz}{y^2}=\dfrac{4x}{y}\cdot\dfrac{y^2}{6xz}=\dfrac{2y}{3z}$

97. $\dfrac{\frac{x-y}{xy}}{\frac{y-x}{x}}=\dfrac{x-y}{xy}\div\dfrac{y-x}{x}=\dfrac{x-y}{xy}\cdot\dfrac{x}{y-x}=\dfrac{x-y}{xy}\cdot\dfrac{x}{-(x-y)}=-\dfrac{1}{y}$

99. $\dfrac{\frac{1}{a}+\frac{1}{b}}{\frac{1}{a}}=\dfrac{\left(\frac{1}{a}+\frac{1}{b}\right)\cdot ab}{\frac{1}{a}\cdot ab}=\dfrac{\frac{1}{a}\cdot ab+\frac{1}{b}\cdot ab}{b}=\dfrac{b+a}{b}$

101. $\dfrac{\frac{y}{x}-\frac{x}{y}}{\frac{1}{x}+\frac{1}{y}}=\dfrac{\left(\frac{y}{x}-\frac{x}{y}\right)\cdot xy}{\left(\frac{1}{x}+\frac{1}{y}\right)\cdot xy}=\dfrac{\frac{y}{x}\cdot xy-\frac{x}{y}\cdot xy}{\frac{1}{x}\cdot xy+\frac{1}{y}\cdot xy}=\dfrac{y^2-x^2}{y+x}=\dfrac{(y+x)(y-x)}{y+x}=y-x$

103. $\dfrac{\frac{1}{a}-\frac{1}{b}}{\frac{a}{b}-\frac{b}{a}}=\dfrac{\left(\frac{1}{a}-\frac{1}{b}\right)\cdot ab}{\left(\frac{a}{b}-\frac{b}{a}\right)\cdot ab}=\dfrac{\frac{1}{a}\cdot ab-\frac{1}{b}\cdot ab}{\frac{a}{b}\cdot ab-\frac{b}{a}\cdot ab}=\dfrac{b-a}{a^2-b^2}=\dfrac{b-a}{(a+b)(a-b)}=-\dfrac{1}{a+b}$

105. $\dfrac{1+\frac{6}{x}+\frac{8}{x^2}}{1+\frac{1}{x}-\frac{12}{x^2}}=\dfrac{\left(1+\frac{6}{x}+\frac{8}{x^2}\right)\cdot x^2}{\left(1+\frac{1}{x}-\frac{12}{x^2}\right)\cdot x^2}=\dfrac{x^2+6x+8}{x^2+x-12}=\dfrac{(x+4)(x+2)}{(x+4)(x-3)}=\dfrac{x+2}{x-3}$

107. $\dfrac{x^{-1}+y^{-1}}{x^{-1}-y^{-1}}=\dfrac{\frac{1}{x}+\frac{1}{y}}{\frac{1}{x}-\frac{1}{y}}=\dfrac{\left(\frac{1}{x}+\frac{1}{y}\right)(xy)}{\left(\frac{1}{x}-\frac{1}{y}\right)(xy)}=\dfrac{y+x}{y-x}$

109. $\dfrac{x-y^{-2}}{y-x^{-2}}=\dfrac{x-\frac{1}{y^2}}{y-\frac{1}{x^2}}=\dfrac{\left(x-\frac{1}{y^2}\right)(x^2y^2)}{\left(y-\frac{1}{x^2}\right)(x^2y^2)}=\dfrac{x^3y^2-x^2}{x^2y^3-y^2}=\dfrac{x^2(xy^2-1)}{y^2(x^2y-1)}$

SECTION 7.5

111. $$\frac{1+\frac{a}{b}}{1-\frac{a}{1-\frac{a}{b}}} = \frac{1+\frac{a}{b}}{1-\frac{a(b)}{\left(1-\frac{a}{b}\right)(b)}} = \frac{1+\frac{a}{b}}{1-\frac{ab}{b-a}} = \frac{\left(1+\frac{a}{b}\right)(b)(b-a)}{\left(1-\frac{ab}{b-a}\right)(b)(b-a)} = \frac{b(b-a)+a(b-a)}{b(b-a)-ab(b)}$$
$$= \frac{b^2-ab+ab-a^2}{b^2-ab-ab^2}$$
$$= \frac{b^2-a^2}{b(b-a-ab)}$$
$$= \frac{(b+a)(b-a)}{b(b-a-ab)}$$

113. $$a+\frac{a}{1+\frac{a}{a+1}} = a+\frac{a(a+1)}{\left(1+\frac{a}{a+1}\right)(a+1)} = a+\frac{a(a+1)}{a+1+a} = a+\frac{a^2+a}{2a+1} = \frac{a(2a+1)}{2a+1}+\frac{a^2+a}{2a+1}$$
$$= \frac{2a^2+a+a^2+a}{2a+1}$$
$$= \frac{3a^2+2a}{2a+1}$$

115. $$\frac{m^2-n^2}{2x^2+3x-2}\cdot\frac{2x^2+5x-3}{n^2-m^2} = \frac{(m+n)(m-n)}{(2x-1)(x+2)}\cdot\frac{(2x-1)(x+3)}{(n+m)(n-m)} = -\frac{x+3}{x+2}$$

117. $$\frac{ax+ay+bx+by}{x^3-27}\cdot\frac{x^2+3x+9}{xc+xd+yc+yd} = \frac{a(x+y)+b(x+y)}{(x-3)(x^2+3x+9)}\cdot\frac{x^2+3x+9}{x(c+d)+y(c+d)}$$
$$= \frac{(x+y)(a+b)}{(x-3)(x^2+3x+9)}\cdot\frac{x^2+3x+9}{(c+d)(x+y)}$$
$$= \frac{a+b}{(x-3)(c+d)}$$

119. $$\frac{x^3+y^3}{x^3-y^3}\div\frac{x^2-xy+y^2}{x^2+xy+y^2} = \frac{x^3+y^3}{x^3-y^3}\cdot\frac{x^2+xy+y^2}{x^2-xy+y^2} = \frac{(x+y)(x^2-xy+y^2)}{(x-y)(x^2+xy+y^2)}\cdot\frac{x^2+xy+y^2}{x^2-xy+y^2}$$
$$= \frac{x+y}{x-y}$$

121. $$\frac{2x^2-7x-4}{20-x-x^2}\div\frac{2x^2-9x-5}{x^2-25} = -\frac{2x^2-7x-4}{x^2+x-20}\cdot\frac{x^2-25}{2x^2-9x-5}$$
$$= -\frac{(2x+1)(x-4)}{(x+5)(x-4)}\cdot\frac{(x+5)(x-5)}{(2x+1)(x-5)} = -1$$

123. $$\frac{x^2-x-6}{x^2-4}\cdot\frac{x^2-x-2}{9-x^2} = \frac{(x-3)(x+2)}{(x+2)(x-2)}\cdot\frac{(x-2)(x+1)}{(3+x)(3-x)} = -\frac{x+1}{x+3}$$

125. $$\frac{3n^2+5n-2}{12n^2-13n+3}\div\frac{n^2+3n+2}{4n^2+5n-6} = \frac{3n^2+5n-2}{12n^2-13n+3}\cdot\frac{4n^2+5n-6}{n^2+3n+2}$$
$$= \frac{(3n-1)(n+2)}{(3n-1)(4n-3)}\cdot\frac{(4n-3)(n+2)}{(n+2)(n+1)} = \frac{n+2}{n+1}$$

127. $\frac{x^2-x-12}{x^2+x-2} \div \frac{x^2-6x+8}{x^2-3x-10} \cdot \frac{x^2-3x+2}{x^2-2x-15} = \frac{x^2-x-12}{x^2+x-2} \cdot \frac{x^2-3x-10}{x^2-6x+8} \cdot \frac{x^2-3x+2}{x^2-2x-15}$
$= \frac{(x-4)(x+3)}{(x+2)(x-1)} \cdot \frac{(x-5)(x+2)}{(x-4)(x-2)} \cdot \frac{(x-2)(x-1)}{(x-5)(x+3)} = 1$

129. $\frac{x+8}{x-3} - \frac{x-14}{3-x} = \frac{x+8}{x-3} - \frac{-x+14}{x-3} = \frac{2x-6}{x-3} = \frac{2(x-3)}{x-3} = 2$

131. $\frac{x-2}{x^2-3x} + \frac{2x-1}{x^2+3x} - \frac{2}{x^2-9} = \frac{x-2}{x(x-3)} + \frac{2x-1}{x(x+3)} - \frac{2}{(x+3)(x-3)}$
$= \frac{(x-2)(x+3)}{x(x-3)(x+3)} + \frac{(2x-1)(x-3)}{x(x-3)(x+3)} - \frac{2x}{x(x-3)(x+3)}$
$= \frac{x^2+x-6}{x(x-3)(x+3)} + \frac{2x^2-7x+3}{x(x-3)(x+3)} - \frac{2x}{x(x-3)(x+3)}$
$= \frac{3x^2-8x-3}{x(x-3)(x+3)} = \frac{(3x+1)(x-3)}{x(x-3)(x+3)} = \frac{3x+1}{x(x+3)}$

133. $\frac{\frac{1}{a+1}+1}{\frac{3}{a-1}+1} = \frac{\left(\frac{1}{a+1}+1\right)(a+1)(a-1)}{\left(\frac{3}{a-1}+1\right)(a+1)(a-1)} = \frac{1(a-1)+1(a+1)(a-1)}{3(a+1)+1(a+1)(a-1)} = \frac{a-1+a^2-1}{3a+3+a^2-1}$
$= \frac{a^2+a-2}{a^2+3a+2}$
$= \frac{(a+2)(a-1)}{(a+2)(a+1)}$
$= \frac{a-1}{a+1}$

135. $\frac{x+y}{x^{-1}+y^{-1}} = \frac{x+y}{\frac{1}{x}+\frac{1}{y}} = \frac{(x+y)(xy)}{\left(\frac{1}{x}+\frac{1}{y}\right)(xy)} = \frac{(x+y)xy}{y+x} = xy$

137. Answers may vary.

139. Answers may vary.

141. $\frac{a-3b}{2b-a} = \frac{-(3b-a)}{-(a-2b)} = \frac{3b-a}{a-2b}$.
The two answers are the same.

143. You can divide out the 4's in parts a and d.

Exercise 7.6 (page 520)

1.
2	1	2	1
		2	8
	1	4	[9]

3.
2	1	−2	1	−2	
		2	0	2	
	1	0	1	0	factor

5. $f(1) = 3(1)^2 + 2(1) - 1 = 4$

7. $f(2a) = 3(2a)^2 + 2(2a) - 1$
$= 12a^2 + 4a - 1$

SECTION 7.6

9. $2(x^2+4x-1)+3(2x^2-2x+2)=2x^2+8x-2+6x^2-6x+6=8x^2+2x+4$

11. $x-r$

13. $P(r)$

15.
$$\begin{array}{r|rrr} 1 & 1 & 1 & -2 \\ & & 1 & 2 \\ \hline & 1 & 2 & 0 \end{array} \Rightarrow \boxed{x+2}$$

17.
$$\begin{array}{r|rrr} 4 & 1 & -7 & 12 \\ & & 4 & -12 \\ \hline & 1 & -3 & 0 \end{array} \Rightarrow \boxed{x-3}$$

19.
$$\begin{array}{r|rrr} -2 & 1 & -5 & 14 \\ & & -2 & 14 \\ \hline & 1 & -7 & 28 \end{array} \Rightarrow \boxed{x-7+\frac{28}{x+2}}$$

21.
$$\begin{array}{r|rrrr} 3 & 3 & -10 & 5 & -6 \\ & & 9 & -3 & 6 \\ \hline & 3 & -1 & 2 & 0 \end{array} \Rightarrow \boxed{3x^2-x+2}$$

23.
$$\begin{array}{r|rrrr} 2 & 2 & 0 & -5 & -6 \\ & & 4 & 8 & 6 \\ \hline & 2 & 4 & 3 & 0 \end{array} \Rightarrow \boxed{2x^2+4x+3}$$

25.
$$\begin{array}{r|rrr} -4 & 1 & 6 & 8 \\ & & -4 & -8 \\ \hline & 1 & 2 & 0 \end{array} \Rightarrow \boxed{x+2}$$

27.
$$\begin{array}{r|rrrr} 2 & 4 & -6 & -5 & 2 \\ & & 8 & 4 & -2 \\ \hline & 4 & 2 & -1 & 0 \end{array}$$
$$\Rightarrow \boxed{4x^2+2x-1}$$

29.
$$\begin{array}{r|rrrr} -1 & 6 & 5 & 0 & 4 \\ & & -6 & 1 & -1 \\ \hline & 6 & -1 & 1 & 3 \end{array}$$
$$\Rightarrow \boxed{6x^2-x+1+\frac{3}{x+1}}$$

31. $P(-2)=2(-2)^3-4(-2)^2+2(-2)-1$
$=\boxed{-37}$
$$\begin{array}{r|rrrr} -2 & 2 & -4 & 2 & -1 \\ & & -4 & 16 & -36 \\ \hline & 2 & -8 & 18 & \boxed{-37} \end{array}$$

33. $P(3)=2(3)^3-4(3)^2+2(3)-1=\boxed{23}$
$$\begin{array}{r|rrrr} 3 & 2 & -4 & 2 & -1 \\ & & 6 & 6 & 24 \\ \hline & 2 & 2 & 8 & \boxed{23} \end{array}$$

35. $P(0)=2(0)^3-4(0)^2+2(0)-1=\boxed{-1}$
$$\begin{array}{r|rrrr} 0 & 2 & -4 & 2 & -1 \\ & & 0 & 0 & 0 \\ \hline & 2 & -4 & 2 & \boxed{-1} \end{array}$$

37. $P(1)=2(1)^3-4(1)^2+2(1)-1=\boxed{-1}$
$$\begin{array}{r|rrrr} 1 & 2 & -4 & 2 & -1 \\ & & 2 & -2 & 0 \\ \hline & 2 & -2 & 0 & \boxed{-1} \end{array}$$

39. $Q(2)=(2)^4-3(2)^3+2(2)^2+(2)-3=\boxed{-1}$
$$\begin{array}{r|rrrrr} 2 & 1 & -3 & 2 & 1 & -3 \\ & & 2 & -2 & 0 & 2 \\ \hline & 1 & -1 & 0 & 1 & \boxed{-1} \end{array}$$

41. $Q(3)=(3)^4-3(3)^3+2(3)^2+(3)-3=\boxed{18}$
$$\begin{array}{r|rrrrr} 3 & 1 & -3 & 2 & 1 & -3 \\ & & 3 & 0 & 6 & 21 \\ \hline & 1 & 0 & 2 & 7 & \boxed{18} \end{array}$$

43. $Q(-3) = (-3)^4 - 3(-3)^3 + 2(-3)^2 + (-3) - 3 = \boxed{174}$

$$\begin{array}{r|rrrrr} -3 & 1 & -3 & 2 & 1 & -3 \\ & & -3 & 18 & -60 & 177 \\ \hline & 1 & -6 & 20 & -59 & \boxed{174} \end{array}$$

45. $Q(-1) = (-1)^4 - 3(-1)^3 + 2(-1)^2 + (-1) - 3 = \boxed{2}$

$$\begin{array}{r|rrrrr} -1 & 1 & -3 & 2 & 1 & -3 \\ & & -1 & 4 & -6 & 5 \\ \hline & 1 & -4 & 6 & -5 & \boxed{2} \end{array}$$

47. $$\begin{array}{r|rrrr} 3 & 1 & -3 & 5 & -15 \\ & & 3 & 0 & 15 \\ \hline & 1 & 0 & 5 & \boxed{0} \end{array} \Rightarrow \text{factor}$$

49. $$\begin{array}{r|rrr} -2 & 3 & -7 & 4 \\ & & -6 & 26 \\ \hline & 3 & -13 & \boxed{30} \end{array} \Rightarrow \text{not a factor}$$

51. $$\begin{array}{r|rrrr} 2 & 1 & -4 & 1 & -2 \\ & & 2 & -4 & -6 \\ \hline & 1 & -2 & -3 & \boxed{-8} \end{array}$$

53. $$\begin{array}{r|rrrr} 3 & 2 & 0 & 1 & 2 \\ & & 6 & 18 & 57 \\ \hline & 2 & 6 & 19 & \boxed{59} \end{array}$$

55. $$\begin{array}{r|rrrrr} -2 & 1 & -2 & 1 & -3 & 2 \\ & & -2 & 8 & -18 & 42 \\ \hline & 1 & -4 & 9 & -21 & \boxed{44} \end{array}$$

57. $$\begin{array}{r|rrrrrr} -\frac{1}{2} & 3 & 0 & 0 & 0 & 0 & 1 \\ & & -\frac{3}{2} & \frac{3}{4} & -\frac{3}{8} & \frac{3}{16} & -\frac{3}{32} \\ \hline & 3 & -\frac{3}{2} & \frac{3}{4} & -\frac{3}{8} & \frac{3}{16} & \boxed{\frac{29}{32}} \end{array}$$

59. $$\begin{array}{r|rrr} 0.2 & 7.2 & -2.1 & 0.5 \\ & & 1.44 & -0.132 \\ \hline & 7.2 & -0.66 & 0.368 \end{array}$$

$\Rightarrow \boxed{7.2x - 0.66 + \frac{0.368}{x-0.2}}$

61. $$\begin{array}{r|rrr} -1.7 & 2.7 & 1.0 & -5.2 \\ & & -4.59 & 6.103 \\ \hline & 2.7 & -3.59 & 0.903 \end{array}$$

$\Rightarrow \boxed{2.7x - 3.59 + \frac{0.903}{x+1.7}}$

63. $$\begin{array}{r|rrrr} -57 & 9 & 0 & 0 & -25 \\ & & -513 & 29241 & -1666737 \\ \hline & 9 & -513 & 29241 & -1666762 \end{array}$$

$\Rightarrow \boxed{9x^2 - 513x + 29{,}241 - \frac{1{,}666{,}762}{x+57}}$

65. $$\begin{array}{r|rrrrrrr} 2 & 1 & 0 & 0 & 0 & 0 & 0 & 0 \\ & & 2 & 4 & 8 & 16 & 32 & 64 \\ \hline & 1 & 2 & 4 & 8 & 16 & 32 & \boxed{64} \end{array}$$

67. **Answers may vary.**

69. remainder $= P(1) = 1^{100} - 1^{99} + 1^{98} - 1^{97} + \cdots + 1^2 - 1 + 1 = 1$

Chapter 7 Review (page 525)

1.
$$\begin{aligned} 4(y-1) &= 28 \\ 4y - 4 &= 28 \\ 4y &= 32 \\ y &= 8 \end{aligned}$$

2.
$$\begin{aligned} 3(x+7) &= 42 \\ 3x + 21 &= 42 \\ 3x &= 21 \\ x &= 7 \end{aligned}$$

3.
$$\begin{aligned} 13(x-9) - 2 &= 7x - 5 \\ 13x - 117 - 2 &= 7x - 5 \\ 13x - 119 &= 7x - 5 \\ 13x - 7x - 119 &= 7x - 7x - 5 \\ 6x - 119 + 119 &= -5 + 119 \\ 6x &= 114 \\ \frac{6x}{6} &= \frac{114}{6} \\ x &= 19 \end{aligned}$$

4.
$$\begin{aligned} \frac{8(x-5)}{3} &= 2(x-4) \\ 3 \cdot \frac{8(x-5)}{3} &= 3 \cdot 2(x-4) \\ 8(x-5) &= 6(x-4) \\ 8x - 40 &= 6x - 24 \\ 8x - 6x - 40 &= 6x - 6x - 24 \\ 2x - 40 + 40 &= -24 + 40 \\ 2x &= 16 \\ \frac{2x}{2} &= \frac{16}{2} \\ x &= 8 \end{aligned}$$

5.
$$\begin{aligned} 2x + 4 &= 2(x+3) - 2 \\ 2x + 4 &= 2x + 6 - 2 \\ 2x + 4 &= 2x + 4 \end{aligned}$$
identity; $\mathbb{R}$

6.
$$\begin{aligned} (3x-2) - x &= 2(x-4) \\ 3x - 2 - x &= 2x - 8 \\ 2x - 2 &= 2x - 8 \\ 2x - 2x - 2 &= 2x - 2x - 8 \\ -2 &\neq -8 \end{aligned}$$
contradiction; $\emptyset$

7.
$$\begin{aligned} V &= \frac{1}{3}\pi r^2 h \\ 3V &= 3\left(\frac{1}{3}\pi r^2 h\right) \\ 3V &= \pi r^2 h \\ \frac{3V}{\pi r^2} &= \frac{\pi r^2 h}{\pi r^2} \\ \frac{3V}{\pi r^2} &= h, \text{ or } h = \frac{3V}{\pi r^2} \end{aligned}$$

8.
$$\begin{aligned} V &= \frac{1}{6}ab(x+y) \\ 6V &= 6 \cdot \frac{1}{6}ab(x+y) \\ 6V &= ab(x+y) \\ \frac{6V}{ab} &= \frac{ab(x+y)}{ab} \\ \frac{6V}{ab} &= x + y \\ \frac{6V}{ab} - y &= x + y - y \\ \frac{6V}{ab} - y &= x, \text{ or } x = \frac{6V}{ab} - y \end{aligned}$$

9. Let $x =$ the length of the first piece.
Then $3x =$ the length of the other piece.

$$\boxed{\text{Sum of lengths}} = \boxed{\text{Total length}}$$
$$x + 3x = 20$$
$$4x = 20$$
$$x = 5$$

He should cut the board 5 feet from one end.

10. Let $w =$ the width of the rectangle. Then $w + 4 =$ the length of the rectangle.

$$\boxed{\text{Perimeter}} = 28$$
$$w + w + w + 4 + w + 4 = 28$$
$$4w + 8 = 28$$
$$4w = 20$$
$$w = 5$$

The dimensions are 5 meters by 9 meters, for an area of 45 m^2.

11.
$$\frac{1}{3}y - 2 \geq \frac{1}{2}y + 2$$
$$6\left(\frac{1}{3}y - 2\right) \geq 6\left(\frac{1}{2}y + 2\right)$$
$$2y - 12 + 12 \geq 3y + 12 + 12$$
$$2y - 3y \geq 3y - 3y + 24$$
$$-y \geq 24$$
$$y \leq -24$$

solution set: $(-\infty, -24]$

−24

12.
$$\frac{7}{4}(x + 3) < \frac{3}{8}(x - 3)$$
$$8 \cdot \frac{7}{4}(x + 3) < 8 \cdot \frac{3}{8}(x - 3)$$
$$14(x + 3) < 3(x - 3)$$
$$14x + 42 < 3x - 9$$
$$14x < 3x - 51$$
$$11x < -51$$
$$x < -\tfrac{51}{11}$$

solution set: $\left(-\infty, -\frac{51}{11}\right)$

$-\frac{51}{11}$

13.
$$3 < 3x + 4 < 10$$
$$3 - 4 < 3x + 4 - 4 < 10 - 4$$
$$-1 < 3x < 6$$
$$-\tfrac{1}{3} < \tfrac{3x}{3} < \tfrac{6}{3}$$

$-\frac{1}{3} < x < 2 \Rightarrow$ solution set: $\left(-\frac{1}{3}, 2\right) \Rightarrow$

$-\frac{1}{3}$ 2

14.
$$4x > 3x + 2 > x - 3$$

$4x > 3x + 2$	and	$3x + 2 > x - 3$
$4x - 3x > 3x - 3x + 2$		$3x + 2 - 2 > x - 3 - 2$
$x > 2$		$3x - x > x - x - 5$
$x > 2$		$2x > -5$
		$x > -\frac{5}{2}$

If $x > 2$ **and** $x > -\frac{5}{2}$, $x > 2$. The solution set is $(2, \infty)$.

2

CHAPTER 7 REVIEW

15.
$$|3x+1| = 10$$
$$3x+1=10 \quad \textbf{or} \quad 3x+1=-10$$
$$3x = 9 \qquad\qquad 3x=-11$$
$$x = 3 \qquad\qquad x = -\tfrac{11}{3}$$

16.
$$\left|\tfrac{3}{2}x-4\right| = 9$$
$$\tfrac{3}{2}x-4=9 \quad \textbf{or} \quad \tfrac{3}{2}x-4=-9$$
$$\tfrac{3}{2}x = 13 \qquad\qquad \tfrac{3}{2}x=-5$$
$$x = \tfrac{26}{3} \qquad\qquad x = -\tfrac{10}{3}$$

17.
$$|3x+2| = |2x-3|$$
$$3x+2=2x-3 \quad \textbf{or} \quad 3x+2=-(2x-3)$$
$$x=-5 \qquad\qquad 3x+2=-2x+3$$
$$5x = 1$$
$$x = \tfrac{1}{5}$$

18.
$$|5x-4| = |4x-5|$$
$$5x-4=4x-5 \quad \textbf{or} \quad 5x-4=-(4x-5)$$
$$x=-1 \qquad\qquad 5x-4=-4x+5$$
$$9x = 9$$
$$x = 1$$

19.
$$|2x+7| < 3$$
$$-3 < 2x+7 < 3$$
$$-3-7 < 2x+7-7 < 3-7$$
$$-10 < 2x < -4$$
$$-5 < x < -2$$

solution set: $(-5,-2)$

$-5 \qquad -2$

20.
$$|3x-8| \geq 4$$
$$3x-8 \leq -4 \quad \textbf{or} \quad 3x-8 \geq 4$$
$$3x \leq 4 \qquad\qquad 3x \geq 12$$
$$x \leq \tfrac{4}{3} \qquad\qquad x \geq 4$$

solution set: $\left(-\infty, \tfrac{4}{3}\right] \cup [4,\infty)$

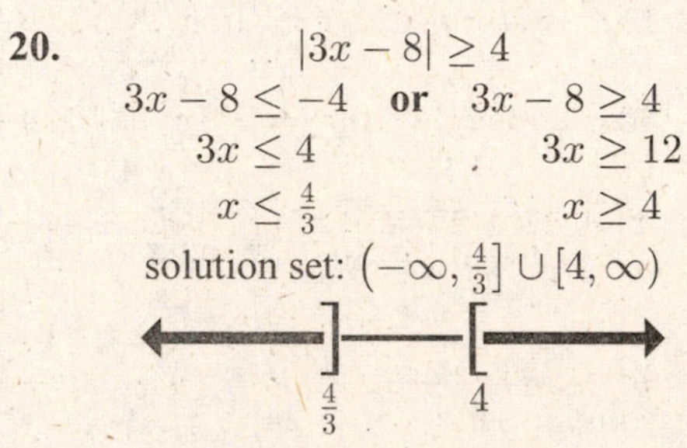

21.
$$\left|\tfrac{3}{2}x-14\right| \geq 0$$

Since an absolute value is always at least 0, this inequality is true for all real #s. Solution set: $(-\infty,\infty)$

0

22. $\left|\tfrac{2}{3}x+14\right| < 0$

Since an absolute value can never be negative, there is no solution. $\emptyset$

23. $4x+8 = 4(x+2)$

24. $5x^2y^3 - 10xy^2 = 5xy^2(xy-2)$

25. $-8x^2y^3z^4 - 12x^4y^3z^2 = -4x^2y^3z^2(2z^2+3x^2)$

26. $12a^6b^4c^2 + 15a^2b^4c^6 = 3a^2b^4c^2(4a^4+5c^4)$

27. $xy+2y+4x+8 = y(x+2)+4(x+2)$
$\qquad = (x+2)(y+4)$

28. $ac+bc+3a+3b = c(a+b)+3(a+b)$
$\qquad = (a+b)(c+3)$

29. $x^{2n}+x^n = x^n(x^{2n-n}+1) = x^n(x^n+1)$

30. $y^{2n}-y^{3n} = y^{2n}(1-y^{3n-2n}) = y^{2n}(1-y^n)$

31. $x^4+4y+4x^2+x^2y = x^4+x^2y+4x^2+4y = x^2(x^2+y)+4(x^2+y) = (x^2+y)(x^2+4)$

32. $a^5+b^2c+a^2c+a^3b^2 = a^5+a^3b^2+a^2c+b^2c = a^3(a^2+b^2)+c(a^2+b^2) = (a^2+b^2)(a^3+c)$

33. $z^2-16 = z^2-4^2 = (z+4)(z-4)$

34. $y^2-121 = y^2-11^2 = (y+11)(y-11)$

35. $2x^4 - 98 = 2(x^4 - 49) = 2(x^2 + 7)(x^2 - 7)$

36. $3x^6 - 300x^2 = 3x^2(x^4 - 100)$
$= 3x^2(x^2 + 10)(x^2 - 10)$

37. $y^2 + 21y + 20 = (y + 20)(y + 1)$

38. $z^2 - 11z + 30 = (z - 5)(z - 6)$

39. $-x^2 - 3x + 28 = -(x^2 + 3x - 28)$
$= -(x + 7)(x - 4)$

40. $-y^2 + 5y + 24 = -\left(y^2 - 5y - 24\right)$
$= -(y - 8)(y + 3)$

41. $y^3 + y^2 - 2y = y(y^2 + y - 2)$
$= y(y + 2)(y - 1)$

42. $2a^4 + 4a^3 - 6a^2 = 2a^2(a^2 + 2a - 3)$
$= 2a^2(a + 3)(a - 1)$

43. $15x^2 - 57xy - 12y^2 = 3(5x^2 - 19xy - 4y^2) = 3(5x + y)(x - 4y)$

44. $30x^2 + 65xy + 10y^2 = 5(6x^2 + 13xy + 2y^2) = 5(6x + y)(x + 2y)$

45. $x^2 + 4x + 4 - 4p^4 = (x^2 + 4x + 4) - 4p^4 = (x + 2)^2 - (2p^2)^2$
$= (x + 2 + 2p^2)(x + 2 - 2p^2)$

46. $y^2 + 3y + 2 + 2x + xy = (y^2 + 3y + 2) + 2x + xy = (y + 1)(y + 2) + x(2 + y)$
$= (y + 2)(y + 1 + x)$

47. $x^3 + 343 = x^3 + 7^3$
$= (x + 7)(x^2 - 7x + 49)$

48. $a^3 - 125 = a^3 - 5^3$
$= (a - 5)(a^2 + 5a + 25)$

49. $8y^3 - 512 = 8(y^3 - 64) = 8(y - 4)(y^2 + 4y + 16)$

50. $4x^3y + 108yz^3 = 4y(x^3 + 27z^3) = 4y(x + 3z)(x^2 - 3xz + 9z^2)$

51. $\dfrac{248x^2y}{576xy^2} = \dfrac{8 \cdot 31x^2y}{8 \cdot 72xy^2} = \dfrac{31x}{72y}$

52. $\dfrac{x^2 - 49}{x^2 + 14x + 49} = \dfrac{(x + 7)(x - 7)}{(x + 7)(x + 7)} = \dfrac{x - 7}{x + 7}$

53. $\dfrac{x^2 + 4x + 4}{x^2 - x - 6} \cdot \dfrac{x^2 - 9}{x^2 + 5x + 6} = \dfrac{(x + 2)(x + 2)}{(x + 2)(x - 3)} \cdot \dfrac{(x + 3)(x - 3)}{(x + 2)(x + 3)} = 1$

54. $\dfrac{x^3 - 64}{x^2 + 4x + 16} \div \dfrac{x^2 - 16}{x + 4} = \dfrac{x^3 - 64}{x^2 + 4x + 16} \cdot \dfrac{x + 4}{x^2 - 16} = \dfrac{(x - 4)(x^2 + 4x + 16)}{x^2 + 4x + 16} \cdot \dfrac{x + 4}{(x + 4)(x - 4)}$
$= 1$

55. $\dfrac{5y}{x - y} - \dfrac{3}{x - y} = \dfrac{5y - 3}{x - y}$

56. $\dfrac{3x - 1}{x^2 + 2} + \dfrac{3(x - 2)}{x^2 + 2} = \dfrac{3x - 1 + 3x - 6}{x^2 + 2}$
$= \dfrac{6x - 7}{x^2 + 2}$

57. $\dfrac{3}{x + 2} + \dfrac{2}{x + 3} = \dfrac{3(x + 3)}{(x + 2)(x + 3)} + \dfrac{2(x + 2)}{(x + 2)(x + 3)} = \dfrac{3x + 9 + 2x + 4}{(x + 2)(x + 3)} = \dfrac{5x + 13}{(x + 2)(x + 3)}$

58. $$\frac{4x}{x-4}-\frac{3}{x+3}=\frac{4x(x+3)}{(x-4)(x+3)}-\frac{3(x-4)}{(x-4)(x+3)}=\frac{4x^2+12x-3x+12}{(x-4)(x+3)}=\frac{4x^2+9x+12}{(x-4)(x+3)}$$

59. $$\begin{aligned}\frac{x^2+3x+2}{x^2-x-6}\cdot\frac{3x^2-3x}{x^2-3x-4}\div\frac{x^2+3x+2}{x^2-2x-8}&=\frac{x^2+3x+2}{x^2-x-6}\cdot\frac{3x^2-3x}{x^2-3x-4}\cdot\frac{x^2-2x-8}{x^2+3x+2}\\&=\frac{(x+2)(x+1)}{(x-3)(x+2)}\cdot\frac{3x(x-1)}{(x-4)(x+1)}\cdot\frac{(x-4)(x+2)}{(x+2)(x+1)}\\&=\frac{3x(x-1)}{(x-3)(x+1)}\end{aligned}$$

60. $$\begin{aligned}\frac{x^2-x-6}{x^2-3x-10}\div\frac{x^2-x}{x^2-5x}\cdot\frac{x^2-4x+3}{x^2-6x+9}&=\frac{x^2-x-6}{x^2-3x-10}\cdot\frac{x^2-5x}{x^2-x}\cdot\frac{x^2-4x+3}{x^2-6x+9}\\&=\frac{(x-3)(x+2)}{(x-5)(x+2)}\cdot\frac{x(x-5)}{x(x-1)}\cdot\frac{(x-3)(x-1)}{(x-3)(x-3)}=1\end{aligned}$$

61. $$\begin{aligned}\frac{2x}{x+1}+\frac{3x}{x+2}+\frac{4x}{x^2+3x+2}&=\frac{2x}{x+1}+\frac{3x}{x+2}+\frac{4x}{(x+2)(x+1)}\\&=\frac{2x(x+2)}{(x+1)(x+2)}+\frac{3x(x+1)}{(x+1)(x+2)}+\frac{4x}{(x+2)(x+1)}\\&=\frac{2x^2+4x+3x^2+3x+4x}{(x+1)(x+2)}=\frac{5x^2+11x}{(x+1)(x+2)}\end{aligned}$$

62. $$\begin{aligned}\frac{5x}{x-3}+\frac{5}{x^2-5x+6}+\frac{x+3}{x-2}&=\frac{5x}{x-3}+\frac{5}{(x-3)(x-2)}+\frac{x+3}{x-2}\\&=\frac{5x(x-2)}{(x-3)(x-2)}+\frac{5}{(x-3)(x-2)}+\frac{(x+3)(x-3)}{(x-3)(x-2)}\\&=\frac{5x^2-10x+5+x^2-9}{(x-3)(x-2)}\\&=\frac{6x^2-10x-4}{(x-3)(x-2)}=\frac{2(3x+1)(x-2)}{(x-3)(x-2)}=\frac{2(3x+1)}{x-3}\end{aligned}$$

63. $$\begin{aligned}\frac{3(x+2)}{x^2-1}-\frac{2}{x+1}+\frac{4(x+3)}{x^2-2x+1}&=\frac{3(x+2)}{(x+1)(x-1)}-\frac{2}{(x+1)}+\frac{4(x+3)}{(x-1)(x-1)}\\&=\frac{3(x+2)(x-1)}{(x+1)(x-1)(x-1)}-\frac{2(x-1)(x-1)}{(x+1)(x-1)(x-1)}+\frac{4(x+3)(x+1)}{(x+1)(x-1)(x-1)}\\&=\frac{3(x^2+x-2)-2(x^2-2x+1)+4(x^2+4x+3)}{(x+1)(x-1)(x-1)}\\&=\frac{3x^2+3x-6-2x^2+4x-2+4x^2+16x+12}{(x+1)(x-1)(x-1)}=\frac{5x^2+23x+4}{(x+1)(x-1)(x-1)}\end{aligned}$$

64. $\dfrac{x}{x^2+4x+4}+\dfrac{2x}{x^2-4}-\dfrac{x^2-4}{x-2}$

$$= \frac{x}{(x+2)(x+2)}+\frac{2x}{(x+2)(x-2)}-\frac{x^2-4}{x-2}$$

$$= \frac{x(x-2)}{(x+2)(x+2)(x-2)}+\frac{2x(x+2)}{(x+2)(x-2)(x+2)}-\frac{(x^2-4)(x+2)(x+2)}{(x-2)(x+2)(x+2)}$$

$$= \frac{x^2-2x+2x^2+4x-(x^2-4)(x^2+4x+4)}{(x+2)(x+2)(x-2)}$$

$$= \frac{3x^2+2x-(x^4+4x^3-16x-16)}{(x+2)(x+2)(x-2)}$$

$$= \frac{3x^2+2x-x^4-4x^3+16x+16}{(x+2)(x+2)(x-2)} = \frac{-x^4-4x^3+3x^2+18x+16}{(x+2)(x+2)(x-2)}$$

65. $\dfrac{\frac{3}{x}-\frac{2}{y}}{xy}=\dfrac{\left(\frac{3}{x}-\frac{2}{y}\right)xy}{xy(xy)}=\dfrac{3y-2x}{x^2y^2}$

66. $\dfrac{\frac{1}{x}+\frac{2}{y}}{\frac{2}{x}-\frac{1}{y}}=\dfrac{\left(\frac{1}{x}+\frac{2}{y}\right)xy}{\left(\frac{2}{x}-\frac{1}{y}\right)xy}=\dfrac{y+2x}{2y-x}$

67. $\dfrac{2x+3+\frac{1}{x}}{x+2+\frac{1}{x}}=\dfrac{\left(2x+3+\frac{1}{x}\right)x}{\left(x+2+\frac{1}{x}\right)x}=\dfrac{2x^2+3x+1}{x^2+2x+1}=\dfrac{(2x+1)(x+1)}{(x+1)(x+1)}=\dfrac{2x+1}{x+1}$

68. $\dfrac{x^{-1}-y^{-1}}{x^{-1}+y^{-1}}=\dfrac{\frac{1}{x}-\frac{1}{y}}{\frac{1}{x}+\frac{1}{y}}=\dfrac{\left(\frac{1}{x}-\frac{1}{y}\right)xy}{\left(\frac{1}{x}+\frac{1}{y}\right)xy}=\dfrac{y-x}{y+x}$

69.

$$\begin{array}{r|rrrr} 2 & 3 & 2 & -7 & 2 \\ & & 6 & 16 & 18 \\ \hline & 3 & 8 & 9 & \boxed{20} \end{array}$$

70.

$$\begin{array}{r|rrrr} -2 & 2 & -4 & -14 & 3 \\ & & -4 & 16 & -4 \\ \hline & 2 & -8 & 2 & \boxed{-1} \end{array}$$

71.

$$\begin{array}{r|rrrr} 5 & 1 & -3 & -8 & -10 \\ & & 5 & 10 & 10 \\ \hline & 1 & 2 & 2 & \boxed{0} \end{array}$$

factor

72.

$$\begin{array}{r|rrrr} -5 & 1 & 4 & -5 & 5 \\ & & -5 & 5 & 0 \\ \hline & 1 & -1 & 0 & \boxed{5} \end{array}$$

not a factor

Chapter 7 Test (page 531)

1.
$$\begin{aligned}9(x+4)+4&=4(x-5)\\9x+36+4&=4x-20\\9x+40&=4x-20\\9x-4x+40-40&=4x-4x-20-40\\5x&=-60\\\frac{5x}{5}&=\frac{-60}{5}\\x&=-12\end{aligned}$$

2.
$$\begin{aligned}\frac{y-1}{5}+2&=\frac{2y-3}{3}\\15\left(\frac{y-1}{5}+2\right)&=15\left(\frac{2y-3}{3}\right)\\3(y-1)+30&=10y-15\\3y-3+30&=10y-15\\3y+27&=10y-15\\3y-10y+27-27&=10y-10y-15-27\\-7y&=-42\\y&=6\end{aligned}$$

3.
$$\begin{aligned}P&=L+\frac{s}{f}i\\P-L&=\frac{s}{f}i\\f(P-L)&=f\cdot\frac{s}{f}i\\f(P-L)&=si\\\frac{f(P-L)}{s}&=\frac{si}{s}\\\frac{f(P-L)}{s}&=i,\text{ or } i=\frac{f(P-L)}{s}\end{aligned}$$

4.
$$\begin{aligned}n&=\frac{360}{180-a}\\(180-a)n&=(180-a)\cdot\frac{360}{180-a}\\180n-an&=360\\180n-360&=an\\\frac{180n-360}{n}&=\frac{an}{n}\\\frac{180n-360}{n}&=a,\text{ or } a=\frac{180n-360}{n}\end{aligned}$$

5. Let x = the length of the shortest piece. Then $2x$ and $6x$ are the other lengths.

$$\begin{aligned}\boxed{\text{Total length}}&=20\\x+2x+6x&=20\\9x&=20\\x&=\frac{20}{9}\text{ feet}\end{aligned}$$

$$\text{Longest}=6\left(\frac{20}{9}\right)=\frac{\not{3}\cdot 2}{1}\cdot\frac{20}{\not{3}\cdot 3}=\frac{40}{3}=13\frac{1}{3}\text{ ft}$$

The longest piece is $13\frac{1}{3}$ ft. long.

6. Let w = the width of the rectangle. Then $w+5$ = the length of the rectangle.

$$\begin{aligned}\boxed{\text{Perimeter}}&=26\\w+w+w+5+w+5&=26\\4w+10&=26\\4w&=16\\w&=4\end{aligned}$$

The dimensions are 4 cm by 9 cm, for an area of 36 cm^2.

CHAPTER 7 TEST

7.
$$\begin{aligned} -2(2x+3) &\geq 14 \\ -4x-6+6 &\geq 14+6 \\ -4x &\geq 20 \\ \frac{-4x}{-4} &\leq \frac{20}{-4} \\ x &\leq -5 \end{aligned}$$

solution set: $(-\infty, -5]$

8.
$$\begin{aligned} &-2 < \frac{x-4}{3} < 4 \\ &-6 < x-4 < 12 \\ &-6+4 < x-4+4 \leq 12+4 \\ &-2 < x < 16 \end{aligned}$$

solution set: $(-2, 16)$

9.
$$|2x+3| = 11$$
$$2x+3 = 11 \quad \textbf{or} \quad 2x+3 = -11$$
$$2x = 8 \qquad\qquad 2x = -14$$
$$x = 4 \qquad\qquad x = -7$$

10.
$$|3x+4| = |x+12|$$
$$3x+4 = x+12 \quad \textbf{or} \quad 3x+4 = -(x+12)$$
$$2x = 8 \qquad\qquad 3x+4 = -x-12$$
$$x = 4 \qquad\qquad 4x = -16$$
$$x = -4$$

11.
$$\begin{aligned} &|x+3| \leq 4 \\ &-4 \leq x+3 \leq 4 \\ &-4-3 \leq x+3-3 \leq 4-3 \\ &-7 \leq x \leq 1 \end{aligned}$$

solution set: $[-7, 1]$

12.
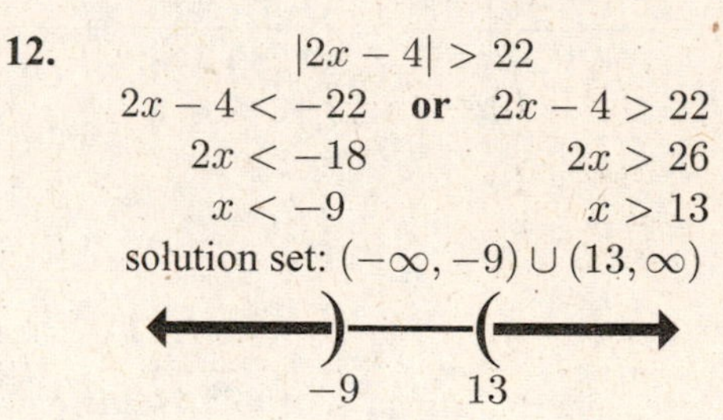
$$|2x-4| > 22$$
$$2x-4 < -22 \quad \textbf{or} \quad 2x-4 > 22$$
$$2x < -18 \qquad\qquad 2x > 26$$
$$x < -9 \qquad\qquad x > 13$$

solution set: $(-\infty, -9) \cup (13, \infty)$

13. $3xy^2 + 6x^2y = 3xy(y+2x)$

14. $12a^3b^2c - 3a^2b^2c^2 + 6abc^3 = 3abc(4a^2b - abc + 2c^2)$

15. $ax - xy + ay - y^2 = x(a-y) + y(a-y) = (a-y)(x+y)$

16. $ax + ay + bx + by - cx - cy = a(x+y) + b(x+y) - c(x+y) = (x+y)(a+b-c)$

17. $x^2 - 49 = x^2 - 7^2 = (x+7)(x-7)$

18. $2x^2 - 32 = 2(x^2 - 16) = 2(x+4)(x-4)$

19. $4y^4 - 64 = 4(y^4 - 16) = 4(y^2+4)(y^2-4) = 4(y^2+4)(y+2)(y-2)$

20. $b^3 + 125 = (b+5)(b^2 - 5b + 25)$

21. $b^3 - 27 = (b-3)(b^2 + 3b + 9)$

22.
$$\begin{aligned} 3u^3 - 24 &= 3(u^3 - 8) \\ &= 3(u-2)(u^2 + 2u + 4) \end{aligned}$$

23. $x^2 + 8x + 15 = (x+3)(x+5)$

24. $6b^2 + b - 2 = (3b+2)(2b-1)$

25.
$$\begin{aligned} 6u^2 + 9u - 6 &= 3(2u^2 + 3u - 2) \\ &= 3(2u-1)(u+2) \end{aligned}$$

26. $x^2+6x+9-y^2=(x^2+6x+9)-y^2=(x+3)^2-y^2=(x+3+y)(x+3-y)$

27. $\dfrac{-12x^2y^3z^2}{18x^3y^4z^2}=-\dfrac{2}{3xy}$

28. $\dfrac{2x^2+7x+3}{4x+12}=\dfrac{(2x+1)(x+3)}{4(x+3)}=\dfrac{2x+1}{4}$

29. $\dfrac{x^2y^{-2}}{x^3z^2}\cdot\dfrac{x^2z^4}{y^2z}=\dfrac{x^4y^{-2}z^4}{x^3y^2z^3}=\dfrac{xz}{y^4}$

30. $\dfrac{u^2+5u+6}{u^2-4}\cdot\dfrac{u^2-5u+6}{u^2-9}=\dfrac{(u+2)(u+3)}{(u+2)(u-2)}\cdot\dfrac{(u-2)(u-3)}{(u+3)(u-3)}=1$

31. $$\dfrac{x^3+y^3}{4}\div\dfrac{x^2-xy+y^2}{2x+2y}=\dfrac{x^3+y^3}{4}\cdot\dfrac{2x+2y}{x^2-xy+y^2}=\dfrac{(x+y)(x^2-xy+y^2)}{4}\cdot\dfrac{2(x+y)}{x^2-xy+y^2}$$
$$=\dfrac{(x+y)^2}{2}$$

32. $$\dfrac{x+2}{x+1}-\dfrac{x+1}{x+2}=\dfrac{(x+2)(x+2)}{(x+1)(x+2)}-\dfrac{(x+1)(x+1)}{(x+1)(x+2)}=\dfrac{x^2+4x+4-(x^2+2x+1)}{(x+1)(x+2)}$$
$$=\dfrac{x^2+4x+4-x^2-2x-1}{(x+1)(x+2)}$$
$$=\dfrac{2x+3}{(x+1)(x+2)}$$

33. $\dfrac{\frac{2u^2w^3}{v^2}}{\frac{4uw^4}{uv}}=\dfrac{2u^2w^3}{v^2}\div\dfrac{4uw^4}{uv}=\dfrac{2u^2w^3}{v^2}\cdot\dfrac{uv}{4uw^4}=\dfrac{2u^3w^3v}{4uv^2w^4}=\dfrac{u^2}{2vw}$

34. $\dfrac{\frac{x}{y}+\frac{1}{2}}{\frac{x}{2}-\frac{1}{y}}=\dfrac{\left(\frac{x}{y}+\frac{1}{2}\right)2y}{\left(\frac{x}{2}-\frac{1}{y}\right)2y}=\dfrac{2x+y}{xy-2}$

35.
$$\begin{array}{r|l} & x^2-5x+10 \\ \hline x+1 & x^3-4x^2+5x+3 \\ & \underline{x^3+x^2} \\ & -5x^2+5x \\ & \underline{-5x^2-5x} \\ & 10x+3 \\ & \underline{10x+10} \\ & -7 \end{array}$$

36.
$$\begin{array}{r|rrrr} 2 & 4 & 3 & 2 & -1 \\ & & 8 & 22 & 48 \\ \hline & 4 & 11 & 24 & \boxed{47} \end{array}$$

Exercise 8.1 (page 540)

1. $x + y = 3$; $x + 0 = 3$; $x = 3$; x-intercept: $(3, 0)$

$x + y = 3$; $0 + y = 3$; $y = 3$; y-intercept: $(0, 3)$

3. $x + 4y = 8$; $x + 4(0) = 8$; $x + 0 = 8$; $x = 8$; x-intercept: $(8, 0)$

$x + 4y = 8$; $0 + 4y = 8$; $4y = 8$; $y = 2$; y-intercept: $(0, 2)$

5. $x = \frac{x_1+x_2}{2} = \frac{2+6}{2} = \frac{8}{2} = 4$

$y = \frac{y_1+y_2}{2} = \frac{4+8}{2} = \frac{12}{2} = 6$

midpoint: $(4, 6)$

7. $-2 \quad 2$

9. $x^2 - x = x(x - 1)$

11. $x^3 - 1 = (x - 1)(x^2 + x + 1)$

13. origin

15. y-coordinate

17. x-axis

19. horizontal

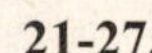

21-27.

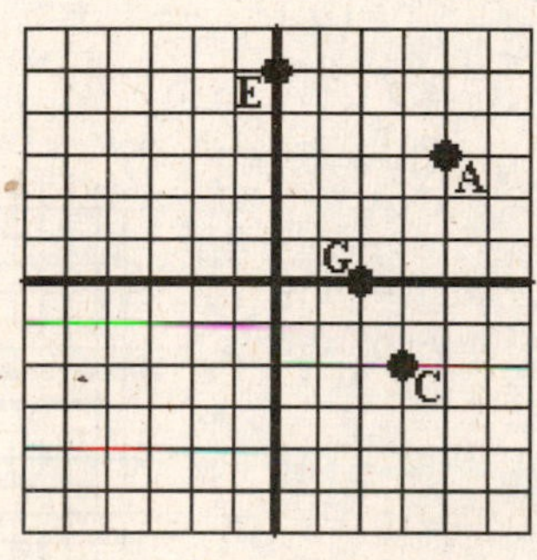

29. $(2, 4)$

31. $(-2, -1)$

33. $(4, 0)$

35. $(0, 0)$

37. $x + y = 4$

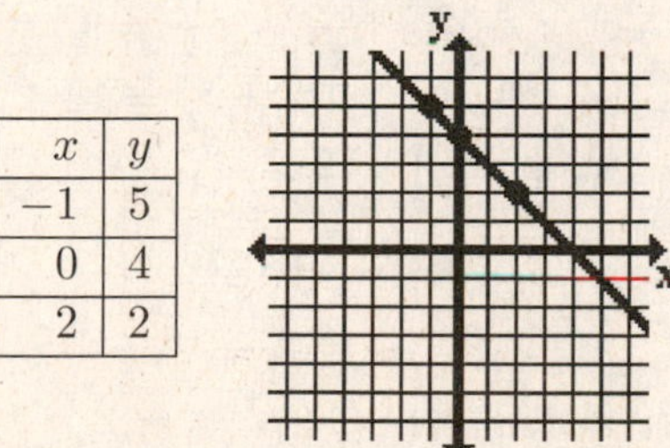

x	y
-1	5
0	4
2	2

39. $2x - y = 3$

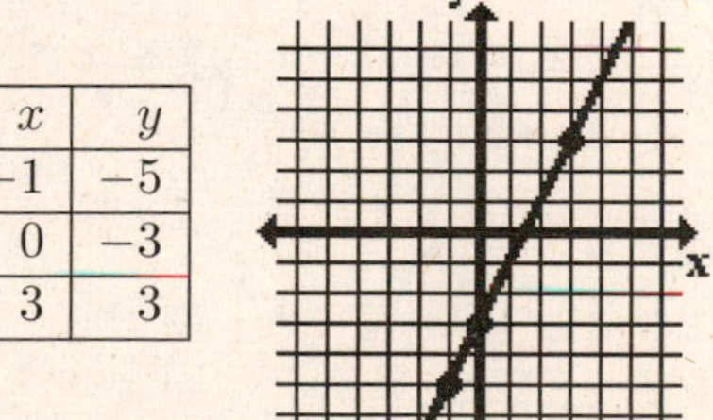

x	y
-1	-5
0	-3
3	3

41.
$$\begin{aligned} 3x + 4y &= 12 \\ 3x + 4(0) &= 12 \\ 3x &= 12 \\ x &= 4 \end{aligned}$$
x-intercept: $(4, 0)$

$$\begin{aligned} 3x + 4y &= 12 \\ 3(0) + 4y &= 12 \\ 4y &= 12 \\ y &= 3 \end{aligned}$$
y-intercept: $(0, 3)$

$3x + 4y = 12$

43.
$$\begin{aligned} y &= -3x + 2 \\ 0 &= -3x + 2 \\ 3x &= 2 \\ x &= \tfrac{2}{3} \end{aligned}$$
x-intercept: $\left(\frac{2}{3}, 0\right)$

$$\begin{aligned} y &= -3x + 2 \\ y &= -3(0) + 2 \\ y &= 2 \end{aligned}$$
y-intercept: $(0, 2)$

$y = -3x + 2$

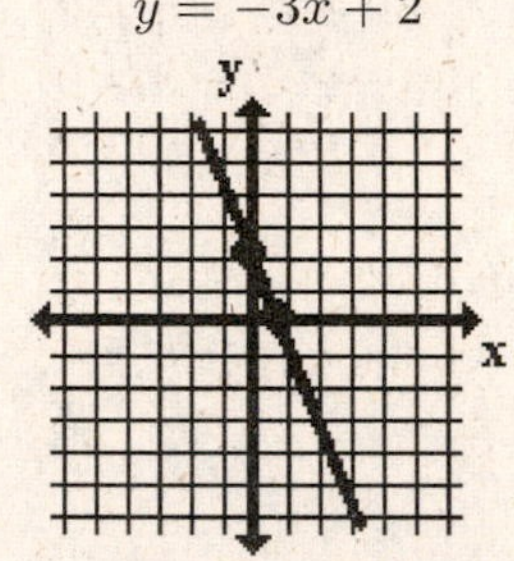

45. $x = 3$
vertical line with
x-coordinate of 3

$x = 3$

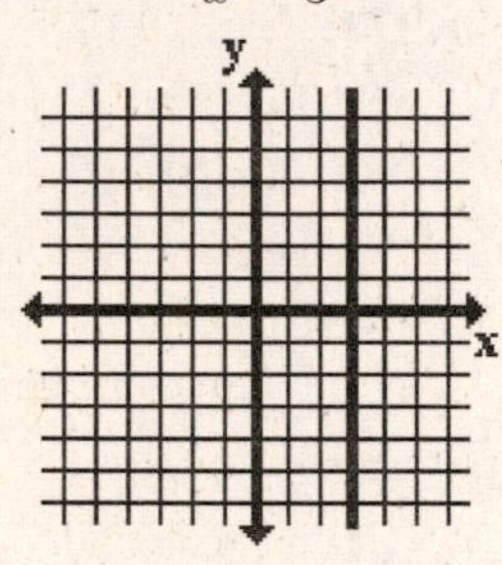

47.
$$\begin{aligned} -3y + 2 &= 5 \\ -3y &= 3 \\ y &= -1 \end{aligned}$$
horizontal line
with y-coordinate
of -1

$y = -1$

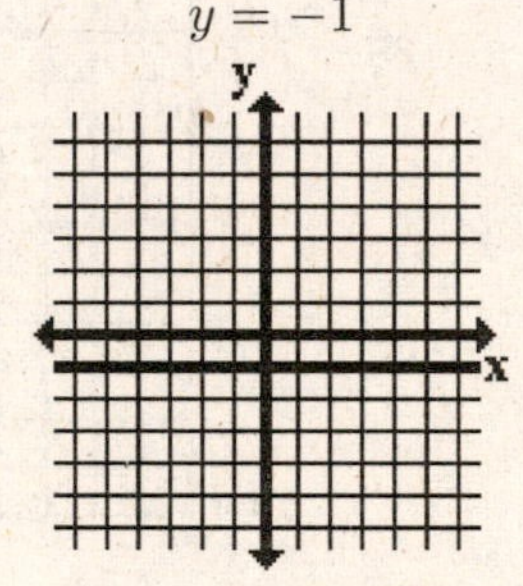

49. $x = \frac{x_1+x_2}{2} = \frac{0+6}{2} = \frac{6}{2} = 3$
$y = \frac{y_1+y_2}{2} = \frac{0+8}{2} = \frac{8}{2} = 4$
midpoint: $(3, 4)$

51. $x = \frac{x_1+x_2}{2} = \frac{6+12}{2} = \frac{18}{2} = 9$
$y = \frac{y_1+y_2}{2} = \frac{8+16}{2} = \frac{24}{2} = 12$
midpoint: $(9, 12)$

53. $x = \frac{x_1+x_2}{2} = \frac{2+5}{2} = \frac{7}{2}$
$y = \frac{y_1+y_2}{2} = \frac{4+8}{2} = \frac{12}{2} = 6$
midpoint: $\left(\frac{7}{2}, 6\right)$

55. $x = \frac{x_1+x_2}{2} = \frac{-2+3}{2} = \frac{1}{2}$
$y = \frac{y_1+y_2}{2} = \frac{-8+4}{2} = \frac{-4}{2} = -2$
midpoint: $\left(\frac{1}{2}, -2\right)$

57.
$$\begin{aligned} 3y &= 6x - 9 \\ 3(0) &= 6x - 9 \\ 0 &= 6x - 9 \\ -6x &= -9 \\ x &= \tfrac{-9}{-6} = \tfrac{3}{2} \end{aligned}$$
x-intercept: $\left(\frac{3}{2}, 0\right)$

$$\begin{aligned} 3y &= 6x - 9 \\ 3y &= 6(0) - 9 \\ 3y &= 0 - 9 \\ 3y &= -9 \\ y &= \tfrac{-9}{3} = -3 \end{aligned}$$
y-intercept: $(0, -3)$

$3y = 6x - 9$

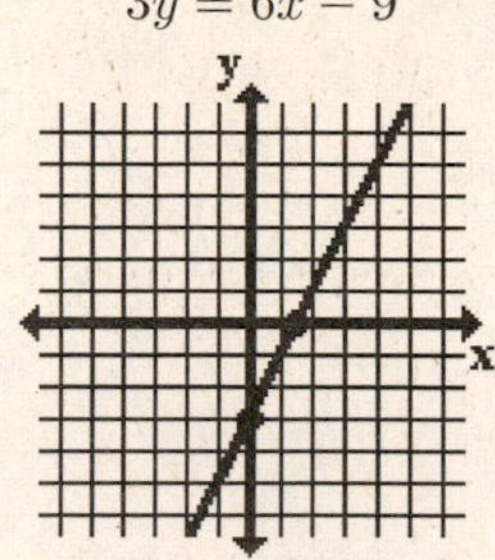

59.
$$\begin{aligned} 3x + 4y - 8 &= 0 \\ 3x + 4(0) - 8 &= 0 \\ 3x - 8 &= 0 \\ 3x &= 8 \\ x &= \tfrac{8}{3} \end{aligned}$$
x-intercept: $\left(\frac{8}{3}, 0\right)$

$$\begin{aligned} 3x + 4y - 8 &= 0 \\ 3(0) + 4y - 8 &= 0 \\ 4y - 8 &= 0 \\ 4y &= 8 \\ y &= 2 \end{aligned}$$
y-intercept: $(0, 2)$

$3x + 4y - 8 = 0$

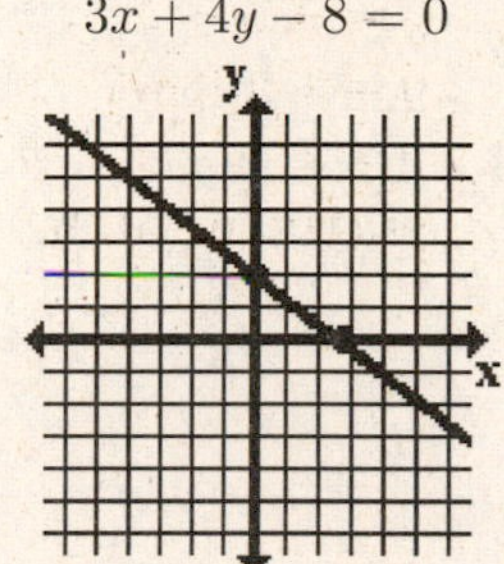

61. $x = \frac{x_1+x_2}{2} = \frac{-3+(-5)}{2} = \frac{-8}{2} = -4$

$y = \frac{y_1+y_2}{2} = \frac{5+(-5)}{2} = \frac{0}{2} = 0$

midpoint: $(-4, 0)$

63. $x = \frac{x_1+x_2}{2} = \frac{a+4a}{2} = \frac{5a}{2}$

$y = \frac{y_1+y_2}{2} = \frac{b+3b}{2} = \frac{4b}{2} = 2b$

midpoint: $\left(\frac{5a}{2}, 2b\right)$

65. $x = \frac{x_1+x_2}{2} = \frac{(a-b)+(a+b)}{2} = \frac{2a}{2} = a$

$y = \frac{y_1+y_2}{2} = \frac{b+3b}{2} = \frac{4b}{2} = 2b$

midpoint: $(a, 2b)$

67.
$$\begin{aligned} x &= \tfrac{x_1+x_2}{2} \\ -2 &= \tfrac{-8+x_2}{2} \\ -4 &= -8 + x_2 \\ 4 &= x_2 \end{aligned} \qquad \begin{aligned} y &= \tfrac{y_1+y_2}{2} \\ 3 &= \tfrac{5+y_2}{2} \\ 6 &= 5 + y_2 \\ 1 &= y_2 \end{aligned}$$
Q has coordinates $(4, 1)$.

69. To find the value of the house after 5 years, let $x = 5$:
$$\begin{aligned} y &= 7500x + 125{,}000 \\ &= 7500(5) + 125{,}000 \\ &= 37{,}500 + 125{,}000 \\ &= 162{,}500 \end{aligned}$$
It will be worth \$162,500 after 5 years.

To find the value of the house after 10 years, let $x = 10$:
$$\begin{aligned} y &= 7500x + 125{,}000 \\ &= 7500(10) + 125{,}000 \\ &= 75{,}000 + 125{,}000 \\ &= 200{,}000 \end{aligned}$$
It will be worth \$200,000 after 10 years.

71. To find the number of TVs sold, let $p = 150$:

$$p = -\frac{1}{10}q + 170$$
$$150 = -\frac{1}{10}q + 170$$
$$\frac{1}{10}q = 20$$
$$q = 200$$

200 TVs will be sold at a price of \$150.

73. Let $V = 60$ and find v.

$$V = \frac{nv}{N}$$
$$60 = \frac{12v}{20}$$
$$1200 = 12v$$
$$100 = v$$

The smaller gear has a speed of 100 revolutions per minute.

75. **Answers may vary.**

77. If the line $y = ax + b$ passes through only quadrants I and II, then $a = 0$ and $b > 0$.

Exercise 8.2 (page 551)

1. $m = \frac{\Delta y}{\Delta x} = \frac{3-0}{1-0} = \frac{3}{1} = 3$

3. $-2 = -\frac{8}{4}$, so parallel

5. $-2\left(\frac{1}{2}\right) = -1$, so perpendicular

7. $\left(\frac{x^5}{x^3}\right)^3 = \left(x^2\right)^3 = x^6$

9. $\left(\frac{x^{-6}}{y^3}\right)^{-4} = \left(\frac{y^3}{x^{-6}}\right)^4 = \left(x^6y^3\right)^4 = x^{24}y^{12}$

11. $\left(\frac{x^3x^{-7}y^{-6}}{x^4y^{-3}y^{-2}}\right)^{-2} = \left(\frac{x^{-4}y^{-6}}{x^4y^{-5}}\right)^{-2} = \left(\frac{x^4y^{-5}}{x^{-4}y^{-6}}\right)^2 = \left(x^8y^1\right)^2 = x^{16}y^2$

13. change

15. rise

17. horizontal

19. positive

21. perpendicular; reciprocals

23. $m = \frac{\Delta y}{\Delta x} = \frac{5-(-3)}{2-(-2)} = \frac{8}{4} = 2$

25. $m = \frac{\Delta y}{\Delta x} = \frac{9-0}{3-0} = \frac{9}{3} = 3$

27. $m = \frac{\Delta y}{\Delta x} = \frac{1-8}{6-(-1)} = \frac{-7}{7} = -1$

29. $m = \frac{\Delta y}{\Delta x} = \frac{2-(-1)}{-6-3} = \frac{3}{-9} = -\frac{1}{3}$

31. $m = \frac{\Delta y}{\Delta x} = \frac{5-5}{-9-7} = \frac{0}{-16} = 0$

33. $m = \frac{\Delta y}{\Delta x} = \frac{-2-(-5)}{-7-(-7)} = \frac{3}{0}$: not defined

35. Find two points on the line:

Let $x = 0$:
$$3x + 2y = 12$$
$$3(0) + 2y = 12$$
$$2y = 12$$
$$y = 6$$
$$(0, 6)$$

Let $y = 0$:
$$3x + 2y = 12$$
$$3x + 2(0) = 12$$
$$3x = 12$$
$$x = 4$$
$$(4, 0)$$

$$m = \frac{\Delta y}{\Delta x} = \frac{6-0}{0-4} = \frac{6}{-4} = -\frac{3}{2}$$

37. Find two points on the line:

Let $x = 0$:
$$3x = 4y - 2$$
$$3(0) = 4y - 2$$
$$0 = 4y - 2$$
$$2 = 4y$$
$$\tfrac{1}{2} = y$$
$$(0, \tfrac{1}{2})$$

Let $y = 0$:
$$3x = 4y - 2$$
$$3x = 4(0) - 2$$
$$3x = -2$$
$$x = -\tfrac{2}{3}$$
$$(-\tfrac{2}{3}, 0)$$

$$m = \frac{\Delta y}{\Delta x} = \frac{\frac{1}{2} - 0}{0 - (-\frac{2}{3})} = \frac{\frac{1}{2}}{\frac{2}{3}} = \frac{3}{4}$$

39. Find two points on the line:

Let $x = 0$:
$$y = \frac{x-4}{2}$$
$$y = \frac{0-4}{2}$$
$$y = \frac{-4}{2}$$
$$y = -2$$
$$(0, -2)$$

Let $y = 0$:
$$y = \frac{x-4}{2}$$
$$0 = \frac{x-4}{2}$$
$$0 = x - 4$$
$$4 = x$$
$$(4, 0)$$

$$m = \frac{\Delta y}{\Delta x} = \frac{-2-0}{0-4} = \frac{-2}{-4} = \frac{1}{2}$$

41.
$$4y = 3(y + 2)$$
$$4y = 3y + 6$$
$$y = 6$$
horizontal line $\Rightarrow m = 0$

43. negative

45. positive

47. not defined

49. 0

51. $m_1 \neq m_2 \Rightarrow$ not parallel

$m_1 \cdot m_2 = 3\left(-\frac{1}{3}\right) = -1 \Rightarrow$ perpendicular

53. $m_1 \neq m_2 \Rightarrow$ not parallel

$m_1 \cdot m_2 = 4(0.25) = 1 \Rightarrow$ not perpendicular

55. $m_{\overline{PQ}} = \frac{\Delta y}{\Delta x} = \frac{2-4}{4-3} = \frac{-2}{1} = -2$

same slope $\Rightarrow$ parallel

57. $m_{\overline{PQ}} = \frac{\Delta y}{\Delta x} = \frac{5-1}{6-(-2)} = \frac{4}{8} = \frac{1}{2}$

opposite reciprocal slope $\Rightarrow$ perpendicular

59. $m_1 = \frac{\Delta y}{\Delta x} = \frac{2.5 - 3.7}{3.7 - 2.5} = \frac{-1.2}{1.2} = -1$

$m_2 = \frac{\Delta y}{\Delta x} = \frac{-1.7 - (-2.3)}{2.3 - 1.7} = \frac{0.6}{0.6} = 1$

opposite reciprocal slope $\Rightarrow$ perpendicular

61. $m_1 \neq m_2 \Rightarrow$ not parallel

$m_1 \cdot m_2 = \frac{3.2}{-9.1} \cdot \frac{-9.1}{3.2} = 1$

$\Rightarrow$ not perpendicular

63. $m_{\overline{PQ}} = \frac{\Delta y}{\Delta x} = \frac{-9-3}{4-(-2)} = \frac{-12}{6} = -2$

same slope $\Rightarrow$ parallel

65. $m_{\overline{PQ}} = \dfrac{\Delta y}{\Delta x} = \dfrac{6-10}{0-6} = \dfrac{-4}{-6} = \dfrac{2}{3}$

$m_{\overline{PR}} = \dfrac{\Delta y}{\Delta x} = \dfrac{8-10}{3-6} = \dfrac{-2}{-3} = \dfrac{2}{3}$

Since they have the same slope and a common point, the three points are on the same line.

67. $m_{\overline{PQ}} = \dfrac{\Delta y}{\Delta x} = \dfrac{-10-(-13)}{-8-(-10)} = \dfrac{3}{2}$

$m_{\overline{PR}} = \dfrac{\Delta y}{\Delta x} = \dfrac{-16-(-13)}{-12-(-10)} = \dfrac{-3}{-2} = \dfrac{3}{2}$

Since they have the same slope and a common point, the three points are on the same line.

69. $m_{\overline{PQ}} = \dfrac{\Delta y}{\Delta x} = \dfrac{-12-(-4)}{0-8} = \dfrac{-8}{-8} = 1$

$m_{\overline{PR}} = \dfrac{\Delta y}{\Delta x} = \dfrac{-20-(-4)}{8-8} = \dfrac{-16}{0} \Rightarrow$ und.

Since they do not have the same slope, the three points are not on the same line.

71. On the y-axis, all x-coordinates are 0. The equation is $x = 0$, and the slope is undefined.

73. Call the points $A(0,0)$, $B(12,0)$ and $C(13,12)$. Compute these slopes:

$m_{\overline{AB}} = \dfrac{\Delta y}{\Delta x} = \dfrac{0-0}{12-0} = \dfrac{0}{12} = 0$

$m_{\overline{AC}} = \dfrac{\Delta y}{\Delta x} = \dfrac{12-0}{13-0} = \dfrac{12}{13}$

$m_{\overline{BC}} = \dfrac{\Delta y}{\Delta x} = \dfrac{12-0}{13-12} = \dfrac{12}{1} = 12$

Since none of the sides are perpendicular, it is not a right triangle.

75. Call the points $A(2b,a)$, $B(b,b)$ and $C(a,0)$. Compute these slopes:

$m_{\overline{AB}} = \dfrac{\Delta y}{\Delta x} = \dfrac{b-a}{b-2b} = \dfrac{b-a}{-b} = -\dfrac{b-a}{b}$

$m_{\overline{AC}} = \dfrac{\Delta y}{\Delta x} = \dfrac{0-a}{a-2b} = \dfrac{-a}{a-2b} = -\dfrac{a}{a-2b}$

$m_{\overline{BC}} = \dfrac{\Delta y}{\Delta x} = \dfrac{b-0}{b-a} = \dfrac{b}{b-a}$

Since $\overline{AB}$ and $\overline{BC}$ are perpendicular, it is a right triangle.

77. Call the points $A(0,0)$, $B(0,b)$, $C(8,b+2)$ and $D(12,3)$. Compute these slopes:

$m_{\overline{AB}} = \dfrac{\Delta y}{\Delta x} = \dfrac{0-b}{0-0} = \dfrac{-b}{0} \Rightarrow$ undefined

$m_{\overline{BC}} = \dfrac{\Delta y}{\Delta x} = \dfrac{b-(b+2)}{0-8} = \dfrac{b-b-2}{-8} = \dfrac{-2}{-8} = \dfrac{1}{4}$

$m_{\overline{CD}} = \dfrac{\Delta y}{\Delta x} = \dfrac{b+2-3}{8-12} = \dfrac{b-1}{-4} = -\dfrac{b-1}{4}$

$m_{\overline{DA}} = \dfrac{\Delta y}{\Delta x} = \dfrac{0-3}{0-12} = \dfrac{-3}{-12} = \dfrac{1}{4}$

Thus, $\overline{BC} \parallel \overline{DA}$ and the figure is a trapezoid.

79. slope $= \dfrac{\text{rise}}{\text{run}} = \dfrac{3 \text{ ft}}{12 \text{ ft}} = \dfrac{1}{4}$

81. $m = \dfrac{\Delta y}{\Delta x} = \dfrac{2}{50} = \dfrac{1}{25}$; $m = \dfrac{\Delta y}{\Delta x} = \dfrac{5}{50} = \dfrac{1}{10}$; $m = \dfrac{\Delta y}{\Delta x} = \dfrac{8}{50} = \dfrac{4}{25}$

83. $m = \dfrac{\Delta y}{\Delta x} = \dfrac{0.7}{50} = \dfrac{7}{500}$; $\dfrac{7}{500}$ degree per year

85. **a.** $m = \dfrac{\Delta y}{\Delta x} = \dfrac{2}{25}$ **b.** $m = \dfrac{\Delta y}{\Delta x} = \dfrac{1}{20}$ (each part)

c. Design #1 has just one level, but the slope is steep. Design #2 has slopes which are less steep, but there are two levels.

87. Let x represent the number of years, and let y represent the price. Then we know two points on the line: $(-10, 5700)$ and $(-2, 1499)$. Find the slope:

$m = \dfrac{\Delta y}{\Delta x} = \dfrac{5700 - 1499}{-10 - (-2)} = \dfrac{4201}{-8} \approx -\$525.13 \Rightarrow$ The cost is decreasing about \$525.13 per year.

89. **Answers may vary.**

91. Find two points on the line:

Let $x = 0$:
$y = mx + b$
$y = m(0) + b$
$y = b$
$(0, b)$ is on the line.

Let $y = 0$:
$y = mx + b$
$0 = mx + b$
$-b = mx$
$\frac{-b}{m} = x$
$\left(-\frac{b}{m}, 0\right)$ is on the line.

$m = \dfrac{\Delta y}{\Delta x} = \dfrac{b - 0}{0 - \left(-\frac{b}{m}\right)} = \dfrac{b}{\frac{b}{m}} = m$

93. $m_{\overline{AB}} = \dfrac{\Delta y}{\Delta x} = \dfrac{3 - 7}{1 - (-2)} = \dfrac{-4}{3} = -\dfrac{4}{3}$. Then $m_{\overline{CD}}$ must equal $\dfrac{3}{4}$.

$$m_{\overline{CD}} = \frac{\Delta y}{\Delta x} = \frac{b - (-1)}{4 - 8} = \frac{b + 1}{-4} = \frac{3}{4}$$
$$-4 \cdot \frac{b + 1}{-4} = -4 \cdot \frac{3}{4}$$
$$b + 1 = -3$$
$$b = -4$$

Exercise 8.3 (page 564)

1. $y - y_1 = m(x - x_1)$
$y - 3 = 2(x - 2)$

3. $y = mx + b$
$y = -3x + 5$

SECTION 8.3

5. $y = 3x - 4 \quad y = 3x + 5$

$m = 3 \quad m = 3$

parallel

7.
$$\begin{aligned} 3(x+2) + x &= 5x \\ 3x + 6 + x &= 5x \\ 4x + 6 &= 5x \\ 6 &= x \end{aligned}$$

9.
$$\begin{aligned} \frac{5(2-x)}{3} - 1 &= x + 5 \\ \frac{10-5x}{3} &= x + 6 \\ 3 \cdot \frac{10-5x}{3} &= 3(x+6) \\ 10 - 5x &= 3x + 18 \\ 10 &= 8x + 18 \\ -8 &= 8x \\ -1 &= x \end{aligned}$$

11. We want the alloy to contain 25% gold. Let $x =$ the amount of copper added to the alloy.

$\boxed{\text{Gold at start}} + \boxed{\text{Gold added}} = \boxed{\text{Gold at end}}$

$$\begin{aligned} 20 + 0 &= 0.25(60 + x) \\ 20 &= 15 + 0.25x \\ 5 &= 0.25x \\ 20 &= x \end{aligned}$$

20 oz of copper should be added to the alloy.

13. $y - y_1 = m(x - x_1)$

15. $Ax + By = C$

17. perpendicular

19.
$$\begin{aligned} y - y_1 &= m(x - x_1) \\ y - 7 &= 5(x - 0) \\ y - 7 &= 5x \end{aligned}$$

21.
$$\begin{aligned} y - y_1 &= m(x - x_1) \\ y - 0 &= -3(x - 2) \\ y &= -3(x - 2) \end{aligned}$$

23. Find the slope of the line:

$$m = \frac{\Delta y}{\Delta x} = \frac{4-0}{4-0} = \frac{4}{4} = 1$$

Use point-slope form to find the equation:

$$\begin{aligned} y - y_1 &= m(x - x_1) \\ y - 0 &= 1(x - 0) \\ y &= x \end{aligned}$$

25. Find the slope of the line:

$$m = \frac{\Delta y}{\Delta x} = \frac{4-(-3)}{3-0} = \frac{7}{3}$$

Use point-slope form to find the equation:

$$\begin{aligned} y - y_1 &= m(x - x_1) \\ y - 4 &= \tfrac{7}{3}(x - 3) \\ y - 4 &= \tfrac{7}{3}x - 7 \\ y &= \tfrac{7}{3}x - 3 \end{aligned}$$

27.
$$\begin{aligned} y &= mx + b \\ y &= 3x + 17 \end{aligned}$$

29.
$$\begin{aligned} y &= mx + b \\ 5 &= -7(7) + b \\ 5 &= -49 + b \\ 54 &= b \\ y &= -7x + 54 \end{aligned}$$

31.
$$\begin{aligned} y &= mx + b \\ -4 &= 0(2) + b \\ -4 &= b \\ y &= 0x + (-4) \\ y &= -4 \end{aligned}$$

SECTION 8.3

33. Find the slope of the line:

$$m = \frac{\Delta y}{\Delta x} = \frac{8-10}{6-2} = \frac{-2}{4} = -\frac{1}{2}$$

$$y = mx + b$$
$$8 = -\frac{1}{2}(6) + b$$
$$8 = -3 + b$$
$$11 = b$$
$$y = -\frac{1}{2}x + 11$$

35. $3x - 2y = 8$

$$-2y = -3x + 8$$
$$y = \tfrac{3}{2}x - 4$$
$$m = \tfrac{3}{2};\ b = -4 \Rightarrow (0, -4)$$

37. $-2(x + 3y) = 5$

$$-2x - 6y = 5$$
$$-6y = 2x + 5$$
$$y = -\tfrac{1}{3}x - \tfrac{5}{6}$$
$$m = -\tfrac{1}{3};\ b = -\tfrac{5}{6} \Rightarrow \left(0, -\tfrac{5}{6}\right)$$

39. 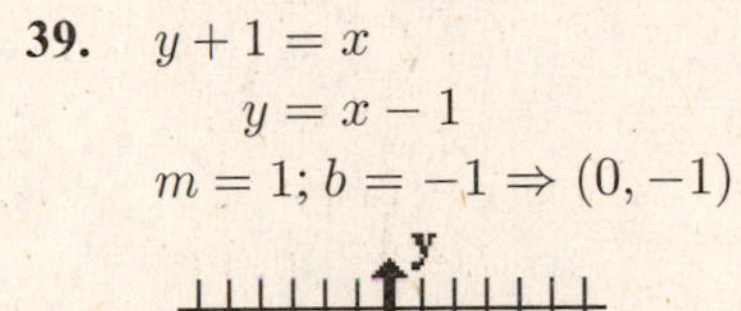

$y + 1 = x$

$$y = x - 1$$
$$m = 1;\ b = -1 \Rightarrow (0, -1)$$

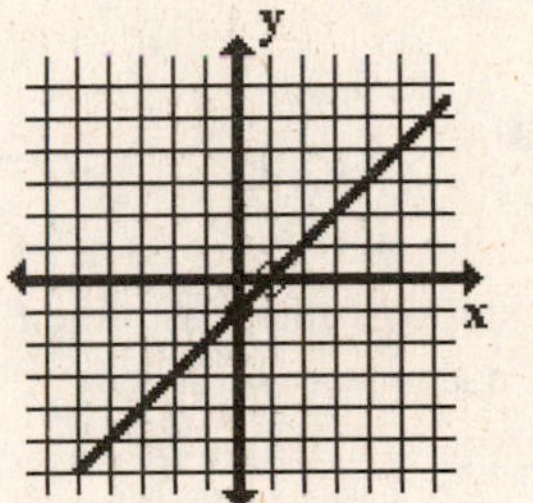

41. $x = \tfrac{3}{2}y - 3$

$$2x = 3y - 6$$
$$-3y = -2x + 6$$
$$y = \tfrac{2}{3}x + 2$$
$$m = \tfrac{2}{3};\ b = 2 \Rightarrow (0, 2)$$

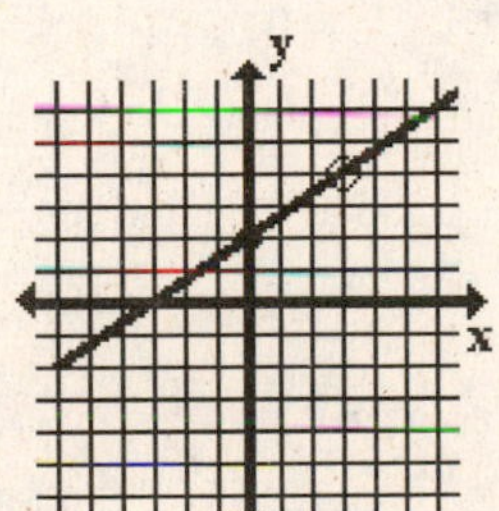

43. $y = 3x + 4 \qquad y = 3x - 7$

$$m = 3 \qquad m = 3$$

parallel

45. $x + y = 2 \qquad y = x + 5$

$$y = -x + 2 \qquad m = 1$$
$$m = -1$$

perpendicular

47. $y = 3x + 7 \qquad 2y = 6x - 9$

$$m = 3 \qquad y = 3x - \frac{9}{2}$$
$$m = 3$$

parallel

49. $x = 3y + 4 \qquad y = -3x + 7$

$$-3y = -x + 4 \qquad m = -3$$
$$y = \tfrac{1}{3}x - \tfrac{4}{3}$$
$$m = \tfrac{1}{3}$$

perpendicular

51. $y = 3$ $x = 4$
horizontal line vertical line
perpendicular

53.
$x = \frac{y-2}{3}$
$3x = y - 2$
$-y = -3x - 2$
$y = 3x + 2$
$m = 3$

$3(y-3) + x = 0$
$3y - 9 + x = 0$
$3y = -x + 9$
$y = -\frac{1}{3}x + 3$
$m = -\frac{1}{3}$

perpendicular

55. Find the slope of the given line:
$y = 4x - 7 \Rightarrow m = 4$
Use the parallel slope.
$y - y_1 = m(x - x_1)$
$y - 0 = 4(x - 0)$
$y = 4x$

57. Find the slope of the given line:
$4x - y = 7$
$-y = -4x + 7$
$y = 4x - 7 \Rightarrow m = 4$
Use the parallel slope.
$y - y_1 = m(x - x_1)$
$y - 5 = 4(x - 2)$
$y - 5 = 4x - 8$
$y = 4x - 3$

59. Find the slope of the given line:
$y = 4x - 7 \Rightarrow m = 4$
Use the perpendicular slope.
$y - y_1 = m(x - x_1)$
$y - 0 = -\frac{1}{4}(x - 0)$
$y = -\frac{1}{4}x$

61. Find the slope of the given line:
$4x - y = 7$
$-y = -4x + 7$
$y = 4x - 7 \Rightarrow m = 4$
Use the perpendicular slope.
$y - y_1 = m(x - x_1)$
$y - 5 = -\frac{1}{4}(x - 2)$
$y - 5 = -\frac{1}{4}x + \frac{1}{2}$
$y = -\frac{1}{4}x + \frac{1}{2} + 5$
$y = -\frac{1}{4}x + \frac{11}{2}$

63. Find the slope of the given line:
$x = \frac{5}{4}y - 2$
$4x = 5y - 8$
$-5y = -4x - 8$
$y = \frac{4}{5}x + \frac{8}{5} \Rightarrow m = \frac{4}{5}$
Use the parallel slope.
$y - y_1 = m(x - x_1)$
$y - (-2) = \frac{4}{5}(x - 4)$
$y + 2 = \frac{4}{5}x - \frac{16}{5}$
$y = \frac{4}{5}x - \frac{16}{5} - 2$
$y = \frac{4}{5}x - \frac{26}{5}$

65. Find the slope of the given line:
$x = \frac{5}{4}y - 2$
$4x = 5y - 8$
$-5y = -4x - 8$
$y = \frac{4}{5}x + \frac{8}{5} \Rightarrow m = \frac{4}{5}$
Use the perpendicular slope.
$y - y_1 = m(x - x_1)$
$y - (-2) = -\frac{5}{4}(x - 4)$
$y + 2 = -\frac{5}{4}x + 5$
$y = -\frac{5}{4}x + 3$

SECTION 8.3

67. Find the slope of the 1st line. $m = -\frac{A}{B} = -\frac{4}{5}$ Find the slope of the 2nd line. $m = -\frac{A}{B} = -\frac{5}{-4} = \frac{5}{4}$ perpendicular

69. Find the slope of the 1st line. $m = -\frac{A}{B} = -\frac{2}{3}$ Find the slope of the 2nd line. $m = -\frac{A}{B} = -\frac{6}{9} = -\frac{2}{3}$ parallel

71.
$$x = \frac{2y-4}{7}$$
$$7x = 2y - 4$$
$$-2y = -7x - 4$$
$$y = \tfrac{7}{2}x + 2$$
$$m = \tfrac{7}{2};\ b = 2 \Rightarrow (0, 2)$$

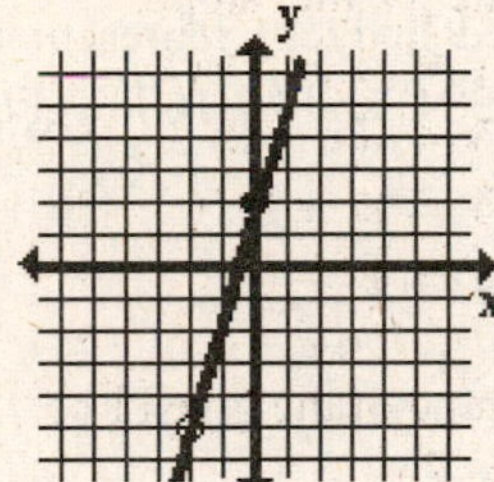

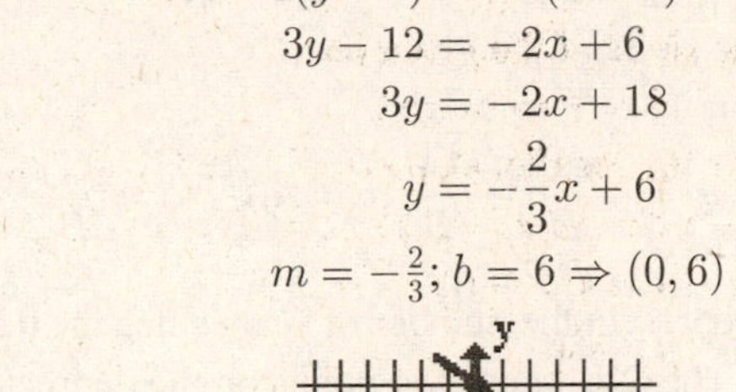

73.
$$3(y-4) = -2(x-3)$$
$$3y - 12 = -2x + 6$$
$$3y = -2x + 18$$
$$y = -\frac{2}{3}x + 6$$
$$m = -\tfrac{2}{3};\ b = 6 \Rightarrow (0, 6)$$

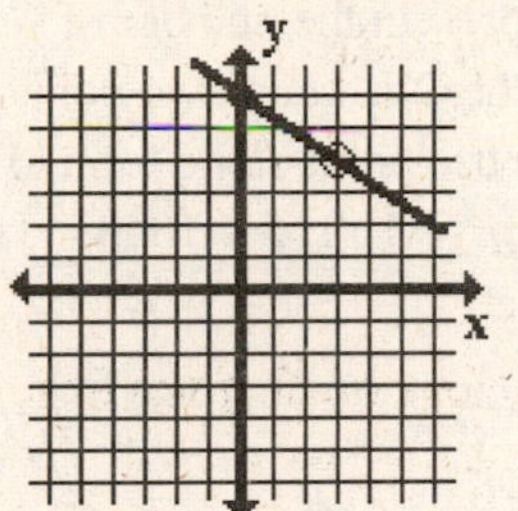

75. Notice that the line goes through $P(2, 5)$ and the point $(-1, 3)$. Find the slope:
$$m = \frac{\Delta y}{\Delta x} = \frac{5-3}{2-(-1)} = \frac{2}{3}$$
Use point-slope form to find the equation:
$$y - y_1 = m(x - x_1)$$
$$y - 5 = \frac{2}{3}(x-2)$$
$$y - 5 = \frac{2}{3}x - \frac{4}{3}$$
$$y = \frac{2}{3}x - \frac{4}{3} + 5$$
$$y = \frac{2}{3}x + \frac{11}{3}$$

77. Find the slope of the line:
$$m = \frac{\Delta y}{\Delta x} = \frac{4-(-3)}{-2-2} = \frac{7}{-4} = -\frac{7}{4}$$
Use point-slope form to find the equation:
$$y - y_1 = m(x - x_1)$$
$$y - 4 = -\tfrac{7}{4}(x - (-2))$$
$$y - 4 = -\tfrac{7}{4}(x + 2)$$
$$y - 4 = -\tfrac{7}{4}x - \tfrac{7}{2}$$
$$y = -\tfrac{7}{4}x - \tfrac{7}{2} + 4$$
$$y = -\tfrac{7}{4}x + \tfrac{1}{2}$$

79. The line $y = 3$ is horizontal, so any perpendicular line will be vertical. Find the midpoint of the described segment:
$$x = \frac{x_1 + x_2}{2} = \frac{2+(-6)}{2} = \frac{-4}{2} = -2;\quad y = \frac{y_1+y_2}{2} = \frac{4+10}{2} = \frac{14}{2} = 7$$
The vertical line through the point $(-2, 7)$ is $x = -2$.

SECTION 8.3

81. The line $x = 3$ is vertical, so any parallel line will be vertical.
Find the midpoint of the described segment:

$$x = \frac{x_1 + x_2}{2} = \frac{2+8}{2} = \frac{10}{2} = 5; \; y = \frac{y_1 + y_2}{2} = \frac{-4+12}{2} = \frac{8}{2} = 4$$

The vertical line through the point $(5, 4)$ is $x = 5$.

83.
$$Ax + By = C$$
$$By = -Ax + C$$
$$\frac{By}{B} = \frac{-Ax + C}{B}$$
$$y = -\frac{A}{B}x + \frac{C}{B}; \; m = -\frac{A}{B}; \; b = \frac{C}{B}$$

85. Let x represent the number of years since the truck was purchased, and let y represent the value of the truck. Then we know two points on the depreciation line: $(0, 19984)$ and $(8, 1600)$. Find the slope and then use point-slope form to find the depreciation equation.

$$m = \frac{\Delta y}{\Delta x} = \frac{19984 - 1600}{0 - 8} = \frac{18384}{-8} = -2298$$
$$y - y_1 = m(x - x_1)$$
$$y - 19984 = -2298(x - 0)$$
$$y = -2298x + 19984$$

87. Let x represent the time in years after 1987, and let y represent the value of the painting. Then we know two points on the appreciation line: $(0, 36225000)$ and $(20, 72450000)$. Find the slope and then use point-slope form to find the appreciation equation.

$$m = \frac{\Delta y}{\Delta x} = \frac{72450000 - 36225000}{20 - 0} = \frac{36225000}{20} = 1811250$$
$$y - y_1 = m(x - x_1)$$
$$y - 36225000 = 1811250(x - 0)$$
$$y = 1{,}811{,}250x + 36{,}225{,}000$$

89. Let x represent the number of years since the painting was purchased, and let y represent the value of the painting. Then we know two points on the appreciation line: $(0, 250000)$ and $(5, 500000)$. Find the slope and then use point-slope form to find the appreciation equation.

$$m = \frac{\Delta y}{\Delta x} = \frac{500000 - 250000}{5 - 0} = \frac{250000}{5} = 50000$$
$$y - y_1 = m(x - x_1)$$
$$y - 250000 = 50000(x - 0)$$
$$y = 50{,}000x + 250{,}000$$

91. Let x represent the number of years since the TV was purchased, and let y represent the value of the TV. Then we know two points on the depreciation line: $(0, 1750)$ and $(3, 800)$. Find the slope and then use point-slope form to find the depreciation equation.

$$m = \frac{\Delta y}{\Delta x} = \frac{1750 - 800}{0 - 3} = \frac{950}{-3} = -\frac{950}{3}$$
$$y - y_1 = m(x - x_1)$$
$$y - 1750 = -\tfrac{950}{3}(x - 0)$$
$$y = -\tfrac{950}{3}x + 1750$$

93. Let x represent the number of years since the copy machine was purchased, and let y represent its value. Since the value decreases \$180 per year, the slope of the depreciation line is -180. Since the copier is worth \$1750 new, the y-intercept is 1750. Thus the depreciation equation can be found:

$$y = mx + b$$
$$y = -180x + 1750$$

Now let $x = 7$: $y = -180x + 1750 = -180(7) + 1750 = -1260 + 1750 = \$490$

95. Let x represent the number of years since the house was purchased, and let y represent its value. Since the value increases \$4000 per year, the slope of the depreciation line is 4000. We also know a point on the line: $(2, 122000)$. Use point-slope form to find the appreciation equation.

$$y - y_1 = m(x - x_1)$$
$$y - 122000 = 4000(x - 2)$$
$$y - 122000 = 4000x - 8000$$
$$y = 4000x + 114000$$

Substitute $x = 10$ and find its value:

$$y = 4000x + 114000 = 4000(10) + 114000 = 40000 + 114000 = \$154{,}000$$

97. The relationship between number of copies and the total cost is linear. Let x represent the number of copies (in hundreds), and let y represent the total cost. The slope of the line will be the charge per one hundred copies, or 15. We also know a point on the line: $(3, 75)$. Use point-slope form to find the equation of the line:

$$y - y_1 = m(x - x_1)$$
$$y - 75 = 15(x - 3)$$
$$y - 75 = 15x - 45$$
$$y = 15x + 30$$

Let $x = 10$ and find the cost:

$$y = 15x + 30 = 15(10) + 30 = 150 + 30 = \$180$$

99-105. Answers may vary.

107. To pass through II and IV, the slope must be negative. To pass through II and not III, the y-intercept must be positive. Thus $a < 0$ and $b > 0$.

Exercise 8.4 (page 574)

1. $y = 2x - 1$ **is a function**, since each value of x corresponds to exactly one value of y.

3. $y^2 = x$ **is not a function**, since $x = 4$ corresponds to both $y = 2$ and $y = -2$.

5. $f(1) = 2(1) + 1 = 2 + 1 = 3$

7.

$$\frac{y+2}{2} = 4(y+2)$$
$$2 \cdot \frac{y+2}{2} = 2 \cdot 4(y+2)$$
$$y + 2 = 8(y+2)$$
$$y + 2 = 8y + 16$$
$$-7y = 14$$
$$y = -2$$

9.

$$\frac{2a}{3} + \frac{1}{2} = \frac{6a-1}{6}$$
$$6\left(\frac{2a}{3} + \frac{1}{2}\right) = 6 \cdot \frac{6a-1}{6}$$
$$2(2a) + 3 = 6a - 1$$
$$4a + 3 = 6a - 1$$
$$-2a = -4$$
$$a = 2$$

SECTION 8.4

11. relation **13.** domain **15.** 0 **17.** cannot

19. slope; y-intercept **21.** y **23.** y

25. domain $= \{3, 5, -4, 0\}$; range $= \{-2, 0, -5\}$. Each element in the domain is paired with exactly one value in the range, so the relation is a function.

27. domain $= \{-2, 6, 5\}$; range $= \{3, 8, 5, 4\}$. The domain value of -2 is paired with range values of 3 and 5, so the relation is not a function.

29. The domain is the set of all x-coordinates on the graph. The domain is $(-\infty, 1]$.
The range is the set of all y-coordinates on the graph. The range is $(-\infty, \infty)$.
Since some vertical lines pass through the graph more than once, it is not the graph of a function.

31. The domain is the set of all x-coordinates on the graph. The domain is $(-\infty, \infty)$.
The range is the set of all y-coordinates on the graph. The range is $(-\infty, \infty)$.
Since each vertical line passes through the graph at most once, it is the graph of a function.

33. $f(x) = 3x$
$f(3) = 3(3) = 9$
$f(-1) = 3(-1) = -3$

$f(x) = 0$
$3x = 0$
$x = 0$

35. $f(x) = 2x - 3$
$f(3) = 2(3) - 3 = 3$
$f(-1) = 2(-1) - 3 = -5$

$f(x) = 0$
$2x - 3 = 0$
$2x = 3$
$x = \frac{3}{2}$

37. $f(x) = x^2$
$f(2) = 2^2 = 4$
$f(3) = 3^2 = 9$

39. $f(x) = x^3 - 1$
$f(2) = 2^3 - 1 = 8 - 1 = 7$
$f(3) = 3^3 - 1 = 27 - 1 = 26$

41. $f(x) = |x| + 2$
$f(2) = |2| + 2 = 2 + 2 = 4$
$f(-2) = |-2| + 2 = 2 + 2 = 4$

43. $f(x) = x^2 - 2$
$f(2) = 2^2 - 2 = 4 - 2 = 2$
$f(-2) = (-2)^2 - 2 = 4 - 2 = 2$

45. $g(x) = 2x$
$g(w) = 2w$
$g(w + 1) = 2(w + 1) = 2w + 2$

47. $g(x) = 3x - 5$
$g(w) = 3w - 5$
$g(w + 1) = 3(w + 1) - 5 = 3w + 3 - 5$
$= 3w - 2$

49. $f(3) + f(2) = 2(3) + 1 + 2(2) + 1$
$= 6 + 1 + 4 + 1 = 12$

51. $f(b) - f(a) = 2b + 1 - (2a + 1)$
$= 2b + 1 - 2a - 1 = 2b - 2a$

53. domain $= \{-2, 4, 6\}$

55. The denominator cannot equal 0, so $x \neq 4$.
domain $= (-\infty, 4) \cup (4, \infty)$

57. The denominator cannot equal 0, so $x \neq -3$.
domain $= (-\infty, -3) \cup (-3, \infty)$

59. All values of x are valid subsitutions.
domain $= (-\infty, \infty)$

SECTION 8.4

61. $f(x) = 2x - 1$
D $= (-\infty, \infty)$
R $= (-\infty, \infty)$

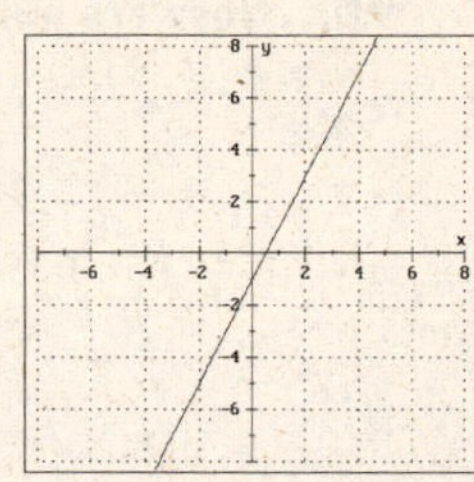

63. $2x - 3y = 6$
D $= (-\infty, \infty)$
R $= (-\infty, \infty)$

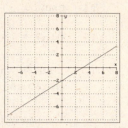

65. $y = 3x^2 + 2 \Rightarrow$ This is not a linear function because the exponent on x is 2.

67. $x = 3y - 4 \Rightarrow y = \frac{1}{3}x + \frac{4}{3} \Rightarrow$ This is a linear function $\left(m = \frac{1}{3}, b = \frac{4}{3}\right)$.

69. $f(x) = (x+1)^2$
$f(2) = (2+1)^2 = 3^2 = 9$
$f(3) = (3+1)^2 = 4^2 = 16$

71. $f(x) = 2x^2 - x$
$f(2) = 2(2)^2 - 2 = 2(4) - 2 = 6$
$f(3) = 2(3)^2 - 3 = 2(9) - 3 = 15$

73. $f(x) = 7 + 5x$
$f(3) = 7 + 5(3) = 22$
$f(-1) = 7 + 5(-1) = 2$

75. $f(x) = 9 - 2x$
$f(3) = 9 - 2(3) = 3$
$f(-1) = 9 - 2(-1) = 11$

77. $f(b) - 1 = 2b + 1 - 1 = 2b$

79. $f(0) + f(-\frac{1}{2}) = 2(0) + 1 + 2(-\frac{1}{2}) + 1 = 0 + 1 - 1 + 1 = 1$

81. Set $R(x)$ equal to $C(x)$ and solve for x.
$R(x) = C(x)$
$120x = 57.50x + 12000$
$62.50x = 12000$
$x = \frac{12000}{62.50}$
$x = 192$
The company must sell 192 DVD players.

83. **a.** $I(h) = 0.10h + 50$

b. $I(115) = 0.10(115) + 50$
$= 11.5 + 50 = \$61.50$

85. Let $t = 3$ and find s.
$s = f(t) = -16t^2 + 256t$
$= -16(3)^2 + 256(3)$
$= -16(9) + 768$
$= -144 + 768$
$= 624$
The bullet will have a height of 624 ft.

87. Let $t = 1.5$ and find h.
$h = -16t^2 + 32t$
$= -16(1.5)^2 + 32(1.5)$
$= -16(2.25) + 48$
$= -36 + 48 = 12$
The dolphin will jump 12 ft above.

89. Let $C = 25$ and find F.

$$\begin{aligned} F(C) &= \frac{9}{5}C + 32 \\ &= \frac{9}{5}(25) + 32 \\ &= 45 + 32 \\ &= 77 \end{aligned}$$

The temperature is $77°F$.

91. **Answers may vary.**

93. $f(x) + g(x) = (2x + 1) + x^2 = 2x + 1 + x^2$; $g(x) + f(x) = x^2 + (2x + 1) = x^2 + 2x + 1$. They are equal.

Exercise 8.5 (page 588)

1. **Answers may vary.**

3. the graph of $f(x) = x^3$ shifted down 4

5. 41, 43, 47

7. $a \cdot b = b \cdot a$

9. 1

11. squaring

13. absolute value

15. horizontal

17. 2; down

19. 4; to the left

21. rational

23. $f(x) = x^2 - 3$

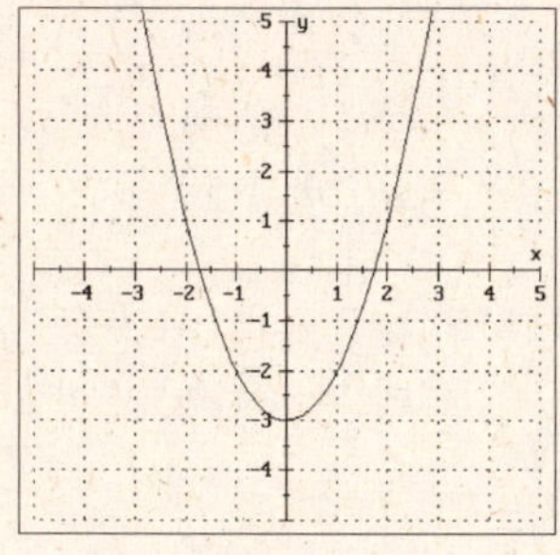

25. $f(x) = (x - 1)^3$

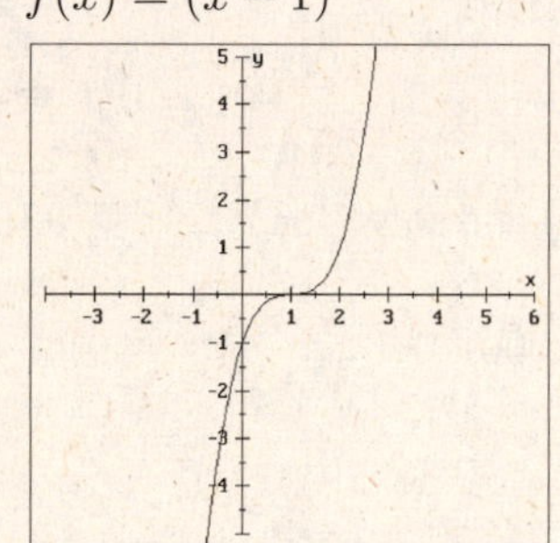

27. $f(x) = |x| - 2$

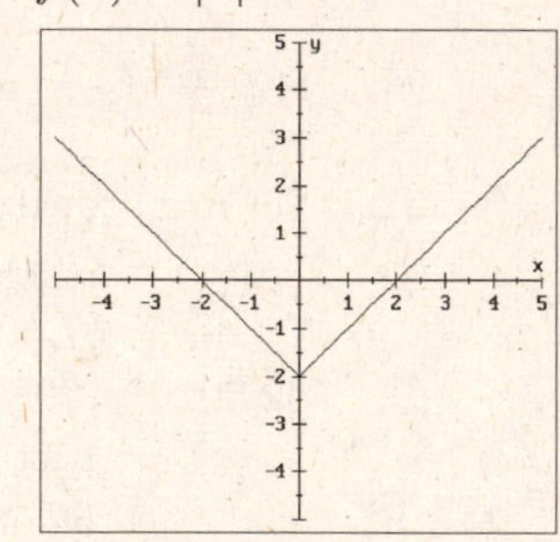

29. $f(x) = |x - 1|$

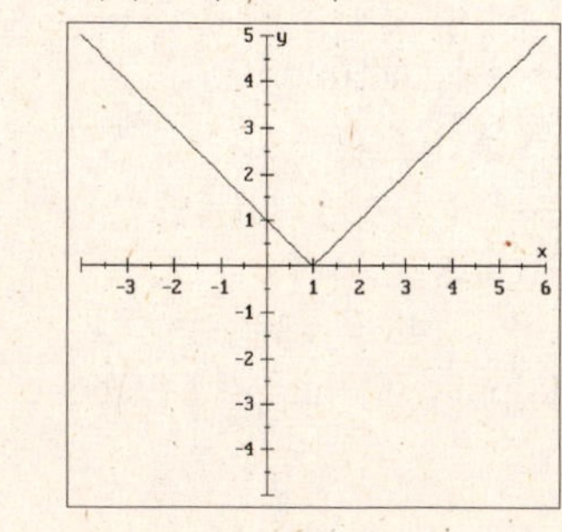

31. $f(x) = x^2 - 5$
Shift $f(x) = x^2$ D 5.

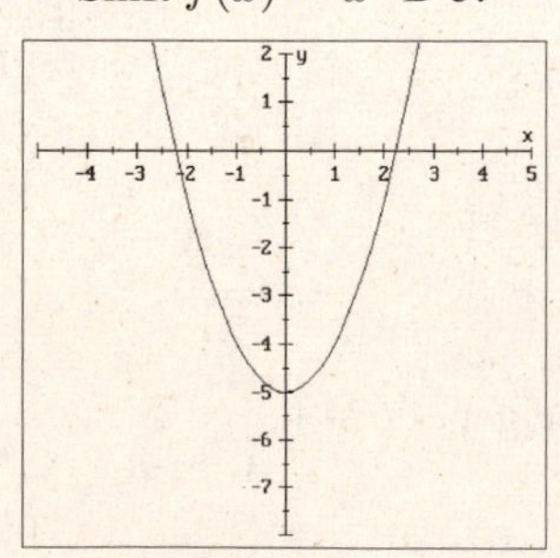

33. $f(x) = (x - 1)^3$

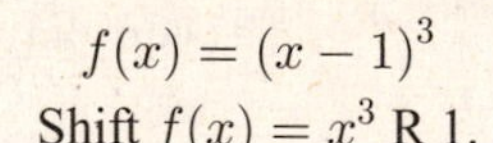
Shift $f(x) = x^3$ R 1.

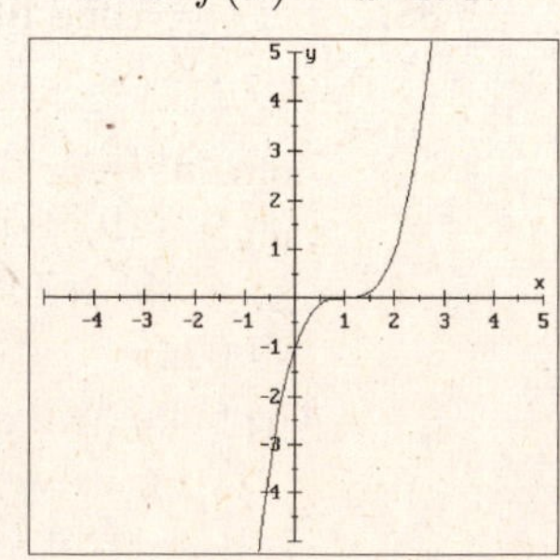

SECTION 8.5

35. $f(x) = |x - 2| - 1$
Shift $f(x) = |x|$ D 1, R 2.

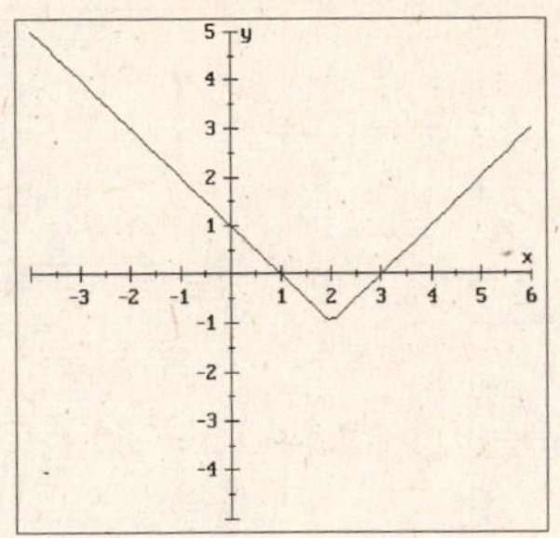

37. $f(x) = (x + 1)^3 - 2$
Shift $f(x) = x^3$ D 2, L 1.

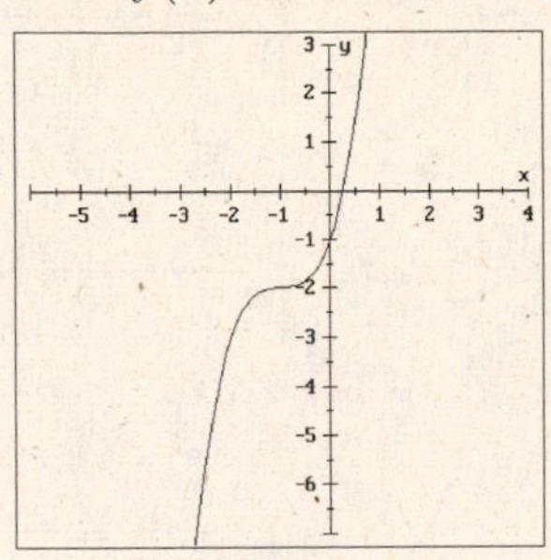

39. $f(x) = -x^2$
Reflect $f(x) = x^2$ about x-axis.

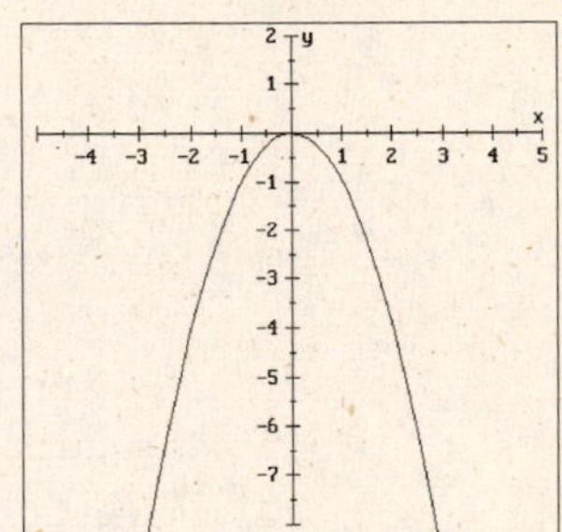

41. $f(x) = -(x - 2)^2 - 3$
Reflect $f(x) = x^2$ about x-axis, shift D 3, R 2.

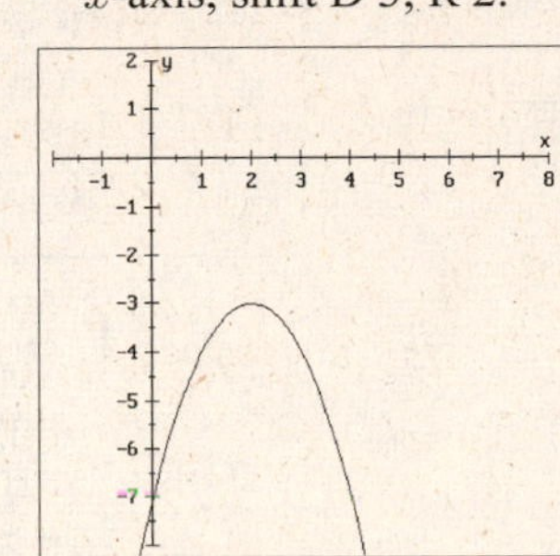

43. $t = f(r) = \dfrac{600}{r}$
$f(30) = \frac{600}{30}$
$= 20$ hrs

45. $t = f(r) = \dfrac{600}{r}$
$f(50) = \frac{600}{50}$
$= 12$ hrs

47. $f(x) = \frac{x}{x-2}$

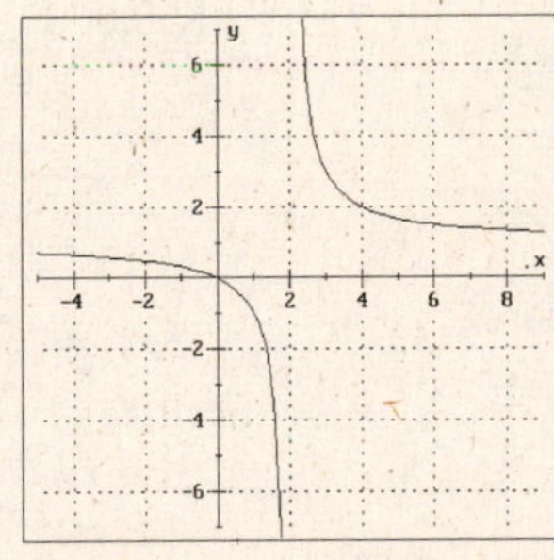

domain: $(-\infty, 2) \cup (2, \infty)$
range: $(-\infty, 1) \cup (1, \infty)$

49. $f(x) = \frac{x+1}{x^2-4}$

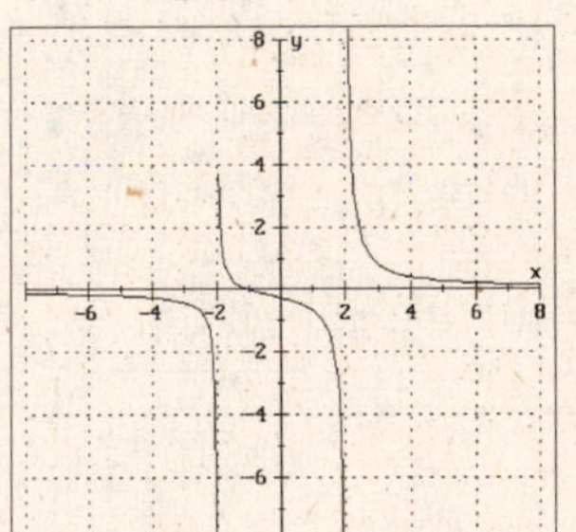

domain: $(-\infty, -2) \cup (-2, 2) \cup (2, \infty)$
range: $(-\infty, \infty)$

SECTION 8.5

51. $f(x) = x^2 + 8$
x: $[-4, 4]$, y: $[-4, 4]$

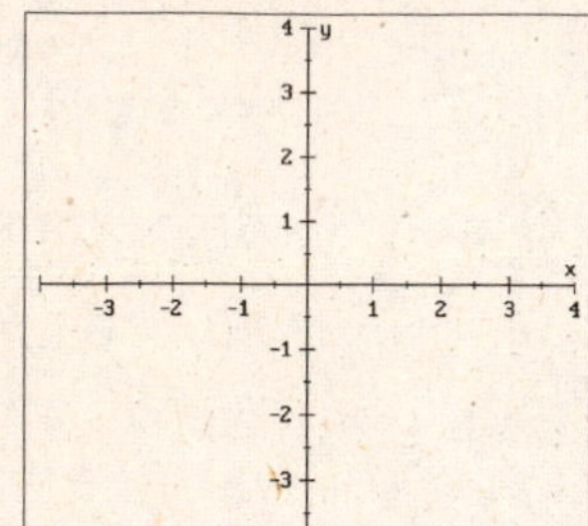

x: $[-7, 7]$, y: $[-2, 12]$

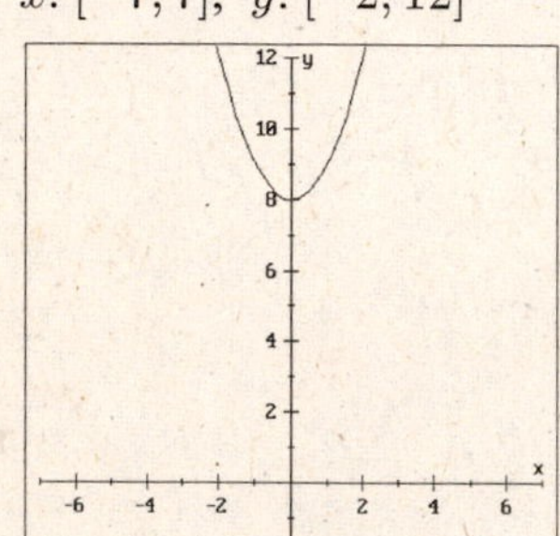

53. $f(x) = |x + 5|$
x: $[-4, 4]$, y: $[-4, 4]$

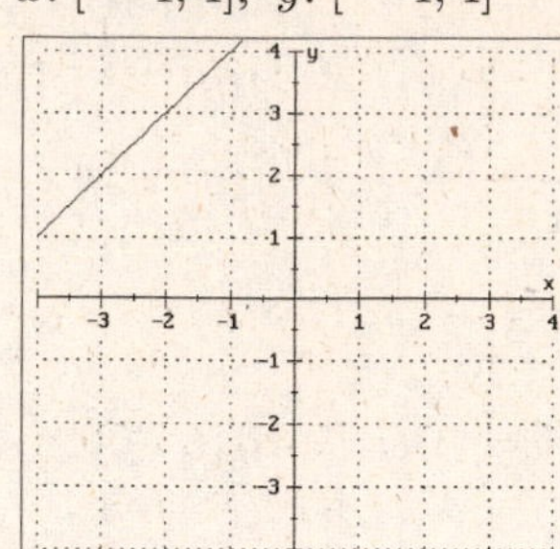

x: $[-12, 2]$, y: $[-4, 10]$

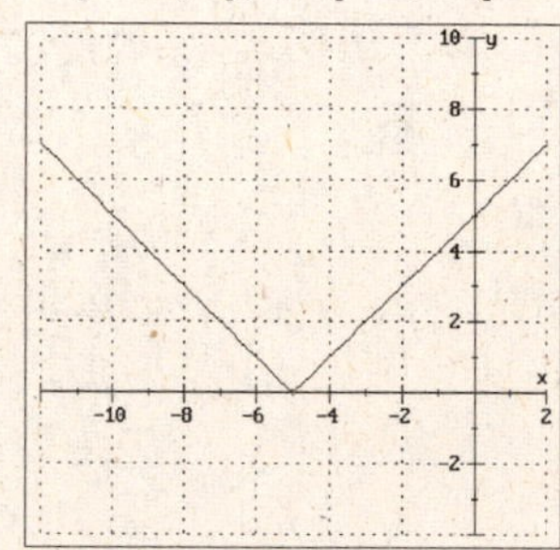

55. $f(x) = (x - 6)^2$
x: $[-4, 4]$, y: $[-4, 4]$

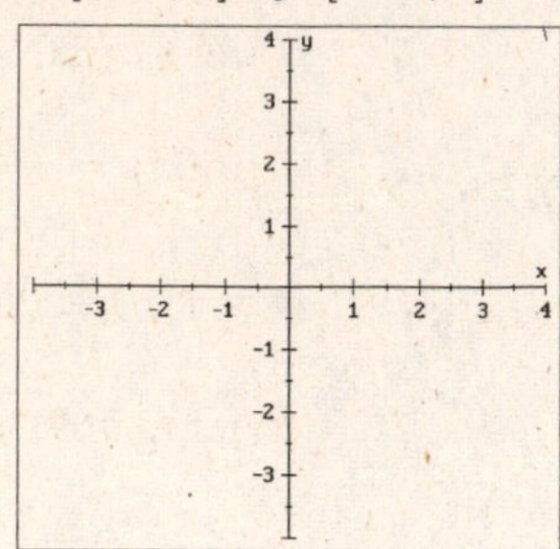

x: $[-2, 12]$, y: $[-2, 12]$

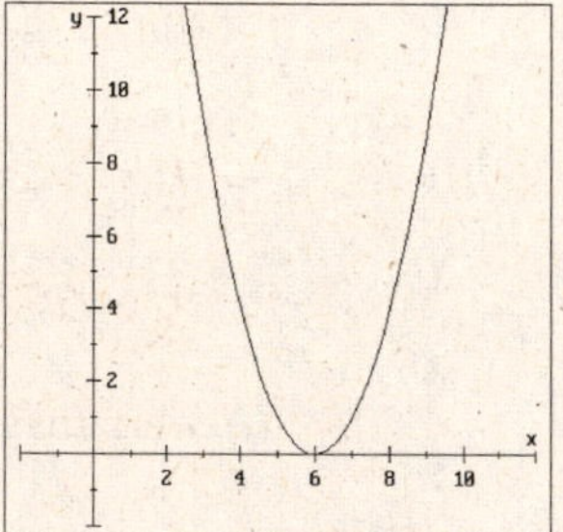

57. $f(x) = x^3 + 8$
x: $[-4, 4]$, y: $[-4, 4]$

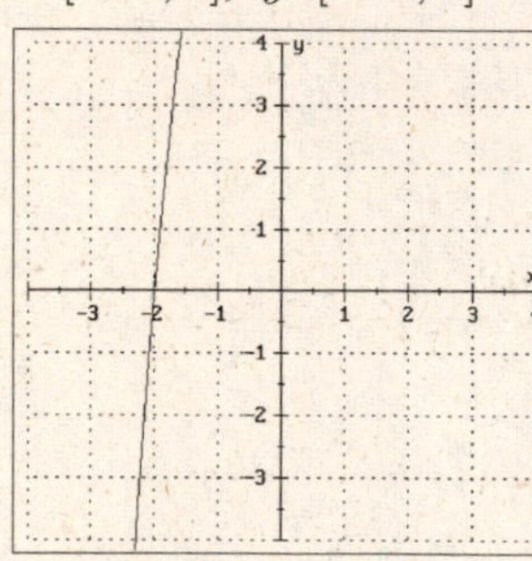

x: $[-10, 10]$, y: $[-4, 16]$

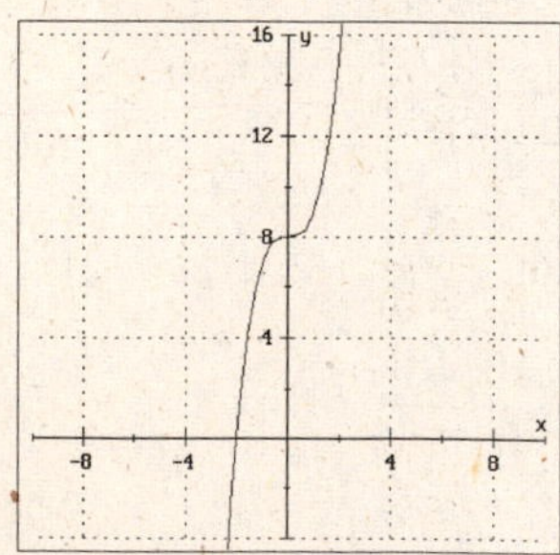

59. $C = f(p) = \dfrac{50{,}000p}{100 - p}$

$$f(10) = \frac{50{,}000(10)}{100 - 10} = \frac{500{,}000}{90} = 5555.56$$

$C = \$5555.56$

61. $C = f(p) = \dfrac{50{,}000p}{100 - p}$

$$f(50) = \frac{50{,}000(50)}{100 - 50} = \frac{2{,}500{,}000}{50} = 50{,}000.00$$

$C = \$50{,}000$

63. $c = f(x) = 1.25x + 700$

65. Refer to **#63**. $c = f(500) = 1.25(500) + 700 = \$1325$

67. Refer to **#63**. The mean cost $= \overline{c} = f(x) = \dfrac{1.25x + 700}{x}$.

$$f(1000) = \frac{1.25(1000) + 700}{1000} = \frac{1950}{1000} = \$1.95$$

69. $c = f(n) = 0.09n + 7.50$

71. Refer to **#69**. $c = f(775) = 0.09(775) + 7.50 = 69.75 + 7.50 = \$77.25$

73. Refer to **#69**. The mean cost $= \overline{c} = f(x) = \dfrac{0.09n + 7.50}{n}$.

$$f(1000) = \frac{0.09(1000) + 7.50}{1000} = \frac{97.50}{1000} = \$0.0975 = 9.75¢$$

75. $c = f(x) = 350x + 5000$

77. Refer to **#75**. $c = f(120) = 350(120) + 5000 = \$47,000$

79. **Answers may vary.**

81. No.

Exercise 8.6 (page 597)

1.
$$\frac{x}{2} = \frac{3}{6}$$
$$6x = 6$$
$$x = 1$$

3. $a = kb$

5. $a = kbc$

7. $(x^2x^3)^2 = (x^2)^2(x^3)^2 = x^4x^6 = x^{10}$

9. $\dfrac{b^0 - 2b^0}{b^0} = \dfrac{1-2}{1} = \dfrac{-1}{1} = -1$

11. $35,000 = 3.5 \times 10^4$

13. $2.5 \times 10^{-3} = 0.0025$

15. proportion

17. direct

19. rational

21. joint

23. direct

25. neither

27.
$$\frac{x}{5} = \frac{15}{25}$$
$$25x = 75$$
$$x = 3$$

29.
$$\frac{r-2}{3} = \frac{r}{5}$$
$$5(r-2) = 3r$$
$$5r - 10 = 3r$$
$$2r = 10$$
$$r = 5$$

SECTION 8.6

31. $$\frac{3}{n} = \frac{2}{n+1}$$
$$3(n+1) = 2n$$
$$3n + 3 = 2n$$
$$n = -3$$

33. $$\frac{5}{5z+3} = \frac{2z}{2z^2+6}$$
$$5(2z^2+6) = 2z(5z+3)$$
$$10z^2 + 30 = 10z^2 + 6z$$
$$30 = 6z$$
$$5 = z$$

35. $A = kp^2$

37. $v = \dfrac{k}{r^3}$

39. $B = kmn$

41. $P = \dfrac{ka^2}{j^3}$

43. L varies jointly with m and n.

45. E varies jointly with a and the square of b.

47. X varies directly with x^2 and inversely with y^2.

49. R varies directly with L and inversely with d^2.

51. $$\frac{2}{c} = \frac{c-3}{2}$$
$$4 = c^2 - 3c$$
$$0 = c^2 - 3c - 4$$
$$0 = (c+1)(c-4)$$
$c + 1 = 0$ **or** $c - 4 = 0$
$c = -1$ $\quad c = 4$

53. $$\frac{2}{3x} = \frac{6x}{36}$$
$$72 = 18x^2$$
$$0 = 18x^2 - 72$$
$$0 = 18(x+2)(x-2)$$
$x + 2 = 0$ **or** $x - 2 = 0$
$x = -2$ $\quad x = 2$

55. $$\frac{2(x+3)}{3} = \frac{4(x-4)}{5}$$
$$10(x+3) = 12(x-4)$$
$$10x + 30 = 12x - 48$$
$$-2x = -78$$
$$x = 39$$

57. $$\frac{1}{x+3} = \frac{-2x}{x+5}$$
$$x + 5 = -2x^2 - 6x$$
$$2x^2 + 7x + 5 = 0$$
$$(2x+5)(x+1)$$
$2x + 5 = 0$ **or** $x + 1 = 0$
$x = -\frac{5}{2}$ $\quad x = -1$

59. $$A = kr^2$$
$$A = \pi r^2$$
$$A = \pi(6 \text{ in.})^2$$
$$\boxed{A = 36\pi \text{ in.}^2}$$

61. $$d = kg$$
$$288 = k(12)$$
$$24 = k$$
$$d = 24g$$
$$d = 24(18)$$
$$\boxed{d = 432 \text{ mi}}$$

63. $$t = \frac{k}{n}$$
$$10 = \frac{k}{25}$$
$$250 = k$$
$$t = \frac{250}{n}$$
$$t = \frac{250}{10}$$
$$\boxed{t = 25 \text{ days}}$$

65. $$V = \frac{k}{P}$$
$$20 = \frac{k}{6}$$
$$120 = k$$
$$V = \frac{120}{P}$$
$$V = \frac{120}{10}$$
$$\boxed{V = 12 \text{ in.}^3}$$

67. $f = \frac{k}{l}$ $256 = \frac{k}{2}$ $512 = k$

$f = \frac{512}{l}$ $f = \frac{512}{6}$ $\boxed{f = 85\frac{1}{3}}$

69. $V_1 = klwh$
$V_2 = k(2l)(3w)(2h) = 12klwh = 12V_1$
The volume is multiplied by 12.

71. $g = khr^2$
$g = 23.5hr^2$
$g = 23.5(20)(7.5)^2$
$\boxed{g = 26{,}437.5 \text{ gallons}}$

73. $V = kC$
$6 = k(2)$
$3 = k$
The resistance is 3 ohms.

75. $D = \frac{k}{wd^3}$ $1.1 = \frac{k}{4(4)^3}$ $1.1 = \frac{k}{256}$ $281.6 = k$

$D = \frac{281.6}{wd^3}$ $D = \frac{281.6}{2(8)^3}$ $D = \frac{281.6}{1024}$ $\boxed{D = 0.275 \text{ in.}}$

77. $P = \frac{kT}{V}$ $1 = \frac{k(273)}{1}$ $\frac{1}{273} = k$

$P = \frac{\frac{1}{273}T}{V}$ $1 = \frac{\frac{1}{273}T}{2}$ $2 = \frac{1}{273}T$ $\boxed{546 \text{ K} = T}$

79. Answers may vary.

81. Answers may vary.

83. This is not direct variation. For this to be direct variation, one temperature would have to be a constant multiple of the other.

85. No. Answers may vary.

Chapter 8 Review (page 602)

1. $x + y = 4$

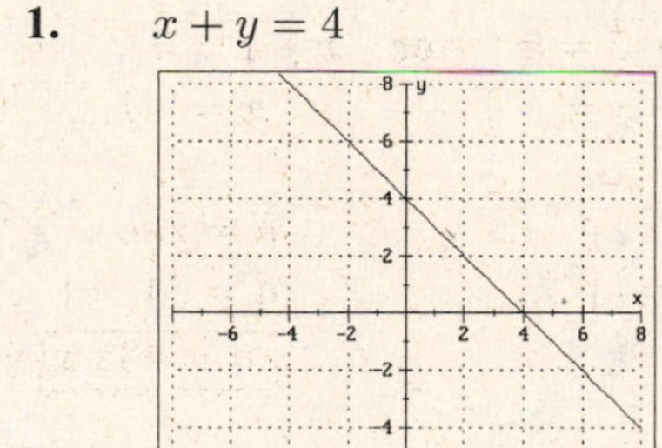

2. $2x - y = 8$

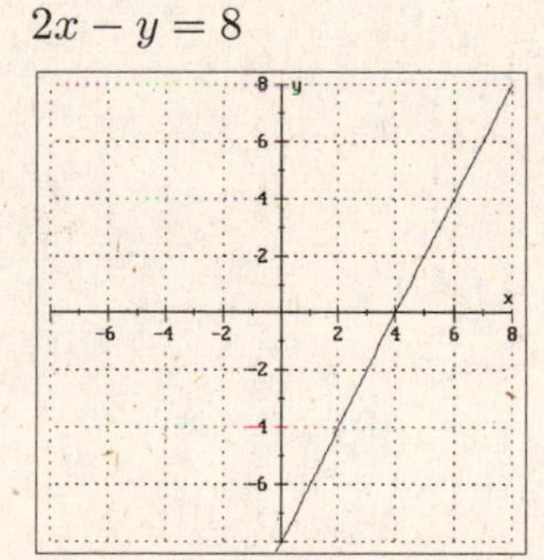

3. $y = 3x + 4$

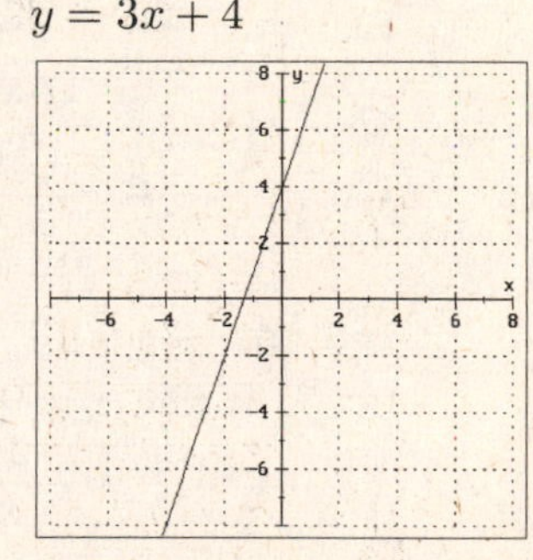

CHAPTER 8 REVIEW

4. $x = 4 - 2y$

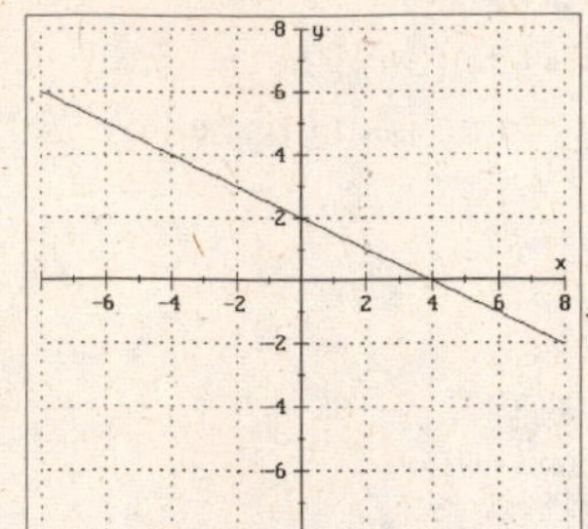

5. $y = 4$

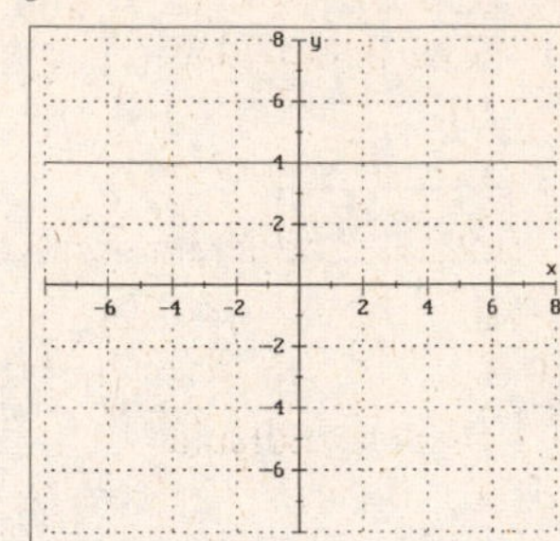

6. $x = -2$

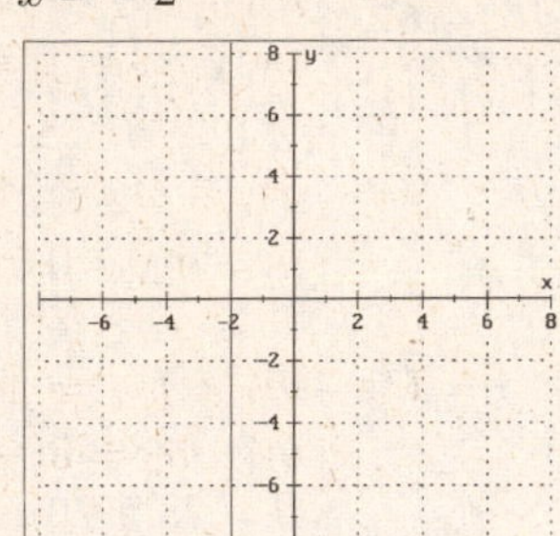

7. $2(x + 3) = x + 2$

$x = -4$

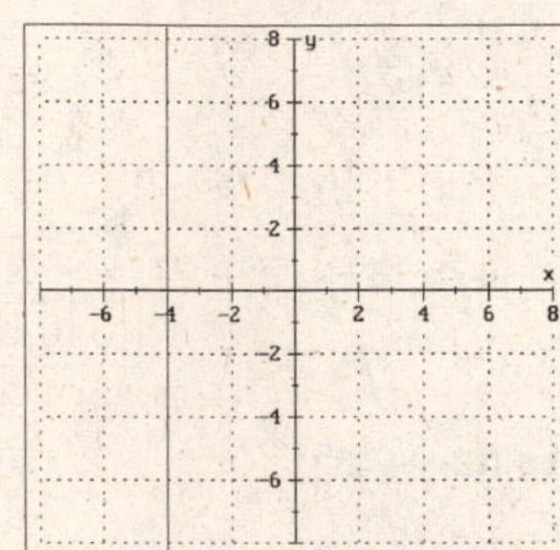

8. $3y = 2(y - 1)$

$y = -2$

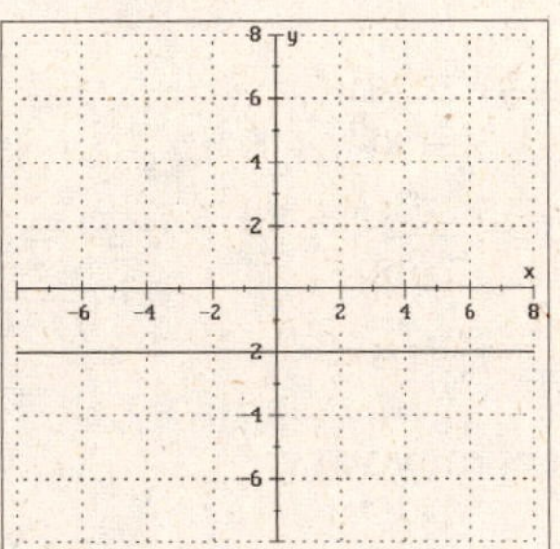

9. $x = \frac{x_1+x_2}{2} = \frac{-3+6}{2} = \frac{3}{2}$; $y = \frac{y_1+y_2}{2} = \frac{5+11}{2} = \frac{16}{2} = 8$; midpoint: $\left(\frac{3}{2}, 8\right)$

10. $m = \dfrac{\Delta y}{\Delta x} = \dfrac{5-8}{2-5} = \dfrac{-3}{-3} = 1$

11. $m = \dfrac{\Delta y}{\Delta x} = \dfrac{-2-12}{-3-6} = \dfrac{-14}{-9} = \dfrac{14}{9}$

12. $m = \dfrac{\Delta y}{\Delta x} = \dfrac{4-(-6)}{-3-(-5)} = \dfrac{10}{2} = 5$

13. $m = \dfrac{\Delta y}{\Delta x} = \dfrac{-4-(-9)}{5-(-6)} = \dfrac{5}{11}$

14. $m = \dfrac{\Delta y}{\Delta x} = \dfrac{4-4}{8-(-2)} = \dfrac{0}{10} = 0$

15. $m = \dfrac{\Delta y}{\Delta x} = \dfrac{8-(-4)}{-5-(-5)} = \dfrac{12}{0} \Rightarrow$ not defined

16.
$$2x - 3y = 18$$
$$-3y = -2x + 18$$
$$\frac{-3y}{-3} = \frac{-2x}{-3} + \frac{18}{-3}$$
$$y = \tfrac{2}{3}x - 6 \Rightarrow m = \tfrac{2}{3}$$

17.
$$2x + y = 8$$
$$y = -2x + 8 \Rightarrow m = -2$$

18.
$$-2(x - 3) = 10$$
$$-2x + 6 = 10$$
$$-2x = 4$$
$$x = -2$$
The slope is not defined (vertical).

19.
$$3y + 1 = 7$$
$$3y = 6$$
$$y = 2$$
$m = 0$ (horizontal)

20. $m_1 \neq m_2 \Rightarrow$ not parallel
$m_1 \cdot m_2 = 4\left(-\frac{1}{4}\right) = -1 \Rightarrow$ perpendicular

21. $m_1 = m_2 \Rightarrow$ parallel

22. $m_1 \neq m_2 \Rightarrow$ not parallel
$m_1 \cdot m_2 = 0.5\left(-\frac{1}{2}\right) = -0.25 \Rightarrow$ not perpendicular

23. $m_1 \neq m_2 \Rightarrow$ not parallel
$m_1 \cdot m_2 = 5(-0.2) = -1 \Rightarrow$ perpendicular

24. Let $y =$ sales and let $x =$ year of business.
$$m = \frac{\Delta y}{\Delta x} = \frac{y_2 - y_1}{x_1 - x_1} = \frac{130{,}000 - 65{,}000}{4 - 1} = \frac{65{,}000}{3} \approx \$21{,}666.67 \text{ per year}$$

25.
$$\begin{aligned} y - y_1 &= m(x - x_1) \\ y - 5 &= 3(x + 8) \\ y - 5 &= 3x + 24 \\ -3x + y &= 29 \\ 3x - y &= -29 \end{aligned}$$

26. $m = \frac{\Delta y}{\Delta x} = \frac{4 - (-9)}{-2 - 6} = \frac{13}{-8} = -\frac{13}{8}$
$$\begin{aligned} y - y_1 &= m(x - x_1) \\ y - 4 &= -\frac{13}{8}(x + 2) \\ 8(y - 4) &= 8\left(-\frac{13}{8}\right)(x + 2) \\ 8y - 32 &= -13(x + 2) \\ 8y - 32 &= -13x - 26 \\ 13x + 8y &= 6 \end{aligned}$$

27. Find the slope of the given line:
$$\begin{aligned} 3x - 2y &= 7 \\ -2y &= -3x + 7 \\ y &= \tfrac{3}{2}x - \tfrac{7}{2} \Rightarrow m = \tfrac{3}{2} \end{aligned}$$
Use the parallel slope:
$$\begin{aligned} y - y_1 &= m(x - x_1) \\ y + 5 &= \tfrac{3}{2}(x + 3) \\ 2(y + 5) &= 2 \cdot \tfrac{3}{2}(x + 3) \\ 2y + 10 &= 3(x + 3) \\ 2y + 10 &= 3x + 9 \\ -3x + 2y &= -1 \Rightarrow 3x - 2y = 1 \end{aligned}$$

28. Find the slope of the given line:
$$\begin{aligned} 3x - 2y &= 7 \\ -2y &= -3x + 7 \\ y &= \tfrac{3}{2}x - \tfrac{7}{2} \Rightarrow m = \tfrac{3}{2} \end{aligned}$$
Use the perpendicular slope:
$$\begin{aligned} y - y_1 &= m(x - x_1) \\ y + 5 &= -\tfrac{2}{3}(x + 3) \\ 3(y + 5) &= 3\left(-\tfrac{2}{3}\right)(x + 3) \\ 3y + 15 &= -2(x + 3) \\ 3y + 15 &= -2x - 6 \\ 2x + 3y &= -21 \end{aligned}$$

29. $y = \frac{2}{3}x + 1;\ m = \frac{2}{3};\ b = 1$

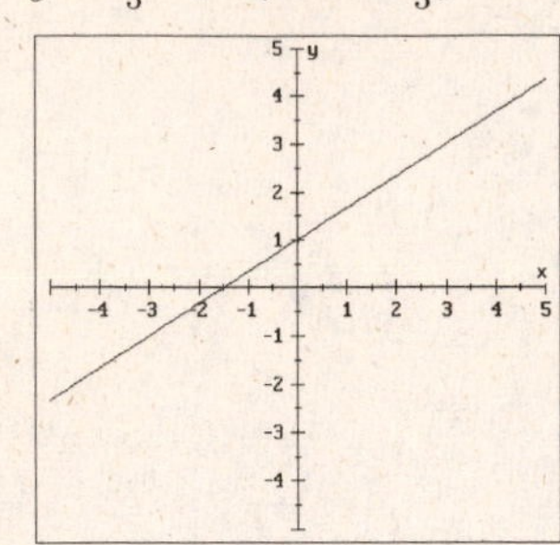

30. $2x + 3y = 8 \Rightarrow m = -\frac{A}{B} = -\frac{2}{3}$

$3x - 2y = 10 \Rightarrow m = -\frac{A}{B} = -\frac{3}{-2} = \frac{3}{2}$

perpendicular

31. Let x represent the number of years since the copy machine was purchased, and let y represent its value. We know two points on the line: $(0, 8700)$ and $(5, 100)$.

Find the slope:

$$m = \frac{\Delta y}{\Delta x} = \frac{8700 - 100}{0 - 5} = \frac{8600}{-5} = -1720$$

Find the equation of the line:

$$y - y_1 = m(x - x_1)$$
$$y - 8700 = -1720(x - 0)$$
$$y = -1720x + 8700$$

32. $y = 6x - 4$ **is a function**, since each value of x corresponds to exactly one value of y.

33. $y = 4 - x$ **is a function**, since each value of x corresponds to exactly one value of y.

34. $y^2 = x$ **is not a function**, since $x = 9$ corresponds to both $y = 3$ and $y = -3$.

35. $|y| = x^2$ **is not a function**, since $x = 2$ corresponds to both $y = 4$ and $y = -4$.

36. $f(x) = 3x + 2$
$f(-3) = 3(-3) + 2 = -9 + 2 = -7$

37. $g(x) = x^2 - 4$
$g(8) = 8^2 - 4 = 64 - 4 = 60$

38. $g(x) = x^2 - 4$
$g(-2) = (-2)^2 - 4 = 4 - 4 = 0$

39. $f(x) = 3x + 2$
$f(5) = 3(5) + 2 = 15 + 2 = 17$

40. $f(x) = 4x - 1$

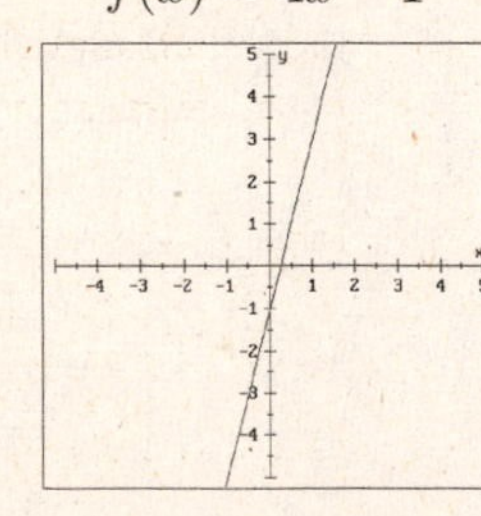

$\text{D} = (-\infty, \infty)$;
$\text{R} = (-\infty, \infty)$

41. $f(x) = 3x - 10$

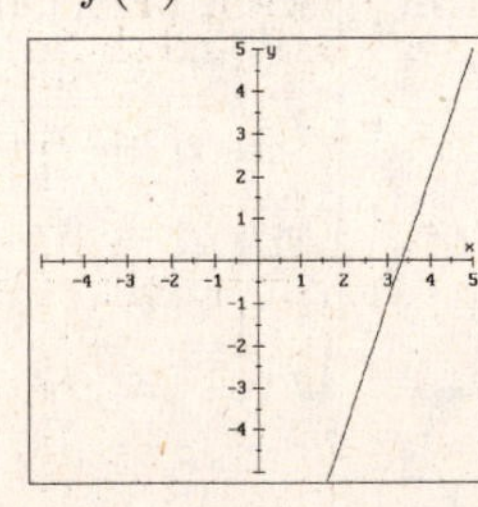

$\text{D} = (-\infty, \infty)$;
$\text{R} = (-\infty, \infty)$

42. $f(x) = x^2 + 1$

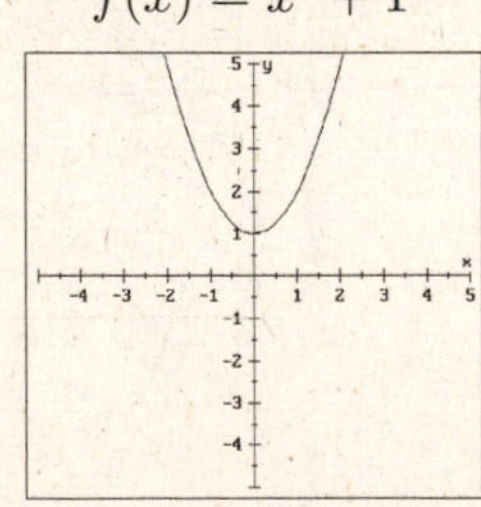

$\text{D} = (-\infty, \infty)$;
$\text{R} = [1, \infty)$

43. 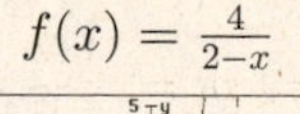

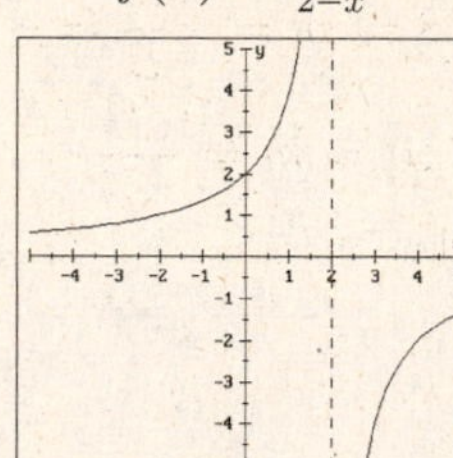

D $= (-\infty, 2) \cup (2, \infty)$; R $= (-\infty, 0) \cup (0, \infty)$

44. $f(x) = \frac{7}{x-3}$

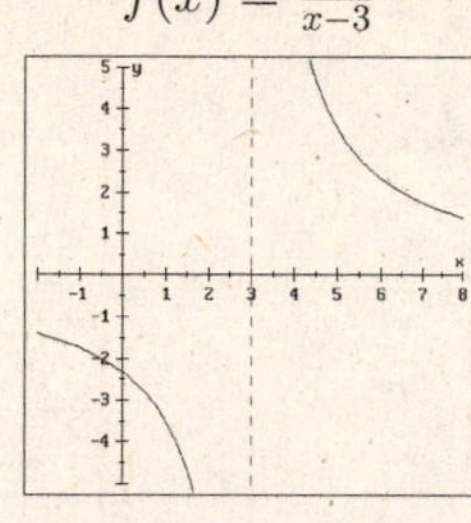

D $= (-\infty, 3) \cup (3, \infty)$; R $= (-\infty, 0) \cup (0, \infty)$

45. $f(x) = 7$

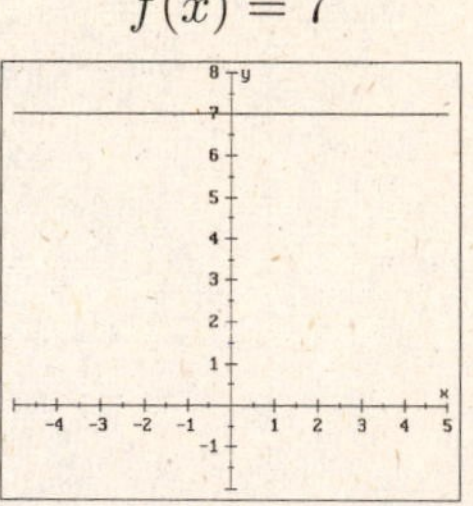

D $= (-\infty, \infty)$; R $= \{7\}$

46. Since each vertical line passes through the graph at most once, it is a function.

47. Since a vertical line can pass through the graph more than once, it is not a function.

48. Since a vertical line can pass through the graph more than once, it is not a function.

49. Since each vertical line passes through the graph at most once, it is a function.

50. $f(x) = x^2 - 3$: Shift $f(x) = x^2$ D 3.

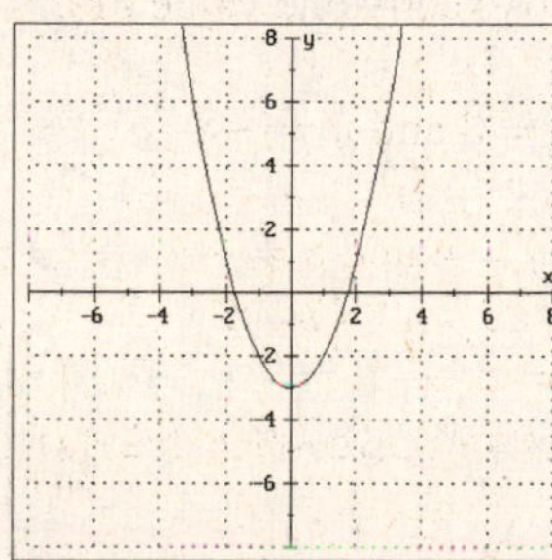

51. $f(x) = |x| - 4$: Shift $f(x) = |x|$ D 4.

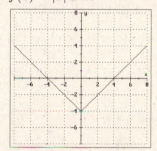

52. $f(x) = (x-2)^3$: Shift $f(x) = x^3$ R 2.

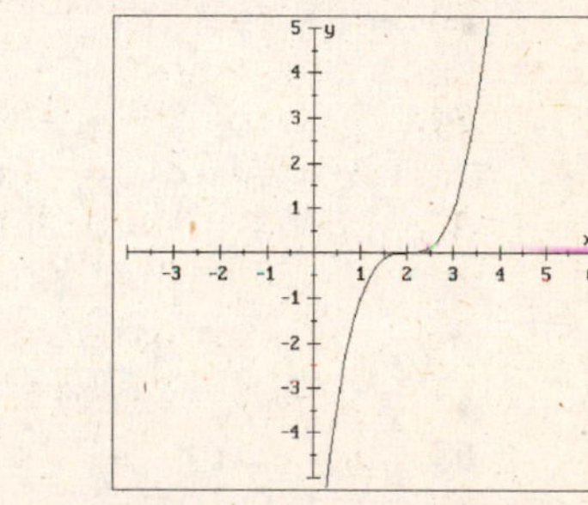

53. $f(x) = (x+4)^2 - 3$: Shift $f(x) = x^2$ D 3 and L 4.

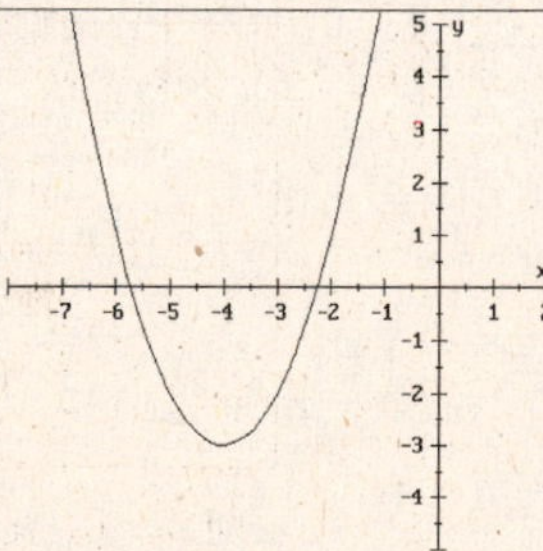

54. $f(x) = -x^3 - 2$: Reflect $f(x) = x^3$ about the x-axis and shift D 2.

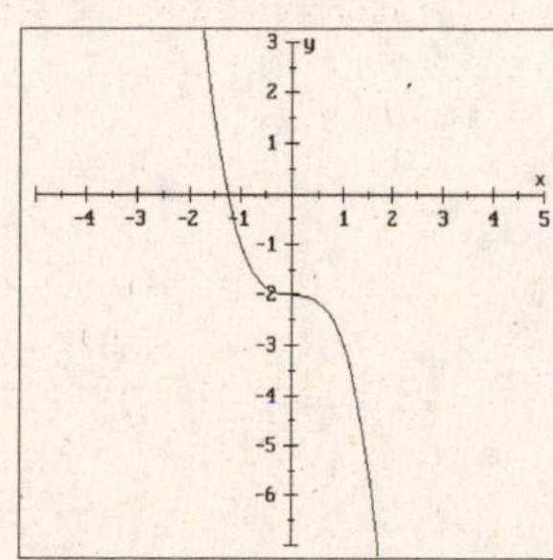

55. $f(x) = -|x-1| + 2$ Reflect $f(x) = |x|$ about the x-axis and shift U 2.

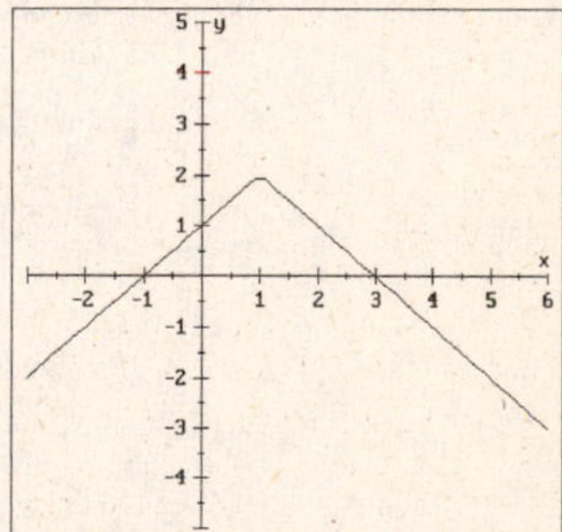

56-61. Compare your graphs to the graphs in numbers **50-55.**

62. $f(x) = \frac{2}{x-2}$

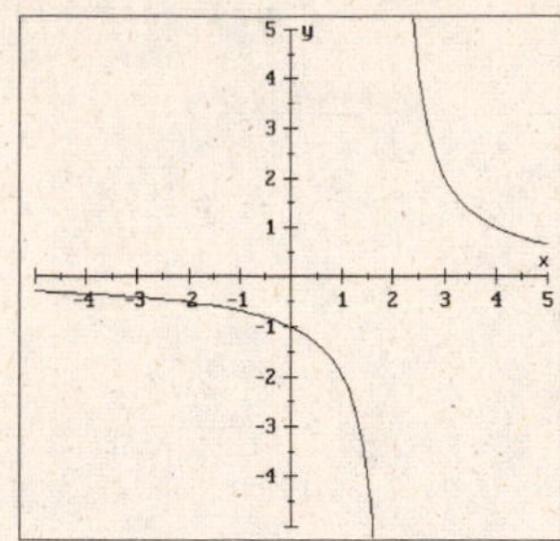

domain $= (-\infty, 2) \cup (2, \infty)$
range $= (-\infty, 0) \cup (0, \infty)$

63. $f(x) = \frac{x}{x+3}$

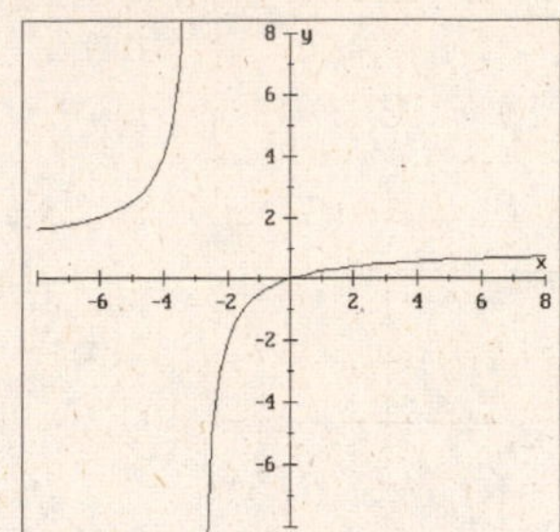

domain $= (-\infty, -3) \cup (-3, \infty)$
range $= (-\infty, 1) \cup (1, \infty)$

64.
$$\begin{aligned} \frac{x+1}{8} &= \frac{4x-2}{23} \\ 23(x+1) &= 8(4x-2) \\ 23x+23 &= 32x-16 \\ -9x &= -39 \\ x &= \tfrac{-39}{-9} = \tfrac{13}{3} \end{aligned}$$

65.
$$\begin{aligned} \frac{1}{x+6} &= \frac{x+10}{12} \\ 12 &= (x+6)(x+10) \\ 12 &= x^2+16x+60 \\ 0 &= x^2+16x+48 \\ 0 &= (x+12)(x+4) \end{aligned}$$
$x+12=0$ **or** $x+4=0$
$x=-12$ $\quad$ $x=-4$

66.
$y = kx$ $\quad$ $y = \frac{1}{6}x$
$2 = k(12)$ $\quad$ $y = \frac{1}{6}(12)$
$\frac{1}{6} = k$ $\quad$ $\boxed{y = 2}$

67.
$y = \dfrac{k}{x}$ $\quad$ $y = \dfrac{72}{x}$
$3 = \dfrac{k}{24}$ $\quad$ $y = \dfrac{72}{12}$
$72 = k$ $\quad$ $\boxed{y = 6}$

68.
$$\begin{aligned} y &= kxz \\ 3 &= k(24)(4) \\ 3 &= 96k \end{aligned}$$
$\boxed{\frac{3}{96} = \frac{1}{32} = k}$

69.
$$\begin{aligned} y &= \frac{kt}{x} \\ 64 &= \frac{k(8)}{2} \\ 128 &= 8k \end{aligned}$$
$\boxed{16 = k}$

Chapter 8 Test (page 608)

1. $2x - 5y = 10$

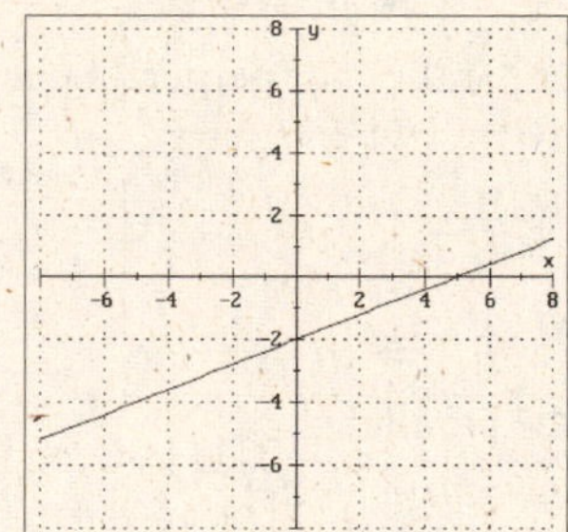

2. $x = \frac{x_1+x_2}{2} = \frac{-3+4}{2} = \frac{1}{2}$

$y = \frac{y_1+y_2}{2} = \frac{3+(-2)}{2} = \frac{1}{2}$

midpoint: $\left(\frac{1}{2}, \frac{1}{2}\right)$

3.

$$y = \frac{x-3}{5}$$
$$0 = \frac{x-3}{5}$$
$$5(0) = 5 \cdot \frac{x-3}{5}$$
$$0 = x - 3$$
$$3 = x$$

x-intercept: $(3, 0)$

$$y = \frac{x-3}{5}$$
$$y = \frac{0-3}{5}$$
$$y = -\frac{3}{5}$$

y-intercept: $\left(0, -\frac{3}{5}\right)$

4. $x - 7 = 0 \Rightarrow x = 7 \Rightarrow$ vertical line

5. $m = \frac{\Delta y}{\Delta x} = \frac{4-8}{-2-6} = \frac{-4}{-8} = \frac{1}{2}$

6.

$$2x - 3y = 8$$
$$-3y = -2x + 8$$
$$y = \frac{2}{3}x - \frac{8}{3} \Rightarrow m = \frac{2}{3}$$

7. The graph of $x = 12$ is a vertical line, so the slope is not defined.

8. The graph of $y = 12$ is a horizontal line, so the slope is 0.

9.

$$y - y_1 = m(x - x_1)$$
$$y + 5 = \frac{2}{3}(x - 4)$$
$$y + 5 = \frac{2}{3}x - \frac{8}{3}$$
$$y = \frac{2}{3}x - \frac{8}{3} - 5$$
$$y = \frac{2}{3}x - \frac{23}{3}$$

10. $m = \frac{\Delta y}{\Delta x} = \frac{6-(-10)}{-2-(-4)} = \frac{16}{2} = 8$

$$y - y_1 = m(x - x_1)$$
$$y - 6 = 8(x + 2)$$
$$y - 6 = 8x + 16$$
$$-8x + y = 22$$
$$8x - y = -22$$

11. $-2(x-3) = 3(2y+5)$
$-2x+6 = 6y+15$
$6y+15 = -2x+6$
$6y = -2x-9$
$y = \frac{-2}{6}x - \frac{9}{6}$
$y = -\frac{1}{3}x - \frac{3}{2}$
$m = -\frac{1}{3}, b = -\frac{3}{2} \Rightarrow \left(0, -\frac{3}{2}\right)$

12. $4x - y = 12$
$-y = -4x + 12$
$y = 4x - 12 \Rightarrow m = 4$
$y = \frac{1}{4}x + 3 \Rightarrow m = \frac{1}{4}$
neither parallel nor perpendicular

13. $y = -\frac{2}{3}x + 4 \Rightarrow m = -\frac{2}{3}$
$2y = 3x - 3$
$y = \frac{3}{2}x - \frac{3}{2} \Rightarrow m = \frac{3}{2}$
perpendicular

14. $y = \frac{3}{2}x - 7 \Rightarrow m = \frac{3}{2}$
Use the parallel slope:
$y - y_1 = m(x - x_1)$
$y - 0 = \frac{3}{2}(x - 0)$
$y = \frac{3}{2}x$

15. $y = -\frac{2}{3}x - 7 \Rightarrow m = -\frac{2}{3}$
Use the perpendicular slope:
$y - y_1 = m(x - x_1)$
$y - 6 = \frac{3}{2}(x + 3)$
$y - 6 = \frac{3}{2}x + \frac{9}{2}$
$y = \frac{3}{2}x + \frac{21}{2}$

16. $|y| = x$ **is not a function**, since $x = 2$ corresponds to both $y = 2$ and $y = -2$.

17. $f(x) = |x|$

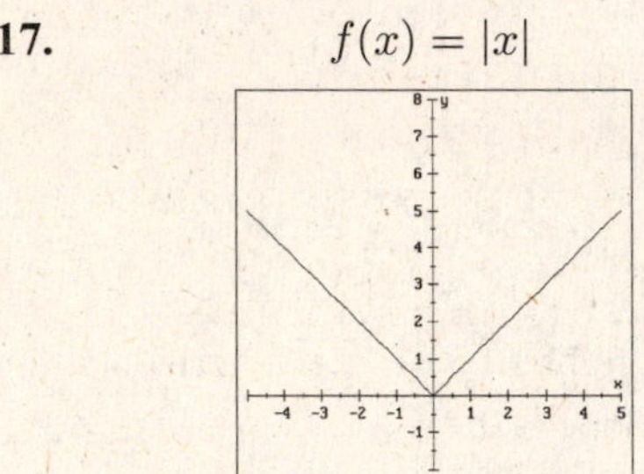

Domain: $(-\infty, \infty)$; Range: $[0, \infty)$

18. $f(x) = x^3$

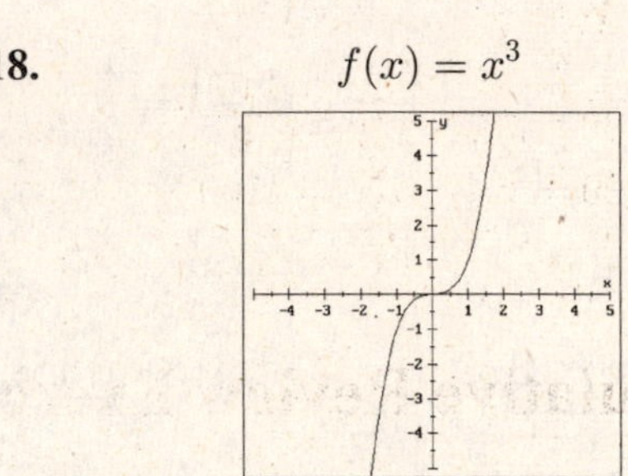

Domain: $(-\infty, \infty)$; Range: $(-\infty, \infty)$

19. $f(x) = 3x + 1$
$f(3) = 3(3) + 1 = 9 + 1 = 10$

20. $g(x) = x^2 - 2$
$g(0) = 0^2 - 2 = 0 - 2 = -2$

21. $f(x) = 3x + 1$
$f(a) = 3a + 1$

22. $g(x) = x^2 - 2$
$g(-x) = (-x)^2 - 2 = x^2 - 2$

23. function

24. not a function

25. $f(x) = x^2 - 1$: Shift $f(x) = x^2$ D 1.

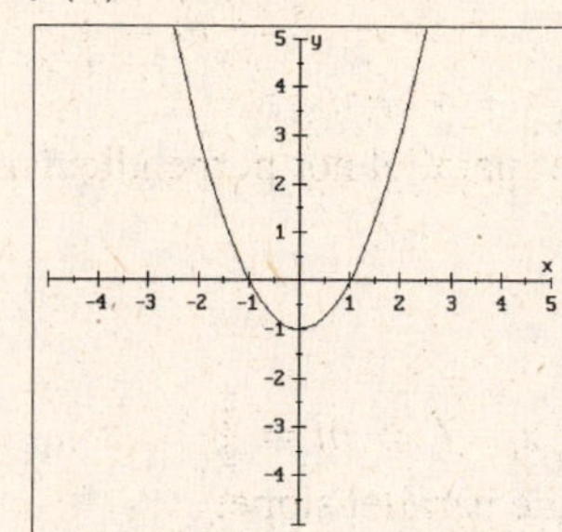

26. $f(x) = -|x + 2|$: Reflect $f(x) = |x|$ about the x-axis and shift L 2.

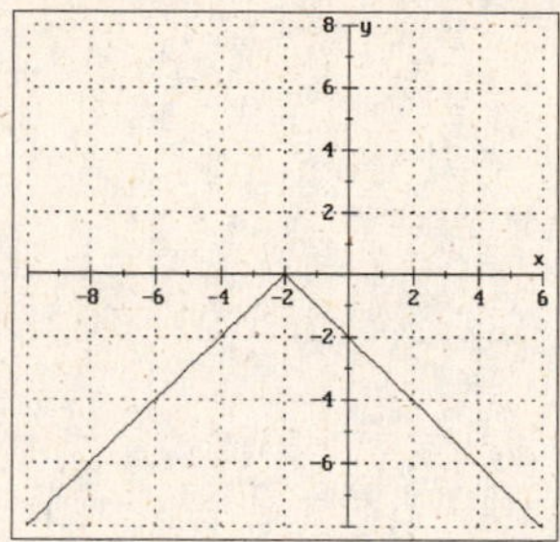

27.
$$\frac{3}{x-2} = \frac{x+3}{2x}$$
$$6x = (x+3)(x-2)$$
$$6x = x^2 + x - 6$$
$$0 = x^2 - 5x - 6$$
$$0 = (x-6)(x+1)$$
$x - 6 = 0$ **or** $x + 1 = 0$
$x = 6$ $\quad\quad x = -1$

28.
$y = kx$ $\quad$ $y = \frac{2}{15}x$
$4 = k(30)$ $\quad$ $y = \frac{2}{15}(9)$
$\frac{4}{30} = k$ $\quad$ $y = \frac{18}{15}$
$\frac{2}{15} = k$ $\quad$ $\boxed{y = \frac{6}{5}}$

29.
$V = \frac{k}{t}$ $\quad$ $V = \frac{1100}{t}$
$55 = \frac{k}{20}$ $\quad$ $75 = \frac{1100}{t}$
$1100 = k$ $\quad$ $75t = 1100$
$\boxed{t = \frac{44}{3}}$

30. function

Cumulative Review Exercises (page 609)

1. natural numbers: $1, 2, 6, 7$

2. whole numbers: $0, 1, 2, 6, 7$

3. rational numbers: $-2, 0, 1, 2, \frac{13}{12}, 6, 7$

4. irrational numbers: $\sqrt{5}, \pi$

5. negative numbers: -2

6. real numbers: $-2, 0, 1, 2, \frac{13}{12}, 6, 7, \sqrt{5}, \pi$

7. prime numbers: $2, 7$

8. composite numbers: 6

9. even numbers: $-2, 0, 2, 6$

10. odd numbers: $1, 7$

11. $-2 < x \leq 5 \Rightarrow$ ←(———]→ with endpoints -2 and 5

12. $[-5, 0) \cup [3, 6] \Rightarrow$ 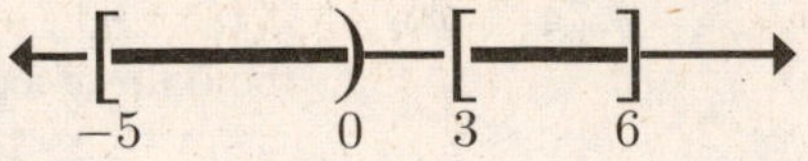

13. $-|5| + |-3| = -5 + 3 = -2$

14. $\dfrac{|-5| + |-3|}{-|4|} = \dfrac{5+3}{-4} = \dfrac{8}{-4} = -2$

15. $2 + 4 \cdot 5 = 2 + 20 = 22$

16. $\dfrac{8-4}{2-4} = \dfrac{4}{-2} = -2$

17. $20 \div (-10 \div 2) = 20 \div (-5) = -4$

18. $\dfrac{6+3(6+4)}{2(3-9)} = \dfrac{6+3(10)}{2(-6)} = \dfrac{6+30}{-12}$
$= \dfrac{36}{-12} = -3$

19. $-x - 2y = -2 - 2(-3) = -2 + 6 = 4$

20. $\dfrac{x^2 - y^2}{2x + y} = \dfrac{2^2 - (-3)^2}{2(2) + (-3)} = \dfrac{4-9}{4-3} = \dfrac{-5}{1}$
$= -5$

21. associative property of addition

22. distributive property

23. commutative property of addition

24. associative property of multiplication

25. $(x^2y^3)^4 = (x^2)^4(y^3)^4 = x^8y^{12}$

26. $\dfrac{c^4c^8}{(c^5)^2} = \dfrac{c^{12}}{c^{10}} = c^2$

27. $\left(-\dfrac{a^3b^{-2}}{ab}\right)^{-1} = \left(-\dfrac{a^2}{b^3}\right)^{-1} = -\dfrac{b^3}{a^2}$

28. $\left(\dfrac{-3a^3b^{-2}}{6a^{-2}b^3}\right)^0 = 1$

29. $0.00000497 = 4.97 \times 10^{-6}$

30. $9.32 \times 10^8 = 932{,}000{,}000$

Exercise 9.1 (page 621)

1. $\sqrt{9} = \sqrt{3^2} = 3$

3. $\sqrt[3]{-8} = \sqrt[3]{(-2)^3} = -2$

5. $\sqrt{64x^2} = \sqrt{(8x)^2} = 8|x|$

7. $\sqrt{-3} \Rightarrow$ not a real number

9. $\dfrac{x^2 + 7x + 12}{x^2 - 16} = \dfrac{(x+4)(x+3)}{(x+4)(x-4)} = \dfrac{x+3}{x-4}$

11. $\dfrac{x^2 - x - 6}{x^2 - 2x - 3} \cdot \dfrac{x^2 - 1}{x^2 + x - 2} = \dfrac{(x-3)(x+2)}{(x-3)(x+1)} \cdot \dfrac{(x+1)(x-1)}{(x+2)(x-1)} = 1$

13. $\dfrac{3}{m+1} + \dfrac{3m}{m-1} = \dfrac{3(m-1)}{(m+1)(m-1)} + \dfrac{3m(m+1)}{(m+1)(m-1)} = \dfrac{3m^2 + 6m - 3}{(m+1)(m-1)} = \dfrac{3(m^2 + 2m - 1)}{(m+1)(m-1)}$

SECTION 9.1

15. $(5x^2)^2$; $6^2 = 36$

17. positive

19. 5; left

21. radical; index; radicand

23. $|x|$

25. x

27. even

29. $3x^2$

31. $a^2 + b^3$

33. $\sqrt{121} = \sqrt{11^2} = 11$

35. $-\sqrt{64} = -\sqrt{8^2} = -8$

37. $\sqrt{\frac{1}{9}} = \sqrt{\left(\frac{1}{3}\right)^2} = \frac{1}{3}$

39. $-\sqrt{\frac{25}{49}} = -\sqrt{\left(\frac{5}{7}\right)^2} = -\frac{5}{7}$

41. $\sqrt{-25}$: not a real number

43. $\sqrt{0.16} = 0.4$

45. $\sqrt{4x^2} = \sqrt{(2x)^2} = |2x| = 2|x|$

47. $\sqrt{9a^4} = \sqrt{(3a^2)^2} = |3a^2| = 3a^2$

49. $\sqrt{(t+5)^2} = |t+5|$

51. $\sqrt{a^2 + 6a + 9} = \sqrt{(a+3)^2} = |a+3|$

53. $\sqrt[3]{1} = \sqrt[3]{1^3} = 1$

55. $\sqrt[3]{-125} = \sqrt[3]{(-5)^3} = -5$

57. $\sqrt[3]{-\frac{8}{27}} = \sqrt[3]{\left(-\frac{2}{3}\right)^3} = -\frac{2}{3}$

59. $\sqrt[3]{0.064} = 0.4$

61. $\sqrt[3]{8a^3} = \sqrt[3]{(2a)^3} = 2a$

63. $\sqrt[3]{-1000p^3q^3} = \sqrt[3]{(-10pq)^3} = -10pq$

65. $\sqrt[4]{81} = \sqrt[4]{3^4} = 3$

67. $-\sqrt[5]{243} = -\sqrt[5]{3^5} = -3$

69. $\sqrt[5]{-32} = \sqrt[5]{(-2)^5} = -2$

71. $\sqrt[4]{\frac{16}{625}} = \sqrt[4]{\left(\frac{2}{5}\right)^4} = \frac{2}{5}$

73. $-\sqrt[5]{-\frac{1}{32}} = -\sqrt[5]{\left(-\frac{1}{2}\right)^5} = -\left(-\frac{1}{2}\right) = \frac{1}{2}$

75. $\sqrt[4]{-256} \Rightarrow$ not a real number

77. $\sqrt[4]{16x^4} = \sqrt[4]{(2x)^4} = |2x| = 2|x|$

79. $\sqrt[3]{8a^3} = \sqrt[3]{(2a)^3} = 2a$

81. $\sqrt[4]{\frac{1}{16}x^4} = \sqrt[4]{\left(\frac{1}{2}x\right)^4} = \left|\frac{1}{2}x\right| = \frac{1}{2}|x|$

83. $\sqrt[4]{x^{12}} = \sqrt[4]{(x^3)^4} = |x^3|$

85. $\sqrt[5]{-x^5} = \sqrt[5]{(-x)^5} = -x$

87. $\sqrt[3]{-27a^6} = \sqrt[3]{(-3a^2)^3} = -3a^2$

89. $f(4) = \sqrt{4-4} = \sqrt{0} = 0$

91. $f(20) = \sqrt{20-4} = \sqrt{16} = 4$

93. $g(9) = \sqrt{9-8} = \sqrt{1} = 1$

95. $g(8.25) = \sqrt{8.25-8} = \sqrt{0.25} = 0.5$

97. $f(x) = \sqrt{x+4}$; Shift $y = \sqrt{x}$ left 4.

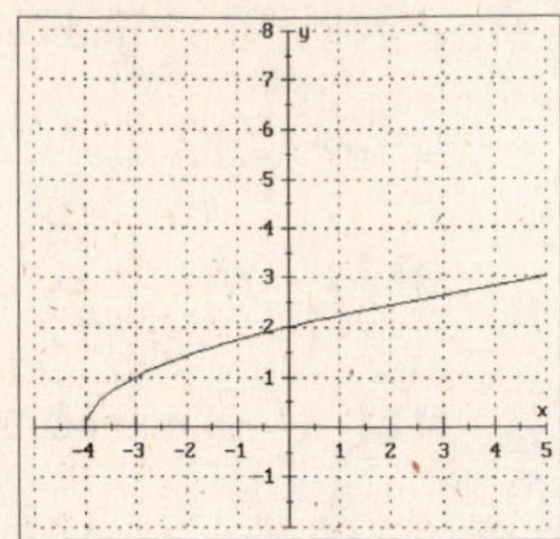

$D = [-4, \infty); R = [0, \infty)$

99. $f(x) = \sqrt[3]{x} - 1$; Shift $y = \sqrt[3]{x}$ down 1.

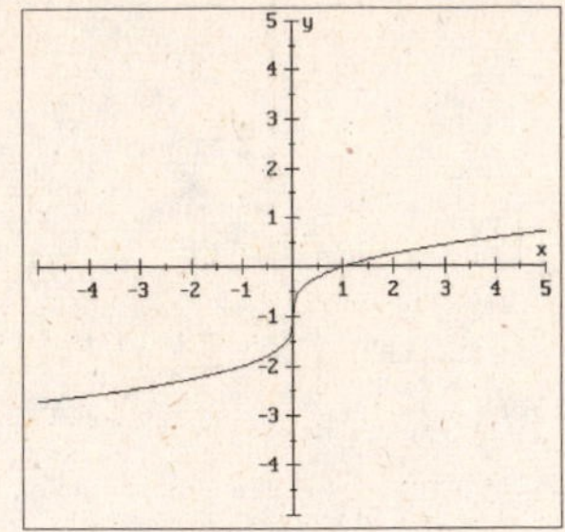

$D = (-\infty, \infty); R = (-\infty, \infty)$

101. $\sqrt{(-4)^2} = \sqrt{16} = 4$

103. $\sqrt{-36}$: not a real number

105. $\sqrt{(-5b)^2} = |-5b| = 5|b|$

107. $\sqrt{t^2 + 24t + 144} = \sqrt{(t+12)^2} = |t+12|$

109. $\sqrt[3]{-\frac{1}{8}m^6n^3} = \sqrt[3]{\left(-\frac{1}{2}m^2n\right)^3} = -\frac{1}{2}m^2n$

111. $\sqrt[3]{0.008z^9} = \sqrt[3]{(0.2z^3)^3} = 0.2z^3$

113. $\sqrt[25]{(x+2)^{25}} = x + 2$

115. $\sqrt[8]{0.00000001x^{16}y^8} = \sqrt[8]{(0.1x^2y)^8} = |0.1x^2y| = 0.1x^2|y|$

117. $\sqrt{12} \approx 3.4641$

119. $\sqrt{679.25} \approx 26.0624$

121. $f(4) = \sqrt{4^2+1} = \sqrt{17} \approx 4.1231$

123. $f(2.35) = \sqrt{(2.35)^2+1} = \sqrt{6.5225} \approx 2.5539$

125. $\text{mean} = \frac{2+5+5+6+7}{5} = \frac{25}{5} = 5$

Original term	Mean	Difference (term−mean)	Square of difference
2	5	−3	9
5	5	0	0
5	5	0	0
6	5	1	1
7	5	2	4

$\text{st. dev.} = \sqrt{\frac{9+0+0+1+4}{5}} \approx 1.67$

127. $s_{\overline{x}} = \frac{s}{\sqrt{N}} = \frac{65}{\sqrt{30}} \approx 11.8673$

129. $r = \sqrt{\frac{A}{\pi}} = \sqrt{\frac{9\pi}{\pi}} = \sqrt{9} = 3$ units

131. $t = \frac{\sqrt{s}}{4} = \frac{\sqrt{256}}{4} = \frac{16}{4} = 4$ seconds

133. $I = \sqrt{\frac{P}{18}} = \sqrt{\frac{980}{18}} \approx \sqrt{54.44} \approx 7.4$ amps

135. **Answers may vary.**

137. $\sqrt{x^2 - 4x + 4} = \sqrt{(x-2)^2} = |x-2|$.
$|x-2| = x - 2$ when $x - 2 \geq 0$, or $x \geq 2$.

Exercise 9.2 (page 629)

1. $\sqrt{25} = 5$

3. $\sqrt{3^2 + 4^2} = \sqrt{9 + 16} = \sqrt{25} = 5$

5. $\sqrt{5^2 - 3^2} = \sqrt{25 - 9} = \sqrt{16} = 4$

7. $(4x + 2)(3x - 5) = 12x^2 - 14x - 10$

9. $(5t + 4s)(3t - 2s) = 15t^2 + 2ts - 8s^2$

11. hypotenuse

13. $a^2 + b^2 = c^2$

15. positive

17.
$$\begin{aligned} c^2 &= a^2 + b^2 \\ c^2 &= 6^2 + 8^2 \\ c^2 &= 36 + 64 \\ c^2 &= 100 \\ c &= \sqrt{100} \\ c &= 10 \text{ ft} \end{aligned}$$

19.
$$\begin{aligned} c^2 &= a^2 + b^2 \\ 82^2 &= a^2 + 18^2 \\ 6724 &= a^2 + 324 \\ 6400 &= a^2 \\ \sqrt{6400} &= a \\ 80 \text{ m} &= a \end{aligned}$$

21.
$$\begin{aligned} c^2 &= a^2 + b^2 \\ 50^2 &= 14^2 + b^2 \\ 2500 &= 196 + b^2 \\ 2304 &= b^2 \\ \sqrt{2304} &= b \\ 48 \text{ in.} &= b \end{aligned}$$

23.
$$\begin{aligned} c^2 &= a^2 + b^2 \\ c^2 &= \left(\sqrt{6}\right)^2 + \left(\sqrt{3}\right)^2 \\ c^2 &= 6 + 3 \\ c^2 &= 9 \\ c &= \sqrt{9} \\ c &= 3 \text{ mi} \end{aligned}$$

25.
$$\begin{aligned} d &= \sqrt{(x_2 - x_1)^2 + (y_2 - y_1)^2} \\ &= \sqrt{(0 - 3)^2 + [0 - (-4)]^2} \\ &= \sqrt{(-3)^2 + (4)^2} \\ &= \sqrt{9 + 16} \\ &= \sqrt{25} = 5 \end{aligned}$$

27.
$$\begin{aligned} d = \sqrt{(x_2 - x_1)^2 + (y_2 - y_1)^2} = \sqrt{(2 - 5)^2 + (4 - 8)^2} = \sqrt{(-3)^2 + (-4)^2} &= \sqrt{9 + 16} \\ &= \sqrt{25} = 5 \end{aligned}$$

29.
$$\begin{aligned} d = \sqrt{(x_2 - x_1)^2 + (y_2 - y_1)^2} = \sqrt{(-2 - 3)^2 + (-8 - 4)^2} = \sqrt{(-5)^2 + (-12)^2} &= \sqrt{25 + 144} \\ &= \sqrt{169} = 13 \end{aligned}$$

31.
$$\begin{aligned} d = \sqrt{(x_2 - x_1)^2 + (y_2 - y_1)^2} = \sqrt{(6 - 12)^2 + (8 - 16)^2} = \sqrt{(-6)^2 + (-8)^2} &= \sqrt{36 + 64} \\ &= \sqrt{100} = 10 \end{aligned}$$

33.
$$\begin{aligned} d = \sqrt{(x_2 - x_1)^2 + (y_2 - y_1)^2} = \sqrt{[-3 - (-5)]^2 + [5 - (-5)]^2} = \sqrt{2^2 + 10^2} &= \sqrt{4 + 100} \\ &= \sqrt{104} \approx 10.2 \end{aligned}$$

35.
$$\begin{aligned} d = \sqrt{(x_2 - x_1)^2 + (y_2 - y_1)^2} = \sqrt{(-9 - 4)^2 + (3 - 7)^2} = \sqrt{(-13)^2 + (-4)^2} &= \sqrt{169 + 16} \\ &= \sqrt{185} \approx 13.6 \end{aligned}$$

SECTION 9.2

37. Let $x =$ distance to 2nd.

$$x^2 = 90^2 + 90^2$$
$$x^2 = 8100 + 8100$$
$$x^2 = 16200$$
$$x \approx 127 \text{ ft}$$

39. Refer to the diagram provided. The 3rd baseman is at B, so $\overline{BC}$ has a length of 10 ft. Let $\overline{AC}$ and $\overline{AB}$ both have a length x.

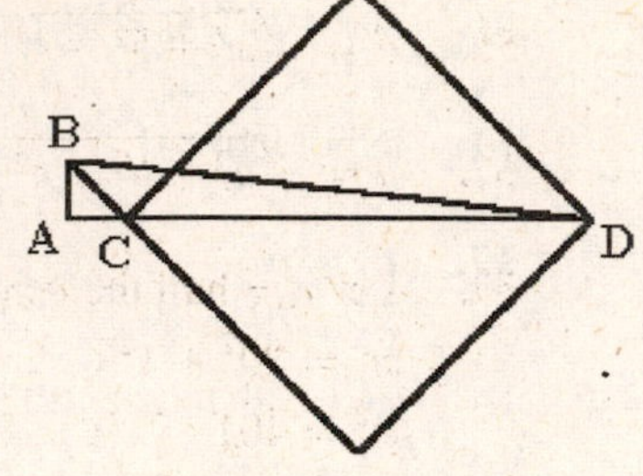

$$x^2 + x^2 = 10^2$$
$$2x^2 = 100$$
$$x^2 = 50$$
$$x = \sqrt{50} \approx 7.1 \text{ ft}$$

From **#37**, the length of $\overline{CD} \approx 127.3$ ft, so $\overline{AD}$ has a length of about $127.3 + 7.1 = 134.4$ ft. Let $y =$ the length of $\overline{BD}$.

$$y^2 = (7.1)^2 + (134.4)^2$$
$$y^2 = 18113.77$$
$$y = \sqrt{18113.77} \approx 135 \text{ ft}$$

41. Let $x =$ the length of the diagonal.

$$7^2 + 7^2 = x^2$$
$$49 + 49 = x^2$$
$$98 = x^2$$
$$\sqrt{98} = x$$
$$9.9 \text{ cm} \approx x$$

43. Let the points be represented by $A(5,1)$, $B(7,0)$ and $C(3,0)$. Find the length of $\overline{AB}$ and $\overline{AC}$:

$\overline{AB}: \sqrt{(5-7)^2 + (1-0)^2} = \sqrt{(-2)^2 + 1^2} = \sqrt{5}$

$\overline{AC}: \sqrt{(5-3)^2 + (1-0)^2} = \sqrt{(2)^2 + 1^2} = \sqrt{5}$

Since $\overline{AB}$ and $\overline{AC}$ have the same length, $(5,1)$ is equidistant from $(7,0)$ and $(3,0)$.

45. Let the points be represented by $A(-2,4)$, $B(2,8)$ and $C(6,4)$. Find the length of each side:

$\overline{AB}: \sqrt{(-2-2)^2 + (4-8)^2} = \sqrt{(-4)^2 + (-4)^2} = \sqrt{32}$

$\overline{AC}: \sqrt{(-2-6)^2 + (4-4)^2} = \sqrt{(-8)^2 + 0^2} = \sqrt{64}$

$\overline{BC}: \sqrt{(2-6)^2 + (8-4)^2} = \sqrt{(-4)^2 + 4^2} = \sqrt{32}$

Since $\overline{AB}$ and $\overline{BC}$ have the same length, the triangle is isosceles.

47.

$$d^2 = 5^2 + 12^2$$
$$d^2 = 25 + 144$$
$$d^2 = 169$$
$$d = 13 \text{ ft}$$

49.

$$37^2 = 9^2 + h^2$$
$$1369 = 81 + h^2$$
$$1288 = h^2$$
$$35.9 = h$$

The ladder will reach.

51. Let x = direct distance from A to E.
$x^2 = (42+26)^2 + (300-15)^2$
$x^2 = 68^2 + 285^2$
$x^2 = 85849$
$x = 293$ yd; savings $= 383 - 293 = 90$ yd

53. $A = 6\sqrt[3]{V^2} = 6\sqrt[3]{8^2} = 6\sqrt[3]{64} = 6(4)$
$= 24\text{ cm}^2$

55. $d = \sqrt{a^2+b^2+c^2} = \sqrt{12^2+24^2+17^2} = \sqrt{1009} = 31.76 \Rightarrow$ The racket will not fit.

57. $d = \sqrt{a^2+b^2+c^2} = \sqrt{21^2+21^2+3^2} = \sqrt{891} = 29.8 \Rightarrow$ The femur will not fit.

59. Let x = half the length of the stretched wire.
$x^2 = 20^2 + 1^2$
$x^2 = 401$
$x = 20.025$ ft
The stretched wire has a length of 40.05 ft. It has been stretched by 0.05 ft.

61. **Answers may vary.**

63. $I = \dfrac{703w}{h^2} = \dfrac{703(104)}{(54.1)^2} = \dfrac{73112}{2926.81} \approx 25$

Exercise 9.3 (page 638)

1. $4^{1/2} = \sqrt{4} = 2$

3. $27^{1/3} = \sqrt[3]{27} = 3$

5. $4^{3/2} = \left(4^{1/2}\right)^3 = 2^3 = 8$

7. $\left(\frac{1}{4}\right)^{1/2} = \sqrt{\frac{1}{4}} = \frac{1}{2}$

9. $(8x^3)^{1/3} = \sqrt[3]{8x^3} = 2x$

11.
$5x - 4 < 11$
$5x < 15$
$x < 3$

13.
$\frac{4}{5}(r-3) > \frac{2}{3}(r+2)$
$15 \cdot \frac{4}{5}(r-3) > 15 \cdot \frac{2}{3}(r+2)$
$12(r-3) > 10(r+2)$
$2r > 56$
$r > 28$

15. Let x = pints of water added (0% alcohol).

Alcohol at start	+	Alcohol added	=	Alcohol at end

$0.20(5) + 0(x) = 0.15(5+x)$
$1 + 0 = 0.75 + 0.15x$
$0.25 = 0.15x$
$x = \dfrac{0.25}{0.15} = \dfrac{5}{3}$
$1\frac{2}{3}$ pints of water should be added.

17. $a \cdot a \cdot a \cdot a$

19. a^{mn}

21. $\frac{a^n}{b^n}$

23. $\frac{1}{a^n}$; 0

25. $\left(\dfrac{b}{a}\right)^n$ **27.** $|x|$ **29.** $7^{1/3} = \sqrt[3]{7}$ **31.** $8^{1/5} = \sqrt[5]{8}$

33. $(3x)^{1/4} = \sqrt[4]{3x}$ **35.** $\left(\frac{1}{2}x^3y\right)^{1/4} = \sqrt[4]{\frac{1}{2}x^3y}$ **37.** $(4a^2b^3)^{1/5} = \sqrt[5]{4a^2b^3}$

39. $(x^2+y^2)^{1/2} = \sqrt{x^2+y^2}$ **41.** $4^{1/2} = \sqrt{4} = 2$ **43.** $27^{1/3} = \sqrt[3]{27} = 3$

45. $\left(\frac{1}{4}\right)^{1/2} = \sqrt{\frac{1}{4}} = \frac{1}{2}$ **47.** $\left(\frac{1}{8}\right)^{1/3} = \sqrt[3]{\frac{1}{8}} = \frac{1}{2}$ **49.** $-16^{1/4} = -\sqrt[4]{16} = -2$

51. $(-64)^{1/2}=\sqrt{-64}$; not a real number

53. $\sqrt{11} = 11^{1/2}$ **55.** $\sqrt[4]{3a} = (3a)^{1/4}$ **57.** $3\sqrt[5]{a} = 3a^{1/5}$

59. $\sqrt[6]{\frac{1}{7}abc} = \left(\frac{1}{7}abc\right)^{1/6}$ **61.** $\sqrt[5]{\frac{1}{2}mn} = \left(\frac{1}{2}mn\right)^{1/5}$ **63.** $\sqrt[3]{a^2-b^2} = (a^2-b^2)^{1/3}$

65. $(25y^2)^{1/2} = \left[(5y)^2\right]^{1/2} = |5y| = 5|y|$ **67.** $(243x^5)^{1/5} = \left[(3x)^5\right]^{1/5} = 3x$

69. $\left[(x+1)^4\right]^{1/4} = |x+1|$ **71.** $(-64x^8)^{1/4} \Rightarrow$ not a real number

73. $36^{3/2} = \left(36^{1/2}\right)^3 = 6^3 = 216$ **75.** $81^{3/4} = \left(81^{1/4}\right)^3 = 3^3 = 27$

77. $144^{3/2} = \left(144^{1/2}\right)^3 = 12^3 = 1728$ **79.** $\left(\frac{1}{8}\right)^{2/3} = \left[\left(\frac{1}{8}\right)^{1/3}\right]^2 = \left(\frac{1}{2}\right)^2 = \frac{1}{4}$

81. $4^{-1/2} = \dfrac{1}{4^{1/2}} = \dfrac{1}{2}$ **83.** $4^{-3/2} = \dfrac{1}{4^{3/2}} = \dfrac{1}{\left(4^{1/2}\right)^3} = \dfrac{1}{2^3} = \dfrac{1}{8}$

85. $(16x^2)^{-3/2} = \dfrac{1}{(16x^2)^{3/2}} = \dfrac{1}{\left[(16x^2)^{1/2}\right]^3} = \dfrac{1}{(4x)^3} = \dfrac{1}{64x^3}$

87. $\left(-27y^3\right)^{-2/3} = \dfrac{1}{(-27y^3)^{2/3}} = \dfrac{1}{\left[(-27y^3)^{1/3}\right]^2} = \dfrac{1}{(-3y)^2} = \dfrac{1}{9y^2}$

89. $\left(\dfrac{1}{4}\right)^{-3/2} = \left(\dfrac{4}{1}\right)^{3/2} = 4^{3/2} = \left(4^{1/2}\right)^3 = 2^3 = 8$

91. $\left(\dfrac{27}{8}\right)^{-4/3} = \left(\dfrac{8}{27}\right)^{4/3} = \left[\left(\dfrac{8}{27}\right)^{1/3}\right]^4 = \left(\dfrac{2}{3}\right)^4 = \dfrac{16}{81}$

93. $5^{4/9}5^{4/9} = 5^{4/9+4/9} = 5^{8/9}$ **95.** $\left(4^{1/5}\right)^3 = 4^{(1/5)\cdot 3} = 4^{3/5}$

SECTION 9.3

97. $6^{-2/3}6^{-4/3} = 6^{-6/3} = 6^{-2} = \frac{1}{36}$

99. $\dfrac{9^{4/5}}{9^{3/5}} = 9^{4/5-3/5} = 9^{1/5}$

101. $\frac{7^{1/2}}{7^0} = 7^{1/2-0} = 7^{1/2}$

103. $\dfrac{2^{5/6}2^{1/3}}{2^{1/2}} = \dfrac{2^{7/6}}{2^{1/2}} = 2^{4/6} = 2^{2/3}$

105. $a^{2/3}a^{1/3} = a^{3/3} = a^1 = a$

107. $\left(a^{2/3}\right)^{1/3} = a^{(2/3)(1/3)} = a^{2/9}$

109. $y^{1/3}\left(y^{2/3} + y^{5/3}\right) = y^{3/3} + y^{6/3} = y + y^2$

111. $x^{3/5}\left(x^{7/5} - x^{2/5} + 1\right) = x^{10/5} - x^{5/5} + x^{3/5}$
$= x^2 - x + x^{3/5}$

113. $\left(x^{1/2} + y^{1/2}\right)\left(x^{1/2} - y^{1/2}\right) = x^{2/2} - x^{1/2}y^{1/2} + x^{1/2}y^{1/2} - y^{2/2} = x - y$

115. $\left(x^{2/3} + y^{2/3}\right)^2 = \left(x^{2/3} + y^{2/3}\right)\left(x^{2/3} + y^{2/3}\right) = x^{4/3} + x^{2/3}y^{2/3} + x^{2/3}y^{2/3} + y^{4/3}$
$= x^{4/3} + 2x^{2/3}y^{2/3} + y^{4/3}$

117. $\sqrt[6]{p^3} = \left(p^3\right)^{1/6} = p^{3/6} = p^{1/2} = \sqrt{p}$

119. $\sqrt[4]{25b^2} = \left(5^2b^2\right)^{1/4} = 5^{1/2}b^{1/2} = \sqrt{5b}$

121. $16^{1/4} = \sqrt[4]{16} = 2$

123. $32^{1/5} = \sqrt[5]{32} = 2$

125. $0^{1/3} = \sqrt[3]{0} = 0$

127. $(-27)^{1/3} = \sqrt[3]{-27} = -3$

129. $\left(25x^4\right)^{3/2} = \left[\left(25x^4\right)^{1/2}\right]^3 = \left(5x^2\right)^3$
$= 125x^6$

131. $\left(\dfrac{8x^3}{27}\right)^{2/3} = \left[\left(\dfrac{8x^3}{27}\right)^{1/3}\right]^2 = \left(\dfrac{2x}{3}\right)^2$
$= \dfrac{4x^2}{9}$

133. $\left(-32p^5\right)^{-2/5} = \dfrac{1}{\left(-32p^5\right)^{2/5}} = \dfrac{1}{\left[\left(-32p^5\right)^{1/5}\right]^2} = \dfrac{1}{(-2p)^2} = \dfrac{1}{4p^2}$

135. $\left(-\dfrac{8x^3}{27}\right)^{-1/3} = \left(-\dfrac{27}{8x^3}\right)^{1/3} = -\dfrac{3}{2x}$

137. $\left(a^{1/2}b^{1/3}\right)^{3/2} = a^{3/4}b^{1/2}$

139. $\left(mn^{-2/3}\right)^{-3/5} = m^{-3/5}n^{2/5} = \dfrac{n^{2/5}}{m^{3/5}}$

141. $\dfrac{\left(4x^3y\right)^{1/2}}{(9xy)^{1/2}} = \dfrac{2x^{3/2}y^{1/2}}{3x^{1/2}y^{1/2}} = \dfrac{2x}{3}$

143. $\left(27x^{-3}\right)^{-1/3} = (27)^{-1/3}x = \dfrac{1}{3}x$

145. $x^{4/3}\left(x^{2/3} + 3x^{5/3} - 4\right) = x^{6/3} + 3x^{9/3} - 4x^{4/3} = x^2 + 3x^3 - 4x^{4/3}$

147. $\left(x^{-1/2}-x^{1/2}\right)^2=\left(x^{-1/2}-x^{1/2}\right)\left(x^{-1/2}-x^{1/2}\right)=x^{-2/2}-x^{1/2}x^{-1/2}-x^{1/2}x^{-1/2}+x^{2/2}$
$=x^{-1}-1-1+x=\dfrac{1}{x}-2+x$

Problems 149-155 are to be solved using a calculator. The keystrokes needed to solve each problem using a TI-83 graphing calculator appear in each solution. There may be other solutions. Keystrokes for other calculators may be slightly different.

149. [1] [5] [^] [(] [1] [÷] [3] [)] [ENTER] {2.4662...} ⇒ 2.47

151. [1] [.] [0] [4] [5] [^] [(] [1] [÷] [5] [)] [ENTER] {1.0088...} ⇒ 1.01

153. [1] [7] [^] [(] [(-)] [1] [÷] [2] [)] [ENTER] {0.2425...} ⇒ 0.24

155. [(] [(-)] [.] [2] [5] [)] [^] [(] [(-)] [1] [÷] [5] [)] [ENTER] {−1.3195...} ⇒ −1.32

157. Answers may vary.

159. $16^{2/4}=2^2=4; 16^{1/2}=4$

Exercise 9.4 (page 648)

1. $\sqrt{7}\sqrt{7}=\sqrt{49}=7$

3. $\dfrac{\sqrt[3]{54}}{\sqrt[3]{2}}=\sqrt[3]{\dfrac{54}{2}}=\sqrt[3]{27}=3$

5. $\sqrt[3]{16}=\sqrt[3]{8\cdot 2}=\sqrt[3]{8}\sqrt[3]{2}=2\sqrt[3]{2}$

7. $3\sqrt{3}+4\sqrt{3}=7\sqrt{3}$

9. $2\sqrt[3]{9}+3\sqrt[3]{9}=5\sqrt[3]{9}$

11. $3x^2y^2(-5x^3y^{-3})=-15x^5y^{-1}=\dfrac{-15x^5}{y}$

13. $(3t+2)^2=(3t+2)(3t+2)$
$=9t^2+12t+4$

15.
$$\begin{array}{r|l} & 3p+\quad 4+\frac{-5}{2p-5} \\ 2p-5 & 6p^2-\ \ 7p-\ \ 25 \\ & \underline{6p^2-15p} \\ & 8p-\ \ 25 \\ & \underline{8p-\ \ 20} \\ & -5 \end{array}$$

17. $\sqrt[n]{a}\sqrt[n]{b}$

19. like

21. $\sqrt{6}\sqrt{6}=\sqrt{36}=6$

23. $\sqrt{t}\sqrt{t}=\sqrt{t^2}=t$

25. $\sqrt[3]{5x^2}\sqrt[3]{25x}=\sqrt[3]{125x^3}=5x$

27. $\dfrac{\sqrt{500}}{\sqrt{5}}=\sqrt{\dfrac{500}{5}}=\sqrt{100}=10$

29. $\dfrac{\sqrt{98x^3}}{\sqrt{2x}}=\sqrt{\dfrac{98x^3}{2x}}=\sqrt{49x^2}=7x$

31. $\dfrac{\sqrt{180ab^4}}{\sqrt{5ab^2}}=\sqrt{\dfrac{180ab^4}{5ab^2}}=\sqrt{36b^2}=6b$

33. $\dfrac{\sqrt[3]{48}}{\sqrt[3]{6}}=\sqrt[3]{\dfrac{48}{6}}=\sqrt[3]{8}=2$

SECTION 9.4

35. $\dfrac{\sqrt[3]{189a^4}}{\sqrt[3]{7a}} = \sqrt[3]{\dfrac{189a^4}{7a}} = \sqrt[3]{27a^3} = 3a$

37. $\sqrt{20} = \sqrt{4 \cdot 5} = \sqrt{4}\sqrt{5} = 2\sqrt{5}$

39. $-\sqrt{200} = -\sqrt{100 \cdot 2} = -\sqrt{100}\sqrt{2}$
$= -10\sqrt{2}$

41. $\sqrt[3]{80} = \sqrt[3]{8 \cdot 10} = \sqrt[3]{8}\sqrt[3]{10} = 2\sqrt[3]{10}$

43. $\sqrt[3]{-81} = \sqrt[3]{-27 \cdot 3} = \sqrt[3]{-27}\sqrt[3]{3} = -3\sqrt[3]{3}$

45. $\sqrt[4]{32} = \sqrt[4]{16 \cdot 2} = \sqrt[4]{16}\sqrt[4]{2} = 2\sqrt[4]{2}$

47. $\sqrt[5]{96} = \sqrt[5]{32 \cdot 3} = \sqrt[5]{32}\sqrt[5]{3} = 2\sqrt[5]{3}$

49. $\sqrt{\dfrac{7}{9x^2}} = \dfrac{\sqrt{7}}{\sqrt{9x^2}} = \dfrac{\sqrt{7}}{3x}$

51. $\sqrt[3]{\dfrac{7a^3}{64}} = \dfrac{\sqrt[3]{7a^3}}{\sqrt[3]{64}} = \dfrac{\sqrt[3]{a^3}\sqrt[3]{7}}{4} = \dfrac{a\sqrt[3]{7}}{4}$

53. $\sqrt[4]{\dfrac{3p^4}{10{,}000q^4}} = \dfrac{\sqrt[4]{3p^4}}{\sqrt[4]{10{,}000q^4}} = \dfrac{\sqrt[4]{p^4}\sqrt[4]{3}}{10q}$
$= \dfrac{p\sqrt[4]{3}}{10q}$

55. $\sqrt[5]{\dfrac{3m^{15}}{32n^{10}}} = \dfrac{\sqrt[5]{3m^{15}}}{\sqrt[5]{32n^{10}}} = \dfrac{\sqrt[5]{m^{15}}\sqrt[5]{3}}{2n^2}$
$= \dfrac{m^3\sqrt[5]{3}}{2n^2}$

57. $\sqrt{50x^2} = \sqrt{25x^2 \cdot 2} = \sqrt{25x^2}\sqrt{2} = 5x\sqrt{2}$

59. $\sqrt{32b} = \sqrt{16 \cdot 2b} = \sqrt{16}\sqrt{2b} = 4\sqrt{2b}$

61. $-\sqrt{112a^3} = -\sqrt{16a^2 \cdot 7a} = -\sqrt{16a^2}\sqrt{7a}$
$= -4a\sqrt{7a}$

63. $\sqrt{175a^2b^3} = \sqrt{25a^2b^2 \cdot 7b} = \sqrt{25a^2b^2}\sqrt{7b} = 5ab\sqrt{7b}$

65. $-\sqrt{300xy} = -\sqrt{100 \cdot 3xy} = -\sqrt{100}\sqrt{3xy} = -10\sqrt{3xy}$

67. $\sqrt[3]{-54x^6} = \sqrt[3]{-27x^6 \cdot 2} = \sqrt[3]{-27x^6}\sqrt[3]{2}$
$= -3x^2\sqrt[3]{2}$

69. $\sqrt[3]{16x^{12}y^3} = \sqrt[3]{8x^{12}y^3 \cdot 2} = \sqrt[3]{8x^{12}y^3}\sqrt[3]{2}$
$= 2x^4y\sqrt[3]{2}$

71. $\sqrt{\dfrac{z^2}{16x^2}} = \dfrac{\sqrt{z^2}}{\sqrt{16x^2}} = \dfrac{z}{4x}$

73. $\sqrt{3} + \sqrt{27} = \sqrt{3} + \sqrt{9}\sqrt{3}$
$= \sqrt{3} + 3\sqrt{3} = 4\sqrt{3}$

75. $\sqrt{2} - \sqrt{8} = \sqrt{2} - \sqrt{4}\sqrt{2}$
$= \sqrt{2} - 2\sqrt{2} = -\sqrt{2}$

77. $\sqrt{98} - \sqrt{50} = \sqrt{49}\sqrt{2} - \sqrt{25}\sqrt{2}$
$= 7\sqrt{2} - 5\sqrt{2} = 2\sqrt{2}$

79. $3\sqrt{24} + \sqrt{54} = 3\sqrt{4}\sqrt{6} + \sqrt{9}\sqrt{6} = 3(2)\sqrt{6} + 3\sqrt{6} = 6\sqrt{6} + 3\sqrt{6} = 9\sqrt{6}$

81. $\sqrt{18} + \sqrt{300} - \sqrt{243} = \sqrt{9}\sqrt{2} + \sqrt{100}\sqrt{3} - \sqrt{81}\sqrt{3} = 3\sqrt{2} + 10\sqrt{3} - 9\sqrt{3} = 3\sqrt{2} + \sqrt{3}$

SECTION 9.4

83. $\sqrt[3]{24}+\sqrt[3]{3}=\sqrt[3]{8}\sqrt[3]{3}+\sqrt[3]{3}=2\sqrt[3]{3}+\sqrt[3]{3}=3\sqrt[3]{3}$

85. $\sqrt[3]{32}-\sqrt[3]{108}=\sqrt[3]{8}\sqrt[3]{4}-\sqrt[3]{27}\sqrt[3]{4}=2\sqrt[3]{4}-3\sqrt[3]{4}=-\sqrt[3]{4}$

87. $2\sqrt[3]{125}-5\sqrt[3]{64}=2(5)-5(4)=10-20=-10$

89. $2\sqrt[3]{16}-\sqrt[3]{54}-3\sqrt[3]{128}=2\sqrt[3]{8}\sqrt[3]{2}-\sqrt[3]{27}\sqrt[3]{2}-3\sqrt[3]{64}\sqrt[3]{2}=2(2)\sqrt[3]{2}-3\sqrt[3]{2}-3(4)\sqrt[3]{2}$
$=4\sqrt[3]{2}-3\sqrt[3]{2}-12\sqrt[3]{2}=-11\sqrt[3]{2}$

91. $\sqrt[3]{3x^5}-\sqrt[3]{24x^5}=\sqrt[3]{x^3}\sqrt[3]{3x^2}-\sqrt[3]{8x^3}\sqrt{3x^2}=x\sqrt[3]{3x^2}-2x\sqrt[3]{3x^2}=-x\sqrt[3]{3x^2}$

93. $\sqrt{25yz^2}+\sqrt{9yz^2}=\sqrt{25z^2}\sqrt{y}+\sqrt{9z^2}\sqrt{y}=5z\sqrt{y}+3z\sqrt{y}=8z\sqrt{y}$

95. $\sqrt{y^5}-\sqrt{9y^5}-\sqrt{25y^5}=\sqrt{y^4}\sqrt{y}-\sqrt{9y^4}\sqrt{y}-\sqrt{25y^4}\sqrt{y}=y^2\sqrt{y}-3y^2\sqrt{y}-5y^2\sqrt{y}$
$=-7y^2\sqrt{y}$

97. $3\sqrt[3]{2x}-\sqrt[3]{54x}=3\sqrt[3]{2x}-\sqrt[3]{27}\sqrt[3]{2x}=3\sqrt[3]{2x}-3\sqrt[3]{2x}=0$

99. $b=a=3; c=3\sqrt{2}$

101. $a=b=\frac{2}{3}; c=\frac{2}{3}\sqrt{2}$

103. $b=a=5\sqrt{2};$
$c=5\sqrt{2}\left(\sqrt{2}\right)=5(2)=10$

105. $a=\frac{7\sqrt{2}}{\sqrt{2}}=7; b=a=7$

107. $b=5\sqrt{3}; c=2(5)=10$

109. $a=\frac{9\sqrt{3}}{\sqrt{3}}=9; c=2(9)=18$

111. $a=\frac{24}{2}=12; b=12\sqrt{3}$

113. $a=\frac{15}{2}; b=\frac{15}{2}\sqrt{3}$

115. $4\sqrt{2x}+6\sqrt{2x}=10\sqrt{2x}$

117. $\sqrt[4]{32x^{12}y^4}=\sqrt[4]{16x^{12}y^4\cdot 2}$
$=\sqrt[4]{16x^{12}y^4}\sqrt[4]{2}=2x^3y\sqrt[4]{2}$

119. $\sqrt[4]{\frac{5x}{16z^4}}=\frac{\sqrt[4]{5x}}{\sqrt[4]{16z^4}}=\frac{\sqrt[4]{5x}}{2z}$

121. $\sqrt{98}-\sqrt{50}-\sqrt{72}=\sqrt{49}\sqrt{2}-\sqrt{25}\sqrt{2}-\sqrt{36}\sqrt{2}=7\sqrt{2}-5\sqrt{2}-6\sqrt{2}=-4\sqrt{2}$

123. $3\sqrt[3]{27}+12\sqrt[3]{216}=3(3)+12(6)=9+72=81$

125. $23\sqrt[4]{768}+\sqrt[4]{48}=23\sqrt[4]{256}\sqrt[4]{3}+\sqrt[4]{16}\sqrt[4]{3}=23(4)\sqrt[4]{3}+2\sqrt[4]{3}=92\sqrt[4]{3}+2\sqrt[4]{3}=94\sqrt[4]{3}$

127. $4\sqrt[4]{243}-\sqrt[4]{48}=4\sqrt[4]{81}\sqrt[4]{3}-\sqrt[4]{16}\sqrt[4]{3}=4(3)\sqrt[4]{3}-2\sqrt[4]{3}=12\sqrt[4]{3}-2\sqrt[4]{3}=10\sqrt[4]{3}$

129. $6\sqrt[3]{5y} + 3\sqrt[3]{5y} = 9\sqrt[3]{5y}$

131. $10\sqrt[6]{12xyz} - \sqrt[6]{12xyz} = 9\sqrt[6]{12xyz}$

133. $\sqrt[5]{x^6y^2} + \sqrt[5]{32x^6y^2} + \sqrt[5]{x^6y^2} = \sqrt[5]{x^5}\sqrt[5]{xy^2} + \sqrt[5]{32x^5}\sqrt[5]{xy^2} + \sqrt[5]{x^5}\sqrt[5]{xy^2}$
$= x\sqrt[5]{xy^2} + 2x\sqrt[5]{xy^2} + x\sqrt[5]{xy^2} = 4x\sqrt[5]{xy^2}$

135. $\sqrt{x^2+2x+1} + \sqrt{x^2+2x+1} = \sqrt{(x+1)^2} + \sqrt{(x+1)^2} = x+1+x+1 = 2x+2$

137. $x = 2.00;\ h = 2\sqrt{2} \approx 2.83$

139. $h = 2(5) = 10.00;\ x = 5\sqrt{3} \approx 8.66$

141. $x = \frac{9.37}{2} \approx 4.69;\ y \approx 4.69\sqrt{3} \approx 8.11$

143. $x = y = \frac{17.12}{\sqrt{2}} \approx 12.11$

145. $x = 5\sqrt{3} \approx 8.66$ mm
$h = 2x = 2\left(5\sqrt{3}\right) = 10\sqrt{3} \approx 17.32$ mm

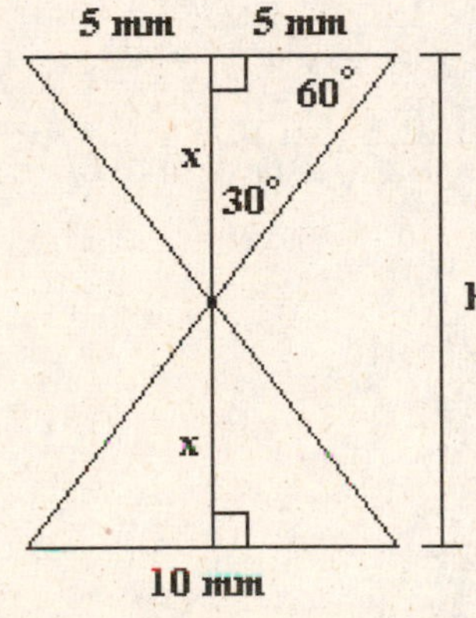

147. Answers may vary.

149. $\sqrt{a+b} = \sqrt{a} + \sqrt{b}$ if either a or b equals 0 and the other is nonnegative.

Exercise 9.5 (page 658)

1. $\sqrt{3}\sqrt{3} = \sqrt{9} = 3$

3. $\sqrt{3}\sqrt{9} = \sqrt{3}\cdot 3 = 3\sqrt{3}$

5. $3\sqrt{2}\left(\sqrt{2}+1\right) = 3\sqrt{4} + 3\sqrt{2} = 3(2) + 3\sqrt{2} = 6 + 3\sqrt{2}$

7. $\dfrac{1}{\sqrt{2}} = \dfrac{1\sqrt{2}}{\sqrt{2}\sqrt{2}} = \dfrac{\sqrt{2}}{2}$

9. $\dfrac{2}{3-a} = 1$
$2 = 3 - a$
$a = 1$

SECTION 9.5

11. $$\frac{8}{b-2}+\frac{3}{2-b}=-\frac{1}{b}$$
$$\frac{8}{b-2}+\frac{-3}{b-2}=-\frac{1}{b}$$
$$\frac{5}{b-2}=\frac{-1}{b}$$
$$5b=-b+2$$
$$6b=2$$
$$b=\frac{2}{6}=\frac{1}{3}$$

13. $2;\ \sqrt{7};\ \sqrt{5}$

15. $\sqrt{x}-1$

17. conjugate

19. $\sqrt{2}\sqrt{8}=\sqrt{16}=4$

21. $\sqrt{5}\sqrt{10}=\sqrt{50}=\sqrt{25}\sqrt{2}=5\sqrt{2}$

23. $2\sqrt{3}\sqrt{6}=2\sqrt{18}=2\sqrt{9}\sqrt{2}=2(3)\sqrt{2}$
$=6\sqrt{2}$

25. $\sqrt[3]{5}\sqrt[3]{25}=\sqrt[3]{125}=5$

27. $\sqrt[3]{2}\sqrt[3]{12}=\sqrt[3]{24}=\sqrt[3]{8}\sqrt[3]{3}=2\sqrt[3]{3}$

29. $\sqrt{ab^3}\sqrt{ab}=\sqrt{a^2b^4}=ab^2$

31. $\sqrt[3]{5r^2s}\sqrt[3]{2r}=\sqrt[3]{10r^3s}=\sqrt[3]{r^3}\sqrt[3]{10s}$
$=r\sqrt[3]{10s}$

33. $\sqrt{x(x+3)}\sqrt{x^3(x+3)}=\sqrt{x^4(x+3)^2}$
$=x^2(x+3)$

35. $3\sqrt{5}\left(4-\sqrt{5}\right)=12\sqrt{5}-3\sqrt{25}=12\sqrt{5}-3(5)=12\sqrt{5}-15$

37. $3\sqrt{2}\left(4\sqrt{3}+2\sqrt{7}\right)=12\sqrt{6}+6\sqrt{14}$

39. $\left(\sqrt{2}+1\right)\left(\sqrt{2}-3\right)=\sqrt{4}-3\sqrt{2}+\sqrt{2}-3=2-2\sqrt{2}-3=-1-2\sqrt{2}$

41. $(4\sqrt{x}+3)(2\sqrt{x}-5)=8\sqrt{x^2}-20\sqrt{x}+6\sqrt{x}-15=8x-14\sqrt{x}-15$

43. $\left(\sqrt{5z}+\sqrt{3}\right)\left(\sqrt{5z}+\sqrt{3}\right)=\sqrt{25z^2}+\sqrt{15z}+\sqrt{15z}+\sqrt{9}=5z+2\sqrt{15z}+3$

45. $\left(2\sqrt{3a}-\sqrt{b}\right)\left(\sqrt{3a}+3\sqrt{b}\right)=2\sqrt{9a^2}+6\sqrt{3ab}-\sqrt{3ab}-3\sqrt{b^2}=6a+5\sqrt{3ab}-3b$

47. $\left(3\sqrt{2r}-2\right)^2=\left(3\sqrt{2r}-2\right)\left(3\sqrt{2r}-2\right)=9\sqrt{4r^2}-6\sqrt{2r}-6\sqrt{2r}+4=18r-12\sqrt{2r}+4$

49. $-2\left(\sqrt{3x}+\sqrt{3}\right)^2=-2\left(\sqrt{3x}+\sqrt{3}\right)\left(\sqrt{3x}+\sqrt{3}\right)$
$=-2\left(\sqrt{9x^2}+\sqrt{9x}+\sqrt{9x}+\sqrt{9}\right)$
$=-2(3x+3\sqrt{x}+3\sqrt{x}+3)=-2(3x+6\sqrt{x}+3)=-6x-12\sqrt{x}-6$

SECTION 9.5

51. $\sqrt{\frac{1}{7}} = \frac{\sqrt{1}}{\sqrt{7}} = \frac{1\sqrt{7}}{\sqrt{7}\sqrt{7}} = \frac{\sqrt{7}}{7}$

53. $\sqrt{\frac{2}{3}} = \frac{\sqrt{2}}{\sqrt{3}} = \frac{\sqrt{2}\sqrt{3}}{\sqrt{3}\sqrt{3}} = \frac{\sqrt{6}}{3}$

55. $\frac{\sqrt{5}}{\sqrt{8}} = \frac{\sqrt{5}\sqrt{2}}{\sqrt{8}\sqrt{2}} = \frac{\sqrt{10}}{\sqrt{16}} = \frac{\sqrt{10}}{4}$

57. $\frac{\sqrt{8}}{\sqrt{2}} = \sqrt{\frac{8}{2}} = \sqrt{4} = 2$

59. $\frac{1}{\sqrt[3]{2}} = \frac{1\sqrt[3]{4}}{\sqrt[3]{2}\sqrt[3]{4}} = \frac{\sqrt[3]{4}}{\sqrt[3]{8}} = \frac{\sqrt[3]{4}}{2}$

61. $\frac{\sqrt[3]{2}}{\sqrt[3]{9}} = \frac{\sqrt[3]{2}\sqrt[3]{3}}{\sqrt[3]{9}\sqrt[3]{3}} = \frac{\sqrt[3]{6}}{\sqrt[3]{27}} = \frac{\sqrt[3]{6}}{3}$

63. $\frac{\sqrt{8x^2y}}{\sqrt{xy}} = \sqrt{\frac{8x^2y}{xy}} = \sqrt{8x} = 2\sqrt{2x}$

65. $\frac{\sqrt{10xy^2}}{\sqrt{2xy^3}} = \sqrt{\frac{10xy^2}{2xy^3}} = \sqrt{\frac{5}{y}} = \frac{\sqrt{5}\sqrt{y}}{\sqrt{y}\sqrt{y}}$

$= \frac{\sqrt{5y}}{y}$

67. $\frac{1}{\sqrt{2}-1} = \frac{1\left(\sqrt{2}+1\right)}{\left(\sqrt{2}-1\right)\left(\sqrt{2}+1\right)} = \frac{\sqrt{2}+1}{\sqrt{4}-1} = \frac{\sqrt{2}+1}{2-1} = \frac{\sqrt{2}+1}{1} = \sqrt{2}+1$

69. $\frac{\sqrt{2}}{\sqrt{5}+3} = \frac{\sqrt{2}\left(\sqrt{5}-3\right)}{\left(\sqrt{5}+3\right)\left(\sqrt{5}-3\right)} = \frac{\sqrt{2}\left(\sqrt{5}-3\right)}{\sqrt{25}-9} = \frac{\sqrt{2}\left(\sqrt{5}-3\right)}{5-9} = \frac{\sqrt{2}\left(\sqrt{5}-3\right)}{-4}$

$= \frac{\sqrt{10}-3\sqrt{2}}{-4}$

$= \frac{3\sqrt{2}-\sqrt{10}}{4}$

71. $\frac{\sqrt{3}+1}{\sqrt{3}-1} = \frac{\left(\sqrt{3}+1\right)\left(\sqrt{3}+1\right)}{\left(\sqrt{3}-1\right)\left(\sqrt{3}+1\right)} = \frac{\sqrt{9}+\sqrt{3}+\sqrt{3}+1}{\sqrt{9}-1} = \frac{4+2\sqrt{3}}{2} = \frac{2\left(2+\sqrt{3}\right)}{2}$

$= 2+\sqrt{3}$

73. $\frac{\sqrt{7}-\sqrt{2}}{\sqrt{2}+\sqrt{7}} = \frac{\left(\sqrt{7}-\sqrt{2}\right)\left(\sqrt{2}-\sqrt{7}\right)}{\left(\sqrt{2}+\sqrt{7}\right)\left(\sqrt{2}-\sqrt{7}\right)} = \frac{\sqrt{14}-7-2+\sqrt{14}}{\sqrt{4}-\sqrt{49}} = \frac{2\sqrt{14}-9}{-5} = \frac{9-2\sqrt{14}}{5}$

75. $\frac{2}{\sqrt{x}+1} = \frac{2(\sqrt{x}-1)}{(\sqrt{x}+1)(\sqrt{x}-1)} = \frac{2(\sqrt{x}-1)}{\sqrt{x^2}-1} = \frac{2(\sqrt{x}-1)}{x-1}$

77. $\frac{x}{\sqrt{x}-4} = \frac{x(\sqrt{x}+4)}{(\sqrt{x}-4)(\sqrt{x}+4)} = \frac{x(\sqrt{x}+4)}{\sqrt{x^2}-16} = \frac{x(\sqrt{x}+4)}{x-16}$

SECTION 9.5

79. $\dfrac{\sqrt{x}-\sqrt{y}}{\sqrt{x}+\sqrt{y}}=\dfrac{(\sqrt{x}-\sqrt{y})(\sqrt{x}-\sqrt{y})}{(\sqrt{x}+\sqrt{y})(\sqrt{x}-\sqrt{y})}=\dfrac{\sqrt{x^2}-\sqrt{xy}-\sqrt{xy}+\sqrt{y^2}}{\sqrt{x^2}-\sqrt{y^2}}=\dfrac{x-2\sqrt{xy}+y}{x-y}$

81. $\dfrac{\sqrt{3}+1}{2}=\dfrac{\left(\sqrt{3}+1\right)\left(\sqrt{3}-1\right)}{2\left(\sqrt{3}-1\right)}=\dfrac{\sqrt{9}-1}{2\left(\sqrt{3}-1\right)}=\dfrac{2}{2\left(\sqrt{3}-1\right)}=\dfrac{1}{\sqrt{3}-1}$

83. $\dfrac{\sqrt{x}+3}{x}=\dfrac{(\sqrt{x}+3)(\sqrt{x}-3)}{x(\sqrt{x}-3)}=\dfrac{\sqrt{x^2}-9}{x(\sqrt{x}-3)}=\dfrac{x-9}{x(\sqrt{x}-3)}$

85. $\left(3\sqrt[3]{9}\right)\left(2\sqrt[3]{3}\right)=6\sqrt[3]{27}=6(3)=18$

87. $\sqrt{5ab}\sqrt{5a}=\sqrt{25a^2b}=\sqrt{25a^2}\sqrt{b}$
$=5a\sqrt{b}$

89. $\sqrt[3]{a^5b}\sqrt[3]{16ab^5}=\sqrt[3]{16a^6b^6}=\sqrt[3]{8a^6b^6}\sqrt[3]{2}=2a^2b^2\sqrt[3]{2}$

91. $\sqrt[3]{6x^2(y+z)^2}\sqrt[3]{18x(y+z)}=\sqrt[3]{108x^3(y+z)^3}=\sqrt[3]{27x^3(y+z)^3}\sqrt[3]{4}=3x(y+z)\sqrt[3]{4}$

93. $-2\sqrt{5x}\left(4\sqrt{2x}-3\sqrt{3}\right)=-8\sqrt{10x^2}+6\sqrt{15x}=-8x\sqrt{10}+6\sqrt{15x}$

95. $\dfrac{3}{\sqrt[3]{9}}=\dfrac{3\sqrt[3]{3}}{\sqrt[3]{9}\sqrt[3]{3}}=\dfrac{3\sqrt[3]{3}}{\sqrt[3]{27}}=\dfrac{3\sqrt[3]{3}}{3}=\sqrt[3]{3}$

97. $\dfrac{1}{\sqrt[4]{4}}=\dfrac{1\sqrt[4]{4}}{\sqrt[4]{4}\sqrt[4]{4}}=\dfrac{\sqrt[4]{4}}{\sqrt[4]{16}}=\dfrac{\sqrt[4]{4}}{2}$

99. $\dfrac{1}{\sqrt[5]{16}}=\dfrac{1\sqrt[5]{2}}{\sqrt[5]{16}\sqrt[5]{2}}=\dfrac{\sqrt[5]{2}}{\sqrt[5]{32}}=\dfrac{\sqrt[5]{2}}{2}$

101. $\dfrac{\sqrt[3]{4a^2}}{\sqrt[3]{2ab}}=\sqrt[3]{\dfrac{4a^2}{2ab}}=\sqrt[3]{\dfrac{2a}{b}}=\dfrac{\sqrt[3]{2a}\sqrt[3]{b^2}}{\sqrt[3]{b}\sqrt[3]{b^2}}$
$=\dfrac{\sqrt[3]{2ab^2}}{b}$

103. $\dfrac{2z-1}{\sqrt{2z}-1}=\dfrac{(2z-1)\left(\sqrt{2z}+1\right)}{\left(\sqrt{2z}-1\right)\left(\sqrt{2z}+1\right)}=\dfrac{(2z-1)\left(\sqrt{2z}+1\right)}{\sqrt{4z^2}-1}=\dfrac{(2z-1)\left(\sqrt{2z}+1\right)}{2z-1}$
$=\sqrt{2z}+1$

105. $\dfrac{\sqrt{x}+\sqrt{y}}{\sqrt{x}}=\dfrac{(\sqrt{x}+\sqrt{y})(\sqrt{x}-\sqrt{y})}{\sqrt{x}(\sqrt{x}-\sqrt{y})}=\dfrac{\sqrt{x^2}-\sqrt{y^2}}{\sqrt{x^2}-\sqrt{xy}}=\dfrac{x-y}{x-\sqrt{xy}}$

107. If the area of the aperture is again cut in half, the area will equal $9\pi/4$ cm^2.

$$A = \pi r^2$$
$$\frac{9\pi}{4} = \pi\left(\frac{d}{2}\right)^2$$
$$\frac{9\pi}{4} = \frac{\pi d^2}{4}$$
$$36\pi = 4\pi d^2$$
$$\tfrac{36\pi}{4\pi} = d^2$$
$$9 = d^2$$
$$3 = d$$

$$f\text{-number} = \frac{f}{d} = \frac{12}{3} = 4$$

$\boxed{f/4}$

109. $r = \sqrt{\dfrac{A}{\pi}} = \dfrac{\sqrt{A}}{\sqrt{\pi}} = \dfrac{\sqrt{A}\sqrt{\pi}}{\sqrt{\pi}\sqrt{\pi}} = \dfrac{\sqrt{A\pi}}{\pi}$

111. $\dfrac{8}{\sqrt{2}} = \dfrac{8\sqrt{2}}{\sqrt{2}\sqrt{2}} = \dfrac{8\sqrt{2}}{2} = 4\sqrt{2}$ cm

113. shorter leg: $\dfrac{6}{\sqrt{3}} = \dfrac{6\sqrt{3}}{\sqrt{3}\sqrt{3}} = \dfrac{6\sqrt{3}}{3} = 2\sqrt{3}$ ft; hypotenuse: $2\left(2\sqrt{3}\right) = 4\sqrt{3}$ ft

115. Answers may vary.

117. $\dfrac{\sqrt{x}-3}{4} = \dfrac{(\sqrt{x}-3)(\sqrt{x}+3)}{4(\sqrt{x}+3)} = \dfrac{x-9}{4(\sqrt{x}+3)}$

Exercise 9.6 (page 667)

1.
$$\sqrt{x+2} = 3$$
$$\left(\sqrt{x+2}\right)^2 = 3^2$$
$$x+2 = 9$$
$$x = 7$$
The answer checks.

3.
$$\sqrt{x+1} = 1$$
$$\left(\sqrt{x+1}\right)^2 = 1^2$$
$$x+1 = 1$$
$$x = 0$$
The answer checks.

5.
$$\sqrt[4]{x-1} = 2$$
$$\left(\sqrt[4]{x-1}\right)^4 = 2^4$$
$$x-1 = 16$$
$$x = 17$$
The answer checks.

7. $f(0) = 3(0)^2 - 4(0) + 2 = 2$

9. $f(2) = 3(2)^2 - 4(2) + 2 = 6$

11. $x^n = y^n$; power rule

13. square

15. extraneous

17.
$$\sqrt{5x-6} = 2$$
$$\left(\sqrt{5x-6}\right)^2 = 2^2$$
$$5x-6 = 4$$
$$5x = 10$$
$$x = 2$$
The answer checks.

19.
$$\sqrt{6x+1}+2 = 7$$
$$\sqrt{6x+1} = 5$$
$$\left(\sqrt{6x+1}\right)^2 = 5^2$$
$$6x+1 = 25$$
$$6x = 24$$
$$x = 4$$
The answer checks.

21.
$$\sqrt[3]{7n-1} = 3$$
$$\left(\sqrt[3]{7n-1}\right)^3 = 3^3$$
$$7n - 1 = 27$$
$$7n = 28$$
$$n = 4$$
The answer checks.

23.
$$x = \frac{\sqrt{12x-5}}{2}$$
$$2x = \sqrt{12x-5}$$
$$(2x)^2 = \left(\sqrt{12x-5}\right)^2$$
$$4x^2 = 12x - 5$$
$$4x^2 - 12x + 5 = 0$$
$$(2x-1)(2x-5)$$
$$2x - 1 = 0 \quad \textbf{or} \quad 2x - 5 = 0$$
$$2x = 1 \qquad 2x = 5$$
$$x = \tfrac{1}{2} \qquad x = \tfrac{5}{2}$$
Both answers check.

25.
$$r - 9 = \sqrt{2r-3}$$
$$(r-9)^2 = \left(\sqrt{2r-3}\right)^2$$
$$r^2 - 18r + 81 = 2r - 3$$
$$r^2 - 20r + 84 = 0$$
$$(r-14)(r-6) = 0$$
$$r - 14 = 0 \quad \textbf{or} \quad r - 6 = 0$$
$$r = 14 \qquad r = 6$$
solution extraneous

27.
$$\sqrt{-5x+24} = 6 - x$$
$$\left(\sqrt{-5x+24}\right)^2 = (6-x)^2$$
$$-5x + 24 = 36 - 12x + x^2$$
$$0 = x^2 - 7x + 12$$
$$0 = (x-3)(x-4)$$
$$x - 3 = 0 \quad \textbf{or} \quad x - 4 = 0$$
$$x = 3 \qquad x = 4$$
solution solution

29.
$$\sqrt{y+2} = 4 - y$$
$$\left(\sqrt{y+2}\right) = (4-y)^2$$
$$y + 2 = 16 - 8y + y^2$$
$$0 = y^2 - 9y + 14$$
$$0 = (y-2)(y-7)$$
$$y - 2 = 0 \quad \textbf{or} \quad y - 7 = 0$$
$$y = 2 \qquad y = 7$$
solution extraneous

31.
$$\sqrt[3]{x^3-7} = x - 1$$
$$\left(\sqrt[3]{x^3-7}\right)^3 = (x-1)^3$$
$$x^3 - 7 = x^3 - 3x^2 + 3x - 1$$
$$3x^2 - 3x - 6 = 0$$
$$3(x-2)(x+1) = 0$$
$$x - 2 = 0 \quad \textbf{or} \quad x + 1 = 0$$
$$x = 2 \qquad x = -1$$
solution solution

33.
$$2\sqrt{4x+1} = \sqrt{x+4}$$
$$\left(2\sqrt{4x+1}\right)^2 = \left(\sqrt{x+4}\right)^2$$
$$4(4x+1) = x + 4$$
$$16x + 4 = x + 4$$
$$15x = 0$$
$$x = 0$$
The answer checks.

35.
$$\sqrt{x+2} = \sqrt{4-x}$$
$$\left(\sqrt{x+2}\right)^2 = \left(\sqrt{4-x}\right)^2$$
$$x + 2 = 4 - x$$
$$2x = 2$$
$$x = 1$$
The answer checks.

37.
$$\begin{aligned} 2\sqrt{x} &= \sqrt{5x-16} \\ \left(2\sqrt{x}\right)^2 &= \left(\sqrt{5x-16}\right)^2 \\ 4x &= 5x-16 \\ -x &= -16 \\ x &= 16 \end{aligned}$$
The answer checks.

39.
$$\begin{aligned} \sqrt{2y+1} &= 1-2\sqrt{y} \\ \left(\sqrt{2y+1}\right)^2 &= \left(1-2\sqrt{y}\right)^2 \\ 2y+1 &= 1-4\sqrt{y}+4y \\ 4\sqrt{y} &= 2y \\ \left(4\sqrt{y}\right)^2 &= (2y)^2 \\ 16y &= 4y^2 \\ 0 &= 4y^2-16y \\ 0 &= 4y(y-4) \end{aligned}$$
$4y = 0$ **or** $y-4=0$

$y=0$ $\quad y=4$

solution $\quad$ extraneous

41.
$$\begin{aligned} 1+\sqrt{z} &= \sqrt{z+3} \\ \left(1+\sqrt{z}\right)^2 &= \left(\sqrt{z+3}\right)^2 \\ 1+2\sqrt{z}+z &= z+3 \\ 2\sqrt{z} &= 2 \\ \left(2\sqrt{z}\right)^2 &= 2^2 \\ 4z &= 4 \\ z &= 1 \end{aligned}$$
The answer checks.

43.
$$\begin{aligned} \sqrt{4s+1}-\sqrt{6s} &= -1 \\ \sqrt{4s+1} &= \sqrt{6s}-1 \\ \left(\sqrt{4s+1}\right)^2 &= \left(\sqrt{6s}-1\right)^2 \\ 4s+1 &= 6s-2\sqrt{6s}+1 \\ 2\sqrt{6s} &= 2s \\ \left(2\sqrt{6s}\right)^2 &= (2s)^2 \\ 24s &= 4s^2 \\ 0 &= 4s^2-24s \\ 0 &= 4s(s-6) \end{aligned}$$
$4s=0$ **or** $s-6=0$

$s=0$ $\quad s=6$

extraneous $\quad$ solution

45.
$$\begin{aligned} \sqrt{2x+5}+\sqrt{x+2} &= 5 \\ \sqrt{2x+5} &= 5-\sqrt{x+2} \\ \left(\sqrt{2x+5}\right)^2 &= \left(5-\sqrt{x+2}\right)^2 \\ 2x+5 &= 25-10\sqrt{x+2}+x+2 \\ 10\sqrt{x+2} &= -x+22 \\ \left(10\sqrt{x+2}\right)^2 &= (-x+22)^2 \end{aligned}$$

$$\begin{aligned} \left(10\sqrt{x+2}\right)^2 &= (-x+22)^2 \\ 100(x+2) &= x^2-44x+484 \\ 0 &= x^2-144x+284 \\ 0 &= (x-142)(x-2) \end{aligned}$$
$x-142=0$ **or** $x-2=0$

$x=142$ $\quad x=2$

extraneous $\quad$ solution

47.

$$\begin{aligned} \sqrt{v}+\sqrt{3} &= \sqrt{v+3} \\ \left(\sqrt{v}+\sqrt{3}\right)^2 &= \left(\sqrt{v+3}\right)^2 \\ v+2\sqrt{3v}+3 &= v+3 \\ 2\sqrt{3v} &= 0 \end{aligned}$$

$$\begin{aligned} 2\sqrt{3v} &= 0 \\ \left(2\sqrt{3v}\right)^2 &= 0^2 \\ 12v &= 0 \\ v &= 0 \end{aligned}$$

The answer checks.

49.

$$\begin{aligned} \sqrt{3x}-\sqrt{x+1} &= \sqrt{x-2} \\ \left(\sqrt{3x}-\sqrt{x+1}\right)^2 &= \left(\sqrt{x-2}\right)^2 \\ 3x-2\sqrt{3x(x+1)}+x+1 &= x-2 \\ 3x+3 &= 2\sqrt{3x^2+3x} \\ (3x+3)^2 &= \left(2\sqrt{3x^2+3x}\right)^2 \\ 9x^2+18x+9 &= 12x^2+12x \\ 0 &= 3x^2-6x-9 \\ 0 &= 3(x-3)(x+1) \end{aligned}$$

$x-3=0$ **or** $x+1=0$

$x=3$ $\qquad x=-1$

solution $\qquad$ extraneous

51.

$$\begin{aligned} v &= \sqrt{2gh} \\ v^2 &= \left(\sqrt{2gh}\right)^2 \\ v^2 &= 2gh \\ \frac{v^2}{2g} &= \frac{2gh}{2g} \\ \frac{v^2}{2g} &= h, \text{ or } h = \frac{v^2}{2g} \end{aligned}$$

53.

$$\begin{aligned} T &= 2\pi\sqrt{\frac{l}{32}} \\ T^2 &= \left(2\pi\sqrt{\frac{l}{32}}\right)^2 \\ T^2 &= 4\pi^2\cdot\frac{l}{32} \\ 32(T^2) &= 32\left(4\pi^2\cdot\frac{l}{32}\right) \\ 32T^2 &= 4\pi^2 l \\ \frac{32T^2}{4\pi^2} &= \frac{4\pi^2 l}{4\pi^2} \\ \frac{8T^2}{\pi^2} &= l, \text{ or } l = \frac{8T^2}{\pi^2} \end{aligned}$$

55.

$$\begin{aligned} r &= \sqrt[3]{\frac{A}{P}} - 1 \\ r + 1 &= \sqrt[3]{\frac{A}{P}} \\ (r+1)^3 &= \left(\sqrt[3]{\frac{A}{P}}\right)^3 \\ (r+1)^3 &= \frac{A}{P} \\ P(r+1)^3 &= P\left(\frac{A}{P}\right) \\ P(r+1)^3 &= A, \text{ or } A = P(r+1)^3 \end{aligned}$$

57.

$$\begin{aligned} L_A &= L_B\sqrt{1 - \frac{v^2}{c^2}} \\ L_A^2 &= \left(L_B\sqrt{1 - \frac{v^2}{c^2}}\right)^2 \\ L_A^2 &= L_B^2\left(1 - \frac{v^2}{c^2}\right) \\ \frac{L_A^2}{L_B^2} &= 1 - \frac{v^2}{c^2} \\ \frac{v^2}{c^2} &= 1 - \frac{L_A^2}{L_B^2} \\ c^2\left(\frac{v^2}{c^2}\right) &= c^2\left(1 - \frac{L_A^2}{L_B^2}\right) \\ v^2 &= c^2\left(1 - \frac{L_A^2}{L_B^2}\right) \end{aligned}$$

59.

$$\begin{aligned} 5r + 4 &= \sqrt{5r + 20} + 4r \\ r + 4 &= \sqrt{5r + 20} \\ (r+4)^2 &= \left(\sqrt{5r + 20}\right)^2 \\ r^2 + 8r + 16 &= 5r + 20 \\ r^2 + 3r - 4 &= 0 \\ (r+4)(r-1) &= 0 \end{aligned}$$

$r + 4 = 0$ **or** $r - 1 = 0$
$r = -4$ $\quad$ $r = 1$
solution $\quad$ solution

61.

$$\begin{aligned} \sqrt{x}\sqrt{x+6} &= 4 \\ \left(\sqrt{x}\sqrt{x+6}\right)^2 &= 4^2 \\ x(x+6) &= 16 \\ x^2 + 6x - 16 &= 0 \\ (x-2)(x+8) &= 0 \end{aligned}$$

$x - 2 = 0$ **or** $x + 8 = 0$
$x = 2$ $\quad$ $x = -8$
solution $\quad$ extraneous

63.

$$\begin{aligned} \sqrt[4]{x^4 + 4x^2 - 4} &= -x \\ \left(\sqrt[4]{x^4 + 4x^2 - 4}\right)^4 &= (-x)^4 \\ x^4 + 4x^2 - 4 &= x^4 \\ 4x^2 - 4 &= 0 \\ 4(x+1)(x-1) &= 0 \end{aligned}$$

$x + 1 = 0$ **or** $x - 1 = 0$
$x = -1$ $\quad$ $x = 1$
solution $\quad$ extraneous

65.

$$\begin{aligned} 2 + \sqrt{u} &= \sqrt{2u + 7} \\ \left(2 + \sqrt{u}\right)^2 &= \left(\sqrt{2u + 7}\right)^2 \\ 4 + 4\sqrt{u} + u &= 2u + 7 \\ 4\sqrt{u} &= u + 3 \\ \left(4\sqrt{u}\right)^2 &= (u+3)^2 \\ 16u &= u^2 + 6u + 9 \\ 0 &= u^2 - 10u + 9 \\ 0 &= (u-9)(u-1) \end{aligned}$$

$u - 9 = 0$ **or** $u - 1 = 0$
$u = 9$ $\quad$ $u = 1$
solution $\quad$ solution

67.

$$\begin{aligned} u &= \sqrt[4]{u^4 - 6u^2 + 24} \\ u^4 &= \left(\sqrt[4]{u^4 - 6u^2 + 24}\right) \\ u^4 &= u^4 - 6u^2 + 24 \\ 6u^2 - 24 &= 0 \\ 6(u+2)(u-2) &= 0 \end{aligned}$$

$u + 2 = 0$ **or** $u - 2 = 0$

$u = -2$ $\quad$ $u = 2$

extraneous $\quad$ solution

69.

$$\begin{aligned} \sqrt{x-5} - \sqrt{x+3} &= 4 \\ \sqrt{x-5} &= \sqrt{x+3} + 4 \\ \left(\sqrt{x-5}\right)^2 &= \left(\sqrt{x+3} + 4\right)^2 \\ x - 5 &= x + 3 + 8\sqrt{x+3} + 16 \\ -24 &= 8\sqrt{x+3} \\ -3 &= \sqrt{x+3} \\ (-3)^2 &= \left(\sqrt{x+3}\right)^2 \\ 9 &= x + 3 \\ 6 &= x \end{aligned}$$

extraneous $\Rightarrow \emptyset$

71.

$$\begin{aligned} \sqrt{x+8} - \sqrt{x-4} &= -2 \\ \sqrt{x+8} &= \sqrt{x-4} - 2 \\ \left(\sqrt{x+8}\right)^2 &= \left(\sqrt{x-4} - 2\right)^2 \\ x + 8 &= x - 4 - 4\sqrt{x-4} + 4 \\ 4\sqrt{x-4} &= -8 \\ \sqrt{x-4} &= -2 \\ \left(\sqrt{x-4}\right)^2 &= (-2)^2 \\ x - 4 &= 4 \\ x &= 8 \end{aligned}$$

extraneous $\Rightarrow \emptyset$

73.

$$\begin{aligned} \sqrt{z-1} + \sqrt{z+2} &= 3 \\ \sqrt{z-1} &= 3 - \sqrt{z+2} \\ \left(\sqrt{z-1}\right)^2 &= \left(3 - \sqrt{z+2}\right)^2 \\ z - 1 &= 9 - 6\sqrt{z+2} + z + 2 \\ 6\sqrt{z+2} &= 12 \\ \sqrt{z+2} &= 2 \\ \left(\sqrt{z+2}\right)^2 &= 2^2 \\ z + 2 &= 4 \end{aligned}$$

$z = 2$: The answer checks.

75.

$$\begin{aligned} \sqrt{\sqrt{a} + \sqrt{a+8}} &= 2 \\ \left(\sqrt{\sqrt{a} + \sqrt{a+8}}\right)^2 &= 2^2 \\ \sqrt{a} + \sqrt{a+8} &= 4 \\ \left(\sqrt{a} + \sqrt{a+8}\right)^2 &= 4^2 \\ a + 2\sqrt{a(a+8)} + a + 8 &= 16 \\ 2\sqrt{a^2 + 8a} &= -2a + 8 \\ \left(2\sqrt{a^2 + 8a}\right)^2 &= (-2a + 8)^2 \\ 4a^2 + 32a &= 4a^2 - 32a + 64 \\ 64a &= 64 \\ a &= 1 \end{aligned}$$

The answer checks.

77.

$$\begin{aligned} \frac{\sqrt{2x}}{\sqrt{x+2}} &= \sqrt{x-1} \\ \left(\frac{\sqrt{2x}}{\sqrt{x+2}}\right)^2 &= \left(\sqrt{x-1}\right)^2 \\ \frac{2x}{x+2} &= x - 1 \\ 2x &= x^2 + x - 2 \\ 0 &= x^2 - x - 2 \\ 0 &= (x-2)(x+1) \end{aligned}$$

$x - 2 = 0$ **or** $x + 1 = 0$

$x = 2$ $\quad$ $x = -1$

solution $\quad$ extraneous

79.
$$\begin{aligned} s &= 1.45\sqrt{r} \\ 65 &= 1.45\sqrt{r} \\ 65^2 &= \left(1.45\sqrt{r}\right)^2 \\ 4225 &= 2.1025r \\ 2010 \text{ ft} &\approx r \end{aligned}$$

81.
$$\begin{aligned} v &= \sqrt[3]{\frac{P}{0.02}} \\ v &= \sqrt[3]{\frac{500}{0.02}} \\ v &= \sqrt[3]{25000} \\ v &\approx 29 \text{ mph} \end{aligned}$$

83. $r = 1 - \sqrt[n]{\dfrac{T}{C}} = 1 - \sqrt[5]{\dfrac{9000}{22000}} \approx 1 - \sqrt[5]{0.40909} \approx 1 - 0.836 \approx 0.164 \approx 16\%$

85. Graph $y = \sqrt{5x}$ and $y = \sqrt{100 - 3x^2}$:

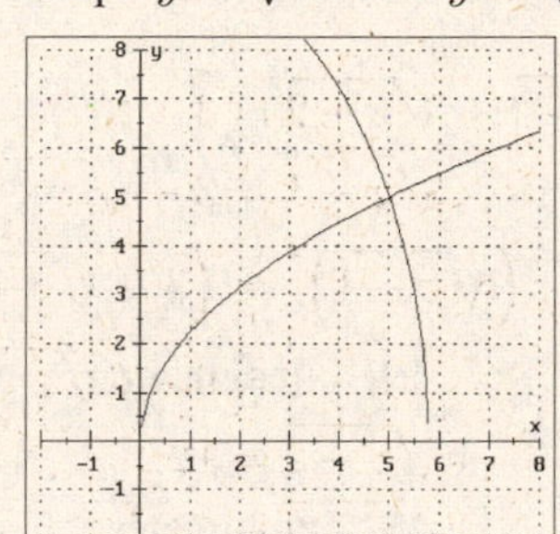

equilibrium price: $x = \$5$

87.

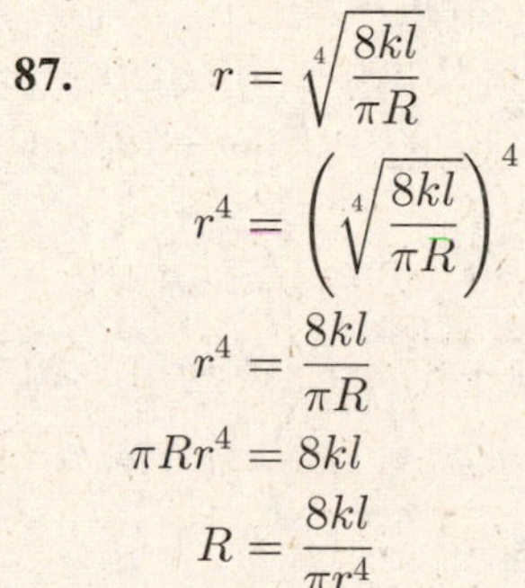

$$\begin{aligned} r &= \sqrt[4]{\frac{8kl}{\pi R}} \\ r^4 &= \left(\sqrt[4]{\frac{8kl}{\pi R}}\right)^4 \\ r^4 &= \frac{8kl}{\pi R} \\ \pi R r^4 &= 8kl \\ R &= \frac{8kl}{\pi r^4} \end{aligned}$$

89. **Answers may vary.**

91.
$$\begin{aligned} \sqrt[3]{2x} &= \sqrt{x} \\ \left(\sqrt[3]{2x}\right)^2 &= \left(\sqrt{x}\right)^2 \\ \left[\left(\sqrt[3]{2x}\right)^2\right]^3 &= x^3 \\ 4x^2 &= x^3 \\ 0 &= x^3 - 4x^2 \\ 0 &= x^2(x-4) \\ x = 0 &\text{ or } x = 4 \end{aligned}$$

Exercise 9.7 (page 678)

1. $\sqrt{-49} = \sqrt{49(-1)} = \sqrt{49}\sqrt{-1} = 7i$

3. $\sqrt{-100} = \sqrt{100(-1)} = \sqrt{100}\sqrt{-1} = 10i$

5. $i^3 = -i$

7. $i^4 = 1$

9. $|-3 + 4i| = \sqrt{(-3)^2 + 4^2} = \sqrt{9 + 16} = \sqrt{25} = 5$

SECTION 9.7

11. $\frac{x^2-x-6}{9-x^2}\cdot\frac{x^2+x-6}{x^2-4}=\frac{(x-3)(x+2)}{(3+x)(3-x)}\cdot\frac{(x+3)(x-2)}{(x+2)(x-2)}=-1$

13. Let w = the speed of the wind.

	Rate	Time	Dist.
Downwind	$200+w$	$\frac{330}{200+w}$	330
Upwind	$200-w$	$\frac{330}{200-w}$	330

$$\frac{330}{200+w}+\frac{330}{200-w}=\frac{10}{3}$$
$$\left(\frac{330}{200+w}+\frac{330}{200-w}\right)3(200+w)(200-w)=\frac{10}{3}\cdot 3(200+w)(200-w)$$
$$330(3)(200-w)+330(3)(200+w)=10(40{,}000-w^2)$$
$$198{,}000-990w+198{,}000+990w=400{,}000-10w^2$$
$$10w^2=4000$$
$$w^2=400$$
$$w=20\text{ mph}$$

15. i

17. $-i$

19. imaginary

21. $\frac{\sqrt{a}}{\sqrt{b}}$

23. 5; 7

25. conjugates

27. $\sqrt{-9}=\sqrt{9(-1)}=\sqrt{9}\sqrt{-1}=3i$

29. $\sqrt{-36}=\sqrt{36(-1)}=\sqrt{36}\sqrt{-1}=6i$

31. $\sqrt{-7}=\sqrt{7(-1)}=\sqrt{7}\sqrt{-1}=i\sqrt{7}$

33. $\sqrt{-8}=\sqrt{8(-1)}=\sqrt{8}\sqrt{-1}=\sqrt{4}\sqrt{2}\cdot i$
$=2i\sqrt{2}$

35. $3+7i\stackrel{?}{=}\sqrt{9}+(5+2)i$
$3+7i\stackrel{?}{=}3+7i$
They are equal.

37. $\sqrt{4}+\sqrt{-4}\stackrel{?}{=}2-2i$
$2+\sqrt{-1\cdot 4}\stackrel{?}{=}2-2i$
$2+2i=2-2i$
They are not equal.

39. $(3+4i)+(5-6i)=3+4i+5-6i$
$=8-2i$

41. $(7-3i)-(4+2i)=7-3i-4-2i$
$=3-5i$

43. $(8+5i)+(7+2i)=8+5i+7+2i$
$=15+7i$

45. $(1+i)-2i+(5-7i)=1+i-2i+5-7i$
$=6-8i$

47. $3i(2-i)=6i-3i^2=6i-3(-1)=6i+3=3+6i$

49. $-5i(5-5i)=-25i+25i^2=-25i+25(-1)=-25i-25=-25-25i$

51. $(2+i)(3-i)=6-2i+3i-i^2=6+i-(-1)=6+i+1=7+i$

53. $(2-4i)(3+2i) = 6+4i-12i-8i^2 = 6-8i-8(-1) = 6-8i+8 = 14-8i$

55. $(2+\sqrt{2}i)(3-\sqrt{2}i) = 6-2\sqrt{2}i+3\sqrt{2}i-2i^2 = 6+\sqrt{2}i-2(-1) = 8+\sqrt{2}i$

57. $(2+i)^2 = (2+i)(2+i) = 4+2i+2i+i^2 = 4+4i-1 = 3+4i$

59. $(2+3i)^2 = (2+3i)(2+3i) = 4+6i+6i+9i^2 = 4+12i-9 = -5+12i$

61. $i(5+i)(3-2i) = i(15-10i+3i-2i^2) = i(15-7i+2) = i(17-7i) = 17i-7i^2 = 7+17i$

63. $(8-\sqrt{-1})(-2-\sqrt{-16}) = (8-i)(-2-4i) = -16-32i+2i+4i^2 = -16-30i-4$
$= -20-30i$

65. $(6-5i)(6+5i) = 36+30i-30i-25i^2 = 36-25(-1) = 36+25 = 61$

67. $\dfrac{5}{2-i} = \dfrac{5(2+i)}{(2-i)(2+i)} = \dfrac{5(2+i)}{4-i^2} = \dfrac{5(2+i)}{5} = 2+i$

69. $\dfrac{13i}{5+i} = \dfrac{13i(5-i)}{(5+i)(5-i)} = \dfrac{65i-13i^2}{25-i^2} = \dfrac{13+65i}{26} = \dfrac{13}{26}+\dfrac{65}{26}i = \dfrac{1}{2}+\dfrac{5}{2}i$

71. $\dfrac{-12}{7-\sqrt{-1}} = \dfrac{-12}{7-i} = \dfrac{-12(7+i)}{(7-i)(7+i)} = \dfrac{-84-12i}{49-i^2} = \dfrac{-84-12i}{50} = \dfrac{-84}{50}-\dfrac{12}{50}i = -\dfrac{42}{25}-\dfrac{6}{25}i$

73. $\dfrac{5i}{6+2i} = \dfrac{5i(6-2i)}{(6+2i)(6-2i)} = \dfrac{30i-10i^2}{36-4i^2} = \dfrac{10+30i}{40} = \dfrac{10}{40}+\dfrac{30}{40}i = \dfrac{1}{4}+\dfrac{3}{4}i$

75. $\dfrac{3-2i}{3+2i} = \dfrac{(3-2i)(3-2i)}{(3+2i)(3-2i)} = \dfrac{9-6i-6i+4i^2}{9-4i^2} = \dfrac{5-12i}{13} = \dfrac{5}{13}-\dfrac{12}{13}i$

77. $\dfrac{3+2i}{3+i} = \dfrac{(3+2i)(3-i)}{(3+i)(3-i)} = \dfrac{9-3i+6i-2i^2}{9-i^2} = \dfrac{11+3i}{10} = \dfrac{11}{10}+\dfrac{3}{10}i$

79. $\dfrac{\sqrt{5}-\sqrt{3}i}{\sqrt{5}+\sqrt{3}i} = \dfrac{\left(\sqrt{5}-\sqrt{3}i\right)\left(\sqrt{5}-\sqrt{3}i\right)}{\left(\sqrt{5}+\sqrt{3}i\right)\left(\sqrt{5}-\sqrt{3}i\right)} = \dfrac{5-\sqrt{15}i-\sqrt{15}i+3i^2}{5-3i^2} = \dfrac{2-2\sqrt{15}i}{8}$

$= \dfrac{2}{8}-\dfrac{2\sqrt{15}}{8}i$

$= \dfrac{1}{4}-\dfrac{\sqrt{15}}{4}i$

SECTION 9.7

81. $\left(\dfrac{i}{3+2i}\right)^2 = \dfrac{i^2}{(3+2i)^2} = \dfrac{-1}{(3+2i)(3+2i)} = \dfrac{-1}{9+12i+4i^2} = \dfrac{-1}{5+12i} = \dfrac{-1(5-12i)}{(5+12i)(5-12i)}$

$= \dfrac{-5+12i}{25-144i^2}$

$= -\dfrac{5}{169} + \dfrac{12}{169}i$

83. $i^{21} = i^{20}i^1 = (i^4)^5 i = 1^5 i = i$

85. $i^{27} = i^{24}i^3 = (i^4)^6 i^3 = 1^6 i^3 = i^3 = -i$

87. $i^{100} = (i^4)^{25} = 1^{25} = 1$

89. $i^{97} = i^{96}i^1 = (i^4)^{24} i = 1^{24} i = i$

91. $3i^3 + i^2 = 3(-i) + (-1) = -1 - 3i$

93. $\dfrac{1}{i} = \dfrac{1i^3}{ii^3} = \dfrac{i^3}{i^4} = \dfrac{i^3}{1} = i^3 = -i = 0 - i$

95. $\dfrac{4}{5i^3} = \dfrac{4i}{5i^3 i} = \dfrac{4i}{5i^4} = \dfrac{4i}{5(1)} = \dfrac{4}{5}i = 0 + \dfrac{4}{5}i$

97. $\dfrac{3i}{8\sqrt{-9}} = \dfrac{3i}{8(3i)} = \dfrac{1}{8} = \dfrac{1}{8} + 0i$

99. $\dfrac{-3}{5i^5} = \dfrac{-3i^3}{5i^5 i^3} = \dfrac{-3(-i)}{5i^8} = \dfrac{3i}{5} = 0 + \dfrac{3}{5}i$

101. $|6+8i| = \sqrt{6^2+8^2} = \sqrt{36+64} = \sqrt{100}$

$= 10$

103. $|12-5i| = \sqrt{12^2+(-5)^2} = \sqrt{144+25} = \sqrt{169} = 13$

105. $|5+7i| = \sqrt{5^2+7^2} = \sqrt{25+49} = \sqrt{74}$

107. $\left|\dfrac{3}{5} - \dfrac{4}{5}i\right| = \sqrt{\left(\dfrac{3}{5}\right)^2 + \left(\dfrac{4}{5}\right)^2} = \sqrt{\dfrac{9}{25} + \dfrac{16}{25}} = \sqrt{\dfrac{25}{25}} = \sqrt{1} = 1$

109. $8 + 5i \stackrel{?}{=} 2^3 + \sqrt{25}i^3$

$8 + 5i \stackrel{?}{=} 8 + 5(-i)$

$8 + 5i \stackrel{?}{=} 8 - 5i \Rightarrow$ They are not equal.

111. $(5+3i) - (3-5i) + \sqrt{-1} = 5 + 3i - 3 + 5i + i = 2 + 9i$

113. $(-8 - \sqrt{3}i) - (7 - 3\sqrt{3}i) = -8 - \sqrt{3}i - 7 + 3\sqrt{3}i = -15 + 2\sqrt{3}i$

115. $(2+i)(2-i)(1+i) = (4 - 2i + 2i - i^2)(1+i) = 5(1+i) = 5 + 5i$

117. $(3+i)[(3-2i)+(2+i)] = (3+i)(5-i) = 15 - 3i + 5i - i^2 = 16 + 2i$

119. $\dfrac{i(3-i)}{3+i} = \dfrac{(3i-i^2)(3-i)}{(3+i)(3-i)} = \dfrac{(1+3i)(3-i)}{9-i^2} = \dfrac{3 - i + 9i - 3i^2}{10} = \dfrac{6+8i}{10} = \dfrac{3}{5} + \dfrac{4}{5}i$

121. $$\frac{(2-5i)-(5-2i)}{5-i} = \frac{2-5i-5+2i}{5-i} = \frac{-3-3i}{5-i} = \frac{(-3-3i)(5+i)}{(5-i)(5+i)} = \frac{-15-3i-15i-3i^2}{25-i^2}$$
$$= \frac{-12-18i}{26}$$
$$= \frac{-12}{26} - \frac{18}{26}i$$
$$= -\frac{6}{13} - \frac{9}{13}i$$

123.
$$\begin{aligned} x^2-2x+26 &= 0 \\ (1-5i)^2-2(1-5i)+26 &= 0 \\ (1-5i)(1-5i)-2+10i+26 &= 0 \\ 1-10i+25i^2+24+10i &= 0 \\ 1-10i-25+24+10i &= 0 \\ 0 &= 0 \end{aligned}$$

125.
$$\begin{aligned} x^4-3x^2-4 &= 0 \\ i^4-3i^2-4 &= 0 \\ 1-3(-1)-4 &= 0 \\ 1+3-4 &= 0 \\ 0 &= 0 \end{aligned}$$

127. $V = IR = (2-3i)(2+i) = 4+2i-6i-3i^2 = 4-4i-3(-1) = 4-4i+3 = 7-4i$ volts

129. $Z = \frac{V}{I} = \frac{1.7+0.5i}{0.5i} = \frac{(1.7+0.5i)i}{(0.5i)i} = \frac{1.7i+0.5i^2}{0.5i^2} = \frac{-0.5+1.7i}{-0.5} = 1-3.4i$

131. Answers may vary.

133. $\frac{3-i}{2} = \frac{(3-i)(3+i)}{2(3+i)} = \frac{9-i^2}{2(3+i)} = \frac{10}{2(3+i)} = \frac{5}{3+i}$

Chapter 9 Review (page 682)

1. $\sqrt{49} = \sqrt{7^2} = 7$

2. $-\sqrt{121} = -\sqrt{11^2} = -11$

3. $-\sqrt{36} = -\sqrt{6^2} = -6$

4. $\sqrt{225} = \sqrt{15^2} = 15$

5. $\sqrt[3]{-27} = \sqrt[3]{(-3)^3} = -3$

6. $-\sqrt[3]{216} = -\sqrt[3]{6^3} = -6$

7. $\sqrt[4]{625} = \sqrt[4]{5^4} = 5$

8. $\sqrt[5]{-32} = \sqrt[5]{(-2)^5} = -2$

9. $\sqrt{25x^2} = \sqrt{5^2x^2} = |5x| = 5|x|$

10. $\sqrt{x^2+4x+4} = \sqrt{(x+2)^2} = |x+2|$

11. $\sqrt[3]{27a^6b^3} = 3a^2b$

12. $\sqrt[4]{256x^8y^4} = |4x^2y| = 4x^2|y|$

13. $y = f(x) = \sqrt{x+2}$; Shift $y = \sqrt{x}$ left 2.

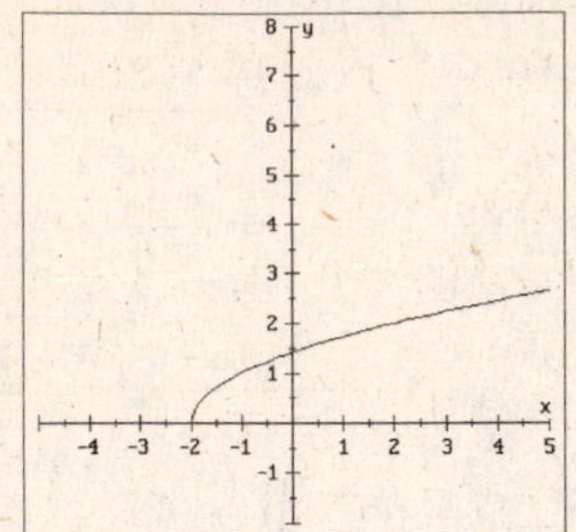

14. $y = f(x) = -\sqrt{x-1}$; Reflect $y = \sqrt{x}$ about the x-axis and shift right 1.

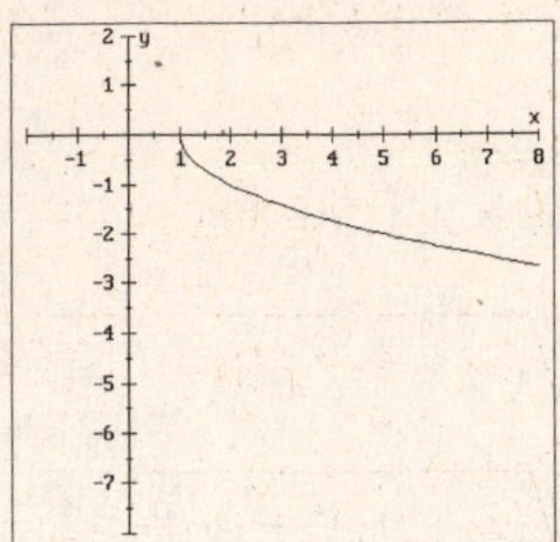

15. $y = f(x) = -\sqrt{x} + 2$; Reflect $y = \sqrt{x}$ about the x-axis and shift up 2.

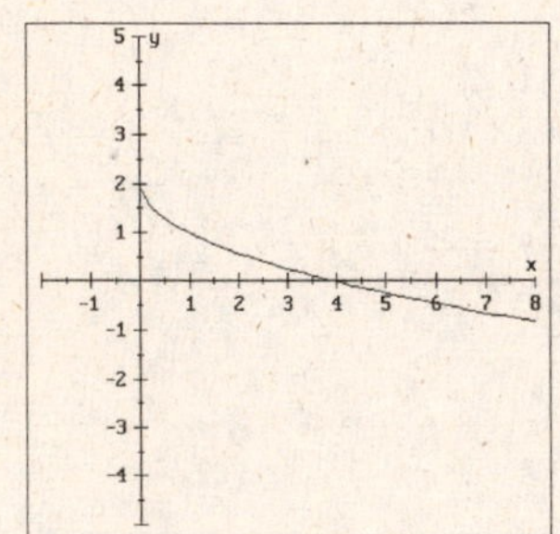

16. $y = f(x) = -\sqrt[3]{x} + 3$; Reflect $y = \sqrt[3]{x}$ about the x-axis and shift up 3.

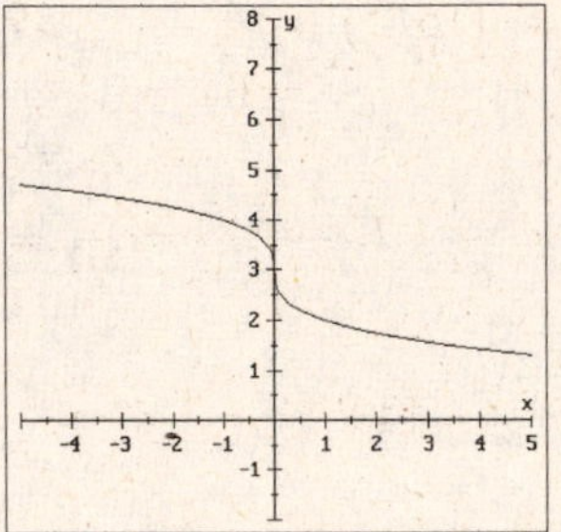

17. mean $= \frac{4+8+12+16+20}{5} = \frac{60}{5} = 12$

18.

Original term	Mean	Difference (term−mean)	Square of difference
4	12	−8	64
8	12	−4	16
12	12	0	0
16	12	4	16
20	12	8	64

$$\begin{aligned} s &= \sqrt{\frac{64+16+0+16+64}{5}} \\ &= \sqrt{\frac{160}{5}} \\ &= \sqrt{32} \\ &\approx 5.7 \end{aligned}$$

19. $d = 1.4\sqrt{h}$

$d = 1.4\sqrt{4.7} \approx 3.0$ miles

20.

$$\begin{aligned} d &= 1.4\sqrt{h} \\ 4 &= 1.4\sqrt{h} \\ 16 &= 1.96h \\ 8.2 \text{ ft} &= h \end{aligned}$$

21. Let x = one-half of d.
$$125^2 = x^2 + 117^2$$
$$15625 = x^2 + 13689$$
$$1936 = x^2$$
$$44 = x$$
$$d = 2x = 2(44) = 88 \text{ yd}$$

22. Let x = one-half of d.
$$8900^2 = x^2 + 3900^2$$
$$79{,}210{,}000 = x^2 + 15{,}210{,}000$$
$$64{,}000{,}000 = x^2$$
$$8000 = x$$
$$d = 2x = 2(8000) = 16{,}000 \text{ yd } (\approx 3 \text{ mi})$$

23. $d = \sqrt{(x_2 - x_1)^2 + (y_2 - y_1)^2} = \sqrt{(0-5)^2 + [0-(-12)]^2} = \sqrt{(-5)^2 + 12^2} = \sqrt{169} = 13$

24. $d = \sqrt{(x_2 - x_1)^2 + (y_2 - y_1)^2} = \sqrt{(-4-(-2))^2 + (6-8)^2} = \sqrt{(-2)^2 + (-2)^2} = \sqrt{8} \approx 2.83$

25. $25^{1/2} = 5$

26. $-36^{1/2} = -6$

27. $9^{3/2} = \left(9^{1/2}\right)^3 = 3^3 = 27$

28. $16^{3/2} = \left(16^{1/2}\right)^3 = 4^3 = 64$

29. $(-8)^{1/3} = -2$

30. $-8^{2/3} = -(8^{1/3})^2 = -2^2 = -4$

31. $8^{-2/3} = \dfrac{1}{8^{2/3}} = \dfrac{1}{(8^{1/3})^2} = \dfrac{1}{2^2} = \dfrac{1}{4}$

32. $8^{-1/3} = \dfrac{1}{8^{1/3}} = \dfrac{1}{2}$

33. $-49^{5/2} = -(49^{1/2})^5 = -7^5 = -16{,}807$

34. $\dfrac{1}{25^{5/2}} = \dfrac{1}{(25^{1/2})^5} = \dfrac{1}{5^5} = \dfrac{1}{3125}$

35. $\left(\dfrac{1}{4}\right)^{-3/2} = 4^{3/2} = (4^{1/2})^3 = 2^3 = 8$

36. $\left(\dfrac{4}{9}\right)^{-3/2} = \left(\dfrac{9}{4}\right)^{3/2} = \left[\left(\dfrac{9}{4}\right)^{1/2}\right]^3 = \left(\dfrac{3}{2}\right)^3 = \dfrac{27}{8}$

37. $(27x^3y)^{1/3} = 3xy^{1/3}$

38. $(81x^4y^2)^{1/4} = 3xy^{1/2}$

39. $(25x^3y^4)^{3/2} = 125x^{9/2}y^6$

40. $(8u^2v^3)^{-2/3} = \dfrac{1}{4u^{4/3}v^2}$

41. $5^{1/4}5^{1/2} = 5^{1/4+1/2} = 5^{3/4}$

42. $a^{3/7}a^{2/7} = a^{3/7+2/7} = a^{5/7}$

43. $u^{1/2}\left(u^{1/2} - u^{-1/2}\right) = u^{2/2} - u^0 = u - 1$

44. $v^{2/3}\left(v^{1/3} + v^{4/3}\right) = v^{3/3} + v^{6/3} = v + v^2$

45. $\left(x^{1/2} + y^{1/2}\right)^2 = \left(x^{1/2} + y^{1/2}\right)\left(x^{1/2} + y^{1/2}\right) = x^{2/2} + x^{1/2}y^{1/2} + x^{1/2}y^{1/2} + y^{2/2} = x + 2x^{1/2}y^{1/2} + y$

46. $\left(a^{2/3} + b^{2/3}\right)\left(a^{2/3} - b^{2/3}\right) = a^{4/3} - a^{2/3}b^{2/3} + a^{2/3}b^{2/3} - b^{4/3} = a^{4/3} - b^{4/3}$

47. $\sqrt[6]{5^2} = 5^{2/6} = 5^{1/3} = \sqrt[3]{5}$

48. $\sqrt[8]{x^4} = x^{4/8} = x^{1/2} = \sqrt{x}$

49. $\sqrt[9]{27a^3b^6} = (3^3a^3b^6)^{1/9} = 3^{3/9}a^{3/9}b^{6/9} = 3^{1/3}a^{1/3}b^{2/3} = \sqrt[3]{3ab^2}$

50. $\sqrt[4]{25a^2b^2} = 5^{2/4}a^{2/4}b^{2/4} = 5^{1/2}a^{1/2}b^{1/2} = \sqrt{5ab}$

51. $\sqrt{240} = \sqrt{16 \cdot 15} = \sqrt{16}\sqrt{15} = 4\sqrt{15}$

52. $\sqrt[3]{54} = \sqrt[3]{27 \cdot 2} = \sqrt[3]{27}\sqrt[3]{2} = 3\sqrt[3]{2}$

53. $\sqrt[4]{32} = \sqrt[4]{16 \cdot 2} = \sqrt[4]{16}\sqrt[4]{2} = 2\sqrt[4]{2}$

54. $\sqrt[5]{96} = \sqrt[5]{32 \cdot 3} = \sqrt[5]{32}\sqrt[5]{3} = 2\sqrt[5]{3}$

55. $\sqrt{8x^3} = \sqrt{4x^2}\sqrt{2x} = 2x\sqrt{2x}$

56. $\sqrt{18x^4y^3} = \sqrt{9x^4y^2}\sqrt{2y} = 3x^2y\sqrt{2y}$

57. $\sqrt[3]{16x^5y^4} = \sqrt[3]{8x^3y^3}\sqrt[3]{2x^2y} = 2xy\sqrt[3]{2x^2y}$

58. $\sqrt[3]{54x^7y^3} = \sqrt[3]{27x^6y^3}\sqrt[3]{2x} = 3x^2y\sqrt[3]{2x}$

59. $\dfrac{\sqrt{32x^3}}{\sqrt{2x}} = \sqrt{\dfrac{32x^3}{2x}} = \sqrt{16x^2} = 4x$

60. $\dfrac{\sqrt[3]{16x^5}}{\sqrt[3]{2x^2}} = \sqrt[3]{\dfrac{16x^5}{2x^2}} = \sqrt[3]{8x^3} = 2x$

61. $\sqrt[3]{\dfrac{2a^2b}{27x^3}} = \dfrac{\sqrt[3]{2a^2b}}{3x}$

62. $\sqrt{\dfrac{17xy}{64a^4}} = \dfrac{\sqrt{17xy}}{8a^2}$

63. $\sqrt{2} + \sqrt{8} = \sqrt{2} + \sqrt{4}\sqrt{2} = \sqrt{2} + 2\sqrt{2}$
$= 3\sqrt{2}$

64. $\sqrt{20} - \sqrt{5} = \sqrt{4}\sqrt{5} - \sqrt{5} = 2\sqrt{5} - \sqrt{5}$
$= \sqrt{5}$

65. $2\sqrt[3]{3} - \sqrt[3]{24} = 2\sqrt[3]{3} - \sqrt[3]{8}\sqrt[3]{3} = 2\sqrt[3]{3} - 2\sqrt[3]{3} = 0$

66. $\sqrt[4]{32} + 2\sqrt[4]{162} = \sqrt[4]{16}\sqrt[4]{2} + 2\sqrt[4]{81}\sqrt[4]{2} = 2\sqrt[4]{2} + 2(3)\sqrt[4]{2} = 2\sqrt[4]{2} + 6\sqrt[4]{2} = 8\sqrt[4]{2}$

67. $2x\sqrt{8} + 2\sqrt{200x^2} + \sqrt{50x^2} = 2x\sqrt{4}\sqrt{2} + 2\sqrt{100x^2}\sqrt{2} + \sqrt{25x^2}\sqrt{2}$
$= 2x(2)\sqrt{2} + 2(10x)\sqrt{2} + 5x\sqrt{2}$
$= 4x\sqrt{2} + 20x\sqrt{2} + 5x\sqrt{2} = 29x\sqrt{2}$

68. $3\sqrt{27a^3} - 2a\sqrt{3a} + 5\sqrt{75a^3} = 3\sqrt{9a^2}\sqrt{3a} - 2a\sqrt{3a} + 5\sqrt{25a^2}\sqrt{3a}$
$= 3(3a)\sqrt{3a} - 2a\sqrt{3a} + 5(5a)\sqrt{3a}$
$= 9a\sqrt{3a} - 2a\sqrt{3a} + 25a\sqrt{3a} = 32a\sqrt{3a}$

69. $\sqrt[3]{54} - 3\sqrt[3]{16} + 4\sqrt[3]{128} = \sqrt[3]{27}\sqrt[3]{2} - 3\sqrt[3]{8}\sqrt[3]{2} + 4\sqrt[3]{64}\sqrt[3]{2}$
$= 3\sqrt[3]{2} - 3(2)\sqrt[3]{2} + 4(4)\sqrt[3]{2} = 3\sqrt[3]{2} - 6\sqrt[3]{2} + 16\sqrt[3]{2} = 13\sqrt[3]{2}$

70. $2\sqrt[4]{32x^5} + 4\sqrt[4]{162x^5} - 5x\sqrt[4]{512x} = 2\sqrt[4]{16x^4}\sqrt[4]{2x} + 4\sqrt[4]{81x^4}\sqrt[4]{2x} - 5x\sqrt[4]{256}\sqrt[4]{2x}$
$= 2(2x)\sqrt[4]{2x} + 4(3x)\sqrt[4]{2x} - 5x(4)\sqrt[4]{2x}$
$= 4x\sqrt[4]{2x} + 12x\sqrt[4]{2x} - 20x\sqrt[4]{2x} = -4x\sqrt[4]{2x}$

71. hypotenuse $= 7\sqrt{2}$ m

72. shorter leg $= \dfrac{1}{2}\left(12\sqrt{3}\right) = 6\sqrt{3}$ cm
longer leg $= \sqrt{3}\left(6\sqrt{3}\right) = 6(3) = 18$ cm

73. $x = 5\sqrt{2} \approx 7.07$ in.

74. $x = \sqrt{3}\left(\frac{1}{2} \cdot 10\right) = 5\sqrt{3} \approx 8.66$ cm

75. $\left(2\sqrt{5}\right)\left(3\sqrt{2}\right) = 6\sqrt{10}$

76. $2\sqrt{6}\sqrt{216} = 2\sqrt{6}\sqrt{36}\sqrt{6} = 2(6)(6) = 72$

77. $\sqrt{9x}\sqrt{x} = \sqrt{9x^2} = 3x$

78. $\sqrt[3]{3}\sqrt[3]{9} = \sqrt[3]{27} = 3$

79. $-\sqrt[3]{2x^2}\sqrt[3]{4x} = -\sqrt[3]{8x^3}$
$= -2x$

80. $-\sqrt[4]{256x^5y^{11}}\sqrt[4]{625x^9y^3} = -\sqrt[4]{256x^4y^8}\sqrt[4]{xy^3}\sqrt[4]{625x^8}\sqrt[4]{xy^3} = -4xy^2(5x^2)\sqrt[4]{x^2y^6}$
$= -20x^3y^2\sqrt[4]{y^4}\sqrt[4]{x^2y^2}$
$= -20x^3y^3\sqrt[4]{x^2y^2} = -20x^3y^3\sqrt{xy}$

81. $\sqrt{2}\left(\sqrt{8} - 3\right) = \sqrt{16} - 3\sqrt{2} = 4 - 3\sqrt{2}$

82. $\sqrt{2}\left(\sqrt{2} + 3\right) = \sqrt{4} + 3\sqrt{2} = 2 + 3\sqrt{2}$

83. $\sqrt{5}\left(\sqrt{2} - 1\right) = \sqrt{10} - \sqrt{5}$

84. $\sqrt{3}\left(\sqrt{3} + \sqrt{2}\right) = \sqrt{9} + \sqrt{6} = 3 + \sqrt{6}$

85. $\left(\sqrt{2} + 1\right)\left(\sqrt{2} - 1\right) = \sqrt{4} - \sqrt{2} + \sqrt{2} - 1 = 1$

86. $\left(\sqrt{3} + \sqrt{2}\right)\left(\sqrt{3} + \sqrt{2}\right) = \sqrt{9} + \sqrt{6} + \sqrt{6} + \sqrt{4} = 5 + 2\sqrt{6}$

87. $(\sqrt{x} + \sqrt{y})(\sqrt{x} - \sqrt{y}) = \sqrt{x^2} - \sqrt{xy} + \sqrt{xy} - \sqrt{y^2} = x - y$

88. $(2\sqrt{u} + 3)(3\sqrt{u} - 4) = 6\sqrt{u^2} - 8\sqrt{u} + 9\sqrt{u} - 12 = 6u + \sqrt{u} - 12$

89. $\dfrac{1}{\sqrt{3}} = \dfrac{1\sqrt{3}}{\sqrt{3}\sqrt{3}} = \dfrac{\sqrt{3}}{3}$

90. $\dfrac{\sqrt{3}}{\sqrt{5}} = \dfrac{\sqrt{3}\sqrt{5}}{\sqrt{5}\sqrt{5}} = \dfrac{\sqrt{15}}{5}$

91. $\dfrac{x}{\sqrt{xy}} = \dfrac{x\sqrt{xy}}{\sqrt{xy}\sqrt{xy}} = \dfrac{x\sqrt{xy}}{xy} = \dfrac{\sqrt{xy}}{y}$

92. $\dfrac{\sqrt[3]{uv}}{\sqrt[3]{u^5v^7}} = \dfrac{\sqrt[3]{uv}\sqrt[3]{uv^2}}{\sqrt[3]{u^5v^7}\sqrt[3]{uv^2}} = \dfrac{\sqrt[3]{u^2v^3}}{\sqrt[3]{u^6v^9}} = \dfrac{v\sqrt[3]{u^2}}{u^2v^3} = \dfrac{\sqrt[3]{u^2}}{u^2v^2}$

93. $\dfrac{2}{\sqrt{2} - 1} = \dfrac{2\left(\sqrt{2} + 1\right)}{\left(\sqrt{2} - 1\right)\left(\sqrt{2} + 1\right)} = \dfrac{2\left(\sqrt{2} + 1\right)}{\sqrt{4} - 1} = \dfrac{2\left(\sqrt{2} + 1\right)}{2 - 1} = \dfrac{2\left(\sqrt{2} + 1\right)}{1} = 2\left(\sqrt{2} + 1\right)$

94. $\dfrac{\sqrt{2}}{\sqrt{3}-1}=\dfrac{\sqrt{2}\left(\sqrt{3}+1\right)}{\left(\sqrt{3}-1\right)\left(\sqrt{3}+1\right)}=\dfrac{\sqrt{2}\left(\sqrt{3}+1\right)}{\sqrt{9}-1}=\dfrac{\sqrt{2}\left(\sqrt{3}+1\right)}{3-1}=\dfrac{\sqrt{2}\left(\sqrt{3}+1\right)}{2}$

$=\dfrac{\sqrt{6}+\sqrt{2}}{2}$

95. $\dfrac{2x-32}{\sqrt{x}+4}=\dfrac{(2x-32)\left(\sqrt{x}-4\right)}{\left(\sqrt{x}+4\right)\left(\sqrt{x}-4\right)}=\dfrac{2(x-16)\left(\sqrt{x}-4\right)}{\sqrt{x^2}-16}=\dfrac{2(x-16)\left(\sqrt{x}-4\right)}{x-16}=2\left(\sqrt{x}-4\right)$

96. $\dfrac{\sqrt{a}+1}{\sqrt{a}-1}=\dfrac{\left(\sqrt{a}+1\right)\left(\sqrt{a}+1\right)}{\left(\sqrt{a}-1\right)\left(\sqrt{a}+1\right)}=\dfrac{\sqrt{a^2}+\sqrt{a}+\sqrt{a}+1}{\sqrt{a^2}-1}=\dfrac{a+2\sqrt{a}+1}{a-1}$

97. $\dfrac{\sqrt{3}}{5}=\dfrac{\sqrt{3}\sqrt{3}}{5\sqrt{3}}=\dfrac{3}{5\sqrt{3}}$

98. $\dfrac{\sqrt[3]{9}}{3}=\dfrac{\sqrt[3]{9}\sqrt[3]{3}}{3\sqrt[3]{3}}=\dfrac{\sqrt[3]{27}}{3\sqrt[3]{3}}=\dfrac{3}{3\sqrt[3]{3}}=\dfrac{1}{\sqrt[3]{3}}$

99. $\dfrac{3-\sqrt{x}}{2}=\dfrac{(3-\sqrt{x})(3+\sqrt{x})}{2(3+\sqrt{x})}=\dfrac{9-\sqrt{x^2}}{2(3+\sqrt{x})}=\dfrac{9-x}{2(3+\sqrt{x})}$

100. $\dfrac{\sqrt{a}-\sqrt{b}}{\sqrt{a}}=\dfrac{\left(\sqrt{a}-\sqrt{b}\right)\left(\sqrt{a}+\sqrt{b}\right)}{\sqrt{a}\left(\sqrt{a}+\sqrt{b}\right)}=\dfrac{\sqrt{a^2}-\sqrt{b^2}}{\sqrt{a^2}+\sqrt{ab}}=\dfrac{a-b}{a+\sqrt{ab}}$

101.
$$\begin{aligned}\sqrt{y+3}&=\sqrt{2y-19}\\ \left(\sqrt{y+3}\right)^2&=\left(\sqrt{2y-19}\right)^2\\ y+3&=2y-19\\ -y&=-22\\ y&=22\end{aligned}$$
The answer checks.

102.
$$\begin{aligned}u&=\sqrt{25u-144}\\ u^2&=\left(\sqrt{25u-144}\right)^2\\ u^2&=25u-144\end{aligned}$$
$u^2-25u+144=0$

$(u-9)(u-16)=0$

$u=9$ or $u=16$; Both answers check.

103.
$$\begin{aligned}r&=\sqrt{12r-27}\\ r^2&=\left(\sqrt{12r-27}\right)^2\\ r^2&=12r-27\end{aligned}$$
$r^2-12r+27=0$

$(r-9)(r-3)=0$

$r=9$ or $r=3$; Both answers check.

104.
$$\begin{aligned}\sqrt{z+1}+\sqrt{z}&=2\\ \sqrt{z+1}&=2-\sqrt{z}\\ \left(\sqrt{z+1}\right)^2&=\left(2-\sqrt{z}\right)^2\\ z+1&=4-4\sqrt{z}+z\\ 4\sqrt{z}&=3\\ \left(4\sqrt{z}\right)^2&=3^2\\ 16z&=9\end{aligned}$$
$z=\frac{9}{16}$; The answer checks.

105. $\sqrt{2x+5}-\sqrt{2x}=1$
$$\sqrt{2x+5}=1+\sqrt{2x}$$
$$\left(\sqrt{2x+5}\right)^2=\left(1+\sqrt{2x}\right)^2$$
$$2x+5=1+2\sqrt{2x}+2x$$
$$4=2\sqrt{2x}$$
$$4^2=\left(2\sqrt{2x}\right)^2$$
$$16=8x$$
$2=x$; The answer checks.

106. $\sqrt[3]{x^3+8}=x+2$
$$\left(\sqrt[3]{x^3+8}\right)^3=(x+2)^3$$
$$x^3+8=x^3+6x^2+12x+8$$
$$0=6x^2+12x$$
$$0=6x(x+2)$$
$x=0$ or $x=-2$; Both answers check.

107. $(5+4i)+(7-12i)=5+4i+7-12i=12-8i$

108. $(-6-40i)-(-8+28i)=-6-40i+8-28i=2-68i$

109. $(-32+\sqrt{-144})-(64+\sqrt{-81})=-32+\sqrt{144i^2}-64-\sqrt{81i^2}=-32+12i-64-9i$
$$=-96+3i$$

110. $(-8+\sqrt{-8})+(6-\sqrt{-32})=-8+\sqrt{4i^2}\sqrt{2}+6-\sqrt{16i^2}\sqrt{2}$
$$=-8+2i\sqrt{2}+6-4i\sqrt{2}=-2-2\sqrt{2}i$$

111. $(2-7i)(-3+4i)=-6+8i+21i-28i^2=-6+29i+28=22+29i$

112. $(-5+6i)(2+i)=-10-5i+12i+6i^2=-10+7i-6=-16+7i$

113. $(5-\sqrt{-27})(-6+\sqrt{-12})=(5-3i\sqrt{3})(-6+2i\sqrt{3})=-30+10i\sqrt{3}+18i\sqrt{3}-6i^2(3)$
$$=-30+28i\sqrt{3}+18=-12+28\sqrt{3}i$$

114. $(2+\sqrt{-128})(3-\sqrt{-98})=(2+8i\sqrt{2})(3-7i\sqrt{2})=6-14i\sqrt{2}+24i\sqrt{2}-56i^2(2)$
$$=6+10i\sqrt{2}+112=118+10\sqrt{2}i$$

115. $\frac{3}{4i}=\frac{3}{4i}\cdot\frac{i}{i}=\frac{3i}{4i^2}=\frac{3i}{-4}=-\frac{3}{4}i=0-\frac{3}{4}i$

116. $\frac{-2}{5i^3}=\frac{-2}{5i^3}\cdot\frac{i}{i}=\frac{-2i}{5i^4}=\frac{-2i}{5}=0-\frac{2}{5}i$

117. $\frac{6}{2+i}=\frac{6}{2+i}\cdot\frac{2-i}{2-i}=\frac{6(2-i)}{4-i^2}=\frac{6(2-i)}{5}=\frac{12-6i}{5}=\frac{12}{5}-\frac{6}{5}i$

118. $\frac{7}{3-i}=\frac{7}{3-i}\cdot\frac{3+i}{3+i}=\frac{7(3+i)}{9-i^2}=\frac{7(3+i)}{10}=\frac{21+7i}{10}=\frac{21}{10}+\frac{7}{10}i$

119. $\frac{4+i}{4-i}=\frac{4+i}{4-i}\cdot\frac{4+i}{4+i}=\frac{16+8i+i^2}{16-i^2}=\frac{15+8i}{17}=\frac{15}{17}+\frac{8}{17}i$

120. $\frac{3-i}{3+i} = \frac{3-i}{3+i} \cdot \frac{3-i}{3-i} = \frac{9-6i+i^2}{9-i^2} = \frac{8-6i}{10} = \frac{8}{10} - \frac{6}{10}i = \frac{4}{5} - \frac{3}{5}i$

121. $\frac{3}{5+\sqrt{-4}} = \frac{3}{5+2i} = \frac{3}{5+2i} \cdot \frac{5-2i}{5-2i} = \frac{3(5-2i)}{25-4i^2} = \frac{3(5-2i)}{29} = \frac{15-6i}{29} = \frac{15}{29} - \frac{6}{29}i$

122. $\frac{2}{3-\sqrt{-9}} = \frac{2}{3-3i} = \frac{2}{3-3i} \cdot \frac{3+3i}{3+3i} = \frac{2(3+3i)}{9-9i^2} = \frac{2(3+3i)}{18} = \frac{3+3i}{9} = \frac{3}{9} + \frac{3}{9}i = \frac{1}{3} + \frac{1}{3}i$

123. $|9+12i| = \sqrt{9^2+12^2} = \sqrt{81+144} = \sqrt{225} = 15$

124. $|24-10i| = \sqrt{24^2+(-10)^2} = \sqrt{576+100} = \sqrt{676} = 26$

125. $i^{12} = (i^4)^3 = 1^3 = 1$

126. $i^{583} = i^{580}i^3 = (i^4)^{145}i^3 = 1^{145}(-i) = -i$

Chapter 9 Test (page 689)

1. $\sqrt{49} = \sqrt{7^2} = 7$

2. $\sqrt[3]{64} = \sqrt[3]{4^3} = 4$

3. $\sqrt{4x^2} = \sqrt{(2x)^2} = |2x| = 2|x|$

4. $\sqrt[3]{8x^3} = \sqrt[3]{(2x)^3} = 2x$

5. $f(x) = \sqrt{x-2}$; Shift $y = \sqrt{x}$ right 2.

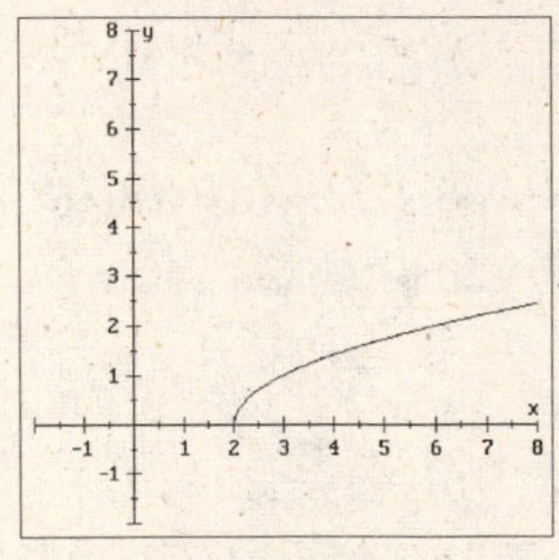

domain $= [2, \infty)$, range $= [0, \infty)$

6. $f(x) = \sqrt[3]{x} + 3$; Shift $y = \sqrt[3]{x}$ up 3.

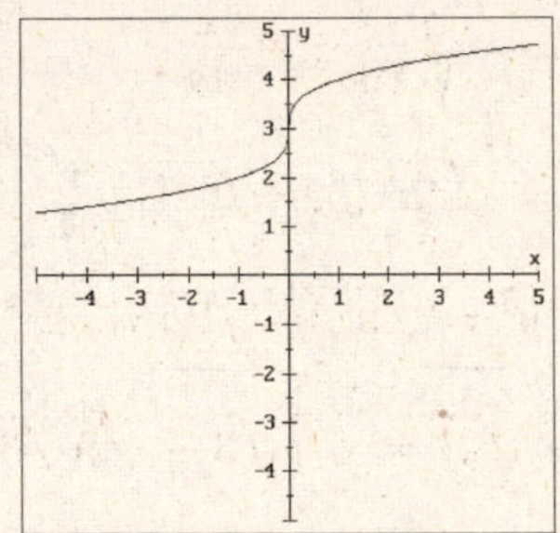

domain $= (-\infty, \infty)$, range $= (-\infty, \infty)$

7.
$$\begin{aligned} 53^2 &= 45^2 + h^2 \\ 2809 &= 2025 + h^2 \\ 784 &= h^2 \\ 28 \text{ in.} &= h \end{aligned}$$

8.
$$\begin{aligned} 2^2 &= \left(\frac{w}{2}\right)^2 + (1.9)^2 \\ 4 &= \frac{w^2}{4} + 3.61 \\ 0.39 &= \frac{w^2}{4} \\ 1.56 &= w^2 \\ 1.25 \text{ meters} &= w \end{aligned}$$

9. $d = \sqrt{(x_2-x_1)^2 + (y_2-y_1)^2} = \sqrt{(6-0)^2+(8-0)^2} = \sqrt{6^2+8^2} = \sqrt{100} = 10$

CHAPTER 9 TEST

10. $d = \sqrt{(x_2 - x_1)^2 + (y_2 - y_1)^2} = \sqrt{(-2-22)^2 + (5-12)^2} = \sqrt{(-24)^2 + (-7)^2} = \sqrt{625} = 25$

11. $16^{1/4} = 2$

12. $27^{2/3} = \left(27^{1/3}\right)^2 = 3^2 = 9$

13. $36^{-3/2} = \dfrac{1}{36^{3/2}} = \dfrac{1}{\left(36^{1/2}\right)^3} = \dfrac{1}{6^3} = \dfrac{1}{216}$

14. $\left(-\dfrac{8}{27}\right)^{-2/3} = \left(-\dfrac{27}{8}\right)^{2/3} = \left[\left(-\dfrac{27}{8}\right)^{1/3}\right]^2 = \left(-\dfrac{3}{2}\right)^2 = \dfrac{9}{4}$

15. $\dfrac{2^{5/3}2^{1/6}}{2^{1/2}} = \dfrac{2^{10/6}2^{1/6}}{2^{3/6}} = 2^{10/6+1/6-3/6} = 2^{8/6} = 2^{4/3}$

16. $\dfrac{\left(8x^3y\right)^{1/2}\left(8xy^5\right)^{1/2}}{\left(x^3y^6\right)^{1/3}} = \dfrac{8^{1/2}x^{3/2}y^{1/2}8^{1/2}x^{1/2}y^{5/2}}{x^{3/3}y^{6/3}} = \dfrac{8^{2/2}x^{4/2}y^{6/2}}{xy^2} = \dfrac{8x^2y^3}{xy^2} = 8xy$

17. $\sqrt{48} = \sqrt{16}\sqrt{3} = 4\sqrt{3}$

18. $\sqrt{250x^3y^5} = \sqrt{25x^2y^4}\sqrt{10xy} = 5xy^2\sqrt{10xy}$

19. $\dfrac{\sqrt[3]{24x^{15}y^4}}{\sqrt[3]{y}} = \sqrt[3]{\dfrac{24x^{15}y^4}{y}} = \sqrt[3]{24x^{15}y^3} = \sqrt[3]{8x^{15}y^3}\sqrt[3]{3} = 2x^5y\sqrt[3]{3}$

20. $\sqrt{\dfrac{3a^5}{48a^7}} = \sqrt{\dfrac{1}{16a^2}} = \dfrac{1}{4a}$

21. $\sqrt{12x^2} = \sqrt{4x^2}\sqrt{3} = 2|x|\sqrt{3}$

22. $\sqrt{8x^6} = \sqrt{4x^6}\sqrt{2} = 2|x^3|\sqrt{2}$

23. $\sqrt[3]{81x^3} = \sqrt[3]{27x^3}\sqrt[3]{3} = 3x\sqrt[3]{3}$

24. $\sqrt{18x^4y^9} = \sqrt{9x^4y^8}\sqrt{2y} = |3x^2y^4|\sqrt{2y} = 3x^2y^4\sqrt{2y}$

25. $\sqrt{12} - \sqrt{27} = \sqrt{4}\sqrt{3} - \sqrt{9}\sqrt{3} = 2\sqrt{3} - 3\sqrt{3} = -\sqrt{3}$

26. $2\sqrt[3]{40} - \sqrt[3]{5000} + 4\sqrt[3]{625} = 2\sqrt[3]{8}\sqrt[3]{5} - \sqrt[3]{1000}\sqrt[3]{5} + 4\sqrt[3]{125}\sqrt[3]{5} = 2(2)\sqrt[3]{5} - 10\sqrt[3]{5} + 4(5)\sqrt[3]{5}$

$$= 4\sqrt[3]{5} - 10\sqrt[3]{5} + 20\sqrt[3]{5}$$

$$= 14\sqrt[3]{5}$$

27. $2\sqrt{48y^5} - 3y\sqrt{12y^3} = 2\sqrt{16y^4}\sqrt{3y} - 3y\sqrt{4y^2}\sqrt{3y} = 2(4y^2)\sqrt{3y} - 3y(2y)\sqrt{3y}$

$$= 8y^2\sqrt{3y} - 6y^2\sqrt{3y} = 2y^2\sqrt{3y}$$

28. $\sqrt[4]{768z^5} + z\sqrt[4]{48z} = \sqrt[4]{256z^4}\sqrt[4]{3z} + z\sqrt[4]{16}\sqrt[4]{3z} = 4z\sqrt[4]{3z} + 2z\sqrt[4]{3z} = 6z\sqrt[4]{3z}$

29. $-2\sqrt{xy}\left(3\sqrt{x} + \sqrt{xy^3}\right) = -6\sqrt{x^2y} - 2\sqrt{x^2y^4} = -6x\sqrt{y} - 2xy^2$

CHAPTER 9 TEST

30. $\left(3\sqrt{2}+\sqrt{3}\right)\left(2\sqrt{2}-3\sqrt{3}\right)=6\sqrt{4}-9\sqrt{6}+2\sqrt{6}-3\sqrt{9}=12-7\sqrt{6}-9=3-7\sqrt{6}$

31. $\dfrac{1}{\sqrt{5}}=\dfrac{1\sqrt{5}}{\sqrt{5}\sqrt{5}}=\dfrac{\sqrt{5}}{5}$

32. $\dfrac{3t-1}{\sqrt{3t}-1}=\dfrac{(3t-1)\left(\sqrt{3t}+1\right)}{\left(\sqrt{3t}-1\right)\left(\sqrt{3t}+1\right)}=\dfrac{(3t-1)\left(\sqrt{3t}+1\right)}{\sqrt{9t^2}-1}=\dfrac{(3t-1)\left(\sqrt{3t}+1\right)}{3t-1}=\sqrt{3t}+1$

33. $\dfrac{\sqrt{3}}{\sqrt{7}}=\dfrac{\sqrt{3}\sqrt{3}}{\sqrt{7}\sqrt{3}}=\dfrac{3}{\sqrt{21}}$

34. $\dfrac{\sqrt{a}+\sqrt{b}}{\sqrt{a}-\sqrt{b}}=\dfrac{\left(\sqrt{a}+\sqrt{b}\right)\left(\sqrt{a}-\sqrt{b}\right)}{\left(\sqrt{a}-\sqrt{b}\right)\left(\sqrt{a}-\sqrt{b}\right)}=\dfrac{\sqrt{a^2}-\sqrt{b^2}}{\sqrt{a^2}-\sqrt{ab}-\sqrt{ab}+\sqrt{b^2}}=\dfrac{a-b}{a-2\sqrt{ab}+b}$

35.
$$\begin{aligned}\sqrt[3]{6n+4}-4&=0\\ \sqrt[3]{6n+4}&=4\\ \left(\sqrt[3]{6n+4}\right)^3&=4^3\\ 6n+4&=64\\ 6n&=60\\ n&=10\end{aligned}$$
The answer checks.

36.
$$\begin{aligned}1-\sqrt{u}&=\sqrt{u-3}\\ \left(1-\sqrt{u}\right)^2&=\left(\sqrt{u-3}\right)^2\\ 1-2\sqrt{u}+u&=u-3\\ 4&=2\sqrt{u}\\ 4^2&=\left(2\sqrt{u}\right)^2\\ 16&=4u\\ 4&=u\end{aligned}$$
extraneous $\Rightarrow \emptyset$

37. $(2+4i)+(-3+7i)=2+4i-3+7i=-1+11i$

38. $(3-\sqrt{-9})-(-1+\sqrt{-16})=3-3i+1-4i=4-7i$

39. $2i(3-4i)=6i-8i^2=6i+8=8+6i$

40. $(3+2i)(-4-i)=-12-3i-8i-2i^2=-12-11i+2=-10-11i$

41. $\dfrac{1}{i\sqrt{2}}=\dfrac{1}{i\sqrt{2}}\cdot\dfrac{i\sqrt{2}}{i\sqrt{2}}=\dfrac{i\sqrt{2}}{2i^2}=-\dfrac{\sqrt{2}i}{2}=0-\dfrac{\sqrt{2}}{2}i$

42. $\dfrac{2+i}{3-i}=\dfrac{2+i}{3-i}\cdot\dfrac{3+i}{3+i}=\dfrac{6+5i+i^2}{9-i^2}=\dfrac{5+5i}{10}=\dfrac{1}{2}+\dfrac{1}{2}i$

Exercise 10.1 (page 701)

1. $x^2 = 49$
$x = \pm\sqrt{49} = \pm 7$

3. $\left[\frac{1}{2}(4)\right]^2 = 2^2 = 4$

5. $\left[\frac{1}{2}(-3)\right]^2 = \left[-\frac{3}{2}\right]^2 = \frac{9}{4}$

7.
$$\begin{aligned} \frac{t+9}{2} + \frac{t+2}{5} &= \frac{8}{5} + 4t \\ 5(t+9) + 2(t+2) &= 2(8) + 40t \\ 5t + 45 + 2t + 4 &= 16 + 40t \\ 33 &= 33t \\ 1 &= t \end{aligned}$$

9.
$$\begin{aligned} 3(t-3) + 3t &\le 2(t+1) + t + 1 \\ 3t - 9 + 3t &\le 2t + 2 + t + 1 \\ 3t &\le 12 \\ t &\le 4 \end{aligned}$$

11. $x = \sqrt{c}; x = -\sqrt{c}$

13. positive or negative

15.
$$\begin{aligned} 2y^2 - 50 &= 0 \\ 2(y+5)(y-5) &= 0 \end{aligned}$$
$y + 5 = 0$ **or** $y - 5 = 0$
$y = -5$ $\quad$ $y = 5$

17.
$$\begin{aligned} 6x^2 + 12x &= 0 \\ 6x(x+2) &= 0 \end{aligned}$$
$6x = 0$ **or** $x + 2 = 0$
$x = 0$ $\quad$ $x = -2$

19.
$$\begin{aligned} r^2 + 6r + 8 &= 0 \\ (r+2)(r+4) &= 0 \end{aligned}$$
$r + 2 = 0$ **or** $r + 4 = 0$
$r = -2$ $\quad$ $r = -4$

21.
$$\begin{aligned} 6s^2 + 11s - 10 &= 0 \\ (2s+5)(3s-2) &= 0 \end{aligned}$$
$2s + 5 = 0$ **or** $3s - 2 = 0$
$s = -\frac{5}{2}$ $\quad$ $s = \frac{2}{3}$

23. $x^2 = 36$
$x = \pm\sqrt{36} = \pm 6$

25. $z^2 = 5$
$z = \pm\sqrt{5}$

27.
$$\begin{aligned} (y+1)^2 &= 1 \\ y + 1 &= \pm\sqrt{1} \\ y + 1 &= \pm 1 \\ y &= -1 \pm 1 \end{aligned}$$
$y = 0$ or $y = -2$

29.
$$\begin{aligned} (x-2)^2 - 5 &= 0 \\ (x-2)^2 &= 5 \\ x - 2 &= \pm\sqrt{5} \\ x &= 2 \pm \sqrt{5} \end{aligned}$$

31.
$$\begin{aligned} p^2 + 16 &= 0 \\ p^2 &= -16 \\ p &= \pm\sqrt{-16} \\ p &= \pm 4i \end{aligned}$$

33.
$$\begin{aligned} 4m^2 + 81 &= 0 \\ 4m^2 &= -81 \\ m^2 &= -\frac{81}{4} \\ m &= \pm\sqrt{-\frac{81}{4}} \\ m &= \pm\frac{9}{2}i \end{aligned}$$

35.
$$\begin{aligned} x^2 + 2x - 8 &= 0 \\ x^2 + 2x &= 8 \\ x^2 + 2x + 1 &= 8 + 1 \\ (x+1)^2 &= 9 \\ x + 1 &= \pm 3 \\ x &= -1 \pm 3 \end{aligned}$$
$x = 2$ or $x = -4$

37.
$$\begin{aligned} x^2 - 6x + 8 &= 0 \\ x^2 - 6x &= -8 \\ x^2 - 6x + 9 &= -8 + 9 \\ (x-3)^2 &= 1 \\ x - 3 &= \pm 1 \\ x &= 3 \pm 1 \end{aligned}$$
$x = 4$ or $x = 2$

39.
$$\begin{aligned} x^2+5x+4&=0\\ x^2+5x&=-4\\ x^2+5x+\frac{25}{4}&=-4+\frac{25}{4}\\ \left(x+\frac{5}{2}\right)^2&=\frac{9}{4}\\ x+\frac{5}{2}&=\pm\frac{3}{2}\\ x&=-\frac{5}{2}\pm\frac{3}{2}\end{aligned}$$
$x=-\frac{2}{2}=-1$ or $x=-\frac{8}{2}=-4$

41.
$$\begin{aligned} x^2-9x-10&=0\\ x^2-9x&=10\\ x^2-9x+\frac{81}{4}&=10+\frac{81}{4}\\ \left(x-\frac{9}{2}\right)^2&=\frac{121}{4}\\ x-\frac{9}{2}&=\pm\frac{11}{2}\\ x&=\frac{9}{2}\pm\frac{11}{2}\end{aligned}$$
$x=\frac{20}{2}=10$ or $x=\frac{-2}{2}=-1$

43.
$$\begin{aligned} 6x^2+11x+3&=0\\ 6x^2+11x&=-3\\ x^2+\frac{11}{6}x&=-\frac{1}{2}\\ x^2+\frac{11}{6}x+\frac{121}{144}&=-\frac{1}{2}+\frac{121}{144}\\ \left(x+\frac{11}{12}\right)^2&=\frac{49}{144}\\ x+\frac{11}{12}&=\pm\frac{7}{12}\\ x&=-\frac{11}{12}\pm\frac{7}{12}\end{aligned}$$
$x=-\frac{4}{12}=-\frac{1}{3}$ or $x=-\frac{18}{12}=-\frac{3}{2}$

45.
$$\begin{aligned} 6x^2-7x-5&=0\\ 6x^2-7x&=5\\ x^2-\frac{7}{6}x&=\frac{5}{6}\\ x^2-\frac{7}{6}x+\frac{49}{144}&=\frac{5}{6}+\frac{49}{144}\\ \left(x-\frac{7}{12}\right)^2&=\frac{169}{144}\\ x-\frac{7}{12}&=\pm\frac{13}{12}\\ x&=\frac{7}{12}\pm\frac{13}{12}\end{aligned}$$
$x=\frac{20}{12}=\frac{5}{3}$ or $x=-\frac{6}{12}=-\frac{1}{2}$

47.
$$\begin{aligned} 9-6r&=8r^2\\ -8r^2-6r&=-9\\ r^2+\frac{3}{4}r&=\frac{9}{8}\\ r^2+\frac{3}{4}r+\frac{9}{64}&=\frac{9}{8}+\frac{9}{64}\\ \left(r+\frac{3}{8}\right)^2&=\frac{81}{64}\\ r+\frac{3}{8}&=\pm\frac{9}{8}\\ r&=-\frac{3}{8}\pm\frac{9}{8}\end{aligned}$$
$r=\frac{6}{8}=\frac{3}{4}$ or $r=-\frac{12}{8}=-\frac{3}{2}$

49.
$$\begin{aligned} x+1&=2x^2\\ -2x^2+x&=-1\\ x^2-\frac{1}{2}x&=\frac{1}{2}\\ x^2-\frac{1}{2}x+\frac{1}{16}&=\frac{1}{2}+\frac{1}{16}\\ \left(x-\frac{1}{4}\right)^2&=\frac{9}{16}\\ x-\frac{1}{4}&=\pm\frac{3}{4}\\ x&=\frac{1}{4}\pm\frac{3}{4}\end{aligned}$$
$x=\frac{4}{4}=1$ or $x=-\frac{2}{4}=-\frac{1}{2}$

51.
$$\begin{aligned} \frac{7x+1}{5} &= -x^2 \\ 7x+1 &= -5x^2 \\ 5x^2+7x &= -1 \\ x^2+\frac{7}{5}x &= -\frac{1}{5} \\ x^2+\frac{7}{5}x+\frac{49}{100} &= -\frac{1}{5}+\frac{49}{100} \\ \left(x+\frac{7}{10}\right)^2 &= \frac{29}{100} \\ x+\frac{7}{10} &= \pm\frac{\sqrt{29}}{10} \\ x &= -\frac{7}{10}\pm\frac{\sqrt{29}}{10} \end{aligned}$$

53.
$$\begin{aligned} p^2+2p+2 &= 0 \\ p^2+2p &= -2 \\ p^2+2p+1 &= -2+1 \\ (p+1)^2 &= -1 \\ p+1 &= \pm\sqrt{-1} \\ p+1 &= \pm i \\ p &= -1\pm i \end{aligned}$$

55.
$$\begin{aligned} y^2+8y+18 &= 0 \\ y^2+8y &= -18 \\ y^2+8y+16 &= -18+16 \\ (y+4)^2 &= -2 \\ y+4 &= \pm\sqrt{-2} \\ y &= -4\pm i\sqrt{2} \end{aligned}$$

57.
$$\begin{aligned} 3m^2-2m+3 &= 0 \\ m^2-\frac{2}{3}m &= -1 \\ m^2-\frac{2}{3}m+\frac{1}{9} &= -1+\frac{1}{9} \\ \left(m-\frac{1}{3}\right)^2 &= -\frac{8}{9} \\ m-\frac{1}{3} &= \pm\sqrt{-\frac{8}{9}} \\ m &= \frac{1}{3}\pm\frac{\sqrt{8}}{3}i \\ m &= \frac{1}{3}\pm\frac{2\sqrt{2}}{3}i \end{aligned}$$

59.
$$\begin{aligned} 7x-6 &= x^2 \\ 0 &= x^2-7x+6 \\ 0 &= (x-6)(x-1) \end{aligned}$$
$x-6=0$ **or** $x-1=0$
$x=6$ $\quad x=1$

61.
$$\begin{aligned} 3x^2-16 &= 0 \\ 3x^2 &= 16 \\ x^2 &= \frac{16}{3} \\ x &= \pm\sqrt{\frac{16}{3}} = \pm\frac{4}{\sqrt{3}} = \pm\frac{4\sqrt{3}}{3} \end{aligned}$$

63.
$$\begin{aligned} (s-7)^2-9 &= 0 \\ (s-7)^2 &= 9 \\ s-7 &= \pm\sqrt{9} \\ s-7 &= \pm 3 \\ s &= 7\pm 3 \end{aligned}$$
$s=10$ or $s=4$

65.
$$\begin{aligned} (x+5)^2-3 &= 0 \\ (x+5)^2 &= 3 \\ x+5 &= \pm\sqrt{3} \\ x &= -5\pm\sqrt{3} \end{aligned}$$

67.
$$\begin{aligned} 2z^2 - 5z + 2 &= 0 \\ (2z-1)(z-2) &= 0 \\ 2z - 1 = 0 \quad \textbf{or} \quad z - 2 &= 0 \\ z = \tfrac{1}{2} \qquad\qquad z &= 2 \end{aligned}$$

69.
$$\begin{aligned} 3x^2 - 6x + 1 &= 0 \\ 3x^2 - 6x &= -1 \\ x^2 - 2x &= -\frac{1}{3} \\ x^2 - 2x + 1 &= -\frac{1}{3} + 1 \\ (x-1)^2 &= \frac{2}{3} \\ x - 1 &= \pm\sqrt{\frac{2}{3}} \\ x - 1 &= \pm\frac{\sqrt{2}}{\sqrt{3}} \cdot \frac{\sqrt{3}}{\sqrt{3}} \\ x - 1 &= \pm\frac{\sqrt{6}}{3} \\ x &= 1 \pm \frac{\sqrt{6}}{3} \end{aligned}$$

71.
$$\begin{aligned} 2x^2 - x + 8 &= 0 \\ 2x^2 - x &= -8 \\ x^2 - \frac{1}{2}x &= -4 \\ x^2 - \frac{1}{2}x + \frac{1}{16} &= -4 + \frac{1}{16} \\ \left(x - \frac{1}{4}\right)^2 &= \frac{-63}{16} \\ x - \frac{1}{4} &= \pm\frac{\sqrt{-63}}{4} \\ x &= \frac{1}{4} \pm \frac{i\sqrt{9}\sqrt{7}}{4} \\ x &= \frac{1}{4} \pm \frac{3\sqrt{7}}{4}i \end{aligned}$$

73.
$$\begin{aligned} 2d^2 &= 3h \\ d^2 &= \frac{3h}{2} \\ d &= \sqrt{\frac{3h}{2}} \\ d &= \frac{\sqrt{3h}}{\sqrt{2}} \cdot \frac{\sqrt{2}}{\sqrt{2}} \\ d &= \frac{\sqrt{6h}}{2} \end{aligned}$$

75.
$$\begin{aligned} E &= mc^2 \\ \frac{E}{m} &= c^2 \\ \sqrt{\frac{E}{m}} &= c \\ \frac{\sqrt{E}}{\sqrt{m}} \cdot \frac{\sqrt{m}}{\sqrt{m}} &= c \\ \frac{\sqrt{Em}}{m} &= c \end{aligned}$$

77.
$$\begin{aligned} f(x) &= 0 \\ 2x^2 + x - 5 &= 0 \\ x^2 + \frac{1}{2}x &= \frac{5}{2} \\ x^2 + \frac{1}{2}x + \frac{1}{16} &= \frac{5}{2} + \frac{1}{16} \\ \left(x + \frac{1}{4}\right)^2 &= \frac{41}{16} \\ x + \frac{1}{4} &= \pm\sqrt{\frac{41}{16}} \\ x + \frac{1}{4} &= \pm\frac{\sqrt{41}}{4} \\ x &= -\frac{1}{4} \pm \frac{\sqrt{41}}{4} \end{aligned}$$

79.
$$\begin{aligned} f(x) &= 0 \\ x^2 + x - 3 &= 0 \\ x^2 + x &= 3 \\ x^2 + x + \frac{1}{4} &= 3 + \frac{1}{4} \\ \left(x + \frac{1}{2}\right)^2 &= \frac{13}{4} \\ x + \frac{1}{2} &= \pm\sqrt{\frac{13}{4}} \\ x &= -\frac{1}{2} \pm \frac{\sqrt{13}}{2} \end{aligned}$$

81.
$$\begin{aligned} s &= 16t^2 \\ 256 &= 16t^2 \\ 16 &= t^2 \\ \pm\sqrt{16} &= t \\ \pm 4 &= t \end{aligned}$$

$t = 4$ is the only answer that makes sense, so it will take 4 seconds.

83.
$$\begin{aligned} s^2 &= 10.5l \\ s^2 &= 10.5(495) \\ s^2 &= 5197.5 \\ s &= \pm\sqrt{5197.5} \\ s &\approx \pm 72.09 \end{aligned}$$

s must be positive, so the speed was about 72 mph.

85.
$$\begin{aligned} A &= P(1 + r)^t \\ 9193.60 &= 8500(1 + r)^2 \\ \frac{9193.60}{8500} &= (1 + r)^2 \\ 1.0816 &= (1 + r)^2 \\ \pm\sqrt{1.0816} &= \sqrt{(1 + r)^2} \\ \pm 1.04 &= 1 + r \\ -1 \pm 1.04 &= r \end{aligned}$$

$r = 0.04$ or $r = -2.04$; r must be positive, so $r = 0.04$, or 4%.

87.
$$\begin{aligned} A &= 100 \\ lw &= 100 \\ (1.9x)(x) &= 100 \\ 1.9x^2 &= 100 \\ x^2 &= \frac{100}{1.9} \\ x &= \pm\sqrt{\frac{100}{1.9}} \approx \pm 7.25 \end{aligned}$$

x must be positive, so $x = 7\frac{1}{4}$ ft.

$1.9x \approx 1.9(7.25) \approx 13.75 = 13\frac{3}{4}$ ft.

89. **Answers may vary.**

91. $\left(\frac{1}{2}\sqrt{3}\right)^2 = \left(\frac{\sqrt{3}}{2}\right)^2 = \frac{3}{4}$

Exercise 10.2 (page 709)

1. $3x^2 - 4x + 7 = 0 \Rightarrow a = 3, b = -4, c = 7$

3. $$\begin{aligned} Ax + By &= C \\ By &= -Ax + C \\ y &= \frac{-Ax + C}{B} \end{aligned}$$

5. $\sqrt{24} = \sqrt{4}\sqrt{6} = 2\sqrt{6}$

7. $\frac{3}{\sqrt{3}} = \frac{3}{\sqrt{3}} \cdot \frac{\sqrt{3}}{\sqrt{3}} = \frac{3\sqrt{3}}{3} = \sqrt{3}$

9. 3; −2; 6

11. $$\begin{aligned} &x^2 + 3x + 2 = 0 \\ &a = 1, b = 3, c = 2 \\ x &= \frac{-b \pm \sqrt{b^2 - 4ac}}{2a} \\ &= \frac{-3 \pm \sqrt{3^2 - 4(1)(2)}}{2(1)} \\ &= \frac{-3 \pm \sqrt{9 - 8}}{2} \\ &= \frac{-3 \pm \sqrt{1}}{2} = \frac{-3 \pm 1}{2} \\ x &= -\tfrac{2}{2} = -1 \text{ or } x = -\tfrac{4}{2} = -2 \end{aligned}$$

13. $$\begin{aligned} &x^2 - 2x - 15 = 0 \\ &a = 1, b = -2, c = -15 \\ x &= \frac{-b \pm \sqrt{b^2 - 4ac}}{2a} \\ &= \frac{2 \pm \sqrt{(-2)^2 - 4(1)(-15)}}{2(1)} \\ &= \frac{2 \pm \sqrt{4 + 60}}{2} \\ &= \frac{2 \pm \sqrt{64}}{2} = \frac{2 \pm 8}{2} \\ x &= \tfrac{10}{2} = 5 \text{ or } x = -\tfrac{6}{2} = -3 \end{aligned}$$

15. $$\begin{aligned} x^2 + 12x &= -36 \\ &x^2 + 12x + 36 = 0 \\ &a = 1, b = 12, c = 36 \\ x &= \frac{-b \pm \sqrt{b^2 - 4ac}}{2a} \\ &= \frac{-12 \pm \sqrt{12^2 - 4(1)(36)}}{2(1)} \\ &= \frac{-12 \pm \sqrt{144 - 144}}{2} \\ &= \frac{-12 \pm \sqrt{0}}{2} = \frac{-12 \pm 0}{2} \\ x &= -\tfrac{12}{2} = -6 \text{ or } x = -\tfrac{12}{2} = -6 \end{aligned}$$

17. $$\begin{aligned} &2x^2 - x - 3 = 0 \\ &a = 2, b = -1, c = -3 \\ x &= \frac{-b \pm \sqrt{b^2 - 4ac}}{2a} \\ &= \frac{1 \pm \sqrt{(-1)^2 - 4(2)(-3)}}{2(2)} \\ &= \frac{1 \pm \sqrt{1 + 24}}{4} \\ &= \frac{1 \pm \sqrt{25}}{4} = \frac{1 \pm 5}{4} \\ x &= \tfrac{6}{4} = \tfrac{3}{2} \text{ or } x = -\tfrac{4}{4} = -1 \end{aligned}$$

19.
$$15x^2 - 14x = 8$$
$$15x^2 - 14x - 8 = 0$$
$$a = 15, b = -14, c = -8$$
$$x = \frac{-b \pm \sqrt{b^2 - 4ac}}{2a}$$
$$= \frac{14 \pm \sqrt{(-14)^2 - 4(15)(-8)}}{2(15)}$$
$$= \frac{14 \pm \sqrt{196 + 480}}{30}$$
$$= \frac{14 \pm \sqrt{676}}{30} = \frac{14 \pm 26}{30}$$
$$x = \tfrac{40}{30} = \tfrac{4}{3} \text{ or } x = -\tfrac{12}{30} = -\tfrac{2}{5}$$

21.
$$8u = -4u^2 - 3$$
$$4u^2 + 8u + 3 = 0$$
$$a = 4, b = 8, c = 3$$
$$u = \frac{-b \pm \sqrt{b^2 - 4ac}}{2a}$$
$$= \frac{-8 \pm \sqrt{8^2 - 4(4)(3)}}{2(4)}$$
$$= \frac{-8 \pm \sqrt{64 - 48}}{8}$$
$$= \frac{-8 \pm \sqrt{16}}{8} = \frac{-8 \pm 4}{8}$$
$$u = \tfrac{-4}{8} = -\tfrac{1}{2} \text{ or } u = \tfrac{-12}{8} = -\tfrac{3}{2}$$

23.
$$16y^2 + 8y - 3 = 0$$
$$a = 16, b = 8, c = -3$$
$$y = \frac{-b \pm \sqrt{b^2 - 4ac}}{2a}$$
$$= \frac{-8 \pm \sqrt{8^2 - 4(16)(-3)}}{2(16)}$$
$$= \frac{-8 \pm \sqrt{64 + 192}}{32}$$
$$= \frac{-8 \pm \sqrt{256}}{32} = \frac{-8 \pm 16}{32}$$
$$y = \tfrac{8}{32} = \tfrac{1}{4} \text{ or } y = \tfrac{-24}{32} = -\tfrac{3}{4}$$

25.
$$5x^2 + 5x + 1 = 0$$
$$a = 5, b = 5, c = 1$$
$$x = \frac{-b \pm \sqrt{b^2 - 4ac}}{2a}$$
$$= \frac{-5 \pm \sqrt{5^2 - 4(5)(1)}}{2(5)}$$
$$= \frac{-5 \pm \sqrt{25 - 20}}{10}$$
$$= \frac{-5 \pm \sqrt{5}}{10}$$
$$= \frac{-5}{10} \pm \frac{\sqrt{5}}{10} = -\frac{1}{2} \pm \frac{\sqrt{5}}{10}$$

27.
$$x^2 + 2x + 2 = 0$$
$$a = 1, b = 2, c = 2$$
$$x = \frac{-b \pm \sqrt{b^2 - 4ac}}{2a}$$
$$= \frac{-2 \pm \sqrt{2^2 - 4(1)(2)}}{2(1)}$$
$$= \frac{-2 \pm \sqrt{4 - 8}}{2}$$
$$= \frac{-2 \pm \sqrt{-4}}{2}$$
$$= \frac{-2 \pm \sqrt{-1 \cdot 4}}{2}$$
$$= \frac{-2 \pm 2i}{2} = \frac{-2}{2} \pm \frac{2i}{2} = -1 \pm i$$

29.
$$2x^2 + x + 1 = 0$$
$$a = 2, b = 1, c = 1$$
$$x = \frac{-b \pm \sqrt{b^2 - 4ac}}{2a}$$
$$= \frac{-1 \pm \sqrt{1^2 - 4(2)(1)}}{2(2)}$$
$$= \frac{-1 \pm \sqrt{1 - 8}}{4}$$
$$= \frac{-1 \pm \sqrt{-7}}{4}$$
$$= \frac{-1 \pm \sqrt{-1 \cdot 7}}{4}$$
$$= \frac{-1 \pm i\sqrt{7}}{4} = -\frac{1}{4} \pm \frac{\sqrt{7}}{4}i$$

31.
$$\begin{aligned} 3x^2 - 4x &= -2 \\ 3x^2 - 4x + 2 &= 0 \\ a = 3, b &= -4, c = 2 \\ x &= \frac{-b \pm \sqrt{b^2 - 4ac}}{2a} \\ &= \frac{4 \pm \sqrt{(-4)^2 - 4(3)(2)}}{2(3)} \\ &= \frac{4 \pm \sqrt{16 - 24}}{6} \\ &= \frac{4 \pm \sqrt{-8}}{6} \\ &= \frac{4 \pm \sqrt{-1 \cdot 4 \cdot 2}}{6} \\ &= \frac{4 \pm 2i\sqrt{2}}{6} = \frac{4}{6} \pm \frac{2\sqrt{2}}{6}i = \frac{2}{3} \pm \frac{\sqrt{2}}{3}i \end{aligned}$$

33.
$$\begin{aligned} 3x^2 - 2x &= -3 \\ 3x^2 - 2x + 3 &= 0 \\ a = 3, b &= -2, c = 3 \\ x &= \frac{-b \pm \sqrt{b^2 - 4ac}}{2a} \\ &= \frac{2 \pm \sqrt{(-2)^2 - 4(3)(3)}}{2(3)} \\ &= \frac{2 \pm \sqrt{4 - 36}}{6} \\ &= \frac{2 \pm \sqrt{-32}}{6} \\ &= \frac{2 \pm \sqrt{-1 \cdot 16 \cdot 2}}{6} \\ &= \frac{2 \pm 4i\sqrt{2}}{6} = \frac{2}{6} \pm \frac{4\sqrt{2}}{6}i = \frac{1}{3} \pm \frac{2\sqrt{2}}{3}i \end{aligned}$$

35.
$$\begin{aligned} C &= \frac{N^2 - N}{2} \\ 2C &= N^2 - N \\ N^2 - N - 2C &= 0 \\ a = 1, b &= -1, c = -2C \\ N &= \frac{-b \pm \sqrt{b^2 - 4ac}}{2a} \\ &= \frac{1 \pm \sqrt{(-1)^2 - 4(1)(-2C)}}{2(1)} \\ &= \frac{1 \pm \sqrt{1 + 8C}}{2} \end{aligned}$$

37.
$$\begin{aligned} x^2 - kx &= -ay \\ x^2 - kx + ay &= 0 \\ a = 1, b &= -k, c = ay \\ x &= \frac{-b \pm \sqrt{b^2 - 4ac}}{2a} \\ &= \frac{k \pm \sqrt{(-k)^2 - 4(1)(ay)}}{2(1)} \\ &= \frac{k \pm \sqrt{k^2 - 4ay}}{2} = \frac{k}{2} \pm \frac{\sqrt{k^2 - 4ay}}{2} \end{aligned}$$

39.
$$\begin{aligned} 6x^2 - x - 1 &= 0 \\ a = 6, b &= -1, c = -1 \\ x &= \frac{-b \pm \sqrt{b^2 - 4ac}}{2a} \\ &= \frac{1 \pm \sqrt{(-1)^2 - 4(6)(-1)}}{2(6)} \\ &= \frac{1 \pm \sqrt{1 + 24}}{12} \\ &= \frac{1 \pm \sqrt{25}}{12} = \frac{1 \pm 5}{12} \end{aligned}$$
$x = \frac{6}{12} = \frac{1}{2}$ or $x = -\frac{4}{12} = -\frac{1}{3}$

41.
$$\begin{aligned} \frac{x^2}{2} + \frac{5}{2}x &= -1 \\ x^2 + 5x &= -2 \\ x^2 + 5x + 2 &= 0 \\ a = 1, b &= 5, c = 2 \\ x &= \frac{-b \pm \sqrt{b^2 - 4ac}}{2a} \\ &= \frac{-5 \pm \sqrt{5^2 - 4(1)(2)}}{2(1)} \\ &= \frac{-5 \pm \sqrt{25 - 8}}{2} \\ &= \frac{-5 \pm \sqrt{17}}{2} = -\frac{5}{2} \pm \frac{\sqrt{17}}{2} \end{aligned}$$

43.
$$\begin{aligned} 2x^2 - 1 &= 3x \\ 2x^2 - 3x - 1 &= 0 \\ a = 2, b &= -3, c = -1 \\ x &= \frac{-b \pm \sqrt{b^2 - 4ac}}{2a} \\ &= \frac{3 \pm \sqrt{(-3)^2 - 4(2)(-1)}}{2(2)} \\ &= \frac{3 \pm \sqrt{9+8}}{4} \\ &= \frac{3 \pm \sqrt{17}}{4} = \frac{3}{4} \pm \frac{\sqrt{17}}{4} \end{aligned}$$

45.
$$\begin{aligned} f(x) &= 0 \\ 4x^2 + 4x - 19 &= 0 \\ a = 4, b &= 4, c = -19 \\ x &= \frac{-b \pm \sqrt{b^2 - 4ac}}{2a} \\ &= \frac{-4 \pm \sqrt{4^2 - 4(4)(-19)}}{2(4)} \\ &= \frac{-4 \pm \sqrt{16+304}}{8} \\ &= \frac{-4 \pm \sqrt{320}}{8} \\ &= -\frac{4}{8} \pm \frac{8\sqrt{5}}{8} = -\frac{1}{2} \pm \sqrt{5} \end{aligned}$$

47.
$$\begin{aligned} f(x) &= 0 \\ 3x^2 + 2x + 2 &= 0 \\ a = 3, b &= 2, c = 2 \\ x &= \frac{-b \pm \sqrt{b^2 - 4ac}}{2a} \\ &= \frac{-2 \pm \sqrt{2^2 - 4(3)(2)}}{2(3)} \\ &= \frac{-2 \pm \sqrt{4-24}}{6} \\ &= \frac{-2 \pm \sqrt{-20}}{6} \\ &= -\frac{2}{6} \pm \frac{2i\sqrt{5}}{6} = -\frac{1}{3} \pm \frac{\sqrt{5}}{3}i \end{aligned}$$

49.
$$\begin{aligned} 0.7x^2 - 3.5x - 25 &= 0 \\ a = 0.7, b &= -3.5, c = -25 \\ x &= \frac{-b \pm \sqrt{b^2 - 4ac}}{2a} \\ &= \frac{3.5 \pm \sqrt{(-3.5)^2 - 4(0.7)(-25)}}{2(0.7)} \\ &= \frac{3.5 \pm \sqrt{12.25+70}}{1.4} \\ &= \frac{3.5 \pm \sqrt{82.25}}{1.4} \\ &= \frac{3.5 \pm 9.069}{1.4} \\ x = 8.98 &\text{ or } x = -3.98 \end{aligned}$$

51.
$$\begin{aligned} (x-3)(x-5) &= 0 \\ x^2 - 8x + 15 &= 0 \end{aligned}$$

53.
$$\begin{aligned} (x-2)(x-3)(x+4) &= 0 \\ (x^2 - 5x + 6)(x+4) &= 0 \\ x^3 - x^2 - 14x + 24 &= 0 \end{aligned}$$

55. length · width = Area

$$\begin{aligned} (x+4)x &= 96 \\ x^2 + 4x - 96 &= 0 \\ (x+12)(x-8) &= 0 \\ x = -12 &\text{ or } x = 8 \end{aligned}$$

Since the width is positive, the dimensions are 8 ft by 12 ft.

57. Let s = the length of a side.

$$\begin{aligned} \text{Area} &= \text{perimeter} \\ s^2 &= 4s \\ s^2 - 4s &= 0 \\ s(s-4) &= 0 \\ s = 0 &\text{ or } s = 4 \end{aligned}$$

Since the length cannot be 0, the length of a side is 4 units.

59. Let b represent the base.
Then $3b+5$ represents the height.

$$\frac{1}{2}\text{base} \cdot \text{height} = \text{Area}$$
$$\frac{1}{2}b(3b+5) = 6$$
$$b(3b+5) = 12$$
$$3b^2+5b-12=0$$
$$(3b-4)(b+3)=0$$
$$b=\frac{4}{3} \text{ or } b=-3$$

Since the base is positive, it must be $\frac{4}{3}$ cm.

61. Let x and $x+2$ represent the integers.

$$x(x+2)=288$$
$$x^2+2x-288=0$$
$$(x+18)(x-16)=0$$
$$x=-18 \text{ or } x=16$$

The integers are 16 & 18, or -18 & -16.

63. Let x and $x+1$ represent the integers.

$$x^2+(x+1)^2=85$$
$$x^2+x^2+2x+1=85$$
$$2x^2+2x-84=0$$
$$2(x+7)(x-6)=0$$

$x=-7$ or $x=6$; The integers are 6 & 7, or -7 & -6.

65. Let $r =$ the slower rate. Then $r+20 =$ the faster rate.

	Rate	Time	Dist.
Slower	r	$\frac{150}{r}$	150
Faster	$r+20$	$\frac{150}{r+20}$	150

$$\boxed{\text{Faster time}} + 2 = \boxed{\text{Slower time}}$$
$$\frac{150}{r+20}+2=\frac{150}{r}$$
$$\left(\frac{150}{r+20}+2\right)(r)(r+20)=\frac{150}{r}\cdot r(r+20)$$
$$150r+2r(r+20)=150(r+20)$$
$$2r^2+40r-3000=0$$
$$2(r+50)(r-30)=0$$

$r=-50$ or $r=30$ $\quad$ $r=30$ is the only answer that makes sense.
Her original speed was 30 mph.

67. Let $x =$ the number of 10¢ increases. Then the ticket price will be $4+0.10x$, while the projected attendance will be $300-5x$, for total receipts of $(4+0.10x)(300-5x)$.

$$\text{Total} = 1248$$
$$(4+0.10x)(300-5x)=1248$$
$$1200+10x-0.5x^2=1248$$
$$-0.5x^2+10x-48=0$$
$$x^2-20x+96=0$$

continued on next page...

67. **continued**

$x^2 - 20x + 96 = 0$

$(x - 12)(x - 8) = 0$

$x = 12$ or $x = 8 \Rightarrow 4 + 0.10(12) = 5.20;\ 4 + 0.10(8) = 4.80$

The ticket price would be either \$5.20 or \$4.80.

69. Let $x =$ the number of additional subscribers. Then the profit per subscriber will be $20 + 0.01x$, for a total profit of $(20 + 0.01x)(3000 + x)$.

$$\begin{aligned}
\text{Total profit} &= 120000 \\
(20 + 0.01x)(3000 + x) &= 120000 \\
60{,}000 + 50x + 0.01x^2 &= 120000 \\
0.01x^2 + 50x - 60{,}000 &= 0 \\
x^2 + 5000x - 6{,}000{,}000 &= 0 \\
(x + 6000)(x - 1000) &= 0
\end{aligned}$$

$x = -6000$ (impossible) or $x = 1000 \Rightarrow$ The total number of subscribers would be 4000.

71. Let $w =$ the constant width.

$$\begin{aligned}
\text{Frame area} &= \text{Picture area} \\
(12 + 2w)(10 + 2w) - 12(10) &= 12(10) \\
120 + 44w + 4w^2 - 240 &= 0 \\
4w^2 + 44w - 120 &= 0 \\
4(w^2 + 11w - 30) &= 0 \\
w^2 + 11w - 30 &= 0 \Rightarrow a = 1, b = 11, c = -30
\end{aligned}$$

$$w = \frac{-11 \pm \sqrt{11^2 - 4(1)(-30)}}{2(1)}$$

$$= \frac{-11 \pm \sqrt{121 + 120}}{2} = \frac{-11 \pm \sqrt{241}}{2} = \frac{-11 \pm 15.52}{2} = \frac{4.52}{2} \text{ or } \frac{-26.52}{2} \text{ (impossible)}$$

$w = 2.26$ in.

73. $P = -0.0072x^2 + 0.4904x + 58.2714$

$65 = -0.0072x^2 + 0.4904x + 58.2714$

$0 = -0.0072x^2 + 0.4904x - 6.7286$

$a = -0.0072,\ b = 0.4904,\ c = -6.7286$

$$x = \frac{-b \pm \sqrt{b^2 - 4ac}}{2a}$$

$$= \frac{-0.4904 \pm \sqrt{(0.4904)^2 - 4(-0.0072)(-6.7286)}}{2(-0.0072)} \approx \frac{-0.4904 \pm \sqrt{0.0467}}{-0.0144} \approx 19.05 \text{ or } 49.06$$

Since $x \leq 42$, the only answer that works is 19.05. The model indicates the desired result happened in 1985.

75. Let [H$^+$] (and then [A$^-$]) $= x$ and [HA] $= 0.1 - x$.

$$\frac{[\text{H}^+][\text{A}^-]}{[\text{HA}]} = 4 \times 10^{-4}$$
$$\frac{x^2}{0.1 - x} = 4 \times 10^{-4}$$
$$x^2 = 4 \times 10^{-5} - (4 \times 10^{-4})x$$
$$x^2 + (4 \times 10^{-4})x - 4 \times 10^{-5} = 0$$
$$x = \frac{-b \pm \sqrt{b^2 - 4ac}}{2a} = \frac{-4 \times 10^{-4} \pm \sqrt{(4 \times 10^{-4})^2 - 4(1)(-4 \times 10^{-5})}}{2(1)}$$
$$\approx \frac{-4 \times 10^{-4} \pm 0.012655}{2} \approx \frac{0.012255}{2} \text{ or } -\frac{0.013055}{2} \text{ (impossible)}$$

The concentration is about 0.00613 M $= 6.13 \times 10^{-3}$ M.

77. **Answers may vary.**

79. $x^2 + 2\sqrt{2}x - 6 = 0$

$a = 1, b = 2\sqrt{2}, c = -6$

$$x = \frac{-b \pm \sqrt{b^2 - 4ac}}{2a}$$
$$= \frac{-2\sqrt{2} \pm \sqrt{\left(2\sqrt{2}\right)^2 - 4(1)(-6)}}{2(1)}$$
$$= \frac{-2\sqrt{2} \pm \sqrt{8 + 24}}{2}$$
$$= \frac{-2\sqrt{2} \pm \sqrt{32}}{2} = \frac{-2\sqrt{2} \pm 4\sqrt{2}}{2}$$
$$x = \tfrac{2\sqrt{2}}{2} = \sqrt{2} \text{ or } x = \tfrac{-6\sqrt{2}}{2} = -3\sqrt{2}$$

81. $x^2 - 3ix - 2 = 0$

$a = 1, b = -3i, c = -2$

$$x = \frac{-b \pm \sqrt{b^2 - 4ac}}{2a}$$
$$= \frac{3i \pm \sqrt{(-3i)^2 - 4(1)(-2)}}{2(1)}$$
$$= \frac{3i \pm \sqrt{9i^2 + 8}}{2}$$
$$= \frac{3i \pm \sqrt{-1}}{2} = \frac{3i \pm i}{2}$$
$$x = \tfrac{4i}{2} = 2i \text{ or } x = \tfrac{2i}{2} = i$$

Exercise 10.3 (page 718)

1. $b^2 - 4ac = (1)^2 - 4(1)(1)$
$= 1 - 4 = -3$

3. $x^2 - 4x + 1 = 0$: $a = 1, b = -4, c = 1$

$b^2 - 4ac = (-4)^2 - 4(1)(1)$
$= 16 - 4 = 12$

irrational and unequal

5.

$x^2 - 7x + 6 = 0$
$1^2 - 7(1) + 6 \stackrel{?}{=} 0$
$1 - 7 + 6 \stackrel{?}{=} 0$
$0 = 0$
1 is a solution.

$x^2 - 7x + 6 = 0$
$5^2 - 7(5) + 6 \stackrel{?}{=} 0$
$25 - 35 + 6 \stackrel{?}{=} 0$
$-4 \neq 0$
5 is a solution.

7.
$$\frac{1}{4} + \frac{1}{t} = \frac{1}{2t}$$
$$\left(\frac{1}{4} + \frac{1}{t}\right)4t = \frac{1}{2t} \cdot 4t$$
$$t + 4 = 2$$
$$t = -2$$

SECTION 10.3

9. $m = \dfrac{\Delta y}{\Delta x} = \dfrac{-4-5}{-2-3} = \dfrac{-9}{-5} = \dfrac{9}{5}$

11. $b^2 - 4ac$

13. rational; unequal

15. $4x^2 - 4x + 1 = 0; a = 4, b = -4, c = 1$

$$\begin{aligned} b^2 - 4ac &= (-4)^2 - 4(4)(1) \\ &= 16 - 16 = 0 \end{aligned}$$

The solutions are rational and equal.

17. $5x^2 + x + 2 = 0; a = 5, b = 1, c = 2$

$$\begin{aligned} b^2 - 4ac &= 1^2 - 4(5)(2) \\ &= 1 - 40 = -39 \end{aligned}$$

The solutions are complex conjugates.

19.

$$\begin{aligned} 2x^2 &= 4x - 1 \\ 2x^2 - 4x + 1 &= 0; a = 2, b = -4, c = 1 \\ b^2 - 4ac &= (-4)^2 - 4(2)(1) \\ &= 16 - 8 = 8 \end{aligned}$$

The solutions are irrational and unequal.

21.

$$\begin{aligned} x(2x - 3) &= 20 \\ 2x^2 - 3x - 20 &= 0; a = 2, b = -3, c = -20 \\ b^2 - 4ac &= (-3)^2 - 4(2)(-20) \\ &= 9 + 160 = 169 \end{aligned}$$

The solutions are rational and unequal.

23. $x^2 + kx + 9 = 0; a = 1, b = k, c = 9$

Set the discriminant equal to 0:

$$\begin{aligned} b^2 - 4ac &= 0 \\ k^2 - 4(1)(9) &= 0 \\ k^2 - 36 &= 0 \\ k^2 &= 36 \\ k &= \pm 6 \end{aligned}$$

25.

$$\begin{aligned} 9x^2 + 4 &= -kx \\ 9x^2 + kx + 4 &= 0; a = 9, b = k, c = 4 \end{aligned}$$

Set the discriminant equal to 0:

$$\begin{aligned} b^2 - 4ac &= 0 \\ k^2 - 4(9)(4) &= 0 \\ k^2 - 144 &= 0 \\ k^2 &= 144 \\ k &= \pm 12 \end{aligned}$$

27.

$$\begin{aligned} x^4 - 17x^2 + 16 &= 0 \\ (x^2 - 16)(x^2 - 1) &= 0 \end{aligned}$$

$x^2 - 16 = 0$ **or** $x^2 - 1 = 0$

$x^2 = 16 \qquad x^2 = 1$

$x = \pm 4 \qquad x = \pm 1$

29.

$$\begin{aligned} x^4 - 3x^2 &= -2 \\ x^4 - 3x^2 + 2 &= 0 \\ (x^2 - 2)(x^2 - 1) &= 0 \end{aligned}$$

$x^2 - 2 = 0$ **or** $x^2 - 1 = 0$

$x^2 = 2 \qquad x^2 = 1$

$x = \pm\sqrt{2} \qquad x = \pm 1$

31.

$$\begin{aligned} x^4 &= 6x^2 - 5 \\ x^4 - 6x^2 + 5 &= 0 \\ (x^2 - 5)(x^2 - 1) &= 0 \end{aligned}$$

$x^2 - 5 = 0$ **or** $x^2 - 1 = 0$

$x^2 = 5 \qquad x^2 = 1$

$x = \pm\sqrt{5} \qquad x = \pm 1$

33.

$$\begin{aligned} 2x^4 - 10x^2 &= -8 \\ 2x^4 - 10x^2 + 8 &= 0 \\ 2(x^2 - 4)(x^2 - 1) &= 0 \end{aligned}$$

$x^2 - 4 = 0$ **or** $x^2 - 1 = 0$

$x^2 = 4 \qquad x^2 = 1$

$x = \pm 2 \qquad x = \pm 1$

35.

$$\begin{aligned} x-6\sqrt{x}+8&=0\\ (\sqrt{x}-2)(\sqrt{x}-4)&=0 \end{aligned}$$

$\sqrt{x}-2=0$ **or** $\sqrt{x}-4=0$

$\sqrt{x}=2$ $\quad\quad$ $\sqrt{x}=4$

$\left(\sqrt{x}\right)^2=2^2$ $\quad\quad$ $\left(\sqrt{x}\right)^2=4^2$

$x=4$ $\quad\quad$ $x=16$

Solution $\quad\quad$ Solution

37.

$$\begin{aligned} 2x-\sqrt{x}&=3\\ 2x-\sqrt{x}-3&=0\\ (2\sqrt{x}-3)(\sqrt{x}+1)&=0 \end{aligned}$$

$2\sqrt{x}-3=0$ **or** $\sqrt{x}+1=0$

$2\sqrt{x}=3$ $\quad\quad$ $\sqrt{x}=-1$

$\left(2\sqrt{x}\right)^2=3^2$ $\quad\quad$ $\left(\sqrt{x}\right)^2=(-1)^2$

$4x=9$ $\quad\quad$ $x=1$

$x=\frac{9}{4}$ $\quad\quad$ Extraneous

Solution

39.

$$\begin{aligned} 2x+x^{1/2}-3&=0\\ \left(2x^{1/2}+3\right)\left(x^{1/2}-1\right)&=0 \end{aligned}$$

$2x^{1/2}+3=0$ **or** $x^{1/2}-1=0$

$2x^{1/2}=-3$ $\quad\quad$ $x^{1/2}=1$

$x^{1/2}=-\frac{3}{2}$ $\quad\quad$ $\left(x^{1/2}\right)^2=1^2$

$\left(x^{1/2}\right)^2=\left(-\frac{3}{2}\right)^2$ $\quad\quad$ $x=1$

$x=\frac{9}{4}$ $\quad\quad$ Solution

Extraneous

41.

$$\begin{aligned} 3x+5x^{1/2}+2&=0\\ \left(3x^{1/2}+2\right)\left(x^{1/2}+1\right)&=0 \end{aligned}$$

$3x^{1/2}+2=0$ **or** $x^{1/2}+1=0$

$3x^{1/2}=-2$ $\quad\quad$ $x^{1/2}=-1$

$x^{1/2}=-\frac{2}{3}$ $\quad\quad$ $\left(x^{1/2}\right)^2=(-1)^2$

$\left(x^{1/2}\right)^2=\left(-\frac{2}{3}\right)^2$ $\quad\quad$ $x=1$

$x=\frac{4}{9}$ $\quad\quad$ Extraneous

Extraneous

solution set: $\emptyset$

43.

$$\begin{aligned} x^{2/3}+5x^{1/3}+6&=0\\ \left(x^{1/3}+2\right)\left(x^{1/3}+3\right)&=0 \end{aligned}$$

$x^{1/3}+2=0$ **or** $x^{1/3}+3=0$

$x^{1/3}=-2$ $\quad\quad$ $x^{1/3}=-3$

$\left(x^{1/3}\right)^3=(-2)^3$ $\quad\quad$ $\left(x^{1/3}\right)^3=(-3)^3$

$x=-8$ $\quad\quad$ $x=-27$

Solution $\quad\quad$ Solution

45.

$$\begin{aligned} x^{2/3}-2x^{1/3}-3&=0\\ \left(x^{1/3}-3\right)\left(x^{1/3}+1\right)&=0 \end{aligned}$$

$x^{1/3}-3=0$ **or** $x^{1/3}+1=0$

$x^{1/3}=3$ $\quad\quad$ $x^{1/3}=-1$

$\left(x^{1/3}\right)^3=(3)^3$ $\quad\quad$ $\left(x^{1/3}\right)^3=(-1)^3$

$x=27$ $\quad\quad$ $x=-1$

Solution $\quad\quad$ Solution

47.

$$\begin{aligned} x+5+\frac{4}{x}&=0\\ x\left(x+5+\frac{4}{x}\right)&=x(0)\\ x^2+5x+4&=0\\ (x+4)(x+1)&=0 \end{aligned}$$

$x+4=0$ **or** $x+1=0$

$x=-4$ $\quad\quad$ $x=-1$

49.

$$\begin{aligned} x+1&=\frac{20}{x}\\ x+1-\frac{20}{x}&=0\\ x\left(x+1-\frac{20}{x}\right)&=x(0)\\ x^2+x-20&=0\\ (x+5)(x-4)&=0 \end{aligned}$$

$x+5=0$ **or** $x-4=0$

$x=-5$ $\quad\quad$ $x=4$

51.
$$\frac{1}{x-1}+\frac{3}{x+1}=2$$
$$\left(\frac{1}{x-1}+\frac{3}{x+1}\right)(x-1)(x+1)=2(x+1)(x-1)$$
$$1(x+1)+3(x-1)=2(x^2-1)$$
$$x+1+3x-3=2x^2-2$$
$$0=2x^2-4x$$
$$0=2x(x-2) \qquad 2x=0 \text{ or } x-2=0$$
$$x=0 \qquad x=2$$

53.
$$\frac{1}{x+2}+\frac{24}{x+3}=13$$
$$\left(\frac{1}{x+2}+\frac{24}{x+3}\right)(x+2)(x+3)=13(x+2)(x+3)$$
$$1(x+3)+24(x+2)=13(x^2+5x+6)$$
$$x+3+24x+48=13x^2+65x+78$$
$$0=13x^2+40x+27$$
$$0=(13x+27)(x+1) \qquad 13x+27=0 \text{ or } x+1=0$$
$$13x=-27 \qquad x=-1$$
$$x=-\tfrac{27}{13}$$

55.
$$x^{-4}-2x^{-2}+1=0$$
$$(x^{-2}-1)(x^{-2}-1)=0$$
$$x^{-2}-1=0 \text{ or } x^{-2}-1=0$$
$$x^{-2}=1 \qquad x^{-2}=1$$
$$\frac{1}{x^2}=1 \qquad \frac{1}{x^2}=1$$
$$1=x^2 \qquad 1=x^2$$
$$\pm 1=x \qquad \pm 1=x$$

57.
$$8a^{-2}-10a^{-1}-3=0$$
$$(2a^{-1}-3)(4a^{-1}+1)=0$$
$$2a^{-1}-3=0 \text{ or } 4a^{-1}+1=0$$
$$2a^{-1}=3 \qquad 4a^{-1}=-1$$
$$\frac{2}{a}=3 \qquad \frac{4}{a}=-1$$
$$2=3a \qquad 4=-a$$
$$\tfrac{2}{3}=a \qquad -4=a$$

59.
$$x^2+y^2=r^2$$
$$x^2=r^2-y^2$$
$$x=\pm\sqrt{r^2-y^2}$$

61. $xy^2+3xy+7=0;\ a=x, b=3x, c=7$
$$y=\frac{-b\pm\sqrt{b^2-4ac}}{2a}$$
$$=\frac{-3x\pm\sqrt{(3x)^2-4(x)(7)}}{2x}$$
$$=\frac{-3x\pm\sqrt{9x^2-28x}}{2x}$$

SECTION 10.3

63. $12x^2 - 5x - 2 = 0;\ a = 12, b = -5, c = -2$

$(4x+1)(3x-2) = 0$

$4x + 1 = 0$ **or** $3x - 2 = 0$

$4x = -1$ $\quad$ $3x = 2$

$x = -\frac{1}{4}$ $\quad$ $x = \frac{2}{3}$

$$-\frac{b}{a} = -\frac{-5}{12} = \frac{5}{12} \qquad \frac{c}{a} = \frac{-2}{12} = -\frac{1}{6}$$

$$-\frac{1}{4} + \frac{2}{3} = -\frac{3}{12} + \frac{8}{12} = \frac{5}{12} \qquad \left(-\frac{1}{4}\right)\left(\frac{2}{3}\right) = -\frac{1}{6}$$

65. $2x^2 + 5x + 1 = 0;\ a = 2, b = 5, c = 1;\ -\frac{b}{a} = -\frac{5}{2};\ \frac{c}{a} = \frac{1}{2}$

$$x = \frac{-b \pm \sqrt{b^2 - 4ac}}{2a} = \frac{-5 \pm \sqrt{5^2 - 4(2)(1)}}{2(2)} = \frac{-5 \pm \sqrt{17}}{4} = -\frac{5}{4} \pm \frac{\sqrt{17}}{4}$$

$$\frac{-5+\sqrt{17}}{4} + \frac{-5-\sqrt{17}}{4} = \frac{-10}{4} = -\frac{5}{2}$$

$$\left(\frac{-5+\sqrt{17}}{4}\right)\left(\frac{-5-\sqrt{17}}{4}\right) = \frac{25 + 5\sqrt{17} - 5\sqrt{17} - 17}{16} = \frac{8}{16} = \frac{1}{2}$$

67. $3x^2 - 2x + 4 = 0;\ a = 3, b = -2, c = 4;\ -\frac{b}{a} = -\frac{-2}{3} = \frac{2}{3};\ \frac{c}{a} = \frac{4}{3}$

$$x = \frac{-b \pm \sqrt{b^2 - 4ac}}{2a} = \frac{-(-2) \pm \sqrt{(-2)^2 - 4(3)(4)}}{2(3)} = \frac{2 \pm \sqrt{-44}}{6} = \frac{1}{3} \pm \frac{i\sqrt{11}}{3}$$

$$\frac{1+i\sqrt{11}}{3} + \frac{1-i\sqrt{11}}{3} = \frac{2}{3}$$

$$\left(\frac{1+i\sqrt{11}}{3}\right)\left(\frac{1-i\sqrt{11}}{3}\right) = \frac{1 - i\sqrt{11} + i\sqrt{11} - 11i^2}{9} = \frac{1+11}{9} = \frac{12}{9} = \frac{4}{3}$$

69. $x^2 + 2x + 5 = 0;\ a = 1, b = 2, c = 5;\ -\frac{b}{a} = -\frac{2}{1} = -2;\ \frac{c}{a} = \frac{5}{1} = 5$

$$x = \frac{-b \pm \sqrt{b^2 - 4ac}}{2a} = \frac{-2 \pm \sqrt{2^2 - 4(1)(5)}}{2(1)} = \frac{-2 \pm \sqrt{-16}}{2} = \frac{-2 \pm 4i}{2} = -1 \pm 2i$$

$(-1+2i) + (-1-2i) = -2;\ (-1+2i)(-1-2i) = 1 + 2i - 2i - 4i^2 = 1 + 4 = 5$

71. $1492x^2 + 1776x - 1984 = 0$

$a = 1492, b = 1776, c = -1984$

$b^2 - 4ac = (1776)^2 - 4(1492)(-1984)$

$= 3{,}154{,}176 + 11{,}840{,}512$

$= 14{,}994{,}688$

The solutions are real numbers.

73. $4x - 5\sqrt{x} - 9 = 0$

$(4\sqrt{x} - 9)(\sqrt{x} + 1) = 0$

$4\sqrt{x} - 9 = 0$ **or** $\sqrt{x} + 1 = 0$

$4\sqrt{x} = 9$ $\quad$ $\sqrt{x} = -1$

$(4\sqrt{x})^2 = 9^2$ $\quad$ $(\sqrt{x})^2 = (-1)^2$

$16x = 81$ $\quad$ $x = 1$

$x = \frac{81}{16}$ $\quad$ Extraneous

Solution

75.
$$\begin{aligned}3x^{2/3}-x^{1/3}-2&=0\\(x^{1/3}-1)(3x^{1/3}+2)&=0\end{aligned}$$
$$\begin{array}{lcl} x^{1/3}-1=0 & \textbf{or} & 3x^{1/3}+2=0\\ x^{1/3}=1 & & 3x^{1/3}=-2\\ \left(x^{1/3}\right)^3=(1)^3 & & \left(3x^{1/3}\right)^3=(-2)^3\\ x=1 & & 27x=-8\\ \text{Solution} & & x=-\frac{8}{27}\\ & & \text{Solution}\end{array}$$

77.
$$\begin{aligned}2x^4+24&=26x^2\\2x^4-26x^2+24&=0\\2(x^2-12)(x^2-1)&=0\end{aligned}$$
$$\begin{array}{lcl} x^2-12=0 & \textbf{or} & x^2-1=0\\ x^2=12 & & x^2=1\\ x=\pm\sqrt{12} & & x=\pm 1\\ =\pm 2\sqrt{3} & & \end{array}$$

79.
$$\begin{aligned}t^4+3t^2&=28\\t^4+3t^2-28&=0\\(t^2+7)(t^2-4)&=-0\end{aligned}$$
$$\begin{array}{lcl} t^2+7=0 & \textbf{or} & t^2-4=0\\ t^2=-7 & & t^2=4\\ t=\pm\sqrt{-7} & & t=\pm 2\\ =\pm i\sqrt{7} & & \end{array}$$

81. Let $y=2x-1$.
$$\begin{aligned}4(2x-1)^2-3(2x-1)-1&=0\\4y^2-3y-1&=0\\(4y+1)(y-1)&=0\end{aligned}$$
$$\begin{array}{lcl} 4y+1=0 & \textbf{or} & y-1=0\\ 4y=-1 & & y=1\\ y=-\frac{1}{4} & & y=1\\ 2x-1=-\frac{1}{4} & & 2x-1=1\\ 2x=\frac{3}{4} & & 2x=2\\ x=\frac{3}{8} & & x=1\end{array}$$

83.
$$\begin{aligned}x^{-2/3}-2x^{-1/3}-3&=0\\(x^{-1/3}-3)(x^{-1/3}+1)&=0\end{aligned}$$
$$\begin{array}{lcl} x^{-1/3}-3=0 & \textbf{or} & x^{-1/3}+1=0\\ x^{-1/3}=3 & & x^{-1/3}=-1\\ \left(x^{-1/3}\right)^{-3}=3^{-3} & & \left(x^{-1/3}\right)^{-3}=(-1)^{-3}\\ x=\frac{1}{3^3} & & x=\frac{1}{(-1)^3}\\ x=\frac{1}{27} & & x=-1\end{array}$$

85.
$$\begin{aligned}x+\frac{2}{x-2}&=0\\\left(x+\frac{2}{x-2}\right)(x-2)&=0(x-2)\\x(x-2)+2&=0\\x^2-2x+2&=0\\x^2-2x&=-2\\x^2-2x+1&=-2+1\\(x-1)^2&=-1\\x-1&=\pm\sqrt{-1}\\x&=1\pm i\end{aligned}$$

87. Let $y = m + 1$.
$$\begin{aligned} 8(m+1)^{-2} - 30(m+1)^{-1} + 7 &= 0 \\ 8y^{-2} - 30y^{-1} + 7 &= 0 \\ (4y^{-1} - 1)(2y^{-1} - 7) &= 0 \end{aligned}$$

$$\begin{aligned} 4y^{-1} - 1 &= 0 & \quad \textbf{or} \quad 2y^{-1} - 7 &= 0 \\ 4y^{-1} &= 1 & 2y^{-1} &= 7 \\ \frac{4}{y} &= 1 & \frac{2}{y} &= 7 \\ 4 &= y & 2 &= 7y \\ & & \tfrac{2}{7} &= y \\ 4 &= m + 1 & \tfrac{2}{7} &= m + 1 \\ 3 &= m & -\tfrac{5}{7} &= m \end{aligned}$$

89.
$$\begin{aligned} I &= \frac{k}{d^2} \\ Id^2 &= k \\ d^2 &= \frac{k}{I} \\ d &= \pm\sqrt{\frac{k}{I}} = \pm\frac{\sqrt{kI}}{I} \end{aligned}$$

91.
$$\begin{aligned} \sigma &= \sqrt{\frac{\Sigma x^2}{N} - \mu^2} \\ \sigma^2 &= \frac{\Sigma x^2}{N} - \mu^2 \\ \mu^2 &= \frac{\Sigma x^2}{N} - \sigma^2 \end{aligned}$$

93. $(k-1)x^2 + (k-1)x + 1 = 0$

$a = k - 1, b = k - 1, c = 1$

Set the discriminant equal to 0:
$$\begin{aligned} b^2 - 4ac &= 0 \\ (k-1)^2 - 4(k-1)(1) &= 0 \\ k^2 - 2k + 1 - 4k + 4 &= 0 \\ k^2 - 6k + 5 &= 0 \\ (k-5)(k-1) &= 0 \end{aligned}$$
$$k - 5 = 0 \quad \textbf{or} \quad k - 1 = 0$$
$$k = 5 \qquad\qquad k = 1$$
doesn't work

95. $(k+4)x^2 + 2kx + 9 = 0$

$a = k + 4, b = 2k, c = 9$

Set the discriminant equal to 0:
$$\begin{aligned} b^2 - 4ac &= 0 \\ (2k)^2 - 4(k+4)(9) &= 0 \\ 4k^2 - 36(k+4) &= 0 \\ 4k^2 - 36k - 144 &= 0 \\ 4(k^2 - 9k - 36) &= 0 \\ 4(k-12)(k+3) &= 0 \end{aligned}$$
$$k - 12 = 0 \quad \textbf{or} \quad k + 3 = 0$$
$$k = 12 \qquad\qquad k = -3$$

97.
$$\begin{aligned} 3x^2 + 4x &= k \\ 3x^2 + 4x - k &= 0 \end{aligned}$$
$a = 3, b = 4, c = -k$

Set the discriminant less than 0:
$$\begin{aligned} b^2 - 4ac &< 0 \\ 4^2 - 4(3)(-k) &< 0 \\ 16 + 12k &< 0 \\ 12k &< -16 \\ k &< -\frac{16}{12}, \text{ or } k < -\frac{4}{3} \end{aligned}$$

99. **Answers may vary.**

101. No

Exercise 10.4 (page 732)

1. $y = -3x^2 + x - 5$; $a = -3$; opens down

3. $y = 2(x-3)^2 - 1$; $a = 2$; opens up

5. $y = 2(x-3)^2 - 1$; vertex $(3, -1)$

7. $3x + 5 = 5x - 15$
$-2x = -20$
$x = 10$

9. Let $t =$ the time of the second train.
Then $t + 3 =$ the time of the first train.

	Rate	Time	Dist.
First	30	$t+3$	$30(t+3)$
Second	55	t	$55t$

1st distance = 2nd distance
$30(t+3) = 55t$
$30t + 90 = 55t$
$-25t = -90$
$t = \frac{-90}{-25} = \frac{18}{5} = 3\frac{3}{5}$ hours

11. $f(x) = ax^2 + bx + c$; $a \neq 0$

13. maximum; minimum; vertex

15. upward

17. to the right

19. upward

21. $f(x) = x^2$
vertex: $(0, 0)$; opens U

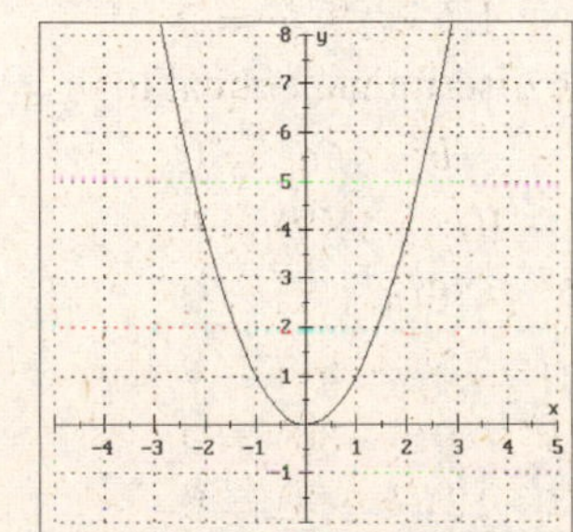

23. $f(x) = x^2 + 2$
vertex: $(0, 2)$; opens U

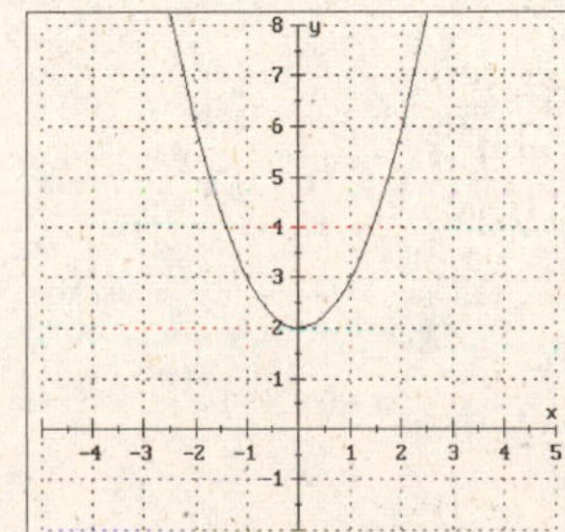

25. $f(x) = -(x-2)^2$
vertex: $(2, 0)$; opens D

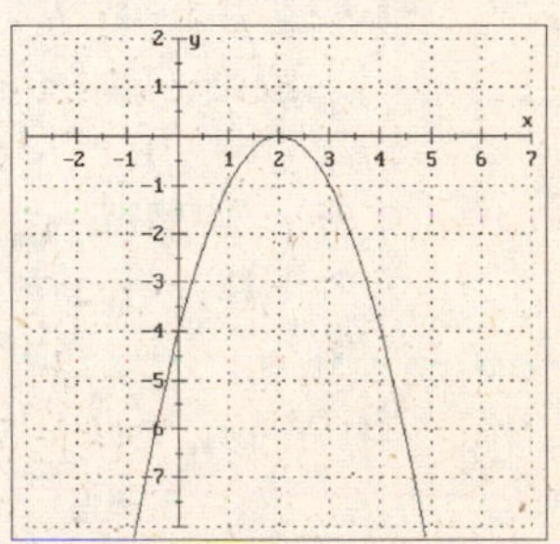

27. $f(x) = (x-3)^2 + 2$
vertex: $(3, 2)$; opens U

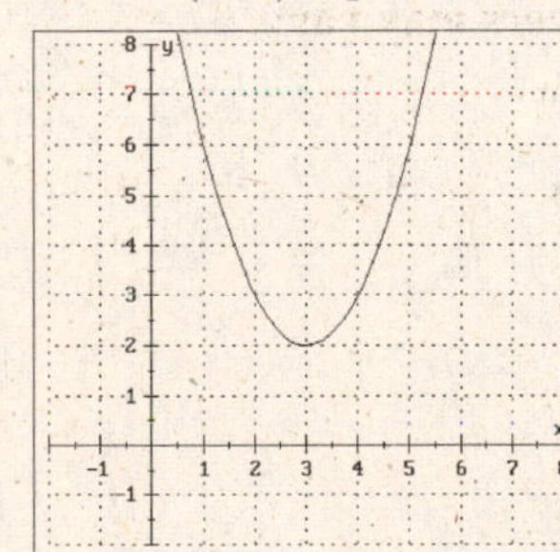

29. $y = (x-1)^2 + 2$; $V(1, 2)$; axis: $x = 1$

31. $y = 2(x+3)^2 - 4$;
$V(-3, -4)$; axis: $x = -3$

33. $y = -3x^2 \Rightarrow y = -3(x-0)^2 + 0$
$V(0,0)$; axis: $x = 0$

35. $y = 2x^2 - 4x$
$y = 2(x^2 - 2x)$
$y = 2(x^2 - 2x + 1) - 2$
$y = 2(x-1)^2 - 2$
$V(1,-2)$; axis: $x = 1$

37. $f(x) = -2x^2 + 4x + 1$
$x = -\frac{b}{2a} = -\frac{4}{2(-2)} = 1$
$y = -2(1)^2 + 4(1) + 1 = 3$
vertex: $(1,3)$; opens D

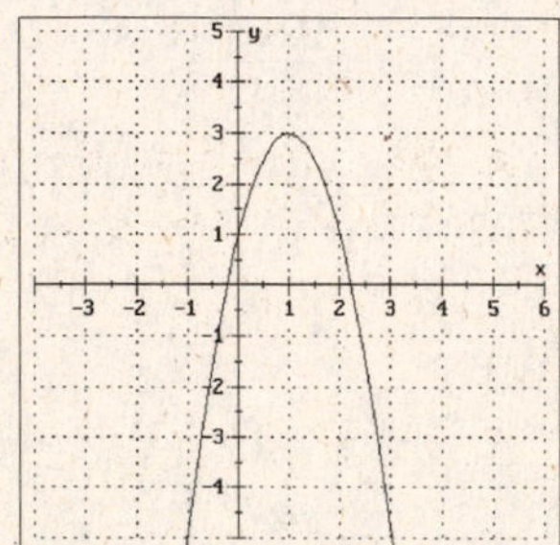

39. $f(x) = 3x^2 - 12x + 10$
$x = -\frac{b}{2a} = -\frac{-12}{2(3)} = 2$
$y = 3(2)^2 - 12(2) + 10 = -2$
vertex: $(2,-2)$; opens U

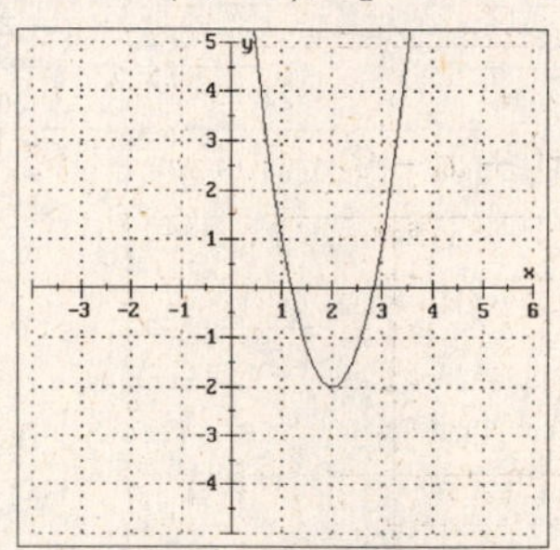

41. $y = -4x^2 + 16x + 5$
$y = -4(x^2 - 4x) + 5$
$y = -4(x^2 - 4x + 4) + 5 + 16$
$y = -4(x-2)^2 + 21$
$V(2,21)$; axis: $x = 2$

43. $y - 7 = 6x^2 - 5x$
$y = 6x^2 - 5x + 7$
$x = -\frac{b}{2a} = -\frac{-5}{2(6)} = \frac{5}{12}$
$y = 6x^2 - 5x + 7$
$= 6\left(\frac{5}{12}\right)^2 - 5\left(\frac{5}{12}\right) + 7$
$= \frac{25}{24} - \frac{25}{12} + 7 = \frac{143}{24}$
$V\left(\frac{5}{12}, \frac{143}{24}\right)$; axis: $x = \frac{5}{12}$

SECTION 10.4

45. $f(x) = x^2 + x - 6$

$f(x) = \left(x + \frac{1}{2}\right)^2 - \frac{25}{4}$

vertex: $\left(-\frac{1}{2}, -\frac{25}{4}\right)$; opens U

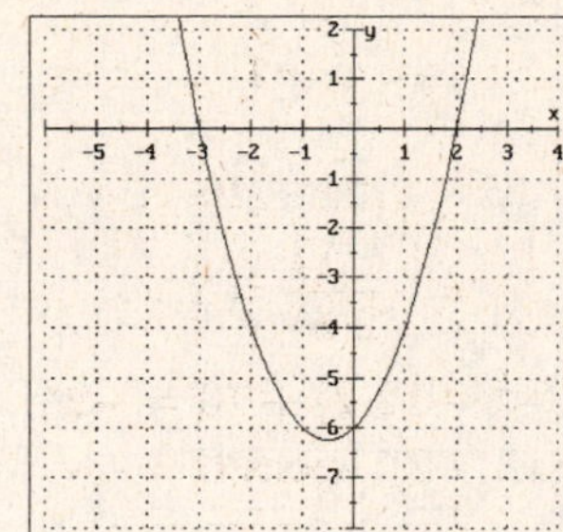

47. $y = 2x^2 - x + 1$

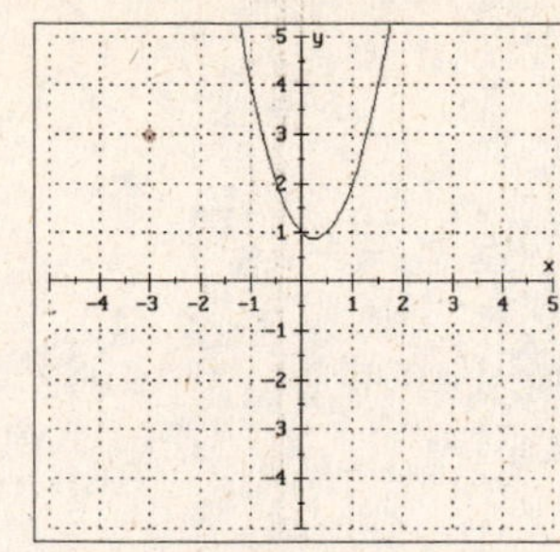

$V(0.25, 0.88)$

49. $y = 7 + x - x^2$

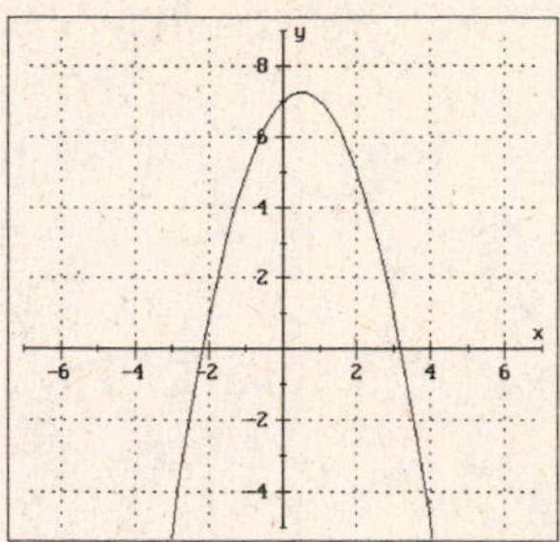

$V(0.5, 7.25)$

51. $y = x^2 + x - 6$

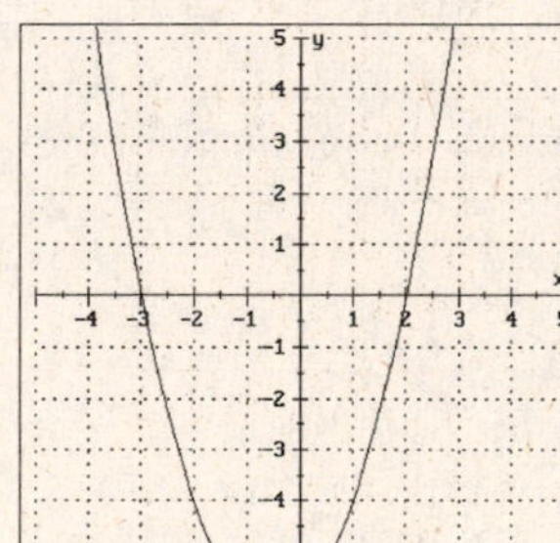

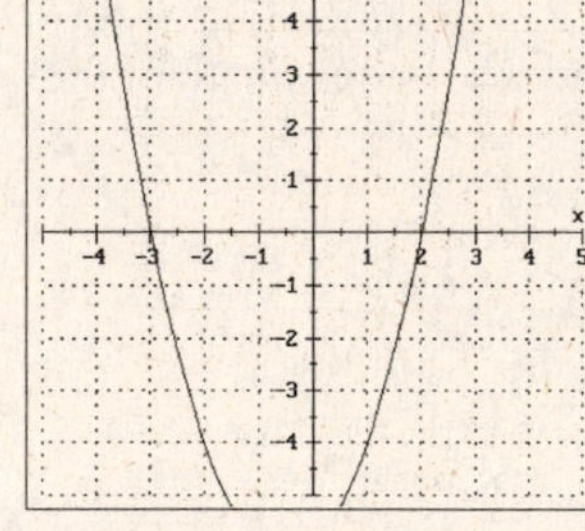

solution set: $\{2, -3\}$

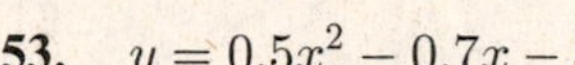

53. $y = 0.5x^2 - 0.7x - 3$

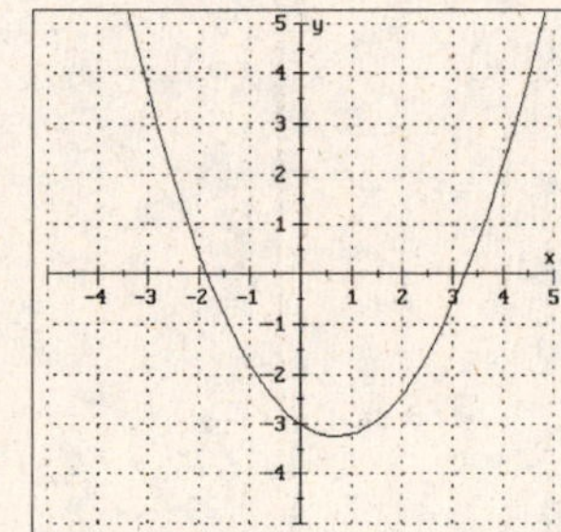

solution set: $\{-1.85, 3.25\}$

55. $y - 2 = (x - 5)^2$

$y = (x - 5)^2 + 2$

$V(5, 2)$

57. Since the graph of the height equation is a parabola, the max. height occurs at the vertex.

$s = 48t - 16t^2$

$s = -16(t^2 - 3t)$

$s = -16\left(t^2 - 3t + \frac{9}{4}\right) + 36$

$s = -16\left(t - \frac{3}{2}\right)^2 + 36$

$V\left(\frac{3}{2}, 36\right) \Rightarrow$ max. height $= 36$ ft

The maximum height is 36 feet, which occurs after $\frac{3}{2}$ seconds (1.5 seconds).

59. Let w = the width of the rectangle.
Then $100 - w$ = the length.
$A = w(100 - w)$
$A = -w^2 + 100w$
$A = -(w^2 - 100w + 2500) + 2500$
$A = -(w - 50)^2 + 2500$
dim: 50 ft by 50 ft; area = 2500 ft^2

61. Replace p with x.
Graph $y = 50x(1 - x)$ and find the x-coordinate(s) when $y = 9.375$.

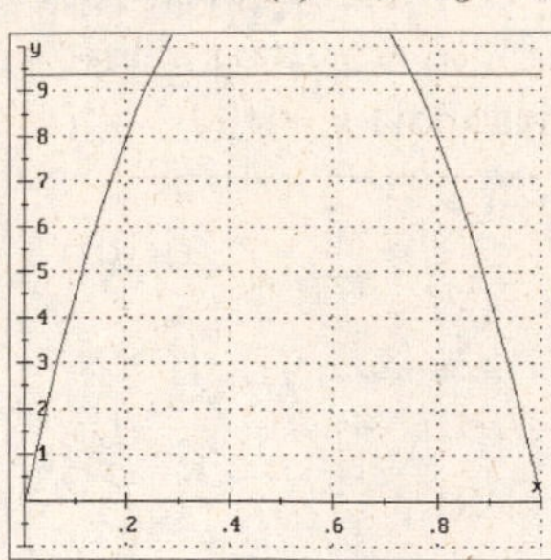

$p = 0.25$ or $p = 0.75$

63. Let w = the width of the rectangle.
Then $150 - w$ = the length.
$A = w(150 - w)$
$A = -w^2 + 150w$
$A = -(w^2 - 150w + 5625) + 5625$
$A = -(w - 75)^2 + 5625$
dim: 75 ft by 75 ft; area = 5625 ft^2

65. Replace t with x.
Graph $y = H = 3.3x^2 - 59.4x + 281.3$ and find the y-coordinate of the vertex:

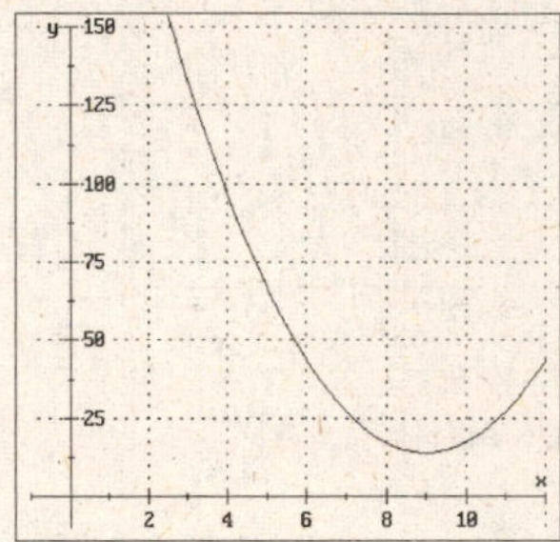

The minimum level was 14 feet.

67. Graph $y = R = -\frac{x^2}{1000} + 10x$ and find the x-coordinate of the vertex:

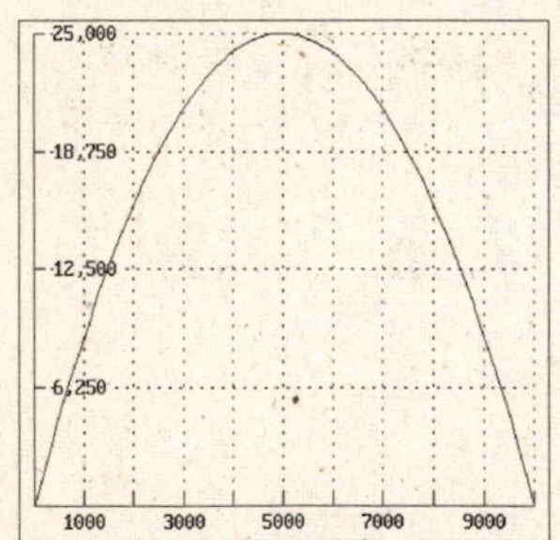

5000 stereos should be sold.

69. Graph $y = R = -\frac{x^2}{728} + 9x$ and find the x- and y- coordinates of the vertex:

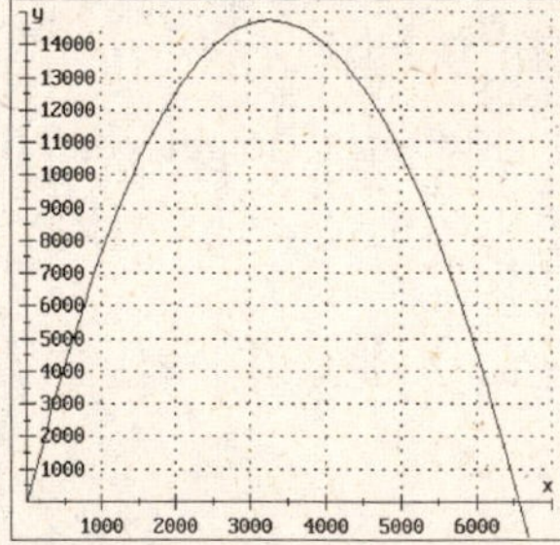

Max. revenue = \$14,742; # radios = 3276

71. Let $x =$ the number of \$1 increases to the price. Then the sales will be $4000 - 100x$, and the revenue will be $(30 + x)(4000 - 100x)$. Find the vertex of the parabola $y = (30 + x)(4000 - 100x)$.

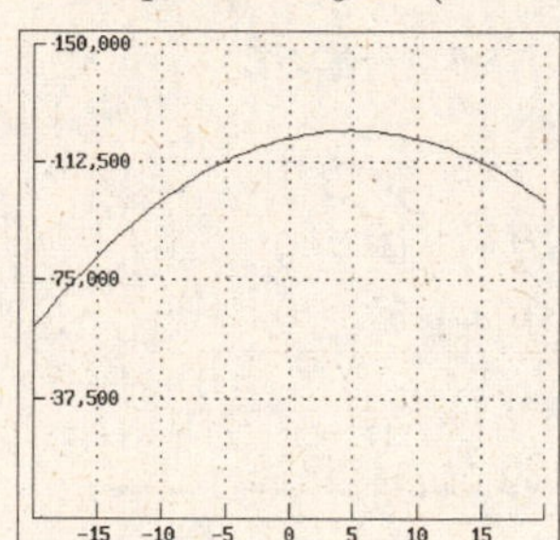

The price should increase \$5, to a total of \$35.

73. Answers may vary.

75. Graph $y = x^2 + x + 1$ to find x-intercept(s):

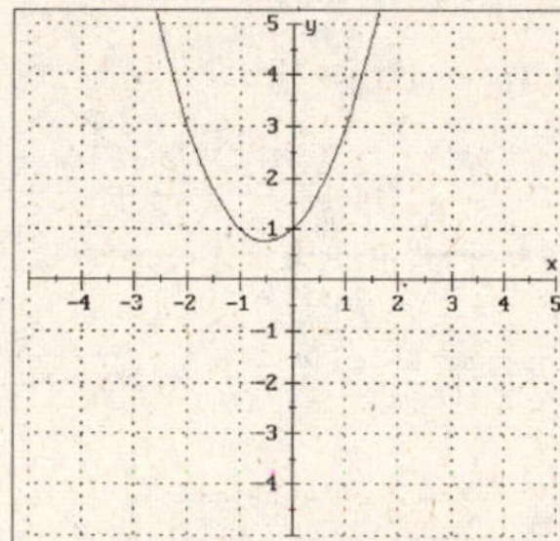

There are no x-intercepts, which means there is no solution to the equation.

Exercise 10.5 (page 743)

1. $x - 2 = 0$
$x = 2$

3. $x - 2 < 0$
$x < 2$

5. $x + 3 > 0$
$x > -3$

7. $\frac{1}{x} < 2$
$x \cdot \frac{1}{x} < x \cdot 2 \ (x > 0)$
$1 < 2x$

9. $y = kx$

11. $t = kxy$

13. $y = 3x - 4$
$m = 3$

15. greater

17. quadratic

19. undefined

21. sign

SECTION 10.5

23. $x^2 - 5x + 4 < 0$

$(x-4)(x-1) < 0$

$x-4$ – – – – – – – – 0 ++++
$x-1$ – – – 0++++++++++

←—(———)—→ (1, 4)

solution set: $(1, 4)$

25. $x^2 - 8x + 15 > 0$

$(x-5)(x-3) > 0$

$x-5$ – – – – – – – – –0++++
$x-3$ – – – 0++++++++++

←—)——(—→ (3, 5)

solution set: $(-\infty, 3) \cup (5, \infty)$

27. $x^2 + x - 12 \le 0$

$(x-3)(x+4) \le 0$

$x-3$ – – – – – – – – – 0++++
$x+4$ – – – 0+++++++++++

←— [———]—→ (−4, 3)

solution set: $[-4, 3]$

29. $x^2 + 2x \ge 15$

$x^2 + 2x - 15 \ge 0$

$(x-3)(x+5) \ge 0$

$x-3$ – – – – – – – – –0++++
$x+5$ – – – 0++++++++++

←—]——[—→ (−5, 3)

solution set: $(-\infty, -5] \cup [3, \infty)$

31. $x^2 + 8x < -16$

$x^2 + 8x + 16 < 0$

$(x+4)(x+4) < 0$

$x+4$ – – – – 0+++++++
$x+4$ – – – – 0+++++++

←———————→ (−4)

Since the product is never negative, there is no solution.

33. $x^2 \ge 9$

$x^2 - 9 \ge 0$

$(x-3)(x+3) \ge 0$

$x-3$ – – – – – – – – –0++++
$x+3$ – – – 0++++++++++

←—]——[—→ (−3, 3)

solution set: $(-\infty, -3] \cup [3, \infty)$

35. $\dfrac{1}{x} < 2$

$\dfrac{1}{x} - 2 < 0$

$\dfrac{1}{x} - \dfrac{2x}{x} < 0$

$\dfrac{1-2x}{x} < 0$

$1-2x$ +++++++++ 0 – – –
x – – – 0++++++++++

←—)——(—→ (0, $\frac{1}{2}$)

solution set: $(-\infty, 0) \cup \left(\frac{1}{2}, \infty\right)$

37. $\dfrac{4}{x} \ge 2$

$\dfrac{4}{x} - 2 \ge 0$

$\dfrac{4}{x} - \dfrac{2x}{x} \ge 0$

$\dfrac{4-2x}{x} \ge 0$

$4-2x$ ++++++++++0 – – –
x – – – 0 +++++++++++

←—(———] —→ (0, 2)

solution set: $(0, 2]$

39. $\dfrac{x^2 - x - 12}{x - 1} < 0$

$\dfrac{(x-4)(x+3)}{x-1} < 0$

$x - 4$ −−−−−−−− −−− 0++++
$x - 1$ −−−−−−−− 0+++ +++++
$x + 3$ −−− 0++++++ ++++ +++++

$-3 \quad 1 \quad 4$

solution set: $(-\infty, -3) \cup (1, 4)$

41. $\dfrac{x^2 + x - 20}{x + 2} \ge 0$

$\dfrac{(x-4)(x+5)}{x+2} \ge 0$

$x - 4$ −−−−−−−− −−−− 0++++
$x + 2$ −−−−−−−− 0++++ +++++
$x + 5$ −−− 0+++++++++++ +++++

$-5 \quad -2 \quad 4$

solution set: $[-5, -2) \cup [4, \infty)$

43. $\dfrac{x^2 - 4x + 4}{x + 4} < 0$

$\dfrac{(x-2)(x-2)}{x+4} < 0$

$x - 2$ −−−−−−−− 0++++
$x - 2$ −−−−−−−− 0++++
$x + 4$ −−− 0++++++ +++++

$-4 \quad 2$

solution set: $(-\infty, -4)$

45. $\dfrac{6x^2 - 5x + 1}{2x + 1} > 0$

$\dfrac{(2x-1)(3x-1)}{2x+1} > 0$

$2x - 1$ −−−−−−−− −−−− 0++++
$3x - 1$ −−−−−−−− 0+++++++++
$2x + 1$ −−− 0+++++++++++++++

$-\frac{1}{2} \quad \frac{1}{3} \quad \frac{1}{2}$

solution set: $\left(-\frac{1}{2}, \frac{1}{3}\right) \cup \left(\frac{1}{2}, \infty\right)$

47. $\dfrac{3}{x-2} < \dfrac{4}{x}$

$\dfrac{3}{x-2} - \dfrac{4}{x} < 0$

$\dfrac{3x}{x(x-2)} - \dfrac{4(x-2)}{x(x-2)} < 0$

$\dfrac{-x+8}{x(x-2)} < 0$

$-x + 8$ +++++++++++++++0−−−
$x - 2$ −−−−−−− 0+++++++++
x −−− 0++++ ++++++++++

$0 \quad 2 \quad 8$

solution set: $(0, 2) \cup (8, \infty)$

49. $\dfrac{-5}{x+2} \ge \dfrac{4}{2-x}$

$\dfrac{-5}{x+2} - \dfrac{4}{2-x} \ge 0$

$\dfrac{-5(2-x)}{(x+2)(2-x)} - \dfrac{4(x+2)}{(x+2)(2-x)} \ge 0$

$\dfrac{x-18}{(x+2)(2-x)} \ge 0$

$x - 18$ −−−−−−−−− −−−− 0++++
$2 - x$ ++++++++++++ 0−−−−−−
$x + 2$ −−− 0++++++++++++++++

$-2 \quad 2 \quad 18$

solution set: $(-\infty, -2) \cup (2, 18]$

51.

$$\frac{7}{x-3} \geq \frac{2}{x+4}$$
$$\frac{7}{x-3} - \frac{2}{x+4} \geq 0$$
$$\frac{7(x+4)}{(x-3)(x+4)} - \frac{2(x-3)}{(x-3)(x+4)} \geq 0$$
$$\frac{5x+34}{(x-3)(x+4)} \geq 0$$

$x-3$ −−−−−−−− −−−− 0++++
$x+4$ −−−−−−−− 0+++++ +++++
$5x+34$ −−− 0+++++++++++ +++++

⟵ [⟶) ⟵(⟶
$-\frac{34}{5}$ -4 3

solution set: $\left[-\frac{34}{5}, -4\right) \cup (3, \infty)$

53.

$$(x+2)^2 > 0$$
$$(x+2)(x+2) > 0$$

$x+2$ −−−−−0+++++++
$x+2$ −−−−−0+++++++

⟵) (⟶
-2

solution set: $(-\infty, -2) \cup (-2, \infty)$

55. $y < x^2 + 1$

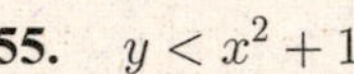

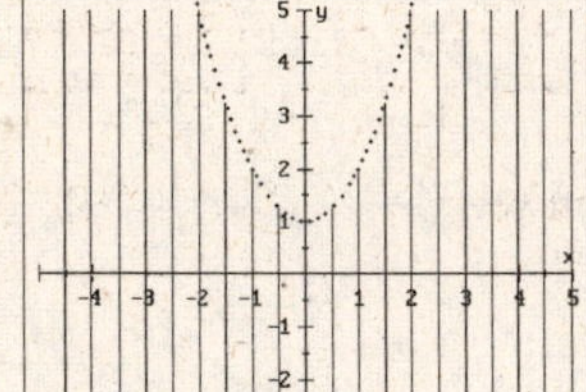

57. $y \leq x^2 + 5x + 6$

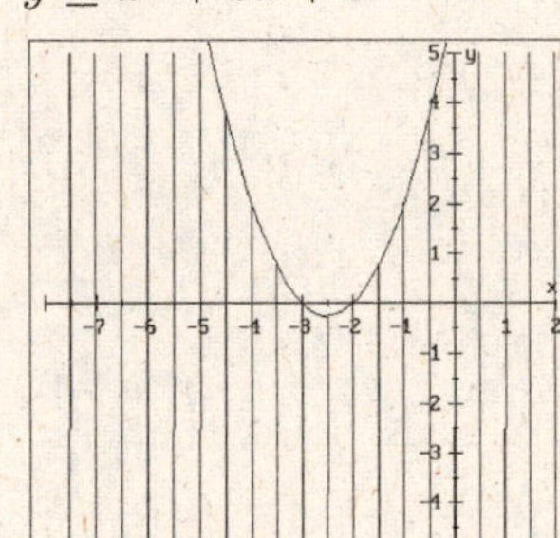

59. $y \geq (x-1)^2$

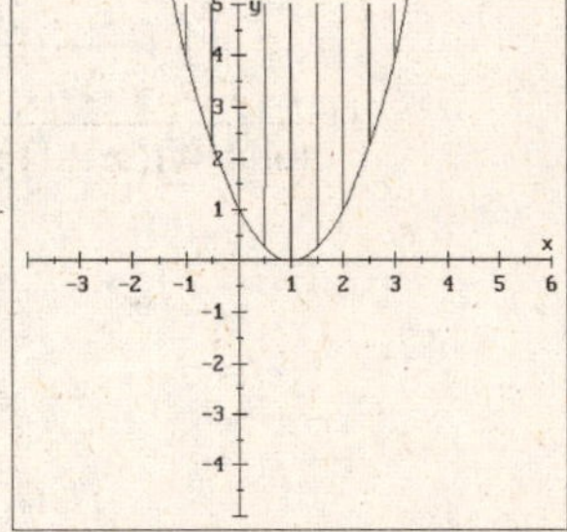

61. $-x^2 - y + 6 > -x$
$-x^2 + x + 6 > y$

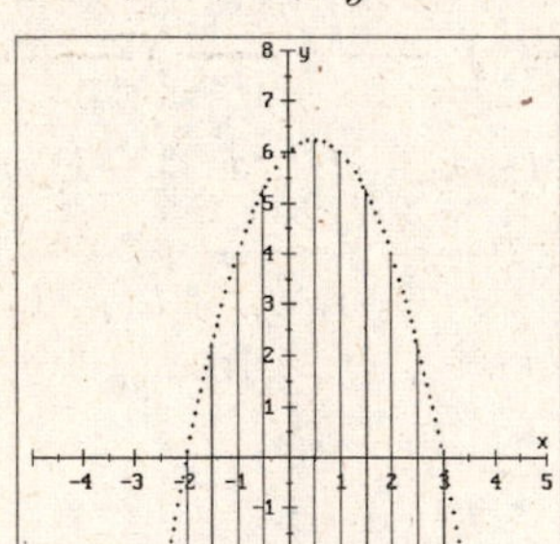

63. $y < |x+4|$

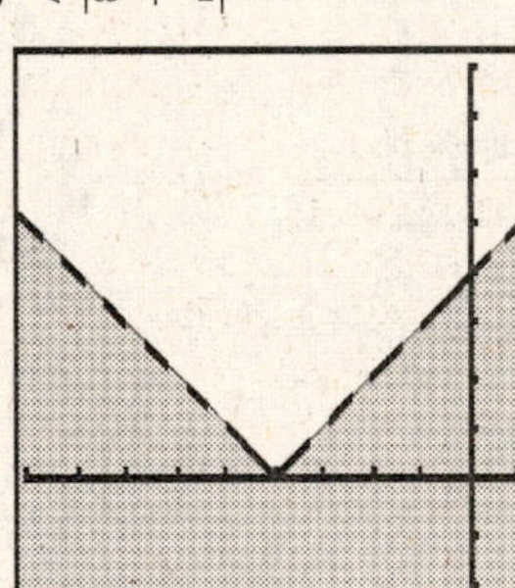

65. $y \leq -|x| + 2$

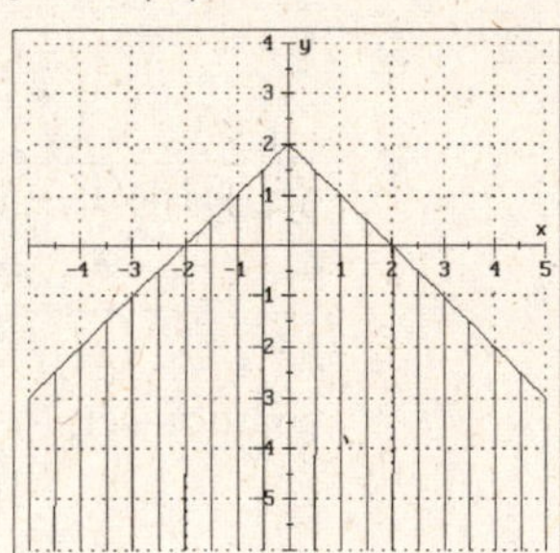

67.
$$2x^2 - 50 < 0$$
$$2(x-5)(x+5) < 0$$

$x-5$ – – – – – – – – – 0++++
$x+5$ – – – 0++++++ ++++

←— (———) —→ at -5, 5

solution set: $(-5, 5)$

69.
$$-\frac{5}{x} < 3$$
$$-\frac{5}{x} - 3 < 0$$
$$\frac{-5}{x} - \frac{3x}{x} < 0$$
$$\frac{-5-3x}{x} < 0$$

$-5-3x$ +++ 0 – – – – – – – –
x – – – – – – – – 0++++

←—) —— (—→ at $-\frac{5}{3}$, 0

solution set: $\left(-\infty, -\frac{5}{3}\right) \cup (0, \infty)$

71.
$$\frac{x}{x+4} \le \frac{1}{x+1}$$
$$\frac{x}{x+4} - \frac{1}{x+1} \le 0$$
$$\frac{x(x+1)}{(x+4)(x+1)} - \frac{1(x+4)}{(x+4)(x+1)} \le 0$$
$$\frac{x^2-4}{(x+4)(x+1)} \le 0$$
$$\frac{(x+2)(x-2)}{(x+4)(x+1)} \le 0$$

$x-2$ – – – – – – – – – – – – – – –0+++
$x+1$ – – – – – – – – – – – 0++++ ++++
$x+2$ – – – – – – 0++++ +++++ ++++
$x+4$ – – 0+++++++++ +++++ ++++

←— (——] —— (——] —→ at -4, -2, -1, 2

solution set: $(-4, -2] \cup (-1, 2]$

73.
$$\frac{x}{x+16} > \frac{1}{x+1}$$
$$\frac{x}{x+16} - \frac{1}{x+1} > 0$$
$$\frac{x(x+1)}{(x+16)(x+1)} - \frac{1(x+16)}{(x+16)(x+1)} > 0$$
$$\frac{x^2-16}{(x+16)(x+1)} > 0$$
$$\frac{(x+4)(x-4)}{(x+16)(x+1)} > 0$$

$x-4$ – – – – – – – – – – – – – – – –0+++
$x+1$ – – – – – – – – – – – – 0++++ ++++
$x+4$ – – – – – – – 0++++ +++++ ++++
$x+16$ – – – 0++++ +++++ +++++ ++++

←—) —— (——) —— (—→ at -16, -4, -1, 4

solution set: $(-\infty, -16) \cup (-4, -1) \cup (4, \infty)$

SECTION 10.5

75. $x^2 - 2x - 3 < 0$
Graph $y = x^2 - 2x - 3$ and find the x-coordinates of points below the x-axis.

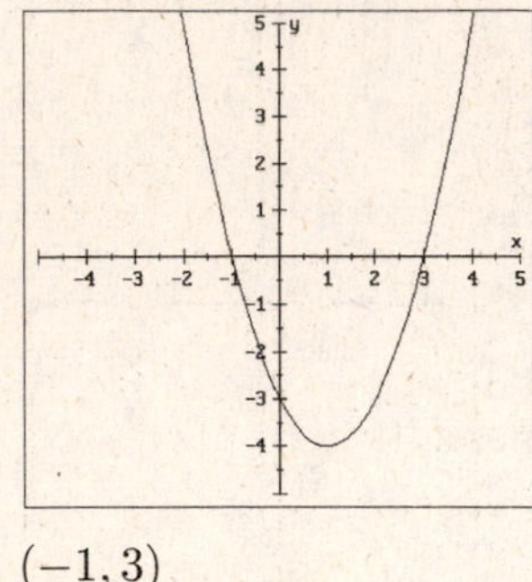

$(-1, 3)$

77. $\frac{x+3}{x-2} > 0$
Graph $y = (x+3)/(x-2)$ and find the x-coordinates of points above the x-axis.

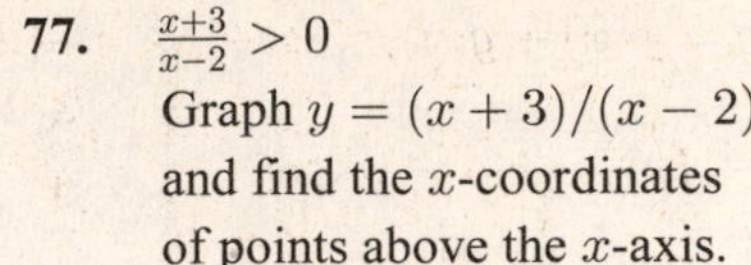

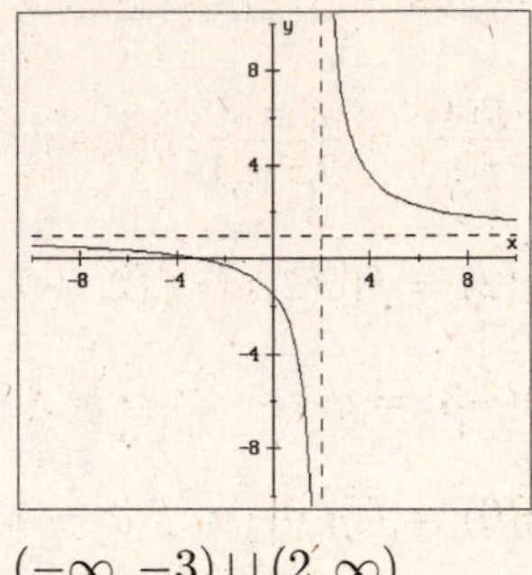

$(-\infty, -3) \cup (2, \infty)$

79. **Answers may vary.**

81. It will be positive if 4, 2 or 0 factors are negative.

Exercise 10.6 (page 752)

1. $f + g = f(x) + g(x) = 2x + 3x = 5x$

3. $f \cdot h = f(x) \cdot h(x) = 2x \cdot 4x = 8x^2$

5. $\frac{h}{f} = \frac{h(x)}{f(x)} = \frac{4x}{2x} = 2$

7. $(f \circ h)(x) = f(h(x)) = f(4x) = 2(4x) = 8x$

9. $\frac{3x^2 + x - 14}{4 - x^2} = \frac{(3x+7)(x-2)}{(2+x)(2-x)} = -\frac{3x+7}{x+2}$

11. $$\frac{8 + 2x - x^2}{12 + x - 3x^2} \div \frac{3x^2 + 5x - 2}{3x - 1} = \frac{x^2 - 2x - 8}{3x^2 - x - 12} \cdot \frac{3x - 1}{3x^2 + 5x - 2} = \frac{(x-4)(x+2)}{3x^2 - x - 12} \cdot \frac{3x - 1}{(3x-1)(x+2)} = \frac{x - 4}{3x^2 - x - 12}$$

13. $f(x) + g(x)$

15. $f(x)g(x)$

17. domain

19. $f(g(x))$

21. $f(x)$

23. $f + g = f(x) + g(x) = 3x + 4x = 7x$
domain $= (-\infty, \infty)$

25. $f \cdot g = f(x) \cdot g(x) = 3x \cdot 4x = 12x^2$
domain $= (-\infty, \infty)$

27. $g - f = g(x) - f(x) = 4x - 3x = x$
domain $= (-\infty, \infty)$

29. $g/f = \frac{g(x)}{f(x)} = \frac{4x}{3x} = \frac{4}{3}$ (for $x \neq 0$); domain $= (-\infty, 0) \cup (0, \infty)$

SECTION 10.6

31. $f + g = f(x) + g(x) = 2x + 1 + x - 3 = 3x - 2$; domain $= (-\infty, \infty)$

33. $f \cdot g = f(x) \cdot g(x) = (2x + 1)(x - 3) = 2x^2 - 5x - 3$; domain $= (-\infty, \infty)$

35. $g - f = g(x) - f(x) = (x - 3) - (2x + 1) = x - 3 - 2x - 1 = -x - 4$; domain $= (-\infty, \infty)$

37. $g/f = \dfrac{g(x)}{f(x)} = \dfrac{x - 3}{2x + 1}$; domain $= \left(-\infty, -\dfrac{1}{2}\right) \cup \left(-\dfrac{1}{2}, \infty\right)$

39. $(f \circ g)(2) = f(g(2)) = f(2^2 - 1) = f(3) = 2(3) + 1 = 7$

41. $(g \circ f)(-3) = g(f(-3)) = g(2(-3) + 1) = g(-5) = (-5)^2 - 1 = 24$

43. $(f \circ g)(0) = f(g(0)) = f(0^2 - 1) = f(-1) = 2(-1) + 1 = -1$

45. $(f \circ g)\left(\dfrac{1}{2}\right) = f\left(g\left(\dfrac{1}{2}\right)\right) = f\left(\left(\dfrac{1}{2}\right)^2 - 1\right) = f\left(-\dfrac{3}{4}\right) = 2\left(-\dfrac{3}{4}\right) + 1 = -\dfrac{3}{2} + 1 = -\dfrac{1}{2}$

47. $(f \circ g)(x) = f(g(x)) = f(x^2 - 1) = 2(x^2 - 1) + 1 = 2x^2 - 2 + 1 = 2x^2 - 1$

49. $(g \circ f)(2x) = g(f(2x)) = g(2(2x) + 1) = g(4x + 1) = (4x + 1)^2 - 1 = 16x^2 + 8x + 1 - 1$
$= 16x^2 + 8x$

51. $\dfrac{f(x + h) - f(x)}{h} = \dfrac{2(x + h) + 3 - (2x + 3)}{h} = \dfrac{2x + 2h + 3 - 2x - 3}{h} = \dfrac{2h}{h} = 2$

53. $\dfrac{f(x + h) - f(x)}{h} = \dfrac{(x + h)^2 - x^2}{h} = \dfrac{x^2 + 2xh + h^2 - x^2}{h} = \dfrac{2xh + h^2}{h} = 2x + h$

55. $\dfrac{f(x + h) - f(x)}{h} = \dfrac{2(x + h)^2 - 1 - (2x^2 - 1)}{h} = \dfrac{2x^2 + 4xh + 2h^2 - 1 - 2x^2 + 1}{h}$
$= \dfrac{4xh + 2h^2}{h} = 4x + 2h$

57. $\dfrac{f(x + h) - f(x)}{h} = \dfrac{(x + h)^2 + (x + h) - (x^2 + x)}{h} = \dfrac{x^2 + 2xh + h^2 + x + h - x^2 - x}{h}$
$= \dfrac{2xh + h^2 + h}{h} = 2x + h + 1$

59. $\dfrac{f(x + h) - f(x)}{h} = \dfrac{(x + h)^2 + 3(x + h) - 4 - (x^2 + 3x - 4)}{h}$
$= \dfrac{x^2 + 2xh + h^2 + 3x + 3h - 4 - x^2 - 3x + 4}{h}$
$= \dfrac{2xh + h^2 + 3h}{h} = 2x + h + 3$

SECTION 10.6

61. $$\frac{f(x+h)-f(x)}{h} = \frac{2(x+h)^2+3(x+h)-7-(2x^2+3x-7)}{h}$$
$$= \frac{2x^2+4xh+2h^2+3x+3h-7-2x^2-3x+7}{h}$$
$$= \frac{4xh+2h^2+3h}{h} = 4x+2h+3$$

63. $f-g = f(x)-g(x) = (3x-2)-(2x^2+1) = 3x-2-2x^2-1 = -2x^2+3x-3$
domain $= (-\infty, \infty)$

65. $f/g = \dfrac{f(x)}{g(x)} = \dfrac{3x-2}{2x^2+1}$; domain $= (-\infty, \infty)$

67. $f-g = f(x)-g(x) = (x^2-1)-(x^2-4) = x^2-1-x^2+4 = 3$; domain $= (-\infty, \infty)$

69. $g/f = \dfrac{g(x)}{f(x)} = \dfrac{x^2-4}{x^2-1} = \dfrac{(x+2)(x-2)}{(x+1)(x-1)}$; domain $= (-\infty, -1) \cup (-1, 1) \cup (1, \infty)$

71. $(f \circ g)(4) = f(g(4)) = f(4^2+4) = f(20) = 3(20)-2 = 58$

73. $(g \circ f)(-3) = g(f(-3)) = g(3(-3)-2) = g(-11) = (-11)^2+(-11) = 110$

75. $(g \circ f)(0) = g(f(0)) = g(3(0)-2) = g(-2) = (-2)^2+(-2) = 2$

77. $(g \circ f)(x) = g(f(x)) = g(3x-2) = (3x-2)^2+3x-2 = 9x^2-12x+4+3x-2$
$= 9x^2-9x+2$

79. $$\frac{f(x)-f(a)}{x-a} = \frac{(2x+3)-(2a+3)}{x-a} = \frac{2x+3-2a-3}{x-a} = \frac{2x-2a}{x-a} = \frac{2(x-a)}{x-a} = 2$$

81. $$\frac{f(x)-f(a)}{x-a} = \frac{x^2-a^2}{x-a} = \frac{(x+a)(x-a)}{x-a} = x+a$$

83. $$\frac{f(x)-f(a)}{x-a} = \frac{(2x^2-1)-(2a^2-1)}{x-a} = \frac{2x^2-1-2a^2+1}{x-a} = \frac{2x^2-2a^2}{x-a}$$
$$= \frac{2(x+a)(x-a)}{x-a}$$
$$= 2(x+a) = 2x+2a$$

85. $$\frac{f(x)-f(a)}{x-a} = \frac{(x^2+x)-(a^2+a)}{x-a} = \frac{x^2+x-a^2-a}{x-a} = \frac{x^2-a^2+x-a}{x-a}$$
$$= \frac{(x+a)(x-a)+1(x-a)}{x-a}$$
$$= \frac{(x-a)(x+a+1)}{x-a} = x+a+1$$

87. $$\frac{f(x)-f(a)}{x-a} = \frac{(x^2+3x-4)-(a^2+3a-4)}{x-a} = \frac{x^2+3x-4-a^2-3a+4}{x-a}$$
$$= \frac{x^2-a^2+3x-3a}{x-a}$$
$$= \frac{(x+a)(x-a)+3(x-a)}{x-a}$$
$$= \frac{(x-a)(x+a+3)}{x-a} = x+a+3$$

89. $$\frac{f(x)-f(a)}{x-a} = \frac{(2x^2+3x-7)-(2a^2+3a-7)}{x-a} = \frac{2x^2+3x-7-2a^2-3a+7}{x-a}$$
$$= \frac{2x^2-2a^2+3x-3a}{x-a}$$
$$= \frac{2(x+a)(x-a)+3(x-a)}{x-a}$$
$$= \frac{(x-a)(2(x+a)+3)}{x-a}$$
$$= 2(x+a)+3 = 2x+2a+3$$

91. $(f \circ g)(x) = f(g(x)) = f(2x-5) = (2x-5)+1 = 2x-4$
$(g \circ f)(x) = g(f(x)) = g(x+1) = 2(x+1)-5 = 2x+2-5 = 2x-3$

93. $f(a) = a^2+2a-3;\ f(h) = h^2+2h-3 \Rightarrow f(a)+f(h) = a^2+h^2+2a+2h-6$
$f(a+h) = (a+h)^2+2(a+h)-3 = a^2+2ah+h^2+2a+2h-3$
$= a^2+h^2+2ah+2a+2h-3$

95. $$\frac{f(x+h)-f(x)}{h} = \frac{(x+h)^3-1-(x^3-1)}{h} = \frac{x^3+3x^2h+3xh^2+h^3-1-x^3+1}{h}$$
$$= \frac{3x^2h+3xh^2+h^3}{h}$$
$$= \frac{h(3x^2+3xh+h^2)}{h} = 3x^2+3xh+h^2$$

97. $F(t) = 2700-200t;\ C(F) = \frac{5}{9}(F-32)$
$C(F(t)) = C(2700-200t) = \frac{5}{9}(2700-200t-32) = \frac{5}{9}(2668-200t)$

99. **Answers may vary.**

101. It is associative. Examples will vary..

Exercise 10.7 (page 761)

1. $\{(2,1),(3,2),(10,5)\}$

3.
$$y = \frac{1}{2}x$$
$$x = \frac{1}{2}y$$
$$2x = 2 \cdot \frac{1}{2}y$$
$$2x = y \Rightarrow f^{-1}(x) = 2x$$

5. $y = 2$ is paired with $x = 2$ and $x = -2$, so the function **is not one-to-one**.

7. $3 - \sqrt{-64} = 3 - \sqrt{64i^2} = 3 - 8i$

9. $(3+4i)(2-3i) = 6 - 9i + 8i - 12i^2 = 6 - i - 12(-1) = 6 - i + 12 = 18 - i$

11. $|6 - 8i| = \sqrt{6^2 + (-8)^2} = \sqrt{36 + 64} = \sqrt{100} = 10$

13. one-to-one

15. 2

17. x

19. Each input has a different output. one-to-one

21. The inputs $x = 2$ and $x = -2$ have the same output. not one-to-one

23. one-to-one

25. one-to-one

27. not one-to-one

29. one-to-one

31. inverse $= \{(2,3), (1,2), (0,1)\}$. Since each x-coordinate is paired with only one y-coordinate, the inverse relation **is a function**.

33. inverse $= \{(2,1), (3,2), (3,1), (5,1)\}$. Since $x = 3$ is paired with more than one y-coordinate, the inverse relation **is not a function**.

35.
$$\begin{aligned} f(x) &= 3x + 1 \\ y &= 3x + 1 \\ x &= 3y + 1 \\ x - 1 &= 3y \\ \frac{x-1}{3} &= y \\ \frac{1}{3}x - \frac{1}{3} &= y \\ f^{-1}(x) &= \tfrac{1}{3}x - \tfrac{1}{3} \end{aligned}$$

$f \circ f^{-1}$
$$\begin{aligned} f\left[f^{-1}(x)\right] &= f\left(\tfrac{1}{3}x - \tfrac{1}{3}\right) \\ &= 3 \cdot \left(\tfrac{1}{3}x - \tfrac{1}{3}\right) + 1 \\ &= x - 1 + 1 \\ &= x \end{aligned}$$

$f^{-1} \circ f$
$$\begin{aligned} f^{-1}[f(x)] &= f^{-1}(3x + 1) \\ &= \tfrac{1}{3}(3x + 1) - \tfrac{1}{3} \\ &= x + \tfrac{1}{3} - \tfrac{1}{3} \\ &= x \end{aligned}$$

37.
$$\begin{aligned} x + 4 &= 5y \\ y = f(x) &= \frac{x+4}{5} \\ x &= \frac{y+4}{5} \\ 5x &= y + 4 \\ 5x - 4 &= y \\ f^{-1}(x) &= 5x - 4 \end{aligned}$$

$f \circ f^{-1}$
$$\begin{aligned} f\left[f^{-1}(x)\right] &= f(5x - 4) \\ &= \frac{(5x-4)+4}{5} \\ &= \frac{5x}{5} \\ &= x \end{aligned}$$

$f^{-1} \circ f$
$$\begin{aligned} f^{-1}[f(x)] &= f^{-1}\left(\frac{x+4}{5}\right) \\ &= 5 \cdot \left(\frac{x+4}{5}\right) - 4 \\ &= x + 4 - 4 \\ &= x \end{aligned}$$

SECTION 10.7

39.

$$\begin{aligned} f(x) &= \frac{x-4}{5} \\ y &= \frac{x-4}{5} \\ x &= \frac{y-4}{5} \\ 5x &= y-4 \\ 5x+4 &= y \\ f^{-1}(x) &= 5x+4 \end{aligned}$$

$f \circ f^{-1}$

$$\begin{aligned} f\left[f^{-1}(x)\right] &= f(5x+4) \\ &= \frac{(5x+4)-4}{5} \\ &= \frac{5x}{5} \\ &= x \end{aligned}$$

$f^{-1} \circ f$

$$\begin{aligned} f^{-1}[f(x)] &= f^{-1}\left(\frac{x-4}{5}\right) \\ &= 5 \cdot \left(\frac{x-4}{5}\right) + 4 \\ &= x - 4 + 4 \\ &= x \end{aligned}$$

41.

$$\begin{aligned} 4x - 5y &= 20 \\ 5y &= 4x - 20 \\ y = f(x) &= \tfrac{4}{5}x - 4 \\ x &= \tfrac{4}{5}y - 4 \\ x + 4 &= \tfrac{4}{5}y \\ \tfrac{5}{4}(x+4) &= y \\ f^{-1}(x) &= \tfrac{5}{4}x + 5 \end{aligned}$$

$f \circ f^{-1}$

$$\begin{aligned} f\left[f^{-1}(x)\right] &= f\left(\tfrac{5}{4}x + 5\right) \\ &= \tfrac{4}{5}\left(\tfrac{5}{4}x + 5\right) - 4 \\ &= x + 4 - 4 \\ &= x \end{aligned}$$

$f^{-1} \circ f$

$$\begin{aligned} f^{-1}[f(x)] &= f^{-1}\left(\tfrac{4}{5}x - 4\right) \\ &= \tfrac{5}{4}\left(\tfrac{4}{5}x - 4\right) + 5 \\ &= x - 5 + 5 \\ &= x \end{aligned}$$

43.

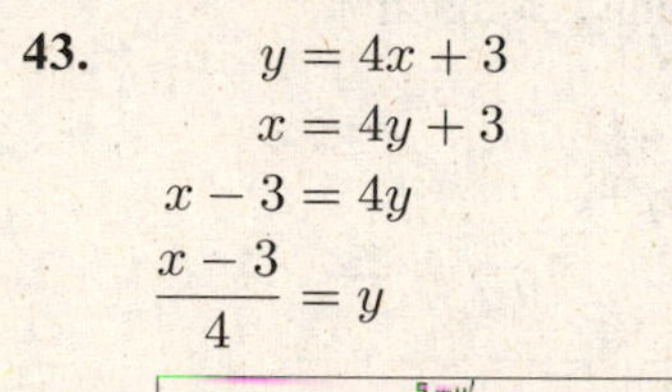

$$\begin{aligned} y &= 4x + 3 \\ x &= 4y + 3 \\ x - 3 &= 4y \\ \frac{x-3}{4} &= y \end{aligned}$$

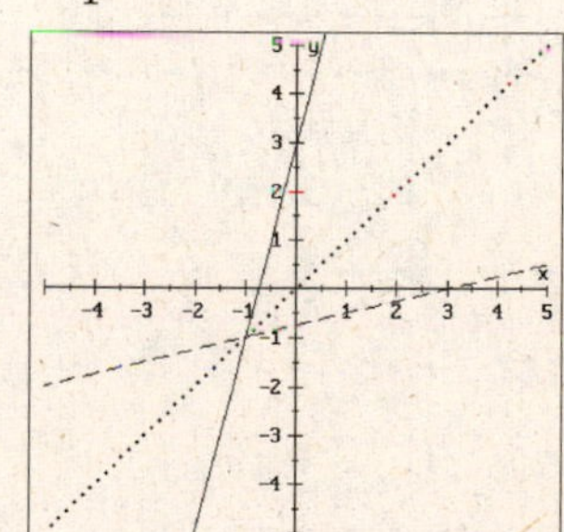

45.

$$\begin{aligned} x &= \frac{y-2}{3} \\ y &= \frac{x-2}{3} \end{aligned}$$

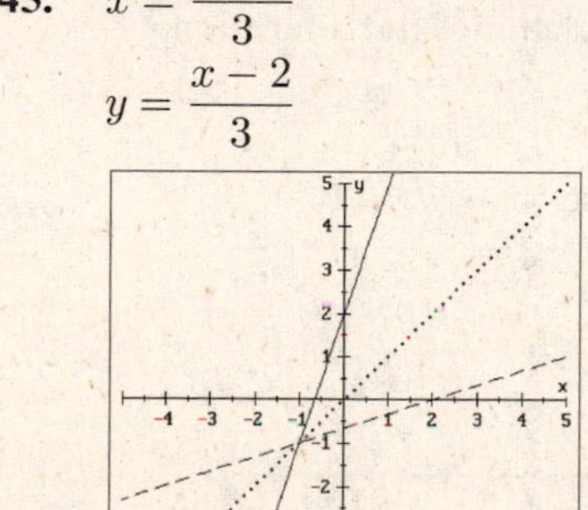

47. $3x - y = 5$

$3y - x = 5$

$3y = x + 5$

$y = \dfrac{x+5}{3}$

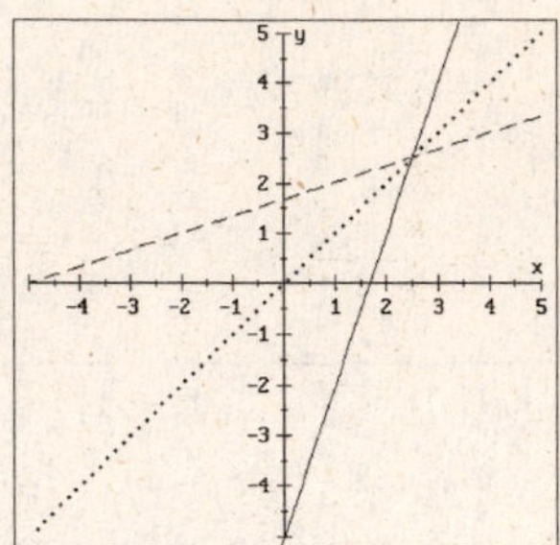

49. $3(x + y) = 2x + 4$

$3(y + x) = 2y + 4$

$3y + 3x = 2y + 4$

$y = 4 - 3x$

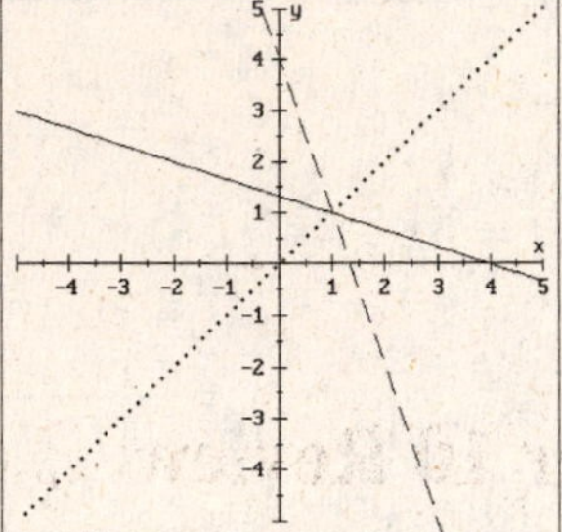

51. $y = x^2 + 4$

$x = y^2 + 4$

$x - 4 = y^2$

$\pm\sqrt{x - 4} = y$

The relation **is not** a function.

53. $y = x^3$

$x = y^3$

$\sqrt[3]{x} = y$

The relation **is** a function.

55. $y = x^2 + 1$

inverse: $x = y^2 + 1$

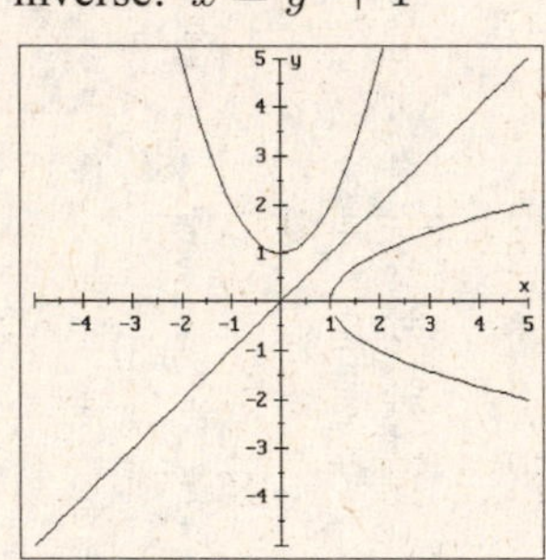

57. $y = \sqrt{x}$

inverse: $x = \sqrt{y}$

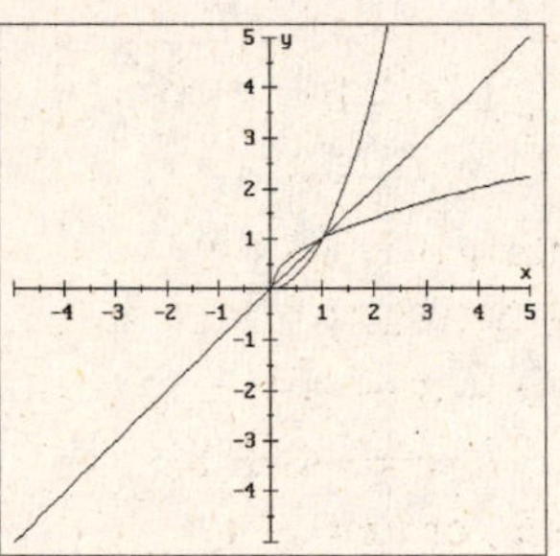

59. inverse $= \{(1,1), (4,2), (9,3), (16,4)\}$. Since each x-coordinate is paired with only one y-coordinate, the inverse relation **is a function**.

61. $y = |x|$

$x = |y|$

63. $y = 2x^3 - 3$

$x = 2y^3 - 3$

$x + 3 = 2y^3$

$\dfrac{x+3}{2} = y^3$

$\sqrt[3]{\dfrac{x+3}{2}} = \sqrt[3]{y^3}$, or $y = f^{-1}(x) = \sqrt[3]{\dfrac{x+3}{2}}$

65. **Answers may vary.**

67.
$$\begin{aligned} y &= \frac{x+1}{x-1} \\ x &= \frac{y+1}{y-1} \\ x(y-1) &= y+1 \\ xy - x &= y+1 \\ xy - y &= x+1 \\ y(x-1) &= x+1 \\ y &= \frac{x+1}{x-1}, \text{ or } f^{-1}(x) = \frac{x+1}{x-1} \end{aligned}$$

Chapter 10 Review (page 765)

1.
$$\begin{aligned} 12x^2 + x - 6 &= 0 \\ (4x+3)(3x-2) &= 0 \\ 4x+3 = 0 \quad &\textbf{or} \quad 3x-2 = 0 \\ x = -\tfrac{3}{4} \quad & \quad x = \tfrac{2}{3} \end{aligned}$$

2.
$$\begin{aligned} 6x^2 + 17x + 5 &= 0 \\ (2x+5)(3x+1) &= 0 \\ 2x+5 = 0 \quad &\textbf{or} \quad 3x+1 = 0 \\ x = -\tfrac{5}{2} \quad & \quad x = -\tfrac{1}{3} \end{aligned}$$

3.
$$\begin{aligned} 15x^2 + 2x - 8 &= 0 \\ (3x-2)(5x+4) &= 0 \\ 3x-2 = 0 \quad &\textbf{or} \quad 5x+4 = 0 \\ x = \tfrac{2}{3} \quad & \quad x = -\tfrac{4}{5} \end{aligned}$$

4.
$$\begin{aligned} (x+2)^2 &= 36 \\ x+2 &= \pm\sqrt{36} \\ x+2 &= \pm 6 \\ x &= -2 \pm 6 \\ x = 4 \quad &\text{or} \quad x = -8 \end{aligned}$$

5.
$$\begin{aligned} x^2 + 6x + 8 &= 0 \\ x^2 + 6x &= -8 \\ x^2 + 6x + 9 &= -8 + 9 \\ (x+3)^2 &= 1 \\ x+3 &= \pm 1 \\ x &= -3 \pm 1 \\ x = -2 \quad &\text{or} \quad x = -4 \end{aligned}$$

6.
$$\begin{aligned} 2x^2 - 9x + 7 &= 0 \\ x^2 - \frac{9}{2}x + \frac{7}{2} &= 0 \\ x^2 - \frac{9}{2}x &= -\frac{7}{2} \\ x^2 - \frac{9}{2}x + \frac{81}{16} &= -\frac{56}{16} + \frac{81}{16} \\ \left(x - \frac{9}{4}\right)^2 &= \frac{25}{16} \\ x - \frac{9}{4} &= \pm\frac{5}{4} \\ x &= \frac{9}{4} \pm \frac{5}{4} \\ x = \frac{7}{2} \quad &\text{or} \quad x = 1 \end{aligned}$$

7.
$$\begin{aligned} 2x^2 - x - 5 &= 0 \\ x^2 - \frac{1}{2}x - \frac{5}{2} &= 0 \\ x^2 - \frac{1}{2}x &= \frac{5}{2} \\ x^2 - \frac{1}{2}x + \frac{1}{16} &= \frac{5}{2} + \frac{1}{16} \\ \left(x - \frac{1}{4}\right)^2 &= \frac{41}{16} \\ x - \frac{1}{4} &= \pm\frac{\sqrt{41}}{4} \\ x &= \frac{1}{4} \pm \frac{\sqrt{41}}{4} \end{aligned}$$

8.
$$\begin{aligned} &x^2 - 8x - 9 = 0 \\ &a = 1, b = -8, c = -9 \\ x &= \frac{-b \pm \sqrt{b^2 - 4ac}}{2a} \\ &= \frac{-(-8) \pm \sqrt{(-8)^2 - 4(1)(-9)}}{2(1)} \\ &= \frac{8 \pm \sqrt{64 + 36}}{2} \\ &= \frac{8 \pm \sqrt{100}}{2} = \frac{8 \pm 10}{2} \\ x &= \frac{18}{2} = 9 \text{ or } x = \frac{-2}{2} = -1 \end{aligned}$$

9.
$$\begin{aligned} &x^2 - 10x = 0 \\ &a = 1, b = -10, c = 0 \\ x &= \frac{-b \pm \sqrt{b^2 - 4ac}}{2a} \\ &= \frac{-(-10) \pm \sqrt{(-10)^2 - 4(1)(0)}}{2(1)} \\ &= \frac{10 \pm \sqrt{100 + 0}}{2} \\ &= \frac{10 \pm \sqrt{100}}{2} = \frac{10 \pm 10}{2} \\ x &= \frac{20}{2} = 10 \text{ or } x = \frac{0}{2} = 0 \end{aligned}$$

10.
$$\begin{aligned} &2x^2 + 13x - 7 = 0 \\ &a = 2, b = 13, c = -7 \\ x &= \frac{-b \pm \sqrt{b^2 - 4ac}}{2a} \\ &= \frac{-(13) \pm \sqrt{13^2 - 4(2)(-7)}}{2(2)} \\ &= \frac{-13 \pm \sqrt{169 + 56}}{4} \\ &= \frac{-13 \pm \sqrt{225}}{4} = \frac{-13 \pm 15}{4} \\ x &= \frac{2}{4} = \frac{1}{2} \text{ or } x = \frac{-28}{4} = -7 \end{aligned}$$

11.
$$\begin{aligned} &3x^2 + 20x - 7 = 0 \\ &a = 3, b = 20, c = -7 \\ x &= \frac{-b \pm \sqrt{b^2 - 4ac}}{2a} \\ &= \frac{-20 \pm \sqrt{(20)^2 - 4(3)(-7)}}{2(3)} \\ &= \frac{-20 \pm \sqrt{400 + 84}}{6} \\ &= \frac{-20 \pm \sqrt{484}}{6} = \frac{-20 \pm 22}{6} \\ x &= \frac{2}{6} = \frac{1}{3} \text{ or } x = \frac{-42}{6} = -7 \end{aligned}$$

12.
$$\begin{aligned} &2x^2 - x - 2 = 0 \\ &a = 2, b = -1, c = -2 \\ x &= \frac{-b \pm \sqrt{b^2 - 4ac}}{2a} \\ &= \frac{-(-1) \pm \sqrt{(-1)^2 - 4(2)(-2)}}{2(2)} \\ &= \frac{1 \pm \sqrt{1 + 16}}{4} \\ &= \frac{1 \pm \sqrt{17}}{4} = \frac{1}{4} \pm \frac{\sqrt{17}}{4} \end{aligned}$$

13. $x^2 + x + 2 = 0$

$a = 1, b = 1, c = 2$

$$\begin{aligned} x &= \frac{-b \pm \sqrt{b^2 - 4ac}}{2a} \\ &= \frac{-1 \pm \sqrt{1^2 - 4(1)(2)}}{2(1)} \\ &= \frac{-1 \pm \sqrt{1-8}}{2} \\ &= \frac{-1 \pm \sqrt{-7}}{2} = -\frac{1}{2} \pm \frac{\sqrt{7}}{2}i \end{aligned}$$

14. Let w represent the original width.
Then $w + 2$ represents the original length.
The new dimensions are then $2w$ and $2(w + 2) = 2w + 4$.

$$\begin{aligned} \text{Old Area} + 72 &= \text{New Area} \\ w(w+2) + 72 &= 2w(2w+4) \\ w^2 + 2w + 72 &= 4w^2 + 8w \\ 0 &= 3w^2 + 6w - 72 \\ 0 &= 3(w+6)(w-4) \end{aligned}$$

$w = -6$ or $w = 4$

Since the width is positive, the dimensions are 4 cm by 6 cm.

15. Let w represent the original width.
Then $w + 1$ represents the original length.
The new dimensions are then $2w$ and $3(w + 1) = 3w + 3$.

$$\begin{aligned} \text{Old Area} + 30 &= \text{New Area} \\ w(w+1) + 30 &= 2w(3w+3) \\ w^2 + w + 30 &= 6w^2 + 6w \\ 0 &= 5w^2 + 5w - 30 \\ 0 &= 5(w+3)(w-2) \end{aligned}$$

$w = -3$ or $w = 2$

Since the width is positive, the dimensions are 2 ft by 3 ft.

16. When the rocket hits the ground, $h = 0$:

$h = 112t - 16t^2$

$0 = 112t - 16t^2$

$0 = 16t(7 - t)$

$t = 0$ or $t = 7$

It hits the ground after 7 seconds.

17. The maximum height occurs at the vertex:

$h = 112t - 16t^2$

$h = -16t^2 + 112t$

$h = -16(t^2 - 7t)$

$h = -16\left(t^2 - 7t + \frac{49}{4}\right) + 196$

$h = -16\left(t^2 - 7t + \frac{49}{4}\right) + 196$

$h = -16\left(t - \frac{7}{2}\right)^2 + 196$

Vertex: $\left(\frac{7}{2}, 196\right) \Rightarrow$ max. height = 196 ft

18. $3x^2 + 4x - 3 = 0$

$a = 3, b = 4, c = -3$

$$\begin{aligned} b^2 - 4ac &= 4^2 - 4(3)(-3) \\ &= 16 + 36 = 52 \end{aligned}$$

irrational unequal solutions

19. $4x^2 - 5x + 7 = 0$

$a = 4, b = -5, c = 7$

$$\begin{aligned} b^2 - 4ac &= (-5)^2 - 4(4)(7) \\ &= 25 - 112 = -87 \end{aligned}$$

complex conjugate solutions

20. $(k-8)x^2+(k+16)x=-49$

$(k-8)x^2+(k+16)x+49=0$

$a=k-8, b=k+16, c=49$

Set the discriminant equal to 0:

$$b^2-4ac=0$$
$$(k+16)^2-4(k-8)(49)=0$$
$$k^2+32k+256-196k+1568=0$$
$$k^2-164k+1824=0$$
$$(k-12)(k-152)=0$$
$$k-12=0 \quad \textbf{or} \quad k-152=0$$
$$k=12 \qquad k=152$$

21. $3x^2+4x=k+1$

$3x^2+4x-k-1=0$

$a=3, b=4, c=-k-1$

Set the discriminant ≥ 0:

$$b^2-4ac\geq 0$$
$$4^2-4(3)(-k-1)\geq 0$$
$$16+12k+12\geq 0$$
$$12k\geq -28$$
$$k\geq -\frac{28}{12}$$
$$k\geq -\frac{7}{3}$$

22.
$$x-13x^{1/2}+12=0$$
$$\left(x^{1/2}-12\right)\left(x^{1/2}-1\right)=0$$
$$x^{1/2}-12=0 \quad \textbf{or} \quad x^{1/2}-1=0$$
$$x^{1/2}=12 \qquad x^{1/2}=1$$
$$\left(x^{1/2}\right)^2=(12)^2 \qquad \left(x^{1/2}\right)^2=1^2$$
$$x=144 \qquad x=1$$

Solution. Solution.

23.
$$a^{2/3}+a^{1/3}-6=0$$
$$\left(a^{1/3}-2\right)\left(a^{1/3}+3\right)=0$$
$$a^{1/3}-2=0 \quad \textbf{or} \quad a^{1/3}+3=0$$
$$a^{1/3}=2 \qquad a^{1/3}=-3$$
$$\left(a^{1/3}\right)^3=(2)^3 \qquad \left(a^{1/3}\right)^3=(-3)^3$$
$$a=8 \qquad a=-27$$

Solution. Solution.

24.
$$\frac{1}{x+1}-\frac{1}{x}=-\frac{1}{x+1}$$
$$\left(\frac{1}{x+1}-\frac{1}{x}\right)(x)(x+1)=-\frac{1}{x+1}(x)(x+1)$$
$$1(x)-1(x+1)=-x$$
$$x-x-1=-x$$
$$-1=-x$$
$$1=x$$

25.
$$\frac{6}{x+2}+\frac{6}{x+1}=5$$
$$\left(\frac{6}{x+2}+\frac{6}{x+1}\right)(x+2)(x+1)=5(x+2)(x+1)$$
$$6(x+1)+6(x+2)=5(x^2+3x+2)$$
$$6x+6+6x+12=5x^2+15x+10$$
$$0=5x^2+3x-8$$
$$0=(5x+8)(x-1) \qquad 5x+8=0 \quad \textbf{or} \quad x-1=0$$
$$x=-\tfrac{8}{5} \qquad x=1$$

26. $3x^2-14x+3=0$

$$\text{sum}=-\frac{b}{a}=-\frac{-14}{3}=\frac{14}{3}$$

27. $3x^2-14x+3=0$

$$\text{product}=\frac{c}{a}=\frac{3}{3}=1$$

28. $y = 2x^2 - 3$
$y = 2(x - 0)^2 - 3$

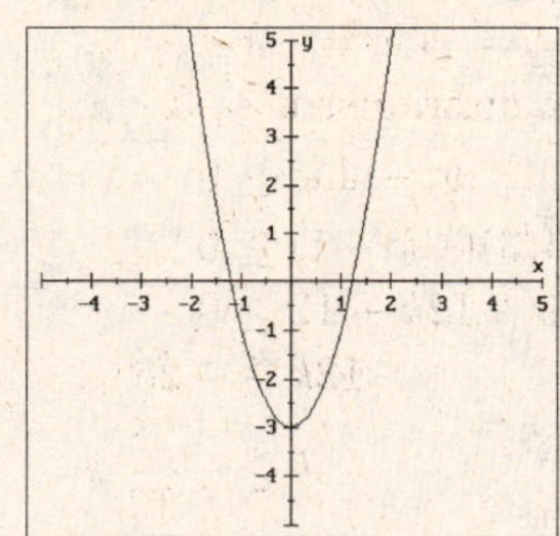

vertex: $(0, -3)$

29. $y = -2x^2 - 1$
$y = -2(x - 0)^2 - 1$

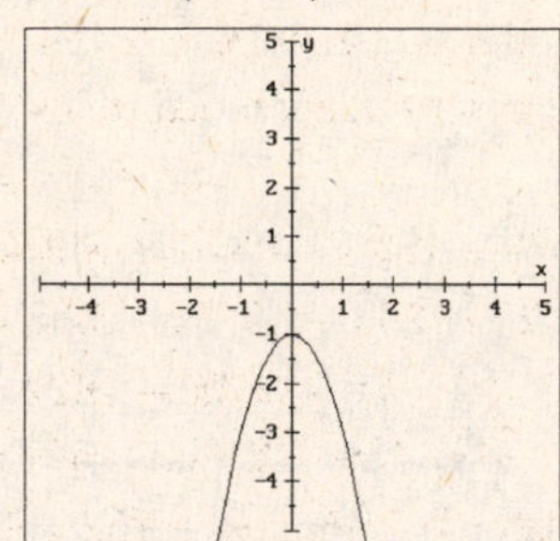

vertex: $(0, -1)$

30. $y = -4(x - 2)^2 + 1$

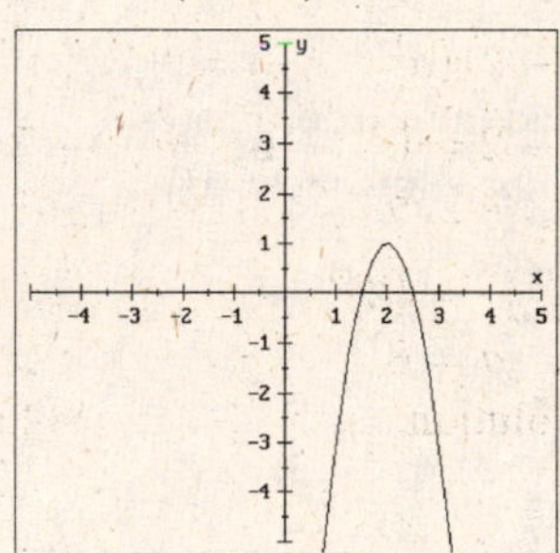

vertex: $(2, 1)$

31. $y = 5x^2 + 10x - 1$
$= 5(x^2 + 2x) - 1$
$= 5(x^2 + 2x + 1) - 1 - 5$
$= 5(x + 1)^2 - 6$

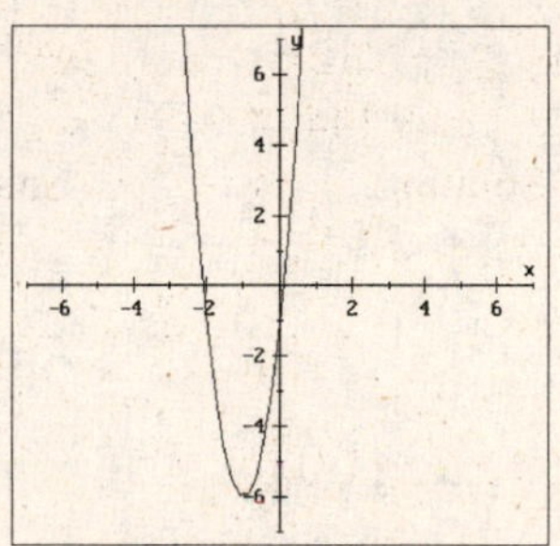

vertex: $(-1, -6)$

32. $y = 3x^2 - 12x - 5 = 3(x^2 - 4x) - 5 = 3(x^2 - 4x + 4) - 5 - 12 = 3(x - 2)^2 - 17$
vertex: $(2, -17)$

33. $x^2 + 2x - 35 > 0$
$(x + 7)(x - 5) > 0$

$x - 5$ ---------0 ++++
$x + 7$ --- 0++++++ ++++
⟵) —— (⟶
−7 5

solution set: $(-\infty, -7) \cup (5, \infty)$

34. $x^2 + 7x - 18 < 0$
$(x - 2)(x + 9) < 0$

$x - 2$ --------- 0++++
$x + 9$ --- 0++++++++++
⟵ (——) ⟶
−9 2

solution set: $(-9, 2)$

35.

$$\frac{3}{x} \le 5$$

$$\frac{3}{x} - 5 \le 0$$

$$\frac{3}{x} - \frac{5x}{x} \le 0$$

$$\frac{3-5x}{x} \le 0$$

$3-5x$ +++++++++ 0 – – –

x – – – 0+++++++++++

⟵) —— [⟶

0 $\frac{3}{5}$

solution set: $(-\infty, 0) \cup \left[\frac{3}{5}, \infty\right)$

36.

$$\frac{2x^2 - x - 28}{x - 1} > 0$$

$$\frac{(2x+7)(x-4)}{x-1} > 0$$

$x-4$ – – – – – – – – – – – – – 0++++

$x-1$ – – – – – – – – – 0+++++++++

$2x+7$ – – – 0+++++++++++++++++

⟵ (——) —— (⟶

$-\frac{7}{2}$ 1 4

solution set: $\left(-\frac{7}{2}, 1\right) \cup (4, \infty)$

37. $x^2 + 2x - 35 > 0$

Graph $y = x^2 + 2x - 35$
and find the x-coordinates
of points above the x-axis.

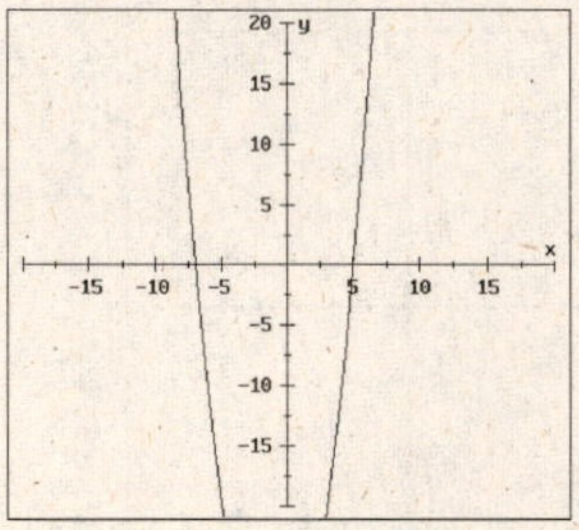

$(-\infty, -7) \cup (5, \infty)$

38. $x^2 + 7x - 18 < 0$

Graph $y = x^2 + 7x - 18$
and find the x-coordinates
of points below the x-axis.

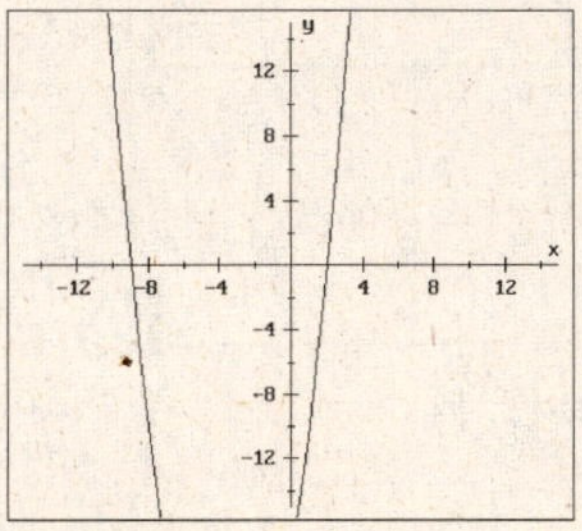

$(-9, 2)$

39. $\frac{3}{x} \le 5 \Rightarrow \frac{3}{x} - 5 \le 0$

Graph $y = (3/x) - 5$
and find the x-coordinates
of points below or on the x-axis.

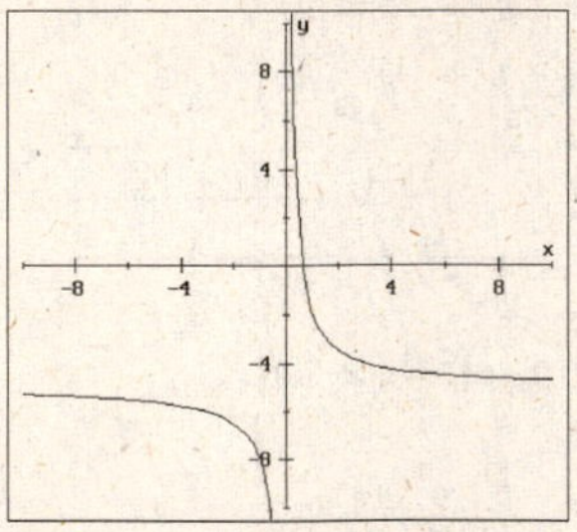

$(-\infty, 0) \cup \left[\frac{3}{5}, \infty\right)$

40. $\frac{2x^2-x-28}{x-1} > 0$

Graph $y = \left(2x^2 - x - 28\right)/(x - 1)$
and find the x-coordinates
of points above the x-axis.

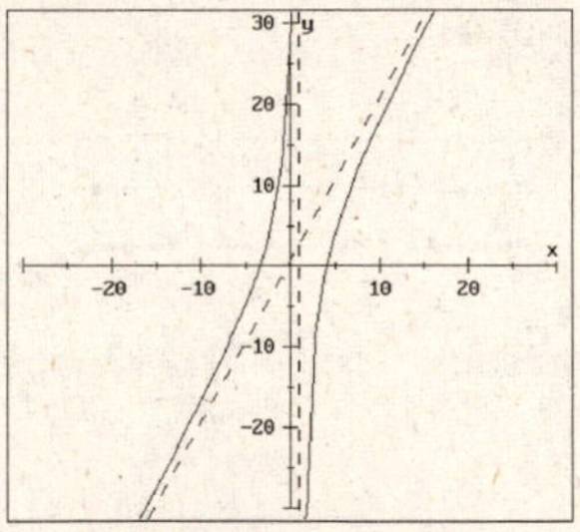

$\left(-\frac{7}{2}, 1\right) \cup (4, \infty)$

41. $y < \frac{1}{2}x^2 - 1$

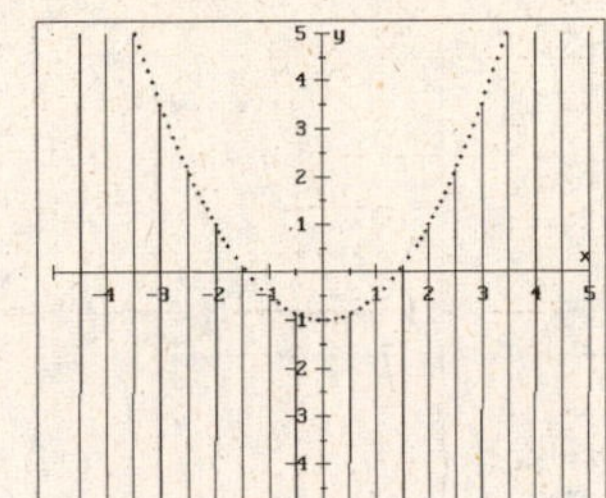

42. $y \geq -|x|$

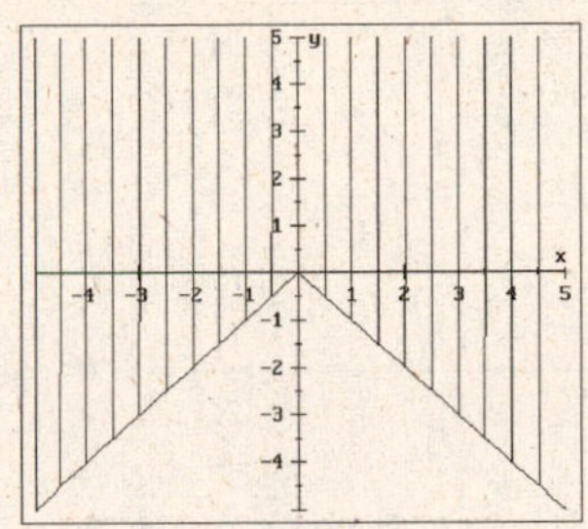

43. $f + g = f(x) + g(x) = 2x + x + 1$
$= 3x + 1$

44. $f - g = f(x) - g(x) = 2x - (x + 1)$
$= x - 1$

45. $f \cdot g = f(x)g(x) = 2x(x + 1) = 2x^2 + 2x$

46. $f/g = \frac{f(x)}{g(x)} = \frac{2x}{x+1}\ (x \neq -1)$

47. $(f \circ g)(2) = f(g(2)) = f(2 + 1)$
$= f(3) = 2(3) = 6$

48. $(g \circ f)(-1) = g(f(-1)) = g(2(-1))$
$= g(-2)$
$= -2 + 1 = -1$

49. $(f \circ g)(x) = f(g(x)) = f(x + 1) = 2(x + 1)$

50. $(g \circ f)(x) = g(f(x)) = g(2x) = 2x + 1$

51. $f(x) = 2(x - 3)$

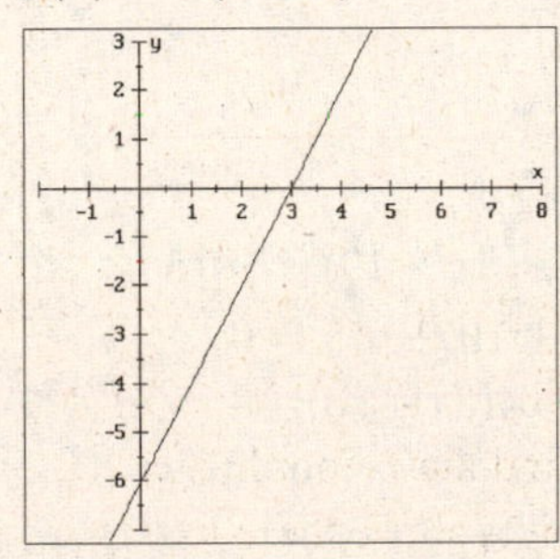

one-to-one

52. $f(x) = x(2x - 3)$

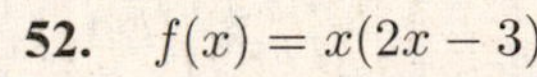

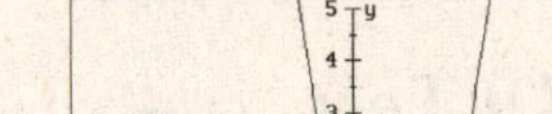

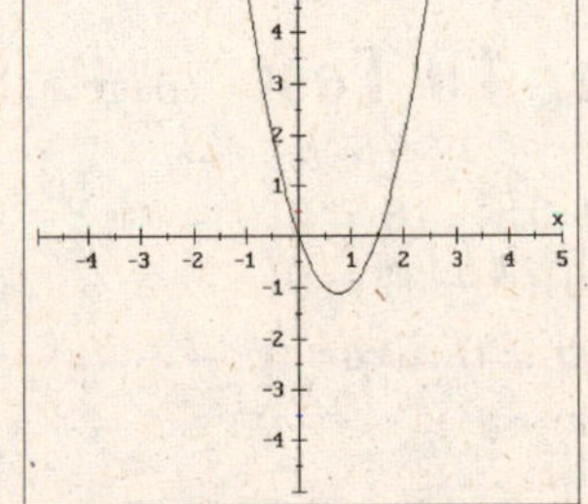

not one-to-one

53. $f(x) = -3(x-2)^2 + 5$

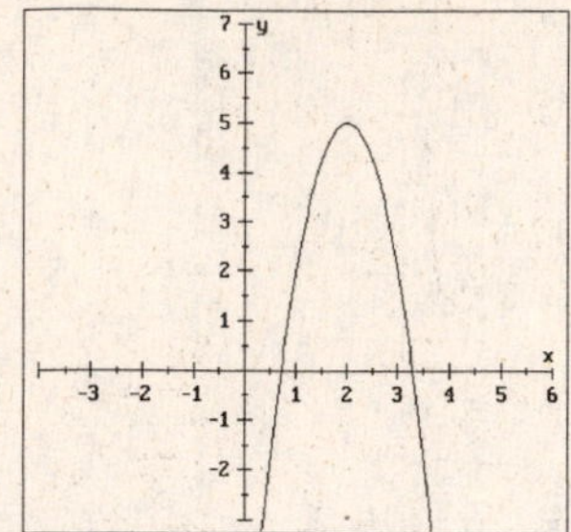

not one-to-one

54. $f(x) = |x|$

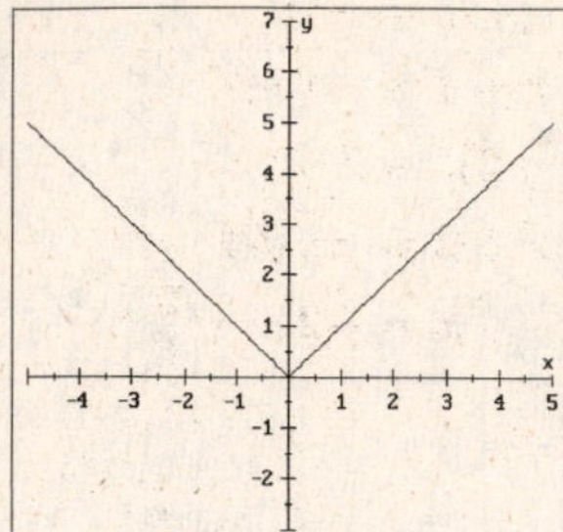

not one-to-one

55.
$$\begin{aligned} y &= 6x - 3 \\ x &= 6y - 3 \\ x + 3 &= 6y \\ \tfrac{x+3}{6} &= y, \text{ or } y = f^{-1}(x) = \tfrac{x+3}{6} \end{aligned}$$

56.
$$\begin{aligned} y &= 4x + 5 \\ x &= 4y + 5 \\ x - 5 &= 4y \\ \tfrac{x-5}{4} &= y, \text{ or } y = f^{-1}(x) = \tfrac{x-5}{4} \end{aligned}$$

57.
$$\begin{aligned} y &= 2x^2 - 1 \\ x &= 2y^2 - 1 \\ x + 1 &= 2y^2 \\ \tfrac{x+1}{2} &= y^2 \\ \sqrt{\tfrac{x+1}{2}} &= y, \text{ or } y = f^{-1}(x) = \sqrt{\tfrac{x+1}{2}} \end{aligned}$$

58.
$$\begin{aligned} y &= |x| \\ x &= |y| \end{aligned}$$

Chapter 10 Test (page 772)

1.
$$\begin{aligned} x^2 + 3x - 18 &= 0 \\ (x+6)(x-3) &= 0 \\ x + 6 = 0 \quad &\textbf{or} \quad x - 3 = 0 \\ x = -6 \quad &\quad\quad\; x = 3 \end{aligned}$$

2.
$$\begin{aligned} x(6x + 19) &= -15 \\ 6x^2 + 19x + 15 &= 0 \\ (2x+3)(3x+5) &= 0 \\ 2x + 3 = 0 \quad &\textbf{or} \quad 3x + 5 = 0 \\ x = -\tfrac{3}{2} \quad &\quad\quad\; x = -\tfrac{5}{3} \end{aligned}$$

3. $\left(\frac{1}{2} \cdot 24\right)^2 = 12^2 = 144$

4. $\left(\frac{1}{2} \cdot (-50)\right)^2 = (-25)^2 = 625$

CHAPTER 10 TEST

5.
$$\begin{aligned} x^2+4x+1&=0\\ x^2+4x&=-1\\ x^2+4x+4&=-1+4\\ (x+2)^2&=3\\ x+2&=\pm\sqrt{3}\\ x&=-2\pm\sqrt{3} \end{aligned}$$

6.
$$\begin{aligned} x^2-5x-3&=0\\ x^2-5x&=3\\ x^2-5x+\frac{25}{4}&=3+\frac{25}{4}\\ \left(x-\frac{5}{2}\right)^2&=\frac{37}{4}\\ x-\frac{5}{2}&=\pm\sqrt{\frac{37}{4}}\\ x&=\frac{5}{2}\pm\frac{\sqrt{37}}{2} \end{aligned}$$

7.
$$2x^2+5x+1=0$$
$$a=2, b=5, c=1$$
$$\begin{aligned} x&=\frac{-b\pm\sqrt{b^2-4ac}}{2a}\\ &=\frac{-5\pm\sqrt{5^2-4(2)(1)}}{2(2)}\\ &=\frac{-5\pm\sqrt{25-8}}{4}\\ &=\frac{-5\pm\sqrt{17}}{4}=-\frac{5}{4}\pm\frac{\sqrt{17}}{4} \end{aligned}$$

8.
$$x^2-x+3=0$$
$$a=1, b=-1, c=3$$
$$\begin{aligned} x&=\frac{-b\pm\sqrt{b^2-4ac}}{2a}\\ &=\frac{-(-1)\pm\sqrt{(-1)^2-4(1)(3)}}{2(1)}\\ &=\frac{1\pm\sqrt{1-12}}{2}\\ &=\frac{1\pm\sqrt{-11}}{2}=\frac{1}{2}\pm\frac{\sqrt{11}}{2}i \end{aligned}$$

9.
$$3x^2+5x+17=0$$
$$a=3, b=5, c=17$$
$$\begin{aligned} b^2-4ac&=5^2-4(3)(17)\\ &=25-208=-183 \end{aligned}$$
nonreal solutions

10.
$$4x^2-2kx+k-1=0$$
$$a=4, b=-2k, c=k-1$$
Set the discriminant equal to 0:
$$\begin{aligned} b^2-4ac&=0\\ (-2k)^2-4(4)(k-1)&=0\\ 4k^2-16k+16&=0\\ 4(k-2)(k-2)&=0 \end{aligned}$$
$$k-2=0 \quad \textbf{or} \quad k-2=0$$
$$k=2 \qquad\qquad k=2$$

11. Let $x=$ the length of the shorter leg.
Then $x+14=$ the other length.
$$\begin{aligned} x^2+(x+14)^2&=26^2\\ x^2+x^2+28x+196&=676\\ 2x^2+28x-480&=0\\ 2(x+24)(x-10)&=0 \end{aligned}$$
$$x=-24 \quad \text{or} \quad x=10$$
The shorter leg is 10 inches long.

12.
$$\begin{aligned} 2y-3y^{1/2}+1&=0\\ \left(2y^{1/2}-1\right)\left(y^{1/2}-1\right)&=0 \end{aligned}$$
$$2y^{1/2}-1=0 \quad \textbf{or} \quad y^{1/2}-1=0$$
$$2y^{1/2}=1 \qquad\qquad y^{1/2}=1$$
$$\left(y^{1/2}\right)^2=\left(\tfrac{1}{2}\right)^2 \qquad\qquad \left(y^{1/2}\right)^2=1^2$$
$$y=\tfrac{1}{4} \qquad\qquad y=1$$
Solution Solution

13. $y = \frac{1}{2}x^2 - 4 = \frac{1}{2}(x-0)^2 - 4$

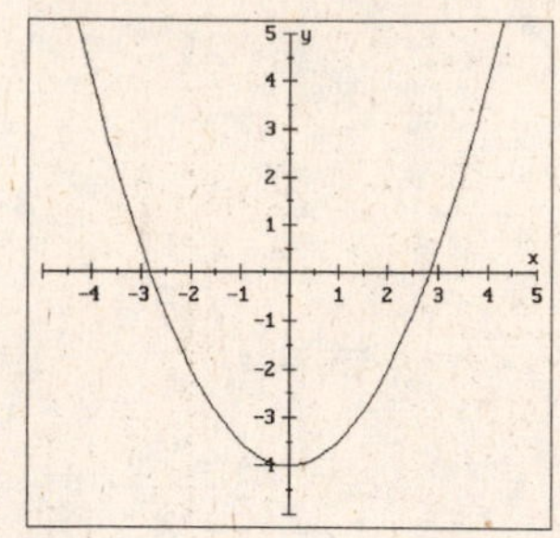

vertex: $(0, -4)$

14. $y = -2x^2 + 8x - 7$
$= -2(x^2 - 4x) - 7$
$= -2(x^2 - 4x + 4) - 7 + 8$
$= -2(x-2)^2 + 1$
Vertex: $(2, 1)$

15. $y \le -x^2 + 3$

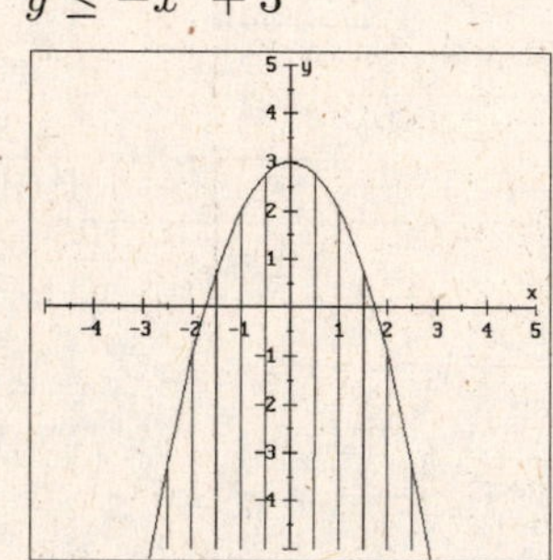

16. $x^2 - 2x - 8 > 0$
$(x+2)(x-4) > 0$
$x - 4$ – – – – – – – – – 0 ++++
$x + 2$ – – – 0+++++++ ++++
⟵) —— (⟶
−2 4
solution set: $(-\infty, -2) \cup (4, \infty)$

17. $\frac{x-2}{x+3} \le 0$
$x - 2$ – – – – – – – – 0 ++++
$x + 3$ – – – 0+++++++ ++++
⟵ (——] ⟶
−3 2
solution set: $(-3, 2]$

18. $g + f = g(x) + f(x) = x - 1 + 4x$
$= 5x - 1$

19. $f - g = f(x) - g(x) = 4x - (x - 1)$
$= 3x + 1$

20. $g \cdot f = g(x)f(x) = (x-1)4x = 4x^2 - 4x$

21. $g/f = \frac{g(x)}{f(x)} = \frac{x-1}{4x}$

22. $(g \circ f)(1) = g(f(1)) = g(4(1))$
$= g(4) = 4 - 1 = 3$

23. $(f \circ g)(0) = f(g(0)) = f(0-1)$
$= f(-1) = 4(-1) = -4$

24. $(f \circ g)(-1) = f(g(-1)) = f(-1-1) = f(-2) = 4(-2) = -8$

25. $(g \circ f)(-2) = g(f(-2)) = g(4(-2)) = g(-8) = -8 - 1 = -9$

26. $(f \circ g)(x) = f(g(x)) = f(x-1) = 4(x-1)$

27. $(g \circ f)(x) = g(f(x)) = g(4x) = 4x - 1$

28.
$$\begin{aligned} 3x + 2y &= 12 \\ 3y + 2x &= 12 \\ 3y &= -2x + 12 \\ y &= \frac{-2x + 12}{3} \end{aligned}$$

29.
$$\begin{aligned} y &= 3x^2 + 4 \\ x &= 3y^2 + 4 \\ x - 4 &= 3y^2 \\ \frac{x-4}{3} &= y^2 \\ -\sqrt{\frac{x-4}{3}} &= y \end{aligned}$$

Cumulative Review Exercises (page 773)

1.
$y = f(x) = 2x^2 - 3$
domain $= (-\infty, \infty)$
range $= [-3, \infty)$

2.
$y = f(x) = -|x - 4|$
domain $= (-\infty, \infty)$
range $= (-\infty, 0]$

3.
$$\begin{aligned} y - y_1 &= m(x - x_1) \\ y + 4 &= 3(x + 2) \\ y &= 3x + 2 \end{aligned}$$

4.
$$\begin{aligned} 2x + 3y &= 6 \\ 3y &= -2x + 6 \\ y &= -\frac{2}{3}x + 2 \\ y - y_1 &= m(x - x_1) \\ y + 2 &= -\frac{2}{3}(x - 0) \\ y &= -\frac{2}{3}x - 2 \end{aligned}$$

5. $(2a^2 + 4a - 7) - 2(3a^2 - 4a) = 2a^2 + 4a - 7 - 6a^2 + 8a = -4a^2 + 12a - 7$

6. $(3x + 2)(2x - 3) = 6x^2 - 9x + 4x - 6 = 6x^2 - 5x - 6$

7. $x^4 - 16y^4 = (x^2 + 4y^2)(x^2 - 4y^2) = (x^2 + 4y^2)(x + 2y)(x - 2y)$

8. $15x^2 - 2x - 8 = (5x - 4)(3x + 2)$

9.
$$\begin{aligned} x^2 - 5x - 6 &= 0 \\ (x - 6)(x + 1) &= 0 \\ x - 6 = 0 \quad &\textbf{or} \quad x + 1 = 0 \\ x = 6 \quad & \quad\;\; x = -1 \end{aligned}$$

10.
$$\begin{aligned} 6a^3 - 2a &= a^2 \\ 6a^3 - a^2 - 2a &= 0 \\ a(6a^2 - a - 2) &= 0 \\ a(3a - 2)(2a + 1) &= 0 \end{aligned}$$
$a = 0$ **or** $3a - 2 = 0$ **or** $2a + 1 = 0$
$a = \frac{2}{3}$ $\quad$ $a = -\frac{1}{2}$

11. $\sqrt{25x^4} = 5x^2$

12. $\sqrt{48t^3} = \sqrt{16t^2}\sqrt{3t} = 4t\sqrt{3t}$

13. $\sqrt[3]{-27x^3} = -3x$

14. $\sqrt[3]{\dfrac{128x^4}{2x}} = \sqrt[3]{64x^3} = 4x$

15. $8^{-1/3} = \dfrac{1}{8^{1/3}} = \dfrac{1}{2}$

16. $64^{2/3} = \left(64^{1/3}\right)^2 = 4^2 = 16$

17. $\dfrac{y^{2/3}y^{5/3}}{y^{1/3}} = \dfrac{y^{7/3}}{y^{1/3}} = y^{6/3} = y^2$

18. $\dfrac{x^{5/3}x^{1/2}}{x^{3/4}} = \dfrac{x^{13/6}}{x^{3/4}} = x^{17/12}$

19. $f(x) = \sqrt{x-2}$; Shift $y = \sqrt{x}$ right 2.

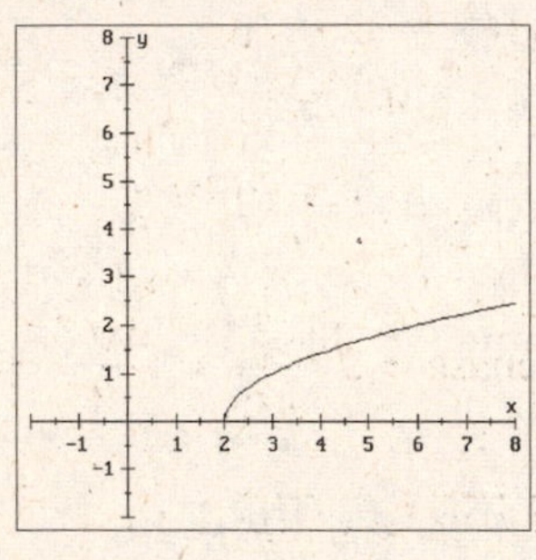

$D = [2, \infty)$, $R = [0, \infty)$

20. $f(x) = -\sqrt{x+2}$; Reflect $y = \sqrt{x}$ about the x-axis and shift left 2.

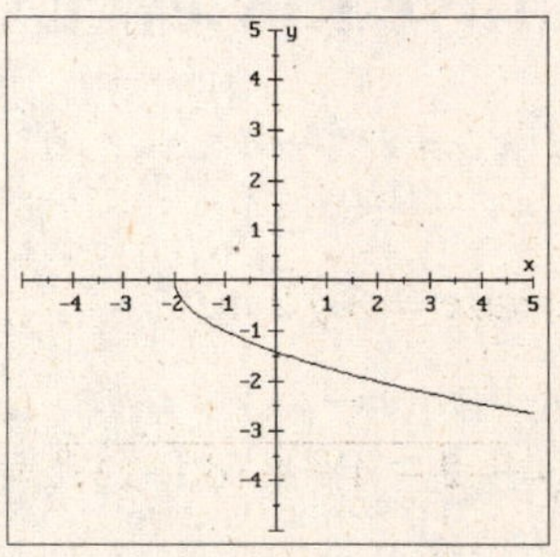

$D = [-2, \infty)$, $R = (-\infty, 0]$

21. $\left(x^{2/3} - x^{1/3}\right)\left(x^{2/3} + x^{1/3}\right) = x^{4/3} + x^{3/3} - x^{3/3} - x^{2/3} = x^{4/3} - x^{2/3}$

22. $\left(x^{-1/2} + x^{1/2}\right)^2 = \left(x^{-1/2} + x^{1/2}\right)\left(x^{-1/2} + x^{1/2}\right) = x^{-2/2} + x^0 + x^0 + x^{2/2} = x + 2 + \frac{1}{x}$

23. $\sqrt{50} - \sqrt{8} + \sqrt{32} = \sqrt{25}\sqrt{2} - \sqrt{4}\sqrt{2} + \sqrt{16}\sqrt{2} = 5\sqrt{2} - 2\sqrt{2} + 4\sqrt{2} = 7\sqrt{2}$

24.
$$\begin{aligned} -3\sqrt[4]{32} - 2\sqrt[4]{162} + 5\sqrt[4]{48} &= -3\sqrt[4]{16}\sqrt[4]{2} - 2\sqrt[4]{81}\sqrt[4]{2} + 5\sqrt[4]{16}\sqrt[4]{3} \\ &= -3(2)\sqrt[4]{2} - 2(3)\sqrt[4]{2} + 5(2)\sqrt[4]{3} \\ &= -6\sqrt[4]{2} - 6\sqrt[4]{2} + 10\sqrt[4]{3} = -12\sqrt[4]{2} + 10\sqrt[4]{3} \end{aligned}$$

25. $3\sqrt{2}(2\sqrt{3} - 4\sqrt{12}) = 6\sqrt{6} - 12\sqrt{24} = 6\sqrt{6} - 12\sqrt{4}\sqrt{6} = 6\sqrt{6} - 24\sqrt{6} = -18\sqrt{6}$

26. $\dfrac{5}{\sqrt[3]{x}} = \dfrac{5}{\sqrt[3]{x}} \cdot \dfrac{\sqrt[3]{x^2}}{\sqrt[3]{x^2}} = \dfrac{5\sqrt[3]{x^2}}{\sqrt[3]{x^3}} = \dfrac{5\sqrt[3]{x^2}}{x}$

27. $\dfrac{\sqrt{x}+2}{\sqrt{x}-1} = \dfrac{\sqrt{x}+2}{\sqrt{x}-1} \cdot \dfrac{\sqrt{x}+1}{\sqrt{x}+1} = \dfrac{x + 3\sqrt{x} + 2}{x - 1}$

28. $\sqrt[6]{x^3y^3} = (x^3y^3)^{1/6} = x^{3/6}y^{3/6} = x^{1/2}y^{1/2} = \sqrt{xy}$

CUMULATIVE REVIEW EXERCISES

29.
$$\begin{aligned}5\sqrt{x+2} &= x+8\\ \left(5\sqrt{x+2}\right)^2 &= (x+8)^2\\ 25(x+2) &= x^2+16x+64\\ 25x+50 &= x^2+16x+64\\ 0 &= x^2-9x+14\\ 0 &= (x-7)(x-2)\end{aligned}$$
$x=7$ or $x=2$ (Both check.)

30.
$$\begin{aligned}\sqrt{x}+\sqrt{x+2} &= 2\\ \sqrt{x} &= 2-\sqrt{x+2}\\ \left(\sqrt{x}\right)^2 &= \left(2-\sqrt{x+2}\right)^2\\ x &= 4-4\sqrt{x+2}+x+2\\ 4\sqrt{x+2} &= 6\\ \left(4\sqrt{x+2}\right)^2 &= 6^2\\ 16(x+2) &= 36\\ 16x+32 &= 36\\ 16x &= 4\\ x &= \frac{4}{16}=\frac{1}{4}\end{aligned}$$

31. hypotenuse $= 3\sqrt{2}$ in.

32. hypotenuse $= 2\cdot\frac{3}{\sqrt{3}}=\frac{6\sqrt{3}}{3}=2\sqrt{3}$ in.

33. $d=\sqrt{(-2-4)^2+(6-14)^2}=\sqrt{(-6)^2+(-8)^2}=\sqrt{36+64}=\sqrt{100}=10$

34. $\left(\frac{1}{2}\cdot 6\right)^2=3^2=9$

35.
$$\begin{aligned}2x^2+x-3 &= 0\\ x^2+\frac{1}{2}x-\frac{3}{2} &= 0\\ x^2+\frac{1}{2}x &= \frac{3}{2}\\ x^2+\frac{1}{2}x+\frac{1}{16} &= \frac{3}{2}+\frac{1}{16}\\ \left(x+\frac{1}{4}\right)^2 &= \frac{25}{16}\\ x+\frac{1}{4} &= \pm\frac{5}{4}\\ x &= -\frac{1}{4}\pm\frac{5}{4}\end{aligned}$$
$x=\frac{4}{4}=1$ or $x=-\frac{6}{4}=-\frac{3}{2}$

36.
$$3x^2+4x-1=0$$
$$a=3, b=4, c=-1$$
$$\begin{aligned}x &= \frac{-b\pm\sqrt{b^2-4ac}}{2a}\\ &= \frac{-4\pm\sqrt{4^2-4(3)(-1)}}{2(3)}\\ &= \frac{-4\pm\sqrt{16+12}}{6}\\ &= \frac{-4\pm\sqrt{28}}{6}\\ &= \frac{-4\pm 2\sqrt{7}}{6}=-\frac{2}{3}\pm\frac{\sqrt{7}}{3}\end{aligned}$$

CUMULATIVE REVIEW EXERCISES

37. $y = \frac{1}{2}x^2 + 5 = \frac{1}{2}(x-0)^2 + 5$

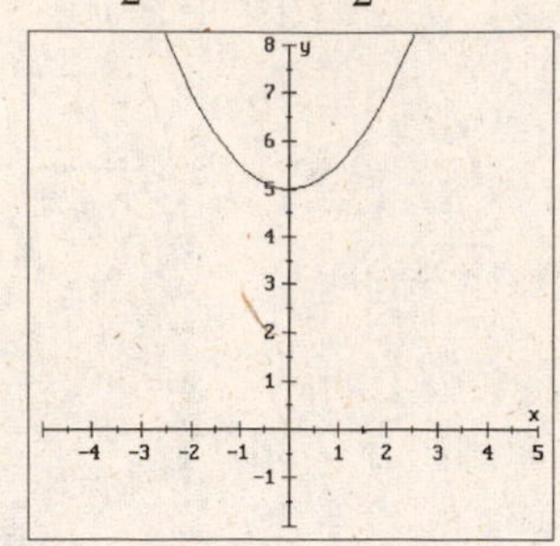

vertex: $(0, 5)$

38. $y \le -x^2 + 3$

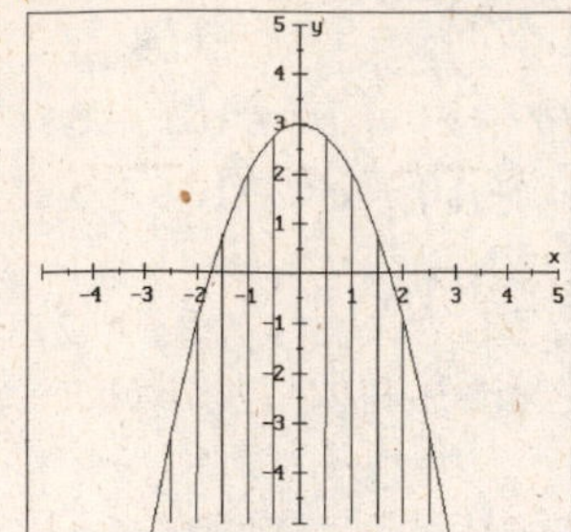

vertex: $(0, 3)$

39. $(3+5i) + (4-3i) = 3 + 5i + 4 - 3i = 7 + 2i$

40. $(7-4i) - (12+3i) = 7 - 4i - 12 - 3i = -5 - 7i$

41. $(2-3i)(2+3i) = 4 + 6i - 6i - 9i^2 = 4 + 9 = 13$

42. $(3+i)(3-3i) = 9 - 9i + 3i - 3i^2 = 9 - 6i + 3 = 12 - 6i$

43. $(3-2i) - (4+i)^2 = 3 - 2i - (16 + 8i + i^2) = 3 - 2i - (15 + 8i) = 3 - 2i - 15 - 8i$
$= -12 - 10i$

44. $\frac{5}{3-i} = \frac{5}{3-i} \cdot \frac{3+i}{3+i} = \frac{5(3+i)}{9-i^2} = \frac{5(3+i)}{10} = \frac{3+i}{2} = \frac{3}{2} + \frac{1}{2}i$

45. $|3+2i| = \sqrt{3^2 + 2^2} = \sqrt{9+4} = \sqrt{13}$

46. $|5-6i| = \sqrt{5^2 + (-6)^2} = \sqrt{25+36} = \sqrt{61}$

47.
$2x^2 + 4x = k$
$2x^2 + 4x - k = 0$
$a = 2, b = 4, c = -k$
Set the discriminant equal to 0:
$b^2 - 4ac = 0$
$4^2 - 4(2)(-k) = 0$
$16 + 8k = 0$
$8k = -16$
$k = -2$

48.
$a - 7a^{1/2} + 12 = 0$
$\left(a^{1/2} - 3\right)\left(a^{1/2} - 4\right) = 0$
$a^{1/2} - 3 = 0$ **or** $a^{1/2} - 4 = 0$
$a^{1/2} = 3$ | $a^{1/2} = 4$
$\left(a^{1/2}\right)^2 = 3^2$ | $\left(a^{1/2}\right)^2 = 4^2$
$a = 9$ | $a = 16$
Solution | Solution

49. $x^2 - x - 6 > 0$
$(x+2)(x-3) > 0$

$x-3$ ---------0 ++++
$x+2$ --- 0++++++ ++++

←——) ——— (——→
−2 3

solution set: $(-\infty, -2) \cup (3, \infty)$

50. $x^2 - x - 6 \le 0$
$(x+2)(x-3) \le 0$

$x-3$ --------0 ++++
$x+2$ ---0++++++ ++++

←—— [———] ——→
−2 3

solution set: $[-2, 3]$

51. $f(-1) = 3(-1)^2 + 2 = 3(1) + 2 = 3 + 2 = 5$

52. $(g \circ f)(2) = g(f(2)) = g(3(2)^2 + 2) = g(14) = 2(14) - 1 = 27$

53. $(f \circ g)(x) = f(g(x)) = f(2x-1) = 3(2x-1)^2 + 2 = 3(4x^2 - 4x + 1) + 2 = 12x^2 - 12x + 5$

54. $(g \circ f)(x) = g(f(x)) = g(3x^2 + 2) = 2(3x^2 + 2) - 1 = 6x^2 + 3$

55.
$$y = 3x + 2$$
$$x = 3y + 2$$
$$x - 2 = 3y$$
$$\frac{x-2}{3} = y, \text{ or } y = f^{-1}(x) = \frac{x-2}{3}$$

56.
$$y = x^3 + 4$$
$$x = y^3 + 4$$
$$x - 4 = y^3$$
$$\sqrt[3]{x-4} = y, \text{ or } y = f^{-1}(x) = \sqrt[3]{x-4}$$

Exercise 11.1 (page 786)

1. $2^x = 2^2 = 4$

3. $2(3^x) = 2(3^2) = 2(9) = 18$

5. $2^x = 2^{-2} = \frac{1}{2^2} = \frac{1}{4}$

7. $2(3^x) = 2(3^{-2}) = 2\left(\frac{1}{3^2}\right) = 2\left(\frac{1}{9}\right) = \frac{2}{9}$

9.
$$3x + 2x - 20 = 180$$
$$5x = 200$$
$$x = 40$$

11. $m(\angle 2) = 3x = 3(40) = 120°$

13. exponential

15. $(0, \infty)$

17. increasing

19. $P\left(1 + \frac{r}{k}\right)^{kt}$

Problems 21-23 are to be solved using a calculator. The keystrokes needed to solve each problem using a TI-84 graphing calculator appear in each solution. There may be other solutions. Keystrokes for other calculators may be slightly different.

21. $2^{\sqrt{2}} \Rightarrow$ [2] [^] [√] [2] [ENTER]
{2.6651}

23. $5^{\sqrt{5}} \Rightarrow$ [5] [^] [√] [5] [ENTER]
{36.5548}

25. $\left(2^{\sqrt{3}}\right)^{\sqrt{3}} = 2^{(\sqrt{3})(\sqrt{3})} = 2^3 = 8$

27. $7^{\sqrt{3}} 7^{\sqrt{12}} = 7^{\sqrt{3}+\sqrt{12}} = 7^{\sqrt{3}+2\sqrt{3}} = 7^{3\sqrt{3}}$

SECTION 11.1

29. $y = f(x) = 3^x$
through $(0, 1)$ and $(1, 3)$

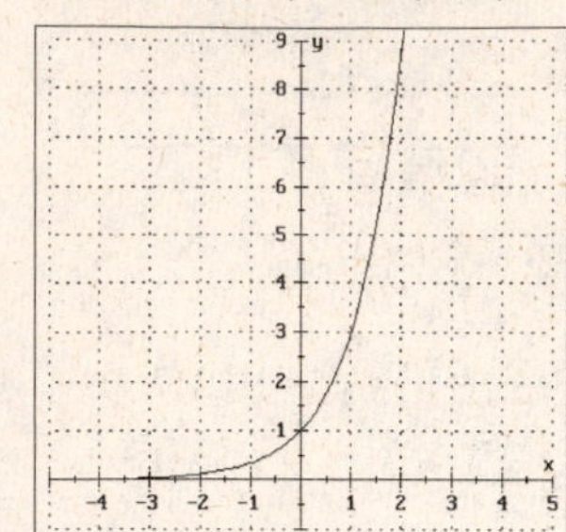

31. $y = f(x) = \left(\frac{1}{3}\right)^x$
through $(0, 1)$ and $\left(1, \frac{1}{3}\right)$

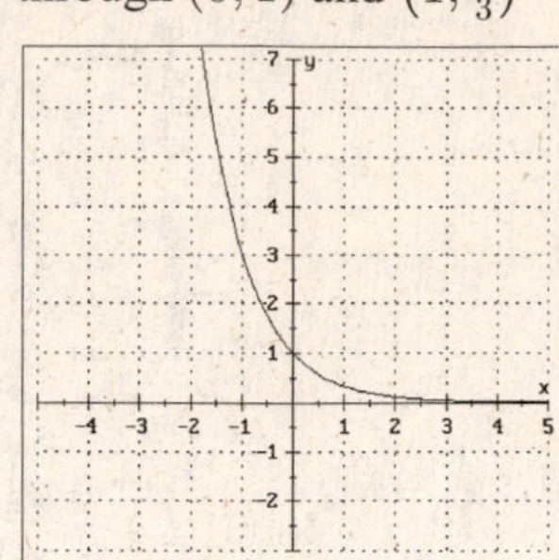

33. $y = b^x$

$\frac{1}{2} = b^1$

$\frac{1}{2} = b$

35. $y = b^x$

$3 = b^1$

$3 = b$

37. $f(x) = 3^x - 2$
Shift $y = 3^x$ down 2.

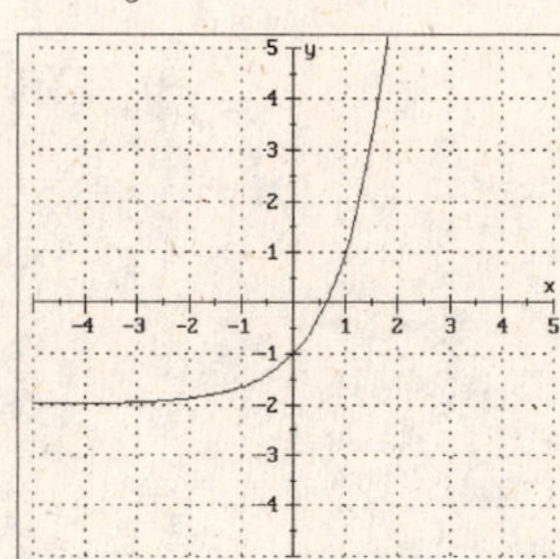

39. $f(x) = 3^{x-1}$
Shift $y = 3^x$ right 1.

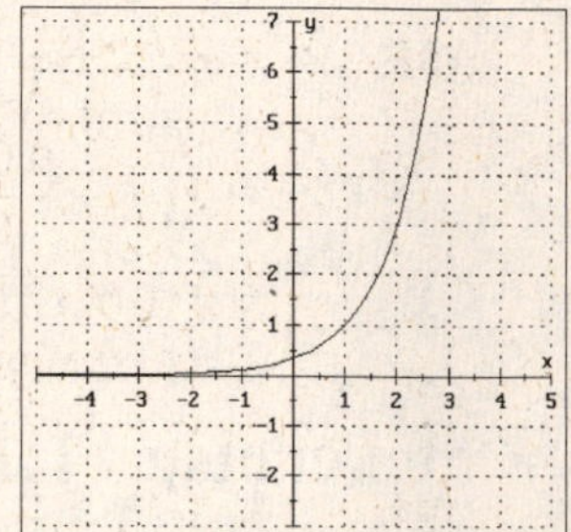

41. $y = b^x$

$2 = b^1$

$2 = b$

43. $y = f(x) = \frac{1}{2}\left(3^{x/2}\right)$

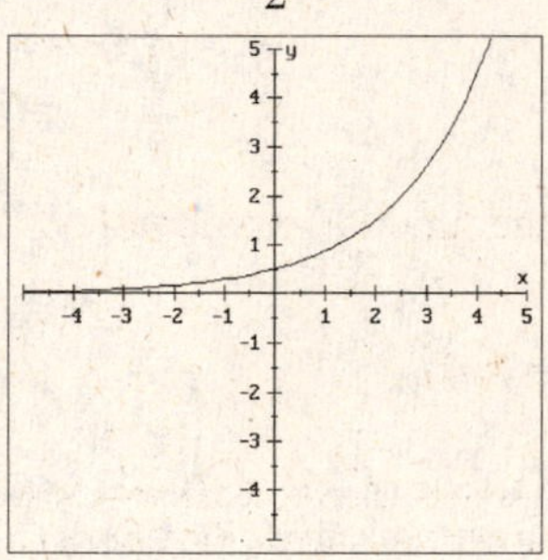

increasing

45. $y = f(x) = 2\left(3^{-x/2}\right)$

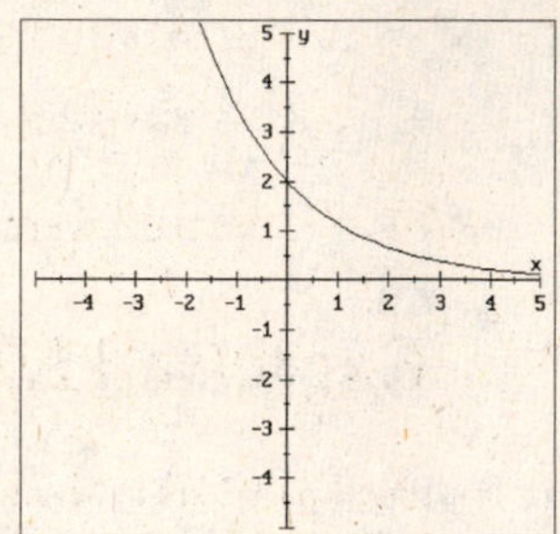

decreasing

47. $S(n) = 5.74(1.39)^n = 5.74(1.39)^5$

≈ 29.8

There were about 29.8 million users in 1995.

49. $A = A_0\left(\frac{2}{3}\right)^t$
$A = A_0\left(\frac{2}{3}\right)^5$
$A = \frac{32}{243}A_0$

51. $A = P\left(1+\frac{r}{k}\right)^{kt}$
$= 10{,}000\left(1+\frac{0.08}{4}\right)^{4(10)}$
$= 10{,}000(1.02)^{40}$
$\approx \$22{,}080.40$

53. $A = P\left(1+\frac{r}{k}\right)^{kt}$
$= 1000\left(1+\frac{0.05}{4}\right)^{4(5)}$
$= 1000(1.0125)^{20}$
$\approx \$1282.040$

$A = P\left(1+\frac{r}{k}\right)^{kt}$
$= 1000\left(1+\frac{0.055}{4}\right)^{4(5)}$
$= 1000(1.01375)^{20}$
$\approx \$1314.07$

difference $= \$1314.07 - \$1282.04$
$= \$32.03$

55. $A = P\left(1+\frac{r}{k}\right)^{kt}$
$= 1\left(1+\frac{0.05}{1}\right)^{1(300)}$
$= 1(1.05)^{300}$
$\approx \$2{,}273{,}996.13$

57. $C = (3\times 10^{-4})(0.7)^t$
$= (3\times 10^{-4})(0.7)^5$
$\approx 5.0421\times 10^{-5}$ coulombs

59. $A = P\left(1+\frac{r}{k}\right)^{kt}$
$= 4700\left(1+\frac{-0.25}{1}\right)^{1(5)}$
$= 4700(0.75)^5$
$\approx \$1115.33$

61. **Answers may vary.**

63. If the base were 0, then the function would not be defined for $x = 0 \Rightarrow y = 0^0$.

Exercise 11.2 (page 794)

Problems 1-3 are to be solved using a calculator. The keystrokes needed to solve each problem using a TI-84 graphing calculator appear in each solution. There may be other solutions. Keystrokes for other calculators may be slightly different.

1. $e^0 \Rightarrow$ [2nd] [LN] [0] [ENTER] {1}

3. $e^2 \Rightarrow$ [2nd] [LN] [2] [ENTER] {7.39}

5. 2

7. $\sqrt{240x^5} = \sqrt{16x^4}\sqrt{15x} = 4x^2\sqrt{15x}$

9. $4\sqrt{48y^3} - 3y\sqrt{12y} = 4\sqrt{16y^2}\sqrt{3y} - 3y\sqrt{4}\sqrt{3y} = 4(4y)\sqrt{3y} - 3y(2)\sqrt{3y}$
$= 16y\sqrt{3y} - 6y\sqrt{3y} = 10y\sqrt{3y}$

11. 2.72

13. increasing

15. $A = Pe^{rt}$

17. $y = f(x) = e^x + 1$
Shift $y = e^x$ up 1.

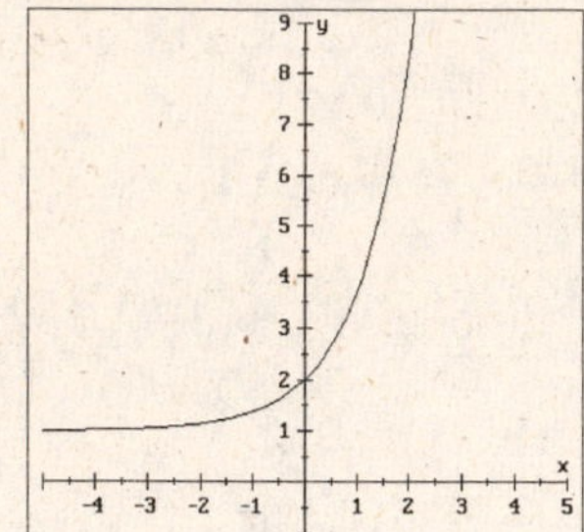

19. $y = f(x) = e^{(x+3)}$
Shift $y = e^x$ left 3.

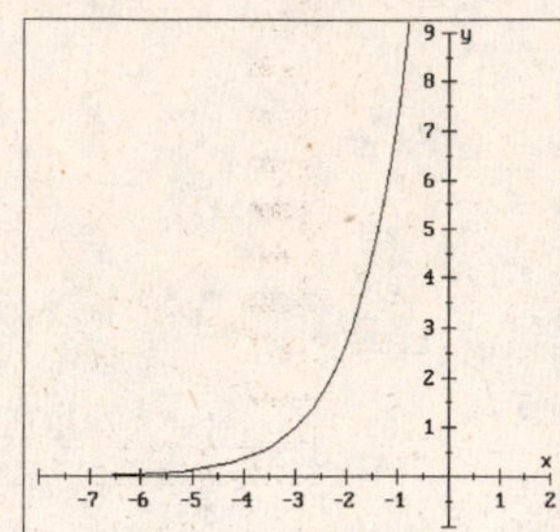

21. $y = f(x) = -e^x$; Reflect $y = e^x$ about the x-axis.

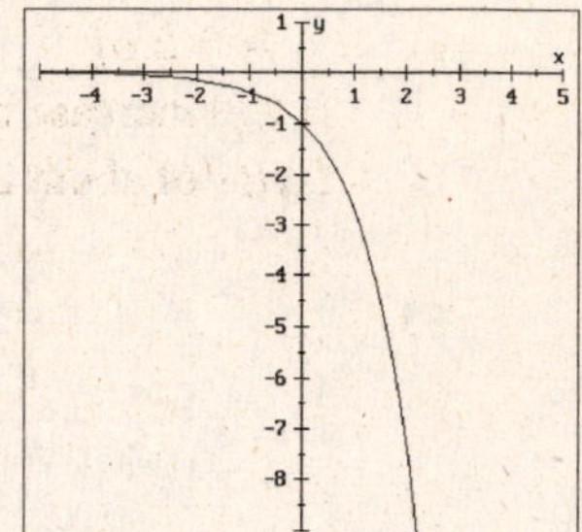

23. $y = f(x) = 2e^x$; Stretch $y = e^x$ vertically by a factor of 2.

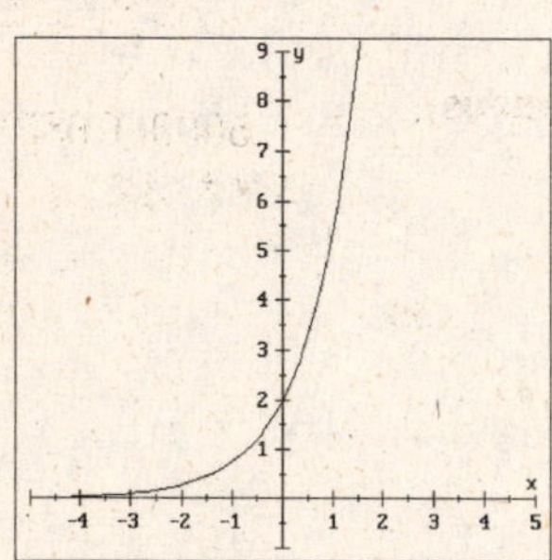

25. The graph should be increasing. The graph could not look like this.

27. The graph should go through the point $(0, 1)$. The graph could not look like this.

29.
$$\begin{aligned} A &= Pe^{rt} \\ &= 5000e^{0.06(12)} \\ &= 5000e^{0.72} \\ &\approx \$10{,}272.17 \end{aligned}$$

31.
$$\begin{aligned} A &= Pe^{rt} \\ 12000 &= Pe^{0.07(9)} \\ 12000 &= Pe^{0.63} \\ 12000 &\approx P(1.8776106) \\ P &\approx \frac{12000}{1.8776106} \\ P &\approx \$6{,}391.10 \end{aligned}$$

33.
$$\begin{aligned} A &= Pe^{rt} \\ &= 6e^{0.019(30)} \\ &= 6e^{0.57} \\ &\approx 10.6 \text{ billion people} \end{aligned}$$

SECTION 11.2

35. $A = Pe^{rt}$
$= 6e^{0.019(50)}$
$= 6e^{0.95}$
$\approx 6(2.6)$
It will increase by a factor of about 2.6.

37. $y = 1000e^{0.02x}$
$y = 31x + 2000$

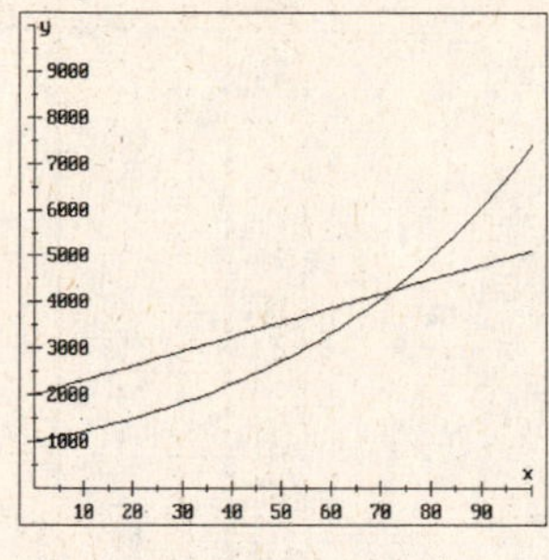

about 72 years

39. $A = A_0e^{-0.087t}$
$= 50e^{-0.087(30)}$
$= 50e^{-2.61}$
≈ 3.68 grams

41. $A = A_0e^{-0.00000693t}$
$= 2500e^{-0.00000693(100)}$
$= 2500e^{-0.000693}$
≈ 2498.27 grams

43. $A = Pe^{rt}$
$= 5000e^{0.085(5)}$
$= 5000e^{0.425}$
$\approx \$7{,}647.95$ (continuous)

$A = P\left(1 + \frac{r}{k}\right)^{kt}$
$= 5000\left(1 + \frac{0.085}{1}\right)^{1(5)}$
$= 5000(1.085)^5$
$\approx \$7{,}518.28$ (annual)

45. $P = 8000e^{-0.008t}$
$= 8000e^{-0.008(20)}$
$= 8000e^{-0.16}$
≈ 6817

47. $x = 0.08\left(1 - e^{-0.1t}\right)$
$= 0.08\left(1 - e^{-0.1(30)}\right)$
$= 0.08\left(1 - e^{-3}\right)$
$\approx 0.08(1 - 0.049787)$
$\approx 0.08(0.950213)$
≈ 0.076

49. $v = 50\left(1 - e^{-0.2t}\right)$
$= 50\left(1 - e^{-0.2(0)}\right)$
$= 50\left(1 - e^{0}\right)$
$= 50(1 - 1)$
$= 50(0) = 0$ mps

51. $v = 50\left(1 - e^{-0.2t}\right)$
$= 50\left(1 - e^{-0.2(2)}\right)$
$= 50\left(1 - e^{-0.4}\right)$
$\approx 50(1 - 0.67032)$
$\approx 50(0.32968) \approx 16.5$ mps

$v = 50\left(1 - e^{-0.3t}\right)$
$= 50\left(1 - e^{-0.3(2)}\right)$
$= 50\left(1 - e^{-0.6}\right)$
$\approx 50(1 - 0.54881)$
$\approx 50(0.45119) \approx 22.6$ mps $\Rightarrow$ faster

53. $A = Pe^{rt}$
$= 4570e^{-0.06(6.5)}$
$\approx \$3094.15$

55. **Answers may vary.**

57. $e \approx 2.7182;\ 1 + 1 + \frac{1}{2} + \frac{1}{2\cdot3} + \frac{1}{2\cdot3\cdot4} + \frac{1}{2\cdot3\cdot4\cdot5} \approx 2.7167$

59. $e^{t+5} = ke^t$
$e^t \cdot e^5 = ke^t$
$e^5e^t = ke^t$
$k = e^5$

Exercise 11.3 (page 804)

1. $\log_2 8 = x \Rightarrow 2^x = 8 \Rightarrow x = 3$

3. $\log_x 125 = 3 \Rightarrow x^3 = 125 \Rightarrow x = 5$

5. $\log_4 16 = x \Rightarrow 4^x = 16 \Rightarrow x = 2$

7. $\log_{1/2} x = 2 \Rightarrow \left(\frac{1}{2}\right)^2 = x \Rightarrow x = \frac{1}{4}$

9. $\log_x \frac{1}{4} = -2 \Rightarrow x^{-2} = \frac{1}{4} \Rightarrow x = 2$

11.
$$\begin{aligned} \sqrt{3x-4} &= \sqrt{-7x+2} \\ 3x-4 &= -7x+2 \\ 10x &= 6 \\ x &= \tfrac{6}{10} = \tfrac{3}{5} \end{aligned}$$
$\frac{3}{5}$ is extraneous $\Rightarrow \emptyset$

13.
$$\begin{aligned} 3 - \sqrt{t-3} &= \sqrt{t} \\ \left(3 - \sqrt{t-3}\right)^2 &= t \\ 9 - 6\sqrt{t-3} + t - 3 &= t \\ 6 &= 6\sqrt{t-3} \\ 1 &= \sqrt{t-3} \\ 1 &= t-3 \\ 4 &= t \end{aligned}$$

15. $(0, \infty)$

17. x

19. exponent

21. $(b, 1); (1, 0)$

23. $20 \log \dfrac{E_O}{E_I}$

25. $\log_3 27 = 3 \Rightarrow 3^3 = 27$

27. $\log_{1/2} \dfrac{1}{4} = 2 \Rightarrow \left(\dfrac{1}{2}\right)^2 = \dfrac{1}{4}$

29. $\log_4 \dfrac{1}{64} = -3 \Rightarrow 4^{-3} = \dfrac{1}{64}$

31. $\log_{1/2} \dfrac{1}{8} = 3 \Rightarrow \left(\dfrac{1}{2}\right)^3 = \dfrac{1}{8}$

33. $6^2 = 36 \Rightarrow \log_6 36 = 2$

35. $5^{-2} = \dfrac{1}{25} \Rightarrow \log_5 \dfrac{1}{25} = -2$

37. $\left(\dfrac{1}{2}\right)^{-5} = 32 \Rightarrow \log_{1/2} 32 = -5$

39. $x^y = z \Rightarrow \log_x z = y$

41. $\log_7 x = 2 \Rightarrow 7^2 = x \Rightarrow x = 49$

43. $\log_6 x = 1 \Rightarrow 6^1 = x \Rightarrow x = 6$

45. $\log_{25} x = \dfrac{1}{2} \Rightarrow 25^{1/2} = x \Rightarrow x = 5$

47. $\log_5 x = -2 \Rightarrow 5^{-2} = x \Rightarrow x = \dfrac{1}{25}$

49. $\log_x 5^3 = 3 \Rightarrow x^3 = 5^3 \Rightarrow x = 5$

51. $\log_x \dfrac{9}{4} = 2 \Rightarrow x^2 = \dfrac{9}{4} \Rightarrow x = \dfrac{3}{2}$

53. $\log_2 16 = x \Rightarrow 2^x = 16 \Rightarrow x = 4$

55. $\log_4 16 = x \Rightarrow 4^x = 16 \Rightarrow x = 2$

SECTION 11.3

57. $\log_{1/2}\dfrac{1}{8} = x \Rightarrow \left(\dfrac{1}{2}\right)^x = \dfrac{1}{8} \Rightarrow x = 3$

59. $\log_9 3 = x \Rightarrow 9^x = 3 \Rightarrow x = \dfrac{1}{2}$

61. $y = f(x) = \log_3 x$
through $(1, 0)$ and $(3, 1)$

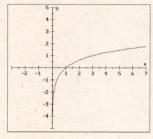

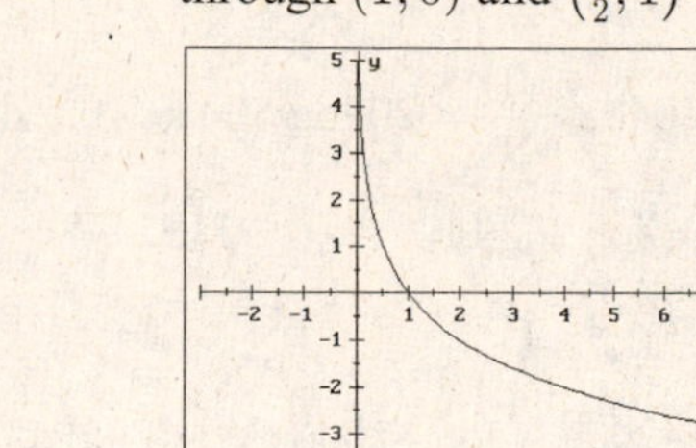

increasing

63. $y = f(x) = \log_{1/2} x$
through $(1, 0)$ and $(\frac{1}{2}, 1)$

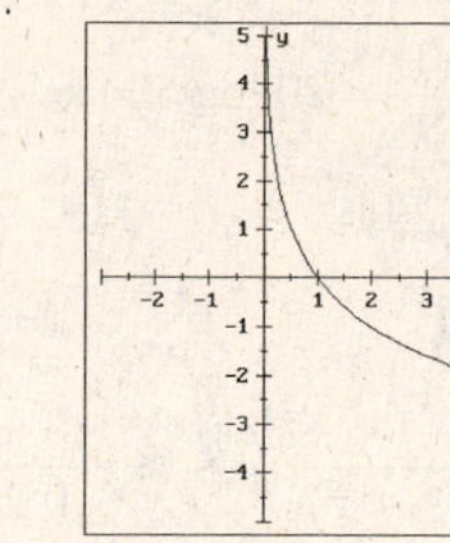

decreasing

65. $y = f(x) = 2^x$
$y = g(x) = \log_2 x$

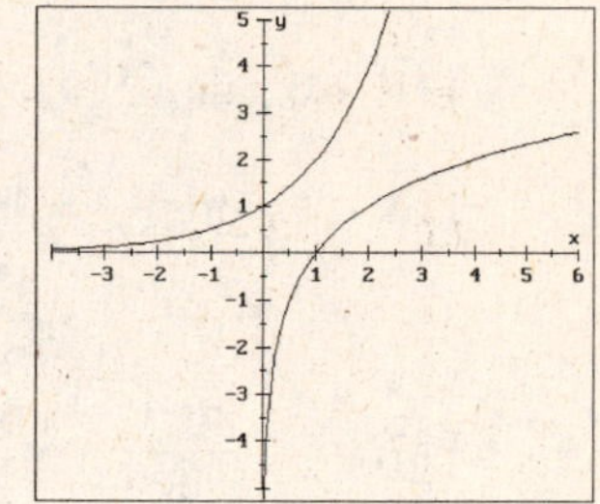

67. $y = f(x) = \left(\dfrac{1}{4}\right)^x$
$y = g(x) = \log_{1/4} x$

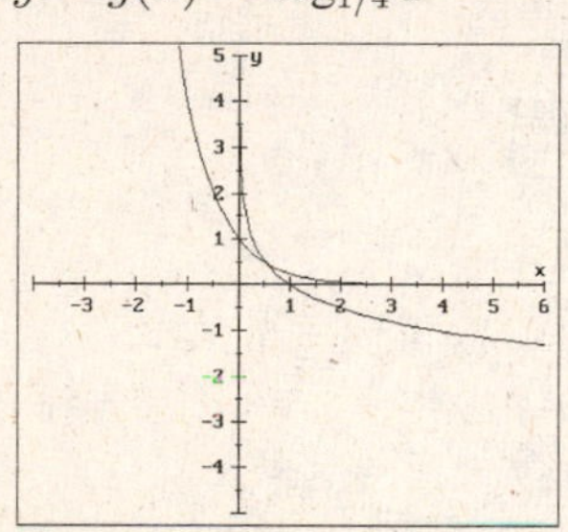

69. $y = f(x) = 3 + \log_3 x$
Shift $y = \log_3 x$ up 3.

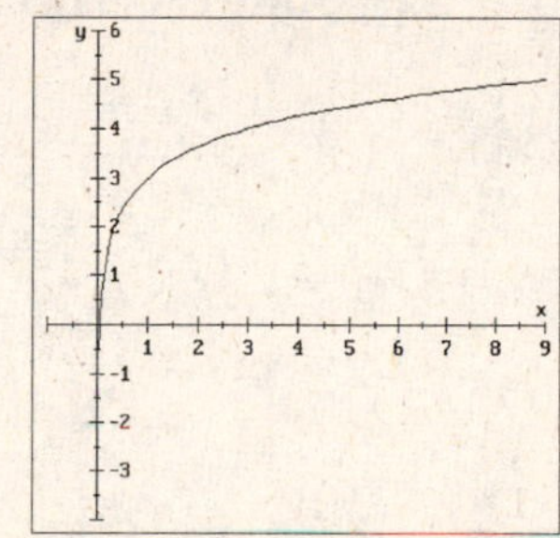

71. $y = f(x) = \log_{1/2}(x - 2)$
Shift $y = \log_{1/2} x$ right 2.

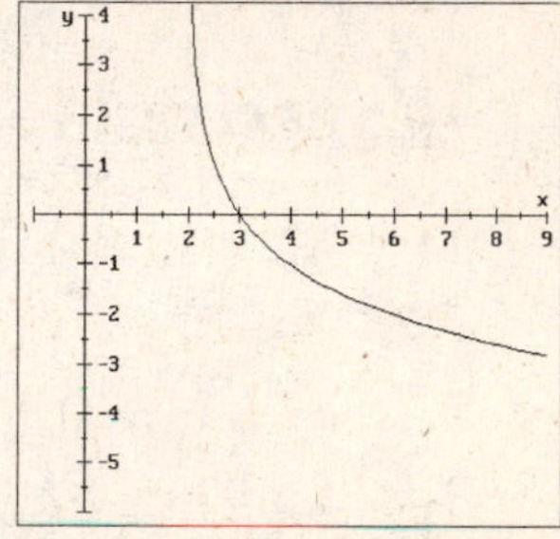

73. $\log 8.25 \approx 0.9165$

75. $\log 0.00867 \approx -2.0620$

77. $\log y = 4.24 \Rightarrow y = 17{,}378.01$

79. $\log y = -3.71 \Rightarrow y = 0.00$

81. $\log_{36} x = -\dfrac{1}{2} \Rightarrow 36^{-1/2} = x \Rightarrow x = \dfrac{1}{6}$

83. $\log_{1/2} 8 = x \Rightarrow \left(\dfrac{1}{2}\right)^x = 8 \Rightarrow x = -3$

85. $\log_{100} \frac{1}{1000} = x \Rightarrow 100^x = \frac{1}{1000} \Rightarrow x = -\frac{3}{2}$

87. $\log_{27} 9 = x \Rightarrow 27^x = 9 \Rightarrow x = \frac{2}{3}$

89. $\log_{2\sqrt{2}} x = 2 \Rightarrow (2\sqrt{2})^2 = x \Rightarrow x = 8$

91. $\log_x \dfrac{1}{64} = -3 \Rightarrow x^{-3} = \dfrac{1}{64} \Rightarrow x = 4$

93. $2^{\log_2 4} = x \Rightarrow x = 4$

95. $x^{\log_4 6} = 6 \Rightarrow x = 4$

97. $\log 10^3 = x \Rightarrow 10^x = 10^3 \Rightarrow x = 3$

99. $10^{\log x} = 100 \Rightarrow \log x = 2 \Rightarrow x = 100$

101. $\log y = 1.4023 \Rightarrow y = 25.25$

103. $\log y = \log 8 \Rightarrow \log y = 0.9030 \Rightarrow y = 8$

105. $\log_b 9 = 2 \Rightarrow b^2 = 9 \Rightarrow b = 3$

107. $\log_b 2 = 0 \Rightarrow b^0 = 2$
No such value exists.

109. $\text{dB gain} = 20 \log \frac{E_O}{E_I} = 20 \log \frac{20}{0.71} = 20 \log 28.169 \approx 29.0 \text{ dB}$

111. $\text{dB gain} = 20 \log \frac{E_O}{E_I} = 20 \log \frac{30}{0.1} = 20 \log 300 \approx 49.5 \text{ dB}$

113. $R = \log \frac{A}{P} = \log \frac{5000}{0.2} = \log 25{,}000 \approx 4.4$

115. $R = \log \frac{A}{P} = \log \frac{2500}{0.25} = \log 10000 = 4$

117. $n = \frac{\log V - \log C}{\log\left(1 - \frac{2}{N}\right)} = \frac{\log 2000 - \log 17000}{\log\left(1 - \frac{2}{5}\right)} \approx \frac{-0.929419}{-0.221849} \approx 4.2 \text{ years old}$

119. $n = \frac{\log\left[\frac{Ar}{P} + 1\right]}{\log(1 + r)} = \frac{\log\left[\frac{20{,}000(0.12)}{1000} + 1\right]}{\log(1 + 0.12)} = \frac{\log 3.4}{\log 1.12} \approx 10.8 \text{ years}$

121. Answers may vary. **123. Answers may vary.** **125. Answers may vary.**

Exercise 11.4 (page 811)

1. $y = \ln x \Rightarrow y = \log_e x \Rightarrow e^y = x$

3. $t = \frac{\ln 2}{r}$

5.
$$y = mx + b$$
$$y = 9x + 5$$

7.
$$3x + 2y = 9$$
$$2y = -3x + 9$$
$$y = -\frac{3}{2}x + \frac{9}{2} \Rightarrow m = -\frac{3}{2}$$
Use the parallel slope:
$$y - y_1 = m(x - x_1)$$
$$y - 5 = -\frac{3}{2}(x - (-3))$$
$$y - 5 = -\frac{3}{2}x - \frac{9}{2}$$
$$y = -\frac{3}{2}x + \frac{1}{2}$$

9. $y = 5$

11. $\frac{x+1}{x} + \frac{x-1}{x+1} = \frac{(x+1)(x+1)}{x(x+1)} + \frac{(x-1)x}{(x+1)x} = \frac{x^2+2x+1}{x(x+1)} + \frac{x^2-x}{x(x+1)} = \frac{2x^2+x+1}{x(x+1)}$

13. $\frac{1 + \frac{y}{x}}{\frac{y}{x} - 1} = \frac{\left(1 + \frac{y}{x}\right)x}{\left(\frac{y}{x} - 1\right)x} = \frac{x+y}{y-x}$

15. $(0, \infty); (-\infty, \infty)$

17. 10

19. $\dfrac{\ln 2}{r}$

21. $\ln 25.25 \approx 3.2288$

23. $\ln 9.89 \approx 2.2915$

25. $\log(\ln 2) \approx \log(0.6931) \approx -0.1592$

27. $\ln(\log 0.5) = \ln(-0.3010) \Rightarrow$ impossible

29. $\ln y = 2.3015 \Rightarrow y = 9.9892$

31. $\ln y = 3.17 \Rightarrow y = 23.8075$

33. $\ln y = -4.72 \Rightarrow y = 0.0089$

35. $\log y = \ln 6 \Rightarrow \log y \approx 1.7918 \Rightarrow y \approx 61.9098$
(The answer will vary if rounding is used on the calculator.)

37. $y = -\ln x$

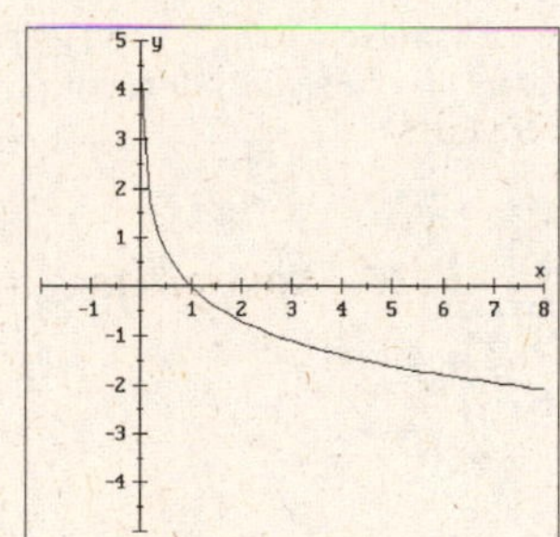

39. $y = \ln(-x)$

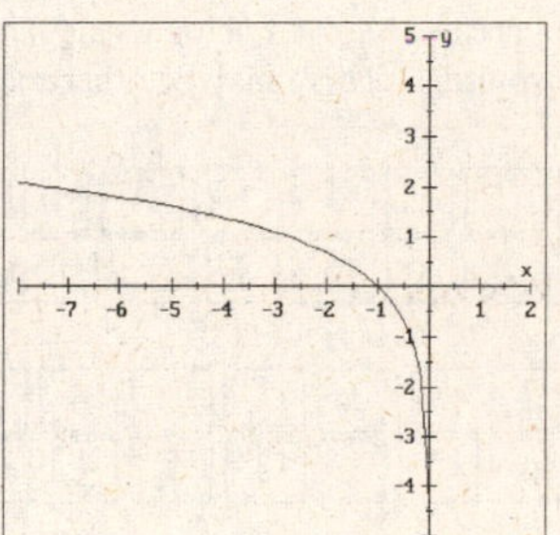

41. The graph must be increasing. The graph could not look like this.

43. The graph must go through $(1, 0)$. The graph could not look like this.

45. $t = \dfrac{\ln 2}{r} = \dfrac{\ln 2}{0.12} \approx 5.8$ years

47. $t = \dfrac{\ln 2}{r} = \dfrac{\ln 2}{0.05} \approx 13.9$ years

49. $t = -\dfrac{1}{0.9} \ln \dfrac{50 - T_r}{200 - T_r} = -\dfrac{1}{0.9} \ln \dfrac{50 - 38}{200 - 38} = -\dfrac{1}{0.9} \ln \dfrac{12}{162} \approx -\dfrac{1}{0.9}(-2.6027) \approx 2.9$ hours

51. **Answers may vary.**

53. $P = P_0 e^{rt} = P_0 e^{r\frac{\ln 3}{r}} = P_0 e^{\ln 3} = 3P_0$

55. Let $t = \dfrac{\ln 5}{r} \Rightarrow P = P_0 e^{rt} = P_0 e^{r\frac{\ln 5}{r}} = P_0 e^{\ln 5} = 5P_0$

Exercise 11.5 (page 821)

1. $\log_3 9 = x \Rightarrow 3^x = 9 \Rightarrow x = 2$

3. $\log_7 x = 3 \Rightarrow 7^3 = x \Rightarrow x = 343$

5. $\log_4 x = \frac{1}{2} \Rightarrow 4^{1/2} = x \Rightarrow x = 2$

7. $\log_{1/2} x = 2 \Rightarrow \left(\frac{1}{2}\right)^2 = x \Rightarrow x = \frac{1}{4}$

9. $\log_x \frac{1}{4} = -2 \Rightarrow x^{-2} = \frac{1}{4} \Rightarrow x = 2$

SECTION 11.5

11. $d = \sqrt{(-2-4)^2 + (3-(-4))^2}$
$= \sqrt{(-6)^2 + 7^2}$
$= \sqrt{36+49}$
$= \sqrt{85}$

13. Use the slope $m = -\frac{7}{6}$ from **#10**.
$y - y_1 = m(x - x_1)$
$y - 3 = -\frac{7}{6}(x+2)$
$y - 3 = -\frac{7}{6}x - \frac{7}{3}$
$y = -\frac{7}{6}x + \frac{2}{3}$

15. 1 **17.** x **19.** $-$ **21.** x

23. $=$ **25.** 0 **27.** 7 **29.** 10

31. 1 **33.** 0 **35.** 7 **37.** 10

39. 1

Problems 41-43 are to be solved using a calculator. The keystrokes needed to solve each problem using a TI-84 graphing calculator appear in each solution. There may be other solutions. Keystrokes for other calculators may be slightly different.

41. [log] [2] [.] [5] [×] [3] [.] [7] [ENTER] {0.96614}
[log] [2] [.] [5] [)] [+] [log] [3] [.] [7] [)] [ENTER] {0.96614}

43. [ln] [2] [.] [2] [5] [^] [4] [ENTER] {3.24372}
[4] [ln] [2] [.] [2] [5] [ENTER] {3.24372}

45. $\log_b xyz = \log_b x + \log_b y + \log_b z$

47. $\log_b \frac{2x}{y} = \log_b 2x - \log_b y$
$= \log_b 2 + \log_b x - \log_b y$

49. $\log_b x^3y^2 = \log_b x^3 + \log_b y^2 = 3\log_b x + 2\log_b y$

51. $\log_b (xy)^{1/2} = \frac{1}{2}\log_b xy = \frac{1}{2}(\log_b x + \log_b y) = \frac{1}{2}\log_b x + \frac{1}{2}\log_b y$

53. $\log_b x\sqrt{z} = \log_b xz^{1/2} = \log_b x + \log_b z^{1/2} = \log_b x + \frac{1}{2}\log_b z$

55. $\log_b \frac{\sqrt[3]{x}}{\sqrt[4]{yz}} = \log_b \frac{x^{1/3}}{(yz)^{1/4}} = \log_b x^{1/3} - \log_b (yz)^{1/4} = \frac{1}{3}\log_b x - \frac{1}{4}\log_b yz$
$= \frac{1}{3}\log_b x - \frac{1}{4}(\log_b y + \log_b z)$
$= \frac{1}{3}\log_b x - \frac{1}{4}\log_b y - \frac{1}{4}\log_b z$

57. $\log_b (x+1) - \log_b x = \log_b \frac{x+1}{x}$

59. $2\log_b x + \frac{1}{2}\log_b y = \log_b x^2 + \log_b y^{1/2}$
$= \log_b x^2y^{1/2}$

61. $-3\log_b x - 2\log_b y + \frac{1}{2}\log_b z = \log_b x^{-3} + \log_b y^{-2} + \log_b z^{1/2} = \log_b x^{-3}y^{-2}z^{1/2} = \log_b \frac{z^{1/2}}{x^3y^2}$

63. $\log_b\left(\frac{x}{z} + x\right) - \log_b\left(\frac{y}{z} + y\right) = \log_b \frac{\frac{x}{z} + x}{\frac{y}{z} + y} = \log_b \frac{x + xz}{y + yz} = \log_b \frac{x(1+z)}{y(1+z)} = \log_b \frac{x}{y}$

65. $\log 28 = \log 4 \cdot 7 = \log 4 + \log 7 = 0.6021 + 0.8451 = 1.4472$

67. $\log 2.25 = \log \frac{9}{4} = \log 9 - \log 4 = 0.9542 - 0.6021 = 0.3521$

69. $\log \frac{63}{4} = \log \frac{7 \cdot 9}{4} = \log 7 + \log 9 - \log 4 = 0.8451 + 0.9542 - 0.6021 = 1.1972$

71. $\log 252 = \log 4 \cdot 7 \cdot 9 = \log 4 + \log 7 + \log 9 = 0.6021 + 0.8451 + 0.9542 = 2.4014$

73. $\log 112 = \log 4^2 \cdot 7 = \log 4^2 + \log 7 = 2\log 4 + \log 7 = 2(0.6021) + 0.8451 = 2.0493$

75. $$\begin{aligned}\log \frac{144}{49} = \log \frac{16 \cdot 9}{7^2} = \log \frac{4^2 \cdot 9}{7^2} = \log 4^2 + \log 9 - \log 7^2 &= 2\log 4 + \log 9 - 2\log 7 \\ &= 2(0.6021) + 0.9542 - 2(0.8451) \\ &= 0.4682\end{aligned}$$

77. $\log_3 7 = \frac{\log 7}{\log 3} \approx 1.7712$

79. $\log_{1/3} 3 = \frac{\log 3}{\log \frac{1}{3}} \approx -1.0000$

81. $\log_3 8 = \frac{\log 8}{\log 3} \approx 1.8928$

83. $\log_{\sqrt{2}} \sqrt{5} = \frac{\log \sqrt{5}}{\log \sqrt{2}} \approx 2.3219$

85. [log] [√] [2] [4] [.] [3] [)] [ENTER] {0.69280}
[.] [5] [log] [2] [4] [.] [3] [)] [ENTER] {0.69280}

87. $\log_b 0 = 1 \Rightarrow b^1 = 0 \Rightarrow b = 0 \Rightarrow$ FALSE $(b \neq 0)$

89. $\log_b xy = (\log_b x)(\log_b y) \Rightarrow$ FALSE $(\log_b xy = \log_b x + \log_b y)$

91. $\log_7 7^7 = 7 \Rightarrow 7^7 = 7^7 \Rightarrow$ TRUE

93. $\frac{\log_b A}{\log_b B} = \log_b A - \log_b B \Rightarrow$ FALSE $\left(\log_b \frac{A}{B} = \log_b A - \log_b B\right)$

95. $3\log_b \sqrt[3]{a} = 3\log_b a^{1/3} = \frac{1}{3} \cdot 3\log_b a = \log_b a \Rightarrow$ TRUE

97. $\log_b \frac{1}{a} = \log_b 1 - \log_b a = 0 - \log_b a = -\log_b a \Rightarrow$ TRUE

99. $\text{pH} = -\log[\text{H}^+] = -\log(1.7 \times 10^{-5}) \approx 4.77$

101. low pH:

$$\begin{aligned} \text{pH} &= -\log[\text{H}^+] \\ 6.8 &= -\log[\text{H}^+] \\ -6.8 &= \log[\text{H}^+] \\ [\text{H}^+] &= 1.5849 \times 10^{-7} \end{aligned}$$

high pH:

$$\begin{aligned} \text{pH} &= -\log[\text{H}^+] \\ 7.6 &= -\log[\text{H}^+] \\ -7.6 &= \log[\text{H}^+] \\ [\text{H}^+] &= 2.5119 \times 10^{-8} \end{aligned}$$

103. $$\begin{aligned} k \ln 2I &= k(\ln 2 + \ln I) \\ &= k \ln 2 + k \ln I \\ &= k \ln 2 + L \end{aligned}$$

The loudness increases by $k \ln 2$.

105. $$\begin{aligned} L = 3k \ln I &= k \cdot 3 \ln I \\ &= k \ln I^3 \end{aligned}$$

The intensity must be cubed.

107. Answers may vary.

109. $\ln(e^x) = \log_e(e^x) = x$

111. Let $\log_{b^2} x = y$. Then

$$\begin{aligned} (b^2)^y &= x \\ b^{2y} &= x \\ (b^{2y})^{1/2} &= x^{1/2} \\ b^y &= x^{1/2} \\ \log_b x^{1/2} &= y \\ \frac{1}{2}\log_b x &= y \end{aligned}$$

Exercise 11.6 (page 832)

1. $$\begin{aligned} 3^x &= 5 \\ \log 3^x &= \log 5 \\ x \log 3 &= \log 5 \\ x &= \frac{\log 5}{\log 3} \end{aligned}$$

3. $$\begin{aligned} 2^{-x} &= 7 \\ \log 2^{-x} &= \log 7 \\ -x \log 2 &= \log 7 \\ x &= -\frac{\log 7}{\log 2} \end{aligned}$$

5. $$\begin{aligned} \log 2x &= \log(x+2) \\ 2x &= x + 2 \\ x &= 2 \end{aligned}$$

7. $$\begin{aligned} \log x^4 &= 4 \\ 10^4 &= x^4 \\ 10 &= x \end{aligned}$$

9. $$\begin{aligned} 5x^2 - 25x &= 0 \\ 5x(x-5) &= 0 \end{aligned}$$

$5x = 0$ **or** $x - 5 = 0$

$x = 0 \qquad x = 5$

11. $$\begin{aligned} 3p^2 + 10p &= 8 \\ 3p^2 + 10p - 8 &= 0 \\ (3p-2)(p+4) &= 0 \end{aligned}$$

$3p - 2 = 0$ **or** $p + 4 = 0$

$p = \frac{2}{3} \qquad p = -4$

13. exponential

15. $A_0 e^{-kt}$

SECTION 11.6

17.
$$\begin{aligned} 4^x &= 5 \\ \log 4^x &= \log 5 \\ x\log 4 &= \log 5 \\ x &= \frac{\log 5}{\log 4} \\ x &\approx 1.1610 \end{aligned}$$

19.
$$\begin{aligned} e^t &= 50 \\ \ln e^t &= \ln 50 \\ t\ln e &= \ln 50 \\ t &= \ln 50 \\ t &\approx 3.9120 \end{aligned}$$

21.
$$\begin{aligned} 2^x &= 3^x \\ \log 2^x &= \log 3^x \\ x\log 2 &= x\log 3 \\ 0 &= x\log 3 - x\log 2 \\ 0 &= x(\log 3 - \log 2) \\ \frac{0}{\log 3 - \log 2} &= x \\ 0 &= x \end{aligned}$$

23.
$$\begin{aligned} 5 &= 2.1(1.04)^t \\ \frac{5}{2.1} &= (1.04)^t \\ \log\frac{5}{2.1} &= \log(1.04)^t \\ \log\frac{5}{2.1} &= t\log 1.04 \\ \frac{\log\frac{5}{2.1}}{\log 1.04} &= t \\ 22.1184 &\approx t \end{aligned}$$

25.
$$\begin{aligned} 13^{x-1} &= 2 \\ \log 13^{x-1} &= \log 2 \\ (x-1)\log 13 &= \log 2 \\ x - 1 &= \frac{\log 2}{\log 13} \\ x &= \frac{\log 2}{\log 13} + 1 \\ x &\approx 1.2702 \end{aligned}$$

27.
$$\begin{aligned} 2^{x+1} &= 3^x \\ \log 2^{x+1} &= \log 3^x \\ (x+1)\log 2 &= x\log 3 \\ x\log 2 + \log 2 &= x\log 3 \\ \log 2 &= x\log 3 - x\log 2 \\ \log 2 &= x(\log 3 - \log 2) \\ \frac{\log 2}{\log 3 - \log 2} &= x \\ 1.7095 &\approx x \end{aligned}$$

29.
$$\begin{aligned} 2^{x^2-3x} &= 16 \\ 2^{x^2-3x} &= 2^4 \\ x^2 - 3x &= 4 \\ x^2 - 3x - 4 &= 0 \\ (x+1)(x-4) &= 0 \\ x+1 = 0 \quad &\textbf{or} \quad x-4=0 \\ x = -1 \quad &\quad\quad x = 4 \end{aligned}$$

31.
$$\begin{aligned} 3^{x^2+4x} &= \frac{1}{81} \\ 3^{x^2+4x} &= 3^{-4} \\ x^2 + 4x &= -4 \\ x^2 + 4x + 4 &= 0 \\ (x+2)(x+2) &= 0 \\ x+2 = 0 \quad &\textbf{or} \quad x+2=0 \\ x = -2 \quad &\quad\quad x = -2 \end{aligned}$$

SECTION 11.6

33.
$$\begin{aligned} 7^{x^2} &= 10 \\ \log 7^{x^2} &= \log 10 \\ x^2 \log 7 &= \log 10 \\ x^2 &= \frac{\log 10}{\log 7} \\ x^2 &\approx 1.1833 \\ x &\approx \pm 1.0878 \end{aligned}$$

35.
$$\begin{aligned} 8^{x^2} &= 9^x \\ \log 8^{x^2} &= \log 9^x \\ x^2 \log 8 &= x \log 9 \\ x^2 \log 8 - x \log 9 &= 0 \\ x(x \log 8 - \log 9) &= 0 \end{aligned}$$
$$x = 0 \quad \textbf{or} \quad x \log 8 - \log 9 = 0$$
$$\begin{aligned} x \log 8 &= \log 9 \\ x &= \frac{\log 9}{\log 8} \\ x &\approx 1.0566 \end{aligned}$$

37. $2^{x+1} = 7 \Rightarrow$ Graph $y = 2^{x+1}$ and $y = 7$.

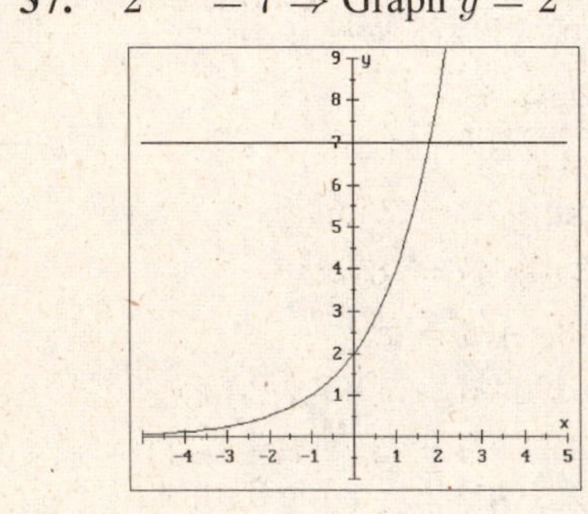

$x \approx 1.8$

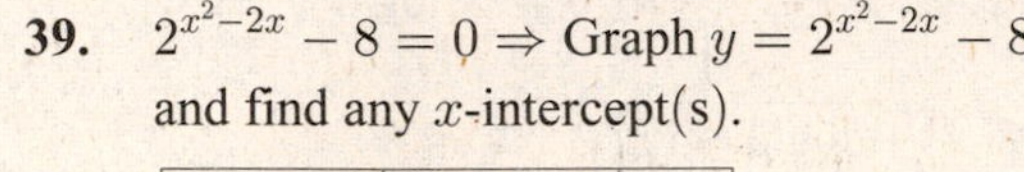

39. $2^{x^2-2x} - 8 = 0 \Rightarrow$ Graph $y = 2^{x^2-2x} - 8$ and find any x-intercept(s).

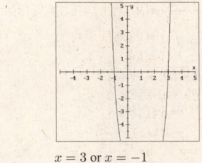

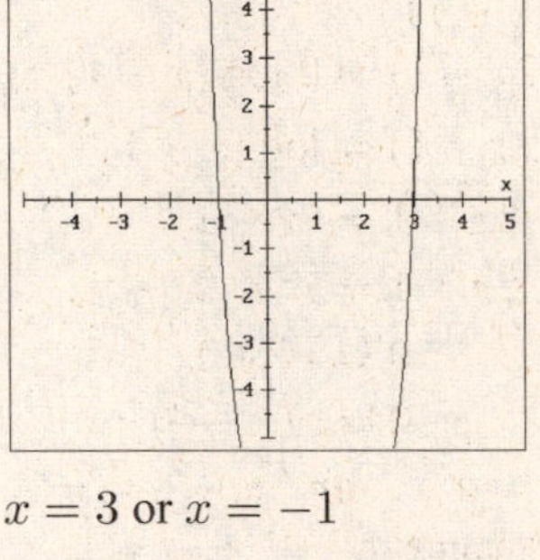

$x = 3$ or $x = -1$

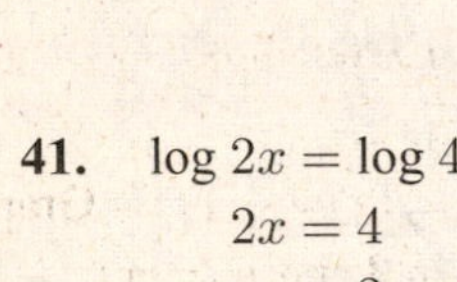

41.
$$\begin{aligned} \log 2x &= \log 4 \\ 2x &= 4 \\ x &= 2 \end{aligned}$$

43.
$$\begin{aligned} \log (3x + 1) &= \log (x + 7) \\ 3x + 1 &= x + 7 \\ 2x &= 6 \\ x &= 3 \end{aligned}$$

45.
$$\begin{aligned} \log (3 - 2x) - \log (x + 24) &= 0 \\ \log (3 - 2x) &= \log (x + 24) \\ 3 - 2x &= x + 24 \\ -21 &= 3x \\ -7 &= x \end{aligned}$$

47.
$$\begin{aligned} \log x^2 &= 2 \\ 10^2 &= x^2 \\ 100 &= x^2 \\ \pm 10 &= x \end{aligned}$$

49.
$$\begin{aligned} \log x + \log (x - 48) &= 2 \\ \log x(x - 48) &= 2 \\ 10^2 &= x(x - 48) \\ 0 &= x^2 - 48x - 100 \\ 0 &= (x - 50)(x + 2) \end{aligned}$$
$$x - 50 = 0 \quad \textbf{or} \quad x + 2 = 0$$
$$x = 50 \qquad x = -2$$
Extraneous

51.
$$\begin{aligned} \log x + \log (x - 15) &= 2 \\ \log x(x - 15) &= 2 \\ 10^2 &= x(x - 15) \\ 0 &= x^2 - 15x - 100 \\ 0 &= (x - 20)(x + 5) \end{aligned}$$
$$x - 20 = 0 \quad \textbf{or} \quad x + 5 = 0$$
$$x = 20 \qquad x = -5$$
Extraneous

53.
$$\begin{aligned} \log (x+90) &= 3 - \log x \\ \log x + \log (x+90) &= 3 \\ \log x(x+90) &= 3 \\ 10^3 &= x(x+90) \\ 0 &= x^2 + 90x - 1000 \\ 0 &= (x-10)(x+100) \end{aligned}$$
$x - 10 = 0$ **or** $x + 100 = 0$
$x = 10$ $\quad x = -100$
Extraneous

55.
$$\begin{aligned} \log (x-6) - \log (x-2) &= \log \frac{5}{x} \\ \log \frac{x-6}{x-2} &= \log \frac{5}{x} \\ \frac{x-6}{x-2} &= \frac{5}{x} \\ x(x-6) &= 5(x-2) \\ x^2 - 6x &= 5x - 10 \\ x^2 - 11x + 10 &= 0 \\ (x-10)(x-1) &= 0 \end{aligned}$$
$x - 10 = 0$ **or** $x - 1 = 0$
$x = 10$ $\quad x = 1$
Extraneous

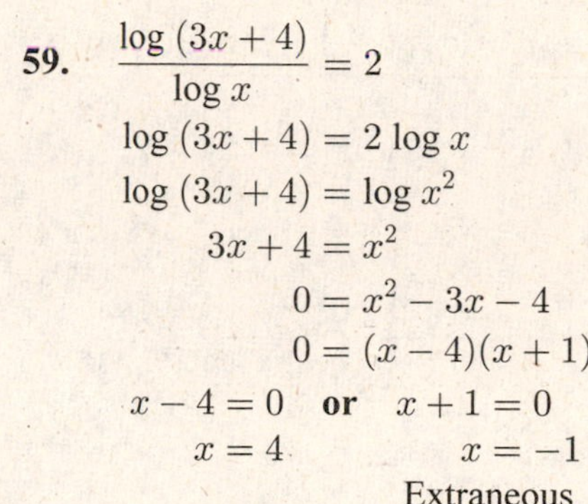

57.
$$\begin{aligned} \frac{\log (2x+1)}{\log (x-1)} &= 2 \\ \log (2x+1) &= 2 \log (x-1) \\ \log (2x+1) &= \log (x-1)^2 \\ 2x+1 &= (x-1)^2 \\ 2x+1 &= x^2 - 2x + 1 \\ 0 &= x^2 - 4x \\ 0 &= x(x-4) \end{aligned}$$
$x = 0$ **or** $x - 4 = 0$
Extraneous $\quad x = 4$

59.
$$\begin{aligned} \frac{\log (3x+4)}{\log x} &= 2 \\ \log (3x+4) &= 2 \log x \\ \log (3x+4) &= \log x^2 \\ 3x+4 &= x^2 \\ 0 &= x^2 - 3x - 4 \\ 0 &= (x-4)(x+1) \end{aligned}$$
$x - 4 = 0$ **or** $x + 1 = 0$
$x = 4$ $\quad x = -1$
Extraneous

61. $\log x + \log (x - 15) = 2$
Graph $y = \log x + \log (x - 15)$ and $y = 2$.

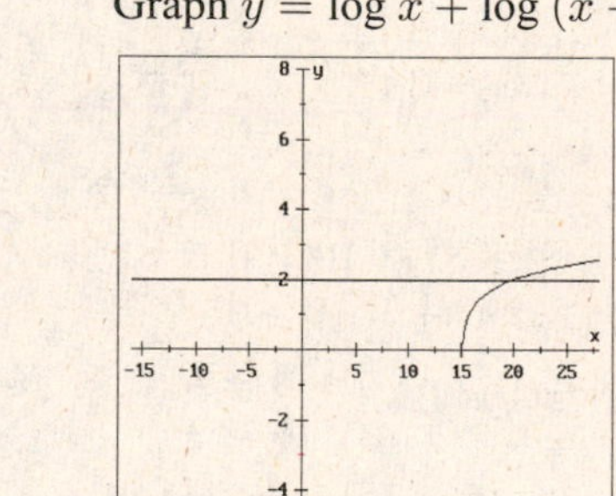

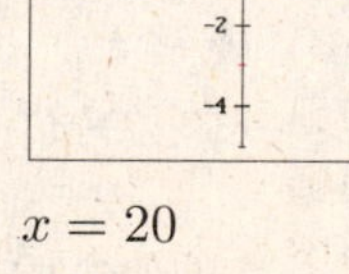

$x = 20$

63. $\ln (2x + 5) - \ln 3 = \ln (x - 1) \Rightarrow$ Graph
$y = \ln (2x + 5) - \ln 3$ and $y = \ln (x - 1)$.

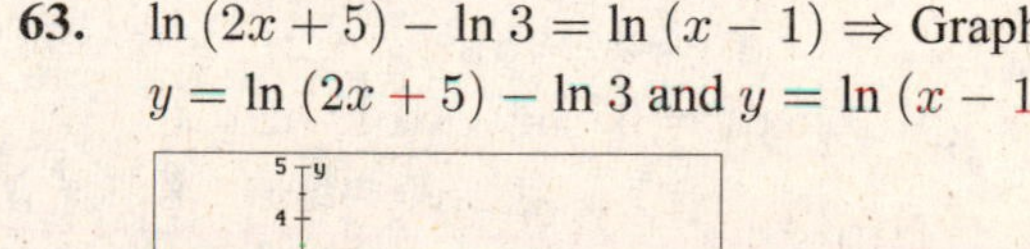

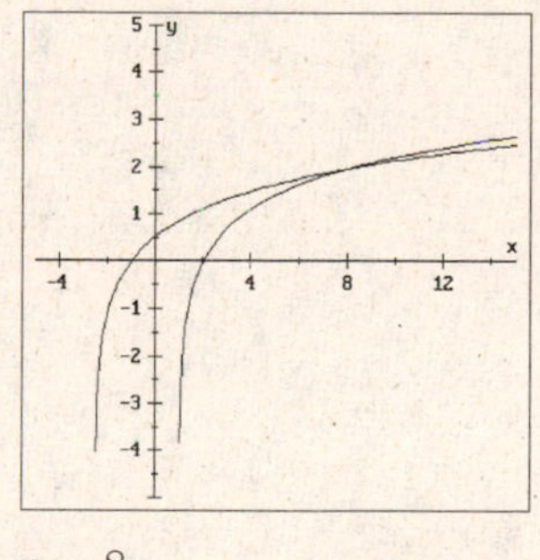

$x = 8$

SECTION 11.6

65.
$$\begin{aligned} 4^{x+2} - 4^x &= 15 \\ 4^x 4^2 - 4^x &= 15 \\ 16 \cdot 4^x - 4^x &= 15 \\ 15 \cdot 4^x &= 15 \\ 4^x &= 1 \\ x &= 0 \end{aligned}$$

67.
$$\begin{aligned} \frac{\log(5x+6)}{2} &= \log x \\ \log(5x+6) &= 2\log x \\ \log(5x+6) &= \log x^2 \\ 5x+6 &= x^2 \\ 0 &= x^2 - 5x - 6 \\ 0 &= (x-6)(x+1) \end{aligned}$$
$x - 6 = 0$ **or** $x + 1 = 0$
$x = 6$ $\quad$ $x = -1$
Extraneous

69.
$$\begin{aligned} \log_3 x &= \log_3\left(\frac{1}{x}\right) + 4 \\ \log_3 x &= \log_3\left(\frac{1}{x}\right) + \log_3 81 \\ \log_3 x &= \log_3\left(\frac{81}{x}\right) \\ x &= \frac{81}{x} \\ x^2 &= 81 \\ x &= 9 \text{ (}-9\text{ is extraneous.)} \end{aligned}$$

71.
$$\begin{aligned} 2(3^x) &= 6^{2x} \\ \log 2(3^x) &= \log 6^{2x} \\ \log 2 + \log 3^x &= 2x\log 6 \\ \log 2 + x\log 3 &= 2x\log 6 \\ \log 2 &= 2x\log 6 - x\log 3 \\ \log 2 &= x(2\log 6 - \log 3) \\ \frac{\log 2}{2\log 6 - \log 3} &= x \\ 0.2789 &\approx x \end{aligned}$$

73.
$$\begin{aligned} \log x^2 &= (\log x)^2 \\ 2\log x &= (\log x)^2 \\ 0 &= (\log x)^2 - 2\log x \\ 0 &= \log x(\log x - 2) \end{aligned}$$
$\log x = 0$ **or** $\log x - 2 = 0$
$x = 1$ $\quad$ $\log x = 2$
$x = 100$

75.
$$\begin{aligned} 2\log_2 x &= 3 + \log_2(x-2) \\ \log_2 x^2 - \log_2(x-2) &= 3 \\ \log_2 \frac{x^2}{x-2} &= 3 \\ \frac{x^2}{x-2} &= 8 \\ x^2 &= 8(x-2) \\ x^2 &= 8x - 16 \\ x^2 - 8x + 16 &= 0 \\ (x-4)(x-4) &= 0 \end{aligned}$$
$x - 4 = 0$ **or** $x - 4 = 0$
$x = 4$ $\quad$ $x = 4$

SECTION 11.6

77.
$$\begin{aligned}\log(7y+1) &= 2\log(y+3) - \log 2\\ \log(7y+1) &= \log(y+3)^2 - \log 2\\ \log(7y+1) &= \log\frac{y^2+6y+9}{2}\\ 7y+1 &= \frac{y^2+6y+9}{2}\\ 2(7y+1) &= y^2+6y+9\\ 14y+2 &= y^2+6y+9\\ 0 &= y^2-8y+7\\ 0 &= (y-7)(y-1)\end{aligned}$$
$y-7=0$ **or** $y-1=0$
$y=7$ $\quad y=1$

79.
$$\begin{aligned}\log\frac{4x+1}{2x+9} &= 0\\ 10^0 &= \frac{4x+1}{2x+9}\\ 1 &= \frac{4x+1}{2x+9}\\ 2x+9 &= 4x+1\\ 8 &= 2x\\ 4 &= x\end{aligned}$$

81.
$$\begin{aligned}A &= A_0e^{-0.013t}\\ 0.5A_0 &= A_0e^{-0.013t}\\ 0.5 &= e^{-0.013t}\\ \ln 0.5 &= \ln e^{-0.013t}\\ \ln 0.5 &= -0.013t\\ \frac{\ln 0.5}{-0.013} &= t\\ 53 \text{ days} &\approx t\end{aligned}$$

83.
$$\begin{aligned}A &= A_0 2^{-t/h}\\ 0.80A_0 &= A_0 2^{-2/h}\\ 0.80 &= 2^{-2/h}\\ \log 0.80 &= \log 2^{-2/h}\\ \log 0.80 &= -\frac{2}{h}\log 2\\ h\log 0.80 &= -2\log 2\\ h &= \frac{-2\log 2}{\log 0.80}\\ h &\approx 6.2 \text{ years}\end{aligned}$$

85.
$$\begin{aligned}A &= A_0 2^{-t/h}\\ 0.60A_0 &= A_0 2^{-t/5700}\\ 0.60 &= 2^{-t/5700}\\ \log 0.60 &= \log 2^{-t/5700}\\ \log 0.60 &= -\frac{t}{5700}\log 2\\ \frac{\log 0.60}{-\log 2} &= \frac{t}{5700}\\ -5700\frac{\log 0.60}{\log 2} &= t\\ 4200 \text{ years} &\approx t\end{aligned}$$

87.
$$\begin{aligned}P &= P_0e^{kt}\\ 2P_0 &= P_0e^{5k}\\ 2 &= e^{5k}\\ \ln 2 &= \ln e^{5k}\\ \ln 2 &= 5k\\ \frac{\ln 2}{5} &= k\end{aligned}$$

$$\begin{aligned}P &= P_0e^{kt}\\ 1{,}000{,}000 &= 30{,}000e^{\frac{\ln 2}{5}t}\\ 33.333 &\approx e^{\frac{\ln 2}{5}t}\\ \ln 33.333 &\approx \ln e^{\frac{\ln 2}{5}t}\\ \ln 33.333 &\approx \frac{\ln 2}{5}t\\ \frac{5\ln 33.333}{\ln 2} &\approx t\\ 25.3 \text{ years} &\approx t\end{aligned}$$

89.

$$\begin{aligned} P &= P_0e^{kt} \\ 2P_0 &= P_0e^{24k} \\ 2 &= e^{24k} \\ \ln 2 &= \ln e^{24k} \\ \ln 2 &= 24k \\ \frac{\ln 2}{24} &= k \end{aligned}$$

$$\begin{aligned} P &= P_0e^{kt} \\ P &= P_0e^{\frac{\ln 2}{24}(36)} \\ P &= P_0e^{\frac{3\ln 2}{2}} \\ P &= P_0(2.828) \end{aligned}$$

It will be about 2.828 timeslarger.

91.

$$\begin{aligned} n &= \frac{1}{\log 2}\left(\log\frac{B}{b}\right) \\ &= \frac{1}{\log 2}\log\frac{5\times 10^6}{500} \\ &= \frac{1}{\log 2}\log 10{,}000 \\ &\approx 13.3 \text{ generations} \end{aligned}$$

93.

$$\begin{aligned} A &= A_0 2^{-t/h} \\ 0.20A_0 &= A_0 2^{-t/18.4} \\ 0.20 &= 2^{-t/18.4} \\ \log 0.20 &= \log 2^{-t/18.4} \\ \log 0.20 &= -\frac{t}{18.4}\log 2 \\ \frac{\log 0.20}{-\log 2} &= \frac{t}{18.4} \\ -18.4\frac{\log 0.20}{\log 2} &= t \\ 42.7 \text{ days} &\approx t \end{aligned}$$

95.

$$\begin{aligned} A &= P\left(1+\frac{r}{k}\right)^{kt} \\ 800 &= 500\left(1+\frac{0.085}{2}\right)^{2t} \\ 1.6 &= (1.0425)^{2t} \\ \log 1.6 &= \log (1.0425)^{2t} \\ \log 1.6 &= 2t\log (1.0425) \\ \frac{\log 1.6}{2\log 1.0425} &= t \\ 5.6 \text{ years} &\approx t \end{aligned}$$

97.

$$\begin{aligned} A &= P\left(1+\frac{r}{k}\right)^{kt} \\ 2100 &= 1300\left(1+\frac{0.09}{4}\right)^{4t} \\ \frac{2100}{1300} &= (1.0225)^{4t} \\ \log\frac{2100}{1300} &= \log (1.0225)^{4t} \\ \log\frac{21}{13} &= 4t\log (1.0225) \\ \frac{\log \frac{21}{13}}{4\log 1.0225} &= t \\ 5.4 \text{ years} &\approx t \end{aligned}$$

99.

$$\begin{aligned} \text{doubling time} &= t \\ &= \frac{\ln 2}{r} \\ &= \frac{100\ln 2}{100r} \\ &\approx \frac{70}{100r} \\ &= \frac{70}{r\text{, written as a }\%} \end{aligned}$$

101. Answers may vary.

103. Since the logarithm of a negative number is not defined (as a real number), the values $x - 3$ and $x^2 + 2$ must be nonnegative. Since $x^2 + 2$ is always greater than 0, the only restriction is that $x - 3 > 0$, or $x > 3$. Thus, x cannot be a solution if $x \leq 3$.

Chapter 11 Review (page 837)

1. $5^{\sqrt{2}} \cdot 5^{\sqrt{2}} = 5^{\sqrt{2}+\sqrt{2}} = 5^{2\sqrt{2}}$

2. $\left(2^{\sqrt{5}}\right)^{\sqrt{2}} = 2^{(\sqrt{5})(\sqrt{2})} = 2^{\sqrt{10}}$

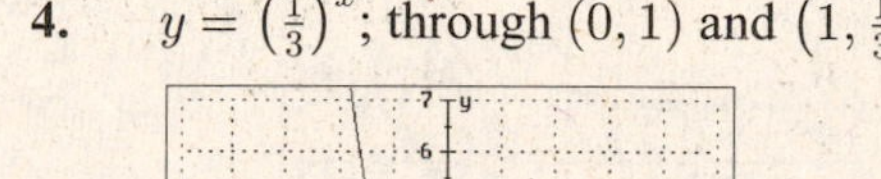

3. $y = 3^x$; through $(0,1)$ and $(1,3)$

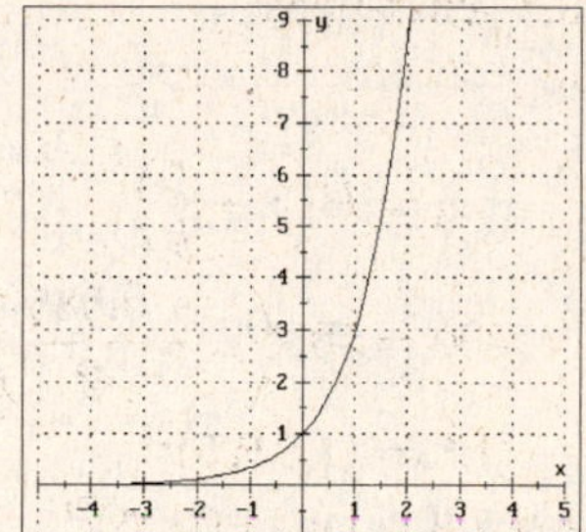

4. $y = \left(\frac{1}{3}\right)^x$; through $(0,1)$ and $\left(1, \frac{1}{3}\right)$

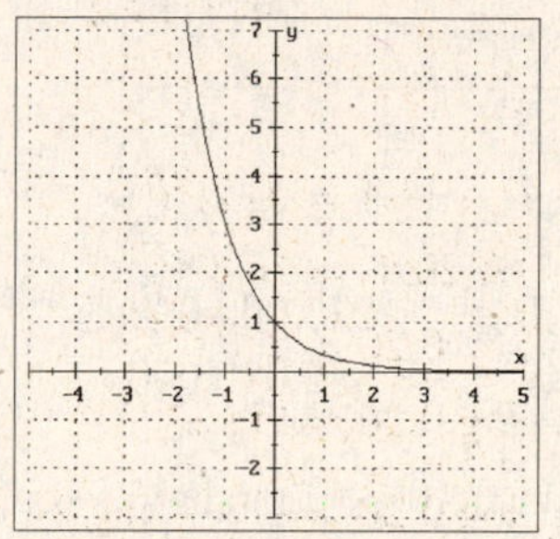

5. The graph will go through $(0,1)$ and $(1,6)$, so $x = 1$ and $y = 6$.

6. domain $= (-\infty, \infty)$; range $= (0, \infty)$

7. $y = f(x) = \left(\frac{1}{2}\right)^x - 2$
Shift $y = \left(\frac{1}{2}\right)^x$ down 2.

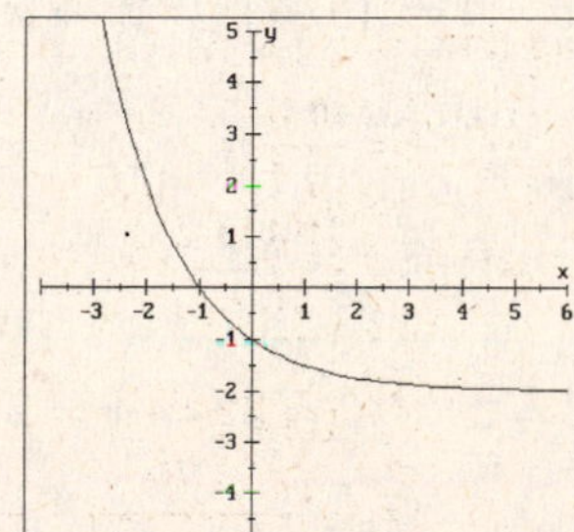

8. $y = f(x) = \left(\frac{1}{2}\right)^{x+2}$
Shift $y = \left(\frac{1}{2}\right)^x$ left 2.

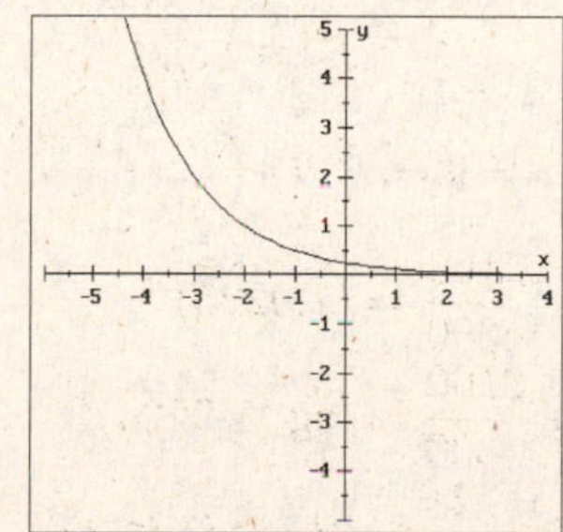

9. $A = P\left(1 + \frac{r}{k}\right)^{kt} = 10500\left(1 + \frac{0.09}{4}\right)^{4 \cdot 60} = 10500(1.0225)^{240} \approx \$2{,}189{,}703.45$

10. $A = Pe^{rt} = 10500e^{0.09(60)} = 10500e^{5.4} \approx \$2{,}324{,}767.37$

11. $y = f(x) = e^x + 1$; Shift $y = e^x$ up 1.

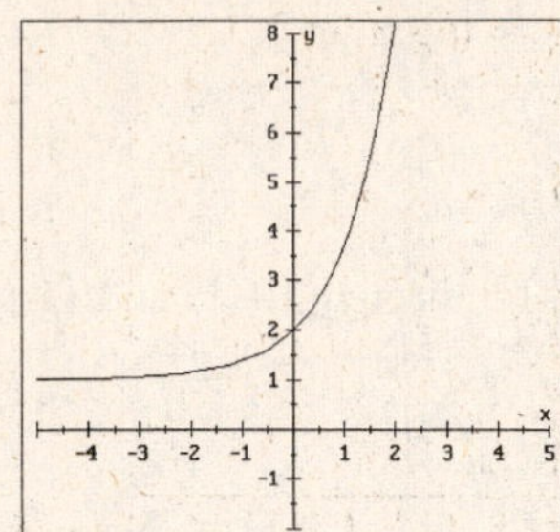

12. $y = f(x) = e^{x-3}$; Shift $y = e^x$ right 3.

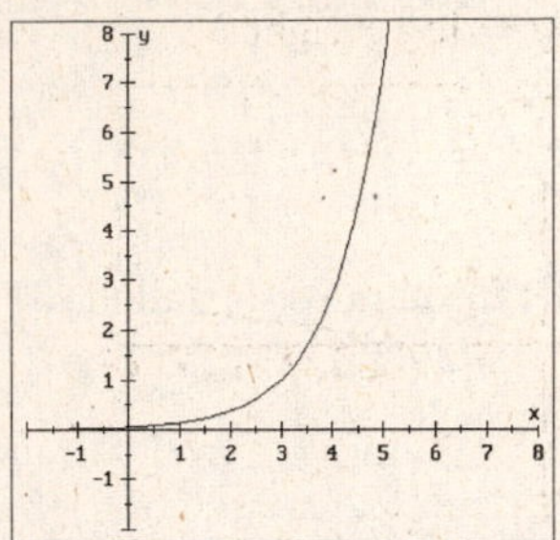

13. $P = P_0e^{kt} = 275{,}000{,}000e^{0.015(50)}$
$\approx 582{,}000{,}000$

14. $A = A_0e^{-0.0244t} = 50e^{-0.0244(20)} = 50e^{-0.488}$
≈ 30.69 g

15. domain $= (0, \infty)$; range $= (-\infty, \infty)$

16. **Answers will vary.**

17. $\log_3 9 = 2$

18. $\log_9 \frac{1}{3} = -\frac{1}{2}$

19. $\log_\pi 1 = 0$

20. $\log_5 0.04 = \log_5 \frac{1}{25} = -2$

21. $\log_a \sqrt{a} = \log_a a^{1/2} = \frac{1}{2}$

22. $\log_a \sqrt[3]{a} = \log_a a^{1/3} = \frac{1}{3}$

23. $\log_2 x = 5 \Rightarrow 2^5 = x \Rightarrow x = 32$

24. $\log_{\sqrt{3}} x = 4 \Rightarrow \left(\sqrt{3}\right)^4 = x \Rightarrow x = 9$

25. $\log_{\sqrt{3}} x = 6 \Rightarrow \left(\sqrt{3}\right)^6 = x \Rightarrow x = 27$

26. $\log_{0.1} 10 = x \Rightarrow (0.1)^x = 10$
$\left(\frac{1}{10}\right)^x = 10 \Rightarrow x = -1$

27. $\log_x 2 = -\frac{1}{3} \Rightarrow x^{-1/3} = 2$
$\left(x^{-1/3}\right)^{-3} = 2^{-3} \Rightarrow x = \frac{1}{8}$

28. $\log_x 32 = 5 \Rightarrow x^5 = 32 \Rightarrow x = 2$

29. $\log_{0.25} x = -1 \Rightarrow (0.25)^{-1} = x$
$\left(\frac{1}{4}\right)^{-1} = x \Rightarrow x = 4$

30. $\log_{0.125} x = -\frac{1}{3} \Rightarrow (0.125)^{-1/3} = x$
$\left(\frac{1}{8}\right)^{-1/3} = x \Rightarrow x = 2$

31. $\log_{\sqrt{2}} 32 = x \Rightarrow \left(\sqrt{2}\right)^x = 32$
$\left(2^{1/2}\right)^x = 2^5 \Rightarrow \frac{1}{2}x = 5 \Rightarrow x = 10$

32. $\log_{\sqrt{5}} x = -4 \Rightarrow \left(\sqrt{5}\right)^{-4} = x$
$\left(5^{1/2}\right)^{-4} = x \Rightarrow 5^{-2} = x \Rightarrow x = \frac{1}{25}$

33. $\log_{\sqrt{3}} 9\sqrt{3} = x \Rightarrow \left(\sqrt{3}\right)^x = 9\sqrt{3}$
$\left(3^{1/2}\right)^x = 3^{5/2} \Rightarrow x = 5$

34. $\log_{\sqrt{5}} 5\sqrt{5} = x \Rightarrow \left(\sqrt{5}\right)^x = 5\sqrt{5}$
$\left(5^{1/2}\right)^x = 5^{3/2} \Rightarrow x = 3$

35. $y = f(x) = \log(x - 2)$
Shift $y = \log x$ right 2.

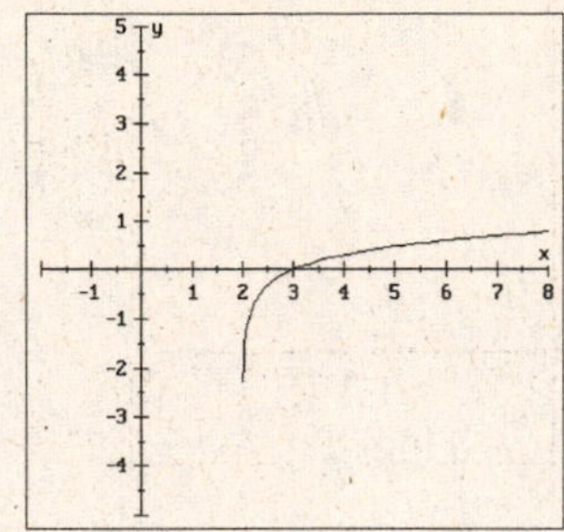

36. $y = f(x) = 3 + \log x$
Shift $y = \log x$ up 3.

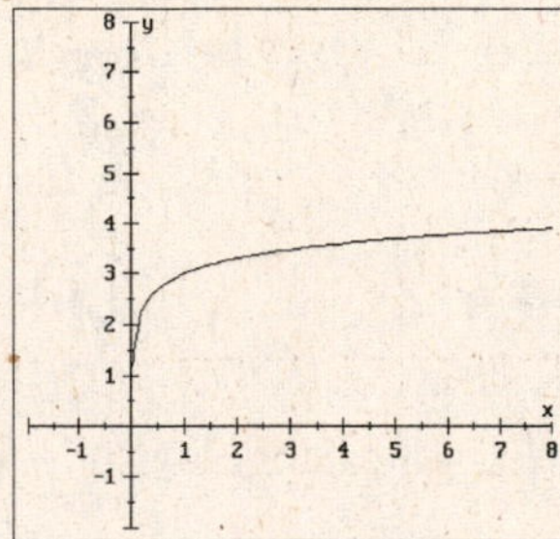

37. $y = 4^x$
$y = \log_4 x$

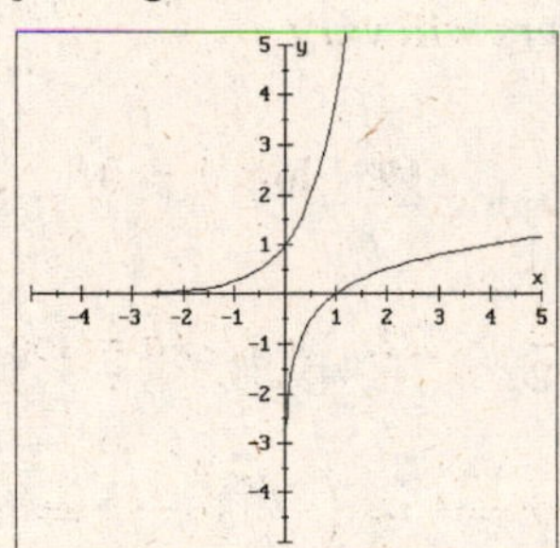

38. $y = \left(\frac{1}{3}\right)^x$
$y = \log_{1/3} x$

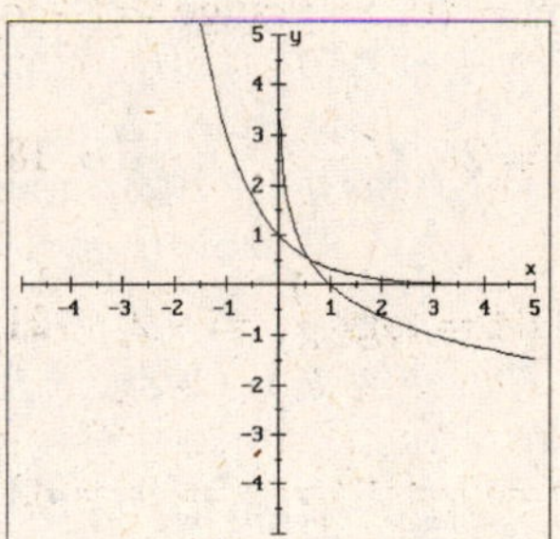

39. dB gain $= 20 \log \frac{E_O}{E_I} = 20 \log \frac{18}{0.04} = 20 \log 450 \approx 53$ dB

40. $R = \log \frac{A}{P} = \log \frac{7500}{0.3} = \log 25{,}000 \approx 4.4$

41. $\ln 452 \approx 6.1137$

42. $\ln(\log 7.85) \approx \ln 0.8949 \approx -0.1111$

43. $\ln x = 2.336 \Rightarrow x = 10.3398$

44. $\ln x = \log 8.8 \Rightarrow x = 2.5715$

45. $y = f(x) = 1 + \ln x$
Shift $y = \ln x$ up 1.

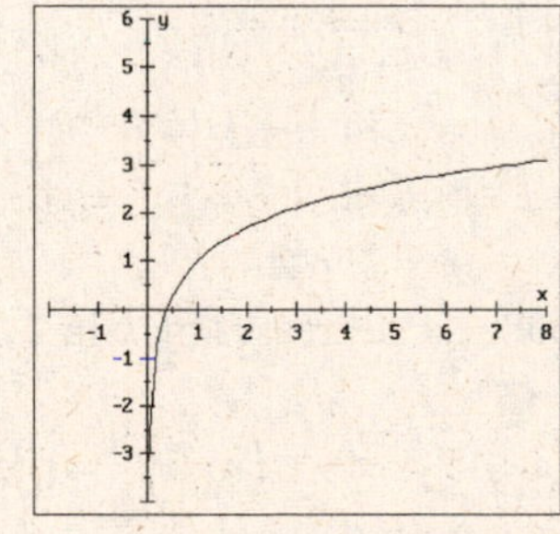

46. $y = f(x) = \ln(x + 1)$
Shift $y = \ln x$ left 1.

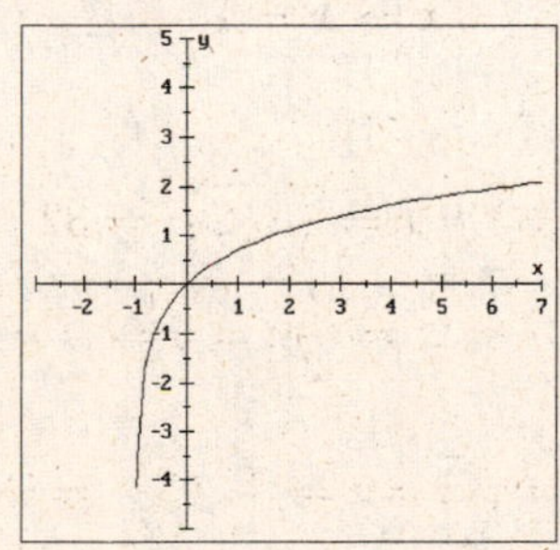

47. $t = \frac{\ln 2}{r} = \frac{\ln 2}{0.03} \approx 23$ years

48. $\log_7 1 = 0$

49. $\log_7 7 = 1$

50. $\log_7 7^3 = 3$

51. $7^{\log_7 4} = 4$

52. $\ln e^4 = 4$

53. $\ln 1 = 0$

54. $10^{\log_{10} 7} = 7$

55. $e^{\ln 3} = 3$

56. $\log_b b^4 = 4$

57. $\ln e^9 = 9$

58. $\log_b \frac{x^2 y^3}{z^4} = \log_b x^2 + \log_b y^3 - \log_b z^4 = 2\log_b x + 3\log_b y - 4\log_b z$

59.
$$\begin{aligned}\log_b \sqrt{\frac{x}{yz^2}} = \log_b \left(\frac{x}{yz^2}\right)^{1/2} = \frac{1}{2}\log_b \frac{x}{yz^2} &= \frac{1}{2}(\log_b x - \log_b y - \log_b z^2)\\ &= \frac{1}{2}(\log_b x - \log_b y - 2\log_b z)\end{aligned}$$

60. $3\log_b x - 5\log_b y + 7\log_b z = \log_b x^3 - \log_b y^5 + \log_b z^7 = \log_b \frac{x^3 z^7}{y^5}$

61. $\frac{1}{2}\log_b x + 3\log_b y - 7\log_b z = \log_b x^{1/2} + \log_b y^3 - \log_b z^7 = \log_b \frac{y^3\sqrt{x}}{z^7}$

62. $\log abc = \log a + \log b + \log c = 0.6 + 0.36 + 2.4 = 3.36$

63. $\log a^2 b = \log a^2 + \log b = 2\log a + \log b = 2(0.6) + 0.36 = 1.56$

64. $\log \frac{ac}{b} = \log a + \log c - \log b = 0.6 + 2.4 - 0.36 = 2.64$

65.
$$\begin{aligned}\log \frac{a^2}{c^3 b^2} = \log a^2 - \log c^3 - \log b^2 &= 2\log a - 3\log c - 2\log b\\ &= 2(0.6) - 3(2.4) - 2(0.36) = -6.72\end{aligned}$$

66. $\log_5 17 = \frac{\log 17}{\log 5} \approx 1.7604$

67.
$$\begin{aligned}\text{pH} &= -\log [\text{H}^+]\\ 3.1 &= -\log [\text{H}^+]\\ -3.1 &= \log [\text{H}^+]\\ [\text{H}^+] &= 7.94 \times 10^{-4} \text{ gram-ions per liter}\end{aligned}$$

68.
$$\begin{aligned}k \ln\left(\tfrac{1}{2}I\right) &= k\left(\ln \tfrac{1}{2} + \ln I\right)\\ &= k \ln \tfrac{1}{2} + k \ln I\\ &= k \ln 2^{-1} + L\\ &= -k\ln 2 + L\end{aligned}$$
The loudness decreases by $k \ln 2$.

69.
$$\begin{aligned} 3^x &= 7 \\ \log 3^x &= \log 7 \\ x\log 3 &= \log 7 \\ x &= \frac{\log 7}{\log 3} \approx 1.7712 \end{aligned}$$

70.
$$\begin{aligned} 5^{x+2} &= 625 \\ 5^{x+2} &= 5^4 \\ x+2 &= 4 \\ x &= 2 \end{aligned}$$

71.
$$\begin{aligned} 25 &= 5.5(1.05)^t \\ \frac{25}{5.5} &= (1.05)^t \\ \log\frac{25}{5.5} &= \log(1.05)^t \\ \log\frac{25}{5.5} &= t\log 1.05 \\ \frac{\log\frac{25}{5.5}}{\log 1.05} &= t \\ 31.0335 &\approx t \end{aligned}$$

72.
$$\begin{aligned} 4^{2t-1} &= 64 \\ 4^{2t-1} &= 4^3 \\ 2t-1 &= 3 \\ 2t &= 4 \\ t &= 2 \end{aligned}$$

73.
$$\begin{aligned} 2^x &= 3^{x-1} \\ \log 2^x &= \log 3^{x-1} \\ x\log 2 &= (x-1)\log 3 \\ x\log 2 &= x\log 3 - \log 3 \\ \log 3 &= x\log 3 - x\log 2 \\ \log 3 &= x(\log 3 - \log 2) \\ \frac{\log 3}{\log 3 - \log 2} &= x \\ 2.7095 &\approx x \end{aligned}$$

74.
$$\begin{aligned} 2^{x^2+4x} &= \frac{1}{8} \\ 2^{x^2+4x} &= 2^{-3} \\ x^2+4x &= -3 \\ x^2+4x+3 &= 0 \\ (x+3)(x+1) &= 0 \\ x+3 = 0 \quad &\textbf{or} \quad x+1=0 \\ x = -3 \quad & \quad\quad x = -1 \end{aligned}$$

75.
$$\begin{aligned} \log x + \log(29-x) &= 2 \\ \log x(29-x) &= 2 \\ 10^2 &= x(29-x) \\ 100 &= 29x - x^2 \\ x^2 - 29x + 100 &= 0 \\ (x-25)(x-4) &= 0 \\ x-25 = 0 \quad &\textbf{or} \quad x-4=0 \\ x = 25 \quad & \quad\quad x = 4 \end{aligned}$$

76.
$$\begin{aligned} \log_2 x + \log_2(x-2) &= 3 \\ \log_2 x(x-2) &= 3 \\ x(x-2) &= 2^3 \\ x^2-2x &= 8 \\ x^2-2x-8 &= 0 \\ (x-4)(x+2) &= 0 \\ x-4=0 \quad &\textbf{or} \quad x+2=0 \\ x=4 \quad & \quad\quad x=-2 \end{aligned}$$

Extraneous

77.
$$\begin{aligned}\log_2(x+2)+\log_2(x-1)&=2\\ \log_2(x+2)(x-1)&=2\\ (x+2)(x-1)&=2^2\\ x^2+x-2&=4\\ x^2+x-6&=0\\ (x-2)(x+3)&=0\end{aligned}$$
$x-2=0$ **or** $x+3=0$

$x=2$ $\qquad x=-3$

Extraneous

78.
$$\begin{aligned}\frac{\log(7x-12)}{\log x}&=2\\ \log(7x-12)&=2\log x\\ \log(7x-12)&=\log x^2\\ 7x-12&=x^2\\ 0&=x^2-7x+12\\ 0&=(x-4)(x-3)\end{aligned}$$
$x-4=0$ **or** $x-3=0$

$x=4$ $\qquad x=3$

79.
$$\begin{aligned}\log x+\log(x-5)&=\log 6\\ \log x(x-5)&=\log 6\\ x(x-5)&=6\\ x^2-5x-6&=0\\ (x-6)(x+1)&=0\end{aligned}$$
$x-6=0$ **or** $x+1=0$

$x=6$ $\qquad x=-1$

Extraneous

80.
$$\begin{aligned}\log 3-\log(x-1)&=-1\\ \log\frac{3}{x-1}&=-1\\ \frac{3}{x-1}&=10^{-1}\\ \frac{3}{x-1}&=\frac{1}{10}\\ 30&=x-1\\ 31&=x\end{aligned}$$

81.
$$\begin{aligned}e^{x\ln 2}&=9\\ e^{\ln 2^x}&=9\\ 2^x&=9\\ \ln 2^x&=\ln 9\\ x\ln 2&=\ln 9\\ x&=\frac{\ln 9}{\ln 2}\approx 3.1699\end{aligned}$$

82.
$$\begin{aligned}\ln x&=\ln(x-1)\\ x&=x-1\\ 0&=-1\end{aligned}$$
There is no solution. $\emptyset$

83.
$$\begin{aligned}\ln x&=\ln(x-1)+1\\ \ln x-\ln(x-1)&=1\\ \ln\frac{x}{x-1}&=1\\ \frac{x}{x-1}&=e^1\\ x&=e(x-1)\\ x&=ex-e\\ e&=ex-x\\ e&=x(e-1)\\ \frac{e}{e-1}&=x\\ 1.5820&\approx x\end{aligned}$$

84.
$$\begin{aligned}\ln x&=\log_{10}x\\ \ln x&=\frac{\ln x}{\ln 10}\\ \ln x\ln 10&=\ln x\\ \ln x\ln 10-\ln x&=0\\ \ln x(\ln 10-1)&=0\\ \ln x&=0\\ x&=1\end{aligned}$$

85.
$$\begin{aligned} A &= A_0 2^{-t/h} \\ \frac{2}{3}A_0 &= A_0 2^{-t/5730} \\ \frac{2}{3} &= 2^{-t/5730} \\ \log\frac{2}{3} &= \log 2^{-t/5730} \\ \log\frac{2}{3} &= -\frac{t}{5730}\log 2 \\ \frac{\log\frac{2}{3}}{-\log 2} &= \frac{t}{5730} \\ -5730\frac{\log\frac{2}{3}}{\log 2} &= t \\ 3352 \text{ years} &\approx t \text{ [or about 3400 years]} \end{aligned}$$

Chapter 11 Test (page 842)

1. $f(x) = 2^x + 1$; Shift $y = 2^x$ up 1.

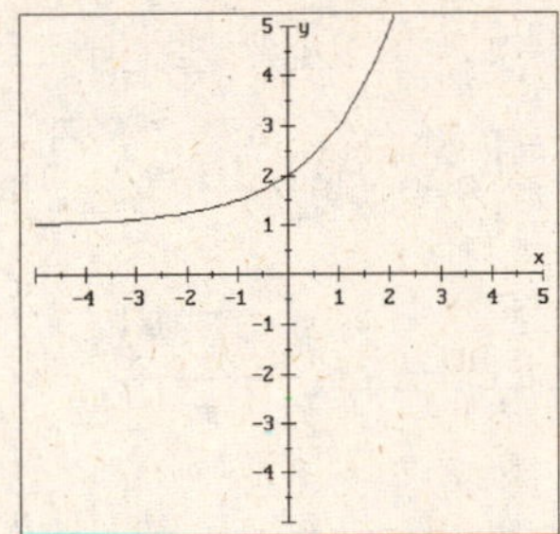

2. $f(x) = 2^{-x}$; Reflect $y = 2^x$ about y-axis.

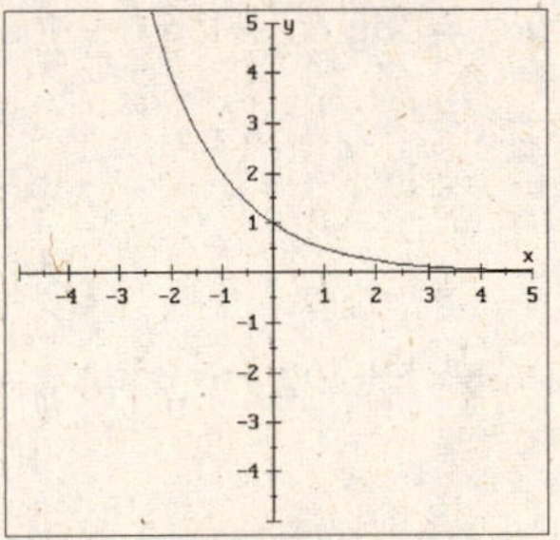

3. $A = A_0(2)^{-t} = 3(2)^{-6} = \frac{3}{2^6} = \frac{3}{64}$ gram

4. $A = A_0\left(1 + \frac{r}{k}\right)^{kt} = 1000\left(1 + \frac{0.06}{2}\right)^{2(1)} = 1000(1.03)^2 \approx \$1060.90$

5. $f(x) = e^x$

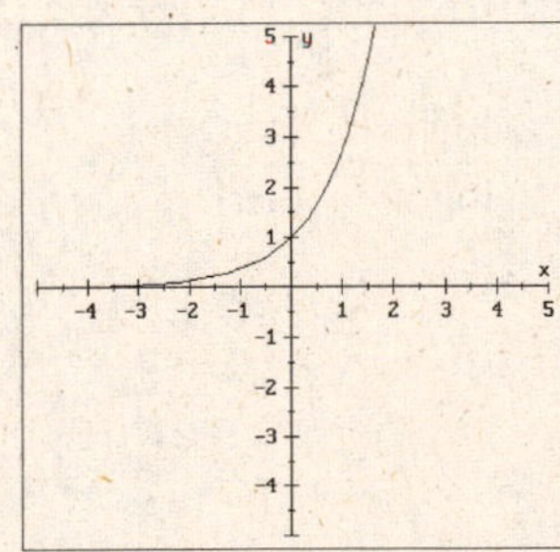

6.
$$\begin{aligned} A = A_0 e^{rt} &= 2000e^{(0.08)10} \\ &= 2000e^{0.8} \\ &\approx \$4451.08 \end{aligned}$$

7. $\log_4 16 = x \Rightarrow 4^x = 16 \Rightarrow x = 2$

8. $\log_x 81 = 4 \Rightarrow x^4 = 81 \Rightarrow x = 3$

9. $\log_3 x = -3 \Rightarrow 3^{-3} = x \Rightarrow x = \dfrac{1}{27}$

10. $\log_x 100 = 2 \Rightarrow x^2 = 100 \Rightarrow x = 10$

11. $\log_{3/2} \dfrac{9}{4} = x \Rightarrow \left(\dfrac{3}{2}\right)^x = \dfrac{9}{4} \Rightarrow x = 2$

12. $\log_{2/3} x = -3 \Rightarrow \left(\dfrac{2}{3}\right)^{-3} = x \Rightarrow x = \dfrac{27}{8}$

13. $f(x) = -\log_3 x$

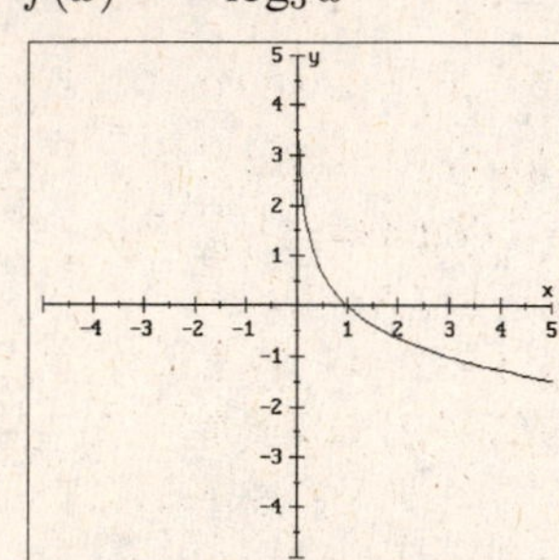

14. $f(x) = \ln x$

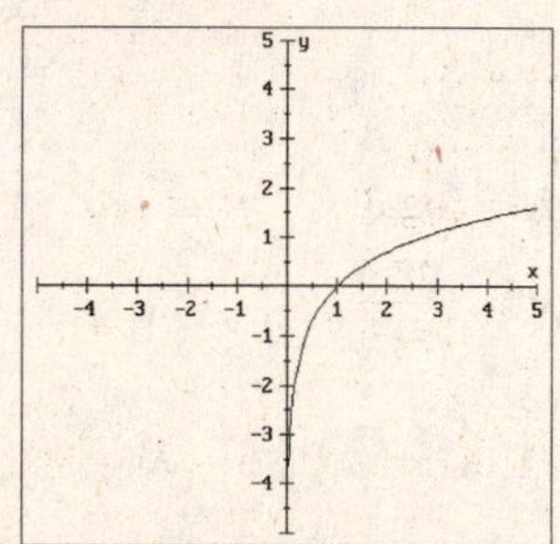

15. $\log a^2bc^3 = \log a^2 + \log b + \log c^3 = 2\log a + \log b + 3\log c$

16. $\ln\sqrt{\dfrac{a}{b^2c}} = \ln\left(\dfrac{a}{b^2c}\right)^{1/2} = \dfrac{1}{2}\ln\dfrac{a}{b^2c} = \dfrac{1}{2}(\ln a - \ln b^2 - \ln c)\ \left\{\text{also } \dfrac{1}{2}\ln a - \ln b - \dfrac{1}{2}\ln c\right\}$

17. $\dfrac{1}{2}\log(a+2) + \log b - 3\log c = \log(a+2)^{1/2} + \log b - \log c^3 = \log\dfrac{b\sqrt{a+2}}{c^3}$

18. $\dfrac{1}{3}(\log a - 2\log b) - \log c = \dfrac{1}{3}(\log a - \log b^2) - \log c = \dfrac{1}{3}\log\dfrac{a}{b^2} - \log c = \log\sqrt[3]{\dfrac{a}{b^2}} - \log c$

$= \log\dfrac{\sqrt[3]{a}}{c\sqrt[3]{b^2}}$

19. $\log 24 = \log 8\cdot 3 = \log 2^3\cdot 3 = \log 2^3 + \log 3 = 3\log 2 + \log 3 = 3(0.3010) + 0.4771 = 1.3801$

20. $\log\dfrac{8}{3} = \log\dfrac{2^3}{3} = \log 2^3 - \log 3 = 3\log 2 - \log 3 = 3(0.3010) - 0.4771 = 0.4259$

21. $\log_7 3 = \dfrac{\log 3}{\log 7}$ or $\dfrac{\ln 3}{\ln 7}$

22. $\log_\pi e = \dfrac{\log e}{\log \pi}$ or $\dfrac{\ln e}{\ln \pi}$

23. $\log_a ab = \log_a a + \log_a b = 1 + \log_a b \Rightarrow$ TRUE

24. $\dfrac{\log a}{\log b} = \log a - \log b \Rightarrow$ FALSE $\left(\log\dfrac{a}{b} = \log a - \log b\right)$

25. $\log a^{-3} = -3 \log a \neq \dfrac{1}{3 \log a} \Rightarrow$ FALSE

26. $\ln(-x) = -\ln x \Rightarrow$ FALSE (This implies one of the logarithms is negative, which is impossible.)

27. $\text{pH} = -\log [\text{H}^+] = -\log (3.7 \times 10^{-7}) \approx 6.4$

28. $\text{dB gain} = 20 \log \dfrac{E_O}{E_I} = 20 \log \dfrac{60}{0.3} = 20 \log 200 \approx 46$

29.
$$\begin{aligned} 5^x &= 3 \\ \log 5^x &= \log 3 \\ x \log 5 &= \log 3 \\ x &= \frac{\log 3}{\log 5} \end{aligned}$$

30.
$$\begin{aligned} 3^{x-1} &= 100^x \\ \log 3^{x-1} &= \log 100^x \\ (x-1)\log 3 &= x \log 100 \\ x \log 3 - \log 3 &= 2x \\ x \log 3 - 2x &= \log 3 \\ x(\log 3 - 2) &= \log 3 \\ x &= \frac{\log 3}{\log 3 - 2} \end{aligned}$$

31.
$$\begin{aligned} \log(5x+2) &= \log(2x+5) \\ 5x+2 &= 2x+5 \\ 3x &= 3 \\ x &= 1 \end{aligned}$$

32.
$$\begin{aligned} \log x + \log(x-9) &= 1 \\ \log x(x-9) &= 1 \\ x(x-9) &= 10 \\ x^2 - 9x - 10 &= 0 \\ (x-10)(x+1) &= 0 \end{aligned}$$
$x - 10 = 0$ **or** $x + 1 = 0$
$x = 10$ $\quad x = -1$
Extraneous

Exercise 12.1 (page 853)

1. $x^2 + y^2 = 144$
C $(0,0)$; $r = \sqrt{144} = 12$

3. $(x-2)^2 + y^2 = 16$
C $(2,0)$; $r = \sqrt{16} = 4$

5. $y = -3x^2 - 2$; down

7. $x = -3y^2$; left

9. $|3x - 4| = 11$
$3x - 4 = 11$ **or** $3x - 4 = -11$
$3x = 15$ $\quad 3x = -7$
$x = 5$ $\quad x = -\frac{7}{3}$

11. $|3x + 4| = |5x - 2|$
$3x + 4 = 5x - 2$ **or** $3x + 4 = -(5x - 2)$
$-2x = -6$ $\quad 3x + 4 = -5x + 2$
$x = 3$ $\quad 8x = -2$
$x = -\frac{1}{4}$

13. conic

15. standard; $(0,3)$; 4

17. circle; general

19. parabola; $(3,2)$; right

21. $x^2 + y^2 = 9$

C $(0,0)$; $r = \sqrt{9} = 3$

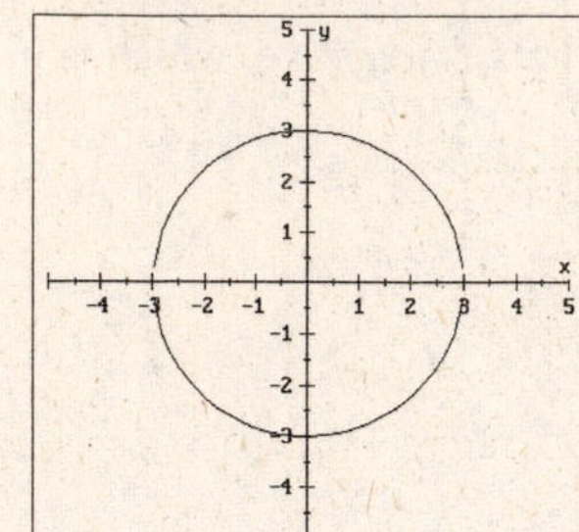

23. $(x-2)^2 + y^2 = 9$

C $(2,0)$; $r = \sqrt{9} = 3$

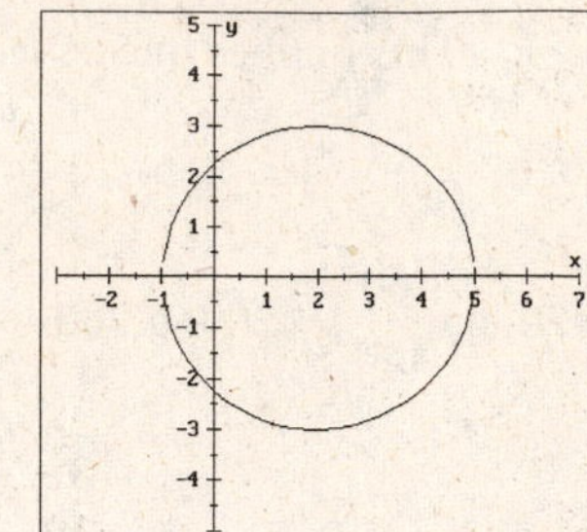

25. $(x-2)^2 + (y-4)^2 = 4$

C $(2,4)$; $r = \sqrt{4} = 2$

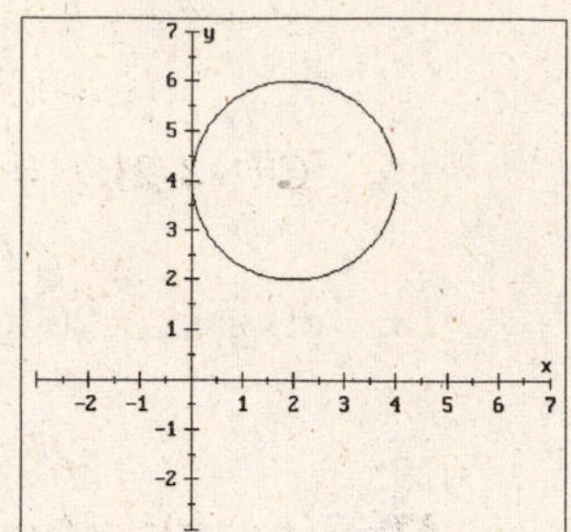

27. $(x+3)^2 + (y-1)^2 = 16$

C $(-3,1)$; $r = \sqrt{16} = 4$

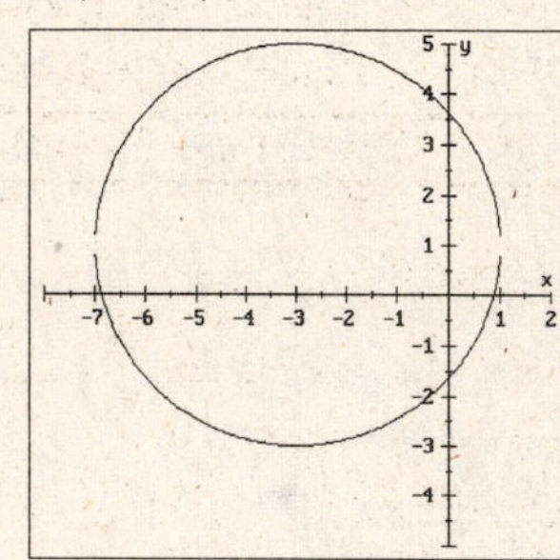

29.

$$x^2 + y^2 + 2x - 8 = 0$$
$$x^2 + 2x + y^2 = 8$$
$$x^2 + 2x + 1 + y^2 = 8 + 1$$
$$(x+1)^2 + y^2 = 9$$

C $(-1,0)$; $r = \sqrt{9} = 3$

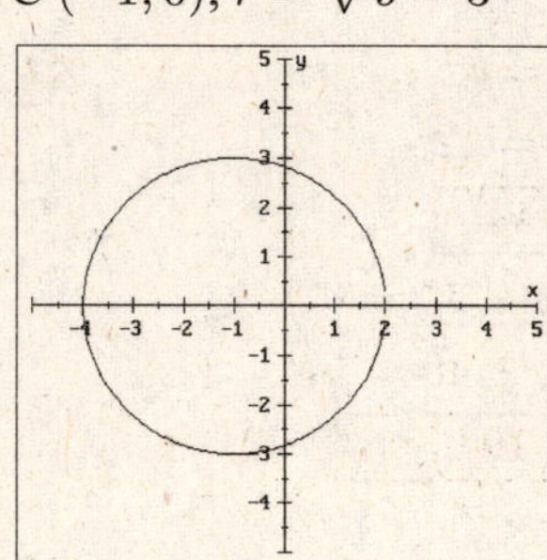

31.

$$9x^2 + 9y^2 - 12y = 5$$
$$x^2 + y^2 - \frac{4}{3}y = \frac{5}{9}$$
$$x^2 + y^2 - \frac{4}{3}y + \frac{4}{9} = \frac{5}{9} + \frac{4}{9}$$
$$x^2 + \left(y - \frac{2}{3}\right)^2 = 1$$

C $\left(0, \frac{2}{3}\right)$; $r = \sqrt{1} = 1$

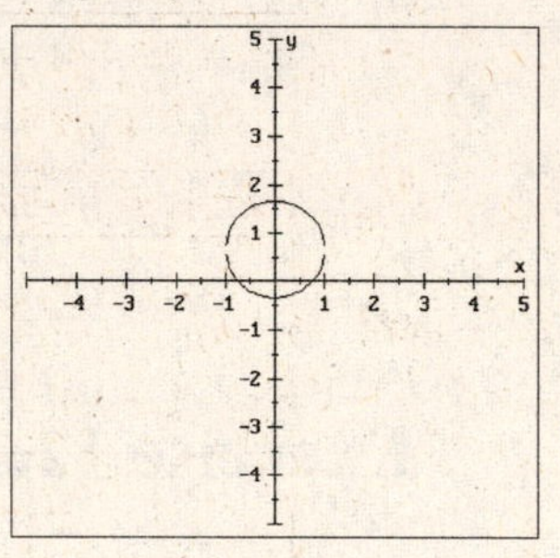

33.

$$x^2 + y^2 - 2x + 4y = -1$$
$$x^2 - 2x + y^2 + 4y = -1$$
$$x^2 - 2x + 1 + y^2 + 4y + 4 = -1 + 1 + 4$$
$$(x-1)^2 + (y+2)^2 = 4$$

C $(1,-2)$; $r = \sqrt{4} = 2$

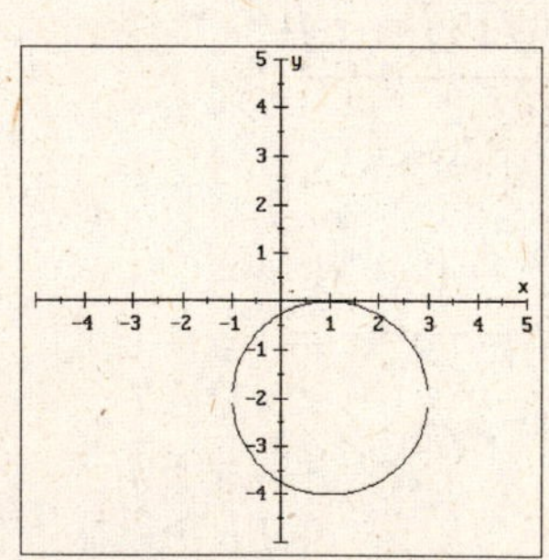

35.
$$x^2 + y^2 + 6x - 4y = -12$$
$$x^2 + 6x + y^2 - 4y = -12$$
$$x^2 + 6x + 9 + y^2 - 4y + 4 = -12 + 9 + 4$$
$$(x+3)^2 + (y-2)^2 = 1$$
C $(-3, 2)$; $r = \sqrt{1} = 1$

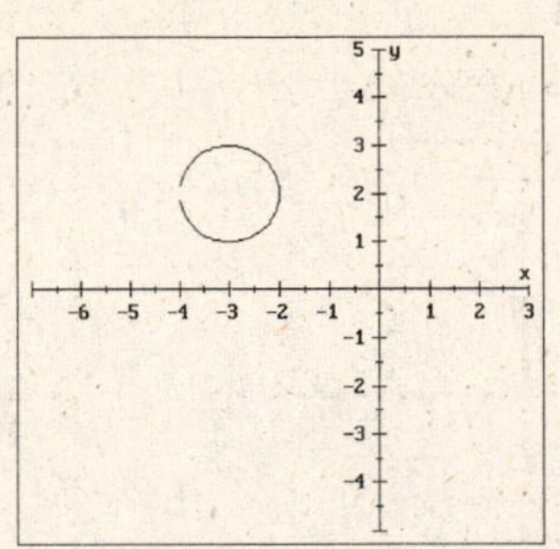

37. $(x-h)^2 + (y-k)^2 = r^2$
$(x-0)^2 + (y-0)^2 = 1^2$
$\boxed{x^2 + y^2 = 1}$
$\boxed{x^2 + y^2 - 1 = 0}$

39. $(x-h)^2 + (y-k)^2 = r^2$
$(x-6)^2 + (y-8)^2 = 5^2$
$\boxed{(x-6)^2 + (y-8)^2 = 25}$
$x^2 - 12x + 36 + y^2 - 16y + 64 = 25$
$x^2 + y^2 - 12x - 16y + 100 = 25$
$\boxed{x^2 + y^2 - 12x - 16y + 75 = 0}$

41. $(x-h)^2 + (y-k)^2 = r^2$
$(x-(-2))^2 + (y-6)^2 = 12^2$
$\boxed{(x+2)^2 + (y-6)^2 = 144}$
$x^2 + 4x + 4 + y^2 - 12y + 36 = 144$
$x^2 + y^2 + 4x - 12y + 40 = 144$
$\boxed{x^2 + y^2 + 4x - 12y - 104 = 0}$

43. $(x-h)^2 + (y-k)^2 = r^2$

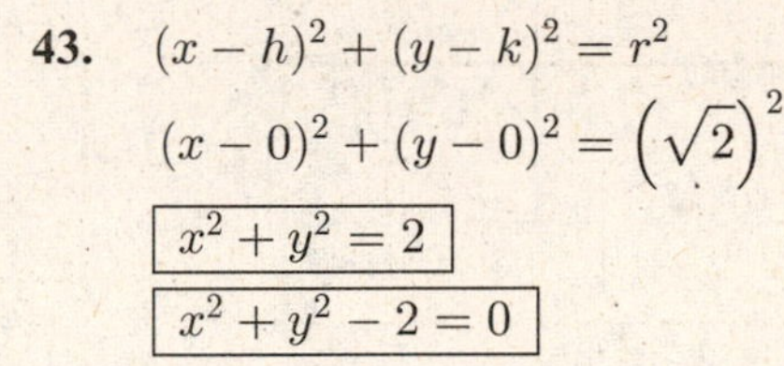

$(x-0)^2 + (y-0)^2 = \left(\sqrt{2}\right)^2$
$\boxed{x^2 + y^2 = 2}$
$\boxed{x^2 + y^2 - 2 = 0}$

45. $x = y^2$
$x = (y-0)^2 + 0$; V $(0, 0)$; opens R

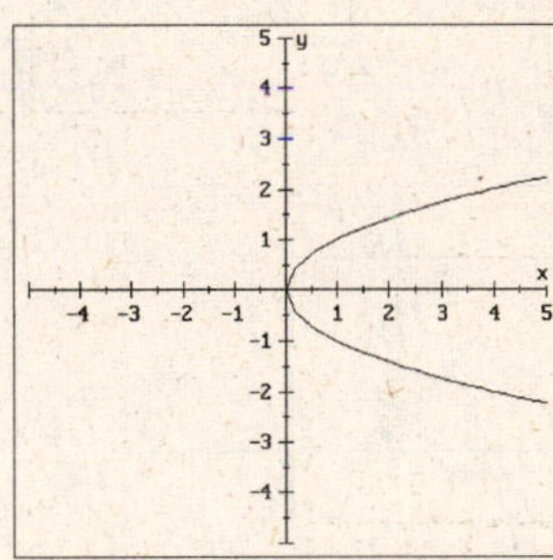

47. $x = -\frac{1}{4}y^2$

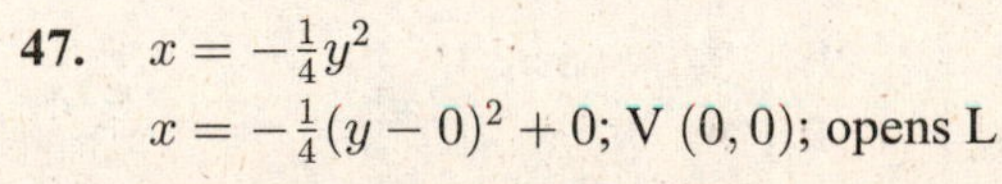

$x = -\frac{1}{4}(y-0)^2 + 0$; V $(0, 0)$; opens L

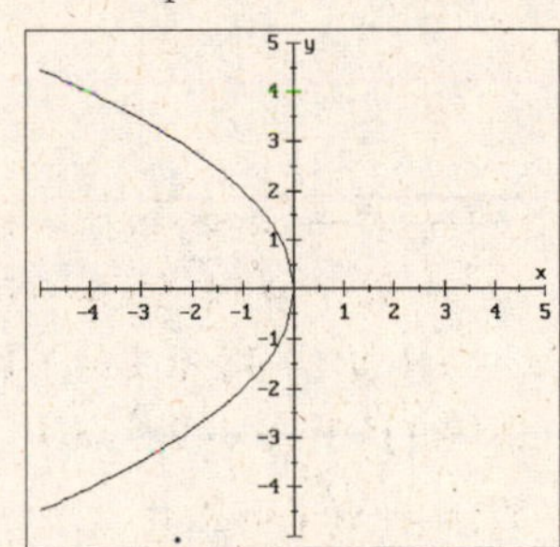

49. $y^2 + 4x - 6y = -1$

$$4x = -y^2 + 6y - 1$$
$$x = -\frac{1}{4}y^2 + \frac{3}{2}y - \frac{1}{4}$$
$$x = -\frac{1}{4}(y^2 - 6y) - \frac{1}{4}$$
$$x = -\frac{1}{4}(y^2 - 6y + 9) - \frac{1}{4} + \frac{9}{4}$$
$$x = -\tfrac{1}{4}(y - 3)^2 + 2$$

V $(2, 3)$; opens L

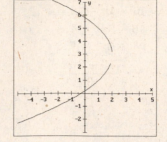

51. $y = 2(x - 1)^2 + 3$

V $(1, 3)$; opens U

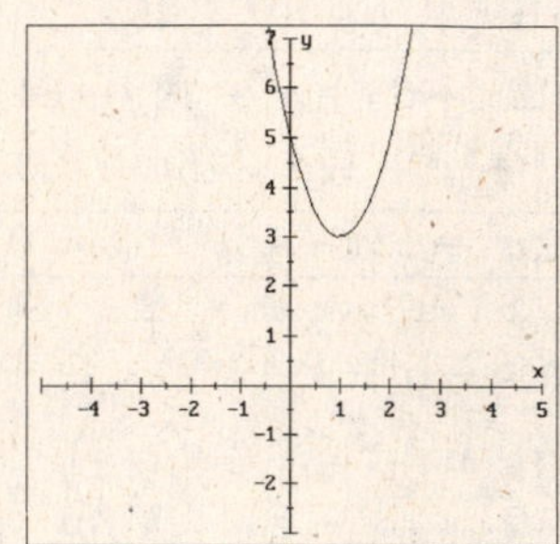

53. $x^2 + (y + 3)^2 = 1$

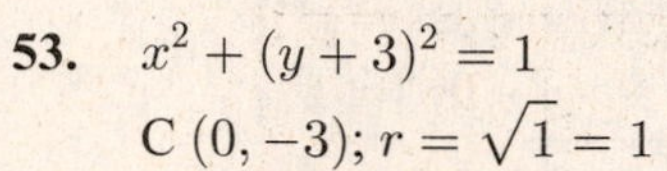

C $(0, -3)$; $r = \sqrt{1} = 1$

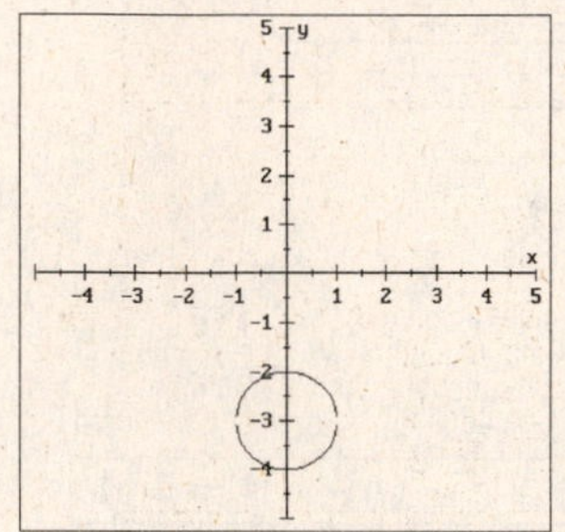

55. $y = x^2 + 4x + 5$

$$y = x^2 + 4x + 4 + 5 - 4$$
$$y = (x + 2)^2 + 1$$

V $(-2, 1)$; opens U

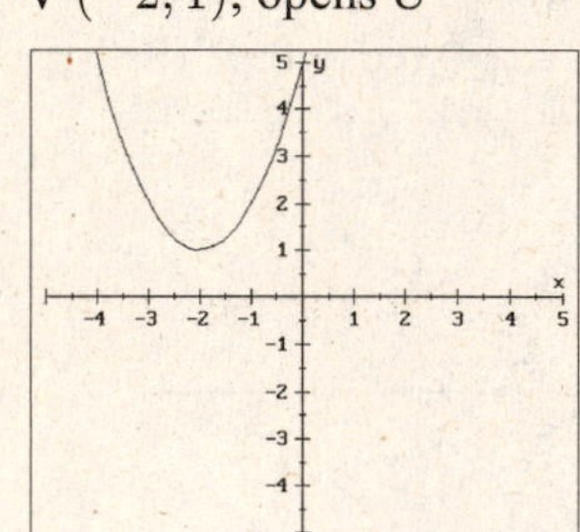

57. $y = -x^2 - x + 1$

$$y = -(x^2 + x) + 1$$

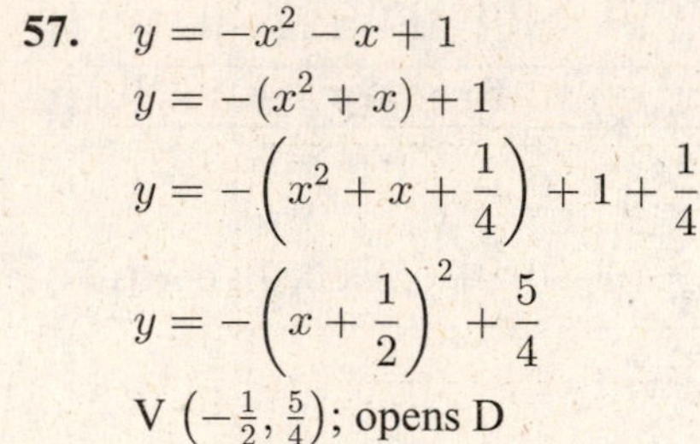

$$y = -\left(x^2 + x + \frac{1}{4}\right) + 1 + \frac{1}{4}$$
$$y = -\left(x + \frac{1}{2}\right)^2 + \frac{5}{4}$$

V $\left(-\frac{1}{2}, \frac{5}{4}\right)$; opens D

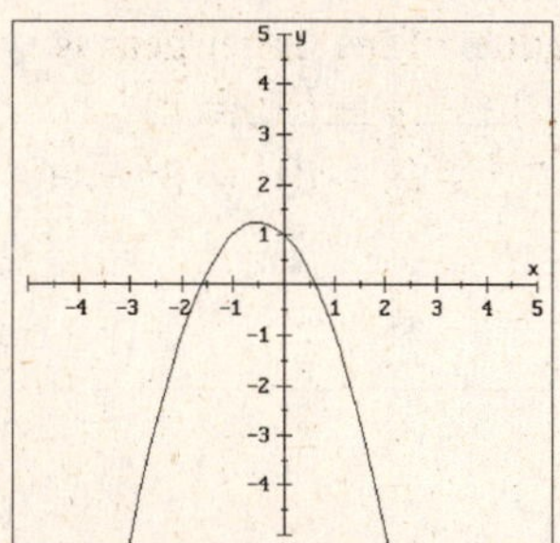

59. $3x^2 + 3y^2 = 16$

$$3y^2 = 16 - 3x^2$$

$$y^2 = \frac{16 - 3x^2}{3}$$

$$y = \pm\sqrt{\frac{16 - 3x^2}{3}}$$

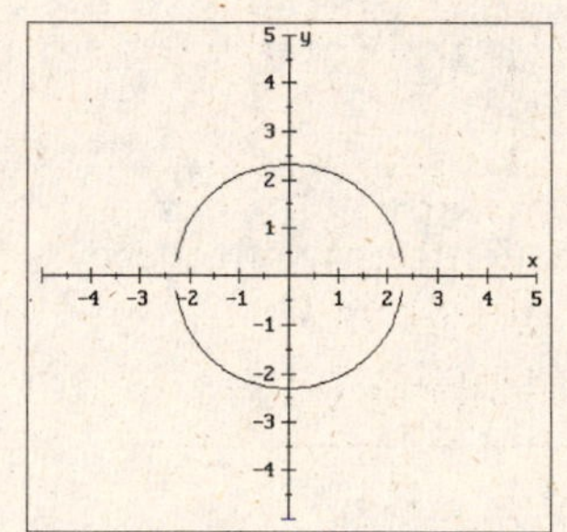

61. $(x + 1)^2 + y^2 = 16$

$$y^2 = 16 - (x + 1)^2$$

$$y = \pm\sqrt{16 - (x + 1)^2}$$

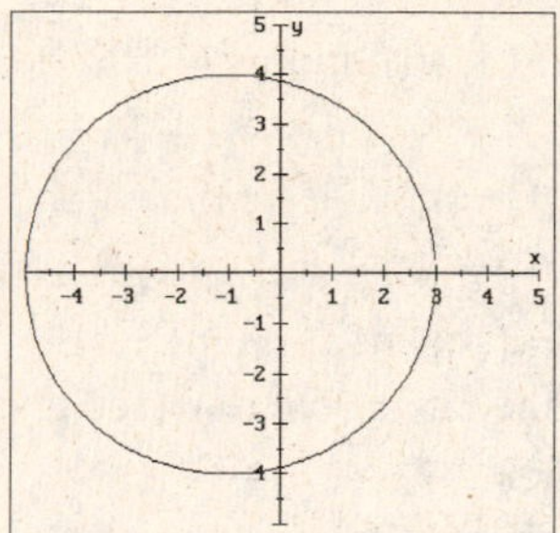

63. $x = 2y^2$

$$y^2 = \frac{x}{2}$$

$$y = \pm\sqrt{\frac{x}{2}}$$

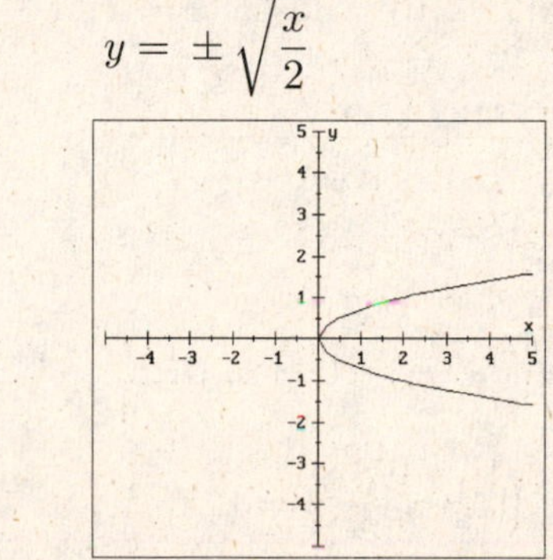

65. $x^2 - 2x + y = 6$

$$y = 6 - x^2 + 2x$$

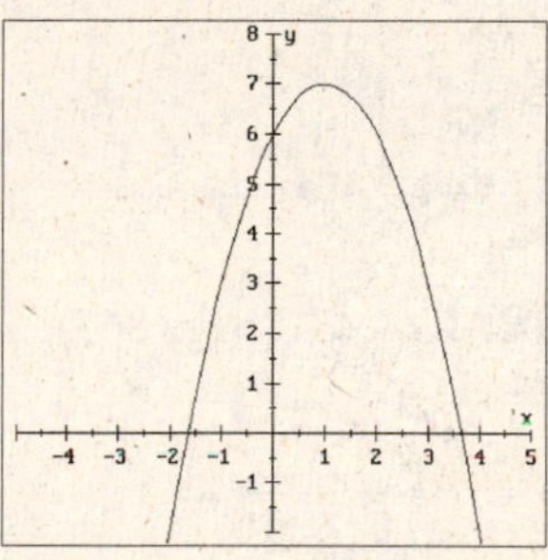

67. The radius of the larger gear is $\sqrt{16} = 4$. Centers: 7 units apart $\Rightarrow$ smaller gear $r = 3$.

$$(x - h)^2 + (y - k)^2 = r^2$$

$$(x - 7)^2 + (y - 0)^2 = 3^2 \Rightarrow (x - 7)^2 + y^2 = 9$$

69.
$$x^2 + y^2 - 8x - 20y + 16 = 0$$
$$x^2 - 8x + y^2 - 20y = -16$$
$$x^2 - 8x + 16 + y^2 - 20y + 100 = -16 + 16 + 100$$
$$(x-4)^2 + (y-10)^2 = 100$$
center: $(4, 10)$; radius $= 10$
$$x^2 + y^2 + 2x + 4y - 11 = 0$$
$$x^2 + 2x + 1 + y^2 + 4y + 4 = 11 + 1 + 4$$
$$(x+1)^2 + (y+2)^2 = 16$$
center: $(-1, -2)$; radius $= 4$

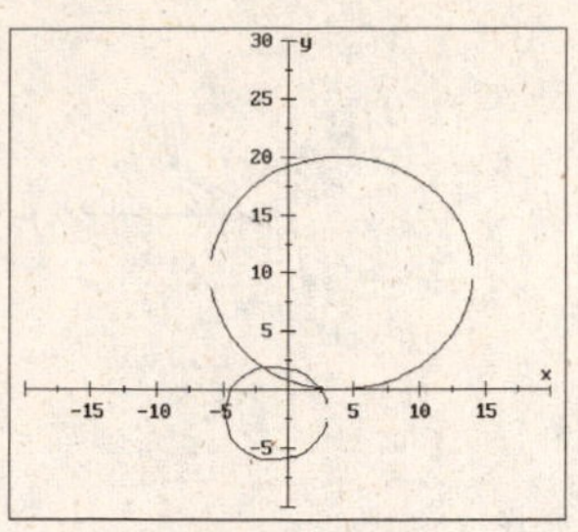

Since the ranges overlap (see graph), they can not be licensed for the same frequency.

71. Set $y = 0$:
$y = 30x - x^2$
$0 = 30x - x^2$
$0 = x(30 - x)$
$x = 0$ or $x = 30$
It lands 30 feet away.

73. Find the vertex:
$2y^2 - 9x = 18$
$$-9x = -2y^2 + 18$$
$$x = \frac{2}{9}y^2 - 2$$
$$x = \frac{2}{9}(y-0)^2 - 2$$
vertex: $(-2, 0) \Rightarrow$ distance $= 2$ AU

75. Answers may vary.

77. Answers may vary.

Exercise 12.2 (page 865)

1.
$$\frac{x^2}{9} + \frac{y^2}{16} = 1$$

x-intercepts	y-intercepts
$\frac{x^2}{9} + \frac{0^2}{16} = 1$	$\frac{0^2}{9} + \frac{y^2}{16} = 1$
$\frac{x^2}{9} = 1$	$\frac{y^2}{16} = 1$
$x^2 = 9$	$y^2 = 16$
$x = \pm 3$	$y = \pm 4$
$(3, 0), (-3, 0)$	$(0, 4), (0, -4)$

3. $\frac{(x-2)^2}{9} + \frac{y^2}{16} = 1$; C $(2, 0)$

5.
$$3x^{-2}y^2(4x^2 + 3y^{-2}) = 12x^0y^2 + 9x^{-2}y^0$$
$$= 12y^2 + \frac{9}{x^2}$$

7.
$$\frac{x^{-2} + y^{-2}}{x^{-2} - y^{-2}} = \frac{x^{-2} + y^{-2}}{x^{-2} - y^{-2}} \cdot \frac{x^2y^2}{x^2y^2} = \frac{y^2 + x^2}{y^2 - x^2}$$

9. ellipse; sum

11. center

13. $(0, 0)$; major axis; $2b$

15. $\frac{x^2}{4}+\frac{y^2}{9}=1$
C $(0,0)$; move 2 horiz. and 3 vert.

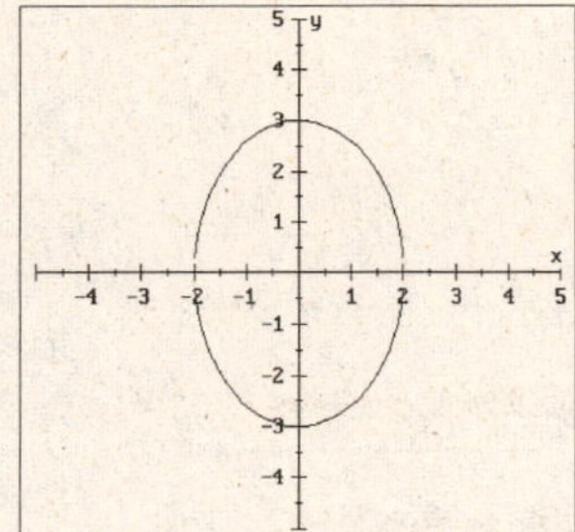

17. $\frac{x^2}{9}+\frac{y^2}{16}=1$
C $(0,0)$; move 3 horiz. and 4 vert.

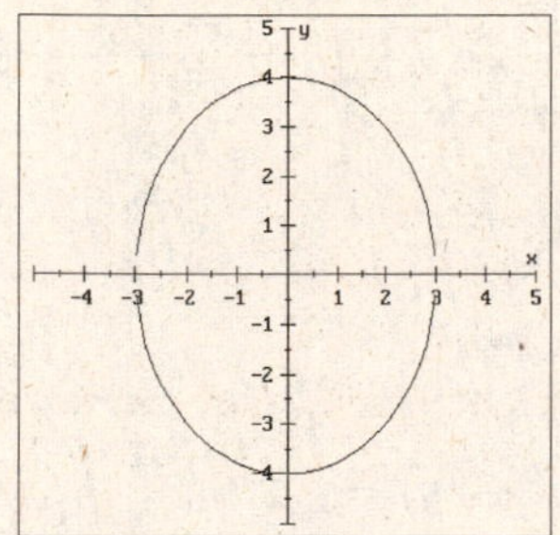

19. $\frac{(x-2)^2}{16}+\frac{y^2}{25}=1$
C $(2,0)$; move 4 horiz. and 5 vert.

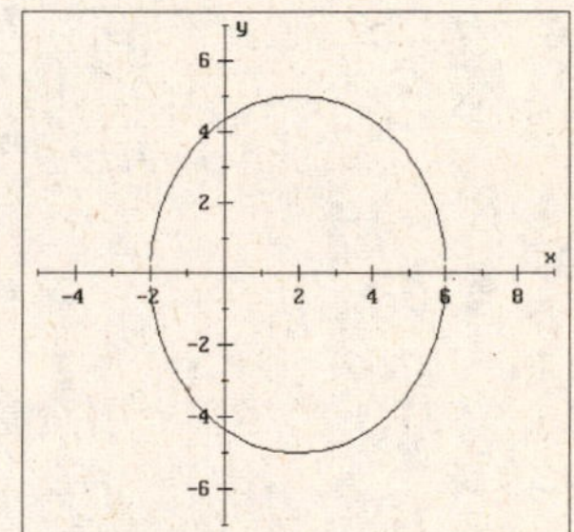

21. $\frac{(x-2)^2}{9}+\frac{(y-1)^2}{4}=1$
C $(2,1)$; move 3 horiz. and 2 vert.

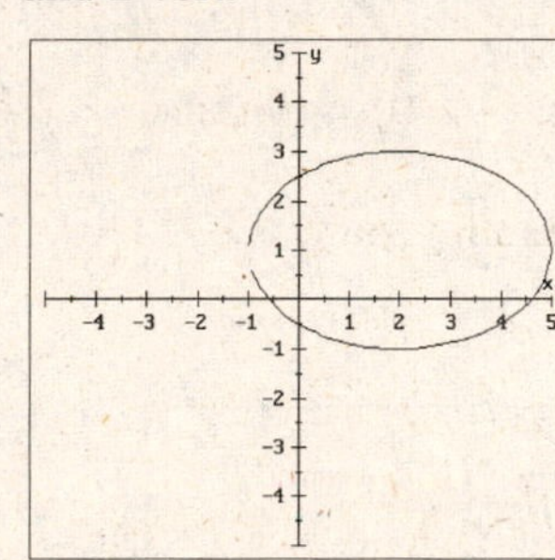

23. $x^2+9y^2=9$

$$\frac{x^2}{9}+\frac{9y^2}{9}=\frac{9}{9}$$

$$\frac{x^2}{9}+\frac{y^2}{1}=1$$

C $(0,0)$; move 3 horiz. and 1 vert.

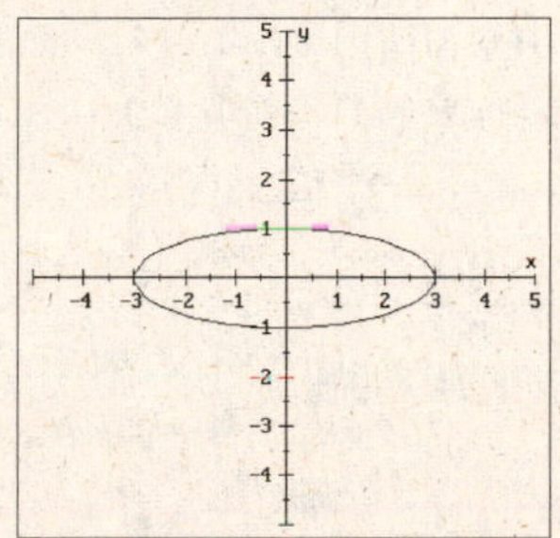

25. $16x^2+4y^2=64$

$$\frac{16x^2}{64}+\frac{4y^2}{64}=\frac{64}{64}$$

$$\frac{x^2}{4}+\frac{y^2}{16}=1$$

C $(0,0)$; move 2 horiz. and 4 vert.

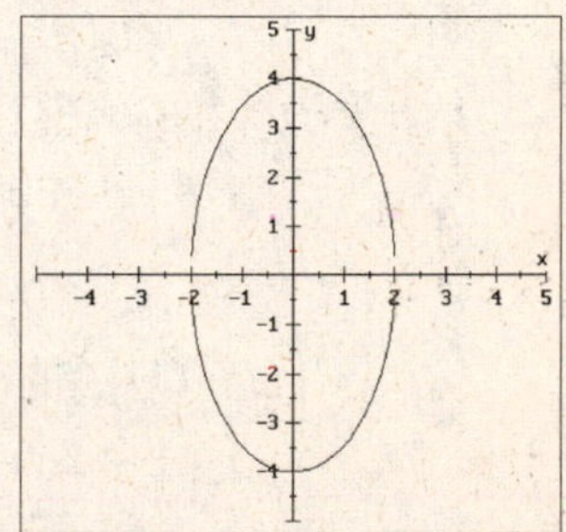

27. $(x+1)^2+4(y+2)^2=4$

$$\frac{(x+1)^2}{4}+\frac{4(y+2)^2}{4}=\frac{4}{4}$$

$$\frac{(x+1)^2}{4}+\frac{(y+2)^2}{1}=1$$

C $(-1,-2)$; move 2 **horiz.** and 1 **vert.**

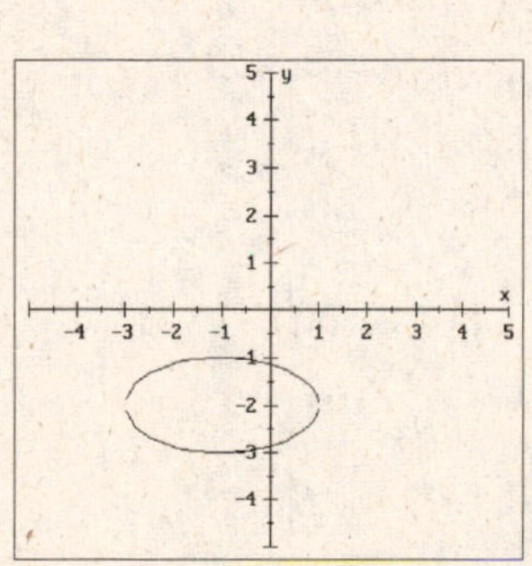

29. $25(x+1)^2 + 9y^2 = 225$

$$\frac{25(x+1)^2}{225} + \frac{9y^2}{225} = \frac{225}{225}$$

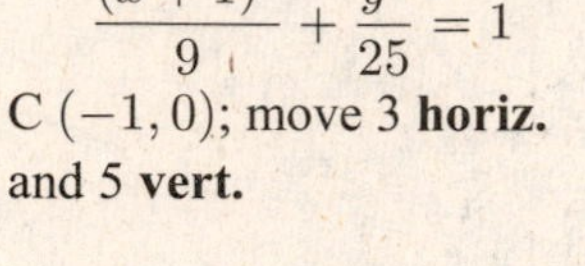

$$\frac{(x+1)^2}{9} + \frac{y^2}{25} = 1$$

C $(-1, 0)$; move 3 **horiz.** and 5 **vert.**

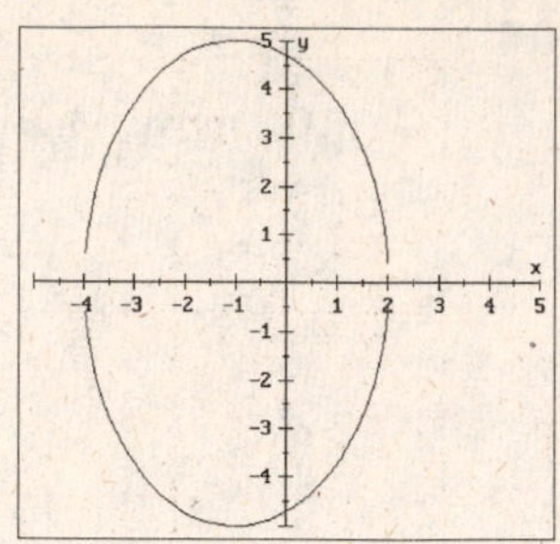

31.

$$x^2 + 4y^2 - 4x + 8y + 4 = 0$$
$$x^2 - 4x + 4(y^2 + 2y) = -4$$
$$x^2 - 4x + 4 + 4(y^2 + 2y + 1) = -4 + 4 + 4$$
$$(x-2)^2 + 4(y+1)^2 = 4$$
$$\frac{(x-2)^2}{4} + \frac{(y+1)^2}{1} = 1$$

C $(2, -1)$; move 2 horiz. and 1 vert.

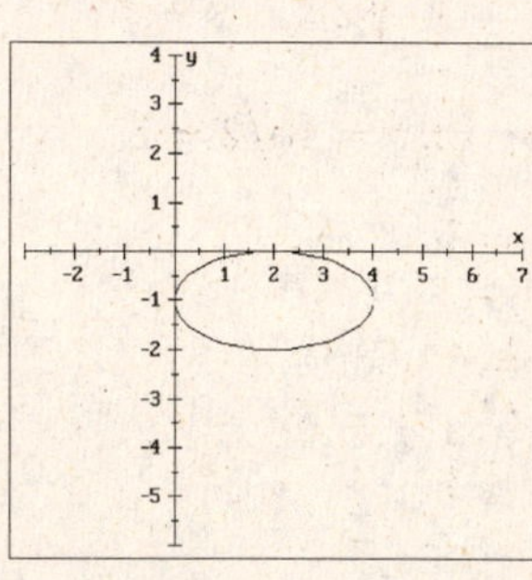

33.

$$9x^2 + 4y^2 - 18x + 16y = 11$$
$$9(x^2 - 2x) + 4(y^2 + 4y) = 11$$
$$9(x^2 - 2x + 1) + 4(y^2 + 4y + 4) = 11 + 9 + 16$$
$$9(x-1)^2 + 4(y+2)^2 = 36$$
$$\frac{(x-1)^2}{4} + \frac{(y+2)^2}{9} = 1$$

C $(1, -2)$; move 2 horiz. and 3 vert.

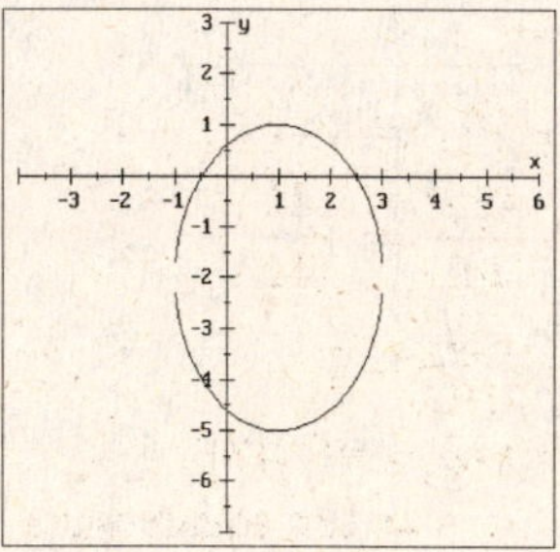

35. $\frac{x^2}{9} + \frac{y^2}{4} = 1$

$$\frac{y^2}{4} = 1 - \frac{x^2}{9}$$
$$y^2 = 4\left(1 - \frac{x^2}{9}\right)$$
$$y = \pm\sqrt{4\left(1 - \frac{x^2}{9}\right)}$$

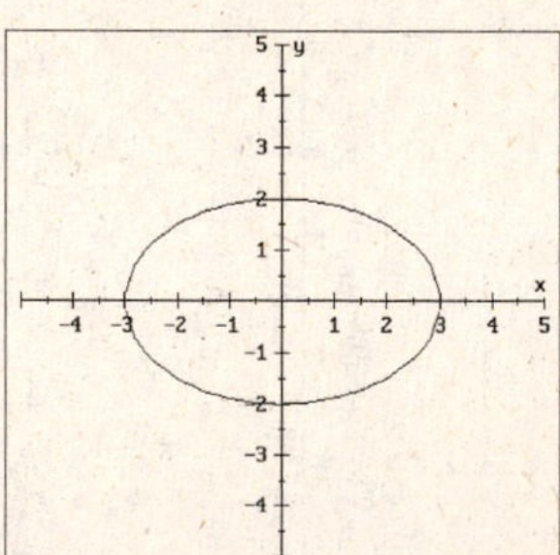

37.
$$\frac{x^2}{4} + \frac{(y-1)^2}{9} = 1$$
$$\frac{(y-1)^2}{9} = 1 - \frac{x^2}{4}$$
$$(y-1)^2 = 9\left(1 - \frac{x^2}{4}\right)$$
$$y - 1 = \pm\sqrt{9\left(1 - \frac{x^2}{4}\right)}$$
$$y = 1 \pm \sqrt{9\left(1 - \frac{x^2}{4}\right)}$$

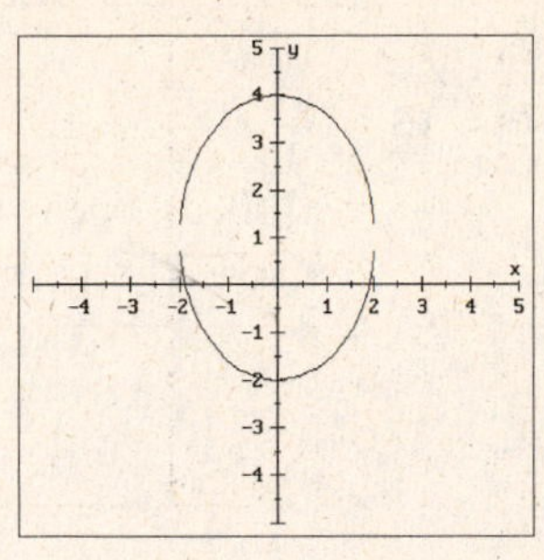

39. $a = 24/2 = 12, b = 10/2 = 5$
$$\frac{x^2}{a^2} + \frac{y^2}{b^2} = 1$$
$$\frac{x^2}{12^2} + \frac{y^2}{5^2} = 1$$
$$\frac{x^2}{144} + \frac{y^2}{25} = 1$$

41. Note: $a = 40/2 = 20, b = 10$
$$\frac{x^2}{a^2} + \frac{y^2}{b^2} = 1$$
$$\frac{x^2}{400} + \frac{y^2}{100} = 1$$
$$x^2 + 4y^2 = 400$$
$$4y^2 = 400 - x^2$$
$$y^2 = \tfrac{1}{4}(400 - x^2)$$
$$y = \tfrac{1}{2}\sqrt{400 - x^2}$$

43.
$$9x^2 + 16y^2 = 144$$
$$\frac{9x^2}{144} + \frac{16y^2}{144} = \frac{144}{144}$$
$$\frac{x^2}{16} + \frac{y^2}{9} = 1$$
$$\text{area} = \pi ab = \pi(4)(3) = 12\pi \text{ square units}$$

45. **Answers may vary.**

47. It is a circle.

Exercise 12.3 (page 877)

1.
$$\frac{x^2}{9} - \frac{y^2}{16} = 1$$

x-intercepts
$$\frac{x^2}{9} - \frac{0^2}{16} = 1$$
$$\frac{x^2}{9} = 1$$
$$x^2 = 9$$
$$x = \pm 3$$
$(3, 0), (-3, 0)$

y-intercepts
$$\frac{0^2}{9} - \frac{y^2}{16} = 1$$
$$-\frac{y^2}{16} = 1$$
$$y^2 = -16$$
no y-intercepts

3. $-6x^4 + 9x^3 - 6x^2 = -3x^2(2x^2 - 3x + 2)$

5. $15a^2 - 4ab - 4b^2 = (3a - 2b)(5a + 2b)$

SECTION 12.3

7. hyperbola; difference

9. center

11. $(\pm a, 0)$; y-intercepts

13. $\frac{x^2}{9} - \frac{y^2}{4} = 1$
C $(0,0)$; open horiz.;
move 3 horiz. and 2 vert.

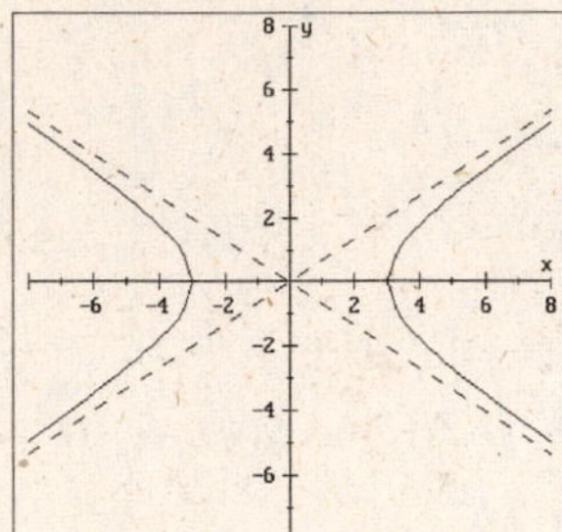

15. $\frac{y^2}{4} - \frac{x^2}{9} = 1$
C $(0,0)$; open vert.;
move 3 horiz. and 2 vert.

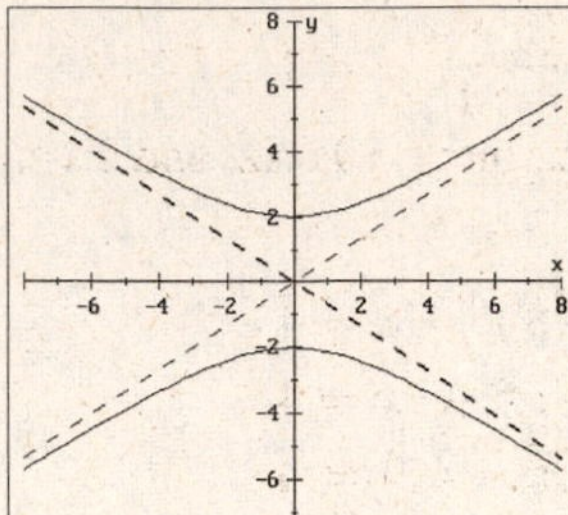

17. $25x^2 - y^2 = 25$
$\frac{25x^2}{25} - \frac{y^2}{25} = 1$
$\frac{x^2}{1} - \frac{y^2}{25} = 1$
C $(0,0)$; open horiz.;
move 1 horiz. and 5 vert.

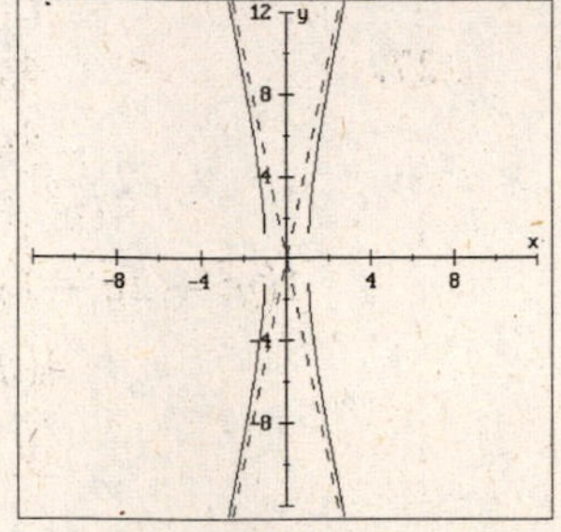

19. $\frac{(x-2)^2}{9} - \frac{y^2}{16} = 1$
C $(2,0)$; open horiz.;
move 3 horiz. and 4 vert.

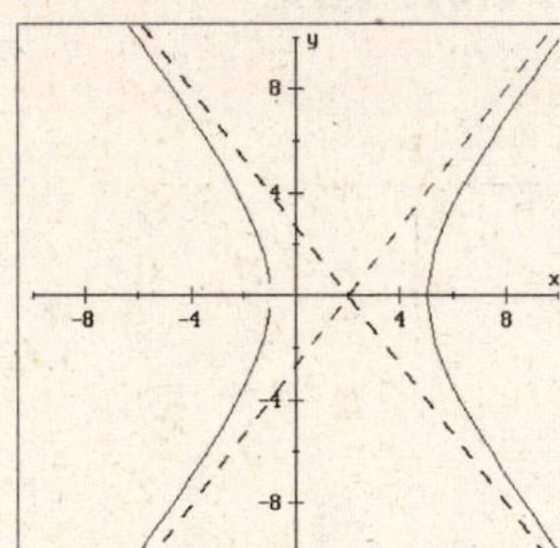

21. $\frac{(y+1)^2}{1} - \frac{(x-2)^2}{4} = 1$
C $(2,-1)$; open vert.;
move 2 horiz. and 1 vert.

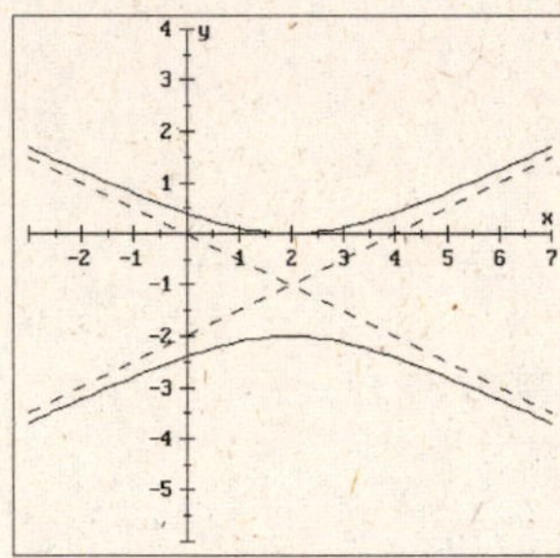

23. $4(x+3)^2 - (y-1)^2 = 4$
$\frac{4(x+3)^2}{4} - \frac{(y-1)^2}{4} = \frac{4}{4}$
$\frac{(x+3)^2}{1} - \frac{(y-1)^2}{4} = 1$
C $(-3,1)$; open horiz.;
move 1 horiz. and 2 vert.

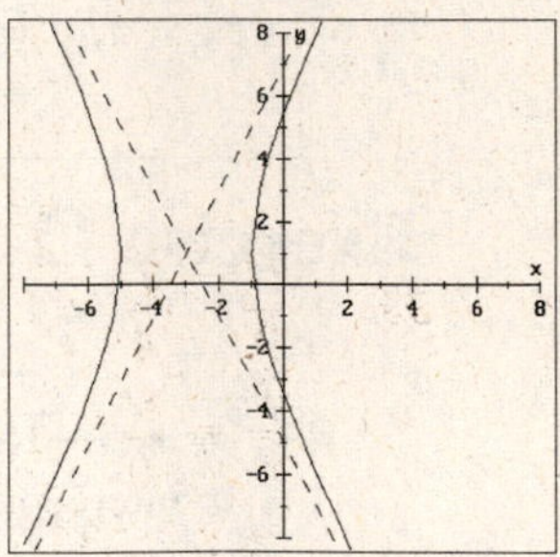

25.

$$\begin{aligned} 4x^2 - y^2 + 8x - 4y &= 4 \\ 4x^2 + 8x - y^2 - 4y &= 4 \\ 4(x^2 + 2x) - (y^2 + 4y) &= 4 \\ 4(x^2 + 2x + 1) - (y^2 + 4y + 4) &= 4 + 4 - 4 \\ 4(x+1)^2 - (y+2)^2 &= 4 \\ \frac{(x+1)^2}{1} - \frac{(y+2)^2}{4} &= 1 \end{aligned}$$

C $(-1, -2)$; opens horiz.; move 1 horiz. and 2 vert.

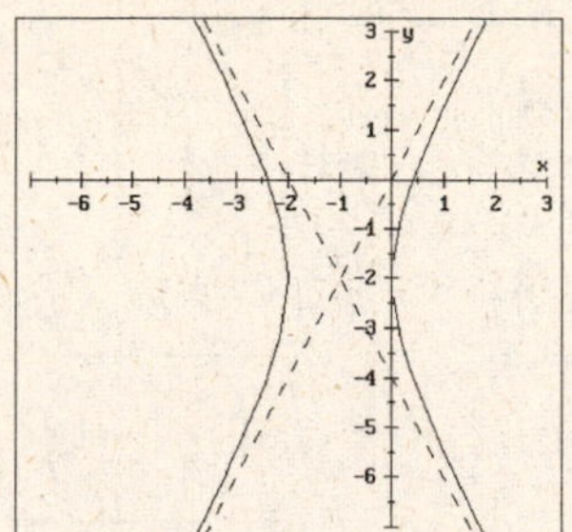

27.

$$\begin{aligned} 4y^2 - x^2 + 8y + 4x &= 4 \\ 4y^2 + 8y - x^2 + 4x &= 4 \\ 4(y^2 + 2y) - (x^2 - 4x) &= 4 \\ 4(y^2 + 2y + 1) - (x^2 - 4x + 4) &= 4 + 4 - 4 \\ 4(y+1)^2 - (x-2)^2 &= 4 \\ \frac{(y+1)^2}{1} - \frac{(x-2)^2}{4} &= 1 \end{aligned}$$

C $(2, -1)$; opens vert.; move 2 horiz. and 1 vert.

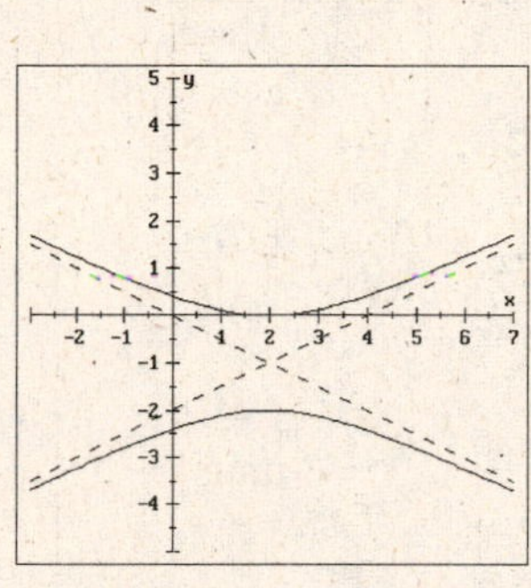

29. $xy = 8$

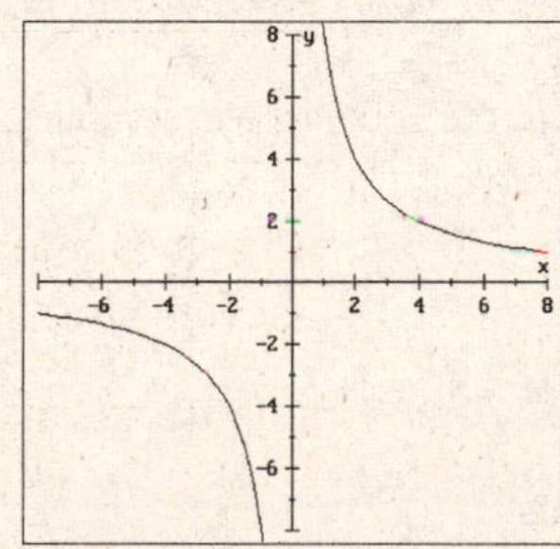

31. $xy = -12$

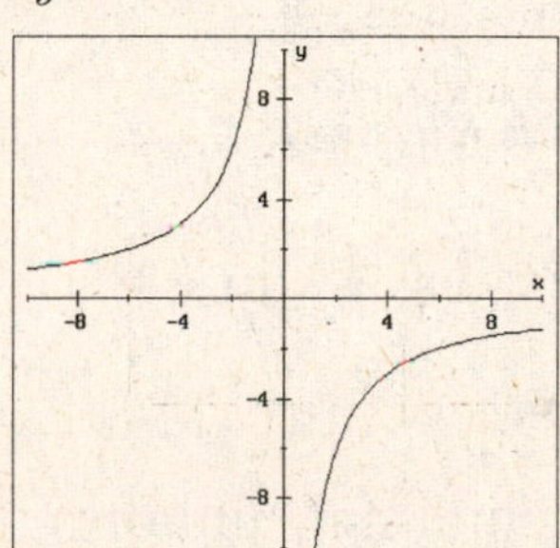

33.

$$\frac{x^2}{9} - \frac{y^2}{4} = 1$$

$$\frac{y^2}{4} = \frac{x^2}{9} - 1$$

$$y^2 = 4\left(\frac{x^2}{9} - 1\right)$$

$$y = \pm\sqrt{4\left(\frac{x^2}{9} - 1\right)}$$

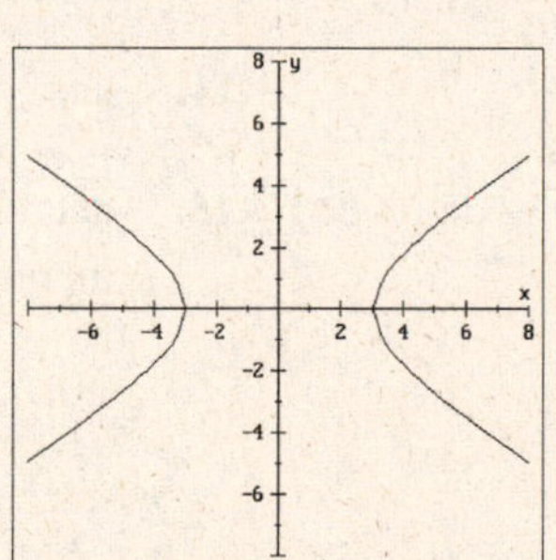

35. $\frac{x^2}{4} - \frac{(y-1)^2}{9} = 1$

$\frac{(y-1)^2}{9} = \frac{x^2}{4} - 1$

$(y-1)^2 = 9\left(\frac{x^2}{4} - 1\right)$

$y - 1 = \pm\sqrt{9\left(\frac{x^2}{4} - 1\right)}$

$y = 1 \pm \sqrt{9\left(\frac{x^2}{4} - 1\right)}$

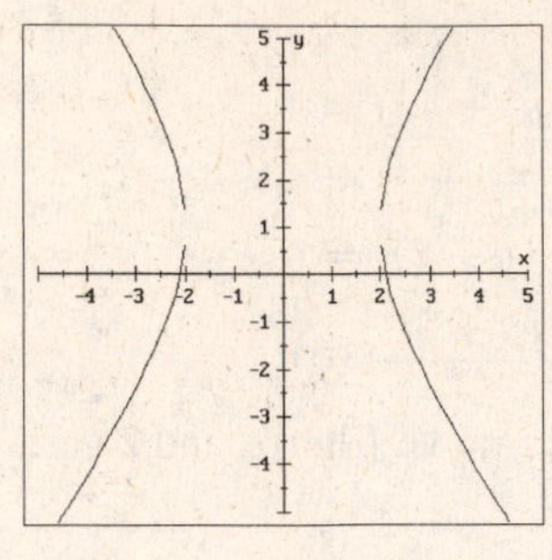

37. $9y^2 - x^2 = 81$

$\frac{9y^2}{81} - \frac{x^2}{81} = \frac{81}{81}$

$\frac{y^2}{9} - \frac{x^2}{81} = 1$

distance $= \sqrt{9} = 3$ units

39. $y^2 - x^2 = 25$

$\frac{y^2}{25} - \frac{x^2}{25} = 1$

vertex: $(0, 5)$

Let $y = 10$:

$10^2 - x^2 = 25$

$-x^2 = -75$

$x = \sqrt{75} = 5\sqrt{3}$

width $= 2(5\sqrt{3}) = 10\sqrt{3}$ miles

41. **Answers may vary.**

43. If $a = b$, the rectangle is a square.

Exercise 12.4 (page 883)

1. increasing

3. constant

5. $(6x - 10) + (3x + 10) = 180$

$9x = 180$

$x = 20$

7. domains

9. constant; $f(x)$

11. step

13. increasing on $(-\infty, 0)$, decreasing on $(0, \infty)$

15. decreasing on $(-\infty, 0)$, constant on $(0, 2)$ increasing on $(2, \infty)$

SECTION 12.4

17. $f(x) = \begin{cases} -1 \text{ if } x \leq 0 \\ x \text{ if } x > 0 \end{cases}$

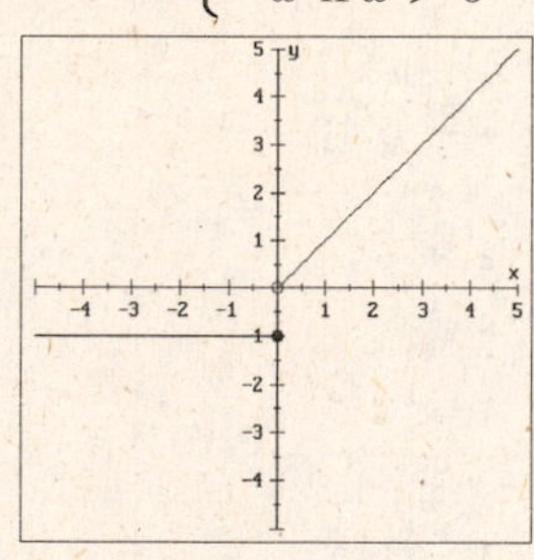

constant on $(-\infty, 0)$
increasing on $(0, \infty)$

19. $f(x) = \begin{cases} -x \text{ if } x \leq 0 \\ x \text{ if } 0 < x < 2 \\ -x \text{ if } x \geq 2 \end{cases}$

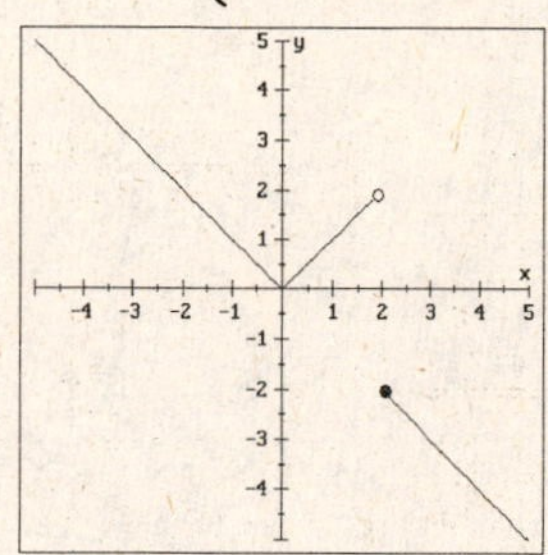

decreasing on $(-\infty, 0)$
increasing on $(0, 2)$
decreasing on $(2, \infty)$

21. $f(x) = -[[x]]$

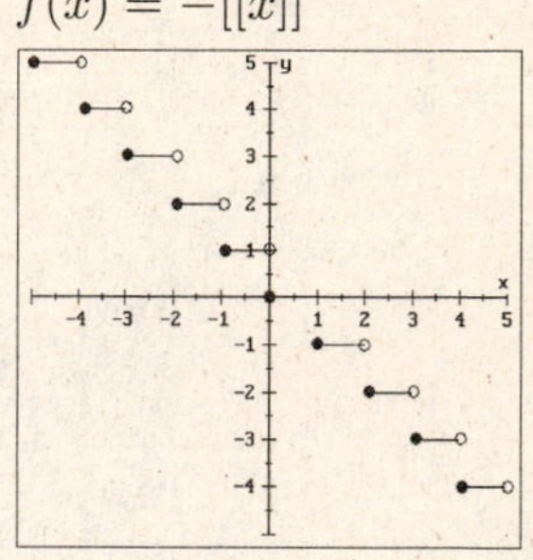

23. $f(x) = 2[[x]]$

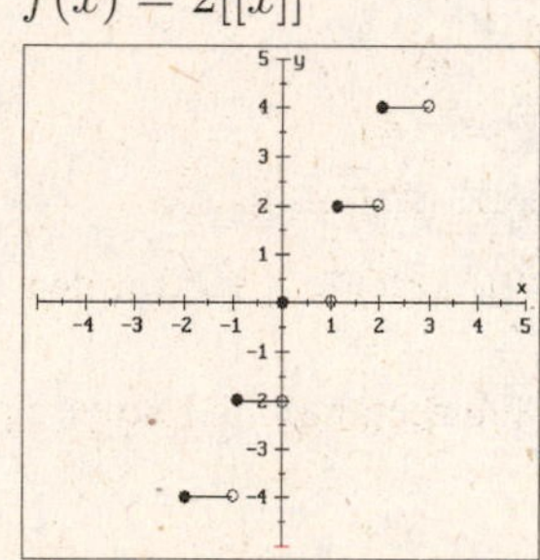

25. $f(x) = \begin{cases} -1 & \text{if } x < 0 \\ 0 & \text{if } x = 0 \\ 1 & \text{if } x > 0 \end{cases}$

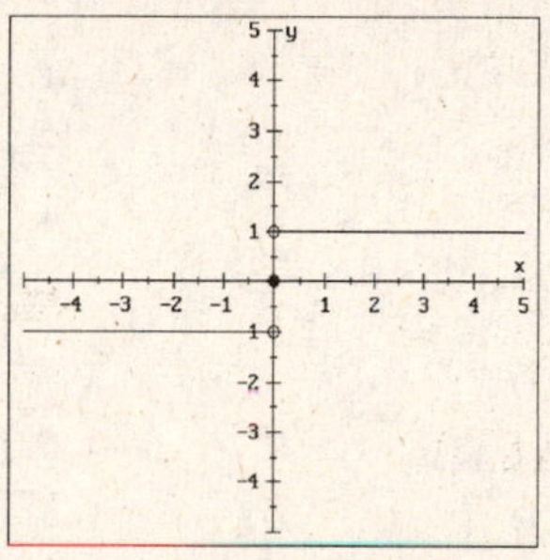

27. Find y when $x = 2.5$.
cost $=$ \$30

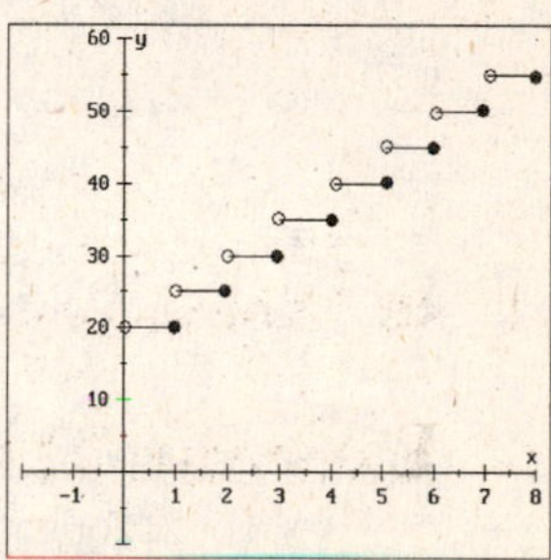

29. After 2 hours, B is cheaper.

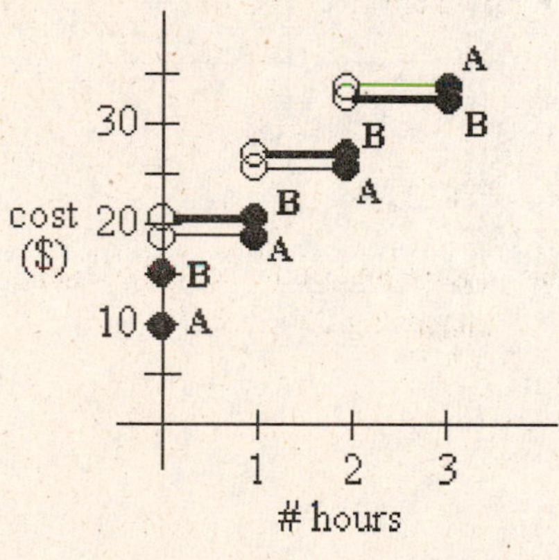

31. **Answers may vary.**

33. $f(x) = \begin{cases} x & \text{if } x < -2 \\ -x & \text{if } x > -2 \end{cases}$

Chapter 12 Review (page 888)

1. $(x-1)^2 + (y+2)^2 = 9$
C $(1, -2)$; $r = \sqrt{9} = 3$

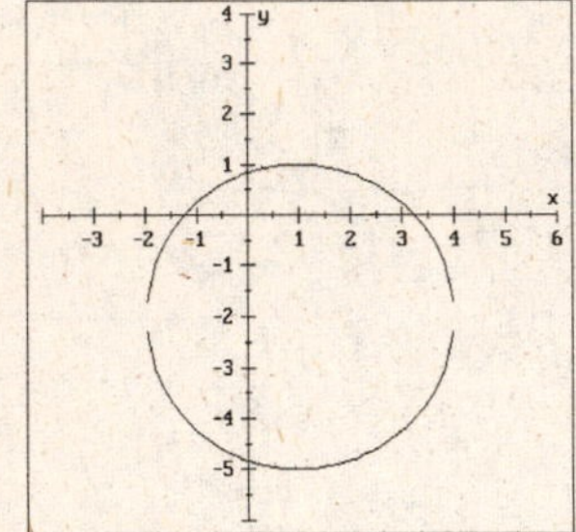

2. $x^2 + y^2 = 16$
C $(0, 0)$; $r = \sqrt{16} = 4$

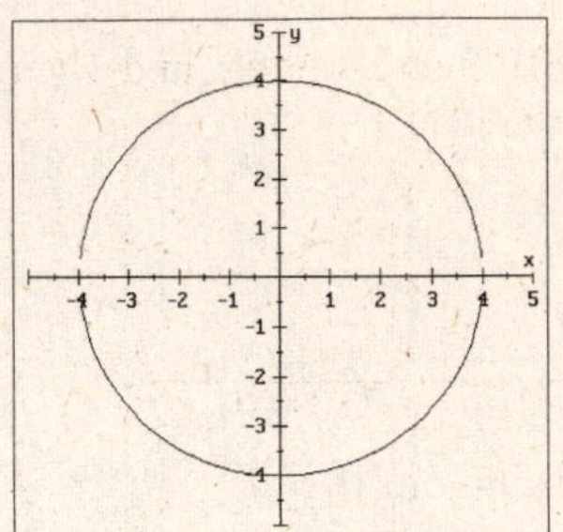

3.
$$x^2 + y^2 + 4x - 2y = 4$$
$$x^2 + 4x + y^2 - 2y = 4$$
$$x^2 + 4x + 4 + y^2 - 2y + 1 = 4 + 4 + 1$$
$$(x+2)^2 + (y-1)^2 = 9$$
C $(-2, 1)$; $r = \sqrt{9} = 3$

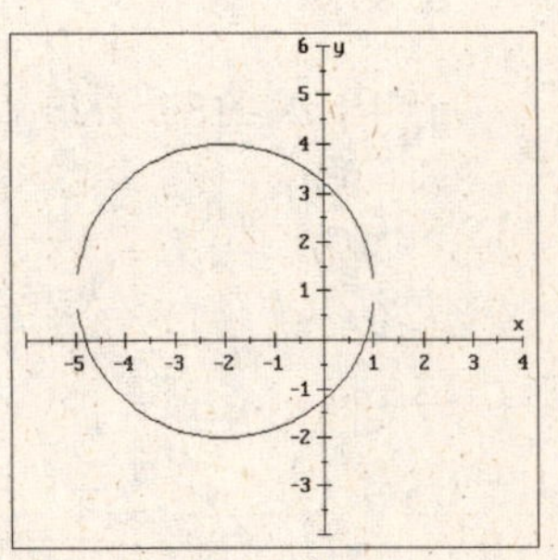

4. $x = -3(y-2)^2 + 5$
V $(5, 2)$; opens L

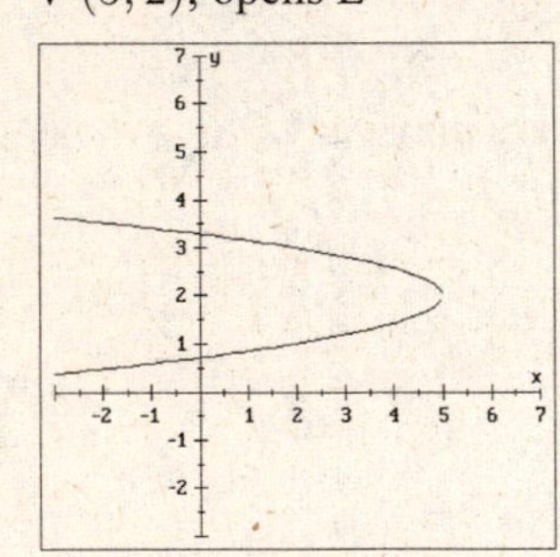

5. $x = 2(y+1)^2 - 2$
V $(-2, -1)$; opens R

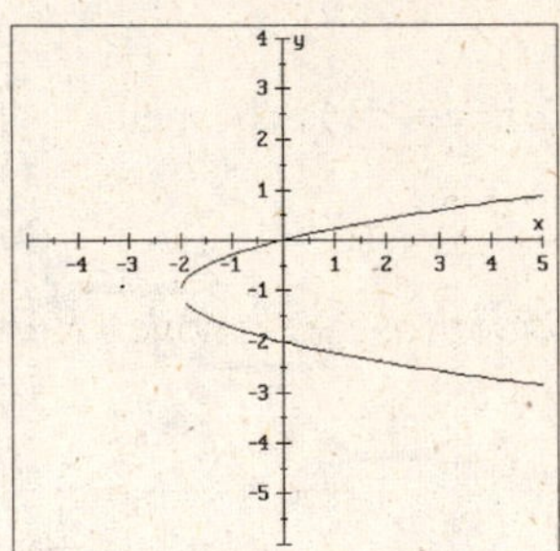

6. $$9x^2 + 16y^2 = 144$$
$$\frac{9x^2}{144} + \frac{16y^2}{144} = \frac{144}{144}$$
$$\frac{x^2}{16} + \frac{y^2}{9} = 1$$
C $(0, 0)$; move 4 horiz. and 3 vert.

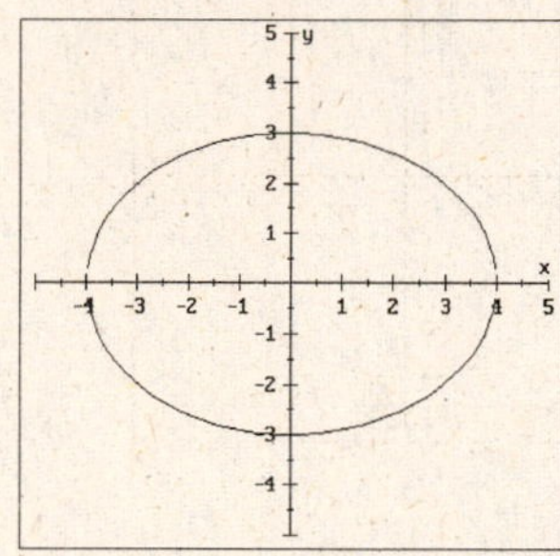

7. $$\frac{(x-2)^2}{4} + \frac{(y-1)^2}{9} = 1$$
C $(2, 1)$; move 2 horiz. and 3 vert.

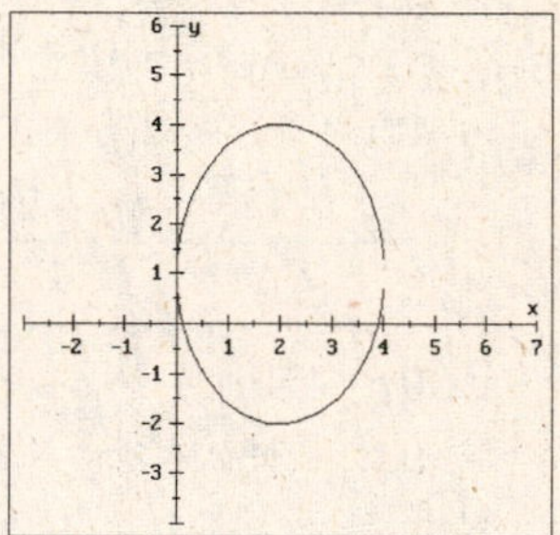

8. $$4x^2 + 9y^2 + 8x - 18y = 23$$
$$4x^2 + 8x + 9y^2 - 18y = 23$$
$$4(x^2 + 2x) + 9(y^2 - 2y) = 23$$
$$4(x^2 + 2x + 1) + 9(y^2 - 2y + 1) = 23 + 4 + 9$$
$$4(x+1)^2 + 9(y-1)^2 = 36$$
$$\frac{(x+1)^2}{9} + \frac{(y-1)^2}{4} = 1$$
C $(-1, 1)$; move 3 horiz. and 2 vert.

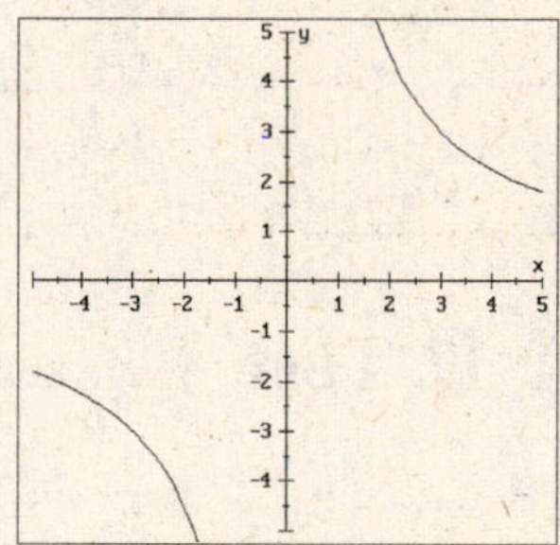

9. $$9x^2 - y^2 = -9$$
$$\frac{9x^2}{-9} - \frac{y^2}{-9} = \frac{-9}{-9}$$
$$\frac{y^2}{9} - \frac{x^2}{1} = 1$$
C $(0, 0)$; opens vert.; move 1 horiz. and 3 vert.

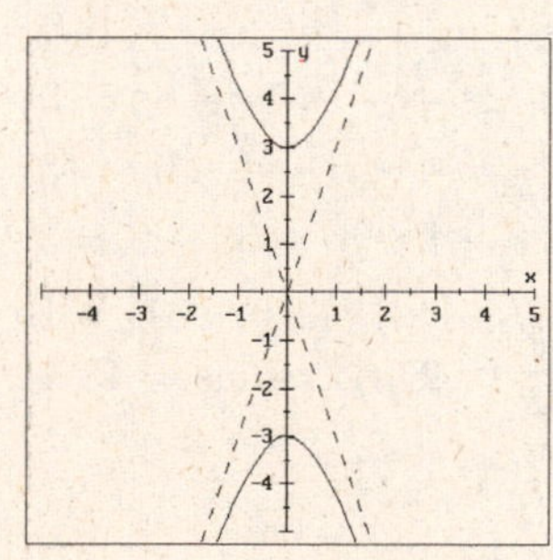

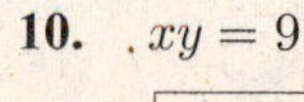

10. $xy = 9$

11.
$$\begin{aligned}
4x^2 - 2y^2 + 8x - 8y &= 8 \\
4x^2 + 8x - 2y^2 - 8y &= 8 \\
4(x^2 + 2x) - 2(y^2 + 4y) &= 8 \\
4(x^2 + 2x + 1) - 2(y^2 + 4y + 4) &= 8 \\
4(x + 1)^2 - 2(y + 2)^2 &= 8 \\
\frac{(x + 1)^2}{2} - \frac{(y + 2)^2}{4} &= 1 \Rightarrow \text{hyperbola}
\end{aligned}$$

12.
$$\begin{aligned}
9x^2 - 4y^2 - 18x - 8y &= 31 \\
9x^2 - 18x - 4y^2 - 8y &= 31 \\
9(x^2 - 2x) - 4(y^2 + 2y) &= 31 \\
9(x^2 - 2x + 1) - 4(y^2 + 2y + 1) &= 31 + 9 - 4 \\
9(x - 1)^2 - 4(y + 1)^2 &= 36 \\
\frac{(x - 1)^2}{4} - \frac{(y + 1)^2}{9} &= 1
\end{aligned}$$
C $(1, -1)$; opens horiz.; move 2 horiz. and 3 vert.

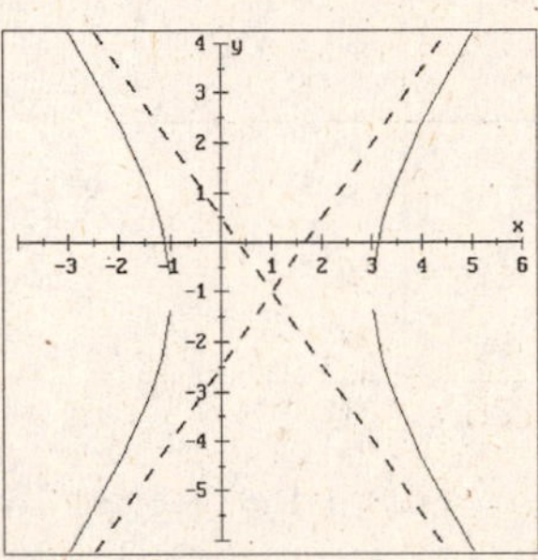

13. increasing on $(-\infty, -2)$; constant on $(-2, 1)$; decreasing on $(1, \infty)$

14. 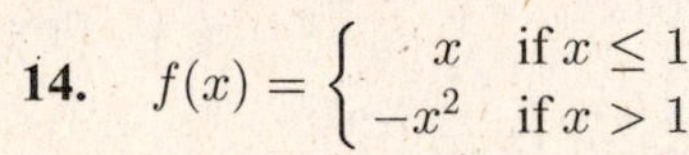
$f(x) = \begin{cases} x & \text{if } x \le 1 \\ -x^2 & \text{if } x > 1 \end{cases}$

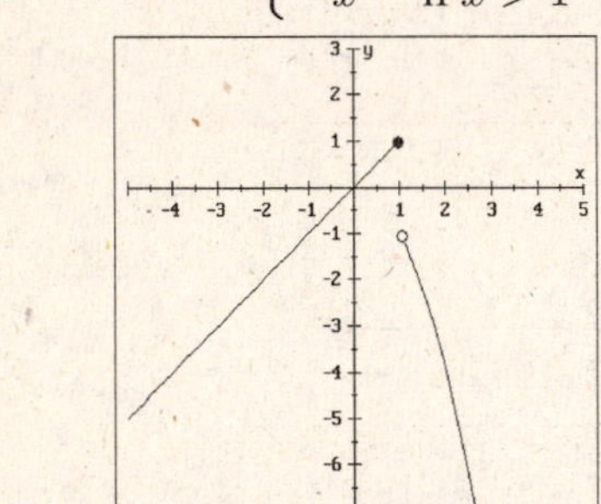

15. $f(x) = 3[[x]]$

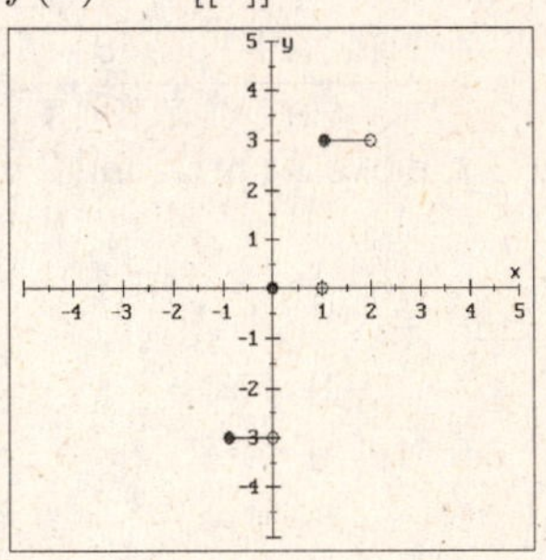

Chapter 12 Test (page 892)

1. $(x - 2)^2 + (y + 3)^2 = 4$
Center: $(2, -3)$; radius $= 2$

2.
$$\begin{aligned}
x^2 + y^2 + 4x - 6y &= 3 \\
x^2 + 4x + y^2 - 6y &= 3 \\
x^2 + 4x + 4 + y^2 - 6y + 9 &= 3 + 4 + 9 \\
(x + 2)^2 + (y - 3)^2 &= 16
\end{aligned}$$
Center: $(-2, 3)$; radius $= 4$

CHAPTER 12 TEST

3. $(x+1)^2+(y-2)^2=9$

C $(-1,2)$; $r=\sqrt{9}=3$

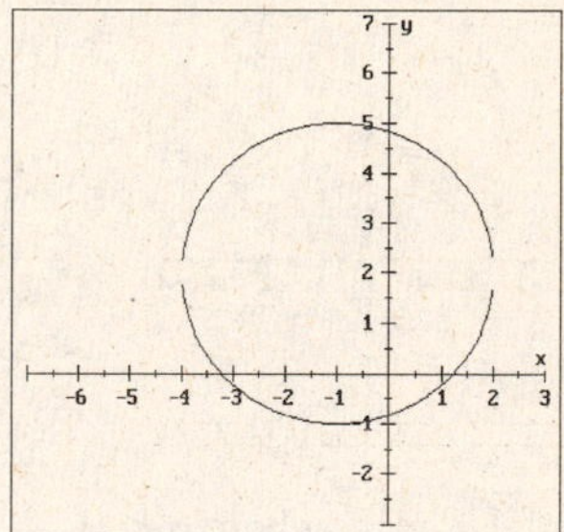

4. $x=(y-2)^2-1$

V $(-1,2)$; opens R

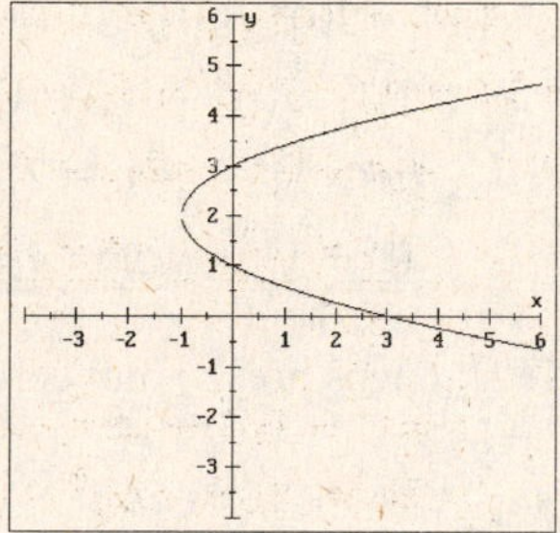

5.
$$9x^2+4y^2=36$$
$$\frac{9x^2}{36}+\frac{4y^2}{36}=\frac{36}{36}$$
$$\frac{x^2}{4}+\frac{y^2}{9}=1$$

C $(0,0)$; move 2 horiz. and 3 vert.

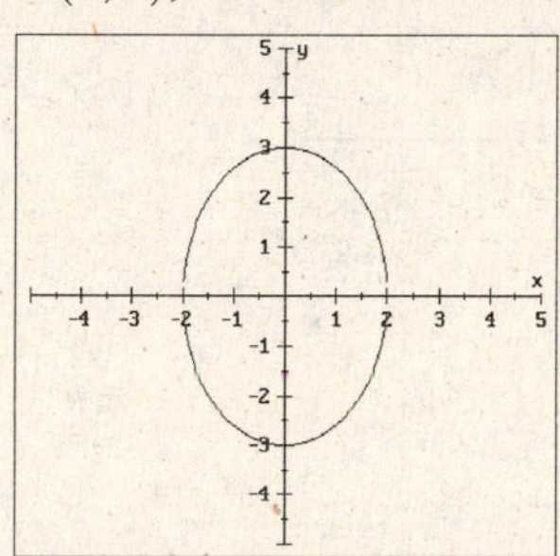

6. $\frac{(x-2)^2}{9}-y^2=1$

C $(2,0)$; opens horiz; move 3 horiz and 1 vert

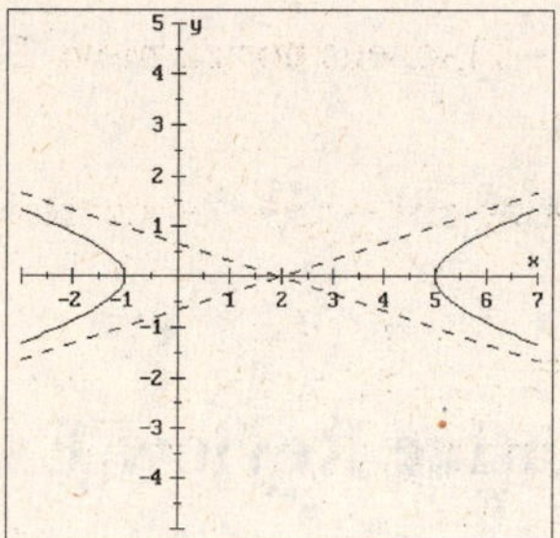

7.
$$4x^2+y^2-24x+2y=-33$$
$$4(x^2-6x)+(y^2+2y)=-33$$
$$4(x^2-6x+9)+(y^2+2y+1)=-33+36+1$$
$$4(x-3)^2+(y+1)^2=4$$
$$\frac{(x-3)^2}{1}+\frac{(y+1)^2}{4}=1$$

C $(3,-1)$; move 1 horiz. and 2 vert.

CHAPTER 12 TEST

8.

$$\begin{aligned}
x^2 - 9y^2 + 2x + 36y &= 44 \\
x^2 + 2x - 9y^2 + 36y &= 44 \\
(x^2 + 2x) - 9(y^2 - 4y) &= 44 \\
(x^2 + 2x + 1) - 9(y^2 - 4y + 4) &= 44 + 1 - 36 \\
(x+1)^2 - 9(y-2)^2 &= 9 \\
\frac{(x+1)^2}{9} - \frac{(y-2)^2}{1} &= 1
\end{aligned}$$

C $(-1, 2)$; opens horiz.; move 3 horiz. and 1 vert.

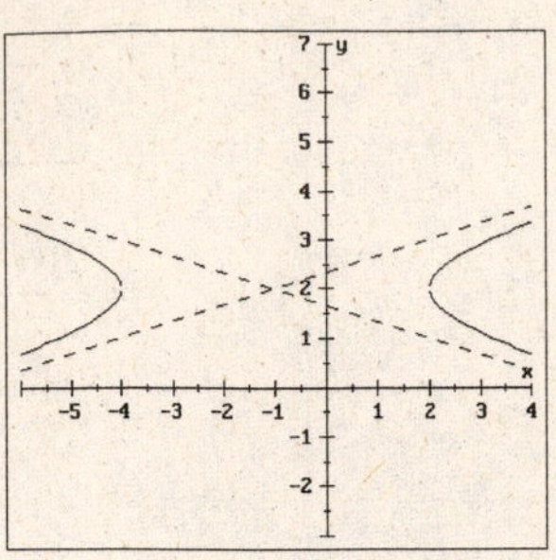

9. increasing: $(-3, 0)$; decreasing: $(0, 3)$

10. 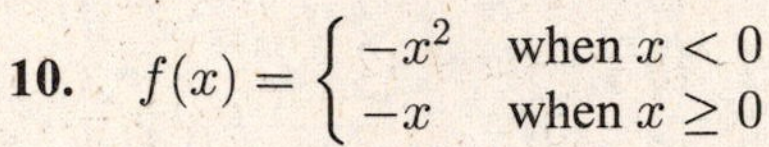

$$f(x) = \begin{cases} -x^2 & \text{when } x < 0 \\ -x & \text{when } x \geq 0 \end{cases}$$

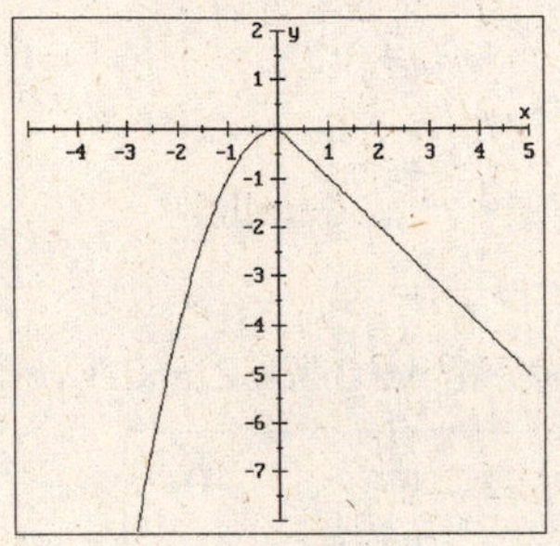

Cumulative Review Exercises (page 893)

1. $(4x - 3y)(3x + y) = 12x^2 + 4xy - 9xy - 3y^2 = 12x^2 - 5xy - 3y^2$

2. $(a^n + 1)(a^n - 3) = a^n a^n - 3a^n + a^n - 3 = a^{2n} - 2a^n - 3$

3. $\dfrac{5a - 10}{a^2 - 4a + 4} = \dfrac{5(a-2)}{(a-2)(a-2)} = \dfrac{5}{a-2}$

4.

$$\begin{aligned}
\frac{a^4 - 5a^2 + 4}{a^2 + 3a + 2} = \frac{(a^2 - 4)(a^2 - 1)}{(a+2)(a+1)} = \frac{(a+2)(a-2)(a+1)(a-1)}{(a+2)(a+1)} &= (a-2)(a-1) \\
&= a^2 - 3a + 2
\end{aligned}$$

5. $\dfrac{a^2 - a - 6}{a^2 - 4} \div \dfrac{a^2 - 9}{a^2 + a - 6} = \dfrac{a^2 - a - 6}{a^2 - 4} \cdot \dfrac{a^2 + a - 6}{a^2 - 9} = \dfrac{(a-3)(a+2)}{(a+2)(a-2)} \cdot \dfrac{(a+3)(a-2)}{(a+3)(a-3)} = 1$

6. $$\frac{2}{a-2}+\frac{3}{a+2}-\frac{a-1}{a^2-4}=\frac{2}{a-2}+\frac{3}{a+2}-\frac{a-1}{(a+2)(a-2)}$$
$$=\frac{2(a+2)}{(a-2)(a+2)}+\frac{3(a-2)}{(a+2)(a-2)}-\frac{a-1}{(a+2)(a-2)}$$
$$=\frac{2(a+2)+3(a-2)-(a-1)}{(a+2)(a-2)}$$
$$=\frac{2a+4+3a-6-a+1}{(a+2)(a-2)}=\frac{4a-1}{(a+2)(a-2)}$$

7. $$\begin{aligned}3x-4y&=12\\ -4y&=-3x+12\\ y&=\frac{3}{4}x-3\\ m&=\frac{3}{4}\end{aligned}\qquad \begin{aligned}y&=\frac{3}{4}x-5\\ m&=\frac{3}{4}\end{aligned}$$

Parallel

8. $$\begin{aligned}y&=3x+4\\ m&=3\end{aligned}\qquad \begin{aligned}x&=-3y+4\\ 3y&=-x+4\\ y&=-\frac{1}{3}x+\frac{4}{3}\\ m&=-\frac{1}{3}\end{aligned}$$

Perpendicular

9. $$\begin{aligned}y-y_1&=m(x-x_1)\\ y-5&=-2(x-0)\\ y-5&=-2x\\ y&=-2x+5\end{aligned}$$

10. $$m=\frac{y_2-y_1}{x_2-x_1}=\frac{4-(-5)}{-5-8}=\frac{9}{-13}=-\frac{9}{13}$$
$$\begin{aligned}y-y_1&=m(x-x_1)\\ y-(-5)&=-\frac{9}{13}(x-8)\\ y+5&=-\frac{9}{13}x+\frac{72}{13}\\ y&=-\frac{9}{13}x+\frac{7}{13}\end{aligned}$$

11. $2x-3y<6$

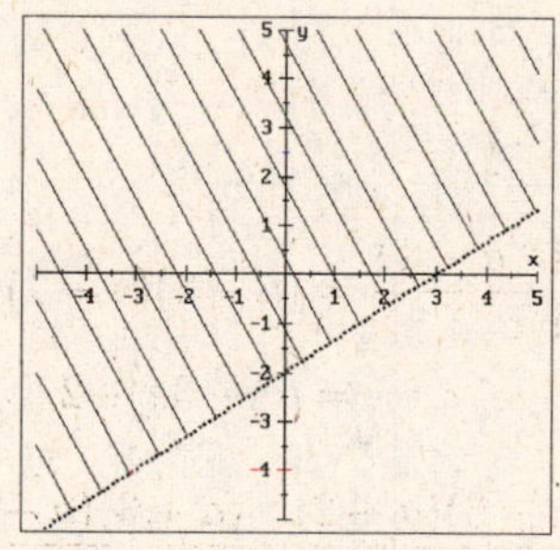

12. $y\geq x^2-4$

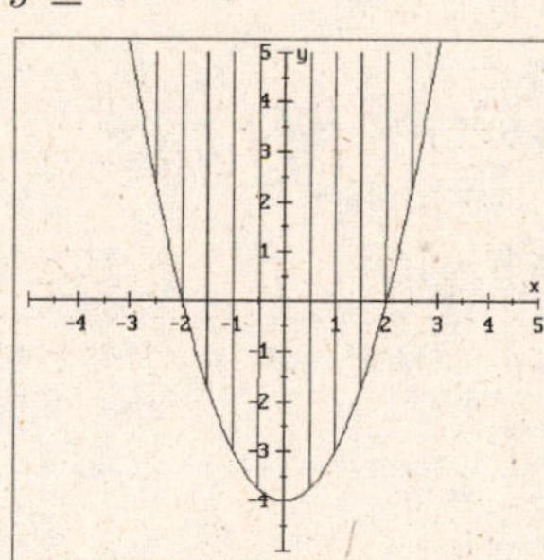

13. $\sqrt{98}+\sqrt{8}-\sqrt{32}=\sqrt{49}\sqrt{2}+\sqrt{4}\sqrt{2}-\sqrt{16}\sqrt{2}=7\sqrt{2}+2\sqrt{2}-4\sqrt{2}=5\sqrt{2}$

14. $$12\sqrt[3]{648x^4}+3\sqrt[3]{81x^4}=12\sqrt[3]{216x^3}\sqrt[3]{3x}+3\sqrt[3]{27x^3}\sqrt[3]{3x}=12(6x)\sqrt[3]{3x}+3(3x)\sqrt[3]{3x}$$
$$=72x\sqrt[3]{3x}+9x\sqrt[3]{3x}=81x\sqrt[3]{3x}$$

15.
$$\begin{aligned}\sqrt{3a+1} &= a-1\\ \left(\sqrt{3a+1}\right)^2 &= (a-1)^2\\ 3a+1 &= a^2-2a+1\\ 0 &= a^2-5a\\ 0 &= a(a-5)\end{aligned}$$
$a=0$ **or** $a-5=0$
extraneous $\quad a=5$

16.
$$\begin{aligned}\sqrt{x+3}-\sqrt{3} &= \sqrt{x}\\ \sqrt{x+3} &= \sqrt{x}+\sqrt{3}\\ \left(\sqrt{x+3}\right)^2 &= \left(\sqrt{x}+\sqrt{3}\right)^2\\ x+3 &= x+2\sqrt{3x}+3\\ 0 &= 2\sqrt{3x}\\ 0^2 &= \left(2\sqrt{3x}\right)^2\\ 0 &= 4(3x)\\ 0 &= 12x\\ 0 &= x\end{aligned}$$

17.
$$\begin{aligned}6a^2+5a-6 &= 0\\ (2a+3)(3a-2) &= 0\end{aligned}$$
$2a+3=0$ **or** $3a-2=0$
$a=-\frac{3}{2} \qquad a=\frac{2}{3}$

18.
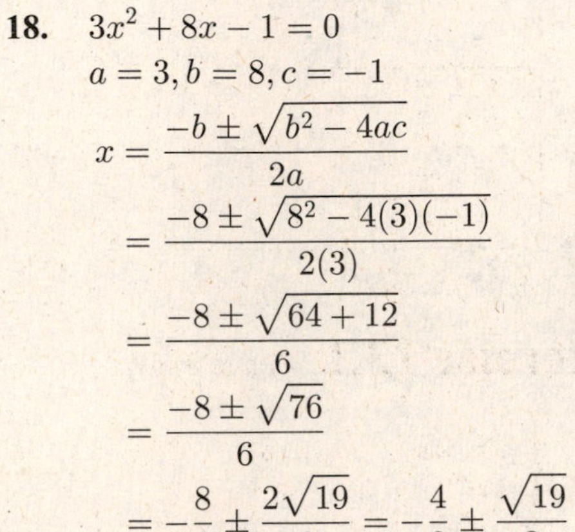
$$3x^2+8x-1=0$$
$$a=3, b=8, c=-1$$
$$\begin{aligned}x &= \frac{-b\pm\sqrt{b^2-4ac}}{2a}\\ &= \frac{-8\pm\sqrt{8^2-4(3)(-1)}}{2(3)}\\ &= \frac{-8\pm\sqrt{64+12}}{6}\\ &= \frac{-8\pm\sqrt{76}}{6}\\ &= -\frac{8}{6}\pm\frac{2\sqrt{19}}{6} = -\frac{4}{3}\pm\frac{\sqrt{19}}{3}\end{aligned}$$

19. $(f\circ g)(x)=f(g(x))=f(2x+1)=(2x+1)^2-2=4x^2+4x+1-2=4x^2+4x-1$

20.
$$\begin{aligned}y &= 2x^3-1\\ x &= 2y^3-1\\ x+1 &= 2y^3\\ \frac{x+1}{2} &= y^3\\ \sqrt[3]{\frac{x+1}{2}} &= y\\ y=f^{-1}(x) &= \sqrt[3]{\frac{x+1}{2}}\end{aligned}$$

21. $y=\left(\frac{1}{2}\right)^x$

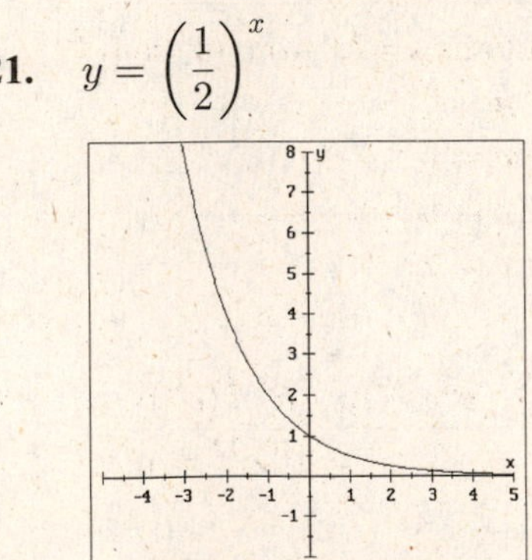

22. $y=\log_2 x \Rightarrow 2^y=x$

23.
$$\begin{aligned} 2^{x+2} &= 3^x \\ \log 2^{x+2} &= \log 3^x \\ (x+2)\log 2 &= x\log 3 \\ x\log 2 + 2\log 2 &= x\log 3 \\ 2\log 2 &= x\log 3 - x\log 2 \\ 2\log 2 &= x(\log 3 - \log 2) \\ \frac{2\log 2}{\log 3 - \log 2} &= x \end{aligned}$$

24.
$$\begin{aligned} 2\log 5 + \log x - \log 4 &= 2 \\ \log 5^2 + \log x - \log 4 &= 2 \\ \log \tfrac{25x}{4} &= 2 \\ 10^2 &= \tfrac{25x}{4} \\ 400 &= 25x \\ 16 &= x \end{aligned}$$

25. $x^2 + (y+1)^2 = 9$

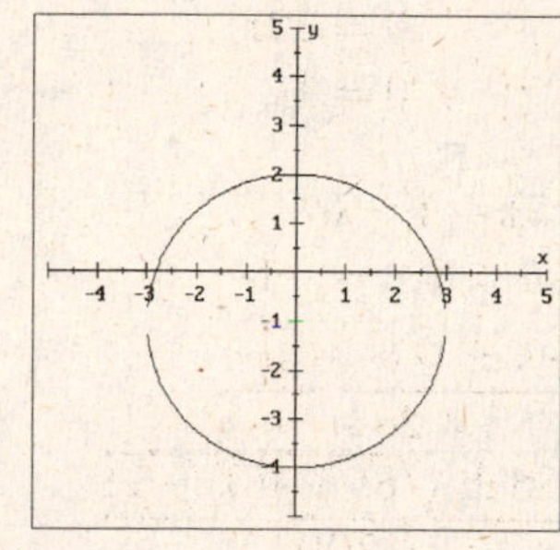

26. $x^2 - 9(y+1)^2 = 9$

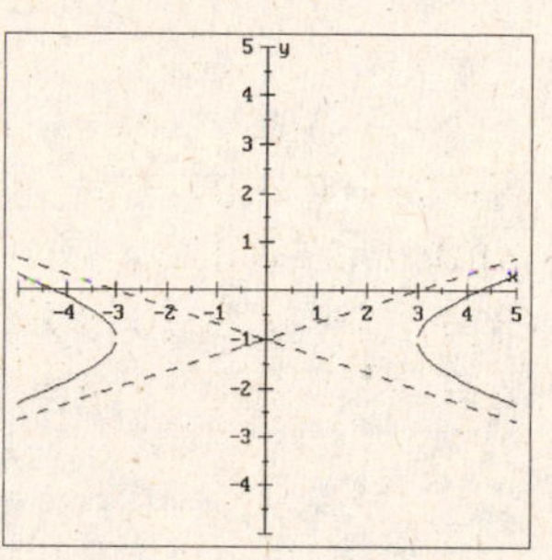

Exercise 13.1 (page 906)

1. $\begin{cases} y = 2x \\ y = 2x + 5 \end{cases}$

parallel lines ⇒ no solution

3. $\begin{cases} y = 2x \\ y = -2x \end{cases}$

intersecting lines ⇒ 1 solution

5. $\begin{cases} (1) \quad y = 2x \\ (2) \quad x + y = 6 \end{cases}$

Substitute $y = 2x$ from (1) into (2):

$$\begin{aligned} x + \boldsymbol{y} &= 6 \\ x + \boldsymbol{2x} &= 6 \\ 3x &= 6 \\ x &= 2 \end{aligned}$$

7.
$$\begin{array}{rl} x - y &= 6 \\ x + y &= 2 \\ \hline 2x \quad &= 8 \\ x \quad &= 4 \end{array}$$

9. $(a^2a^3)^2(a^4a^2)^2 = (a^5)^2(a^6)^2 = a^{10}a^{12} = a^{22}$

11. $\left(\dfrac{-3x^3y^4}{x^{-5}y^3}\right)^{-4} = \left(\dfrac{x^{-5}y^3}{-3x^3y^4}\right)^4 = \left(\dfrac{1}{-3x^8y}\right)^4 = \dfrac{1}{81x^{32}y^4}$

13.
$$\begin{aligned} A &= p + prt \\ A - p &= prt \\ \frac{A-p}{pt} &= r, \text{ or } r = \frac{A-p}{pt} \end{aligned}$$

15.
$$\begin{aligned} \frac{1}{r} &= \frac{1}{r_1} + \frac{1}{r_2} \\ \frac{1}{r} \cdot rr_1r_2 &= \left(\frac{1}{r_1} + \frac{1}{r_2}\right) rr_1r_2 \\ r_1r_2 &= rr_2 + rr_1 \\ r_1r_2 &= r(r_2 + r_1) \\ \frac{r_1r_2}{r_2 + r_1} &= r, \text{ or } r = \frac{r_1r_2}{r_2 + r_1} \end{aligned}$$

17. consistent

19. independent

21. $\begin{cases} x - y = 4 \\ 2x + y = 5 \end{cases}$

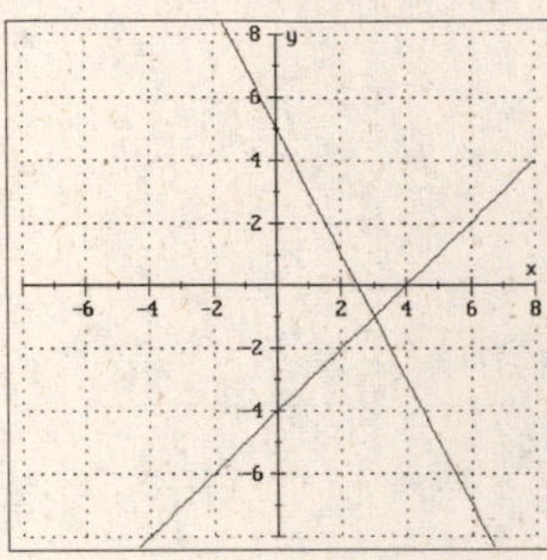

$(3, -1)$ is the solution.

23. $\begin{cases} x = 13 - 4y \\ 3x = 4 + 2y \end{cases}$

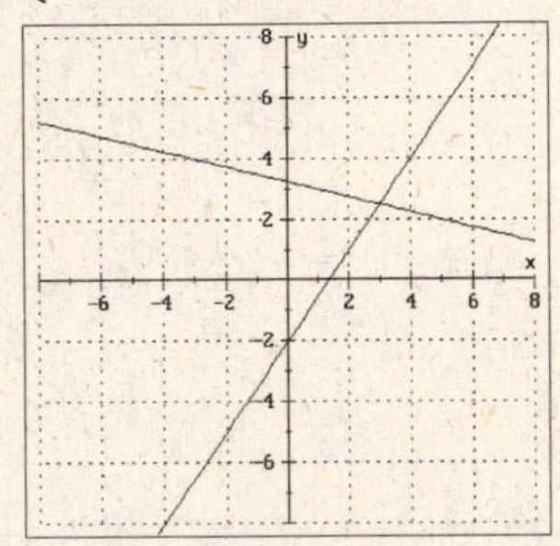

$(3, \frac{5}{2})$ is the solution.

25. $\begin{cases} 2x + 3y = 0 \\ 2x + y = 4 \end{cases}$

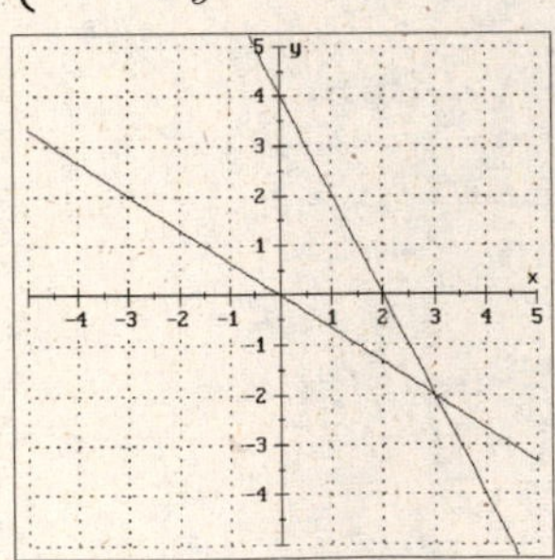

$(3, -2)$ is the solution.

27. $\begin{cases} x = 3 - 2y \\ 2x + 4y = 6 \end{cases}$

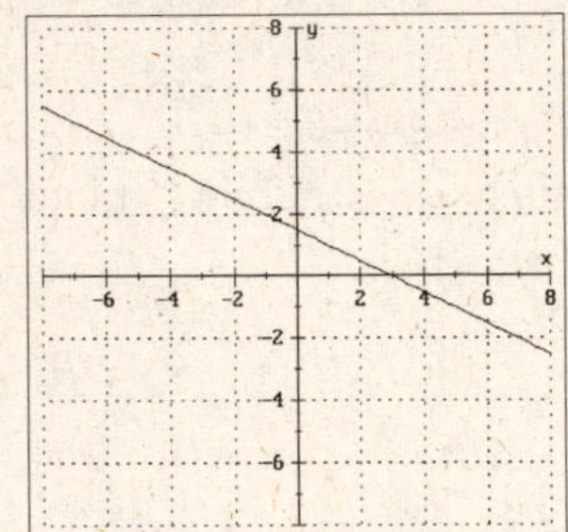

dependent equations
$(x, \frac{3-x}{2})$

29. $\begin{cases} (1) & y = x \\ (2) & x + y = 4 \end{cases}$

Substitute $y = x$ from (1) into (2):

$$x + \boldsymbol{y} = 4$$
$$x + \boldsymbol{x} = 4$$
$$2x = 4$$
$$x = 2$$

Substitute this and solve for y:

$$y = x = 2$$

Solution: $(2, 2)$

31. $\begin{cases} (1) & x - y = 2 \\ (2) & 2x + y = 13 \end{cases}$

Substitute $x = y + 2$ from (1) into (2):

$$2\boldsymbol{x} + y = 13$$
$$2\boldsymbol{(y + 2)} + y = 13$$
$$2y + 4 + y = 13$$
$$3y = 9$$
$$y = 3$$

Substitute this and solve for x:

$$x = y + 2 = 3 + 2 = 5$$

Solution: $(5, 3)$

33. $\begin{cases} (1) & x + 2y = 6 \\ (2) & 3x - y = -10 \end{cases}$

Substitute $x = -2y + 6$ from (1) into (2):

$$3\boldsymbol{x} - y = -10$$
$$3\boldsymbol{(-2y + 6)} - y = -10$$
$$-6y + 18 - y = -10$$
$$-7y = -28$$
$$y = 4$$

Substitute this and solve for x:

$$x = -2y + 6$$
$$x = -2(4) + 6$$
$$x = -8 + 6 = -2$$

Solution: $(-2, 4)$

35. $\begin{cases} (1) & 3x = 2y - 4 \\ (2) & 6x - 4y = -4 \end{cases}$

Substitute $x = \dfrac{2y - 4}{3}$ from (1) into (2):

$$6\boldsymbol{x} - 4y = -4$$
$$6\left(\boldsymbol{\frac{2y - 4}{3}}\right) - 4y = -4$$
$$2(2y - 4) - 4y = -4$$
$$4y - 8 - 4y = -4$$
$$-8 = -4$$

Impossible $\Rightarrow$ no solution $(\emptyset)$

37.

$$\begin{array}{rcr} x - y &=& 3 \\ x + y &=& 7 \\ \hline 2x &=& 10 \\ x &=& 5 \end{array}$$

Substitute and solve for y:

$$x + y = 7$$
$$5 + y = 7$$
$$y = 2$$

The solution is $(5, 2)$.

39.

$$\begin{array}{rcr} 2x + y &=& -10 \\ 2x - y &=& -6 \\ \hline 4x &=& -16 \\ x &=& -4 \end{array}$$

Substitute and solve for y:

$$2x + y = -10$$
$$2(-4) + y = -10$$
$$-8 + y = -10$$
$$y = -2$$

The solution is $(-4, -2)$.

41.

$$\begin{array}{l} 8x - 4y = 16 \Rightarrow 8x - 4y = 16 \Rightarrow \\ 2x - 4 = y \Rightarrow 2x - y = 4 \Rightarrow \times(-4) \end{array} \quad \begin{array}{rcr} 8x - 4y &=& 16 \\ -8x + 4y &=& -16 \\ \hline 0 &=& 0 \end{array} \Rightarrow \boxed{\begin{array}{l}\text{Dependent equations} \\ (x, 2x - 4)\end{array}}$$

43. $x = \frac{3}{2}y + 5 \Rightarrow \times(2)\ 2x = 3y + 10 \Rightarrow 2x - 3y = 10 \Rightarrow \qquad 2x - 3y = 10$

$\underline{2x - 3y = 8} \Rightarrow \qquad \underline{2x - 3y = 8} \Rightarrow \underline{2x - 3y = 8} \Rightarrow \times(-1)\ \underline{-2x + 3y = -8}$

$0 \neq 2 \Rightarrow$ No solution

45. $\begin{cases} y < 3x + 2 \\ y < -2x + 3 \end{cases}$

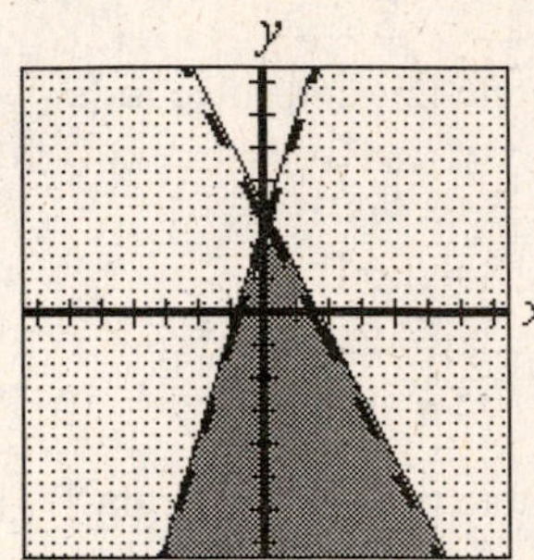

47. $\begin{cases} 3x + 2y > 6 \\ x + 3y \leq 2 \end{cases}$

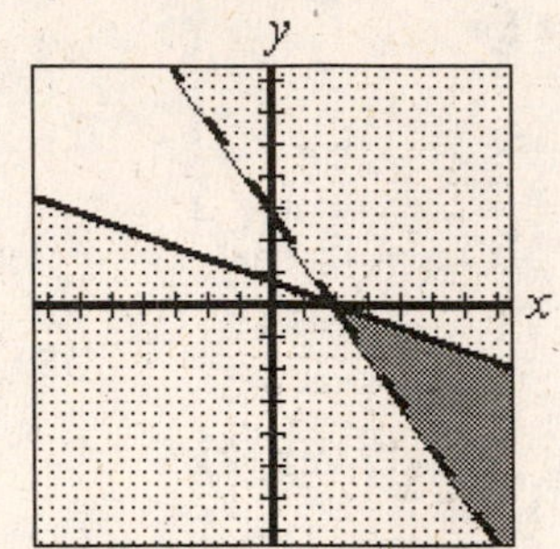

49. $\begin{cases} y = 3 \\ x = 2 \end{cases}$

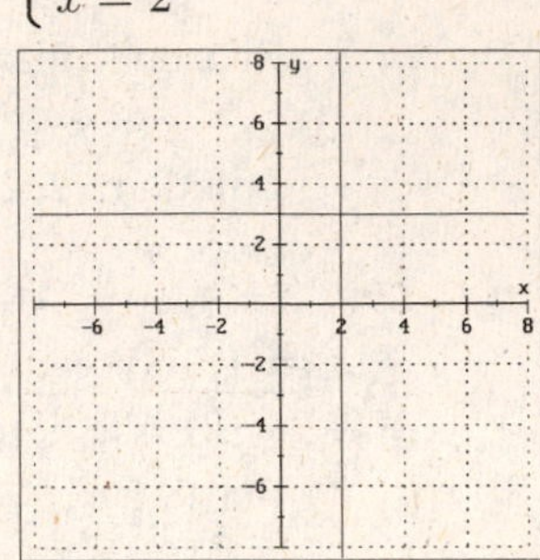

$(2, 3)$ is the solution.

51. $\begin{cases} x = \dfrac{11 - 2y}{3} \\ y = \dfrac{11 - 6x}{4} \end{cases}$

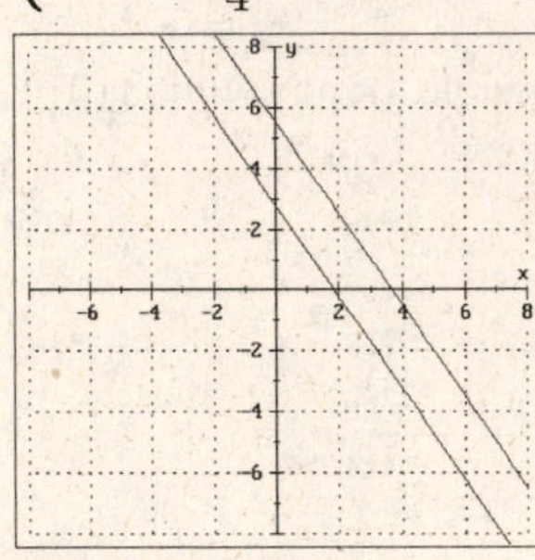

inconsistent system

53. $\begin{cases} \dfrac{5}{2}x + y = \dfrac{1}{2} \\ 2x - \dfrac{3}{2}y = 5 \end{cases}$

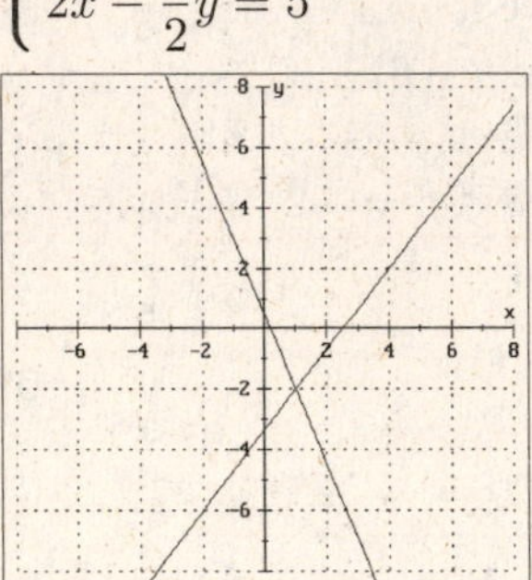

$(1, -2)$ is the solution.

55. $\begin{cases} (1) \quad 3x - 4y = 9 \\ (2) \quad x + 2y = 8 \end{cases}$

Substitute $x = -2y + 8$ from (2) into (1):

$$3\boldsymbol{x} - 4y = 9$$
$$3(\boldsymbol{-2y + 8}) - 4y = 9$$
$$-6y + 24 - 4y = 9$$
$$-10y = -15$$
$$y = \tfrac{3}{2}$$

Substitute this and solve for x:

$$x = -2y + 8$$
$$x = -2\left(\tfrac{3}{2}\right) + 8$$
$$x = -3 + 8 = 5$$

Solution: $\left(5, \dfrac{3}{2}\right)$

57. $\begin{cases} (1) & 2x + 2y = -1 \\ (2) & 3x + 4y = 0 \end{cases}$

Substitute $y = \dfrac{-2x - 1}{2}$ from (1) into (2):

$$\begin{aligned} 3x + 4\boldsymbol{y} &= 0 \\ 3x + 4\left(\frac{\boldsymbol{-2x - 1}}{\boldsymbol{2}}\right) &= 0 \\ 3x + 2(-2x - 1) &= 0 \\ 3x - 4x - 2 &= 0 \\ -x &= 2 \\ x &= -2 \end{aligned}$$

Substitute this and solve for y:

$$\begin{aligned} y &= \frac{-2x - 1}{2} \\ y &= \frac{-2(-2) - 1}{2} \\ y &= \frac{4 - 1}{2} = \frac{3}{2} \end{aligned}$$

Solution: $\left(-2, \dfrac{3}{2}\right)$

59.

$$\begin{aligned} 2x + 3y &= 8 \Rightarrow \times (2) & 4x + 6y &= 16 \\ 3x - 2y &= -1 \Rightarrow \times (3) & 9x - 6y &= -3 \\ & & 13x &= 13 \\ & & x &= 1 \end{aligned}$$

$$\begin{aligned} 2x + 3y &= 8 \\ 2(1) + 3y &= 8 \\ 3y &= 6 \\ y &= 2 \end{aligned}$$

Solution: $\boxed{(1, 2)}$

61.

$$\begin{aligned} 4x + 9y &= 8 & 4x + 9y &= 8 \\ 2x - 6y &= -3 \Rightarrow \times (-2) & -4x + 12y &= 6 \\ & & 21y &= 14 \\ & & y &= \tfrac{14}{21} = \tfrac{2}{3} \end{aligned}$$

$$\begin{aligned} 2x - 6y &= -3 \\ 2x - 6\left(\tfrac{2}{3}\right) &= -3 \\ 2x - 4 &= -3 \\ 2x &= 1 \\ x &= \tfrac{1}{2} \end{aligned}$$

Solution: $\boxed{\left(\tfrac{1}{2}, \tfrac{2}{3}\right)}$

63.

$$\begin{aligned} \tfrac{x}{2} + \tfrac{y}{2} &= 6 \Rightarrow \times 2 & x + y &= 12 \\ \tfrac{x}{2} - \tfrac{y}{2} &= -2 \Rightarrow \times 2 & x - y &= -4 \\ & & 2x &= 8 \\ & & x &= 4 \end{aligned}$$

$$\begin{aligned} x + y &= 12 \\ 4 + y &= 12 \\ y &= 8 \end{aligned}$$

Solution: $\boxed{(4, 8)}$

65.

$$\begin{aligned} \tfrac{3}{4}x + \tfrac{2}{3}y &= 7 \Rightarrow \times 12 & 9x + 8y &= 84 \Rightarrow \times 2 & 18x + 16y &= 168 \\ \tfrac{3}{5}x - \tfrac{1}{2}y &= 18 \Rightarrow \times 10 & 6x - 5y &= 180 \Rightarrow \times (-3) & -18x + 15y &= -540 \\ & & & & 31y &= -372 \\ & & & & y &= -12 \end{aligned}$$

$$\begin{aligned} 6x - 5y &= 180 \\ 6x - 5(-12) &= 180 \\ 6x &= 120 \\ x &= 20 \end{aligned}$$

Solution: $\boxed{(20, -12)}$

67. $\begin{cases} y = 3.2x - 1.5 \\ y = -2.7x - 3.7 \end{cases}$

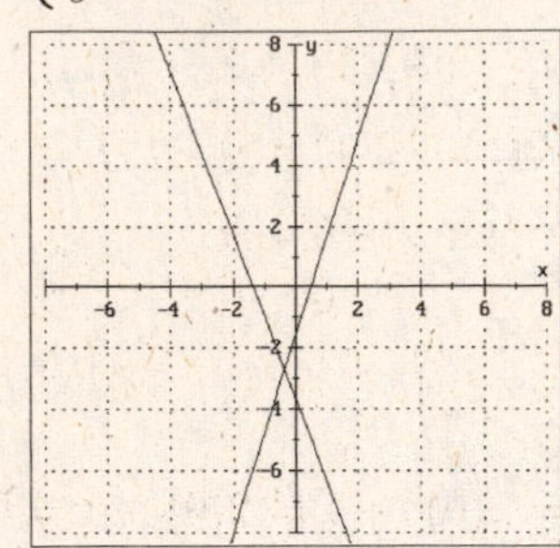

solution: $(-0.37, -2.69)$

69. $\begin{cases} 1.7x + 2.3y = 3.2 \\ y = 0.25x + 8.95 \end{cases}$

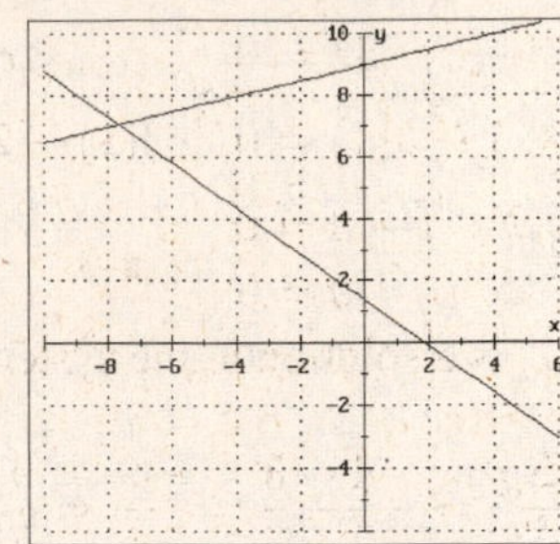

solution: $(-7.64, 7.04)$

71. **a.** The point (15, 2.0) is on the graph of the cost function, so it costs \$2 million to manufacture 15,000 cameras.

b. The point (20, 3.0) is on the graph of the revenue function, so there is a revenue of \$3 million when 20,000 cameras are sold.

c. The graphs of the cost function and the revenue function meet at the point (10, 1.5), so the revenue and cost functions are equal for 10,000 cameras.

73. Let $x =$ the cost of the pair of shoes and $y =$ the cost of the sweater.

$$\begin{array}{lr} (1) & x + y = 98 \\ (2) & y = x + 16 \end{array} \qquad \begin{aligned} x + \mathbf{y} &= 98 \\ x + \mathbf{x + 16} &= 98 \\ 2x &= 82 \\ x &= 41 \end{aligned} \qquad \begin{aligned} y &= x + 16 \\ y &= 41 + 16 \\ y &= 57 \end{aligned}$$

The sweater cost \$57.

75.

$$\begin{array}{lr} (1) & R_1 + R_2 = 1375 \\ (2) & R_1 = R_2 + 125 \end{array} \qquad \begin{aligned} \mathbf{R_1} + R_2 &= 1375 \\ \mathbf{R_2 + 125} + R_2 &= 1375 \\ 2R_2 &= 1250 \\ R_2 &= 625 \end{aligned} \qquad \begin{aligned} R_1 &= R_2 + 125 \\ R_1 &= 625 + 125 \\ R_1 &= 750 \end{aligned}$$

The resistances are $R_1 = 750$ ohms and $R_2 = 625$ ohms.

77. Let $l =$ the length of the field and $w =$ the width of the field.

$$\begin{aligned} 2w + 2l = 72 &\Rightarrow \times(-1) & -2w - 2l &= -72 \\ 3w + 2l = 88 &\Rightarrow & 3w + 2l &= 88 \\ & & w &= 16 \end{aligned} \qquad \begin{aligned} 2w + 2l &= 72 \\ 2(16) + 2l &= 72 \\ 32 + 2l &= 72 \\ 2l &= 40 \\ l &= 20 \end{aligned}$$

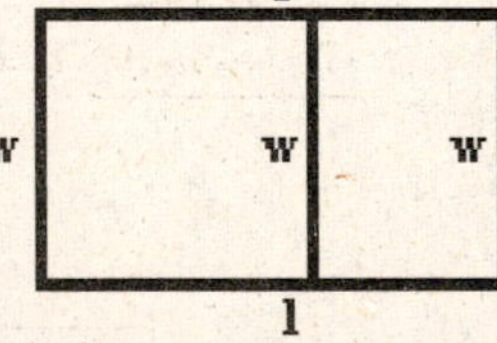

The dimensions of the field are 20 meters by 16 meters.

79-83. Answers may vary.

85. No.

Exercise 13.2 (page 916)

1.

$$\begin{array}{rrr} 2x+y-3z=0 & 3x-2y+4z=5 & 4x+2y-6z=0 \\ 2(1)+1-3(1)\stackrel{?}{=}0 & 3(1)-2(1)+4(1)\stackrel{?}{=}5 & 4(1)+2(1)-6(1)\stackrel{?}{=}0 \\ 2+1-3\stackrel{?}{=}0 & 3-2+4\stackrel{?}{=}5 & 4+2-6\stackrel{?}{=}0 \\ 0=0 & 5=5 & 0=0 \end{array}$$

$(1,1,1)$ is a solution to the system.

3. $m=\dfrac{\Delta y}{\Delta x}=\dfrac{-4-5}{-2-3}=\dfrac{-9}{-5}=\dfrac{9}{5}$

5. $f(0)=2(0)^2+1=2(0)+1=0+1=1$

7. $f(s)=2s^2+1$

9. plane

11. infinitely

13.

$$\begin{array}{rrr} x-y+z=2 & 2x+y-z=4 & 2x-3y+z=2 \\ 2-1+1\stackrel{?}{=}2 & 2(2)+1-1\stackrel{?}{=}4 & 2(2)-3(1)+1\stackrel{?}{=}2 \\ 2=2 & 4+1-1\stackrel{?}{=}4 & 4-3+1\stackrel{?}{=}2 \\ & 4=4 & 2=2 \end{array}$$

$(2,1,1)$ is a solution to the system.

15.

$$\begin{array}{llllll} (1) & x+y+z=4 & (1) & x+y+z=4 & (2) & 2x+y-z=1 \\ (2) & 2x+y-z=1 & (2) & 2x+y-z=1 & (3) & 2x-3y+z=1 \\ (3) & 2x-3y+z=1 & (4) & 3x+2y=5 & (5) & 4x-2y=2 \end{array}$$

$$\begin{array}{lll} (4)\ 3x+2y=5 & 3x+2y=5 & x+y+z=4 \\ (5)\ 4x-2y=2 & 3(1)+2y=5 & 1+1+z=4 \\ 7x=7 & 3+2y=5 & 2+z=4 \\ x=1 & 2y=2 & z=2 \\ & y=1 & \end{array}$$

The solution is $(1,1,2)$.

17.

$$\begin{array}{llllll} (1) & 2x+2y+3z=10 & (1) & 2x+2y+3z=10 & (3) & x+y+2z=6 \\ (2) & 3x+y-z=0 & 3\cdot(2) & 9x+3y-3z=0 & 2\cdot(2) & 6x+2y-2z=0 \\ (3) & x+y+2z=6 & (4) & 11x+5y=10 & (5) & 7x+3y=6 \end{array}$$

$$\begin{array}{llll} 11x+5y=10\Rightarrow\times 3 & 33x+15y=30 & 11x+5y=10 & x+y+2z=6 \\ 7x+3y=6\Rightarrow\times(-5) & -35x-15y=-30 & 11(0)+5y=10 & 0+2+2z=6 \\ & -2x=0 & 0+5y=10 & 2+2z=6 \\ & x=0 & 5y=10 & 2z=4 \\ & & y=2 & z=2 \end{array}$$

Solution: $(0,2,2)$

19.

$$\begin{array}{lrl} (1) & 4x+3z=4 \\ (2) & 2y-6z=-1 \\ (3) & 8x+4y+3z=9 \end{array} \quad \begin{array}{lrl} (2) & 2y-6z=-1 \\ 2\cdot(1) & 8x\quad+6z=\ \ 8 \\ \hline (4) & 8x+2y\quad=\ \ 7 \end{array} \quad \begin{array}{lrl} (2) & 2y-6z=-1 \\ 2\cdot(3) & 16x+\ 8y+6z=\ 18 \\ \hline (5) & 16x+10y\quad=\ 17 \end{array}$$

$$\begin{array}{l} 8x+\ 2y=\ \ 7\Rightarrow\times(-2) \\ 16x+10y=17\Rightarrow \end{array} \quad \begin{array}{r} -16x-\ \ 4y=-14 \\ 16x+10y=\ \ 17 \\ \hline 6y=\ \ 3 \\ y=\ \ \frac{1}{2} \end{array} \quad \begin{array}{r} 8x+2y=7 \\ 8x+2\left(\frac{1}{2}\right)=7 \\ 8x+1=7 \\ 8x=6 \\ x=\frac{3}{4} \end{array} \quad \begin{array}{r} 4x+3z=4 \\ 4\left(\frac{3}{4}\right)+3z=4 \\ 3+3z=4 \\ 3z=1 \\ z=\frac{1}{3} \end{array}$$

Solution: $\boxed{\left(\frac{3}{4},\frac{1}{2},\frac{1}{3}\right)}$

21.

$$\begin{array}{lr} (1) & 2a+3b+c=2 \\ (2) & 4a+6b+2c=5 \\ (3) & a-2b+c=3 \end{array} \quad \begin{array}{lr} (2) & 4a+6b+2c=\ \ 5 \\ -2\cdot(1) & -4a-6b-2c=-4 \\ \hline (4) & 0=\ \ 1 \end{array}$$

Since equation (4) is always false, there is no solution. The system is inconsistent.

23.

$$\begin{array}{lr} (1) & x-2y+3z=9 \\ (2) & -x+3y=-4 \\ (3) & 2x-5y+3z=13 \end{array} \quad \begin{array}{lr} (1) & x-2y+3z=\ \ 9 \\ (2) & -x+3y\quad=-4 \\ \hline (4) & y+3z=\ \ 5 \end{array} \quad \begin{array}{lr} (3) & 2x-5y+3z=\ 13 \\ 2\cdot(2) & -2x+6y\quad=-8 \\ \hline (5) & y+3z=\ \ 5 \end{array}$$

$$\begin{array}{l} y+3z=5\Rightarrow\times(-1) \\ y+3z=5\Rightarrow \end{array} \quad \begin{array}{r} -y-3z\ =-5 \\ y+3z\ =\ \ 5 \\ \hline 0=\ \ 0 \end{array}$$

Dependent equations
If $z=z$, then $y=5-3z$.

$$\begin{array}{r} x-2y+3z=9 \\ x-2(5-3z)+3z=9 \\ x-10+6z+3z=9 \\ x=19-9z \end{array}$$

$\boxed{(19-9z,5-3z,z)}$

25.

$$\begin{array}{lr} (1) & a+b+2c=7 \\ (2) & a+2b+c=8 \\ (3) & 2a+b+c=9 \end{array} \quad \begin{array}{lr} (1) & a+\ \ b+2c=\ \ \ \ 7 \\ -2\cdot(2) & -2a-4b-2c=-16 \\ \hline (4) & -a-3b\quad=\ -9 \end{array} \quad \begin{array}{lr} (1) & a+\ \ b+2c=\ \ \ \ 7 \\ -2\cdot(3) & -4a-2b-2c=-18 \\ \hline (5) & -3a-\ \ b\quad=-11 \end{array}$$

$$\begin{array}{l} -a-3b=\ \ -9\Rightarrow\times(-3) \\ -3a-\ \ b=-11\Rightarrow \end{array} \quad \begin{array}{r} 3a+9b=\ \ 27 \\ -3a-\ \ b=-11 \\ \hline 8b=\ \ 16 \\ b=\ \ \ 2 \end{array} \quad \begin{array}{r} -a-3b=-9 \\ -a-3(2)=-9 \\ -a-6=-9 \\ -a=-3 \\ a=3 \end{array} \quad \begin{array}{r} 2a+b+c=9 \\ 2(3)+2+c=9 \\ 6+2+c=9 \\ 8+c=9 \\ c=1 \end{array}$$

Solution: $\boxed{(3,2,1)}$

27.

$$\begin{array}{lll} x+\frac{1}{3}y+z=13 & \Rightarrow\times3 & (1)\ 3x+y+3z=39 \\ \frac{1}{2}x-y+\frac{1}{3}z=-2 & \Rightarrow\times6 & (2)\ 3x-6y+2z=-12 \\ x+\frac{1}{2}y-\frac{1}{3}z=2 & \Rightarrow\times6 & (3)\ 6x+3y-2z=12 \end{array}$$

continued on next page...

27. continued

$$\begin{array}{rl} (2) & 3x-6y+\ \ 2z=-12 \\ 6\cdot(1) & 18x+6y+18z=\ \ 234 \\ \hline (4) & 21x \qquad +20z=\ \ 222 \end{array} \qquad \begin{array}{rl} (2) & 3x-6y+2z=-12 \\ 2\cdot(3) & 12x+6y-4z=\ \ 24 \\ \hline (5) & 15x \qquad -2z=\ \ 12 \end{array}$$

$$\begin{array}{l} 21x+20z=222 \Rightarrow \\ \underline{15x-\ \ 2z=\ \ 12} \Rightarrow \times 10 \end{array} \quad \begin{array}{rl} 21x+20z&=222 \\ 150x-20z&=120 \\ \hline 171x \qquad &=342 \\ x \qquad &=\ \ 2 \end{array} \quad \begin{array}{r} 15x-2z=12 \\ 15(2)-2z=12 \\ 30-2z=12 \\ -2z=-18 \\ z=9 \end{array} \quad \begin{array}{r} 3x+y+3z=39 \\ 3(2)+y+3(9)=39 \\ 6+y+27=39 \\ y=6 \end{array}$$

Solution: $\boxed{(2,6,9)}$

29. $$\begin{array}{rl} (1) & x-3y+z=1 \\ (2) & 2x-y-2z=2 \\ (3) & x+2y-3z=-1 \end{array} \quad \begin{array}{rr} -2\cdot(1) & -2x+6y-2z=-2 \\ (2) & 2x-\ \ y-2z=\ \ 2 \\ \hline (4) & 5y-4z=\ \ 0 \end{array} \quad \begin{array}{rr} -2\cdot(3) & -2x-4y+6z=2 \\ (2) & 2x-\ \ y-2z=2 \\ \hline (5) & -5y+4z=4 \end{array}$$

$$\begin{array}{rl} 5y-4z &=0 \\ -5y+4z &=4 \\ \hline 0\neq 4 \end{array}$$

Since this equation is always false, there is no solution. $\boxed{\text{The system is inconsistent.}}$

31. $$\begin{array}{rl} (1) & 2x+3y+4z=6 \\ (2) & 2x-3y-4z=-4 \\ (3) & 4x+6y+8z=12 \end{array} \quad \begin{array}{rl} (1) & 2x+3y+4z=\ \ 6 \\ (2) & 2x-3y-4z=-4 \\ \hline (4) & 4x \qquad\qquad =\ \ 2 \\ & x \qquad\qquad =\ \ \frac{1}{2} \end{array}$$

Substitute $x=\frac{1}{2}$ into (1):

$$\begin{array}{r} 2\left(\frac{1}{2}\right)+3y+4z=6 \\ 1+3y+4z=6 \\ y=\frac{5}{3}-\frac{4}{3}z \end{array}$$

Dependent equations: $\boxed{\left(\frac{1}{2},\frac{5}{3}-\frac{4}{3}z,z\right)}$

33. Let $x=$ the first integer, $y=$ the second integer and $z=$ the third integer.

$$\begin{array}{rcl} x+y+z=18 & \Rightarrow & (1)\ x+y+z=18 \\ z=4y & \Rightarrow & (2)\ -4y+z=0 \\ y=x+6 & \Rightarrow & (3)\ -x+y=6 \end{array} \quad \begin{array}{rr} (1) & x+\ \ y+z=18 \\ (3) & -x+\ \ y \qquad =\ \ 6 \\ \hline (4) & 2y+z=24 \end{array}$$

$$\begin{array}{l} 2y+z=24 \Rightarrow \times 2 \\ \underline{-4y+z=\ \ 0 \Rightarrow} \end{array} \quad \begin{array}{r} 4y+2z=48 \\ -4y+\ \ z=\ \ 0 \\ \hline 3z=48 \\ z=16 \end{array} \quad \begin{array}{r} 2y+z=24 \\ 2y+16=24 \\ 2y=8 \\ y=4 \end{array} \quad \begin{array}{r} x+y+z=18 \\ x+4+16=18 \\ x=-2 \end{array}$$

The integers are $-2, 4$ and 16.

35. Let A, B and C represent the measures of the three angles.

$$\begin{array}{rcl} A+B+C=180 & \Rightarrow & (1)\ A+B+C=180 \\ A=B+C-100 & \Rightarrow & (2)\ A-B-C=-100 \\ C=2B-40 & \Rightarrow & (3)\ -2B+C=-40 \end{array} \quad \begin{array}{rr} (1) & A+B+C=\ \ 180 \\ (2) & A-B-C=-100 \\ \hline (4) & 2A \qquad\quad =\ \ 80 \\ & A \qquad\quad =\ \ 40 \end{array}$$

$$\begin{array}{l} A+B+C=180 \\ 40+B+C=180 \\ (5)\ \ B+C=140 \end{array} \quad \begin{array}{rr} -1\cdot(3) & 2B-C=\ \ 40 \\ (5) & B+C=140 \\ \hline & 3B \qquad =180 \\ & B \qquad =\ \ 60 \end{array} \quad \begin{array}{r} B+C=140 \\ 60+C=140 \\ C=80 \end{array}$$

The angles have measures of 40°, 60° and 80°.

SECTION 13.2

37. Let $A =$ the units of food A, $B =$ the units of food B and $C =$ the units of food C.

$$\begin{array}{lll} (1) & A+2B+2C=11 & \text{(fat)} \\ (2) & A+B+C=6 & \text{(carbohydrate)} \\ (3) & 2A+B+2C=10 & \text{(protein)} \end{array} \qquad \begin{array}{rrcl} (1) & A+2B+2C & = & 11 \\ -2\cdot(2) & -2A-2B-2C & = & -12 \\ \hline (4) & -A & = & -1 \\ & A & = & 1 \end{array}$$

$$\begin{array}{rrcl} (1) & A+2B+2C & = & 11 \\ -2\cdot(3) & -4A-2B-4C & = & -20 \\ \hline (5) & -3A \qquad -2C & = & -9 \end{array} \qquad \begin{aligned} -3A-2C &= -9 \\ -3(1)-2C &= -9 \\ -3-2C &= -9 \\ -2C &= -6 \\ C &= 3 \end{aligned} \qquad \begin{aligned} A+B+C &= 6 \\ 1+B+3 &= 6 \\ B+4 &= 6 \\ B &= 2 \end{aligned}$$

1 unit of food A, 2 units of food B and 3 units of food C should be used.

39. Let $x =$ the number of \$5 statues, $y =$ the number of \$4 statues and $z =$ the number of \$3 statues.

$$\begin{array}{lll} (1) & x+y+z=180 & \text{(total number made)} \\ (2) & 5x+4y+3z=650 & \text{(total cost)} \\ (3) & 20x+12y+9z=2100 & \text{(total revenue)} \end{array} \qquad \begin{array}{rrcl} -3\cdot(1) & -3x-3y-3z & = & -540 \\ (2) & 5x+4y+3z & = & 650 \\ \hline (4) & 2x+y & = & 110 \end{array}$$

$$\begin{array}{rrcl} -9\cdot(1) & -9x-9y-9z & = & -1620 \\ (3) & 20x+12y+9z & = & 2100 \\ \hline (5) & 11x+3y & = & 480 \end{array} \qquad \begin{array}{l} 2x+y=110 \Rightarrow \times(-3) \\ \underline{11x+3y=480} \Rightarrow \end{array} \qquad \begin{array}{rcl} -6x-3y & = & -330 \\ 11x+3y & = & 480 \\ \hline 5x & = & 150 \\ x & = & 30 \end{array}$$

$$\begin{aligned} 2x+y &= 110 \\ 2(30)+y &= 110 \\ 60+y &= 110 \\ y &= 50 \end{aligned} \qquad \begin{aligned} x+y+z &= 180 \\ 30+50+z &= 180 \\ z &= 100 \end{aligned}$$

30 of the \$5, 50 of the \$4 and 100 of the \$3 statues should be made.

41. Let $x =$ the number of \$5 tickets, $y =$ the number of \$3 tickets and $z =$ the number of \$2 tickets.

$$\begin{array}{lll} (1) & x+y+z=750 & \text{(total sold)} \\ (2) & x=2z \Rightarrow x-2z=0 & \text{(twice as many)} \\ (3) & 5x+3y+2z=2625 & \text{(total revenue)} \end{array} \qquad \begin{array}{rrcl} 2\cdot(1) & 2x+2y+2z & = & 1500 \\ (2) & x \qquad -2z & = & 0 \\ \hline (4) & 3x+2y & = & 1500 \end{array}$$

$$\begin{array}{rrcl} (2) & x \qquad -2z & = & 0 \\ (3) & 5x+3y+2z & = & 2625 \\ \hline (5) & 6x+3y & = & 2625 \end{array} \qquad \begin{array}{l} 3x+2y=1500 \Rightarrow \times(-2) \\ \underline{6x+3y=2625} \Rightarrow \end{array} \qquad \begin{array}{rcl} -6x-4y & = & -3000 \\ 6x+3y & = & 2625 \\ \hline -y & = & -375 \\ y & = & 375 \end{array}$$

$$\begin{aligned} 3x+2y &= 1500 \\ 3x+2(375) &= 1500 \\ 3x+750 &= 1500 \\ 3x &= 750 \\ x &= 250 \end{aligned} \qquad \begin{aligned} x+y+z &= 750 \\ 250+375+z &= 750 \\ 625+z &= 750 \\ z &= 125 \end{aligned}$$

250 of the \$5, 375 of the \$3 and 125 of the \$2 tickets were sold.

SECTION 13.2

43. Let $x =$ the number of totem poles, $y =$ the number of bears and $z =$ the number of deer.

$$\begin{array}{llll} (1) & 2x + 2y + z = 14 & \text{(carving)} & -2\cdot(1) \quad -4x - 4y - 2z = -28 \\ (2) & x + 2y + 2z = 15 & \text{(sanding)} & (2) \quad\quad x + 2y + 2z = \ \ 15 \\ (3) & 3x + 2y + 2z = 21 & \text{(painting)} & (4) \quad -3x - 2y \quad\quad = -13 \end{array}$$

$$\begin{array}{ll} -2\cdot(1) & -4x - 4y - 2z = -28 \\ (3) & \ \ 3x + 2y + 2z = \ \ 21 \\ (5) & \ \ -x - 2y \quad\quad = -7 \end{array} \qquad \begin{array}{l} -3x - 2y = -13 \Rightarrow \times(-1) \\ -x - 2y = \ -7 \Rightarrow \end{array} \qquad \begin{array}{rl} 3x + 2y &= 13 \\ -x - 2y &= -7 \\ 2x &= 6 \\ x &= 3 \end{array}$$

$$\begin{array}{rr} 3x + 2y = 13 & 2x + 2y + z = 14 \\ 3(3) + 2y = 13 & 2(3) + 2(2) + z = 14 \\ 9 + 2y = 13 & 6 + 4 + z = 14 \\ 2y = 4 & z = 4 \\ y = 2 & \end{array}$$

3 totem poles, 2 bears and 4 deer should be made.

45. Let $x =$ the % of nitrogen, $y =$ the % of oxygen and $z =$ the % of other gases.

$$\begin{array}{lll} (1) & x + y + z = 100 & \\ (2) & x = 3(y + z) + 12 \Rightarrow & x - 3y - 3z = 12 \\ (3) & z = y - 20 \Rightarrow & -y + z = -20 \end{array} \qquad \begin{array}{ll} (1) & x + y + \ z = 100 \\ (3) & \quad -y + \ z = -20 \\ (4) & x \quad\ + 2z = \ \ 80 \end{array}$$

$$\begin{array}{ll} 3\cdot(1) & 3x + 3y + 3z = 300 \\ (2) & \ \ x - 3y - 3z = \ \ 12 \\ (5) & 4x \quad\quad\quad = 312 \\ & x \quad\quad\quad\ = \ \ 78 \end{array} \qquad \begin{array}{r} x + 2z = 80 \\ 78 + 2z = 80 \\ 2z = 2 \\ z = 1 \end{array} \qquad \begin{array}{r} x + y + z = 100 \\ 78 + y + 1 = 100 \\ y + 79 = 100 \\ y = 21 \end{array}$$

The composition is 78% nitrogen, 21% oxygen, and 1% other gases.

47. Substitute the coordinates of each point for x and y in the equation $y = ax^2 + bx + c$.

$$\begin{array}{lll} y = ax^2 + bx + c & y = ax^2 + bx + c & y = ax^2 + bx + c \\ 0 = a(0)^2 + b(0) + c & -4 = a(2)^2 + b(2) + c & 0 = a(4)^2 + b(4) + c \\ 0 = c & -4 = 4a + 2b + c & 0 = 16a + 4b + c \end{array}$$

Solve the system of equations formed from the three equations:

$$\begin{array}{ll} (1) & c = 0 \\ (2) & 4a + 2b + c = -4 \\ (3) & 16a + 4b + c = 0 \end{array} \qquad \begin{array}{l} 4a + 2b = -4 \Rightarrow \times(-2) \\ 16a + 4b = \ \ 0 \Rightarrow \end{array} \qquad \begin{array}{rl} -8a - 4b &= 8 \\ 16a + 4b &= 0 \\ 8a &= 8 \\ a &= 1 \end{array} \qquad \begin{array}{r} 16a + 4b = 0 \\ 16(1) + 4b = 0 \\ 4b = -16 \\ b = -4 \end{array}$$

The equation is $y = x^2 - 4x$.

49. Substitute the coordinates of each point for x and y in the equation $x^2+y^2+cx+dy+e=0$.

$$\begin{aligned} x^2+y^2+cx+dy+e&=0\\ (1)^2+(3)^2+c(1)+d(3)+e&=0\\ 1+9+c+3d+e&=0\\ c+3d+e&=-10 \end{aligned}$$

$$\begin{aligned} x^2+y^2+cx+dy+e&=0\\ (3)^2+(1)^2+c(3)+d(1)+e&=0\\ 9+1+3c+d+e&=0\\ 3c+d+e&=-10 \end{aligned}$$

$$\begin{aligned} x^2+y^2+cx+dy+e&=0\\ (1)^2+(-1)^2+c(1)+d(-1)+e&=0\\ 1+1+c-d+e&=0\\ c-d+e&=-2 \end{aligned}$$

$$\begin{aligned} (1)\quad c+3d+e&=-10\\ (2)\quad 3c+d+e&=-10\\ (3)\quad c-d+e&=-2 \end{aligned}$$

$$\begin{array}{rrl} (1) & c+3d+e & =-10\\ -1\cdot(2) & -3c-\ d-e & =\ 10\\ \hline (4) & -2c+2d & =\ 0 \end{array}$$

$$\begin{array}{rrl} (1) & c+3d+e & =-10\\ -1\cdot(3) & -c+\ d-e & =\ 2\\ \hline (5) & 4d & =-8\\ & d & =-2 \end{array}$$

$$\begin{aligned} -2c+2d&=0\\ -2c+2(-2)&=0\\ -2c-4&=0\\ -2c&=4\\ c&=-2 \end{aligned}$$

$$\begin{aligned} c+3d+e&=-10\\ -2+3(-2)+e&=-10\\ -2-6+e&=-10\\ -8+e&=-10\\ e&=-2 \end{aligned}$$

The equation is $x^2+y^2-2x-2y-2=0$.

51. **Answers may vary.**

53.

$$\begin{aligned} (1)\quad x+y+z+w&=3\\ (2)\quad x-y-z-w&=-1\\ (3)\quad x+y-z-w&=1\\ (4)\quad x+y-z+w&=3 \end{aligned}$$

$$\begin{array}{rl} (1)\ x+y+z+w= & 3\\ (2)\ x-y-z-w= & -1\\ \hline 2x\qquad\qquad = & 2\\ x\qquad\qquad = & 1 \end{array}$$

$$\begin{array}{rl} (1)\ x+\ y+z+w & =3\\ (3)\ x+\ y-z-w & =1\\ \hline 2x+2y & =4 \end{array}$$

$$\begin{array}{rl} (1)\ x+\ y+z+\ w & =3\\ (4)\ x+\ y-z+\ w & =3\\ \hline 2x+2y\quad +2w & =6 \end{array}$$

$$\begin{aligned} 2x+2y&=4\\ 2(1)+2y&=4\\ 2y&=2\\ y&=1 \end{aligned}$$

$$\begin{aligned} 2x+2y+2w&=6\\ 2(1)+2(1)+2w&=6\\ 4+2w&=6\\ 2w&=2\\ w&=1 \end{aligned}$$

$$\begin{aligned} x+y+z+w&=3\\ 1+1+z+1&=3\\ 3+z&=3\\ z&=0 \end{aligned}$$

The solution is $(1,1,0,1)$.

Exercise 13.3 (page 926)

Note: The notation $3R_1+R_3 \Rightarrow R_2$ means to multiply Row #1 of the previous matrix, add that result to Row #3 of the previous matrix, and write the final result in Row #2 of the current matrix.

1. $\begin{bmatrix} 3 & 2\\ 4 & -3 \end{bmatrix}$

3. yes

5. $93{,}000{,}000 = 9.3\times10^7$

7. $63\times10^3 = 6.3\times10^1\times10^3 = 6.3\times10^4$

SECTION 13.3

9. matrix **11.** 3; columns **13.** augmented; coefficient

15. type 1 **17.** nonzero

19. $\begin{bmatrix} 2 & 1 & 1 \\ 5 & 4 & 1 \end{bmatrix} \overset{R_2+(-R_1)\Rightarrow R_2}{\Rightarrow} \begin{bmatrix} 2 & 1 & 1 \\ 3 & 3 & \boxed{0} \end{bmatrix}$

21. $\begin{bmatrix} 3 & -2 & 1 \\ -1 & 2 & 4 \end{bmatrix} \overset{2R_2\Rightarrow R_2}{\Rightarrow} \begin{bmatrix} 3 & -2 & 1 \\ -2 & 4 & \boxed{8} \end{bmatrix}$

23. $\left[\begin{array}{cc|c} 1 & 1 & 2 \\ 1 & -1 & 0 \end{array}\right] \overset{R_1+(-R_2)\Rightarrow R_2}{\Rightarrow} \left[\begin{array}{cc|c} 1 & 1 & 2 \\ 0 & 2 & 2 \end{array}\right] \overset{\frac{1}{2}R_2\Rightarrow R_2}{\Rightarrow} \left[\begin{array}{cc|c} 1 & 1 & 2 \\ 0 & 1 & 1 \end{array}\right]$

From R_2, $y = 1$. From R_1:

$x + y = 2$

$x + 1 = 2 \Rightarrow x = 1$

The solution is $(1, 1)$.

25. $\left[\begin{array}{cc|c} 1 & 2 & -4 \\ 2 & 1 & 1 \end{array}\right] \overset{-2R_1+R_2\Rightarrow R_2}{\Rightarrow} \left[\begin{array}{cc|c} 1 & 2 & -4 \\ 0 & -3 & 9 \end{array}\right] \overset{-\frac{1}{3}R_2\Rightarrow R_2}{\Rightarrow} \left[\begin{array}{cc|c} 1 & 2 & -4 \\ 0 & 1 & -3 \end{array}\right]$

From R_2, $y = -3$. From R_1:

$x + 2y = -4$

$x + 2(-3) = -4$

$x - 6 = -4 \Rightarrow x = 2$

The solution is $(2, -3)$.

27. $\left[\begin{array}{ccc|c} 1 & 1 & 1 & 6 \\ 1 & 2 & 1 & 8 \\ 1 & 1 & 2 & 9 \end{array}\right] \overset{\substack{-R_1+R_2\Rightarrow R_2 \\ -R_1+R_3\Rightarrow R_3}}{\Rightarrow} \left[\begin{array}{ccc|c} 1 & 1 & 1 & 6 \\ 0 & 1 & 0 & 2 \\ 0 & 0 & 1 & 3 \end{array}\right]$

From R_3, $z = 3$. From R_2, $y = 2$. From R_1:

$x + y + z = 6$

$x + 2 + 3 = 6$

$x + 5 = 6$

$x = 1$

The solution is $(1, 2, 3)$.

29. $\left[\begin{array}{ccc|c} 2 & 1 & 3 & 3 \\ -2 & -1 & 1 & 5 \\ 4 & -2 & 2 & 2 \end{array}\right] \overset{\substack{R_1+R_2\Rightarrow R_2 \\ -2R_1+R_3\Rightarrow R_3}}{\Rightarrow} \left[\begin{array}{ccc|c} 2 & 1 & 3 & 3 \\ 0 & 0 & 4 & 8 \\ 0 & -4 & -4 & -4 \end{array}\right] \overset{R_2\Leftrightarrow R_3}{\Rightarrow} \left[\begin{array}{ccc|c} 2 & 1 & 3 & 3 \\ 0 & -4 & -4 & -4 \\ 0 & 0 & 4 & 8 \end{array}\right] \overset{\substack{-\frac{1}{4}R_2\Rightarrow R_2 \\ \frac{1}{4}R_3\Rightarrow R_3}}{\Rightarrow} \left[\begin{array}{ccc|c} 2 & 1 & 3 & 3 \\ 0 & 1 & 1 & 1 \\ 0 & 0 & 1 & 2 \end{array}\right]$

continued on next page...

29. continued

From R_3, $z = 2$.

From R_2:
$$y + z = 1$$
$$y + 2 = 1$$
$$y = -1$$

From R_1:
$$2x + y + 3z = 3$$
$$2x + (-1) + 3(2) = 3$$
$$2x - 1 + 6 = 3$$
$$2x + 5 = 3$$
$$2x = -2$$
$$x = -1$$

The solution is $(-1, -1, 2)$.

31. $$\left[\begin{array}{ccc|c} 1 & 2 & 2 & 2 \\ 2 & 1 & -1 & 1 \\ 4 & 5 & 3 & 3 \end{array}\right] \xRightarrow{\substack{-2R_1 + R_2 \Rightarrow R_2 \\ -4R_1 + R_3 \Rightarrow R_3}} \left[\begin{array}{ccc|c} 1 & 2 & 2 & 2 \\ 0 & -3 & -5 & -3 \\ 0 & -3 & -5 & -5 \end{array}\right] \xRightarrow{-R_2 + R_3 \Rightarrow R_3} \left[\begin{array}{ccc|c} 1 & 2 & 2 & 2 \\ 0 & -3 & -5 & -3 \\ 0 & 0 & 0 & -2 \end{array}\right]$$

The last row indicates that $0x + 0y + 0z = -2$. This is false, so there is no solution. $\emptyset$

33. $$\left[\begin{array}{cc|c} 1 & 1 & 3 \\ 3 & -1 & 1 \\ 2 & 1 & 4 \end{array}\right] \xRightarrow{\substack{-3R_1 + R_2 \Rightarrow R_2 \\ -2R_1 + R_3 \Rightarrow R_3}} \left[\begin{array}{cc|c} 1 & 1 & 3 \\ 0 & -4 & -8 \\ 0 & -1 & -2 \end{array}\right] \xRightarrow{\substack{-\frac{1}{4}R_2 \Rightarrow R_2 \\ -4R_3 + R_2 \Rightarrow R_3}} \left[\begin{array}{cc|c} 1 & 1 & 3 \\ 0 & 1 & 2 \\ 0 & 0 & 0 \end{array}\right]$$

From R_2, $y = 2$.

From R_1:
$$x + y = 3$$
$$x + 2 = 3$$
$$x = 1$$

The solution is $(1, 2)$.

35. $$\left[\begin{array}{cc|c} 2 & -1 & 4 \\ 1 & 3 & 2 \\ -1 & -4 & -2 \end{array}\right] \xRightarrow{\substack{-2R_2 + R_1 \Rightarrow R_2 \\ 2R_3 + R_1 \Rightarrow R_3}} \left[\begin{array}{cc|c} 2 & -1 & 4 \\ 0 & -7 & 0 \\ 0 & -9 & 0 \end{array}\right] \xRightarrow{\substack{-\frac{1}{7}R_2 \Rightarrow R_2 \\ -\frac{7}{9}R_2 + R_3 \Rightarrow R_3}} \left[\begin{array}{cc|c} 2 & -1 & 4 \\ 0 & 1 & 0 \\ 0 & 0 & 0 \end{array}\right]$$

From R_2, $y = 0$.

From R_1:
$$2x - y = 4$$
$$2x - 0 = 4$$
$$2x = 4$$
$$x = 2$$

The solution is $(2, 0)$.

37. $$\left[\begin{array}{cc|c} 2 & 1 & 7 \\ 1 & -1 & 2 \\ -1 & 3 & -2 \end{array}\right] \xRightarrow{\substack{-2R_2 + R_1 \Rightarrow R_2 \\ 2R_3 + R_1 \Rightarrow R_3}} \left[\begin{array}{cc|c} 2 & 1 & 7 \\ 0 & 3 & 3 \\ 0 & 7 & 3 \end{array}\right] \xRightarrow{\substack{\frac{1}{3}R_2 \Rightarrow R_2 \\ -\frac{7}{3}R_2 + R_3 \Rightarrow R_3}} \left[\begin{array}{cc|c} 2 & 1 & 7 \\ 0 & 1 & 1 \\ 0 & 0 & -4 \end{array}\right]$$

R_3: $0x + 0y = -4$, or $0 = -4$, which is an impossible equation. $\emptyset$

39. $\begin{bmatrix} 1 & 2 & 3 & | & -2 \\ -1 & -1 & -2 & | & 4 \end{bmatrix} \overset{R_1 + R_2 \Rightarrow R_2}{\Rightarrow} \begin{bmatrix} 1 & 2 & 3 & | & -2 \\ 0 & 1 & 1 & | & 2 \end{bmatrix}$

From R_2:
$$y + z = 2$$
$$y = 2 - z$$

From R_1:
$$x + 2y + 3z = -2$$
$$x + 2(2 - z) + 3z = -2$$
$$x + 4 - 2z + 3z = -2$$
$$x + z = -6$$
$$x = -6 - z$$

The solution is $(-6 - z, 2 - z, z)$.

41. $\begin{bmatrix} 1 & -1 & 0 & | & 1 \\ 0 & 1 & 1 & | & 1 \\ 1 & 0 & 1 & | & 2 \end{bmatrix} \overset{-R_1 + R_3 \Rightarrow R_3}{\Rightarrow} \begin{bmatrix} 1 & -1 & 0 & | & 1 \\ 0 & 1 & 1 & | & 1 \\ 0 & 1 & 1 & | & 1 \end{bmatrix} \overset{-R_2 + R_3 \Rightarrow R_3}{\Rightarrow} \begin{bmatrix} 1 & -1 & 0 & | & 1 \\ 0 & 1 & 1 & | & 1 \\ 0 & 0 & 0 & | & 0 \end{bmatrix}$

From R_2:
$$y + z = 1$$
$$y = 1 - z$$

From R_1:
$$x - y = 1$$
$$x - (1 - z) = 1$$
$$x - 1 + z = 1$$
$$x + z = 2$$
$$x = 2 - z$$

The solution is $(2 - z, 1 - z, z)$.

43. $\begin{bmatrix} 3 & 4 & | & -12 \\ 9 & -2 & | & 6 \end{bmatrix} \overset{-3R_1 + R_2 \Rightarrow R_2}{\Rightarrow} \begin{bmatrix} 3 & 4 & | & -12 \\ 0 & -14 & | & 42 \end{bmatrix} \overset{-\frac{1}{14}R_2 \Rightarrow R_2}{\Rightarrow} \begin{bmatrix} 3 & 4 & | & -12 \\ 0 & 1 & | & -3 \end{bmatrix}$

From R_2, $y = -3$.

From R_1:
$$3x + 4y = -12$$
$$3x + 4(-3) = -12$$
$$3x - 12 = -12$$
$$3x = 0 \Rightarrow x = 0$$

The solution is $(0, -3)$.

45. $\begin{cases} 5a = 24 + 2b \\ 5b = 3a + 16 \end{cases} \Rightarrow \begin{cases} 5a - 2b = 24 \\ -3a + 5b = 16 \end{cases}$

$\begin{bmatrix} 5 & -2 & | & 24 \\ -3 & 5 & | & 16 \end{bmatrix} \overset{2R_2 + R_1 \Rightarrow R_1}{\Rightarrow} \begin{bmatrix} -1 & 8 & | & 56 \\ -3 & 5 & | & 16 \end{bmatrix} \overset{-R_1 \Rightarrow R_1}{\Rightarrow} \begin{bmatrix} 1 & -8 & | & -56 \\ -3 & 5 & | & 16 \end{bmatrix} \overset{3R_1 + R_2 \Rightarrow R_2}{\Rightarrow} \begin{bmatrix} 1 & -8 & | & -56 \\ 0 & -19 & | & -152 \end{bmatrix}$

$\overset{-\frac{1}{19}R_2 \Rightarrow R_2}{\Rightarrow} \begin{bmatrix} 1 & -8 & | & -56 \\ 0 & 1 & | & 8 \end{bmatrix}$

From R_2, $b = 8$.

From R_1:
$$a - 8b = -56$$
$$a - 8(8) = -56$$
$$a - 64 = -56 \Rightarrow a = 8$$

The solution is $(8, 8)$.

47. $-3R_2 + R_1 \Rightarrow R_2$, $-3R_3 + R_1 \Rightarrow R_3$; $\frac{2}{7}R_2 + R_3 \Rightarrow R_3$

$$\left[\begin{array}{ccc|c} 3 & 1 & -3 & 5 \\ 1 & -2 & 4 & 10 \\ 1 & 1 & 1 & 13 \end{array}\right] \Rightarrow \left[\begin{array}{ccc|c} 3 & 1 & -3 & 5 \\ 0 & 7 & -15 & -25 \\ 0 & -2 & -6 & -34 \end{array}\right] \Rightarrow \left[\begin{array}{ccc|c} 3 & 1 & -3 & 5 \\ 0 & 7 & -15 & -25 \\ 0 & 0 & -\frac{72}{7} & -\frac{288}{7} \end{array}\right]$$

From R_3:

$$\begin{aligned} -\frac{72}{7}c &= -\frac{288}{7} \\ -\frac{7}{72}\left(-\frac{72}{7}c\right) &= -\frac{7}{72}\left(-\frac{288}{7}\right) \\ c &= 4 \end{aligned}$$

From R_2:

$$\begin{aligned} 7b - 15c &= -25 \\ 7b - 15(4) &= -25 \\ 7b &= 35 \\ b &= 5 \end{aligned}$$

From R_1:

$$\begin{aligned} 3a + b - 3c &= 5 \\ 3a + 5 - 3(4) &= 5 \\ 3a - 7 &= 5 \\ 3a &= 12 \\ a &= 4 \end{aligned}$$

The solution is $(4, 5, 4)$.

49. $-R_2 + R_1 \Rightarrow R_2$, $-3R_1 + R_3 \Rightarrow R_3$; $\frac{1}{2}R_2 \Rightarrow R_2$, $4R_2 + R_3 \Rightarrow R_3$

$$\left[\begin{array}{cc|c} 1 & 3 & 7 \\ 1 & 1 & 3 \\ 3 & 1 & 5 \end{array}\right] \Rightarrow \left[\begin{array}{cc|c} 1 & 3 & 7 \\ 0 & 2 & 4 \\ 0 & -8 & -16 \end{array}\right] \Rightarrow \left[\begin{array}{cc|c} 1 & 3 & 7 \\ 0 & 1 & 2 \\ 0 & 0 & 0 \end{array}\right]$$

From R_2, $y = 2$. From R_1:

$$\begin{aligned} x + 3y &= 7 \\ x + 3(2) &= 7 \\ x + 6 = 7 &\Rightarrow x = 1 \end{aligned}$$

The solution is $(1, 2)$.

51. $\frac{1}{2}R_2 \Rightarrow R_2$; $R_1 + (-5R_2) \Rightarrow R_2$; $\frac{1}{8}R_2 \Rightarrow R_2$

$$\left[\begin{array}{cc|c} 5 & -2 & 4 \\ 2 & -4 & -8 \end{array}\right] \Rightarrow \left[\begin{array}{cc|c} 5 & -2 & 4 \\ 1 & -2 & -4 \end{array}\right] \Rightarrow \left[\begin{array}{cc|c} 5 & -2 & 4 \\ 0 & 8 & 24 \end{array}\right] \Rightarrow \left[\begin{array}{cc|c} 5 & -2 & 4 \\ 0 & 1 & 3 \end{array}\right]$$

From R_2, $y = 3$. From R_1:

$$\begin{aligned} 5x - 2y &= 4 \\ 5x - 2(3) &= 4 \\ 5x = 10 &\Rightarrow x = 2 \end{aligned}$$

The solution is $(2, 3)$.

53. $-3R_1 + R_2 \Rightarrow R_2$

$$\left[\begin{array}{cc|c} 2 & 1 & -4 \\ 6 & 3 & 1 \end{array}\right] \Rightarrow \left[\begin{array}{cc|c} 2 & 1 & -4 \\ 0 & 0 & 13 \end{array}\right]$$

The last equation is impossible. $\emptyset$

55. $-3R_2 + R_1 \Rightarrow R_2$, $-2R_1 + R_3 \Rightarrow R_3$; $\frac{2}{5}R_2 + R_3 \Rightarrow R_3$

$$\left[\begin{array}{ccc|c} 3 & -2 & 4 & 4 \\ 1 & 1 & 1 & 3 \\ 6 & -2 & -3 & 10 \end{array}\right] \Rightarrow \left[\begin{array}{ccc|c} 3 & -2 & 4 & 4 \\ 0 & -5 & 1 & -5 \\ 0 & 2 & -11 & 2 \end{array}\right] \Rightarrow \left[\begin{array}{ccc|c} 3 & -2 & 4 & 4 \\ 0 & -5 & 1 & -5 \\ 0 & 0 & -\frac{53}{5} & 0 \end{array}\right]$$

continued on next page...

55. continued

From R_3, $z = 0$.

From R_2:

$$\begin{aligned}-5y + z &= -5\\ -5y + 0 &= -5\\ -5y &= -5\\ y &= 1\end{aligned}$$

From R_1:

$$\begin{aligned}3x - 2y + 4z &= 4\\ 3x - 2(1) + 4(0) &= 4\\ 3x - 2 &= 4\\ 3x &= 6\\ x &= 2\end{aligned}$$

The solution is $(2, 1, 0)$.

57. $$\left[\begin{array}{cc|c} 3 & -1 & 9\\ -6 & 2 & -18\end{array}\right] \overset{2R_1 + R_2 \Rightarrow R_2}{\Rightarrow} \left[\begin{array}{cc|c} 3 & -1 & 9\\ 0 & 0 & 0\end{array}\right]$$

From R_2:

$$\begin{aligned}3x - y &= 9\\ y &= 3x - 9\end{aligned}$$

The solution is $(x, 3x - 9)$.

59. $$\left[\begin{array}{ccc|c} 1 & 1 & 1 & 6\\ 1 & -1 & 1 & 2\end{array}\right] \overset{-R_1 + R_2 \Rightarrow R_2}{\Rightarrow} \left[\begin{array}{ccc|c} 1 & 1 & 1 & 6\\ 0 & -2 & 0 & -4\end{array}\right] \Rightarrow -2y = -4 \Rightarrow y = 2$$

From R_1:

$$\begin{aligned}x + y + z &= 6\\ x + 2 + z &= 6\\ x &= 4 - z\end{aligned}$$

The solution is $(4 - z, 2, z)$.

61. $$\left[\begin{array}{ccc|c} 1 & 2 & 1 & 1\\ 2 & -1 & 2 & 2\\ 3 & 1 & 3 & 3\end{array}\right] \overset{\substack{-2R_1 + R_2 \Rightarrow R_2\\ -3R_1 + R_3 \Rightarrow R_3}}{\Rightarrow} \left[\begin{array}{ccc|c} 1 & 2 & 1 & 1\\ 0 & -5 & 0 & 0\\ 0 & -5 & 0 & 0\end{array}\right] \overset{R_2 + (-R_3) \Rightarrow R_3}{\Rightarrow} \left[\begin{array}{ccc|c} 1 & 2 & 1 & 1\\ 0 & -5 & 0 & 0\\ 0 & 0 & 0 & 0\end{array}\right]$$

From R_2:

$$\begin{aligned}-5y &= 0\\ y &= 0\end{aligned}$$

From R_1:

$$\begin{aligned}x + 2y + z &= 1\\ x + 2(0) + z &= 1\\ z &= 1 - x\end{aligned}$$

The solution is $(x, 0, 1 - x)$.

63. Let $x =$ the measure of the first angle and $y =$ the measure of the second angle. Form and solve this system of equations: $\begin{cases} x + y = 90\\ y = x + 46\end{cases} \Rightarrow \begin{cases} x + y = 90\\ -x + y = 46\end{cases}$

$$\left[\begin{array}{cc|c} 1 & 1 & 90\\ -1 & 1 & 46\end{array}\right] \overset{R_1 + R_2 \Rightarrow R_2}{\Rightarrow} \left[\begin{array}{cc|c} 1 & 1 & 90\\ 0 & 2 & 136\end{array}\right] \overset{\frac{1}{2}R_2 \Rightarrow R_2}{\Rightarrow} \left[\begin{array}{cc|c} 1 & 1 & 90\\ 0 & 1 & 68\end{array}\right]$$

From R_2, $y = 68$.

From R_1:

$$\begin{aligned}x + y &= 90\\ x + 68 &= 90\\ x &= 22\end{aligned}$$

The angles have measures of 22° and 68°.

65. Let A, B and C represent the measures of the three angles.

$$\begin{cases} A+B+C=180 \\ B=A+25 \\ C=2A-5 \end{cases} \Rightarrow \begin{cases} A+B+C=180 \\ -A+B=25 \\ -2A+C=-5 \end{cases}$$

$$\left[\begin{array}{ccc|c} 1 & 1 & 1 & 180 \\ -1 & 1 & 0 & 25 \\ -2 & 0 & 1 & -5 \end{array}\right] \underset{2R_1+R_3 \Rightarrow R_3}{\overset{R_1+R_2 \Rightarrow R_2}{\Rightarrow}} \left[\begin{array}{ccc|c} 1 & 1 & 1 & 180 \\ 0 & 2 & 1 & 205 \\ 0 & 2 & 3 & 355 \end{array}\right] \overset{-R_2+R_3 \Rightarrow R_3}{\Rightarrow} \left[\begin{array}{ccc|c} 1 & 1 & 1 & 180 \\ 0 & 2 & 1 & 205 \\ 0 & 0 & 2 & 150 \end{array}\right] \overset{\frac{1}{2}R_3 \Rightarrow R_3}{\Rightarrow} \left[\begin{array}{ccc|c} 1 & 1 & 1 & 180 \\ 0 & 2 & 1 & 205 \\ 0 & 0 & 1 & 75 \end{array}\right]$$

From R_3, $C=75$.

From R_2:

$$\begin{aligned} 2B+C&=205 \\ 2B+75&=205 \\ 2B&=130 \\ B&=65 \end{aligned}$$

From R_1:

$$\begin{aligned} A+B+C&=180 \\ A+65+75&=180 \\ A+140&=180 \\ A&=40 \end{aligned}$$

The angles have measures of 40°, 65° and 75°.

67. Plug the coordinates of the points into the general equation to form and solve a system of equations.

$$\begin{aligned} y&=ax^2+bx+c \\ 1&=a(0)^2+b(0)+c \\ 1&=c \end{aligned} \qquad \begin{aligned} y&=ax^2+bx+c \\ 2&=a(1)^2+b(1)+c \\ 2&=a+b+c \end{aligned} \qquad \begin{aligned} y&=ax^2+bx+c \\ 4&=a(-1)^2+b(-1)+c \\ 4&=a-b+c \end{aligned}$$

$$\begin{cases} a+b+c=2 \\ a-b+c=4 \\ c=1 \end{cases} \Rightarrow \left[\begin{array}{ccc|c} 1 & 1 & 1 & 2 \\ 1 & -1 & 1 & 4 \\ 0 & 0 & 1 & 1 \end{array}\right] \overset{-R_2+R_1 \Rightarrow R_2}{\Rightarrow} \left[\begin{array}{ccc|c} 1 & 1 & 1 & 2 \\ 0 & 2 & 0 & -2 \\ 0 & 0 & 1 & 1 \end{array}\right] \overset{\frac{1}{2}R_2 \Rightarrow R_2}{\Rightarrow} \left[\begin{array}{ccc|c} 1 & 1 & 1 & 2 \\ 0 & 1 & 0 & -1 \\ 0 & 0 & 1 & 1 \end{array}\right]$$

From R_3, $c=1$. From R_2, $b=-1$. From R_1:

$$\begin{aligned} a+b+c&=2 \\ a+(-1)+1&=2 \\ a&=2 \end{aligned}$$

The equation is $y=2x^2-x+1$.

69. Let $x=$ the measure of the first angle and $y=$ the measure of the second angle. Form and solve this system of equations:

$$\begin{cases} x+y=180 \\ x=y-28 \end{cases} \Rightarrow \begin{cases} x+y=180 \\ x-y=-28 \end{cases}$$

$$\left[\begin{array}{cc|c} 1 & 1 & 180 \\ 1 & -1 & -28 \end{array}\right] \overset{-R_2+R_1 \Rightarrow R_2}{\Rightarrow} \left[\begin{array}{cc|c} 1 & 1 & 180 \\ 0 & 2 & 208 \end{array}\right] \overset{\frac{1}{2}R_2 \Rightarrow R_2}{\Rightarrow} \left[\begin{array}{cc|c} 1 & 1 & 180 \\ 0 & 1 & 104 \end{array}\right]$$

From R_2, $y=104$. From R_1:

$$\begin{aligned} x+y&=180 \\ x+104&=180 \\ x&=76 \end{aligned}$$

The angles have measures of 76° and 104°.

71. Let $x =$ the number of nickels, $y =$ the number of dimes, and $z =$ the number of quarters.
$$\begin{cases} x + y + z = 64 \\ 5x + 10y + 25z = 600 \\ 10x + 5y + 25z = 500 \end{cases}$$

$$\left[\begin{array}{ccc|c} 1 & 1 & 1 & 64 \\ 5 & 10 & 25 & 600 \\ 10 & 5 & 25 & 500 \end{array}\right] \overset{\substack{-5R_1 + R_2 \Rightarrow R_2 \\ -10R_1 + R_3 \Rightarrow R_3}}{\Rightarrow} \left[\begin{array}{ccc|c} 1 & 1 & 1 & 64 \\ 0 & 5 & 20 & 280 \\ 0 & -5 & 15 & -140 \end{array}\right] \overset{R_2 + R_3 \Rightarrow R_3}{\Rightarrow} \left[\begin{array}{ccc|c} 1 & 1 & 1 & 64 \\ 0 & 5 & 20 & 280 \\ 0 & 0 & 35 & 140 \end{array}\right] \overset{\substack{\frac{1}{5}R_2 \Rightarrow R_2 \\ \frac{1}{35}R_3 \Rightarrow R_3}}{\Rightarrow} \left[\begin{array}{ccc|c} 1 & 1 & 1 & 64 \\ 0 & 1 & 4 & 56 \\ 0 & 0 & 1 & 4 \end{array}\right]$$

From R_3, $z = 4$.

From R_2, $y + 4z = 56$
$$\begin{aligned} y + 4(4) &= 56 \\ y + 16 &= 56 \\ y &= 40 \end{aligned}$$

From R_3, $x + y + z = 64$
$$\begin{aligned} x + 40 + 4 &= 64 \\ x + 44 &= 64 \\ x &= 20 \end{aligned}$$

There are 20 nickels, 40 dimes, and 4 quarters.

73. **Answers may vary.**

75. The last equation represents the equation $0x + 0y + 0z = k$, or $0 = k$. If $k = 0$, then the system can be solved. However, if $k \neq 0$, the system will have no solution.

Exercise 13.4 (page 936)

1. $\begin{vmatrix} 2 & 1 \\ 1 & 1 \end{vmatrix} = 2(1) - 1(1) = 2 - 1 = 1$

3. $\begin{vmatrix} 0 & 1 \\ 0 & 1 \end{vmatrix} = 0(1) - 1(0) = 0 - 0 = 0$

5. $\begin{vmatrix} 5 & 2 \\ 4 & -1 \end{vmatrix}$

7.
$$\begin{aligned} 3(x + 2) - (2 - x) &= x - 5 \\ 3x + 6 - 2 + x &= x - 5 \\ 4x + 4 &= x - 5 \\ 3x &= -9 \\ x &= -3 \end{aligned}$$

9.
$$\begin{aligned} \frac{5}{3}(5x + 6) - 10 &= 0 \\ 3 \cdot \frac{5}{3}(5x + 6) - 3 \cdot 10 &= 3 \cdot 0 \\ 5(5x + 6) - 30 &= 0 \\ 25x + 30 - 30 &= 0 \\ 25x &= 0 \\ x &= 0 \end{aligned}$$

11. number; square

13. $\begin{vmatrix} a_2 & c_2 \\ a_3 & c_3 \end{vmatrix}$

15. Cramer's rule

17. consistent; independent

19. $\begin{vmatrix} 2 & 3 \\ -2 & 1 \end{vmatrix} = 2(1) - 3(-2) = 2 + 6 = 8$

21. $\begin{vmatrix} -1 & 2 \\ 3 & -4 \end{vmatrix} = -1(-4) - 2(3) = 4 - 6 = -2$

SECTION 13.4

23. $\begin{vmatrix} 1 & 0 & 1 \\ 0 & 1 & 0 \\ 1 & 1 & 1 \end{vmatrix} = 1\begin{vmatrix} 1 & 0 \\ 1 & 1 \end{vmatrix} - 0\begin{vmatrix} 0 & 0 \\ 1 & 1 \end{vmatrix} + 1\begin{vmatrix} 0 & 1 \\ 1 & 1 \end{vmatrix} = 1(1) - 0(0) + 1(-1) = 1 - 0 - 1 = 0$

25. $\begin{vmatrix} -1 & 2 & 1 \\ 2 & 1 & -3 \\ 1 & 1 & 1 \end{vmatrix} = -1\begin{vmatrix} 1 & -3 \\ 1 & 1 \end{vmatrix} - 2\begin{vmatrix} 2 & -3 \\ 1 & 1 \end{vmatrix} + 1\begin{vmatrix} 2 & 1 \\ 1 & 1 \end{vmatrix} = -1(4) - 2(5) + 1(1) = -4 - 10 + 1$

$= -13$

27. $\begin{vmatrix} 1 & -2 & 3 \\ -2 & 1 & 1 \\ -3 & -2 & 1 \end{vmatrix} = 1\begin{vmatrix} 1 & 1 \\ -2 & 1 \end{vmatrix} - (-2)\begin{vmatrix} -2 & 1 \\ -3 & 1 \end{vmatrix} + 3\begin{vmatrix} -2 & 1 \\ -3 & -2 \end{vmatrix} = 1(3) + 2(1) + 3(7) = 3 + 2 + 21$

$= 26$

29. $\begin{vmatrix} 1 & 2 & 3 \\ 4 & 5 & 6 \\ 7 & 8 & 9 \end{vmatrix} = 1\begin{vmatrix} 5 & 6 \\ 8 & 9 \end{vmatrix} - 2\begin{vmatrix} 4 & 6 \\ 7 & 9 \end{vmatrix} + 3\begin{vmatrix} 4 & 5 \\ 7 & 8 \end{vmatrix} = 1(-3) - 2(-6) + 3(-3) = -3 + 12 - 9 = 0$

31. $x = \dfrac{\begin{vmatrix} 1 & 1 \\ -7 & -2 \end{vmatrix}}{\begin{vmatrix} 2 & 1 \\ 1 & -2 \end{vmatrix}} = \dfrac{-2 - (-7)}{-4 - 1} = \dfrac{5}{-5} = -1;\ y = \dfrac{\begin{vmatrix} 2 & 1 \\ 1 & -7 \end{vmatrix}}{\begin{vmatrix} 2 & 1 \\ 1 & -2 \end{vmatrix}} = \dfrac{-14 - 1}{-5} = \dfrac{-15}{-5} = 3$

solution: $(-1, 3)$

33. $x = \dfrac{\begin{vmatrix} 6 & 1 \\ 2 & -1 \end{vmatrix}}{\begin{vmatrix} 1 & 1 \\ 1 & -1 \end{vmatrix}} = \dfrac{-6 - 2}{-1 - 1} = \dfrac{-8}{-2} = 4;\ y = \dfrac{\begin{vmatrix} 1 & 6 \\ 1 & 2 \end{vmatrix}}{\begin{vmatrix} 1 & 1 \\ 1 & -1 \end{vmatrix}} = \dfrac{2 - 6}{-2} = \dfrac{-4}{-2} = 2$; solution: $(4, 2)$

35. $\begin{cases} 5x = 3y - 7 \\ y = \dfrac{5x - 7}{3} \end{cases} \Rightarrow \begin{cases} 5x = 3y - 7 \\ 3y = 5x - 7 \end{cases} \Rightarrow \begin{cases} 5x - 3y = -7 \\ -5x + 3y = -7 \end{cases}$

$x = \dfrac{\begin{vmatrix} -7 & -3 \\ -7 & 3 \end{vmatrix}}{\begin{vmatrix} 5 & -3 \\ -5 & 3 \end{vmatrix}} = \dfrac{-21 - 21}{15 - 15} = \dfrac{-42}{0} \Rightarrow$ denominator $= 0$, numerator $\neq 0 \Rightarrow$ no solution

37. $\begin{cases} 2x + 3y = 9 \\ y = -\dfrac{2}{3}x + 3 \end{cases} \Rightarrow \begin{cases} 2x + 3y = 9 \\ 3y = -2x + 9 \end{cases} \Rightarrow \begin{cases} 2x + 3y = 9 \\ 2x + 3y = 9 \end{cases}$

$x = \dfrac{\begin{vmatrix} 9 & 3 \\ 9 & 3 \end{vmatrix}}{\begin{vmatrix} 2 & 3 \\ 2 & 3 \end{vmatrix}} = \dfrac{27 - 27}{6 - 6} = \dfrac{0}{0} \Rightarrow$ denominator $= 0$, numerator $= 0 \Rightarrow$ dependent equations

solution: $\left(x, -\frac{2}{3}x + 3\right)$

SECTION 13.4

Note: In the following problems, D stands for the denominator determinant, while N_x, N_y and N_z stand for the numerator determinants for x, y and z, respectively.

39. $D = \begin{vmatrix} 1 & 1 & 1 \\ 1 & 1 & -1 \\ 1 & -1 & 1 \end{vmatrix} = 1\begin{vmatrix} 1 & -1 \\ -1 & 1 \end{vmatrix} - 1\begin{vmatrix} 1 & -1 \\ 1 & 1 \end{vmatrix} + 1\begin{vmatrix} 1 & 1 \\ 1 & -1 \end{vmatrix} = 1(0) - 1(2) + 1(-2) = -4$

$N_x = \begin{vmatrix} 4 & 1 & 1 \\ 0 & 1 & -1 \\ 2 & -1 & 1 \end{vmatrix} = 4\begin{vmatrix} 1 & -1 \\ -1 & 1 \end{vmatrix} - 1\begin{vmatrix} 0 & -1 \\ 2 & 1 \end{vmatrix} + 1\begin{vmatrix} 0 & 1 \\ 2 & -1 \end{vmatrix} = 4(0) - 1(2) + 1(-2) = -4$

$N_y = \begin{vmatrix} 1 & 4 & 1 \\ 1 & 0 & -1 \\ 1 & 2 & 1 \end{vmatrix} = 1\begin{vmatrix} 0 & -1 \\ 2 & 1 \end{vmatrix} - 4\begin{vmatrix} 1 & -1 \\ 1 & 1 \end{vmatrix} + 1\begin{vmatrix} 1 & 0 \\ 1 & 2 \end{vmatrix} = 1(2) - 4(2) + 1(2) = -4$

$N_z = \begin{vmatrix} 1 & 1 & 4 \\ 1 & 1 & 0 \\ 1 & -1 & 2 \end{vmatrix} = 1\begin{vmatrix} 1 & 0 \\ -1 & 2 \end{vmatrix} - 1\begin{vmatrix} 1 & 0 \\ 1 & 2 \end{vmatrix} + 4\begin{vmatrix} 1 & 1 \\ 1 & -1 \end{vmatrix} = 1(2) - 1(2) + 4(-2) = -8$

$x = \frac{N_x}{D} = \frac{-4}{-4} = 1$; $y = \frac{N_y}{D} = \frac{-4}{-4} = 1$; $z = \frac{N_z}{D} = \frac{-8}{-4} = 2 \Rightarrow$ solution: $(1, 1, 2)$

41. $D = \begin{vmatrix} 1 & 1 & 2 \\ 1 & 2 & 1 \\ 2 & 1 & 1 \end{vmatrix} = 1\begin{vmatrix} 2 & 1 \\ 1 & 1 \end{vmatrix} - 1\begin{vmatrix} 1 & 1 \\ 2 & 1 \end{vmatrix} + 2\begin{vmatrix} 1 & 2 \\ 2 & 1 \end{vmatrix} = 1(1) - 1(-1) + 2(-3) = -4$

$N_x = \begin{vmatrix} 7 & 1 & 2 \\ 8 & 2 & 1 \\ 9 & 1 & 1 \end{vmatrix} = 7\begin{vmatrix} 2 & 1 \\ 1 & 1 \end{vmatrix} - 1\begin{vmatrix} 8 & 1 \\ 9 & 1 \end{vmatrix} + 2\begin{vmatrix} 8 & 2 \\ 9 & 1 \end{vmatrix} = 7(1) - 1(-1) + 2(-10) = -12$

$N_y = \begin{vmatrix} 1 & 7 & 2 \\ 1 & 8 & 1 \\ 2 & 9 & 1 \end{vmatrix} = 1\begin{vmatrix} 8 & 1 \\ 9 & 1 \end{vmatrix} - 7\begin{vmatrix} 1 & 1 \\ 2 & 1 \end{vmatrix} + 2\begin{vmatrix} 1 & 8 \\ 2 & 9 \end{vmatrix} = 1(-1) - 7(-1) + 2(-7) = -8$

$N_z = \begin{vmatrix} 1 & 1 & 7 \\ 1 & 2 & 8 \\ 2 & 1 & 9 \end{vmatrix} = 1\begin{vmatrix} 2 & 8 \\ 1 & 9 \end{vmatrix} - 1\begin{vmatrix} 1 & 8 \\ 2 & 9 \end{vmatrix} + 7\begin{vmatrix} 1 & 2 \\ 2 & 1 \end{vmatrix} = 1(10) - 1(-7) + 7(-3) = -4$

$x = \frac{N_x}{D} = \frac{-12}{-4} = 3$; $y = \frac{N_y}{D} = \frac{-8}{-4} = 2$; $z = \frac{N_z}{D} = \frac{-4}{-4} = 1 \Rightarrow$ solution: $(3, 2, 1)$

43. $D = \begin{vmatrix} 2 & 1 & -1 \\ 1 & 2 & 2 \\ 4 & 5 & 3 \end{vmatrix} = 2\begin{vmatrix} 2 & 2 \\ 5 & 3 \end{vmatrix} - 1\begin{vmatrix} 1 & 2 \\ 4 & 3 \end{vmatrix} + (-1)\begin{vmatrix} 1 & 2 \\ 4 & 5 \end{vmatrix} = 2(-4) - 1(-5) - 1(-3) = 0$

$N_x = \begin{vmatrix} 1 & 1 & -1 \\ 2 & 2 & 2 \\ 3 & 5 & 3 \end{vmatrix} = 1\begin{vmatrix} 2 & 2 \\ 5 & 3 \end{vmatrix} - 1\begin{vmatrix} 2 & 2 \\ 3 & 3 \end{vmatrix} + (-1)\begin{vmatrix} 2 & 2 \\ 3 & 5 \end{vmatrix} = 1(-4) - 1(0) - 1(4) = -8$

$x = \frac{N_x}{D} = \frac{-8}{0} \Rightarrow$ denominator $= 0$, numerator $\neq 0 \Rightarrow$ no solution.

45. $D = \begin{vmatrix} 2 & 3 & 4 \\ 2 & -3 & -4 \\ 4 & 6 & 8 \end{vmatrix} = 2\begin{vmatrix} -3 & -4 \\ 6 & 8 \end{vmatrix} - 3\begin{vmatrix} 2 & -4 \\ 4 & 8 \end{vmatrix} + 4\begin{vmatrix} 2 & -3 \\ 4 & 6 \end{vmatrix} = 2(0) - 3(32) + 4(24) = 0$

$N_x = \begin{vmatrix} 6 & 3 & 4 \\ -4 & -3 & -4 \\ 12 & 6 & 8 \end{vmatrix} = 6\begin{vmatrix} -3 & -4 \\ 6 & 8 \end{vmatrix} - 3\begin{vmatrix} -4 & -4 \\ 12 & 8 \end{vmatrix} + 4\begin{vmatrix} -4 & -3 \\ 12 & 6 \end{vmatrix} = 6(0) - 3(16) + 4(12) = 0$

$x = \frac{N_x}{D} = \frac{0}{0} \Rightarrow$ denominator $= 0$, numerator $= 0 \Rightarrow$ dependent equations.

$$\left[\begin{array}{ccc|c} 2 & 3 & 4 & 6 \\ 2 & -3 & -4 & -4 \\ 4 & 6 & 8 & 12 \end{array}\right] \underset{-2R_1 + R_3 \Rightarrow R_3}{\overset{-R_1 + R_2 \Rightarrow R_2}{\Rightarrow}} \left[\begin{array}{ccc|c} 2 & 3 & 4 & 6 \\ 0 & -6 & -8 & -10 \\ 0 & 0 & 0 & 0 \end{array}\right]$$

From R_2:
$-6y - 8z = -10$
$y = \frac{5}{3} - \frac{4}{3}z$

From R_1:
$2x + 3y + 4z = 6$
$2x + 3\left(\frac{5}{3} - \frac{4}{3}z\right) + 4z = 6$
$2x + 5 - 4z + 4z = 6$
$x = \frac{1}{2}$

The solution is $\left(\frac{1}{2}, \frac{5}{3} - \frac{4}{3}z, z\right)$.

47. $\begin{vmatrix} x & y \\ y & x \end{vmatrix} = x(x) - y(y) = x^2 - y^2$

49. $\begin{vmatrix} a & 2a & -a \\ 2 & -1 & 3 \\ 1 & 2 & -3 \end{vmatrix} = a\begin{vmatrix} -1 & 3 \\ 2 & -3 \end{vmatrix} - 2a\begin{vmatrix} 2 & 3 \\ 1 & -3 \end{vmatrix} + (-a)\begin{vmatrix} 2 & -1 \\ 1 & 2 \end{vmatrix} = a(-3) - 2a(-9) - a(5)$

$= -3a + 18a - 5a = 10a$

51. $\begin{vmatrix} 1 & a & b \\ 1 & 2a & 2b \\ 1 & 3a & 3b \end{vmatrix} = 1\begin{vmatrix} 2a & 2b \\ 3a & 3b \end{vmatrix} - a\begin{vmatrix} 1 & 2b \\ 1 & 3b \end{vmatrix} + b\begin{vmatrix} 1 & 2a \\ 1 & 3a \end{vmatrix} = 1(0) - a(b) + b(a) = 0 - ab + ab = 0$

53. $\begin{vmatrix} 2 & -3 & 4 \\ -1 & 2 & 4 \\ 3 & -3 & 1 \end{vmatrix} = -23$

55. $\begin{vmatrix} 2 & 1 & -3 \\ -2 & 2 & 4 \\ 1 & -2 & 2 \end{vmatrix} = 26$

57. $x = \frac{\begin{vmatrix} 0 & 3 \\ -4 & -6 \end{vmatrix}}{\begin{vmatrix} 2 & 3 \\ 4 & -6 \end{vmatrix}} = \frac{0 - (-12)}{-12 - 12} = \frac{12}{-24} = -\frac{1}{2}$; $y = \frac{\begin{vmatrix} 2 & 0 \\ 4 & -4 \end{vmatrix}}{\begin{vmatrix} 2 & 3 \\ 4 & -6 \end{vmatrix}} = \frac{-8 - 0}{-24} = \frac{-8}{-24} = \frac{1}{3}$

solution: $\left(-\frac{1}{2}, \frac{1}{3}\right)$

59. $\begin{cases} y = \dfrac{-2x+1}{3} \\ 3x - 2y = 8 \end{cases} \Rightarrow \begin{cases} 3y = -2x+1 \\ 3x - 2y = 8 \end{cases} \Rightarrow \begin{cases} 2x + 3y = 1 \\ 3x - 2y = 8 \end{cases}$

$x = \dfrac{\begin{vmatrix} 1 & 3 \\ 8 & -2 \end{vmatrix}}{\begin{vmatrix} 2 & 3 \\ 3 & -2 \end{vmatrix}} = \dfrac{-2-24}{-4-9} = \dfrac{-26}{-13} = 2$; $y = \dfrac{\begin{vmatrix} 2 & 1 \\ 3 & 8 \end{vmatrix}}{\begin{vmatrix} 2 & 3 \\ 3 & -2 \end{vmatrix}} = \dfrac{16-3}{-13} = \dfrac{13}{-13} = -1$; solution: $(2, -1)$

61. $\begin{cases} x = \dfrac{5y-4}{2} \\ y = \dfrac{3x-1}{5} \end{cases} \Rightarrow \begin{cases} 2x = 5y - 4 \\ 5y = 3x - 1 \end{cases} \Rightarrow \begin{cases} 2x - 5y = -4 \\ -3x + 5y = -1 \end{cases}$

$x = \dfrac{\begin{vmatrix} -4 & -5 \\ -1 & 5 \end{vmatrix}}{\begin{vmatrix} 2 & -5 \\ -3 & 5 \end{vmatrix}} = \dfrac{-20-5}{10-15} = \dfrac{-25}{-5} = 5$; $y = \dfrac{\begin{vmatrix} 2 & -4 \\ -3 & -1 \end{vmatrix}}{\begin{vmatrix} 2 & -5 \\ -3 & 5 \end{vmatrix}} = \dfrac{-2-12}{-5} = \dfrac{-14}{-5} = \dfrac{14}{5}$

solution: $\left(5, \frac{14}{5}\right)$

63. $D = \begin{vmatrix} 2 & 1 & 1 \\ 1 & -2 & 3 \\ 1 & 1 & -4 \end{vmatrix} = 2\begin{vmatrix} -2 & 3 \\ 1 & -4 \end{vmatrix} - 1\begin{vmatrix} 1 & 3 \\ 1 & -4 \end{vmatrix} + 1\begin{vmatrix} 1 & -2 \\ 1 & 1 \end{vmatrix} = 2(5) - 1(-7) + 1(3) = 20$

$N_x = \begin{vmatrix} 5 & 1 & 1 \\ 10 & -2 & 3 \\ -3 & 1 & -4 \end{vmatrix} = 5\begin{vmatrix} -2 & 3 \\ 1 & -4 \end{vmatrix} - 1\begin{vmatrix} 10 & 3 \\ -3 & -4 \end{vmatrix} + 1\begin{vmatrix} 10 & -2 \\ -3 & 1 \end{vmatrix}$

$= 5(5) - 1(-31) + 1(4) = 60$

$N_y = \begin{vmatrix} 2 & 5 & 1 \\ 1 & 10 & 3 \\ 1 & -3 & -4 \end{vmatrix} = 2\begin{vmatrix} 10 & 3 \\ -3 & -4 \end{vmatrix} - 5\begin{vmatrix} 1 & 3 \\ 1 & -4 \end{vmatrix} + 1\begin{vmatrix} 1 & 10 \\ 1 & -3 \end{vmatrix}$

$= 2(-31) - 5(-7) + 1(-13) = -40$

$N_z = \begin{vmatrix} 2 & 1 & 5 \\ 1 & -2 & 10 \\ 1 & 1 & -3 \end{vmatrix} = 2\begin{vmatrix} -2 & 10 \\ 1 & -3 \end{vmatrix} - 1\begin{vmatrix} 1 & 10 \\ 1 & -3 \end{vmatrix} + 5\begin{vmatrix} 1 & -2 \\ 1 & 1 \end{vmatrix} = 2(-4) - 1(-13) + 5(3) = 20$

$x = \dfrac{N_x}{D} = \dfrac{60}{20} = 3$; $y = \dfrac{N_y}{D} = \dfrac{-40}{20} = -2$; $z = \dfrac{N_z}{D} = \dfrac{20}{20} = 1 \Rightarrow$ solution: $(3, -2, 1)$

65. $D = \begin{vmatrix} 4 & 0 & 3 \\ 0 & 2 & -6 \\ 8 & 4 & 3 \end{vmatrix} = 4\begin{vmatrix} 2 & -6 \\ 4 & 3 \end{vmatrix} - 0\begin{vmatrix} 0 & -6 \\ 8 & 3 \end{vmatrix} + 3\begin{vmatrix} 0 & 2 \\ 8 & 4 \end{vmatrix} = 4(30) - 0(48) + 3(-16) = 72$

$N_x = \begin{vmatrix} 4 & 0 & 3 \\ -1 & 2 & -6 \\ 9 & 4 & 3 \end{vmatrix} = 4\begin{vmatrix} 2 & -6 \\ 4 & 3 \end{vmatrix} - 0\begin{vmatrix} -1 & -6 \\ 9 & 3 \end{vmatrix} + 3\begin{vmatrix} -1 & 2 \\ 9 & 4 \end{vmatrix} = 4(30) - 0(51) + 3(-22) = 54$

$N_y = \begin{vmatrix} 4 & 4 & 3 \\ 0 & -1 & -6 \\ 8 & 9 & 3 \end{vmatrix} = 4\begin{vmatrix} -1 & -6 \\ 9 & 3 \end{vmatrix} - 4\begin{vmatrix} 0 & -6 \\ 8 & 3 \end{vmatrix} + 3\begin{vmatrix} 0 & -1 \\ 8 & 9 \end{vmatrix} = 4(51) - 4(48) + 3(8) = 36$

continued on next page...

65. continued

$$N_z = \begin{vmatrix} 4 & 0 & 4 \\ 0 & 2 & -1 \\ 8 & 4 & 9 \end{vmatrix} = 4\begin{vmatrix} 2 & -1 \\ 4 & 9 \end{vmatrix} - 0\begin{vmatrix} 0 & -1 \\ 8 & 9 \end{vmatrix} + 4\begin{vmatrix} 0 & 2 \\ 8 & 4 \end{vmatrix} = 4(22) - 0(8) + 4(-16) = 24$$

$$x = \frac{N_x}{D} = \frac{54}{72} = \frac{3}{4}; y = \frac{N_y}{D} = \frac{36}{72} = \frac{1}{2}; z = \frac{N_z}{D} = \frac{24}{72} = \frac{1}{3} \Rightarrow \text{solution: } \left(\frac{3}{4}, \frac{1}{2}, \frac{1}{3}\right)$$

67. $\begin{cases} x + y = 1 \Rightarrow & x + y = 1 \\ \frac{1}{2}y + z = \frac{5}{2} \Rightarrow & y + 2z = 5 \\ x - z = -3 \Rightarrow & x - z = -3 \end{cases}$

$$D = \begin{vmatrix} 1 & 1 & 0 \\ 0 & 1 & 2 \\ 1 & 0 & -1 \end{vmatrix} = 1\begin{vmatrix} 1 & 2 \\ 0 & -1 \end{vmatrix} - 1\begin{vmatrix} 0 & 2 \\ 1 & -1 \end{vmatrix} + 0\begin{vmatrix} 0 & 1 \\ 1 & 0 \end{vmatrix} = 1(-1) - 1(-2) + 0(-1) = 1$$

$$N_x = \begin{vmatrix} 1 & 1 & 0 \\ 5 & 1 & 2 \\ -3 & 0 & -1 \end{vmatrix} = 1\begin{vmatrix} 1 & 2 \\ 0 & -1 \end{vmatrix} - 1\begin{vmatrix} 5 & 2 \\ -3 & -1 \end{vmatrix} + 0\begin{vmatrix} 5 & 1 \\ -3 & 0 \end{vmatrix} = 1(-1) - 1(1) + 0(3) = -2$$

$$N_y = \begin{vmatrix} 1 & 1 & 0 \\ 0 & 5 & 2 \\ 1 & -3 & -1 \end{vmatrix} = 1\begin{vmatrix} 5 & 2 \\ -3 & -1 \end{vmatrix} - 1\begin{vmatrix} 0 & 2 \\ 1 & -1 \end{vmatrix} + 0\begin{vmatrix} 0 & 5 \\ 1 & -3 \end{vmatrix} = 1(1) - 1(-2) + 0(-5) = 3$$

$$N_z = \begin{vmatrix} 1 & 1 & 1 \\ 0 & 1 & 5 \\ 1 & 0 & -3 \end{vmatrix} = 1\begin{vmatrix} 1 & 5 \\ 0 & -3 \end{vmatrix} - 1\begin{vmatrix} 0 & 5 \\ 1 & -3 \end{vmatrix} + 1\begin{vmatrix} 0 & 1 \\ 1 & 0 \end{vmatrix} = 1(-3) - 1(-5) + 1(-1) = 1$$

$$x = \frac{N_x}{D} = \frac{-2}{1} = -2; y = \frac{N_y}{D} = \frac{3}{1} = 3; z = \frac{N_z}{D} = \frac{1}{1} = 1 \Rightarrow \text{solution: } (-2, 3, 1)$$

69. $\begin{vmatrix} x & 1 \\ 3 & 2 \end{vmatrix} = 1$

$$\begin{aligned} 2x - 3 &= 1 \\ 2x &= 4 \\ x &= 2 \end{aligned}$$

71. $\begin{vmatrix} x & -2 \\ 3 & 1 \end{vmatrix} = \begin{vmatrix} 4 & 2 \\ x & 3 \end{vmatrix}$

$$\begin{aligned} x - (-6) &= 12 - 2x \\ x + 6 &= 12 - 2x \\ 3x &= 6 \\ x &= 2 \end{aligned}$$

73. $\begin{cases} 2x + y = 180 \\ y = x + 30 \end{cases} \Rightarrow \begin{cases} 2x + y = 180 \\ -x + y = 30 \end{cases}$

$$x = \frac{\begin{vmatrix} 180 & 1 \\ 30 & 1 \end{vmatrix}}{\begin{vmatrix} 2 & 1 \\ -1 & 1 \end{vmatrix}} = \frac{180 - 30}{2 - (-1)} = \frac{150}{3} = 50; y = \frac{\begin{vmatrix} 2 & 180 \\ -1 & 30 \end{vmatrix}}{\begin{vmatrix} 2 & 1 \\ -1 & 1 \end{vmatrix}} = \frac{60 - (-180)}{3} = \frac{240}{3} = 80$$

75. Let x = the amount invested in HiTech, y = the amount invested in SaveTel and z = the amount invested in HiGas. Form and solve the following system of equations:

$$\begin{cases} x+y+z=20000 \Rightarrow & x+y+z=20000 \\ 0.10x+0.05y+0.06z=0.066(20000) \Rightarrow & 10x+5y+6x=132000 \\ y+z=3x \Rightarrow & -3x+y+z=0 \end{cases}$$

$$D=\begin{vmatrix} 1 & 1 & 1 \\ 10 & 5 & 6 \\ -3 & 1 & 1 \end{vmatrix}=1\begin{vmatrix} 5 & 6 \\ 1 & 1 \end{vmatrix}-1\begin{vmatrix} 10 & 6 \\ -3 & 1 \end{vmatrix}+1\begin{vmatrix} 10 & 5 \\ -3 & 1 \end{vmatrix}=1(-1)-1(28)+1(25)=-4$$

$$N_x=\begin{vmatrix} 20000 & 1 & 1 \\ 132000 & 5 & 6 \\ 0 & 1 & 1 \end{vmatrix}=20000\begin{vmatrix} 5 & 6 \\ 1 & 1 \end{vmatrix}-1\begin{vmatrix} 132000 & 6 \\ 0 & 1 \end{vmatrix}+1\begin{vmatrix} 132000 & 5 \\ 0 & 1 \end{vmatrix}$$

$$=20000(-1)-1(132000)+1(132000)=-20000$$

$$N_y=\begin{vmatrix} 1 & 20000 & 1 \\ 10 & 132000 & 6 \\ -3 & 0 & 1 \end{vmatrix}=1\begin{vmatrix} 132000 & 6 \\ 0 & 1 \end{vmatrix}-20000\begin{vmatrix} 10 & 6 \\ -3 & 1 \end{vmatrix}+1\begin{vmatrix} 10 & 132000 \\ -3 & 0 \end{vmatrix}$$

$$=1(132000)-20000(28)+1(396000)=-32000$$

$$N_z=\begin{vmatrix} 1 & 1 & 20000 \\ 10 & 5 & 132000 \\ -3 & 1 & 0 \end{vmatrix}=1\begin{vmatrix} 5 & 132000 \\ 1 & 0 \end{vmatrix}-1\begin{vmatrix} 10 & 132000 \\ -3 & 0 \end{vmatrix}+20000\begin{vmatrix} 10 & 5 \\ -3 & 1 \end{vmatrix}$$

$$=1(-132000)-1(396000)+20000(25)=-28000$$

$$x=\frac{N_x}{D}=\frac{-20000}{-4}=5000;\ y=\frac{N_y}{D}=\frac{-32000}{-4}=8000;\ z=\frac{N_z}{D}=\frac{-28000}{-4}=7000$$

He should invest \$5000 in HiTech, \$8000 in SaveTel and \$7000 in HiGas.

77. **Answers may vary.**

79.
$$\begin{vmatrix} x & y & 1 \\ -2 & 3 & 1 \\ 3 & 5 & 1 \end{vmatrix}=0$$

$$x\begin{vmatrix} 3 & 1 \\ 5 & 1 \end{vmatrix}-y\begin{vmatrix} -2 & 1 \\ 3 & 1 \end{vmatrix}+1\begin{vmatrix} -2 & 3 \\ 3 & 5 \end{vmatrix}=0$$

$$x(3-5)-y(-2-3)+1(-10-9)=0$$

$$-2x+5y-19=0$$

$$2x-5y=-19$$ Verify that both points satisfy this equation.

81.
$$\begin{vmatrix} 1 & 0 & 2 & 1 \\ 2 & 1 & 1 & 3 \\ 1 & 1 & 1 & 1 \\ 2 & 1 & 1 & 1 \end{vmatrix}=1\begin{vmatrix} 1 & 1 & 3 \\ 1 & 1 & 1 \\ 1 & 1 & 1 \end{vmatrix}-0\begin{vmatrix} 2 & 1 & 3 \\ 1 & 1 & 1 \\ 2 & 1 & 1 \end{vmatrix}+2\begin{vmatrix} 2 & 1 & 3 \\ 1 & 1 & 1 \\ 2 & 1 & 1 \end{vmatrix}-1\begin{vmatrix} 2 & 1 & 1 \\ 1 & 1 & 1 \\ 2 & 1 & 1 \end{vmatrix}$$

$$=1(0)-0(???)+2(-2)-1(0)=-4$$

Exercise 13.5 (page 944)

1. 0, 1, 2

3. 0, 1, 2, 3, 4

5. $\sqrt{200x^2} - 3\sqrt{98x^2} = \sqrt{100x^2}\sqrt{2} - 3\sqrt{49x^2}\sqrt{2} = 10x\sqrt{2} - 3(7x)\sqrt{2} = -11x\sqrt{2}$

7. $\dfrac{3t\sqrt{2t} - 2\sqrt{2t^3}}{\sqrt{18t} - \sqrt{2t}} = \dfrac{3t\sqrt{2t} - 2\sqrt{t^2}\sqrt{2t}}{\sqrt{9}\sqrt{2t} - \sqrt{2t}} = \dfrac{3t\sqrt{2t} - 2t\sqrt{2t}}{3\sqrt{2t} - \sqrt{2t}} = \dfrac{t\sqrt{2t}}{2\sqrt{2t}} = \dfrac{t}{2}$

9. graphing; substitution

11. four

13. $\begin{cases} 8x^2 + 32y^2 = 256 \\ x = 2y \end{cases}$

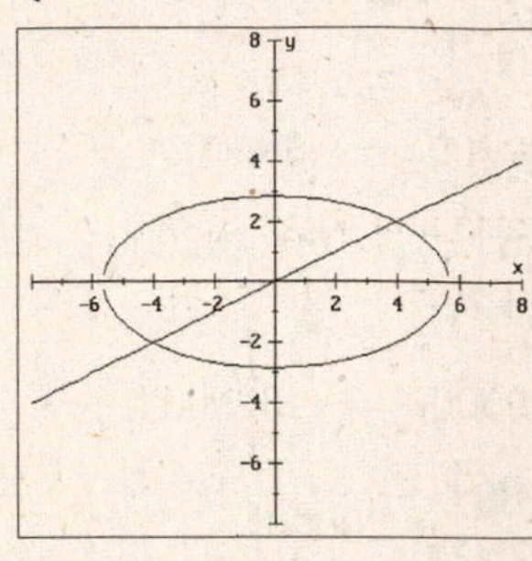

$(-4,-2), (4,2)$

15. $\begin{cases} x^2 + y^2 = 10 \\ y = 3x^2 \end{cases}$

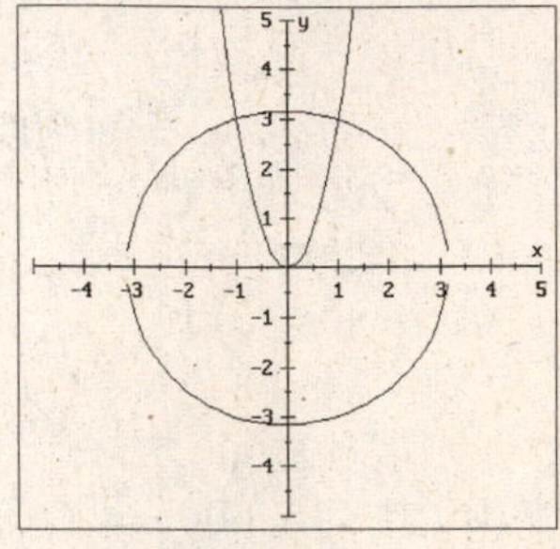

$(-1,3), (1,3)$

17. $\begin{cases} x^2 + y^2 = 25 \\ 12x^2 + 64y^2 = 768 \end{cases}$

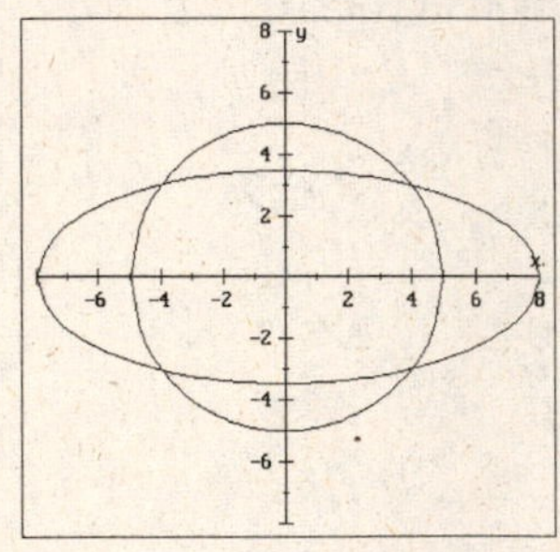

$(-4,3), (4,3), (4,-3)$
$(-4,-3)$

19. $\begin{cases} x^2 - 13 = -y^2 \\ y = 2x - 4 \end{cases}$

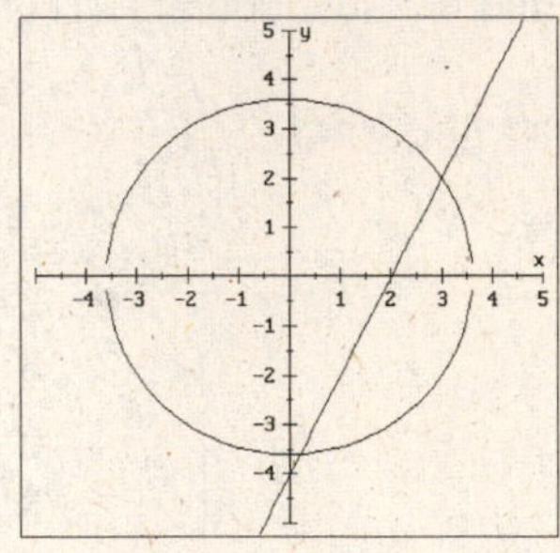

$\left(\frac{1}{5}, -\frac{18}{5}\right), (3,2)$

21. $\begin{cases} (1) \quad 25x^2 + 9y^2 = 225 \\ (2) \quad 5x + 3y = 15 \end{cases}$

Substitute $x = -\frac{3}{5}y + 3$ from (2) into (1):

$$25x^2 + 9y^2 = 225$$
$$25\left(-\tfrac{3}{5}y + 3\right)^2 + 9y^2 = 225$$
$$25\left(\tfrac{9}{25}y^2 - \tfrac{18}{5}y + 9\right) + 9y^2 = 225$$
$$9y^2 - 90y + 225 + 9y^2 = 225$$
$$18y^2 - 90y = 0$$
$$18y(y-5) = 0$$

$18y = 0$ **or** $y - 5 = 0$
$y = 0 \qquad y = 5$

Substitute these and solve for x:

$5x + 3y = 15$	$5x + 3y = 15$
$5x + 3(0) = 15$	$5x + 3(5) = 15$
$5x = 15$	$5x = 0$
$x = 3$	$x = 0$

Solutions: $(3,0), (0,5)$

SECTION 13.5

23. $\begin{cases} (1) & x^2+y^2=2 \\ (2) & x+y=2 \end{cases}$

Substitute $x=2-y$ from (2) into (1):

$$\begin{aligned} x^2+y^2&=2 \\ (2-y)^2+y^2&=2 \\ 4-4y+y^2+y^2&=2 \\ 2y^2-4y+2&=0 \\ 2(y-1)(y-1)&=0 \end{aligned}$$

$y-1=0$ **or** $y-1=0$

$y=1$ $\qquad y=1$

Substitute this and solve for x:

$$\begin{aligned} x&=2-y \\ x&=2-1 \\ x&=1 \end{aligned}$$

Solution: $(1,1)$

25. $\begin{cases} (1) & x^2+y^2=5 \\ (2) & x+y=3 \end{cases}$

Substitute $x=3-y$ from (2) into (1):

$$\begin{aligned} x^2+y^2&=5 \\ (3-y)^2+y^2&=5 \\ 9-6y+y^2+y^2&=5 \\ 2y^2-6y+4&=0 \\ 2(y-2)(y-1)&=0 \end{aligned}$$

$y-2=0$ **or** $y-1=0$

$y=2$ $\qquad y=1$

Substitute these and solve for x:

$x=3-y \qquad x=3-y$

$x=3-2 \qquad x=3-1$

$x=1 \qquad x=2$

Solutions: $(1,2),(2,1)$

27. $\begin{cases} (1) & x^2+y^2=13 \\ (2) & y=x^2-1 \end{cases}$

Substitute $x^2=13-y^2$ from (1) into (2):

$$\begin{aligned} y&=x^2-1 \\ y&=13-y^2-1 \\ y^2+y-12&=0 \\ (y+4)(y-3)&=0 \end{aligned}$$

$y+4=0$ **or** $y-3=0$

$y=-4$ $\qquad y=3$

Substitute these and solve for x:

$x^2=13-y^2 \qquad x^2=13-y^2$

$x^2=13-16 \qquad x^2=13-9$

$x^2=-4 \qquad x^2=4$

complex $\qquad x=\pm 2$

Solutions: $(2,3),(-2,3)$

29. $\begin{cases} (1) & x^2+y^2=30 \\ (2) & y=x^2 \end{cases}$

Substitute $x^2=y$ from (2) into (1):

$$\begin{aligned} x^2+y^2&=30 \\ y+y^2&=30 \\ y^2+y-30&=0 \\ (y-5)(y+6)&=0 \end{aligned}$$

$y-5=0$ **or** $y+6=0$

$y=5$ $\qquad y=-6$

Substitute these and solve for x:

$x^2=y \qquad x^2=y$

$x^2=5 \qquad x^2=-6$

$x=\pm\sqrt{5} \qquad$ complex

Solutions: $\left(\sqrt{5},5\right),\left(-\sqrt{5},5\right)$

31.

$$\begin{aligned} x^2+y^2&=13 \\ x^2-y^2&=5 \\ \hline 2x^2&=18 \\ x^2&=9 \\ x&=\pm 3 \end{aligned}$$

Substitute and solve for y:

$x^2+y^2=13 \qquad x^2+y^2=13$

$3^2+y^2=13 \qquad (-3)^2+y^2=13$

$y^2=4 \qquad y^2=4$

$y=\pm 2 \qquad y=\pm 2$

Solutions: $(3,2),(3,-2),(-3,2),(-3,-2)$

33. $\begin{cases} (1) & y = x^2 - 4 \\ (2) & x^2 - y^2 = -16 \end{cases}$

Substitute $x^2 = y + 4$ from (1) into (2):

$$x^2 - y^2 = -16$$
$$y + 4 - y^2 = -16$$
$$y^2 - y - 20 = 0$$
$$(y+4)(y-5) = 0$$

$y + 4 = 0$ **or** $y - 5 = 0$
$y = -4$ $\quad y = 5$

Substitute these and solve for x:

$x^2 = y + 4 \qquad x^2 = y + 4$
$x^2 = -4 + 4 \qquad x^2 = 5 + 4$
$x^2 = 0 \qquad x^2 = 9$
$x = 0 \qquad x = \pm 3$

Solutions: $(0, -4), (3, 5), (-3, 5)$

35. $x^2 - y^2 = -5 \Rightarrow (\times 2)\ 2x^2 - 2y^2 = -10$
$3x^2 + 2y^2 = 30 \Rightarrow \quad 3x^2 + 2y^2 = 30$

$$5x^2 = 20$$
$$x^2 = 4$$
$$x = \pm 2$$

Substitute and solve for y:

$x^2 - y^2 = -5 \qquad x^2 - y^2 = -5$
$2^2 - y^2 = -5 \qquad (-2)^2 - y^2 = -5$
$-y^2 = -9 \qquad -y^2 = -9$
$y^2 = 9 \qquad y^2 = 9$
$y = \pm 3 \qquad y = \pm 3$

Solutions: $(2, 3), (2, -3), (-2, 3), (-2, -3)$

37. $\begin{cases} 2x - y > 4 \\ y < -x^2 + 2 \end{cases}$

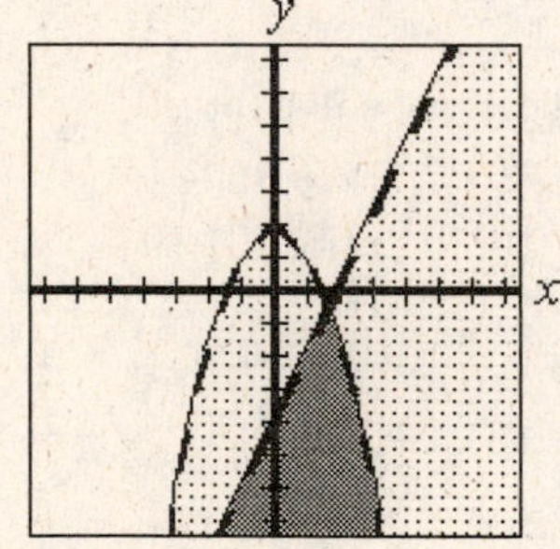

39. $\begin{cases} y > x^2 - 4 \\ y < -x^2 + 4 \end{cases}$

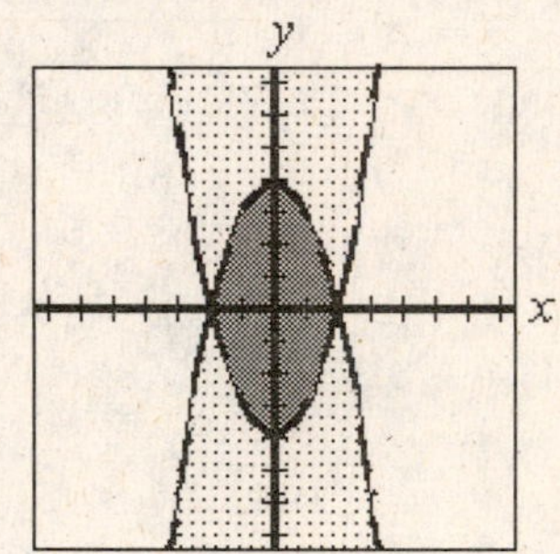

41. $\begin{cases} (1) & x^2 + y^2 = 5 \\ (2) & y = x + 1 \end{cases}$

Substitute $y = x + 1$ from (2) into (1):

$$x^2 + y^2 = 5$$
$$x^2 + (x+1)^2 = 5$$
$$x^2 + x^2 + 2x + 1 = 5$$
$$2x^2 + 2x - 4 = 0$$
$$2(x+2)(x-1) = 0$$

$x + 2 = 0$ **or** $x - 1 = 0$
$x = -2 \qquad x = 1$

Substitute these and solve for x:

$y = x + 1 \qquad y = x + 1$
$y = -2 + 1 \qquad y = 1 + 1$
$y = -1 \qquad y = 2$

Solutions: $(-2, -1), (1, 2)$

43. $\begin{aligned} x^2 + y^2 &= 20 \\ x^2 - y^2 &= -12 \\ \hline 2x^2 &= 8 \\ x^2 &= 4 \\ x &= \pm 2 \end{aligned}$

Substitute and solve for y:

$\begin{aligned} x^2 + y^2 &= 20 \\ 2^2 + y^2 &= 20 \\ y^2 &= 16 \\ y &= \pm 4 \end{aligned}$ $\qquad$ $\begin{aligned} x^2 + y^2 &= 20 \\ (-2)^2 + y^2 &= 20 \\ y^2 &= 16 \\ y &= \pm 4 \end{aligned}$

Solutions: $(2, 4), (2, -4), (-2, 4), (-2, -4)$

45. $\begin{cases} (1) & y^2 = 40 - x^2 \\ (2) & y = x^2 - 10 \end{cases}$

Substitute $x^2 = 40 - y^2$ from (1) into (2):

$$\begin{aligned} y &= x^2 - 10 \\ y &= 40 - y^2 - 10 \\ y^2 + y - 30 &= 0 \\ (y + 6)(y - 5) &= 0 \end{aligned}$$

$\begin{aligned} y + 6 &= 0 \\ y &= -6 \end{aligned}$ **or** $\begin{aligned} y - 5 &= 0 \\ y &= 5 \end{aligned}$

Substitute these and solve for x:

$\begin{aligned} x^2 &= 40 - y^2 \\ x^2 &= 40 - 36 \\ x^2 &= 4 \\ x &= \pm 2 \end{aligned}$ $\qquad$ $\begin{aligned} x^2 &= 40 - y^2 \\ x^2 &= 40 - 25 \\ x^2 &= 15 \\ x &= \pm \sqrt{15} \end{aligned}$

$(2, -6), (-2, -6), \left(\sqrt{15}, 5\right), \left(-\sqrt{15}, 5\right)$

47. $\begin{aligned} \tfrac{1}{x} + \tfrac{2}{y} &= 1 \Rightarrow \\ \tfrac{2}{x} - \tfrac{1}{y} &= \tfrac{1}{3} \Rightarrow (\times 2) \end{aligned}$ $\qquad$ $\begin{aligned} \tfrac{1}{x} + \tfrac{2}{y} &= 1 \\ \tfrac{4}{x} - \tfrac{2}{y} &= \tfrac{2}{3} \\ \hline \tfrac{5}{x} &= \tfrac{5}{3} \\ 15 &= 5x \\ 3 &= x \end{aligned}$

Substitute and solve for y:

$$\begin{aligned} \tfrac{1}{x} + \tfrac{2}{y} &= 1 \\ \tfrac{1}{3} + \tfrac{2}{y} &= 1 \\ \tfrac{2}{y} &= \tfrac{2}{3} \\ 6 &= 2y \\ 3 &= y \end{aligned}$$

Solution: $(3, 3)$

49. $\begin{cases} (1) & 3y^2 = xy \\ (2) & 2x^2 + xy - 84 = 0 \end{cases}$

From (1): $\begin{aligned} 3y^2 - xy &= 0 \\ y(3y - x) &= 0 \end{aligned}$

$y = 0$ or $y = \dfrac{1}{3}x$

Substitute these into (2):

$\begin{aligned} 2x^2 + xy - 84 &= 0 \\ 2x^2 + x(0) - 84 &= 0 \\ 2x^2 &= 84 \\ x^2 &= 42 \\ x &= \pm \sqrt{42} \end{aligned}$ $\qquad$ $\begin{aligned} 2x^2 + xy - 84 &= 0 \\ 2x^2 + x\left(\tfrac{1}{3}x\right) - 84 &= 0 \\ 2x^2 + \tfrac{1}{3}x^2 &= 84 \\ 6x^2 + x^2 &= 252 \\ 7x^2 &= 252 \\ x^2 &= 36 \\ x &= \pm 6 \end{aligned}$

(substitute and solve for y)

Solutions: $\left(\sqrt{42}, 0\right), \left(-\sqrt{42}, 0\right), (6, 2), (-6, -2)$

51. $\begin{cases} (1) \quad xy = \frac{1}{6} \\ (2) \quad y + x = 5xy \end{cases}$

Substitute $x = \frac{1}{6y}$ from (1) into (2):

$$y + \frac{1}{6y} = \frac{5y}{6y}$$
$$6y^2 + 1 = 5y$$
$$6y^2 - 5y + 1 = 0$$
$$(2y - 1)(3y - 1) = 0$$

$2y - 1 = 0$ **or** $3y - 1 = 0$

$y = \frac{1}{2}$ $\qquad y = \frac{1}{3}$

Substitute these and solve for x:

$x = \frac{1}{6y}$ $\qquad x = \frac{1}{6y}$

$x = \dfrac{1}{6\left(\frac{1}{2}\right)}$ $\qquad x = \dfrac{1}{6\left(\frac{1}{3}\right)}$

$x = \frac{1}{3}$ $\qquad x = \frac{1}{2}$

Solutions: $\left(\frac{1}{3}, \frac{1}{2}\right), \left(\frac{1}{2}, \frac{1}{3}\right)$

53. Let the integers be x and y. Then the equations are

$\begin{cases} (1) \quad xy = 32 \\ (2) \quad x + y = 12 \end{cases}$

Substitute $x = \frac{32}{y}$ from (1) into (2):

$$\frac{32}{y} + y = 12$$
$$32 + y^2 = 12y$$
$$y^2 - 12y + 32 = 0$$
$$(y - 4)(y - 8) = 0$$

$y - 4 = 0$ **or** $y - 8 = 0$

$y = 4$ $\qquad y = 8$

Substitute these and solve for x:

$x = \frac{32}{y} = \frac{32}{4} = 8$ $\qquad x = \frac{32}{y} = \frac{32}{8} = 4$

The integers are 8 and 4.

55. Let $l =$ the length of the rectangle, and $w =$ the width of the rectangle. Then the equations are:

$\begin{cases} (1) \quad lw = 63 \\ (2) \quad 2l + 2w = 32 \end{cases}$

Substitute $l = \frac{63}{w}$ from (1) into (2):

$$2\left(\frac{63}{w}\right) + 2w = 32$$
$$\frac{126}{w} + 2w = 32$$
$$126 + 2w^2 = 32w$$
$$2w^2 - 32y + 126 = 0$$
$$2(w - 7)(w - 9) = 0$$

$w - 7 = 0$ **or** $w - 9 = 0$

$w = 7$ $\qquad w = 9$

Substitute these and solve for l:

$l = \frac{63}{w} = \frac{63}{7} = 9$ $\qquad l = \frac{63}{w} = \frac{63}{9} = 7$

The dimensions are 7 cm by 9 cm.

57. Let $r =$ Carol's rate, and let $p =$ the amount Carol invested.
Then Juan invested $p + 150$ at a rate of $r + 0.015$. The equations are

$\begin{cases} (1) \quad pr = 67.50 \Rightarrow p = \frac{67.5}{r} \\ (2) \quad (p + 150)(r + 0.015) = 94.5 \end{cases}$

Substitute $p = \frac{67.5}{r}$ from (1) into (2):

$$\left(\frac{67.5}{r} + 150\right)(r + 0.015) = 94.5$$
$$67.5 + \frac{1.0125}{r} + 150r + 2.25 = 94.5$$
$$67.5r + 1.0125 + 150r^2 + 2.25r = 94.5r$$
$$150r^2 - 24.75r - 1.0125 = 0$$
$$12{,}000r^2 - 1980r + 81 = 0$$
$$(100r - 9)(120r - 9) = 0$$

$100r - 9 = 0$ **or** $120r - 9 = 0$

$r = 0.09$ $\qquad r = 0.075$

Substitute and solve for p:

$p = \frac{67.5}{r} = \frac{67.5}{0.09} = 750$ or $p = \frac{67.5}{r} = \frac{67.5}{0.075} = 900$

Carol invested \$750 at 9% or \$900 at 7.5%.

59. Let r = Jim's rate and t = Jim's time. Then his brother's rate was $r - 17$ and his time was $t + 1.5$.

$$\begin{cases} (1)\ rt = 306 \Rightarrow t = \frac{306}{r} \\ (2)\ (r-17)(t+1.5) = 306 \end{cases}$$

Substitute $t = \frac{306}{r}$ from (1) into (2):

$$(r-17)\left(\tfrac{306}{r} + 1.5\right) = 306$$
$$306 + 1.5r - \tfrac{5202}{r} - 25.5 = 306$$
$$306r + 1.5r^2 - 5202 - 25.5r = 306r$$
$$1.5r^2 - 25.5r - 5202 = 0$$
$$3r^2 - 51r - 10{,}404 = 0$$
$$(3r + 153)(r - 68) = 0$$

$3r + 153 = 0$ **or** $r - 68 = 0$

$r = -153/3$ $\qquad$ $r = 68$

Substitute and solve for t:

$t = \frac{306}{r} = \frac{306}{68} = 4.5$

Jim drove for 4.5 hours at 68 miles per hour.

61. **Answers may vary.**

63. $0, 1, 2, 3, 4$

Chapter 13 Review (page 949)

1. $\begin{cases} 2x + y = 11 \\ -x + 2y = 7 \end{cases}$

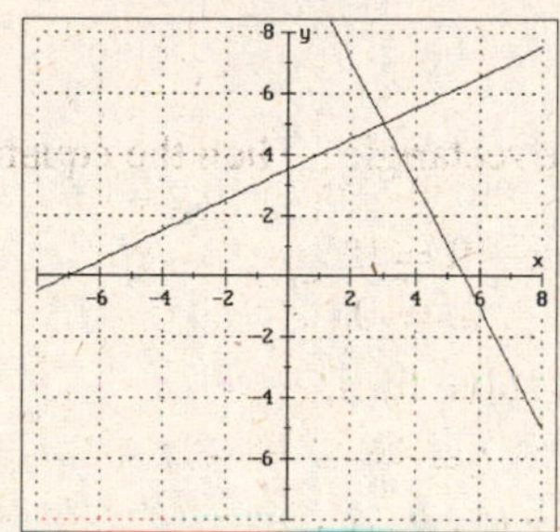

$(3, 5)$ is the solution.

2. $\begin{cases} 3x + 2y = 0 \\ 2x - 3y = -13 \end{cases}$

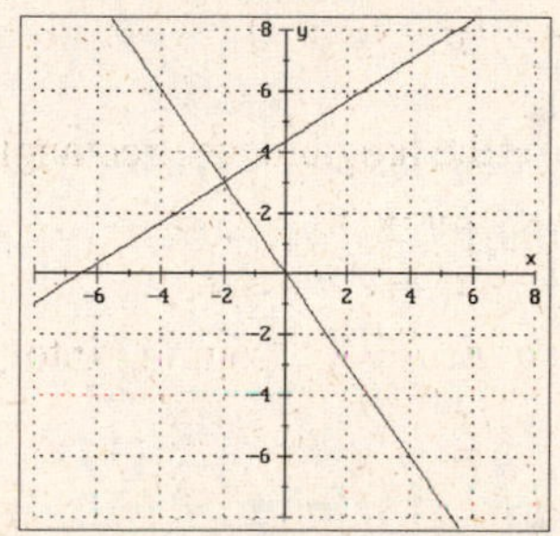

$(-2, 3)$ is the solution.

3. $\begin{cases} \frac{1}{2}x + \frac{1}{3}y = 2 \\ y = 6 - \frac{3}{2}x \end{cases}$

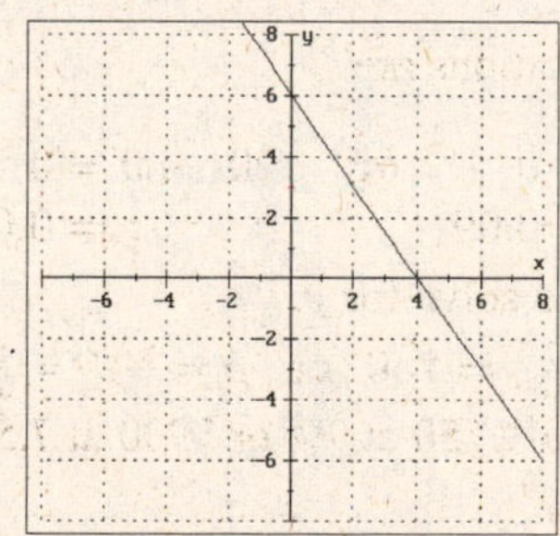

dependent equations

4. $\begin{cases} \frac{1}{3}x - \frac{1}{2}y = 1 \\ 6x - 9y = 2 \end{cases}$

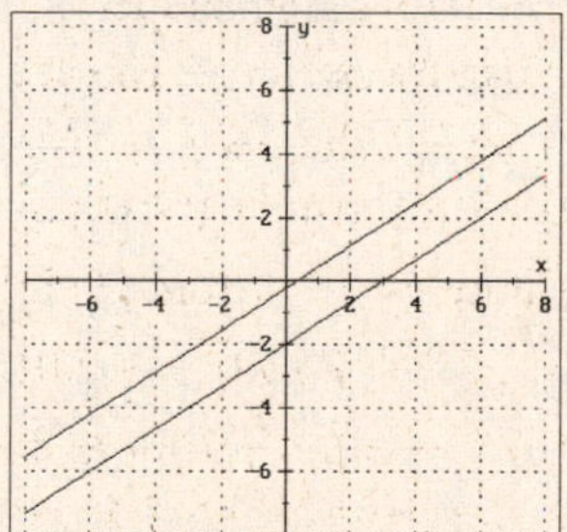

inconsistent system

5. $\begin{cases} (1) & y = x + 4 \\ (2) & 2x + 3y = 7 \end{cases}$

Substitute $y = x + 4$ from (1) into (2):

$$\begin{aligned} 2x + 3\boldsymbol{y} &= 7 \\ 2x + 3(\boldsymbol{x+4}) &= 7 \\ 2x + 3x + 12 &= 7 \\ 5x &= -5 \\ x &= -1 \end{aligned}$$

Substitute this and solve for y:

$y = x + 4 = -1 + 4 = 3$

Solution: $(-1, 3)$

6. $\begin{cases} (1) & y = 2x + 5 \\ (2) & 3x - 5y = -4 \end{cases}$

Substitute $y = 2x + 5$ from (1) into (2):

$$\begin{aligned} 3x - 5\boldsymbol{y} &= -4 \\ 3x - 5(\boldsymbol{2x+5}) &= -4 \\ 3x - 10x - 25 &= -4 \\ -7x &= 21 \\ x &= -3 \end{aligned}$$

Substitute this and solve for y:

$y = 2x + 5 = 2(-3) + 5 = -1$

Solution: $(-3, -1)$

7. $\begin{cases} (1) & x + 2y = 11 \\ (2) & 2x - y = 2 \end{cases}$

Substitute $x = -2y + 11$ from (1) into (2):

$$\begin{aligned} 2\boldsymbol{x} - y &= 2 \\ 2(\boldsymbol{-2y+11}) - y &= 2 \\ -4y + 22 - y &= 2 \\ -5y &= -20 \\ y &= 4 \end{aligned}$$

Substitute this and solve for x:

$x = -2y + 11 = -2(4) + 11 = 3$

Solution: $(3, 4)$

8. $\begin{cases} (1) & 2x + 3y = -2 \\ (2) & 3x + 5y = -2 \end{cases}$

Substitute $x = \frac{-3y-2}{2}$ from (1) into (2):

$$\begin{aligned} 3\boldsymbol{x} + 5y &= -2 \\ 3 \cdot \boldsymbol{\tfrac{-3y-2}{2}} + 5y &= -2 \\ 3(-3y-2) + 10y &= -4 \\ -9y - 6 + 10y &= -4 \\ y &= 2 \end{aligned}$$

Substitute this and solve for y:

$x = \frac{-3y-2}{2} = \frac{-3(2)-2}{2} = \frac{-8}{2} = -4$

Solution: $(-4, 2)$

9. $$\begin{array}{ll} x + y = -2 \Rightarrow \times(-2) & -2x - 2y = 4 \\ \underline{2x + 3y = -3 \Rightarrow} & \underline{2x + 3y = -3} \\ & y = 1 \end{array}$$

Substitute and solve for x:

$x + y = -2$

$x + 1 = -2$

$x = -3$ Solution: $\boxed{(-3, 1)}$

10. $$\begin{array}{ll} 3x + 2y = 1 \Rightarrow \times 3 & 9x + 6y = 3 \\ \underline{2x - 3y = 5 \Rightarrow \times 2} & \underline{4x - 6y = 10} \\ & 13x = 13 \\ & x = 1 \end{array}$$

Substitute and solve for y:

$$\begin{aligned} 3x + 2y &= 1 \\ 3(1) + 2y &= 1 \\ 3 + 2y &= 1 \\ 2y &= -2 \\ y &= -1 \end{aligned}$$ Solution: $\boxed{(1, -1)}$

11. $$\begin{array}{ll} x + \frac{1}{2}y = 7 \Rightarrow \times 2 & 2x + y = 14 \Rightarrow \\ \underline{-2x = 3y - 6 \Rightarrow} & \underline{-2x - 3y = -6 \Rightarrow} \\ & -2y = 8 \\ & y = -4 \end{array}$$

Solve for x:

$$\begin{aligned} 2x + y &= 14 \\ 2x + (-4) &= 14 \\ 2x &= 18 \\ x &= 9 \end{aligned}$$ $\boxed{\text{Solution: } (9, -4)}$

12.
$$\begin{array}{lll} y = \dfrac{x-3}{2} \Rightarrow \times 2 & 2y = x - 3 \Rightarrow & -x + 2y = -3 \\ x = \dfrac{2y+7}{2} \Rightarrow \times 2 & 2x = 2y + 7 \Rightarrow & 2x - 2y = 7 \\ \hline & & x = 4 \end{array}$$

Solve for y:
$$y = \frac{x-3}{2}$$
$$y = \frac{4-3}{2} = \frac{1}{2}$$
Solution: $\boxed{(4, \frac{1}{2})}$

13.
$$\begin{array}{ll} (1) & x + y + z = 6 \\ (2) & x - y - z = -4 \\ (3) & -x + y - z = -2 \end{array} \qquad \begin{array}{ll} (1) & x + y + z = 6 \\ (2) & x - y - z = -4 \\ \hline (4) & 2x = 2 \\ & x = 1 \end{array} \qquad \begin{array}{ll} (1) & x + y + z = 6 \\ (3) & -x + y - z = -2 \\ \hline (5) & 2y = 4 \\ & y = 2 \end{array}$$

$$\begin{aligned} x + y + z &= 6 \\ 1 + 2 + z &= 6 \\ 3 + z &= 6 \\ z &= 3 \end{aligned}$$
Solution: $\boxed{(1, 2, 3)}$

14.
$$\begin{array}{ll} (1) & 2x + 3y + z = -5 \\ (2) & -x + 2y - z = -6 \\ (3) & 3x + y + 2z = 4 \end{array} \qquad \begin{array}{ll} (1) & 2x + 3y + z = -5 \\ (2) & -x + 2y - z = -6 \\ \hline (4) & x + 5y = -11 \end{array} \qquad \begin{array}{ll} (3) & 3x + y + 2z = 4 \\ 2 \cdot (2) & -2x + 4y - 2z = -12 \\ \hline (5) & x + 5y = -8 \end{array}$$

$$\begin{array}{ll} x + 5y = -11 \Rightarrow \times(-1) & -x - 5y = 11 \\ x + 5y = -8 \Rightarrow & x + 5y = -8 \\ \hline & 0 = 3 \end{array}$$

Since this is an impossible equation, there is no solution. It is an inconsistent system.

15.
$$\left[\begin{array}{cc|c} 1 & 2 & 4 \\ 2 & -1 & 3 \end{array}\right] \xRightarrow{-2R_1 + R_2 \Rightarrow R_2} \left[\begin{array}{cc|c} 1 & 2 & 4 \\ 0 & -5 & -5 \end{array}\right] \xRightarrow{-\frac{1}{5}R_2 \Rightarrow R_2} \left[\begin{array}{cc|c} 1 & 2 & 4 \\ 0 & 1 & 1 \end{array}\right]$$

From R_2, $y = 1$. From R_1:
$$\begin{aligned} x + 2y &= 4 \\ x + 2(1) &= 4 \\ x &= 2 \end{aligned}$$
Solution: $\boxed{(2, 1)}$

16.
$$\left[\begin{array}{ccc|c} 1 & 1 & 1 & 6 \\ 2 & -1 & 1 & 1 \\ 4 & 1 & -1 & 5 \end{array}\right] \xRightarrow{\substack{-2R_1 + R_2 \Rightarrow R_2 \\ -4R_1 + R_3 \Rightarrow R_3}} \left[\begin{array}{ccc|c} 1 & 1 & 1 & 6 \\ 0 & -3 & -1 & -11 \\ 0 & -3 & -5 & -19 \end{array}\right] \xRightarrow{\substack{-R_2 \Rightarrow R_2 \\ -R_3 + R_2 \Rightarrow R_3}} \left[\begin{array}{ccc|c} 1 & 1 & 1 & 6 \\ 0 & 3 & 1 & 11 \\ 0 & 0 & 4 & 8 \end{array}\right] \xRightarrow{\frac{1}{4}R_3 \Rightarrow R_3} \left[\begin{array}{ccc|c} 1 & 1 & 1 & 6 \\ 0 & 3 & 1 & 11 \\ 0 & 0 & 1 & 2 \end{array}\right]$$

From R_3, $z = 2$. From R_2:
$$\begin{aligned} 3y + z &= 11 \\ 3y + 2 &= 11 \\ 3y &= 9 \\ y &= 3 \end{aligned}$$
From R_1:
$$\begin{aligned} x + y + z &= 6 \\ x + 3 + 2 &= 6 \\ x &= 1 \end{aligned}$$
Solution: $\boxed{(1, 3, 2)}$

17. $\begin{bmatrix} 1 & 1 & | & 3 \\ 1 & -2 & | & -3 \\ 2 & 1 & | & 4 \end{bmatrix} \Rightarrow \begin{matrix} -R_1 + R_2 \Rightarrow R_2 \\ -2R_1 + R_3 \Rightarrow R_3 \end{matrix} \begin{bmatrix} 1 & 1 & | & 3 \\ 0 & -3 & | & -6 \\ 0 & -1 & | & -2 \end{bmatrix} \Rightarrow \begin{matrix} -\frac{1}{3}R_2 \Rightarrow R_2 \\ -3R_3 + R_2 \Rightarrow R_3 \end{matrix} \begin{bmatrix} 1 & 1 & | & 3 \\ 0 & 1 & | & 2 \\ 0 & 0 & | & 0 \end{bmatrix}$

From R_2, $y = 2$. From R_1: Solution: $\boxed{(1,2)}$

$x + y = 3$
$x + 2 = 3$
$x = 1$

18. $\begin{bmatrix} 1 & 2 & 1 & | & 2 \\ 2 & 5 & 4 & | & 5 \end{bmatrix} \Rightarrow^{-2R_1 + R_2 \Rightarrow R_2} \begin{bmatrix} 1 & 2 & 1 & | & 2 \\ 0 & 1 & 2 & | & 1 \end{bmatrix}$

From R_2: From R_1: Solution: $\boxed{(3z, 1-2z, z)}$

$y + 2z = 1$
$y = 1 - 2z$

$x + 2y + z = 2$
$x + 2(1-2z) + z = 2$
$x + 2 - 4z + z = 2$
$x - 3z = 0$
$x = 3z$

19. $\begin{vmatrix} 2 & 3 \\ -4 & 3 \end{vmatrix} = 2(3) - 3(-4) = 6 + 12 = 18$

20. $\begin{vmatrix} -3 & -4 \\ 5 & -6 \end{vmatrix} = -3(-6) - (-4)(5) = 18 - (-20) = 18 + 20 = 38$

21. $\begin{vmatrix} -1 & 2 & -1 \\ 2 & -1 & 3 \\ 1 & -2 & 2 \end{vmatrix} = -1\begin{vmatrix} -1 & 3 \\ -2 & 2 \end{vmatrix} - 2\begin{vmatrix} 2 & 3 \\ 1 & 2 \end{vmatrix} + (-1)\begin{vmatrix} 2 & -1 \\ 1 & -2 \end{vmatrix} = -1(4) - 2(1) - 1(-3) = -3$

22. $\begin{vmatrix} 3 & -2 & 2 \\ 1 & -2 & -2 \\ 2 & 1 & -1 \end{vmatrix} = 3\begin{vmatrix} -2 & -2 \\ 1 & -1 \end{vmatrix} - (-2)\begin{vmatrix} 1 & -2 \\ 2 & -1 \end{vmatrix} + 2\begin{vmatrix} 1 & -2 \\ 2 & 1 \end{vmatrix} = 3(4) + 2(3) + 2(5) = 28$

23. $x = \dfrac{\begin{vmatrix} 10 & 4 \\ 1 & -3 \end{vmatrix}}{\begin{vmatrix} 3 & 4 \\ 2 & -3 \end{vmatrix}} = \dfrac{-30-4}{-9-8} = \dfrac{-34}{-17} = 2;\ y = \dfrac{\begin{vmatrix} 3 & 10 \\ 2 & 1 \end{vmatrix}}{\begin{vmatrix} 3 & 4 \\ 2 & -3 \end{vmatrix}} = \dfrac{3-20}{-17} = \dfrac{-17}{-17} = 1$

Solution: $\boxed{(2,1)}$

24. $x = \dfrac{\begin{vmatrix} -17 & -5 \\ 3 & 2 \end{vmatrix}}{\begin{vmatrix} 2 & -5 \\ 3 & 2 \end{vmatrix}} = \dfrac{-34+15}{4+15} = \dfrac{-19}{19} = -1;\ y = \dfrac{\begin{vmatrix} 2 & -17 \\ 3 & 3 \end{vmatrix}}{\begin{vmatrix} 2 & -5 \\ 3 & 2 \end{vmatrix}} = \dfrac{6+51}{19} = \dfrac{57}{19} = 3$

Solution: $\boxed{(-1, 3)}$

25. $D = \begin{vmatrix} 1 & 2 & 1 \\ 2 & 1 & 1 \\ 1 & 1 & 2 \end{vmatrix} = 1\begin{vmatrix} 1 & 1 \\ 1 & 2 \end{vmatrix} - 2\begin{vmatrix} 2 & 1 \\ 1 & 2 \end{vmatrix} + 1\begin{vmatrix} 2 & 1 \\ 1 & 1 \end{vmatrix} = 1(1) - 2(3) + 1(1) = -4$

$N_x = \begin{vmatrix} 0 & 2 & 1 \\ 3 & 1 & 1 \\ 5 & 1 & 2 \end{vmatrix} = 0\begin{vmatrix} 1 & 1 \\ 1 & 2 \end{vmatrix} - 2\begin{vmatrix} 3 & 1 \\ 5 & 2 \end{vmatrix} + 1\begin{vmatrix} 3 & 1 \\ 5 & 1 \end{vmatrix} = 0(1) - 2(1) + 1(-2) = -4$

$N_y = \begin{vmatrix} 1 & 0 & 1 \\ 2 & 3 & 1 \\ 1 & 5 & 2 \end{vmatrix} = 1\begin{vmatrix} 3 & 1 \\ 5 & 2 \end{vmatrix} - 0\begin{vmatrix} 2 & 1 \\ 1 & 2 \end{vmatrix} + 1\begin{vmatrix} 2 & 3 \\ 1 & 5 \end{vmatrix} = 1(1) - 0(3) + 1(7) = 8$

$N_z = \begin{vmatrix} 1 & 2 & 0 \\ 2 & 1 & 3 \\ 1 & 1 & 5 \end{vmatrix} = 1\begin{vmatrix} 1 & 3 \\ 1 & 5 \end{vmatrix} - 2\begin{vmatrix} 2 & 3 \\ 1 & 5 \end{vmatrix} + 0\begin{vmatrix} 2 & 1 \\ 1 & 1 \end{vmatrix} = 1(2) - 2(7) + 0(1) = -12$

$x = \dfrac{N_x}{D} = \dfrac{-4}{-4} = 1;\ y = \dfrac{N_y}{D} = \dfrac{8}{-4} = -2;\ z = \dfrac{N_z}{D} = \dfrac{-12}{-4} = 3 \Rightarrow$ solution: $(1, -2, 3)$

26. $D = \begin{vmatrix} 2 & 3 & 1 \\ 1 & 3 & 2 \\ 1 & -1 & -1 \end{vmatrix} = 2\begin{vmatrix} 3 & 2 \\ -1 & -1 \end{vmatrix} - 3\begin{vmatrix} 1 & 2 \\ 1 & -1 \end{vmatrix} + 1\begin{vmatrix} 1 & 3 \\ 1 & -1 \end{vmatrix} = 2(-1) - 3(-3) + 1(-4)$

$= 3$

$N_x = \begin{vmatrix} 2 & 3 & 1 \\ 7 & 3 & 2 \\ -7 & -1 & -1 \end{vmatrix} = 2\begin{vmatrix} 3 & 2 \\ -1 & -1 \end{vmatrix} - 3\begin{vmatrix} 7 & 2 \\ -7 & -1 \end{vmatrix} + 1\begin{vmatrix} 7 & 3 \\ -7 & -1 \end{vmatrix}$

$= 2(-1) - 3(7) + 1(14) = -9$

$N_y = \begin{vmatrix} 2 & 2 & 1 \\ 1 & 7 & 2 \\ 1 & -7 & -1 \end{vmatrix} = 2\begin{vmatrix} 7 & 2 \\ -7 & -1 \end{vmatrix} - 2\begin{vmatrix} 1 & 2 \\ 1 & -1 \end{vmatrix} + 1\begin{vmatrix} 1 & 7 \\ 1 & -7 \end{vmatrix}$

$= 2(7) - 2(-3) + 1(-14) = 6$

$N_z = \begin{vmatrix} 2 & 3 & 2 \\ 1 & 3 & 7 \\ 1 & -1 & -7 \end{vmatrix} = 2\begin{vmatrix} 3 & 7 \\ -1 & -7 \end{vmatrix} - 3\begin{vmatrix} 1 & 7 \\ 1 & -7 \end{vmatrix} + 2\begin{vmatrix} 1 & 3 \\ 1 & -1 \end{vmatrix}$

$= 2(-14) - 3(-14) + 2(-4) = 6$

$x = \dfrac{N_x}{D} = \dfrac{-9}{3} = -3;\ y = \dfrac{N_y}{D} = \dfrac{6}{3} = 2;\ z = \dfrac{N_z}{D} = \dfrac{6}{3} = 2 \Rightarrow$ solution: $(-3, 2, 2)$

27.
$$\begin{aligned} 3x^2 + y^2 &= 52 \\ x^2 - y^2 &= 12 \\ \hline 4x^2 &= 64 \\ x^2 &= 16 \\ x &= \pm 4 \end{aligned}$$

Substitute and solve for y:

$$\begin{aligned} x^2 - y^2 &= 12 & x^2 - y^2 &= 12 \\ 4^2 - y^2 &= 12 & (-4)^2 - y^2 &= 12 \\ y^2 &= 4 & y^2 &= 4 \\ y &= \pm 2 & y &= \pm 2 \end{aligned}$$

Solutions: $(4,2), (4,-2), (-4,2), (-4,-2)$.

28.
$$\begin{aligned} \tfrac{x^2}{16} + \tfrac{y^2}{12} = 1 \Rightarrow \times 48 \quad & 3x^2 + 4y^2 = 48 \\ x^2 - \tfrac{y^2}{3} = 1 \Rightarrow \times(-3) \quad & -3x^2 + y^2 = -3 \\ \hline & 5y^2 = 45 \\ & y^2 = 9 \\ & y = \pm 3 \end{aligned}$$

Substitute and solve for x:

$$\begin{aligned} 3x^2 + 4y^2 &= 48 \\ 3x^2 + 4(9) &= 48 \\ 3x^2 &= 12 \\ x^2 = 4 &\Rightarrow x = \pm 2 \end{aligned}$$

Solutions: $(2,3), (2,-3), (-2,3), (-2,-3)$

29. $\begin{cases} y \geq x^2 - 4 \\ y < x + 3 \end{cases}$

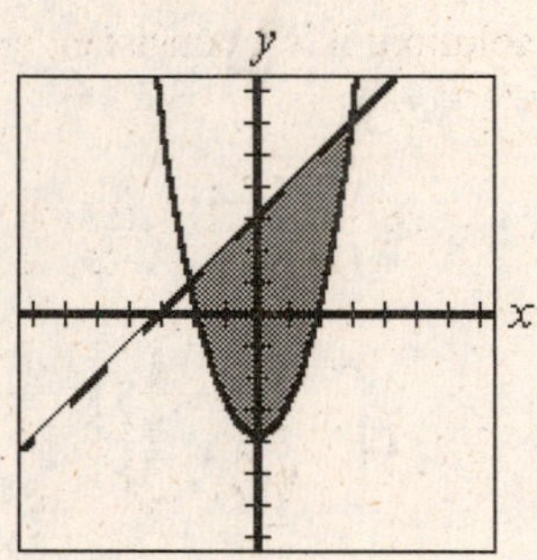

Chapter 13 Test (page 955)

1. $\begin{cases} 2x + y = 5 \\ y = 2x - 3 \end{cases}$

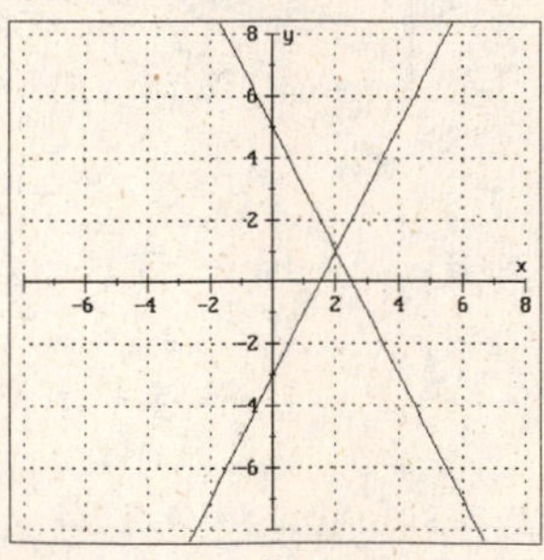

$(2,1)$ is the solution.

2. $\begin{cases} (1) \quad 2x - 4y = 14 \\ (2) \quad x = -2y + 7 \end{cases}$

Substitute $x = -2y + 7$ from (2) into (1):

$$\begin{aligned} 2\boldsymbol{x} - 4y &= 14 \\ 2(\boldsymbol{-2y + 7}) - 4y &= 14 \\ -4y + 14 - 4y &= 14 \\ -8y &= 0 \\ y &= 0 \end{aligned}$$

Substitute this and solve for x:

$$\begin{aligned} x &= -2y + 7 \\ x &= -2(0) + 7 = 7 \end{aligned}$$

Solution: $(7,0)$

3. $2x+3y=-5 \Rightarrow \times 2$ $\quad 4x+6y=-10$ $\quad 2x+3y=-5$ $\quad$ Solution:

$\underline{3x-2y=\ 12} \Rightarrow \times 3$ $\quad \underline{9x-6y\ \ =36}$ $\quad 2(2)+3y=-5$ $\quad \boxed{(2,-3)}$

$$\begin{aligned} 13x &= 26 \\ x &= 2 \end{aligned} \qquad \begin{aligned} 4+3y &= -5 \\ 3y &= -9 \\ y &= -3 \end{aligned}$$

4. $\frac{x}{2}-\frac{y}{4}=-4 \Rightarrow \times 4$ $\quad 2x-y=-16$ $\quad x+y=-2$ $\quad$ Solution:

$\underline{x+\ y=-2} \Rightarrow$ $\quad \underline{x+y=\ -2}$ $\quad -6+y=-2$ $\quad \boxed{(-6,4)}$

$$\begin{aligned} 3x &= -18 \\ x &= -6 \end{aligned} \qquad y=4$$

5. $3(x+y)=x-3 \Rightarrow$ $\quad 2x+3y=-3$ $\qquad 2x+3y=-3 \Rightarrow$ $\qquad 2x+3y=-3$

$-y=\dfrac{2x+3}{3} \Rightarrow$ $\quad 2x+3y=-3$ $\qquad \underline{2x+3y=-3} \Rightarrow \times(-1)$ $\underline{-2x-3y\ \ =3}$

$$0=0$$

The equation is an identity, so the system has infinitely many solutions. ⇒ dependent equations

6. See #5. Since the system has at least one solution, it is a consistent system.

7. $\begin{bmatrix} 1 & 2 & -1 \\ 2 & -2 & 3 \end{bmatrix} \overset{-3R_1+R_2 \Rightarrow R_2}{\Rightarrow} \begin{bmatrix} 1 & 2 & -1 \\ -1 & -8 & \boxed{6} \end{bmatrix}$

8. $\begin{bmatrix} -1 & 3 & 6 \\ 3 & -2 & 4 \end{bmatrix} \overset{-2R_1+R_2 \Rightarrow R_2}{\Rightarrow} \begin{bmatrix} -1 & 3 & 6 \\ 5 & -8 & \boxed{-8} \end{bmatrix}$

9. $\left[\begin{array}{ccc|c} 1 & 1 & 1 & 4 \\ 1 & 1 & -1 & 6 \\ 2 & -3 & 1 & -1 \end{array}\right]$

10. $\begin{bmatrix} 1 & 1 & 1 \\ 1 & 1 & -1 \\ 2 & -3 & 1 \end{bmatrix}$

11. $\left[\begin{array}{cc|c} 1 & 1 & 4 \\ 2 & -1 & 2 \end{array}\right] \overset{-2R_1+R_2 \Rightarrow R_2}{\Rightarrow} \left[\begin{array}{cc|c} 1 & 1 & 4 \\ 0 & -3 & -6 \end{array}\right] \overset{-\frac{1}{3}R_2 \Rightarrow R_2}{\Rightarrow} \left[\begin{array}{cc|c} 1 & 1 & 4 \\ 0 & 1 & 2 \end{array}\right]$

From R_2, $y=2$. From R_1: Solution: $\boxed{(2,2)}$

$$\begin{aligned} x+y &= 4 \\ x+2 &= 4 \\ x &= 2 \end{aligned}$$

12. $\left[\begin{array}{cc|c} 1 & 1 & 2 \\ 1 & -1 & -4 \\ 2 & 1 & 1 \end{array}\right] \overset{\substack{-R_2+R_1 \Rightarrow R_2 \\ -2R_1+R_3 \Rightarrow R_3}}{\Rightarrow} \left[\begin{array}{cc|c} 1 & 1 & 2 \\ 0 & 2 & 6 \\ 0 & -1 & -3 \end{array}\right] \overset{\substack{\frac{1}{2}R_2 \Rightarrow R_2 \\ 2R_3+R_2 \Rightarrow R_3}}{\Rightarrow} \left[\begin{array}{cc|c} 1 & 1 & 2 \\ 0 & 1 & 3 \\ 0 & 0 & 0 \end{array}\right]$

From R_2, $y=3$. From R_1: Solution: $\boxed{(-1,3)}$

$$\begin{aligned} x+y &= 2 \\ x+3 &= 2 \\ x &= -1 \end{aligned}$$

13. $\begin{vmatrix} 2 & -3 \\ 4 & 5 \end{vmatrix} = 2(5) - (-3)(4) = 10 - (-12)$
$= 22$

14. $\begin{vmatrix} -3 & -4 \\ -2 & 3 \end{vmatrix} = -3(3) - (-4)(-2)$
$= -9 - 8 = -17$

15. $\begin{vmatrix} 1 & 2 & 0 \\ 2 & 0 & 3 \\ 1 & -2 & 2 \end{vmatrix} = 1\begin{vmatrix} 0 & 3 \\ -2 & 2 \end{vmatrix} - 2\begin{vmatrix} 2 & 3 \\ 1 & 2 \end{vmatrix} + 0\begin{vmatrix} 2 & 0 \\ 1 & -2 \end{vmatrix} = 1(6) - 2(1) + 0(-4) = 4$

16. $\begin{vmatrix} 2 & -1 & 1 \\ 3 & 1 & 0 \\ 0 & 1 & 2 \end{vmatrix} = 2\begin{vmatrix} 1 & 0 \\ 1 & 2 \end{vmatrix} - (-1)\begin{vmatrix} 3 & 0 \\ 0 & 2 \end{vmatrix} + 1\begin{vmatrix} 3 & 1 \\ 0 & 1 \end{vmatrix} = 2(2) + 1(6) + 1(3) = 13$

17. $\begin{vmatrix} -6 & -1 \\ -6 & 1 \end{vmatrix}$

18. $\begin{vmatrix} 1 & -1 \\ 3 & 1 \end{vmatrix}$

19. $x = \dfrac{\begin{vmatrix} -6 & -1 \\ -6 & 1 \end{vmatrix}}{\begin{vmatrix} 1 & -1 \\ 3 & 1 \end{vmatrix}} = \dfrac{-6-6}{1-(-3)} = \dfrac{-12}{4} = -3$

20. $y = \dfrac{\begin{vmatrix} 1 & -6 \\ 3 & -6 \end{vmatrix}}{\begin{vmatrix} 1 & -1 \\ 3 & 1 \end{vmatrix}} = \dfrac{-6-(-18)}{1-(-3)} = \dfrac{12}{4} = 3$

21. $D = \begin{vmatrix} 1 & 1 & 1 \\ 1 & 1 & -1 \\ 2 & -3 & 1 \end{vmatrix} = 1\begin{vmatrix} 1 & -1 \\ -3 & 1 \end{vmatrix} - 1\begin{vmatrix} 1 & -1 \\ 2 & 1 \end{vmatrix} + 1\begin{vmatrix} 1 & 1 \\ 2 & -3 \end{vmatrix} = 1(-2) - 1(3) + 1(-5) = -10$

$N_x = \begin{vmatrix} 4 & 1 & 1 \\ 6 & 1 & -1 \\ -1 & -3 & 1 \end{vmatrix} = 4\begin{vmatrix} 1 & -1 \\ -3 & 1 \end{vmatrix} - 1\begin{vmatrix} 6 & -1 \\ -1 & 1 \end{vmatrix} + 1\begin{vmatrix} 6 & 1 \\ -1 & -3 \end{vmatrix}$ $\boxed{x = \dfrac{N_x}{D} = \dfrac{-30}{-10} = 3}$

$= 4(-2) - 1(5) + 1(-17) = -30$

22. See #21. $D = -10$.

$N_z = \begin{vmatrix} 1 & 1 & 4 \\ 1 & 1 & 6 \\ 2 & -3 & -1 \end{vmatrix} = 1\begin{vmatrix} 1 & 6 \\ -3 & -1 \end{vmatrix} - 1\begin{vmatrix} 1 & 6 \\ 2 & -1 \end{vmatrix} + 4\begin{vmatrix} 1 & 1 \\ 2 & -3 \end{vmatrix}$ $\boxed{z = \dfrac{N_z}{D} = \dfrac{10}{-10} = -1}$

$= 1(17) - 1(-13) + 4(-5) = 10$

23.
$$\begin{aligned} x^2 + y^2 &= 5 \\ x^2 - y^2 &= 3 \\ \hline 2x^2 &= 8 \\ x^2 &= 4 \\ x &= \pm 2 \end{aligned}$$

Substitute and solve for y:

$$\begin{array}{cc} x^2 + y^2 = 5 & x^2 + y^2 = 5 \\ 2^2 + y^2 = 5 & (-2)^2 + y^2 = 5 \\ y^2 = 1 & y^2 = 1 \\ y = \pm 1 & y = \pm 1 \end{array}$$

Solutions: $(2, 1), (2, -1), (-2, 1), (-2, -1)$.

24. $\begin{cases} (1) \quad x^2 + y^2 = 25 \\ (2) \quad 4x^2 - 9y = 0 \end{cases}$

Substitute $x^2 = 25 - y^2$ from (1) into (2):

$$\begin{aligned} 4x^2 - 9y &= 0 \\ 4(25 - y^2) - 9y &= 0 \\ 100 - 4y^2 - 9y &= 0 \\ -4y^2 - 9y + 100 &= 0 \\ 4y^2 + 9y - 100 &= 0 \\ (y-4)(4y+25) &= 0 \end{aligned}$$

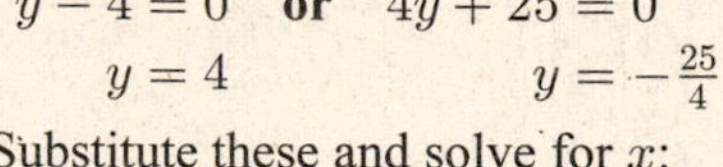

$y - 4 = 0$ **or** $4y + 25 = 0$

$y = 4 \qquad y = -\frac{25}{4}$

Substitute these and solve for x:

$$\begin{aligned} x^2 &= 25 - y^2 & x^2 &= 25 - y^2 \\ x^2 &= 25 - 4^2 & x^2 &= 25 - \left(-\tfrac{25}{4}\right)^2 \\ x^2 &= 25 - 16 = 9 & x^2 &= 25 - \tfrac{625}{16} = -\tfrac{225}{16} \\ x &= \pm 3 & x &\text{ is nonreal.} \end{aligned}$$

Solutions: $(3, 4), (-3, 4)$

25. $\begin{cases} y \geq x^2 \\ y < x + 3 \end{cases}$

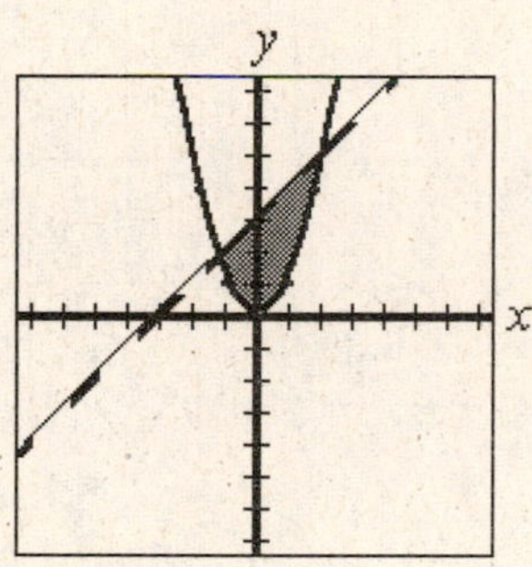

Exercise 14.1 (page 963)

1. $1! = 1$

3. $0! = 1$

5. $(m+n)^2 = (m+n)(m+n)$
$= m^2 + 2mn + n^2$

7. $(p+2q)^2 = (p+2q)(p+2q)$
$= p^2 + 4pq + 4q^2$

9. $\log_4 16 = x \Rightarrow 4^x = 16 \Rightarrow x = 2$

11. $\log_{25} x = \frac{1}{2} \Rightarrow 25^{1/2} = x \Rightarrow x = 5$

13. one

15. Pascal's

17. $6!$

19. 1

21. $(a+b)^3 = 1a^3 + 3a^2b + 3ab^2 + 1b^3 = a^3 + 3a^2b + 3ab^2 + b^3$

23. $(a-b)^4 = 1a^4 + 4a^3(-b) + 6a^2(-b)^2 + 4a(-b)^3 + 1(-b)^4$
$= a^4 - 4a^3b + 6a^2b^2 - 4ab^3 + b^4$

25. $3! = 3 \cdot 2 \cdot 1 = 6$

27. $-5! = -1 \cdot 5! = -1 \cdot 5 \cdot 4 \cdot 3 \cdot 2 \cdot 1$
$= -120$

29. $3! + 4! = 3 \cdot 2 \cdot 1 + 4 \cdot 3 \cdot 2 \cdot 1$
$= 6 + 24 = 30$

31. $3!(4!) = 3 \cdot 2 \cdot 1 \cdot 4 \cdot 3 \cdot 2 \cdot 1 = 144$

SECTION 14.1

33. $\frac{9!}{11!} = \frac{9!}{11 \cdot 10 \cdot 9!} = \frac{1}{11 \cdot 10} = \frac{1}{110}$

35. $\frac{49!}{47!} = \frac{49 \cdot 48 \cdot 47!}{47!} = 49 \cdot 48 = 2352$

37. $\frac{9!}{7!\,0!} = \frac{9 \cdot 8 \cdot 7!}{7! \cdot 1} = 9 \cdot 8 = 72$

39. $\frac{5!}{3!(5-3)!} = \frac{5!}{3!2!} = \frac{5 \cdot 4 \cdot 3!}{3! \cdot 2 \cdot 1} = \frac{5 \cdot 4}{2 \cdot 1} = 10$

41.
$$\begin{aligned}(x+y)^3 = x^3 + \frac{3!}{1!(3-1)!}x^2y + \frac{3!}{2!(3-2)!}xy^2 + y^3 &= x^3 + \frac{3!}{1!2!}x^2y + \frac{3!}{2!1!}xy^2 + y^3 \\ &= x^3 + \frac{3 \cdot 2!}{1!2!}x^2y + \frac{3 \cdot 2!}{2!1!}xy^2 + y^3 \\ &= x^3 + \frac{3}{1}x^2y + \frac{3}{1}xy^2 + y^3 \\ &= x^3 + 3x^2y + 3xy^2 + y^3\end{aligned}$$

43.
$$\begin{aligned}(x-y)^4 &= x^4 + \frac{4!}{1!(4-1)!}x^3(-y) + \frac{4!}{2!(4-2)!}x^2(-y)^2 + \frac{4!}{3!(4-3)!}x(-y)^3 + (-y)^4 \\ &= x^4 + \frac{4!}{1!3!}(-x^3y) + \frac{4!}{2!2!}x^2y^2 + \frac{4!}{3!1!}(-xy^3) + y^4 \\ &= x^4 - \frac{4 \cdot 3!}{1!3!}x^3y + \frac{4 \cdot 3 \cdot 2!}{2! \cdot 2 \cdot 1}x^2y^2 - \frac{4 \cdot 3!}{3!1!}xy^3 + y^4 \\ &= x^4 - \frac{4}{1}x^3y + \frac{12}{2}x^2y^2 - \frac{4}{1}xy^3 + y^4 \\ &= x^4 - 4x^3y + 6x^2y^2 - 4xy^3 + y^4\end{aligned}$$

45.
$$\begin{aligned}(2x+y)^3 &= (2x)^3 + \frac{3!}{1!(3-1)!}(2x)^2y + \frac{3!}{2!(3-2)!}2xy^2 + y^3 \\ &= 8x^3 + \frac{3!}{1!2!} \cdot 4x^2y + \frac{3!}{2!1!} \cdot 2xy^2 + y^3 \\ &= 8x^3 + \frac{3 \cdot 2!}{1!2!} \cdot 4x^2y + \frac{3 \cdot 2!}{2!1!} \cdot 2xy^2 + y^3 \\ &= 8x^3 + \frac{3}{1} \cdot 4x^2y + \frac{3}{1} \cdot 2xy^2 + y^3 \\ &= 8x^3 + 12x^2y + 6xy^2 + y^3\end{aligned}$$

47.
$$\begin{aligned}(x-2y)^3 &= x^3 + \frac{3!}{1!(3-1)!}x^2(-2y) + \frac{3!}{2!(3-2)!}x(-2y)^2 + (-2y)^3 \\ &= x^3 + \frac{3!}{1!2!} \cdot (-2x^2y) + \frac{3!}{2!1!} \cdot 4xy^2 - 8y^3 \\ &= x^3 - \frac{3 \cdot 2!}{1!2!} \cdot 2x^2y + \frac{3 \cdot 2!}{2!1!} \cdot 4xy^2 - 8y^3 \\ &= x^3 - \frac{3}{1} \cdot 2x^2y + \frac{3}{1} \cdot 4xy^2 - 8y^3 \\ &= x^3 - 6x^2y + 12xy^2 - 8y^3\end{aligned}$$

49. $(2x+3y)^3 = (2x)^3 + \frac{3!}{1!(3-1)!}(2x)^2(3y) + \frac{3!}{2!(3-2)!}2x(3y)^2 + (3y)^3$

$$= 8x^3 + \frac{3!}{1!2!}\cdot 4x^2(3y) + \frac{3!}{2!1!}\cdot 2x(9y^2) + 27y^3$$

$$= 8x^3 + \frac{3\cdot 2!}{1!2!}\cdot 12x^2y + \frac{3\cdot 2!}{2!1!}\cdot 18xy^2 + 27y^3$$

$$= 8x^3 + \frac{3}{1}\cdot 12x^2y + \frac{3}{1}\cdot 18xy^2 + 27y^3$$

$$= 8x^3 + 36x^2y + 54xy^2 + 27y^3$$

51. $\left(\frac{x}{2}-\frac{y}{3}\right)^3 = \left(\frac{x}{2}\right)^3 + \frac{3!}{1!(3-1)!}\left(\frac{x}{2}\right)^2\left(-\frac{y}{3}\right) + \frac{3!}{2!(3-2)!}\left(\frac{x}{2}\right)\left(-\frac{y}{3}\right)^2 + \left(-\frac{y}{3}\right)^3$

$$= \frac{x^3}{8} - \frac{3!}{1!2!}\cdot\frac{x^2}{4}\cdot\frac{y}{3} + \frac{3!}{2!1!}\cdot\frac{x}{2}\cdot\frac{y^2}{9} - \frac{y^3}{27}$$

$$= \frac{x^3}{8} - \frac{3\cdot 2!}{1!2!}\cdot\frac{x^2y}{12} + \frac{3\cdot 2!}{2!1!}\cdot\frac{xy^2}{18} - \frac{y^3}{27}$$

$$= \frac{x^3}{8} - \frac{3}{1}\cdot\frac{x^2y}{12} + \frac{3}{1}\cdot\frac{xy^2}{18} - \frac{y^3}{27}$$

$$= \frac{x^3}{8} - \frac{x^2y}{4} + \frac{xy^2}{6} - \frac{y^3}{27}$$

53. $8(7!) = 8\cdot 7\cdot 6\cdot 5\cdot 4\cdot 3\cdot 2\cdot 1 = 40{,}320$

55. $\frac{7!}{5!(7-5)!} = \frac{7!}{5!2!} = \frac{7\cdot 6\cdot 5!}{5!\cdot 2\cdot 1} = \frac{7\cdot 6}{2\cdot 1} = 21$

57. $\frac{5!(8-5)!}{4!7!} = \frac{5!3!}{4!7!} = \frac{5!}{7!}\cdot\frac{3!}{4!} = \frac{5!}{7\cdot 6\cdot 5!}\cdot\frac{3!}{4\cdot 3!} = \frac{1}{7\cdot 6}\cdot\frac{1}{4} = \frac{1}{168}$

59. $11! = 39{,}916{,}800$

61. $20! = 2.432902008\times 10^{18}$

63. $(3+2y)^4 = 3^4 + \frac{4!}{1!(4-1)!}3^3(2y) + \frac{4!}{2!(4-2)!}3^2(2y)^2 + \frac{4!}{3!(4-3)!}3(2y)^3 + (2y)^4$

$$= 81 + \frac{4!}{1!3!}\cdot 27(2y) + \frac{4!}{2!2!}\cdot 9(4y^2) + \frac{4!}{3!1!}\cdot 3(8y^3) + 16y^4$$

$$= 81 + \frac{4\cdot 3!}{1!3!}\cdot 54y + \frac{4\cdot 3\cdot 2!}{2!\cdot 2\cdot 1}\cdot 36y^2 + \frac{4\cdot 3!}{3!1!}\cdot 24y^3 + 16y^4$$

$$= 81 + \frac{4}{1}\cdot 54y + \frac{12}{2}\cdot 36y^2 + \frac{4}{1}\cdot 24y^3 + 16y^4$$

$$= 81 + 216y + 216y^2 + 96y^3 + 16y^4$$

65. $\left(\frac{x}{3}-\frac{y}{2}\right)^4$

$$= \left(\frac{x}{3}\right)^4 + \frac{4!}{1!(4-1)!}\left(\frac{x}{3}\right)^3\left(-\frac{y}{2}\right) + \frac{4!}{2!(4-2)!}\left(\frac{x}{3}\right)^2\left(-\frac{y}{2}\right)^2 + \frac{4!}{3!(4-3)!}\left(\frac{x}{3}\right)\left(-\frac{y}{2}\right)^3 + \left(-\frac{y}{2}\right)^4$$

$$= \frac{x^4}{81} - \frac{4!}{1!3!}\cdot\frac{x^3}{27}\cdot\frac{y}{2} + \frac{4!}{2!2!}\cdot\frac{x^2}{9}\cdot\frac{y^2}{4} - \frac{4!}{3!1!}\cdot\frac{x}{3}\cdot\frac{y^3}{8} + \frac{y^4}{16}$$

$$= \frac{x^4}{81} - \frac{4\cdot 3!}{1!3!}\cdot\frac{x^3y}{54} + \frac{4\cdot 3\cdot 2!}{2!2!}\cdot\frac{x^2y^2}{36} - \frac{4\cdot 3!}{3!1!}\cdot\frac{xy^3}{24} + \frac{y^4}{16}$$

$$= \frac{x^4}{81} - \frac{4}{1}\cdot\frac{x^3y}{54} + \frac{12}{2}\cdot\frac{x^2y^2}{36} - \frac{4}{1}\cdot\frac{xy^3}{24} + \frac{y^4}{16}$$

$$= \frac{x^4}{81} - \frac{2x^3y}{27} + \frac{x^2y^2}{6} - \frac{xy^3}{6} + \frac{y^4}{16}$$

67.

```
                    1
                  1   1
                1   2   1
              1   3   3   1
            1   4   6   4   1
          1   5  10  10   5   1
        1   6  15  20  15   6   1
      1   7  21  35  35  21   7   1
    1   8  28  56  70  56  28   8   1
  1   9  36  84 126 126  84  36   9   1
```

69. **Answers may vary.**

71. $\frac{n!}{0!(n-0)!} = \frac{n!}{0!n!} = \frac{n!}{1\cdot n!} = \frac{n!}{n!} = 1$

73. $1, 1, 2, 3, 5, 8, 13, \ldots$; Beginning with 2, each number is the sum of the previous two numbers.

Exercise 14.2 (page 967)

1. In the 3rd term, the exponent on y is 2.

3. In the 7th term, the exponent on y is 6.

5. In the 4th term, the exponent on y is 3. The exponent on x is $8 - 3 = 5$.

7. $\frac{8!}{0!\,8!} = 1$

9.

$$\begin{aligned} 3x + 2y &= 12 \Rightarrow & 3x + 2y &= 12 \\ 2x - y &= 1 \Rightarrow \times 2 & 4x - 2y &= 2 \\ & & 7x &= 14 \\ & & x &= 2 \end{aligned}$$

Substitute and solve for y:

$$\begin{aligned} 2x - y &= 1 \\ 2(2) - y &= 1 \\ 4 - y &= 1 \\ -y &= -3 \\ y &= 3 \end{aligned}$$

Solution: $(2, 3)$

SECTION 14.2

11. $\begin{vmatrix} 2 & -3 \\ 4 & -2 \end{vmatrix} = 2(-2) - (-3)(4) = -4 - (-12) = -4 + 12 = 8$

13. 3

15. 7

17. In the 2nd term, the exponent on b is 1.
Variables: $a^2b^1 = a^2b$
Coef. $= \dfrac{n!}{r!(n-r)!} = \dfrac{3!}{1!2!} = 3$
Term $= 3a^2b$

19. In the 5th term, the exponent on y is 4.
Variables: x^2y^4
Coef. $= \dfrac{n!}{r!(n-r)!} = \dfrac{6!}{4!2!} = 15$
Term $= 15x^2y^4$

21. In the 4th term, the exponent on $-y$ is 3.
Variables: $x^1(-y)^3 = -xy^3$
Coef. $= \dfrac{n!}{r!(n-r)!} = \dfrac{4!}{3!1!} = 4$
Term $= 4(-xy^3) = -4xy^3$

23. In the 3rd term, the exponent on $-y$ is 2.
Variables: $x^6(-y)^2 = x^6y^2$
Coef. $= \dfrac{n!}{r!(n-r)!} = \dfrac{8!}{2!6!} = 28$
Term $= 28x^6y^2$

25. In the 3rd term, the exponent on y is 2.
Variables: $(4x)^3y^2 = 64x^3y^2$
Coef. $= \dfrac{n!}{r!(n-r)!} = \dfrac{5!}{2!3!} = 10$
Term $= 10(64x^3y^2) = 640x^3y^2$

27. In the 2nd term, the exponent on $-3y$ is 1.
Variables: $x^3(-3y)^1 = -3x^3y$
Coef. $= \dfrac{n!}{r!(n-r)!} = \dfrac{4!}{1!3!} = 4$
Term $= 4(-3x^3y) = -12x^3y$

29. In the 4th term, the exponent on -5 is 3. Variables: $(2x)^4(-5)^3 = (16x^4)(-125) = -2000x^4$
Coef. $= \dfrac{n!}{r!(n-r)!} = \dfrac{7!}{3!4!} = 35$; Term $= 35(-2000x^4) = -70{,}000x^4$

31. In the 5th term, the exponent on $-3y$ is 4. Variables: $(2x)^1(-3y)^4 = 2x(81y^4) = 162xy^4$
Coef. $= \dfrac{n!}{r!(n-r)!} = \dfrac{5!}{4!1!} = 5$; Term $= 5(162xy^4) = 810xy^4$

33. In the 3rd term, the exponent on $\sqrt{3}y$ is 2. Variables: $(\sqrt{2}x)^4(\sqrt{3}y)^2 = (4x^4)(3y^2) = 12x^4y^2$
Coef. $= \dfrac{n!}{r!(n-r)!} = \dfrac{6!}{2!4!} = 15$; Term $= 15(12x^4y^2) = 180x^4y^2$

35. In the 2nd term, the exponent on $-\dfrac{y}{3}$ is 1. Variables: $\left(\dfrac{x}{2}\right)^3\left(-\dfrac{y}{3}\right)^1 = \left(\dfrac{x^3}{8}\right)\left(-\dfrac{y}{3}\right) = -\dfrac{x^3y}{24}$
Coef. $= \dfrac{n!}{r!(n-r)!} = \dfrac{4!}{1!3!} = 4$; Term $= 4\left(-\dfrac{x^3y}{24}\right) = -\dfrac{x^3y}{6} = -\dfrac{1}{6}x^3y$

37. In the 3rd term, the exponent on 3 is 2. Variables: $x^3(3)^2 = 9x^3$
Coef. $= \dfrac{n!}{r!(n-r)!} = \dfrac{5!}{2!3!} = 10$; Term $= 10(9x^3) = 90x^3$

39. In the 4th term, the exponent on b is 3. Variables: $a^{n-3}b^3$

Coef. $= \dfrac{n!}{r!(n-r)!} = \dfrac{n!}{3!(n-3)!}$; Term $= \dfrac{n!}{3!(n-3)!}a^{n-3}b^3$

41. In the 5th term, the exponent on $-b$ is 4. Variables: $a^{n-4}(-b)^4 = a^{n-4}b^4$

Coef. $= \dfrac{n!}{r!(n-r)!} = \dfrac{n!}{4!(n-4)!}$; Term $= \dfrac{n!}{4!(n-4)!}a^{n-4}b^4$

43. In the rth term, the coefficient on b is $r-1$. Variables: $a^{n-(r-1)}b^{r-1} = a^{n-r+1}b^{r-1}$

Coef. $= \dfrac{n!}{r!(n-r)!} = \dfrac{n!}{(r-1)![n-(r-1)]!} = \dfrac{n!}{(r-1)!(n-r+1)!}$

Term $= \dfrac{n!}{(r-1)!(n-r+1)!}a^{n-r+1}b^{r-1}$

45. **Answers may vary.**

47. $\left(x+\dfrac{1}{x}\right)^{10} = \left(x+x^{-1}\right)^{10}$. The constant term occurs when the exponent is 0.

The $(r+1)$th term of $(x+x^{-1})^{10}$ is $\dfrac{10!}{r!(10-r)!}x^{10-r}\left(x^{-1}\right)^r = \dfrac{10!}{r!(10-r)!}x^{10-r}x^{-r}$.

But $\dfrac{10!}{r!(10-r)!}x^{10-r}x^{-r} = \dfrac{10!}{r!(10-r)!}x^{10-2r}$. If $10-2r=0$, then $r=5$.

The term is $\dfrac{10!}{5!(10-5)!}x^{10-5}x^{-5} = \dfrac{10!}{5!5!} = \dfrac{10\cdot 9\cdot 8\cdot 7\cdot 6\cdot 5!}{5!\cdot 5\cdot 4\cdot 3\cdot 2\cdot 1} = 252$

Exercise 14.3 (page 975)

1. $2, 6, 10, \boxed{14}$

3. $-2, 3, 8, \ldots \Rightarrow d = 5$

5. $\sum_{k=1}^{2} k = 1+2=3$

7. $3(2x^2-4x+7)+4(3x^2+5x-6) = 6x^2-12x+21+12x^2+20x-24 = 18x^2+8x-3$

9. $\dfrac{3a+4}{a-2}+\dfrac{3a-4}{a+2} = \dfrac{(3a+4)(a+2)}{(a-2)(a+2)}+\dfrac{(3a-4)(a-2)}{(a+2)(a-2)}$

$= \dfrac{3a^2+10a+8+3a^2-10a+8}{(a+2)(a-2)} = \dfrac{6a^2+16}{(a+2)(a-2)}$

11. sequence; finite; infinite

13. arithmetic; difference

15. arithmetic mean

17. series

19. $1+2+3+4+5$

21. $a_1 = 3(1)-2 = 1$

23. $a_{25} = 3(25)-2 = 73$

25. $3, 5, 7, 9, 11$

27. $-5, -8, -11, -14, -17$

SECTION 14.3

29. $a_n = a_1 + (n-1)d$
$a_5 = a_1 + (5-1)d$
$29 = 5 + 4d$
$24 = 4d$
$6 = d$
$5, 11, 17, 23, 29$

31. $a_n = a_1 + (n-1)d$
$a_6 = a_1 + (6-1)d$
$-39 = -4 + 5d$
$-35 = 5d$
$-7 = d$
$-4, -11, -18, -25, -32$

33. Form an arithmetic sequence with a 1st term of 2 and a 5th term of 11:
$a_n = a_1 + (n-1)d$
$a_5 = a_1 + (5-1)d$
$11 = 2 + 4d$
$9 = 4d$
$\frac{9}{4} = d$
$2, \boxed{\frac{17}{4}, \frac{13}{2}, \frac{35}{4}}, 11$

35. Form an arithmetic sequence with a 1st term of 10 and a 6th term of 20:
$a_n = a_1 + (n-1)d$
$a_6 = a_1 + (6-1)d$
$20 = 10 + 5d$
$10 = 5d$
$2 = d$
$10, \boxed{12, 14, 16, 18}, 20$

37. Form an arithmetic sequence with a 1st term of 10 and a 3rd term of 19:
$a_n = a_1 + (n-1)d$
$a_3 = a_1 + (3-1)d$
$19 = 10 + 2d$
$9 = 2d$
$\frac{9}{2} = d$
$10, \boxed{\frac{29}{2}}, 19$

39. Form an arithmetic sequence with a 1st term of -4.5 and a 3rd term of 7:
$a_n = a_1 + (n-1)d$
$a_3 = a_1 + (3-1)d$
$7 = -4.5 + 2d$
$11.5 = 2d$
$5.75 = d$
$-4.5, \boxed{1.25}, 7$

41. $a_1 = 1, d = 3, n = 30$
$a_n = a_1 + (n-1)d = 1 + 29(3) = 88$
$$S_n = \frac{n(a_1 + a_n)}{2} = \frac{30(1 + 88)}{2} = 1335$$

43. $a_1 = -5, d = 4, n = 17$
$a_n = a_1 + (n-1)d = -5 + 16(4) = 59$
$$S_n = \frac{n(a_1 + a_n)}{2} = \frac{17(-5 + 59)}{2} = 459$$

45. $\sum_{k=1}^{4}(3k) = 3(1) + 3(2) + 3(3) + 3(4) = 3 + 6 + 9 + 12$

47. $\sum_{k=4}^{6} k^2 = 4^2 + 5^2 + 6^2 = 16 + 25 + 36$

49. $\sum_{k=1}^{4} 6k = 6(1) + 6(2) + 6(3) + 6(4) = 6 + 12 + 18 + 24 = 60$

51. $\sum_{k=3}^{4}(k^2 + 3) = (3^2 + 3) + (4^2 + 3) = 9 + 3 + 16 + 3 = 31$

53.
$$\begin{aligned} a_n &= a_1 + (n-1)d \\ a_6 &= a_1 + (6-1)d \\ -83 &= a_1 + 5(7) \\ -83 &= a_1 + 35 \\ -118 &= a_1 \end{aligned}$$
$-118, -111, -104, -97, -90$

55.
$$\begin{aligned} a_n &= a_1 + (n-1)d \\ a_7 &= a_1 + (7-1)d \\ 16 &= a_1 + 6(-3) \\ 16 &= a_1 - 18 \\ 34 &= a_1 \end{aligned}$$
$34, 31, 28, 25, 22$

57.
$$\begin{aligned} a_n &= a_1 + (n-1)d \\ a_{19} &= a_1 + (19-1)d \\ 131 &= a_1 + 18d \\ a_n &= a_1 + (n-1)d \\ a_{20} &= a_1 + (20-1)d \\ 138 &= a_1 + 19d \end{aligned}$$

$$\begin{aligned} a_1 + 18d = 131 &\Rightarrow \times(-1) & -a_1 - 18d &= -131 \\ a_1 + 19d = 138 &\Rightarrow & a_1 + 19d &= 138 \\ & & d &= 7 \end{aligned}$$

Substitute and solve for a_1:
$$\begin{aligned} a_1 + 18d &= 131 \\ a_1 + 18(7) &= 131 \\ a_1 + 126 &= 131 \\ a_1 &= 5 \Rightarrow 5, 12, 19, 26, 33 \end{aligned}$$

59. $a_n = a_1 + (n-1)d \quad a_{30} = a_1 + (30-1)d = 7 + 29(12)$
$$= 7 + 348 = 355$$

61.
$$\begin{aligned} a_n &= a_1 + (n-1)d \\ a_2 &= a_1 + (2-1)d \\ -4 &= a_1 + d \\ a_n &= a_1 + (n-1)d \\ a_3 &= a_1 + (3-1)d \\ -9 &= a_1 + 2d \end{aligned}$$

$$\begin{aligned} a_1 + d = -4 &\Rightarrow \times(-1) & -a_1 - d &= 4 \\ a_1 + 2d = -9 &\Rightarrow & a_1 + 2d &= -9 \\ & & d &= -5 \end{aligned}$$

Substitute and solve for a_1:
$$\begin{aligned} a_1 + d &= -4 \\ a_1 + (-5) &= -4 \\ a_1 &= 1 \end{aligned}$$

Find the desired term:
$$\begin{aligned} a_n &= a_1 + (n-1)d \\ a_{37} &= a_1 + (37-1)d \\ &= 1 + 36(-5) \\ &= 1 - 180 = \boxed{-179} \end{aligned}$$

63.
$$\begin{aligned} a_n &= a_1 + (n-1)d \\ a_{27} &= a_1 + (27-1)d \\ 263 &= a_1 + 26(11) \\ 263 &= a_1 + 286 \\ -23 &= a_1 \end{aligned}$$

65.
$$\begin{aligned} a_n &= a_1 + (n-1)d \\ a_{44} &= a_1 + (44-1)d \\ 556 &= 40 + 43d \\ 516 &= 43d \\ 12 &= d \end{aligned}$$

67.
$$\begin{aligned} a_n &= a_1 + (n-1)d \\ a_2 &= a_1 + (2-1)d \\ 7 &= a_1 + d \\ a_n &= a_1 + (n-1)d \\ a_3 &= a_1 + (3-1)d \\ 12 &= a_1 + 2d \end{aligned}$$

$$\begin{aligned} a_1 + d = 7 &\Rightarrow \times(-1) & -a_1 - d &= -7 \\ a_1 + 2d = 12 &\Rightarrow & a_1 + 2d &= 12 \\ & & d &= 5 \end{aligned}$$

Substitute and solve for a_1:
$$\begin{aligned} a_1 + d &= 7 \\ a_1 + 5 &= 7 \end{aligned}$$
$$a_1 = 2, d = 5, n = 12$$
$$a_n = a_1 + (n-1)d = 2 + 11(5) = 57$$
$$S_n = \frac{n(a_1 + a_n)}{2} = \frac{12(2 + 57)}{2} = 354$$

69. $f(n) = 2n + 1 \Rightarrow f(1) = 3$
$f(n) = 2n + 1 = 31$
$2n = 30$
$n = 15$
$$S_n = \frac{n(a_1 + a_n)}{2} = \frac{15(3 + 31)}{2} = 255$$

71. $a_1 = 1, d = 1, n = 50$
$a_n = a_1 + (n - 1)d = 1 + 49(1) = 50$
$$S_n = \frac{n(a_1 + a_n)}{2} = \frac{50(1 + 50)}{2} = 1275$$

73. $a_1 = 1, d = 2, n = 50$
$a_n = a_1 + (n - 1)d = 1 + 49(2) = 99$
$$S_n = \frac{n(a_1 + a_n)}{2} = \frac{50(1 + 99)}{2} = 2500$$

75. $\sum_{k=4}^{4}(2k + 4) = 2(4) + 4 = 8 + 4 = 12$

77. $a_1 = 60, d = 50 \Rightarrow 60, 110, 160, 210, 260, 310; n = 121$
$a_n = a_1 + (n - 1)d = 60 + (121 - 1)(50) = 60 + 120(50) = \$6060$

79. $a_1 = 1, d = 1, n = 150, a_n = 150 \Rightarrow S_n = \frac{n(a_1+a_n)}{2} = \frac{150(1+150)}{2} = 11{,}325$ bricks

81. After 1 sec.: $s = 16(1)^2 = 16$; After 2 sec.: $s = 16(2)^2 = 64$; After 3 sec.: $s = 16(3)^2 = 144$
During 2nd second $\Rightarrow$ falls $64 - 16 = 48$ ft; During 3rd second $\Rightarrow$ falls $144 - 64 = 80$ ft
The sequence of the amounts fallen during each second is $16, 48, 80 \Rightarrow a_1 = 16, d = 32$
$a_n = a_1 + (n - 1)d = 16 + (12 - 1)(32) = 16 + 11(32) = 368$ ft

83. **Answers may vary.**

85. $\sum_{n=1}^{6}\left(\frac{1}{2}n + 1\right)$: $\frac{3}{2}, 2, \frac{5}{2}, 3, \frac{7}{2}, 4$

87. Form an arithmetic sequence with a 1st term of a and a 3rd term of b:
$a_n = a_1 + (n - 1)d$
$b = a_1 + (3 - 1)d$
$b = a + 2d$
$b - a = 2d$
$$\frac{b - a}{2} = d \Rightarrow \text{mean} = a_1 + \frac{b - a}{2} = a + \frac{b - a}{2} = \frac{2a}{2} + \frac{b - a}{2} = \frac{a + b}{2}$$

89. $\sum_{k=1}^{5} 5k = 5(1) + 5(2) + 5(3) + 5(4) + 5(5) = 5(1 + 2 + 3 + 4 + 5) = 5\sum_{k=1}^{5} k.$

91. $\sum_{k=1}^{n} 3 = \sum_{k=1}^{n} 3k^0 = 3(1)^0 + 3(2)^0 + \cdots + 3(n)^0 = 3 + 3 + \cdots + 3 = 3n$

Exercise 14.4 (page 983)

1. $1, 3, 9, \boxed{27}$

3. $0.2, 0.5, 1.25$; $\frac{0.5}{0.2} = 2.5$; $r = 2.5$

5. $2, \boxed{6}, 18, 54$

SECTION 14.4

7. $x^2 - 5x - 6 \le 0$
$(x-6)(x+1) \le 0$
$x - 6$ - - - - - - - - - 0++++
$x + 1$ - - - 0+++++++++++
⟵ [——] ⟶
−1 6
solution set: $[-1, 6]$

9. $\dfrac{x-4}{x+3} \ge 0$
$x - 4$ - - - - - - - - - 0++++
$x + 3$ - - - 0+++++++++++
⟵)——[⟶
−3 4
solution set: $(-\infty, -3) \cup [4, \infty)$

11. geometric

13. common ratio

15. $S_n = \dfrac{a_1 - a_1 r^n}{1 - r}$

17. $3, 6, 12, 24, 48$
$a_1 r^{8-1} = 3(2)^7 = 3(128) = 384$

19. $-5, -1, -\frac{1}{5}, -\frac{1}{25}, -\frac{1}{125}$
$a_1 r^{8-1} = -5\left(\frac{1}{5}\right)^7 = -\frac{1}{5^6} = -\frac{1}{15{,}625}$

21.
$$\begin{aligned} a_n &= a_1 r^{n-1} \\ 32 &= 2r^{3-1} \\ 32 &= 2r^2 \\ 16 &= r^2 \\ \pm 4 &= r, \text{ so } r = 4\ (r > 0) \end{aligned}$$
$2, 8, 32, 128, 512$

23.
$$\begin{aligned} a_n &= a_1 r^{n-1} \\ -192 &= -3r^{4-1} \\ -192 &= -3r^3 \\ 64 &= r^3 \\ 4 &= r \end{aligned}$$
$-3, -12, -48, -192, -768$

25.
$$\begin{aligned} a_n &= a_1 r^{n-1} \\ -4 &= -64r^{5-1} \\ -4 &= -64r^4 \\ \tfrac{1}{16} &= r^4 \\ \pm \tfrac{1}{2} &= r, \text{ so } r = -\tfrac{1}{2}\ (r < 0) \end{aligned}$$
$-64, 32, -16, 8, -4$

27.
$$\begin{aligned} a_n &= a_1 r^{n-1} \\ -2 &= -64r^{6-1} \\ -2 &= -64r^5 \\ \tfrac{1}{32} &= r^5 \\ \tfrac{1}{2} &= r \end{aligned}$$
$-64, -32, -16, -8, -4$

29. $a_1 = 2, a_5 = 162$
$$\begin{aligned} a_n &= a_1 r^{n-1} \\ a_5 &= a_1 r^{5-1} \\ 162 &= 2r^4 \\ 81 &= r^4 \\ \pm 3 &= r \Rightarrow \text{choose } r = 3 \end{aligned}$$
$2, \boxed{6, 18, 54}, 162$

31. $a_1 = -4, a_6 = -12500$
$$\begin{aligned} a_n &= a_1 r^{n-1} \\ a_6 &= a_1 r^{6-1} \\ -12500 &= -4r^5 \\ 3125 &= r^5 \\ 5 &= r \end{aligned}$$
$-4, \boxed{-20, -100, -500, -2500}, -12500$

33. $a_1 = 2, a_3 = 128$

$$a_n = a_1 r^{n-1}$$
$$a_3 = a_1 r^{3-1}$$
$$128 = 2r^2$$
$$64 = r^2$$
$$\pm 8 = r \Rightarrow \text{choose } r = -8$$
$$2, \boxed{-16}, 128$$

35. $a_1 = 10, a_3 = 20$

$$a_n = a_1 r^{n-1}$$
$$a_3 = a_1 r^{3-1}$$
$$20 = 10r^2$$
$$2 = r^2$$
$$\pm \sqrt{2} = r \Rightarrow \text{choose } r = \sqrt{2}$$
$$10, \boxed{10\sqrt{2}}, 20$$

37. $a_1 = 2, r = 3, n = 6;\ S_n = \dfrac{a_1 - a_1 r^n}{1 - r} = \dfrac{2 - 2(3)^6}{1 - 3} = \dfrac{2 - 2(729)}{-2} = \dfrac{-1456}{-2} = 728$

39. $a_1 = 2, r = -3, n = 5;\ S_n = \dfrac{a_1 - a_1 r^n}{1 - r} = \dfrac{2 - 2(-3)^5}{1 - (-3)} = \dfrac{2 - 2(-243)}{4} = \dfrac{488}{4} = 122$

41. If the 3rd term is $\frac{1}{5}$ and the 2nd term is 1, then the common ratio $r = \frac{1}{5} \div 1 = \frac{1}{5}$.

$a_1 = 1 \div \frac{1}{5} = 5, r = \frac{1}{5}, n = 4;\ S_n = \dfrac{a_1 - a_1 r^n}{1 - r} = \dfrac{5 - 5\left(\frac{1}{5}\right)^4}{1 - \frac{1}{5}} = \dfrac{5 - \frac{1}{125}}{\frac{4}{5}} = \dfrac{\frac{624}{125}}{\frac{4}{5}} = \dfrac{156}{25}$

43. If the 4th term is 1 and the 3rd term is -2, then the common ratio $r = 1 \div (-2) = -\frac{1}{2}$.

$a_2 = -2 \div \left(-\frac{1}{2}\right) = 4;\ a_1 = 4 \div \left(-\frac{1}{2}\right) = -8;\ a_1 = -8, r = -\frac{1}{2}, n = 6$

$$S_n = \frac{a_1 - a_1 r^n}{1 - r} = \frac{-8 - (-8)\left(-\frac{1}{2}\right)^6}{1 - \left(-\frac{1}{2}\right)} = \frac{-8 + \frac{1}{8}}{\frac{3}{2}} = \frac{-\frac{63}{8}}{\frac{3}{2}} = -\frac{21}{4}$$

45. If the 3rd term is 50 and the 2nd term is 10, then the common ratio $r = 50 \div 10 = 5$.

$a_1 = a_2 \div r = 10 \div 5 = 2 \Rightarrow 2, 10, 50, 250, 1250$

47. $a_n = a_1 r^{n-1} = 7 \cdot 2^{10-1} = 7 \cdot 2^9 = 7 \cdot 512 = 3584$

49.

$$a_n = a_1 r^{n-1}$$
$$-81 = a_1(-3)^{8-1}$$
$$-81 = a_1(-3)^7$$
$$-81 = -2187a_1$$
$$\tfrac{1}{27} = a_1$$

51.

$$a_n = a_1 r^{n-1}$$
$$-1944 = -8r^{6-1}$$
$$-1944 = -8r^5$$
$$243 = r^5$$
$$3 = r$$

53. $a_1 = -50, a_3 = 10$

$$a_n = a_1 r^{n-1}$$
$$a_3 = a_1 r^{3-1}$$
$$10 = -50r^2$$
$$-\tfrac{1}{5} = r^2$$

No such mean exists.

55. $a_1 = 3, r = -2, n = 8;\ S_n = \dfrac{a_1 - a_1 r^n}{1 - r} = \dfrac{3 - 3(-2)^8}{1 - (-2)} = \dfrac{3 - 3(256)}{3} = \dfrac{-765}{3} = -255$

57. $a_1 = 3, r = 2, n = 7;\ S_n = \dfrac{a_1 - a_1 r^n}{1 - r} = \dfrac{3 - 3(2)^7}{1 - 2} = \dfrac{3 - 3(128)}{-1} = \dfrac{-381}{-1} = 381$

59. Sequence of population: $500, 500(1.06), 500(1.06)^2, ...$
$a_1 = 500, r = 1.06, n = 6 \Rightarrow a_n = a_1 r^{n-1} = 500(1.06)^5 \approx 669$

61. Sequence of amounts: $10000, 10000(0.88), 10000(0.88)^2, ...$
$a_1 = 10000, r = 0.88, n = 16 \Rightarrow a_n = a_1 r^{n-1} = 10000(0.88)^{15} \approx \$1{,}469.74$

63. Sequence of values: $70000, 70000(1.06), 70000(1.06)^2, ...$
$a_1 = 70000, r = 1.06, n = 13 \Rightarrow a_n = a_1 r^{n-1} = 70000(1.06)^{12} \approx \$140{,}853.75$

65. Sequence of areas: $1, \frac{1}{2}, \frac{1}{4}, ...$
$a_1 = 1, r = \frac{1}{2}, n = 12 \Rightarrow a_n = a_1 r^{n-1} 1 = 1\left(\frac{1}{2}\right)^{11} = \left(\frac{1}{2}\right)^{11} \approx 0.0005$

67. Sequence of amounts: $1000(1.03), 1000(1.03)^2, 1000(1.03)^3, ...$
$$a_1 = 1030, r = 1.03, n = 4 \Rightarrow S_n = \frac{a_1 - a_1 r^n}{1 - r} = \frac{1030 - 1030(1.03)^4}{1 - 1.03} = \frac{-129.2740743}{-0.03} \approx \$4{,}309.14$$

69. **Answers may vary.**

71. **Answers may vary.**

73. arithmetic mean

75. **Answers may vary.**

Exercise 14.5 (page 988)

1. 8

3. $\frac{1}{2}$

5. $\frac{a_1}{1 - r} = \frac{18}{1 - \frac{1}{3}} = \frac{18}{\frac{2}{3}} = 27$

7. $y = 3x^3 - 4$
function

9. $3x = y^2 + 4$; not a function

11. infinite

13 $S_\infty = \frac{a_1}{1 - r}$

15. $a_1 = 8, r = \frac{1}{2}$
$$S_\infty = \frac{a_1}{1 - r} = \frac{8}{1 - \frac{1}{2}} = \frac{8}{\frac{1}{2}} = 16$$

17. $a_1 = 54, r = \frac{1}{3}$
$$S_\infty = \frac{a_1}{1 - r} = \frac{54}{1 - \frac{1}{3}} = \frac{54}{\frac{2}{3}} = 81$$

19. $a_1 = 12, r = -\frac{1}{2}$
$$S_\infty = \frac{a_1}{1 - r} = \frac{12}{1 - \left(-\frac{1}{2}\right)} = \frac{12}{\frac{3}{2}} = 8$$

21. $a_1 = \frac{9}{2}, r = \frac{4}{3} \Rightarrow$ no sum ($|r| > 1$)

SECTION 14.5

23. $0.\overline{1} = \frac{1}{10} + \frac{1}{100} + \frac{1}{1000} + \cdots \Rightarrow a_1 = \frac{1}{10}, r = \frac{1}{10} \Rightarrow S_\infty = \dfrac{a_1}{1-r} = \dfrac{\frac{1}{10}}{1-\frac{1}{10}} = \dfrac{\frac{1}{10}}{\frac{9}{10}} = \dfrac{1}{9}$

25. $-0.\overline{3} = -\frac{3}{10} - \frac{3}{100} - \frac{3}{1000} + \cdots \Rightarrow a_1 = -\frac{3}{10}, r = \frac{1}{10} \Rightarrow S_\infty = \dfrac{a_1}{1-r} = \dfrac{-\frac{3}{10}}{1-\frac{1}{10}} = \dfrac{-\frac{3}{10}}{\frac{9}{10}} = -\dfrac{1}{3}$

27. $0.\overline{12} = \dfrac{12}{100} + \dfrac{12}{10{,}000} + \dfrac{12}{1{,}000{,}000} + \cdots \Rightarrow a_1 = \dfrac{12}{100}, r = \dfrac{1}{100}$

$$S_\infty = \frac{a_1}{1-r} = \frac{\frac{12}{100}}{1-\frac{1}{100}} = \frac{\frac{12}{100}}{\frac{99}{100}} = \frac{12}{99} = \frac{4}{33}$$

29. $0.\overline{75} = \dfrac{75}{100} + \dfrac{75}{10{,}000} + \dfrac{75}{1{,}000{,}000} + \cdots \Rightarrow a_1 = \dfrac{75}{100}, r = \dfrac{1}{100}$

$$S_\infty = \frac{a_1}{1-r} = \frac{\frac{75}{100}}{1-\frac{1}{100}} = \frac{\frac{75}{100}}{\frac{99}{100}} = \frac{75}{99} = \frac{25}{33}$$

31. $a_1 = -54, r = -\dfrac{1}{3}$

$$S_\infty = \frac{a_1}{1-r} = \frac{-54}{1-\left(-\frac{1}{3}\right)} = \frac{-54}{\frac{4}{3}} = -\frac{81}{2}$$

33. $a_1 = -\dfrac{27}{2}, r = \dfrac{2}{3}$

$$S_\infty = \frac{a_1}{1-r} = \frac{-\frac{27}{2}}{1-\frac{2}{3}} = \frac{-\frac{27}{2}}{\frac{1}{3}} = -\frac{81}{2}$$

35. Distance ball travels down $= 10 + 5 + 2.5 + \cdots = \dfrac{a_1}{1-r} = \dfrac{10}{1-\frac{1}{2}} = \dfrac{10}{\frac{1}{2}} = 20$

Distance ball travels up $= 5 + 2.5 + 1.25 + \cdots = \dfrac{a_1}{1-r} = \dfrac{5}{1-\frac{1}{2}} = \dfrac{5}{\frac{1}{2}} = 10$

Total distance $= 20 + 10 = 30$ m

37. $S_\infty = \dfrac{a_1}{1-r} = \dfrac{1000}{1-0.8} = \dfrac{1000}{0.2}$

$= 5{,}000$ moths

39. **Answers may vary.**

41.

$$\begin{aligned} S_\infty &= \frac{a_1}{1-r} \\ 5 &= \frac{1}{1-r} \\ 5(1-r) &= 1 \\ 5 - 5r &= 1 \\ 4 &= 5r \\ \frac{4}{5} &= r \end{aligned}$$

43. $0.\overline{9} = \dfrac{9}{10} + \dfrac{9}{100} + \dfrac{9}{1000} + \cdots$

$\Rightarrow a = \dfrac{9}{10}, r = \dfrac{1}{10}$

$$S_\infty = \frac{a_1}{1-r} = \frac{\frac{9}{10}}{1-\frac{1}{10}} = \frac{\frac{9}{10}}{\frac{9}{10}} = \frac{9}{9} = 1$$

45. No. $0.999999 = \dfrac{999{,}999}{1{,}000{,}000} < 1$

Exercise 14.6 (page 998)

1. $3 \cdot 5 = 15$

3. $P(3,1) = \frac{3!}{(3-1)!} = \frac{3!}{2!} = \frac{6}{2} = 3$

5. $$C(3,0) = \frac{3!}{0!(3-0)!} = \frac{3!}{0!3!} = \frac{6}{1 \cdot 6} = \frac{6}{6} = 1$$

7. $$\begin{aligned} |2x - 3| &= 9 \\ 2x - 3 = 9 \quad &\textbf{or} \quad 2x - 3 = -9 \\ 2x = 12 \quad &\quad\quad 2x = -6 \\ x = 6 \quad &\quad\quad x = -3 \end{aligned}$$

9. $$\begin{aligned} \frac{3}{x-5} &= \frac{8}{x} \\ 3x &= 8(x-5) \\ 3x &= 8x - 40 \\ -5x &= -40 \\ x &= 8 \end{aligned}$$

11. $p \cdot q$

13. permutation

15. $P(n,r) = \dfrac{n!}{(n-r)!}$

17. combination

19. $C(n,r) = \dfrac{n!}{r!(n-r)!}$

21. $P(3,3) = \dfrac{3!}{(3-3)!} = \dfrac{3!}{0!} = \dfrac{6}{1} = 6$

23. $P(5,3) = \dfrac{5!}{(5-3)!} = \dfrac{5!}{2!} = \dfrac{120}{2} = 60$

25. $P(2,2) \cdot P(3,3) = \dfrac{2!}{(2-2)!} \cdot \dfrac{3!}{(3-3)!} = \dfrac{2!}{0!} \cdot \dfrac{3!}{0!} = \dfrac{2}{1} \cdot \dfrac{6}{1} = 12$

27. $$\frac{P(5,3)}{P(4,2)} = \frac{\frac{5!}{(5-3)!}}{\frac{4!}{(4-2)!}} = \frac{\frac{5!}{2!}}{\frac{4!}{2!}} = \frac{\frac{120}{2}}{\frac{24}{2}} = \frac{60}{12} = 5$$

29. $$C(5,3) = \frac{5!}{3!(5-3)!} = \frac{5!}{3!2!} = \frac{120}{6 \cdot 2} = \frac{120}{12} = 10$$

31. $\dbinom{6}{3} = \dfrac{6!}{3!(6-3)!} = \dfrac{6!}{3!3!} = \dfrac{720}{6 \cdot 6} = \dfrac{720}{36} = 20$

33. $\dbinom{5}{4}\dbinom{5}{3} = \dfrac{5!}{4!(5-4)!} \cdot \dfrac{5!}{3!(5-3)!} = \dfrac{5!}{4!1!} \cdot \dfrac{5!}{3!2!} = \dfrac{120}{24 \cdot 1} \cdot \dfrac{120}{6 \cdot 2} = \dfrac{120}{24} \cdot \dfrac{120}{12} = 5 \cdot 10 = 50$

35. $$\frac{C(38,37)}{C(19,18)} = \frac{\frac{38!}{37!(38-37)!}}{\frac{19!}{18!(19-18)!}} = \frac{\frac{38 \cdot 37!}{37!1!}}{\frac{19 \cdot 18!}{18!1!}} = \frac{\frac{38}{1}}{\frac{19}{1}} = \frac{38}{19} = 2$$

37. $$\begin{aligned} (x+y)^4 &= \binom{4}{0}x^4y^0 + \binom{4}{1}x^3y^1 + \binom{4}{2}x^2y^2 + \binom{4}{3}x^1y^3 + \binom{4}{4}x^0y^4 \\ &= \frac{4!}{0!4!}x^4 + \frac{4!}{1!3!}x^3y + \frac{4!}{2!2!}x^2y^2 + \frac{4!}{3!1!}xy^3 + \frac{4!}{4!0!}y^4 = x^4 + 4x^3y + 6x^2y^2 + 4xy^3 + y^4 \end{aligned}$$

39. $(2x+y)^3 = \binom{3}{0}(2x)^3y^0 + \binom{3}{1}(2x)^2y^1 + \binom{3}{2}(2x)^1y^2 + \binom{3}{3}(2x)^0y^3$
$= \frac{3!}{0!3!}\cdot 8x^3 + \frac{3!}{1!2!}\cdot 4x^2y + \frac{3!}{2!1!}\cdot 2xy^2 + \frac{3!}{3!0!}y^3$
$= 8x^3 + 12x^2y + 6xy^2 + y^3$

41. $C(12,0)C(12,12) = \frac{12!}{0!(12-0)!}\cdot\frac{12!}{12!(12-12)!} = \frac{12!}{0!12!}\cdot\frac{12!}{12!0!} = \frac{12!}{12!}\cdot\frac{12!}{12!} = 1\cdot 1 = 1$

43. $\frac{P(6,2)\cdot P(7,3)}{P(5,1)} = \frac{\frac{6!}{(6-2)!}\cdot\frac{7!}{(7-3)!}}{\frac{5!}{(5-1)!}} = \frac{\frac{6!}{4!}\cdot\frac{7!}{4!}}{\frac{5!}{4!}} = \frac{\frac{720}{24}\cdot\frac{5040}{24}}{\frac{120}{24}} = \frac{30\cdot 210}{5} = 1{,}260$

45. $C(n,2) = \frac{n!}{2!(n-2)!}$

47. $(3x-2)^4$
$= \binom{4}{0}(3x)^4(-2)^0 + \binom{4}{1}(3x)^3(-2)^1 + \binom{4}{2}(3x)^2(-2)^2 + \binom{4}{3}(3x)^1(-2)^3$
$+ \binom{4}{4}(3x)^0(-2)^4$
$= \frac{4!}{0!4!}\cdot 81x^4(1) + \frac{4!}{1!3!}\cdot 27x^3(-2) + \frac{4!}{2!2!}\cdot 9x^2(4) + \frac{4!}{3!1!}\cdot 3x(-8) + \frac{4!}{4!0!}\cdot 1(16)$
$= 81x^4 - 216x^3 + 216x^2 - 96x + 16$

49. $\binom{5}{3}x^2(-5y)^3 = \frac{5!}{3!2!}x^2(-125y^3) = 10x^2(-125y^3) = -1{,}250x^2y^3$

51. $\binom{4}{1}(x^2)^3(-y^3)^1 = \frac{4!}{1!3!}x^6(-y^3) = -4x^6y^3$

53. $7\cdot 5 = 35$

55. $7! = 5{,}040$

57. $10\cdot 10\cdot 10\cdot 10\cdot 10\cdot 10 = 1{,}000{,}000$

59. $9\cdot 9\cdot 8\cdot 7\cdot 6\cdot 5 = 136{,}080$

61. $8\cdot 10\cdot 10\cdot 10\cdot 10\cdot 10\cdot 10 = 8{,}000{,}000$

63. $4!\cdot 5! = 24\cdot 120 = 2{,}880$

65. $25\cdot 24\cdot 23 = 13{,}800$

67. $P(10,3) = \frac{10!}{(10-3)!} = \frac{10!}{7!} = 720$

69. $9\cdot 10\cdot 10\cdot 1\cdot 1 = 900$

71. $C(14,3) = \frac{14!}{3!(14-3)!} = \frac{14!}{3!11!} = 364$

73. $C(5,3) = 10 \Rightarrow 5$ persons

75. $C(100,6) = \dfrac{100!}{6!(100-6)!} = \dfrac{100!}{6!94!} = \dfrac{100 \cdot 99 \cdot 98 \cdot 97 \cdot 96 \cdot 95 \cdot 94!}{6!94!} = 1{,}192{,}052{,}400$

77. $C(3,2) \cdot C(4,2) = \dfrac{3!}{2!1!} \cdot \dfrac{4!}{2!2!} = 3 \cdot 6 = 18$

79. $C(12,2) \cdot C(10,3) = \dfrac{12!}{2!10!} \cdot \dfrac{10!}{3!7!} = 66 \cdot 120 = 7{,}920$

81. **Answers may vary.**

83. Consider the two people who insist on standing together as one person. Then there are a total of 4 "persons" to be arranged. This can be done in $4! = 24$ ways. However, the two people who are standing together can be arranged in 2 different ways, so there are $24 \cdot 2 = 48$ arrangements.

Exercise 14.7 (page 1004)

1. $\frac{1}{6}$

3.
$$\begin{aligned} 5^{4x} &= \frac{1}{125} \\ 5^{4x} &= 5^{-3} \\ 4x &= -3 \\ x &= -\frac{3}{4} \end{aligned}$$

5.
$$\begin{aligned} 2^{x^2-2x} &= 8 \\ 2^{x^2-2x} &= 2^3 \\ x^2 - 2x &= 3 \\ x^2 - 2x - 3 &= 0 \\ (x+1)(x-3) &= 0 \end{aligned}$$
$$x + 1 = 0 \quad \textbf{or} \quad x - 3 = 0$$
$$x = -1 \qquad\qquad x = 3$$

7.
$$\begin{aligned} 3^{x^2+4x} &= \frac{1}{81} \\ 3^{x^2+4x} &= 3^{-4} \\ x^2 + 4x &= -4 \\ x^2 + 4x + 4 &= 0 \\ (x+2)(x+2) &= 0 \end{aligned}$$
$$x + 2 = 0 \quad \textbf{or} \quad x + 2 = 0$$
$$x = -2 \qquad\qquad x = -2$$

9. experiment

11. $\dfrac{s}{n}$

13. 0

15. **a.** 6 **b.** 52 **c.** $\dfrac{6}{52}; \dfrac{3}{26}$

17. $\{(1,H), (2,H), (3,H), (4,H), (5,H), (6,H), (1,T), (2,T), (3,T), (4,T), (5,T), (6,T)\}$

19. $\{A, B, C, D, E, F, G, H, I, J, K, L, M, N, O, P, Q, R, S, T, U, V, W, X, Y, Z\}$

21. $\dfrac{1}{6}$

23. $\dfrac{4}{6} = \dfrac{2}{3}$

25. $\dfrac{19}{42}$

27. $\dfrac{13}{42}$

29. $\dfrac{3}{8}$

31. $\dfrac{0}{8} = 0$

33. $\dfrac{\#\text{ diamonds}}{\#\text{ cards}} = \dfrac{13}{52} = \dfrac{1}{4}$

35. $\dfrac{\#\text{ red face cards}}{\#\text{ cards in deck}} = \dfrac{6}{52} = \dfrac{3}{26}$

37. $\dfrac{\#\text{ ace}\cdot\#\text{ ten}}{\underset{\text{get 2 cards}}{\text{3 ways to}}} = \dfrac{4\cdot 4}{52\cdot 52} = \dfrac{16}{2704} = \dfrac{1}{169}$

39. impossible $\Rightarrow 0$

41. rolls of 4: $\{(1, 3), (2, 2), (3, 1)\}$
Probability $= \dfrac{3}{36} = \dfrac{1}{12}$

43. $\dfrac{\#\text{ yellow}}{\#\text{ eggs}} = \dfrac{7}{12}$

45. $FFFF \Rightarrow \dfrac{1}{16}$

47. $SSFF, SFSF, SFFS, FSSF, FSFS, FFSS \Rightarrow \dfrac{6}{16} = \dfrac{3}{8}$

49. $SSSS \Rightarrow \dfrac{1}{16}$

51. $\dfrac{\binom{8}{4}}{\binom{10}{4}} = \dfrac{70}{210} = \dfrac{1}{3}$

53. $\dfrac{176}{282} = \dfrac{88}{141}$

55. $\dfrac{15}{71}$

57. **Answers may vary.**

59. $P(\text{lux and } 2^{\text{nd}} \text{ car}) = P(\text{lux})\cdot P(2^{\text{nd}}|\text{lux}) = 0.2(0.7) = 0.14$

Chapter 14 Review (page 1008)

1. $(4!)(3!) = 4\cdot 3\cdot 2\cdot 1\cdot 3\cdot 2\cdot 1 = 144$

2. $\dfrac{5!}{3!} = \dfrac{5\cdot 4\cdot 3!}{3!} = 5\cdot 4 = 20$

3. $\dfrac{6!}{2!(6-2)!} = \dfrac{6!}{2!4!} = \dfrac{6\cdot 5\cdot 4!}{2\cdot 1\cdot 4!} = \dfrac{30}{2} = 15$

4. $\dfrac{12!}{3!(12-3)!} = \dfrac{12!}{3!9!} = \dfrac{12\cdot 11\cdot 10\cdot 9!}{3\cdot 2\cdot 1\cdot 9!} = \dfrac{1320}{6} = 220$

5. $(n-n)! = 0! = 1$

6. $\dfrac{8!}{7!} = \dfrac{8\cdot 7!}{7!} = 8$

7.
$$\begin{aligned}(x+y)^5 &= x^5 + \frac{5!}{1!(5-1)!}x^4y + \frac{5!}{2!(5-2)!}x^3y^2 + \frac{5!}{3!(5-3)!}x^2y^3 + \frac{5!}{4!(5-4)!}xy^4 + y^5\\
&= x^5 + \frac{5!}{1!4!}x^4y + \frac{5!}{2!3!}x^3y^2 + \frac{5!}{3!2!}x^2y^3 + \frac{5!}{4!1!}xy^4 + y^5\\
&= x^5 + \frac{5\cdot 4!}{1!4!}x^4y + \frac{5\cdot 4\cdot 3!}{2\cdot 1\cdot 3!}x^3y^2 + \frac{5\cdot 4\cdot 3!}{3!\cdot 2\cdot 1}x^2y^3 + \frac{5\cdot 4!}{4!\cdot 1}xy^4 + y^5\\
&= x^5 + \frac{5}{1}x^4y + \frac{20}{2}x^3y^2 + \frac{20}{2}x^2y^3 + \frac{5}{1}xy^4 + y^5\\
&= x^5 + 5x^4y + 10x^3y^2 + 10x^2y^3 + 5xy^4 + y^5\end{aligned}$$

8. $(x-y)^4 = x^4 + \dfrac{4!}{1!(4-1)!}x^3(-y) + \dfrac{4!}{2!(4-2)!}x^2(-y)^2 + \dfrac{4!}{3!(4-3)!}x(-y)^3 + (-y)^4$

$= x^4 + \dfrac{4!}{1!3!}(-x^3y) + \dfrac{4!}{2!2!}x^2y^2 + \dfrac{4!}{3!1!}(-xy^3) + y^4$

$= x^4 - \dfrac{4\cdot 3!}{1!3!}x^3y + \dfrac{4\cdot 3\cdot 2!}{2!\cdot 2\cdot 1}x^2y^2 - \dfrac{4\cdot 3!}{3!1!}xy^3 + y^4$

$= x^4 - \dfrac{4}{1}x^3y + \dfrac{12}{2}x^2y^2 - \dfrac{4}{1}xy^3 + y^4 = x^4 - 4x^3y + 6x^2y^2 - 4xy^3 + y^4$

9. $(4x-y)^3 = (4x)^3 + \dfrac{3!}{1!(3-1)!}(4x)^2(-y) + \dfrac{3!}{2!(3-2)!}(4x)(-y)^2 + (-y)^3$

$= 64x^3 + \dfrac{3!}{1!2!}\cdot(-16x^2y) + \dfrac{3!}{2!1!}\cdot 4xy^2 - y^3$

$= 64x^3 - \dfrac{3\cdot 2!}{1!2!}\cdot 16x^2y + \dfrac{3\cdot 2!}{2!1!}\cdot 4xy^2 - y^3$

$= 64x^3 - \dfrac{3}{1}\cdot 16x^2y + \dfrac{3}{1}\cdot 4xy^2 - y^3 = 64x^3 - 48x^2y + 12xy^2 - y^3$

10. $(x+4y)^3 = x^3 + \dfrac{3!}{1!(3-1)!}x^2(4y) + \dfrac{3!}{2!(3-2)!}x(4y)^2 + (4y)^3$

$= x^3 + \dfrac{3!}{1!2!}\cdot 4x^2y + \dfrac{3!}{2!1!}\cdot 16xy^2 + 64y^3$

$= x^3 + \dfrac{3\cdot 2!}{1!2!}\cdot 4x^2y + \dfrac{3\cdot 2!}{2!1!}\cdot 16xy^2 + 64y^3$

$= x^3 + \dfrac{3}{1}\cdot 4x^2y + \dfrac{3}{1}\cdot 16xy^2 + 64y^3 = x^3 + 12x^2y + 48xy^2 + 64y^3$

11. In the 3rd term, the exponent on 3 is 2.

Variables: x^2y^2

Coef. $= \dfrac{n!}{r!(n-r)!} = \dfrac{4!}{2!2!} = 6$

Term $= 6x^2y^2$

12. In the 4th term, the exponent on $-y$ is 3.

Variables: $x^2(-y)^3 = -x^2y^3$

Coef. $= \dfrac{n!}{r!(n-r)!} = \dfrac{5!}{3!2!} = 10$

Term $= 10(-x^2y^3) = -10x^2y^3$

13. 2nd term: The exponent on $-4y$ is 1. Variables: $(3x)^2(-4y)^1 = (9x^2)(-4y) = -36x^2y$

Coef. $= \dfrac{n!}{r!(n-r)!} = \dfrac{3!}{2!1!} = 3$; Term $= 3(-36x^2y) = -108x^2y$

14. 3rd term: The exponent on $3y$ is 2. Variables: $(4x)^2(3y)^2 = (16x^2)(9y^2) = 144x^2y^2$

Coef. $= \dfrac{n!}{r!(n-r)!} = \dfrac{4!}{2!2!} = 6$; Term $= 6(144x^2y^2) = 864x^2y^2$

15. $a_n = a_1 + (n-1)d = 7 + (8-1)5 = 7 + 7\cdot 5 = 42$

16.

$$a_n = a_1 + (n-1)d \qquad a_n = a_1 + (n-1)d$$
$$242 = a_1 + (9-1)d \qquad 212 = a_1 + (7-1)d$$
$$242 = a_1 + 8d \qquad 212 = a_1 + 6d$$

$$\begin{aligned} a_1 + 8d &= 242 \Rightarrow \times(-1) & -a_1 - 8d &= -242 \\ a_1 + 6d &= 212 \Rightarrow & a_1 + 6d &= 212 \\ & & -2d &= -30 \\ & & d &= 15 \end{aligned}$$

Substitute and solve for a_1:

$$\begin{aligned} a_1 + 6d &= 212 \\ a_1 + 6(15) &= 212 \\ a_1 + 90 &= 212 \\ a_1 &= 122 \Rightarrow 122, 137, 152, 167, 182 \end{aligned}$$

17. Form an arithmetic sequence with a 1st term of 8 and a 4th term of 25:

$$\begin{aligned} a_n &= a_1 + (n-1)d \\ 25 &= 8 + (4-1)d \\ 17 &= 3d \\ \tfrac{17}{3} &= d \Rightarrow 8, \boxed{\tfrac{41}{3}, \tfrac{58}{3}}, 25 \end{aligned}$$

18. $a_1 = 11, d = 7, n = 20$

$$a_n = a_1 + (n-1)d = 11 + 19(7) = 144$$
$$S_n = \frac{n(a_1 + a_n)}{2} = \frac{20(11+144)}{2} = 1{,}550$$

19. $a_1 = 9, d = -\frac{5}{2}, n = 10$

$$a_n = a_1 + (n-1)d = 9 + 9\left(-\tfrac{5}{2}\right) = -\tfrac{27}{2}$$
$$S_n = \frac{n(a_1 + a_n)}{2} = \frac{10\left(9 - \frac{27}{2}\right)}{2} = -\frac{45}{2}$$

20. $\sum_{k=4}^{6} \frac{1}{2}k = \frac{1}{2}(4) + \frac{1}{2}(5) + \frac{1}{2}(6) = 2 + \frac{5}{2} + 3 = \frac{15}{2}$

21. $\sum_{k=2}^{5} 7k^2 = 7(2)^2 + 7(3)^2 + 7(4)^2 + 7(5)^2 = 28 + 63 + 112 + 175 = 378$

22. $\sum_{k=1}^{4} (3k-4) = (3(1)-4) + (3(2)-4) + (3(3)-4) + (3(4)-4) = -1 + 2 + 5 + 8 = 14$

23. $\sum_{k=10}^{10} 36k = 36(10) = 360$

24. If the 5th term is $\frac{3}{2}$ and the 4th term is 3, then the common ratio $r = \frac{3}{2} \div 3 = \frac{1}{2}$.

$$\begin{aligned} a_3 &= a_4 \div r = 3 \div \tfrac{1}{2} = 6 \\ a_2 &= a_3 \div r = 6 \div \tfrac{1}{2} = 12 \\ a_1 &= a_2 \div r = 12 \div \tfrac{1}{2} = 24 \end{aligned}$$

$24, 12, 6, 3, \frac{3}{2}$

25.

$$\begin{aligned} a_n &= a_1 r^{n-1} \\ &= \frac{1}{8}(2)^{6-1} \\ &= \frac{1}{8}(32) = 4 \end{aligned}$$

26. $a_1 = -6, a_4 = 384$

$$\begin{aligned} a_n &= a_1 r^{n-1} \\ 384 &= -6r^{4-1} \\ -64 &= r^3 \\ -4 &= r \Rightarrow -6, \boxed{24, -96}, 384 \end{aligned}$$

27. $a_1 = 162, r = \frac{1}{3}, n = 7;\ S_n = \dfrac{a_1 - a_1 r^n}{1 - r} = \dfrac{162 - 162\left(\frac{1}{3}\right)^7}{1 - \frac{1}{3}} = \dfrac{162 - 162\left(\frac{1}{2187}\right)}{-\frac{2}{3}} = \dfrac{\frac{4372}{27}}{-\frac{2}{3}} = \dfrac{2186}{9}$

28. $a_1 = \frac{1}{8}, r = -2, n = 8;\ S_n = \dfrac{a_1 - a_1 r^n}{1 - r} = \dfrac{\frac{1}{8} - \frac{1}{8}(-2)^8}{1 - (-2)} = \dfrac{\frac{1}{8} - \frac{1}{8}(256)}{3} = \dfrac{-\frac{255}{8}}{3} = -\dfrac{85}{8}$

29. Sequence of amounts: $5000, 5000(0.8), 5000(0.8)^2, \ldots$
$a_1 = 5000, r = 0.8, n = 6 \Rightarrow a_n = a_1 r^{n-1} = 5000(0.8)^5 \approx \$1{,}638.40$

30. Sequence of amounts: $25700, 25700(1.18), 25700(1.18)^2, \ldots$
$a_1 = 25700, r = 1.18, n = 11 \Rightarrow a_n = a_1 r^{n-1} = 25700(1.18)^{10} \approx \$134{,}509.57$

31. $a_1 = 300, d = 75$

$$\begin{aligned} a_n &= a_1 + (n-1)d \\ 1200 &= 300 + (n-1)75 \\ 1200 &= 300 + 75n - 75 \\ 975 &= 75n \\ 13 &= n \end{aligned}$$

$13 = n$; In the 13th year, or after 12 years of increases.

32. $16, 48, 80, \ldots;\ a_1 = 16, d = 32$

$$\begin{aligned} a_n &= a_1 + (n-1)d = 16 + (10-1)(32) \\ &= 16 + 9(32) = 304 \text{ ft} \end{aligned}$$

$$S_n = \frac{n(a_1 + a_n)}{2} = \frac{10(16 + 304)}{2} = 1600 \text{ ft}$$

33. $a_1 = 25, r = \frac{4}{5};\ S_\infty = \dfrac{a_1}{1 - r} = \dfrac{25}{1 - \frac{4}{5}} = \dfrac{25}{\frac{1}{5}} = 125$

34. $0.\overline{05} = \dfrac{5}{100} + \dfrac{5}{10{,}000} + \dfrac{5}{1{,}000{,}000} + \cdots \Rightarrow a_1 = \dfrac{5}{100}, r = \dfrac{1}{100}$

$$S_\infty = \frac{a_1}{1 - r} = \frac{\frac{5}{100}}{1 - \frac{1}{100}} = \frac{\frac{5}{100}}{\frac{99}{100}} = \frac{5}{99}$$

35. $17 \cdot 8 = 136$

36. $P(7,7) = \dfrac{7!}{(7-7)!} = \dfrac{7!}{0!} = 7! = 5{,}040$

37. $P(7,0) = \dfrac{7!}{(7-0)!} = \dfrac{7!}{7!} = 1$

38. $P(8,6) = \dfrac{8!}{(8-6)!} = \dfrac{8!}{2!} = \dfrac{40{,}320}{2} = 20{,}160$

39. $\dfrac{P(9,6)}{P(10,7)} = \dfrac{\frac{9!}{(9-6)!}}{\frac{10!}{(10-7)!}} = \dfrac{\frac{9!}{3!}}{\frac{10!}{3!}} = \dfrac{9!}{3!} \cdot \dfrac{3!}{10!} = \dfrac{9!}{10!} = \dfrac{9!}{10 \cdot 9!} = \dfrac{1}{10}$

CHAPTER 14 TEST

25. $\dfrac{\text{\# ways to get 5 hearts}}{\text{\# ways to get 5 cards}} = \dfrac{\binom{13}{5}}{\binom{52}{5}} = \dfrac{1287}{2{,}598{,}960} = \dfrac{33}{66{,}640}$

26. $\dfrac{\text{\# ways to get 2 heads}}{\text{\# ways to toss 5 times}} = \dfrac{\binom{5}{2}}{2^5} = \dfrac{20}{32} = \dfrac{5}{16}$

Cumulative Review Exercises (page 1013)

1. $\begin{cases} 2x + y = 5 \\ x - 2y = 0 \end{cases}$

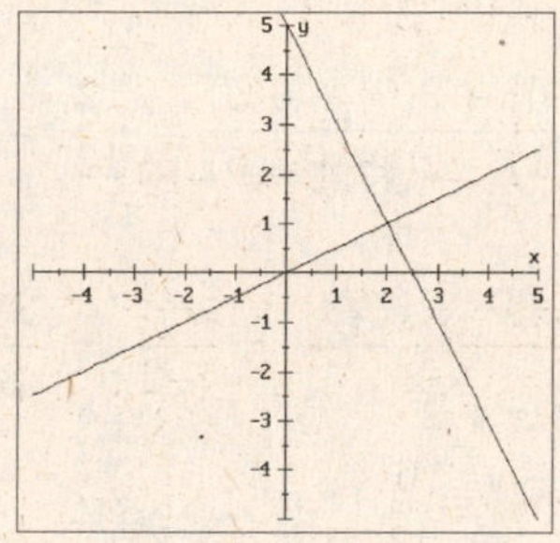

Solution: $(2, 1)$

2. $\begin{cases} (1) & 3x + y = 4 \\ (2) & 2x - 3y = -1 \end{cases}$

Substitute $y = 4 - 3x$ from (1) into (2):

$$\begin{aligned} 2x - 3y &= -1 \\ 2x - 3(4 - 3x) &= -1 \\ 2x - 12 + 9x &= -1 \\ 11x &= 11 \\ x &= 1 \end{aligned}$$

Substitute this and solve for y:

$y = 4 - 3x = 4 - 3(1) = 1$

Solution: $(1, 1)$

3. $\begin{array}{rcl} x + 2y = -2 & & x + 2y = -2 \\ 2x - \ \ y = \ \ 6 & \Rightarrow \times (2) & 4x - 2y = 12 \\ \hline & & 5x \quad = 10 \\ & & x \quad = 2 \end{array}$ $\quad \begin{aligned} x + 2y &= -2 \\ 2 + 2y &= -2 \\ 2y &= -4 \\ y &= -2 \end{aligned}$ $\quad$ Solution: $\boxed{(2, -2)}$

4. $\begin{array}{rcl} \dfrac{x}{10} + \dfrac{y}{5} = \dfrac{1}{2} & \Rightarrow \times 10 & x + 2y = 5 \\ \dfrac{x}{2} - \dfrac{y}{5} = \dfrac{13}{10} & \Rightarrow \times 10 & 5x - 2y = 13 \\ \hline & & 6x \quad = 18 \\ & & x \quad = 3 \end{array}$ $\quad \begin{aligned} x + 2y &= 5 \\ 3 + 2y &= 5 \\ 2y &= 2 \\ y &= 1 \Rightarrow \text{Solution: } \boxed{(3, 1)} \end{aligned}$

5. $\begin{vmatrix} 3 & -2 \\ 1 & -1 \end{vmatrix} = 3(-1) - (-2)(1)$
$= -3 + 2 = -1$

6. $y = \dfrac{\begin{vmatrix} 4 & -1 \\ 3 & -7 \end{vmatrix}}{\begin{vmatrix} 4 & -3 \\ 3 & 4 \end{vmatrix}} = \dfrac{4(-7) - (-1)(3)}{4(4) - (-3)(3)}$
$= \dfrac{-25}{25} = -1$

7.

$$\begin{array}{ll} (1) & x+y+z=1 \\ (2) & 2x-y-z=-4 \\ (3) & x-2y+z=4 \end{array} \qquad \begin{array}{lrcl} (1) & x+y+z & = & 1 \\ (2) & 2x-y-z & = & -4 \\ \hline (4) & 3x & = & -3 \\ & x & = & -1 \end{array} \qquad \begin{array}{lrcl} (2) & 2x-\ \ y-z & = & -4 \\ (3) & x-2y+z & = & 4 \\ \hline (5) & 3x-3y & = & 0 \end{array}$$

$$\begin{aligned} 3x-3y&=0 \\ 3(-1)-3y&=0 \\ -3-3y&=0 \\ -3y&=3 \\ y&=-1 \end{aligned} \qquad \begin{aligned} x+y+z&=1 \\ -1+(-1)+z&=1 \\ -2+z&=1 \\ z&=3 \end{aligned}$$

$\boxed{\text{The solution is } (-1,-1,3).}$

8. $$z=\frac{\begin{vmatrix} 1 & 2 & 6 \\ 3 & 2 & 6 \\ 2 & 3 & 6 \end{vmatrix}}{\begin{vmatrix} 1 & 2 & 3 \\ 3 & 2 & 1 \\ 2 & 3 & 1 \end{vmatrix}}=\frac{1\begin{vmatrix} 2 & 6 \\ 3 & 6 \end{vmatrix}-2\begin{vmatrix} 3 & 6 \\ 2 & 6 \end{vmatrix}+6\begin{vmatrix} 3 & 2 \\ 2 & 3 \end{vmatrix}}{1\begin{vmatrix} 2 & 1 \\ 3 & 1 \end{vmatrix}-2\begin{vmatrix} 3 & 1 \\ 2 & 1 \end{vmatrix}+3\begin{vmatrix} 3 & 2 \\ 2 & 3 \end{vmatrix}}=\frac{1(-6)-2(6)+6(5)}{1(-1)-2(1)+3(5)}=\frac{12}{12}=1$$

9. $\begin{cases} 3x-2y<6 \\ y<-x+2 \end{cases}$

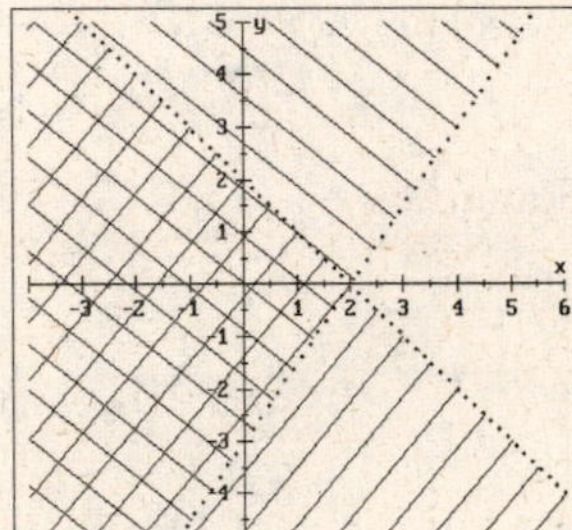

10. $\begin{cases} y<x+2 \\ 3x+y\le 6 \end{cases}$

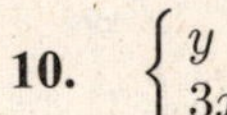

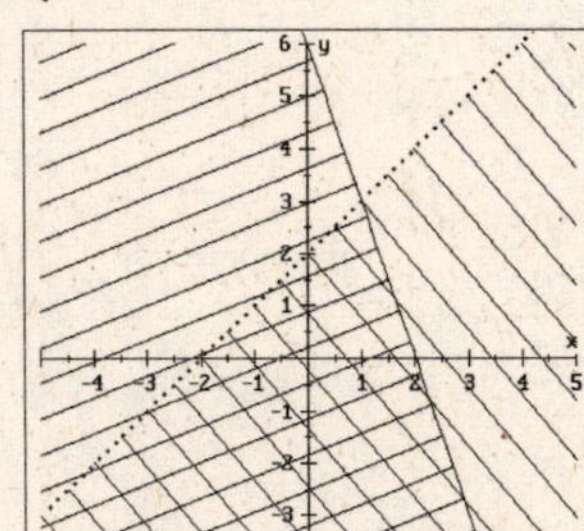

11. $y=\left(\frac{1}{2}\right)^x$

12. $y=\log_2 x \Rightarrow 2^y=x$

13. $\log_x 25=2 \Rightarrow x^2=25 \Rightarrow x=5$

14. $\log_5 125=x \Rightarrow 5^x=125 \Rightarrow x=3$

15. $\log_3 x=-3 \Rightarrow 3^{-3}=x \Rightarrow x=\frac{1}{27}$

16. $\log_5 x=0 \Rightarrow 5^0=x \Rightarrow x=1$

CUMULATIVE REVIEW EXERCISES

17. $y = 2^x$

18. x

19. $\log 98 = \log(14 \cdot 7) = \log 14 + \log 7 = 1.1461 + 0.8451 = 1.9912$

20. $\log 2 = \log \frac{14}{7} = \log 14 - \log 7 = 1.1461 - 0.8451 = 0.3010$

21. $\log 49 = \log 7^2 = 2\log 7 = 2(0.8451) = 1.6902$

22. $\log \frac{7}{5} = \log \frac{14}{10} = \log 14 - \log 10 = 1.1461 - 1 = 0.1461$

23.
$$\begin{aligned} 2^{x+5} &= 3^x \\ \log 2^{x+5} &= \log 3^x \\ (x+5)\log 2 &= x\log 3 \\ x\log 2 + 5\log 2 &= x\log 3 \\ 5\log 2 &= x\log 3 - x\log 2 \\ 5\log 2 &= x(\log 3 - \log 2) \\ \frac{5\log 2}{\log 3 - \log 2} &= x \end{aligned}$$

24.
$$\begin{aligned} \log 5 + \log x - \log 4 &= 1 \\ \log \frac{5x}{4} &= 1 \\ 10^1 &= \frac{5x}{4} \\ 40 &= 5x \\ 8 &= x \end{aligned}$$

25. $A = P\left(1 + \frac{r}{k}\right)^{kt} = 9000\left(1 + \frac{-0.12}{1}\right)^{1(9)} \approx \$2{,}848.31$

26. $\log_6 8 = \frac{\log 8}{\log 6} \approx 1.16056$

27. $\frac{6!7!}{5!} = \frac{6 \cdot 5! \cdot 7!}{5!} = 6 \cdot 7! = 30{,}240$

28. $(3a - b)^4$
$$\begin{aligned} &= (3a)^4 + \frac{4!}{1!(4-1)!}(3a)^3(-b) + \frac{4!}{2!(4-2)!}(3a)^2(-b)^2 + \frac{4!}{3!(4-3)!}(3a)(-b)^3 + (-b)^4 \\ &= 81a^4 + \frac{4!}{1!3!}(-27a^3b) + \frac{4!}{2!2!}(9a^2b^2) + \frac{4!}{3!1!}(-3ab^3) + b^4 \\ &= 81a^4 - \frac{4 \cdot 3!}{1!3!}(27a^3b) + \frac{4 \cdot 3 \cdot 2!}{2! \cdot 2 \cdot 1}(9a^2b^2) - \frac{4 \cdot 3!}{3!1!}(3ab^3) + b^4 \\ &= 81a^4 - \frac{4}{1}(27a^3b) + \frac{12}{2}(9a^2b^2) - \frac{4}{1}(3ab^3) + b^4 \\ &= 81a^4 - 108a^3b + 54a^2b^2 - 12ab^3 + b^4 \end{aligned}$$

29. In the 7th term, the exponent on $-y$ is 6.
Variables: $(2x)^2(-y)^6 = 4x^2y^6$
Coef. $= \frac{n!}{r!(n-r)!} = \frac{8!}{6!2!} = 28$
Term $= 28(4x^2y^6) = 112x^2y^6$

30. $a_1 = -11, d = 6, n = 20$
$$\begin{aligned} a_n &= a_1 + (n-1)d \\ &= -11 + (19)(6) \\ &= -11 + 114 = 103 \end{aligned}$$

CUMULATIVE REVIEW EXERCISES

31. $a_1 = 6, d = 3, n = 20; a_n = a_1 + (n-1)d = 6 + (20-1)(3) = 6 + 19(3) = 63$

$$S_n = \frac{n(a_1 + a_n)}{2} = \frac{20(6+63)}{2} = \frac{20(69)}{2} = 690$$

32. $a_1 = -3; a_4 = 30$:

$a_n = a_1 + (n-1)d$

$30 = -3 + (4-1)d$

$33 = 3d$

$11 = d \Rightarrow -3, \boxed{8, 19}, 30$

33. $\sum_{k=1}^{3} 3k^2 = 3(1)^2 + 3(2)^2 + 3(3)^2$

$= 3 + 12 + 27 = 42$

34. $\sum_{k=3}^{5} (2k+1) = (2(3)+1) + (2(4)+1) + (2(5)+1) = 7 + 9 + 11 = 27$

35. $a_1 = \frac{1}{27}, r = 3, n = 7; a_n = a_1 r^{n-1} = \frac{1}{27}(3)^{7-1} = \frac{1}{27}(3)^6 = \frac{1}{27}(729) = 27$

36. $a_1 = \frac{1}{64}, r = 2, n = 10; S_n = \dfrac{a_1 - a_1 r^n}{1-r} = \dfrac{\frac{1}{64} - \frac{1}{64}(2)^{10}}{1-2} = \dfrac{\frac{1}{64} - \frac{1}{64}(1024)}{-1} = \dfrac{-\frac{1023}{64}}{-1} = \dfrac{1023}{64}$

37. $a_1 = -3, a_4 = 192$

$a_n = a_1 r^{n-1}$

$192 = -3r^{4-1}$

$-64 = r^3$

$-4 = r \Rightarrow -3, \boxed{12, -48}, 192$

38. $a_1 = 9, r = \dfrac{1}{3}$

$$S_\infty = \frac{a_1}{1-r} = \frac{9}{1-\frac{1}{3}} = \frac{9}{\frac{2}{3}} = \frac{27}{2}$$

39. $P(9,3) = \dfrac{9!}{(9-3)!} = \dfrac{9!}{6!} = \dfrac{9 \cdot 8 \cdot 7 \cdot 6!}{6!} = 9 \cdot 8 \cdot 7 = 504$

40. $C(7,4) = \dfrac{7!}{4!(7-4)!} = \dfrac{7!}{4!3!} = \dfrac{7 \cdot 6 \cdot 5 \cdot 4!}{4! \cdot 3 \cdot 2 \cdot 1} = \dfrac{210}{6} = 35$

41. $\dfrac{C(8,4)C(8,0)}{P(6,2)} = \dfrac{\frac{8!}{4!4!} \cdot \frac{8!}{0!8!}}{\frac{6!}{4!}} = \dfrac{70 \cdot 1}{30} = \dfrac{7}{3}$

42. $C(n,n) = 1$ is smaller than $P(n,n) = n!$.

43. $7! = 5{,}040$

44. $\binom{9}{3} = \frac{9!}{3!6!} = 84$

45. $\dfrac{6}{52} = \dfrac{3}{26}$

Appendix 1 (page A-4)

1. $y = x^2 - 1$

x-axis	y-axis	origin
$-y = x^2 - 1$	$y = (-x)^2 - 1$	$-y = (-x)^2 - 1$
not equivalent: no symmetry	$y = x^2 - 1$	$-y = x^2 - 1$
	equivalent: **symmetry**	not equivalent: no symmetry

3. $y = x^5$

x-axis	y-axis	origin
$-y = x^5$	$y = (-x)^3$	$-y = (-x)^3$
not equivalent: no symmetry	$y = -x^5$	$-y = -x^5$
	not equivalent: no symmetry	$y = x^5$
		equivalent: **symmetry**

5. $y = -x^2 + 2$

x-axis	y-axis	origin
$-y = -x^2 + 2$	$y = -(-x)^2 + 2$	$-y = (-x)^2 + 2$
not equivalent: no symmetry	$y = -x^2 + 2$	$-y = x^2 + 2$
	equivalent: **symmetry**	not equivalent: no symmetry

7. $y = x^2 - x$

x-axis	y-axis	origin
$-y = x^2 - x$	$y = (-x)^2 - (-x)$	$-y = (-x)^2 - (-x)$
not equivalent: no symmetry	$y = x^2 + x$	$-y = x^2 + x$
	not equivalent: no symmetry	not equivalent: no symmetry

9. $y = -|x + 2|$

x-axis	y-axis	origin
$-y = -|x + 2|$	$y = -|-x + 2|$	$-y = -|-x + 2|$
not equivalent: no symmetry	not equivalent: no symmetry	not equivalent: no symmetry

11. $|y| = x$

x-axis	y-axis	origin
$|-y| = x$	$|y| = -x$	$|-y| = -x$
$|-1||y| = x$	not equivalent: no symmetry	$|-1||y| = -x$
$|y| = x$		$|y| = -x$
equivalent: **symmetry**		not equivalent: no symmetry

13. $y = x^4 - 4$

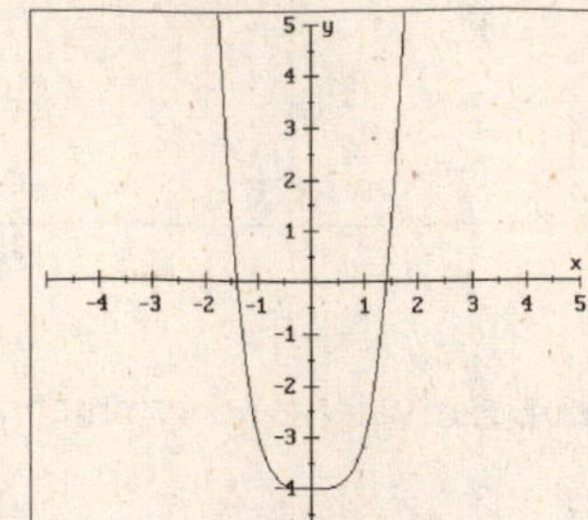

D $(-\infty, \infty)$; R $[-4, \infty)$

15. $y = -x^3$

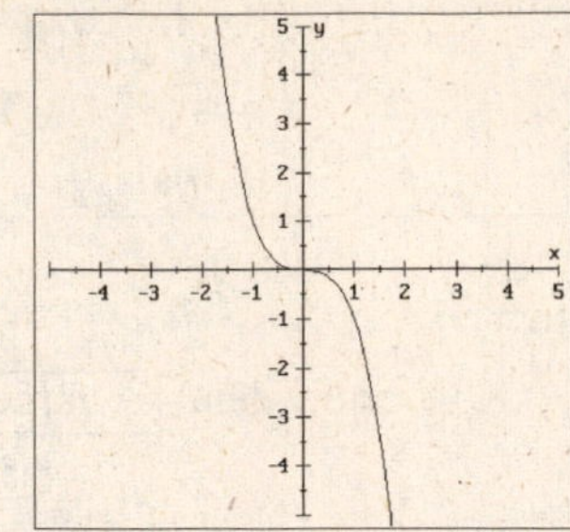

D $(-\infty, \infty)$; R $(-\infty, \infty)$

17. $y = x^4 + x^2$

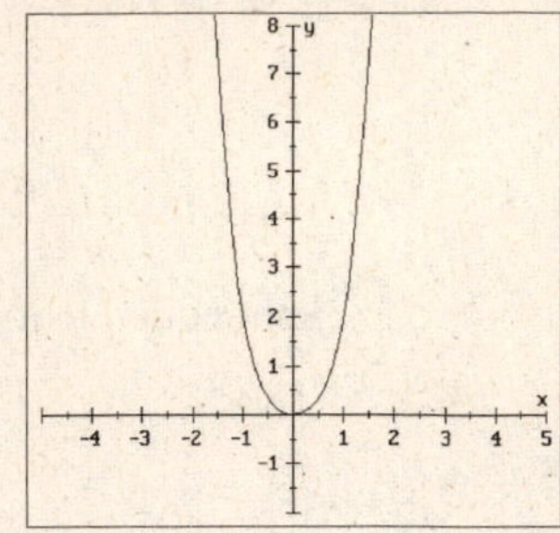

D $(-\infty, \infty)$; R $[0, \infty)$

19. $y = x^3 - x$

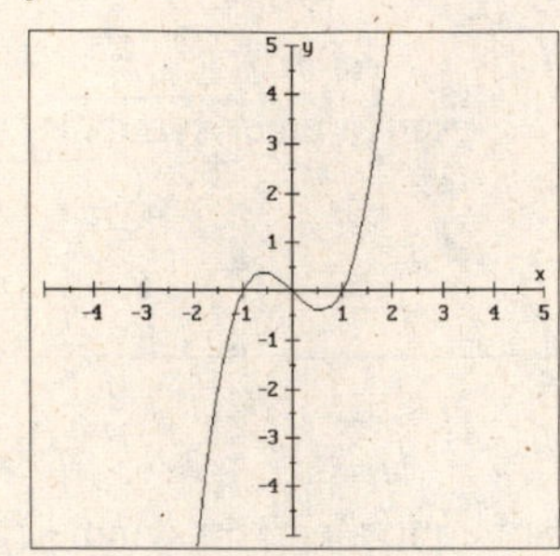

D $(-\infty, \infty)$; R $(-\infty, \infty)$

21. $y = \frac{1}{2}|x| - 1$

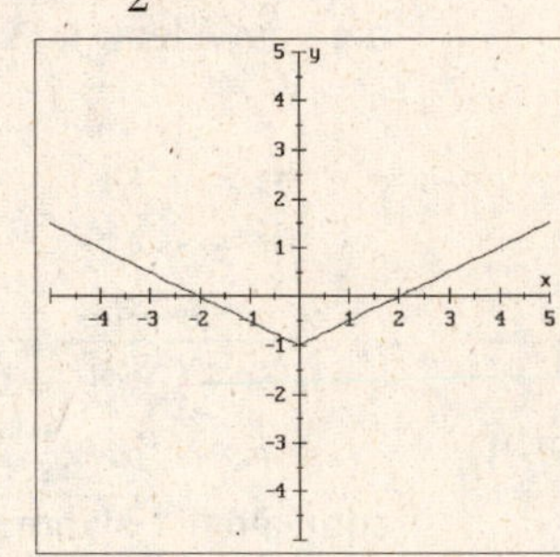

D $(-\infty, \infty)$; R $[-1, \infty)$

23. $y = -|x + 2|$

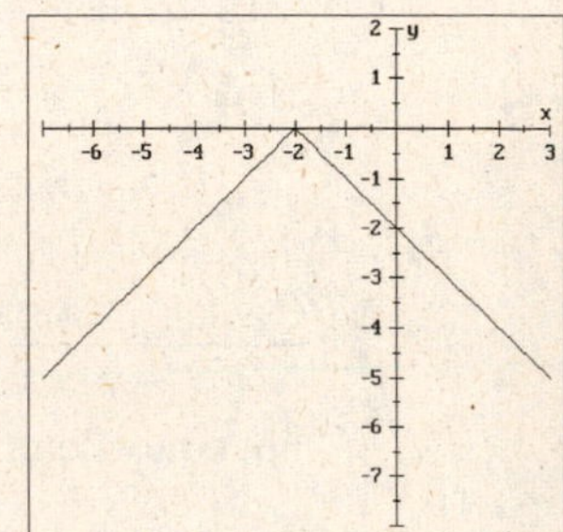

D $(-\infty, \infty)$; R $(-\infty, 0]$